Nutrition and Metabolism

The Nutrition Society Textbook Series

Sport and Exercise Nutrition, 2nd Edition
Susan A. Lanham-New, Samantha Stear, Susan Shirreffs, Adam Collins, Richard Budgett

Clinical Nutrition, 2nd Edition
Marinos Elia, Olle Ljungqvist, Rebecca J. Stratton, Susan A. Lanham-New

Nutrition Research Methodologies, 1st Edition
Julie A. Lovegrove, Leanne Hodson, Sangita Sharma, Susan A. Lanham-New, Lord John Krebs

Public Health Nutrition, 2nd Edition
Judith L. Buttriss, Ailsa A. Welch, John M. Kearney, Susan A. Lanham-New

Introduction to Human Nutrition, 3rd Edition
Susan A. Lanham-New, Thomas R. Hill, Alison M. Gallagher, Hester H. Vorster

Nutrition and Metabolism, 3rd Edition
Helen M. Roche, Ian A. MacDonald, Annemie M.W.J. Schols, Susan A. Lanham-New

Nutrition and Metabolism

Third Edition

Edited on behalf of The Nutrition Society by

Helen M. Roche
Full Professor of Nutrigenomics
Director, UCD Conway Institute
University College Dublin
Dublin, Ireland

Ian A. Macdonald
Emeritus Professor of Metabolic Physiology
School of Life Sciences, Faculty of Medicine and Health Sciences
University of Nottingham
Nottingham, UK

Annemie M.W.J. Schols
Professor of Nutrition and Metabolism in Chronic Diseases
Dean Faculty of Health, Medicine and Lifesciences, Maastricht University
Vice-president Executive Board, Maastricht University Medical Centre
Maastricht University
Maastricht, The Netherlands

Susan A. Lanham-New
Professor of Human Nutrition & Head, Nutritional Sciences Department
School of Biosciences & Medicine, Faculty of Health & Medical Sciences
University of Surrey
Guildford, UK

This edition first published 2024

Edition History
1e, 2002; 2e, 2009 The Nutrition Society

Registered Offices
John Wiley & Sons, Inc., 111 River Street, Hoboken, NJ 07030, USA
John Wiley & Sons Ltd, The Atrium, Southern Gate, Chichester, West Sussex, PO19 8SQ, UK

Editorial Office
9600 Garsington Road, Oxford, OX4 2DQ, UK

For details of our global editorial offices, customer services, and more information about Wiley products visit us at www.wiley.com.

Wiley also publishes its books in a variety of electronic formats and by print-on-demand. Some content that appears in standard print versions of this book may not be available in other formats.

Library of Congress Cataloging-in-Publication Data
Names: Roche, Helen M., editor. | MacDonald, Ian A., editor. | Schols, Annemie M.W.J., editor. | Lanham-New, Susan A., editor.
Title: Nutrition and metabolism / edited on behalf of The Nutrition Society by Helen M. Roche, Full Professor of Nutrigenomics, Director, UCD Conway Institute, University College Dublin, Dublin, Ireland, Ian A. Macdonald, Emeritus Professor of Metabolic Physiology, School of Life Sciences, Faculty of Medicine and Health Sciences, University of Nottingham, Nottingham, UK, Annemie M.W.J. Schols, Professor of Nutrition and Metabolism in chronic diseases, Dean, Faculty of Health, Medicine and Lifesciences, Maastricht University, Vice-president Executive Board, Maastricht University Medical Centre, Maastricht University, Maastricht, The Netherlands, Susan A. Lanham-New, Professor of Human Nutrition & Head, Nutritional Sciences Department, School of Biosciences & Medicine, Faculty of Health & Medical Sciences, University of Surrey, Guildford, UK.
Description: Third edition. | Hoboken, NJ, USA : John Wiley & Sons Ltd, 2024. | Series: The nutrition society textbook series | Revised edition of: Nutrition and metabolism / edited on behalf of the Nutrition Society by Susan A. Lanham-New, Helen M. Roche, Ian MacDonald. 2nd ed. 2011. | Includes bibliographical references and index.
Identifiers: LCCN 2022054800 (print) | LCCN 2022054801 (ebook) | ISBN 9781119840237 (Paperback) | ISBN 9781119840244 (ePDF) | ISBN 9781119840251 (ePUB)
Subjects: LCSH: Nutrition. | Metabolism.
Classification: LCC QP141 .N7768 2023 (print) | LCC QP141 (ebook) | DDC 613.2--dc23/eng/20221215
LC record available at https://lccn.loc.gov/2022054800
LC ebook record available at https://lccn.loc.gov/2022054801

Cover Design: Wiley
Cover Images: © jxfzsy/Getty Images

Set in 10/12pt Minion by 9.5/12.5pt STIXTwoText by Integra Software Services Pvt. Ltd, Pondicherry, India
SKY10081853_081324

Contents

Contributors

Abayomi O. Akanji
Quinnipiac University, USA

Ellen E. Blaak
Maastricht University, the Netherlands

Nicholas A. Burd
University of Illinois at Urbana-Champaign, USA

Louise M. Burke
Australian Catholic University, Australia

Philip C. Calder
University of Southampton, UK

Tyler A. Churchward-Venne
McGill University, Quebec, Canada

Emilio Martínez de Victoria
Universidad de Granada, Spain

Connor M. Delahunty
CSIRO, Canberra, Austrailia

Emily J. Ferguson
Queen's University, Canada

Graham Finlayson
University of Leeds, UK

Albert Flynn
University College Cork, Ireland

Keith N. Frayn
University of Oxford, UK

Ángel Gil
Universidad de Granada, Spain

Gijs H. Goossens
Maastricht University, the Netherlands

Wendy L. Hall
King's College London, UK

Remco C. Havermans
Maastricht University, the Netherlands

Phoebe Hodges
Queen Mary University of London, UK

Mark Hopkins
University of Leeds, UK

Asker E. Jeukendrup
Loughborough University, UK

Paul Kelly
Queen Mary University of London, UK

Mara Loana Iesanu
Carol Davila University of Medicine and Pharmacy, Romania

J.A. Lovegrove
University of Reading, UK

Ansuyah Magan
University of the Witwatersrand, Johannesburg, South Africa

Mariano Mañas
Universidad de Granada, Spain

John C. Mathers
University of Newcastle, UK

Chris McGlory
Queen's University, Canada

Michelle K. McGuire
University of Idaho, USA

Ronald P. Mensink
Maastricht University, the Netherlands

Jennifer L. Miles-Chan
University of Auckland – Waipapa Taumata Rau, New Zealand

Siobhain M. O'Mahony
University College Cork, Ireland

Sharleen L. O'Reilly
University College Dublin, Ireland

John M. Pettifor
University of the Witwatersrand, New Zealand

Herman E. Popeijus
Maastricht University, the Netherlands

Ann Prentice
University of Cambridge, UK

Harriët Schellekens
University College Cork, Ireland

Jeremy P.E. Spencer
University of Reading, UK

Mark L. Wahlqvist
Monash University, Australia

Kate A. Ward
University of Southampton, UK

Gary Williamson
Monash University, Australia

Jayne V. Woodside
Queen's University Belfast, UK

María D. Yago
Universidad de Granada, Spain

Parveen Yaqoob
University of Reading, UK

Series Foreword – Nutrition and Metabolism 3rd Edition

In 1941, a group of leading physiologists, biochemists and medical scientists recognised that the emerging discipline of nutrition needed its own Learned Society and The Nutrition Society was established. The original mission remains *"to advance the scientific study of nutrition and its application to the maintenance of human and animal health"*. The Society is now one of the world's largest Learned Societies for nutrition, with over 2,500 members. You can learn more about it and how to become a member by visiting www.nutritionsociety.org. The Society's first journal, *The Proceedings of the Nutrition Society*, published in 1944, records the scientific presentations made at the time. Shortly afterwards, 1947, scientists worldwide established the *British Journal of Nutrition* to publish primary research on human and animal nutrition. Recognising the needs of students and their teachers for authoritative reviews on topical issues in nutrition, the Society began publishing *Nutrition Research Reviews* in 1988. *Public Health Nutrition*, the first international journal dedicated to this critical and growing area, was launched in 1998 and made open access in January 2022. The Society's first open-access journal, the *Journal of Nutritional Science*, was launched in 2012. The Society is constantly evolving in response to emerging areas of nutritional science. It has most recently launched the journal *Gut Microbiome*, an open-access journal published in partnership with Cambridge University Press.

Now 25 years old, the Nutrition Society Textbook Series was first established by Professor Michael Gibney (University College Dublin) in 1998. Under the direction of the second Editor-in-Chief, Professor Susan Lanham-New (University of Surrey), the Series continues to be extraordinarily successful for the Society. This series of Nutrition textbooks is designed for use worldwide. This has been achieved by translating the Textbook Series into many different languages, including Spanish, Greek, Portuguese, Italian and Indonesian. The success of the Textbook Series is a tribute to the quality of the authorship and the value placed on them in the UK and Worldwide as a core educational tool and a resource for practitioners. *Nutrition and Metabolism* is an important textbook, with more than 10,000 copies sold to date and with it now transcending into its 3rd Edition.

Therefore, writing the Foreword for this book gives me great pleasure. As President of the Royal College of General Practitioners, I know how important it is to understand the scientific basics of nutrition in the context of a systems and health approach. This Textbook brings together science and the practical application of methodologies in nutrition and is a most valuable resource to all those working in the field.

Professor Dame Clare Gerada, DBE, PRCGP FRCPsych FRCP (Hons) FRCGP

President of the Royal College of General Practitioners

Series Editor's Preface

I am absolutely delighted in my capacity as Editor-in-Chief (E-i-C) of the Nutrition Society Textbook Series to introduce the 3rd Edition of *Nutrition and Metabolism (N&M3e)*. The production of this Third Edition represents a significant milestone for the Textbook Series, given that it is now exactly 25 years on since the production of the first textbook, *Introduction to Human Nutrition*, and 22 years since the production of the 1st Edition of N&M.

The Team of *Nutrition and Metabolism* 3rd Edition, namely Professor Helen Roche (University College Dublin), Professor Ian Macdonald (University of Nottingham) and Professor Annemie Schols (Maastricht University) have been meticulous in ensuring that each chapter is updated & accurate, and to ensuring that new aspects of N&M3e are also brought into the book. N&M3e comprises of a total of 19 chapters, each with their own unique summary of the take home messages. How indebted we are to have so many experts in the field who have written chapters to make N&M3e a complete and thorough review of the area of Metabolic Science - a must read!

N&M3e is intended for those with an interest in nutrition and metabolic science whether they are nutritionists, food scientists, dietitians, medics, nursing staff or other allied health professionals. We hope that both undergraduate and postgraduate students will find the book of great help with their respective studies and that the book will really put nutrition and metabolic science as a *discipline* into context.

It is a great honour for our 3rd Edition of N&M to have the Foreword written by Professor Dame Clare Gerada, DBE, PRCGP FRCPsych FRCP (Hons) FRCGP, and we are most grateful for her support of our work at the Society, particularly in her role as President of the Royal College of General Practitioners. We are also most grateful to the following individuals for their support and most generous Forewords in *Introduction to Human Nutrition3e, Public Health Nutrition2e, Sport and Exercise Nutrition1e, Clinical Nutrition2e* and *Nutrition Research Methodologies1e*; namely – the late The Earl of Selbourne, Her Royal Highness The Princess Royal; Professor Richard Budgett OBE, Chief Medical Officer for the London 2012 Olympic and Paralympic Games and now Medical and Scientific Director at the International Olympic Committee (IOC); Dame Sally Davies, former Chief Medical Officer (CMO) for England, and the UK Government's Principal Medical Adviser; Professor Lord John Krebs, Principal, Jesus College, University of Oxford and our first Chairman of the UK Food Standards Agency. We are now planning ahead with respect to the production of the 3rd Edition of *Clinical Nutrition* and 2nd Edition of *Sport and Exercise Nutrition* as well as bringing a *seventh* book to the Textbook Series, *Animal Nutrition*, published in collaboration with the British Society of Animal Sciences.

The Society is most grateful to the Textbook publishers, Wiley-Blackwell, for their continued help with the production of the textbook and in particular: Tom Marriott - Commissioning Editor; Charlie Hamlyn and Vallikannu Narayanan - Managing Editors; Durgadevi Shanmugasundaram - Content Refinement Specialist. In addition, I would like to acknowledge formally my great personal appreciation to: Professor G.Q. Max Lu AO, DL, FREng, FAA, FTSE, FIChemE, FRSC, FCAS, FNAI, Vice-Chancellor & President of the University of Surrey; Professor Tim Dunne, Provost & Executive Vice-President of the University of Surrey; Professor Paul Townsend, Pro-Vice-Chancellor and Executive Dean of the Faculty of Health and Medical Sciences and Professor Roberto La Ragione BSc (Hons) MSc PhD FRSB CBiol FIBMS CSci AECVM FRCPath HonAssocRCVS, Head of the School of Biosciences, University of Surrey, for their respective great encouragement of the nutritional sciences field in general, especially in light of Surrey's success in the 2017/2018 Queen's Anniversary Prize for our work in *Food and Nutrition for Health*, and for their support of the Textbook Series production in particular.

Sincerest appreciation indeed to the Nutrition Society President, Professor Mary Ward (Ulster

University), and our past President, Professor Julie Lovegrove (University of Reading) for their great support and belief in the Textbook Series. With special thanks to Honorary Publications Officer Professor Jayne Woodside (Queen's University Belfast) for her support of the Textbook Series and for being such a great sounding board, and to past-Honorary Publications Officer, Professor Paul Trayhurn for his wise counsel to me during the six years we worked together on the Textbooks. And finally an enormous thank you indeed to: Mark Hollingsworth MBA, FInstLM, Chief Executive Officer of the Nutrition Society for his unstinting support of the Textbook Series, to Cassandra Ellis, Science Director and Deputy Editor, for her pivotal continued contribution to the development of the Textbook Series, and to Caroline Roberts, (formerly Science Communications Manager), whose excellent support made this latest edition possible.

Finally, as I always write and mean with absolute sincerity, how the Textbook Series is indebted to the forward thinking focus that Professor Michael Gibney (University College Dublin) had at that time of the Textbook Series development. It shall remain for always, such a tremendous privilege for me to continue to follow in his footsteps as the second E-i-C.

I really hope that you will find the textbook a great resource of information and inspiration ......please enjoy, and with so many grateful thanks to all those who made it happen!

With my warmest of wishes indeed

Professor Susan A. Lanham-New RNutr, FAfN
E-i-C, Nutrition Society Textbook Series
Professor of Human Nutrition and Head, Department of Nutrition, Food & Exercise Sciences
School of Biosciences,
Faculty of Health and Medical Sciences
University of Surrey

About the Companion Website

www.wiley.com/go/nutrition/metabolism3e

- Multiple choice questions
- Short answer questions
- Essay questions

1 Core Concepts of Nutrition

Ian A. Macdonald and Annemie M.W.J. Schols

Key messages

- The change in body reserves or stores of a nutrient is the difference between the intake of that nutrient and the body's utilisation of that nutrient. The time-frame necessary to assess the body's balance of a particular nutrient varies from one nutrient to another.
- The concept of turnover can be applied at various levels within the body (molecular, cellular, tissue/organs, whole body).
- The flux of a nutrient through a metabolic pathway is a measure of the rate of activity of the pathway. Flux is not necessarily related to the size of the pool or pathway through which the nutrient or metabolite flows.
- Nutrients and metabolites are present in several pools in the body. The size of these metabolic pools varies substantially for different nutrients/metabolites, and a knowledge of how these pools are interconnected greatly helps us to understand nutrition and metabolism in health and disease.
- The Darwinian theory of evolution implies a capacity to adapt to adverse conditions, including adverse dietary conditions. Many such examples can be cited. Some allow for long-term adaptation and others buy time until better conditions arrive.

1.1 Introduction

Nutrition and metabolism are addressed in this textbook in an integrated fashion. Thus, rather than considering nutrients separately, this book brings together information on macronutrients, energy and substrate metabolism in relation to specific nutritional or disease states or topics (e.g. undernutrition, overnutrition, cardiovascular disease). Before considering these topics in detail, it is necessary to outline the core concepts of nutritional metabolism. The core concepts covered in this chapter are nutrient balance, turnover and flux, metabolic pools and adaptation to altered nutrient supply.

1.2 Balance

As discussed in Chapters 3 and 4, nutrient balance must be considered separately from the concepts of metabolic equilibrium or steady state. In this chapter, the concept of balance is considered in the context of the classic meaning of that term, i.e. the long-term sum of all the forces of metabolic equilibrium for a given nutrient.

The concept of nutrient balance essentially restates the law of conservation of mass in terms of nutrient exchange in the body. It has become common practice to refer to the content of the nutrient within the body as a 'store' but in many cases this is not appropriate and the term 'reserves' is better. Thus, the idea of nutrient balance is summarised by the equation:

$$\begin{bmatrix}\text{nutrient}\\ \text{intake}\end{bmatrix} - \begin{bmatrix}\text{nutrient}\\ \text{utilisation}\end{bmatrix} = \begin{bmatrix}\text{change in body}\\ \text{nutrient reserves}\end{bmatrix}$$

The above equation can have three outcomes:

- *Zero balance* (or nutrient balance): Intake matches utilisation and reserves remain constant
- *Positive balance* (or positive imbalance): Intake exceeds utilisation and reserves expand

Nutrition and Metabolism, Third Edition. Edited on behalf of The Nutrition Society by Helen M. Roche, Ian A. Macdonald, Annemie M.W.J. Schols and Susan A. Lanham-New.

Companion Website: www.wiley.com/go/nutrition/metabolism3e

- *Negative balance* (or negative imbalance): Utilisation exceeds intake and reserves become depleted.

In relation to macronutrient metabolism, the concept of balance is most often applied to protein (nitrogen) and to energy. However, many research studies now subdivide energy into the three macronutrients and consider fat, carbohydrate and protein balance separately. This separation of the macronutrients is valuable in conditions of altered dietary composition (e.g. low-carbohydrate diets) where a state of energy balance might exist over a few days but be the result of negative carbohydrate balance (using the body's glycogen reserves to satisfy the brain's requirement for glucose) matched in energy terms by positive fat balance.

Balance is a function not only of nutrient intake but also of metabolic requirements and metabolically elevated requirements. Fat balance is generally driven by periods where energy intake exceeds energy expenditure (positive energy balance) and by periods when intakes are maintained below energy expenditure, such as in dieting, acute and chronic disease, hypoxia (negative energy balance). However, nutrient balance can also be driven by metabolic regulators through hormones or cytokines. For example, the dominance of growth hormone during childhood ensures positive energy and nutrient balance. In pregnancy, a wide range of hormones lead to a positive balance of all nutrients in the overall placental, fetal and maternal tissues, although this may be associated with a redistribution of some nutrient reserves from the mother to the fetus (see Chapter 6). By contrast, severe trauma or illness will dramatically induce energy and protein losses, an event due to elevated metabolic requirements unrelated to eating patterns.

Balance is not something to be thought of in the short term. Following each meal, there is either storage of absorbed nutrients (triacylglycerol [TAG] in adipose tissue or glucose in glycogen) or a cessation of nutrient losses (breakdown of stored TAG to non-esterified fatty acids or amino acid conversion to glucose via gluconeogenesis). As the period of postprandial metabolism is extended, the recently stored nutrients are drawn upon and the catabolic state commences again. This is best reflected in the high glucagon to insulin ratio in the fasted state before the meal and the opposite high insulin to glucagon ratio during the meal and immediate postprandial period. However, when balance is measured over a sufficient period, which varies from nutrient to nutrient, a stable pattern can be seen: zero, positive or negative (Figure 1.1).

It is critically important with respect to both obesity and malnutrition that the concept of balance is correctly considered. While at some stage energy balance must have been positive to reach an overweight or obese stage, once attained most people sustain a stable weight over quite long periods.

In the context of the present chapter, it is worth reflecting on the reasons why the time taken to assess energy balance correctly varies for different nutrients. Because of differences in capacity and mobilisation as summarised below and explained in Chapters 4 and 12, calcium

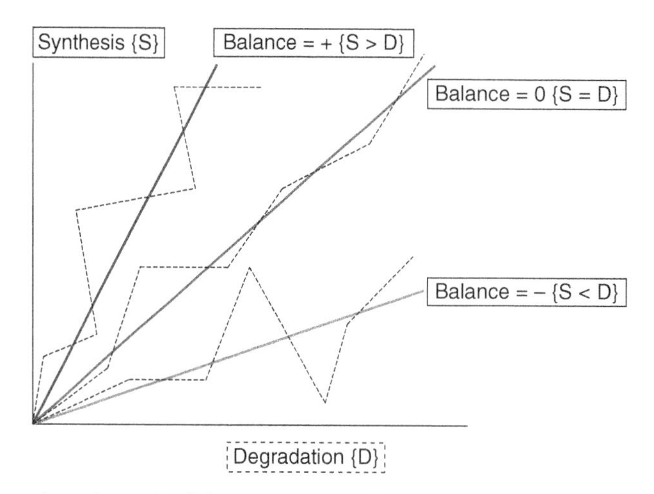

Figure 1.1 Positive, zero and negative nutrient balance over time with fluctuations upwards and downwards within that time.

balance, for example, will require months of equilibrium while fat balance could be equilibrated in days or at most a few weeks.

Fat and adipose tissue (Chapter 4)

- There is a very large capacity to vary the body's pool of adipose tissue. One can double or halve the level of the fat reserves in the body.
- The capacity to vary the level of TAG in blood en route to adipose tissue can vary.
- Almost all of the TAG reserves in adipose tissue are exchangeable.

Calcium and bone (Chapter 12)

- The human being must maintain a large skeleton as the scaffold on which the musculature and organs are held.
- There is a very strict limit to the level of calcium that can be transported in blood. Excess or insufficient plasma calcium levels influence neural function and muscle function, since calcium is a major component of muscle and nerve function.
- Only a small fraction (the miscible pool) of bone is available for movement into plasma.

1.3 Turnover

Although the composition of the body and of the constituents of the blood may appear constant, the component parts are not static. In fact, most metabolic substrates are continually being utilised and replaced (i.e. they turn over). This process of turnover is well illustrated by considering protein metabolism in the body. Daily adult dietary protein intakes are in the region of 50–100 g and the rates of urinary excretion of nitrogen match the protein intake. However, isotopically derived rates of protein degradation indicate that approximately 350 g is broken down per day. This is matched by an equivalent amount of protein synthesis per day, with most of this synthesis representing turnover of substrate (i.e. degradation and resynthesis) rather than being derived de novo from dietary protein (see Chapter 5).

Similar metabolic turnover occurs with other nutrients; glucose is a good example, with a relatively constant blood glucose concentration arising from a match between production by the liver and utilisation by the glucose-dependent metabolic tissues (see Chapter 4).

The concept of turnover can be applied at various levels within the body (molecular, cellular, tissue/organs, whole body). Thus, within a cell, the concentration of adenosine triphosphate (ATP) remains relatively constant, with utilisation being matched by synthesis. Within most tissues and organs, there is a continuous turnover of cells, with degradation and death of some cells matched by the production of new ones. Some cells, such as red blood cells, have a long lifespan (c. 120 days) while others, such as platelets, turn over in a matter of 1–2 days. In the case of proteins, those with very short half-lives have amino acid sequences that favour rapid proteolysis by the range of enzymes designed to hydrolyse proteins. Equally, those with longer half-lives have a more proteolytic-resistant structure.

A major advantage of this process of turnover is that the body is able to respond rapidly to a change in metabolic state by altering both synthesis and degradation to achieve the necessary response. One consequence of this turnover is the high energy cost of continuing synthesis. There is also the potential for nutrient imbalance and metabolic dysfunction if the rates of synthesis and degradation do not match.

The consequences of a reduction in substrate synthesis will vary between the nutrients, depending on the half-life of the nutrient. The half-life is the time taken for half of the substrate to be used up, and is dependent on the rate of utilisation of the nutrient. Thus, if synthesis of a nutrient with a short half-life is stopped, the level of that nutrient will fall quickly. By contrast, a nutrient with a long half-life will disappear more slowly.

Since proteins have the most complex structures undergoing very significant turnover, it is worth dwelling on the mechanism of this turnover. Synthesis is fairly straightforward. Each protein has its own gene and the extent to which that gene is expressed will vary according to metabolic needs. In contrast to synthesis, a reasonably small array of lysosomal enzymes is responsible for protein degradation.

1.4 Flux

The flux of a nutrient through a metabolic pathway is a measure of the activity of the pathway. If one considers the flux of glucose from the blood to the tissues, the rate of utilisation is approximately 2 mg/kg body weight per minute at rest.

However, this does not normally lead to a fall in blood glucose because it is balanced by an equivalent rate of glucose production by the liver, so the net flux is zero. This concept of flux can be applied at the cellular, tissue/organ or whole-body level, and can also relate to the conversion of one substrate/nutrient to another (i.e. the movement between metabolic pathways).

Flux is not necessarily related to the size of the pool or pathway through which the nutrient or metabolite flows. For example, the membrane of a cell will have several phospholipids present and each will have some level of arachidonic acid. The rate at which arachidonic acid enters one of the phospholipid pools and exits from that phospholipid pool is often higher in the smaller pools.

1.5 Metabolic pools

An important aspect of metabolism is that the nutrients and metabolites are present in several pools in the body (Figure 1.2). At the simplest level, for a given metabolite there are three pools, which will be illustrated using the role of dietary essential fatty acids in eicosanoid synthesis.

In the *functional pool*, the nutrient/metabolite has a direct involvement in one or more bodily functions. In the chosen example, intracellular free arachidonic acid, released from membrane-bound stores on stimulation with some extracellular signal, is the functional pool. It will be acted on by the key enzyme in eicosanoid synthesis, cyclo-oxygenase.

The *storage pool* provides a buffer of material that can be made available for the functional pool when required. Membrane phospholipids store arachidonic acid in the sn-2 position at quite high concentrations, simply to release this fatty acid when prostaglandin synthesis is needed. In the case of platelets, the eicosanoid thromboxane A_2 is synthesised from arachidonic acid released into the cytoplasm by stimuli such as collagen.

The *precursor pool* provides the substrate from which the nutrient/metabolite can be synthesised. Linoleic acid represents a good example of a precursor pool. It is elongated and desaturated in the liver to yield arachidonic acid. Thus, the hepatic pool of linoleic acid is the precursor pool in this regard.

The precursor, storage and functional pool model does not apply to all nutrients outlined above. The essential nutrients and the minerals and trace elements do not have a precursor pool. Nevertheless, no nutrient exists in a single homogeneous pool and an awareness of the existence of metabolic pools is essential to understand human metabolism. For example, one might expect that a fasted individual would show a fall in all essential nutrient levels in the plasma pool. In many instances, this is not the case initially because of the existence of storage pools, such as liver stores of iron or vitamin A. In the case of folic acid, fasting causes a rise in blood folic acid levels and this is explained by the concept of metabolic pools. A considerable amount of folic acid enters the gut via the bile duct and is reabsorbed further down the digestive tract. Thus, there is an equilibrium between the blood folate pool and the gut folate pool. Fasting stops gall bladder contraction and thus the flow of folate to the gut, and hence folate is redistributed from one pool to another.

Another example of how an awareness of metabolic pools helps us to understand nutrition and metabolism is the intracellular free amino acid pool. This is the functional pool from which protein is synthesised. As this pool is depleted in the process of protein synthesis, it must be repleted, otherwise protein synthesis stops. Moreover, it is not just the intracellular pool of amino acids that matters but the intracellular pool of essential amino acids or, more precisely, the intracellular pool of the most limiting essential amino acid. Calculations show that if the pool of the most limiting amino acid in mammalian cells was not

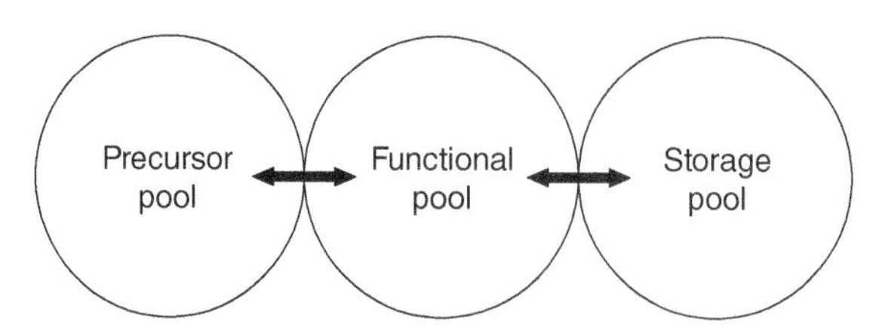

Figure 1.2 The pools in the body in which nutrients and metabolites may exist.

replenished, protein synthesis would cease in under one hour. This highlights the need to transfer the limiting amino acid across the cell membrane, which raises the question of how that pool is repleted. Effectively, it can only be repleted if there is a comparable rate of protein degradation to provide the key amino acid, assuming the balance is zero. Thus, there are links between the protein pool of amino acids and the extra- and intracellular pools of amino acids.

The size of these pools varies substantially for different nutrients and metabolites. When studying the activities of metabolic processes within the body, it is often necessary to measure or estimate the size of the various pools in order to derive quantitative information about the overall rates of the processes. In addition, the actual situation may be more complex than the simple three-pool model described above.

Nutritional assessment often involves some biochemical assessment of nutritional status. Blood is frequently the pool that is sampled and even there, blood can be separated into:

- Erythrocytes, which have a long lifespan and are frequently used to assess folic acid status
- Cells of the immune system, which can be used to measure zinc or ascorbic acid status
- Plasma, which is used to ascertain the levels of many biomarkers
- Fractions of plasma, such as cholesteryl esters used to ascertain long-term intake of polyunsaturated fatty acids.

In addition to sampling blood, nutritionists may take muscle or adipose tissue biopsies, or samples of saliva, buccal cells, hair and even toenails. A knowledge of how a nutrient behaves in different metabolic pools is critically important in choosing the correct tissue to sample to measure or judge nutritional status. For example, the level of folic acid in plasma is determined by the most recent intake pattern and thus is subject to considerable fluctuation. However, since erythrocytes remain in the circulation for about 120 days, a sample of erythrocytes will represent very recently synthesised cells right through to erythrocytes ready for recycling through the turnover mechanism previously described. As erythrocytes do not have a nucleus, they cannot switch on genes that might influence folate levels, and so the cell retains the level of folate that prevailed at the time of synthesis. Thus, erythrocyte folate is a good marker of long-term intake. The free form of many minerals and trace elements is potentially toxic, and for this reason their level in the plasma is strictly regulated. Hence, blood levels are not used to assess long-term intake of selenium, but toenail clippings can be used.

1.6 Adaptation to altered nutrient supply

In many circumstances, the body can respond to altered metabolic and nutritional states in order to minimise the consequences of such alterations. For example, the brain has an obligatory requirement for glucose as a substrate for energy and it accounts for a significant part of resting energy expenditure. During undernutrition, where glucose input does not match glucose needs, the first adaptation to the altered metabolic environment is to increase the process of gluconeogenesis, which involves the diversion of amino acids into glucose synthesis. That means less amino acid entering the protein synthesis cycle of protein turnover. Inevitably, protein reserves begin to fall. Thus, two further adaptations are made. The first is that the brain begins to use less glucose for energy (replacing it by ketones as an alternative metabolic fuel). The second is that overall, resting energy expenditure falls to help sustain a new balance if possible. Stunting in infants and children, reflected in a low height for age, can be regarded as an example of successful adaptation to chronic low energy intake. If the period of energy deprivation is not too long, the child will subsequently exhibit a period of accelerated or catch-up growth (see Chapter 7). If it is protracted, the stunting will lead to a permanent reprogramming of genetic balance.

In many instances, the rate of absorption of nutrients may be enhanced as an adaptive mechanism to low intakes. Some adaptations appear to be unsuccessful but work for a period, effectively buying time in the hope that normal intakes will be resumed. In essential fatty acid deficiency, the normal processes of elongation and desaturation of fatty acids take place but the emphasis is on the wrong fatty acid, that is, the non-essential 18-carbon monounsaturated fatty acid (oleic acid, C18:1 n-9) rather than the deficient dietary essential 18-carbon polyunsaturated fatty acid (linoleic acid, C18:2 n-6). The resultant

20-carbon fatty acid does not produce a functional eicosanoid. However, the body has significant reserves of linoleic acid which are also used for eicosanoid synthesis and so the machinery of this synthesis operates at a lower efficiency than normal. Eventually, if the dietary deficiency continues then pathological consequences ensue. In effect, adaptation to adverse metabolic and nutritional circumstances is a feature of survival until the crisis abates. The greater the capacity to mount adaptations to adverse nutritional circumstances, the greater the capacity to survive.

1.7 Perspectives on the future

These core concepts of nutrition will remain forever, but they will be refined by the emerging concept of nutrigenomics (see Chapter 2). We will develop a greater understanding of how changes in the nutrient content of one pool will alter gene expression to influence events in another pool and how this influences the flux of nutrients between pools. We will better understand how common single nucleotide polymorphisms will determine the level of nutrient intake to achieve nutrient balance in different individuals.

There is growing interest in the role of the gut microbiome in promoting good health in people. Care is needed in the use of only the term 'microbiome' as it covers a wide variety of different locations in the body (e.g. skin, respiratory system, genitourinary system, different stages of the gastrointestinal tract) and not just the large intestine. The potential for the composition of the microbiome to be influenced by the diet has generated great interest in identifying dietary components that may improve or diminish gut health, and in turn have broader effects on general health. The possibility that manipulation of the gut microbiome may influence signalling in the gut–brain axis is likely to be a major topic of investigation going forward, and the influence of dietary components on the composition of the microbiome as well as providing the substrates for bacteria could be major aspects of nutrition and health to be incorporated into the core concepts in the future.

A final topic to consider is the issue of personalisation of diet and whether it is a valid biological concept which requires much greater flexibility in the food supply. Clearly, the genomic differences and variability in epigenetic changes between people imply potential differences in nutrient requirements. The possibility that the gut microbiome varies between people is likely to add to these differences but whether it is possible to truly personalise the diet at the level of the individual based on these factors is likely to need a substantial amount of research in the future.

Further reading

Frayn, K.N. (2003). Metabolic Regulation: A Human Perspective, 2e. Oxford: Blackwell Publishing.

Websites

health.nih.gov/search.asp?category_id=29
http://themedicalbiochemistrypage.org
www.nlm.nih.gov/medlineplus/foodnutritionandmetabolism.html

2
Molecular Aspects of Nutrition

Herman E. Popeijus, Helen M. Roche, and Ronald P. Mensink

Key messages

- The genome forms the information or blueprint to build up an organism and contains the full complement of genes (genotype). An individual's genotype interacts with the environment (of which nutrition is one element), which are collectively expressed as an individual's phenotype.
- The specific order of nucleotides within DNA forms the basic units of genetic information. It is organised into chromosomes. Typically, every human cell contains 46 chromosomes (or 23 chromosome pairs). The chromosomes are a series of individual genes, wherein diploid cells contain two copies of each.
- Genetic variation reflects DNA sequence alterations, due to random mating or damage. This variation is referred to as genetic polymorphisms or mutations. Genetic polymorphisms are common forms of genetic heterogeneity, whereby different forms of a given allele are common in a population. Genetic mutations refer to more uncommon alterations.
- Transcription (or gene expression) refers to the process whereby information encoded in the genes is converted into mRNA sequence. Translation refers to converting mRNA sequence information into a protein, which may in turn affect metabolism. Some, but not all, transcriptional and translational elements are converted into observable phenotypes.
- There are several tools used to investigate molecular aspects of nutrition: animal models, organoids, cell/tissue culture models, molecular cloning, CRISPR/Cas9, RNAi, gene expression analysis (polymerase chain reaction [PCR], DNA microarrays and sequencing), proteomics, stable isotopes, and metabolomics.
- Genetic variability is relatively simple to characterise. Understanding the interaction with the environment, of which food is one complex element, is the challenge. The resultant gene–environment interactions can determine nutrient requirements, the metabolic response to food and nutrient intake and/or susceptibility to diet-related diseases.
- Nutrients and non-nutrient food components can interact with the genome to modulate gene, protein and metabolite expression. These interactions between nutrition and the genome are referred to as *molecular nutrition* or *nutrigenomics*.
- Personalised nutrition seeks to understand if/how food or nutrient intake could be used to manipulate an individual's metabolic response and/or to reduce their predisposition to diet-related diseases, based on their genetic and/or metabolic phenotype.

2.1 Introduction

Our genes determine every characteristic of life: gender, physical characteristics, metabolic functions, life stage and responses to external or environmental factors, which include nutrition. Nutrients have the ability to interact with the human genome to alter gene, protein and metabolite expression, which in turn can affect normal growth, health and disease. The Human Genome Project has provided an enormous amount of genetic information and thus a greater understanding of our genetic background. It is well known that nutrients, and non-nutritive food components, can interact with the genome. This aspect of nutritional science is known as *molecular nutrition or nutrigenomics*.

Molecular nutrition looks at the relationship between the human genome and nutrition from two perspectives. First, the genome determines every individual's genotype (or genetic background), which in turn can determine their

Nutrition and Metabolism, Third Edition. Edited on behalf of The Nutrition Society by Helen M. Roche, Ian A. Macdonald, Annemie M.W.J. Schols and Susan A. Lanham-New.

Companion Website: www.wiley.com/go/nutrition/metabolism3e

nutrient state, metabolic response and/or genetic predisposition to diet-related diseases. Second, nutrients can interact with the genome and alter gene, protein or metabolite expression. Gene expression (or transcription) is only the first stage of the whole-body or metabolic response to a nutrient and a number of post-translational events (e.g. enzyme activity, protein half-life, co-activators, co-repressors), but metabolomic events can also modify the ability of nutrients to alter an individual's phenotype.

This chapter will review the core concepts in molecular biology, introduce the genome and discuss how we can characterise the effect of nutrition on gene, protein and metabolite expression using state-of-the-art transcriptomic, proteomic and metabolomic technologies, identifying some important research tools used to investigate these molecular aspects of nutrition, such as characterising how genetic background can determine nutrition and health. Chapter 13 'Application of "Omics" Technology', in the Nutrition Research Methodologies textbook) provides detailed information in relation to each technology. This chapter provides some examples of how nutrients regulate gene, protein and metabolite expression, within the context of human nutrition and metabolism.

Overall, the principal aim of nutrigenomics/molecular nutrition is to understand how the genome interacts with food, nutrients and non-nutrient food components, within the context of diet-related diseases. It attempts to determine nutrients that enhance the expression of gene, protein and metabolic pathways/networks that are associated with health and suppress those that predispose to disease. While it is unrealistic to assume that food intake and good nutrition can overcome our genetic fate, good nutrition can improve health and quality of life. Therefore, it is essential that we extend our understanding of the molecular interplay between the genome, food and nutrients to facilitate a greater understanding of the mechanistic relationship between diet, health and disease.

2.2 Core concepts in molecular biology

The genome, DNA, and the genetic code

The *genome* refers to the total genetic information carried by a cell or organism. In simple terms, the genome (or DNA sequence) contains the full complement of genes. The expression of each gene leads in general to the formation of a protein which, together with many other proteins coded by other genes, forms tissues, organs and systems, to constitute the whole organism. In complex multicellular organisms, the information carried within the genome gives rise to multiple tissues (muscle, bone, adipose tissue, etc.).

The characteristics of each cell type and tissue are dependent on differential gene expression by the genome, whereby only those genes are expressed that code for specific proteins to confer the individual characteristics of the cells that constitute each organ. For example, gene expression in muscle cells results in the formation of muscle-specific proteins, that are critical for the differentiation, development and maintenance of muscle tissue. These genes are completely different from those expressed in osteoblasts, osteoclasts and osteocytes, which form bone. These differentially expressed proteins can have a wide variety of functions: as structural components of the cell or as regulatory proteins, including enzymes, hormones, receptors and intracellular signalling proteins that confer tissue specificity.

It is very important to understand the molecular basis of cellular metabolism because incorrect expression of genes at the cellular level can disrupt whole-body metabolism and lead to disease. Aberrant gene expression can lead to cellular disease when proteins are produced in the wrong place, at the wrong time, at abnormal levels or as a malfunctioning isoform that can compromise whole-body health. Furthermore, different nutritional states and intervention therapies can modulate the expression of cellular genes and thereby the formation of proteins. The ultimate goal of molecular nutrition is to understand how nutrients interact with the genome to alter the expression of genes, determine the formation and function of proteins and modulate metabolite profiles, all of which play a role in health and disease.

Deoxyribonucleic acid (DNA) is the most basic unit of genetic information, as the DNA sequence codes for the amino acids that form cellular proteins. Two individual DNA molecules are packaged as the chromosomes within the nucleus of animal and plant cells. The basic structure and composition of DNA are illustrated in Figure 2.1. DNA is composed of large polymers, with a linear backbone composed of residues of

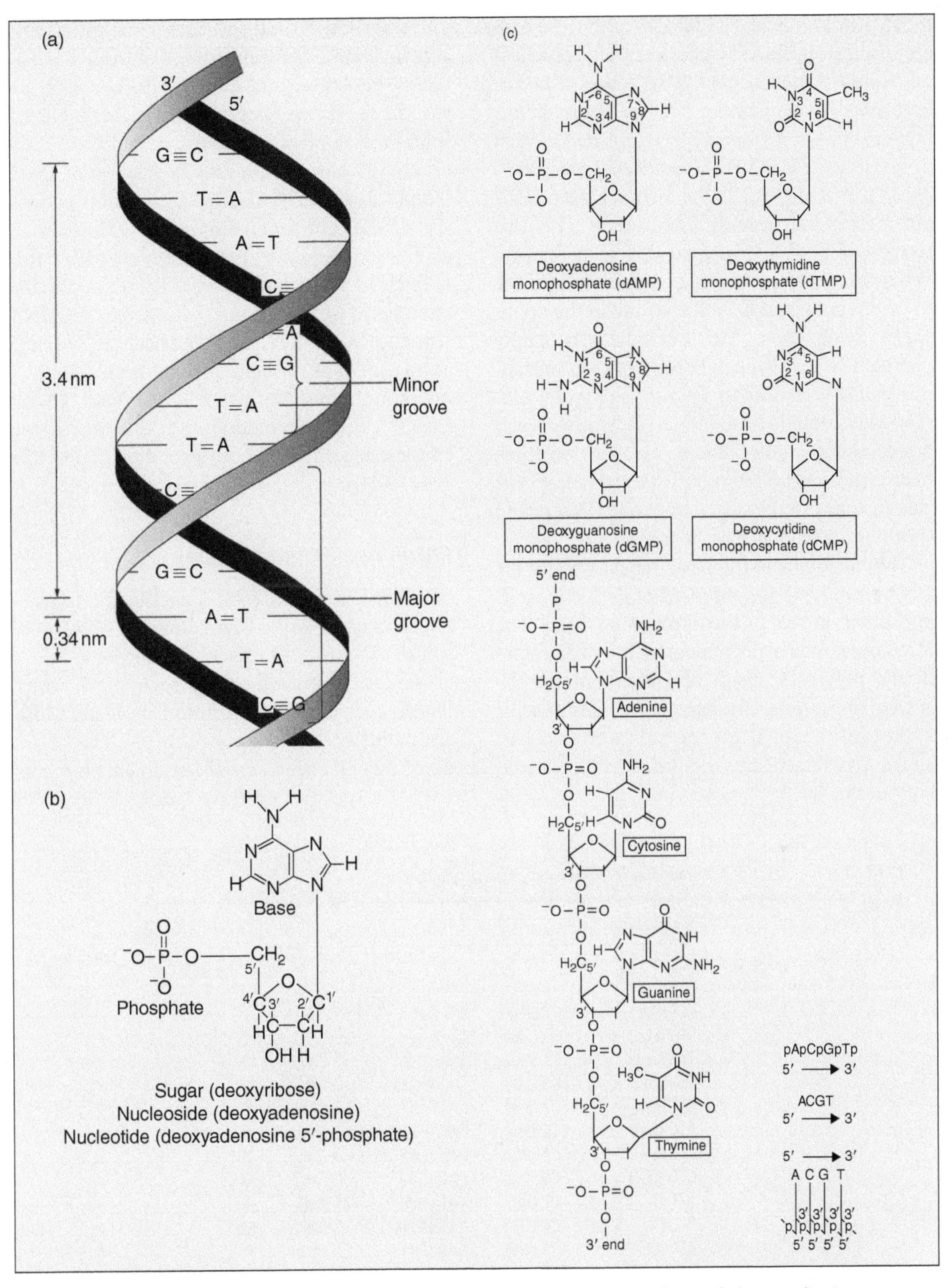

Figure 2.1 Structure and composition of DNA. DNA contains deoxynucleotides consisting of a specific heterocyclic nitrogenous base (adenine, guanine, cytosine or thymine) joined to a deoxyribose phosphate moiety. Adjacent deoxynucleotides are linked through their phosphate groups to form long polynucleotide chains. (a) The DNA double helix; (b) a nucleotide; (c) the purine and pyrimidine bases. *Source:* Cox TM and Sinclair J 1997/With permission from John Wiley & Sons.

the five-carbon sugar residue deoxyribose, which are successively linked by covalent phosphodiester bonds. A nitrogenous base, either a *purine* (*adenine* [A] or *guanine* [G]) or a *pyrimidine* (*cytosine* [C] or *thymine* [T]), is attached to each deoxyribose. DNA forms *a double-stranded helical structure*, in which the two separate DNA polymers wind around each other. The two strands of DNA run *antiparallel*, such that the deoxyribose linkages of one strand runs in the 5′–3′ direction and the other strand in the opposite 3′–5′ direction. The double helix is mainly maintained by hydrogen bonds between nucleotide pairs. According to the *base-pair rules*, adenine always binds to thymine via two hydrogen bonds and guanine binds to cytosine via three hydrogen bonds. This complementary base-pair rule ensures that the sequence of one DNA strand specifies the sequence of the other.

The *nucleotide* is the basic repeat unit of the DNA strand and is composed of deoxyribose, a phosphate group and a nitrogenous base. The 5′–3′ sequential arrangement of the nucleotides in the polymeric chain of DNA contains the *genetic code* for the arrangement of amino acids in proteins. The genetic code is the universal language that translates the information stored within the DNA of genes into proteins. It is universal between all species known so far. The genetic code is read in groups of three nucleotides. These three nucleotides, called a *codon*, are specific for one particular amino acid. Table 2.1 shows the 64 possible codons, of which 61 specify for 22 different amino acids, while three sequences (TAA, TAG, TGA) are stop codons (i.e. do not code for an amino acid). Most amino acids are coded for by more than one codon; this is referred to as *redundancy*. For example, the amino acid isoleucine may be coded by the DNA sequence ATT, ATC or ATA. Each amino acid sequence of a protein always begins with a methionine residue because the start codon (ATG) codes for methionine. The three stop codons signal the end of the coding region of a gene and the resultant polypeptide sequence.

Chromosome (karyotype)

In eukaryotic cells, DNA is packaged into *chromosomes* and every cell contains a set of chromosomes (Figure 2.2). Each chromosome has a narrow waist known as the *centromere*, which divides each chromosome into a short and a long arm, labelled p and q, respectively. The beginning and end of each double-stranded helix consists of a specific repetitive sequence of several

Table 2.1 Conversion of the genetic code into amino acids.

	Second base				
First base	T	C	A	G	Third base
T	TTT (Phe)	TCT (Ser)	TAT (Tyr)	TGT (Cys)	T
	TTC (Phe)	TCC (Ser)	TAC (Tyr)	TGC (Cys)	C
	TTA (Leu)	TCA (Ser)	TAA (Stop)	TGA (Stop)	A
	TTG (Leu)	TCG (Ser)	TAG (Stop)	TGG (Trp)	G
C	CTT (Leu)	CCT (Pro)	CAT (His)	CGT (Arg)	T
	CTC (Leu)	CCC (Pro)	CAC (His)	CGC (Arg)	C
	CTA (Leu)	CCA (Pro)	CAA (Gln)	CGA (Arg)	A
	CTG (Leu)	CCG (Pro)	CAG (Gln)	CGG (Arg)	G
A	ATT (Ile)	ACT (Thr)	AAT (Asn)	AGT (Ser)	T
	ATC (Ile)	ACC (Thr)	AAC (Asn)	AGC (Ser)	C
	ATA (Ile)	ACA (Thr)	AAA (Lys)	AGA (Arg)	A
	ATG (Met)	ACG (Thr)	AAG (Lys)	AGG (Arg)	G
G	GTT (Val)	GCT (Ala)	GAT (Asp)	GGT (Gly)	T
	GTC (Val)	GCC (Ala)	GAC (Asp)	GGC (Gly)	C
	GTA (Val)	GCA (Ala)	GAA (Glu)	GGA (Gly)	A
	GTG (Val)	GCG (Ala)	GAG (Glu)	GGG (Gly)	G

Codon table; from genetic code into amino acids: Ala, alanine; Arg, arginine; Asn, asparagine; Asp, aspartic acid; Cys, cysteine; Gln, glutamine; Glu, glutamic acid; Gly, glycine; His, histidine; Ile, isoleucine; Leu, leucine; Lys, lysine; Met, methionine; Phe, phenylalanine; Pro, proline; Ser, serine; Thr, threonine; Trp, tryptophan; Tyr, tyrosine; Val, valine, and "stop" for no amino acid. Furthermore, there are three Stop codons -TAA; TG; TGA-, a sequence that does not code for an amino acid as there is not tRNA to bind to that codon.

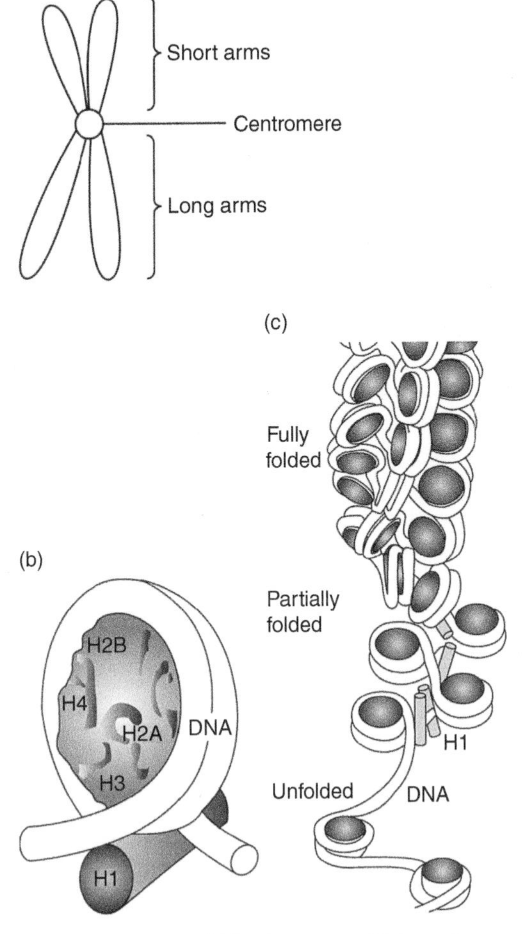

Figure 2.2 Structure of (a) a chromosome, (b) the nucleosome and (c) chromatin. *Source:* Cox TM and Sinclair J 1997/With permission from John Wiley & Sons.

kilobases, also known as the *telomere*. DNA is packaged in a very compact structure within the nucleus. Condensing of DNA is essential because the human cell contains approximately 4×10^9 nucleotide pairs, termed *base pairs* (*bp*) of DNA, whose extended length would approach more than 1 m. The DNA of each chromosome is wrapped approximately 2.5 times around a set of eight histone proteins and stored in the nucleus. The *nucleosome* is the most basic unit of the chromosome. It is composed of a 145 bp linear strand of double-stranded DNA wound around a complex of *histone proteins* (two of each of the four histone proteins H2a, H2b, H3 and H4). Nucleosomes are linked together by the histone protein H1 to form *chromatin*. During cell division, this is then further compacted with the aid of non-histone chromosomal proteins to generate a chromosome. The structure of DNA in chromatin is important because it has profound effects on the ability of DNA to be transcribed.

The chromosomal complement or karyotype refers to the number, size and shape of the chromosomes as seen during mitosis, a specific phase during the cell cycle. The human karyotype is composed of 22 pairs of autosomes and a pair of sex chromosomes: XX in the female and XY in the male. Most human cells contain 46 chromosomes, the *diploid* number. Chromosomal disorders are characterised by abnormalities of chromosomal number or structure. They may involve the autosomes or the sex chromosomes and may be the result of a *germ cell mutation* in the parent (or a more distant ancestor) or a *somatic mutation* in which only a proportion of cells will be affected (mosaicism). The normal chromosome number is an exact multiple of the *haploid* number (23) and is referred to as the diploid number. A chromosomal number that exceeds the diploid number (46) is called *polyploidy*, and one that is not an exact multiple number is *aneuploidy*. Aneuploidy usually occurs when the pair of chromosomes fails to segregate (non-disjunction) during meiosis, which results in an extra copy of a chromosome (*trisomy*) or a missing copy of a chromosome (*monosomy*). Down syndrome is a common example of trisomy – it is due to the presence of three copies of chromosome 21 (trisomy 21).

Structural abnormalities of chromosomes also occur. A *translocation* is the transfer of chromosomal material between chromosomes. Chronic myeloid leukaemia results from the translocation of genetic material between chromosome 8 and chromosome 22. This results in an abnormal chromosome, known as the Philadelphia chromosome, the expression of which results in leukaemia. *Chromosomal deletions* arise from the loss of a portion of the chromosome between two break points. *Inversions* arise from two chromosomal breaks with inversion through 180° of the chromosomal segment between the breaks.

Genotype, phenotype, and allelic expression

The *genotype* of an organism is the total number of genes that make up a cell or organism. The term, however, is also used to refer to alleles present at one locus. Each diploid cell contains two copies of each gene; the individual copies of the gene are called *alleles*. The definition of an allele is one of two (or more) alternative forms of a gene located at the corresponding site (*locus*) on homologous chromosomes. One allele is inherited from the maternal gamete and the other from the paternal gamete, therefore the cell can contain the same or different alleles of every gene. *Homozygous* individuals carry two identical alleles of a particular gene. *Heterozygotes* have two different alleles of a particular gene. The term *haplotype* describes a cluster of alleles that occur together on a DNA segment and/or are inherited together. *Genetic linkage* is the tendency for alleles located close together to be transmitted together through meiosis and hence be inherited together.

Genetic polymorphisms are different forms of the same allele in the population. The 'normal' allele is known as the *wild-type allele*, whereas the variant is known as the polymorphic or mutant allele. A polymorphism differs from a mutation, in that it occurs in a population more frequently than a recurrent mutation. By convention, a polymorphic locus is one at which there are at least two alleles, each of which occurs with frequencies greater than 1%. Alleles with frequencies less than 1% are considered as a recurrent mutation. The alleles of the ABO blood group system are examples of genetic polymorphisms. The *single nucleotide polymorphism* (*SNP*) is a common pattern of inherited genetic variation (or common mutation) that involves a single base change in the DNA. Another common form of genetic variation includes *copy number variations* (*CNV*). It is estimated that about 0.4% of the human genome differs with respect to CNV.

There are several ways to characterise genetic variation. Traditionally, genetic variants were identified using methods including restriction fragment length polymorphism (RFLP), positional cloning and gene sequencing. These approaches tended to identify specific candidate genes or haplotypes. More recently, whole-genome and transcriptome sequencing technologies that allow full genetic sequence determination of all expressed proteins have become available. Sequencing now provides the opportunity to identify multiple genetic variants. The biggest challenge is to translate genetic data into knowledge, wherein genetic data are coupled with physiology data to make sense out of them. Complex bioinformatics using neural networks, machine learning and artificial intelligence facilitate analysis of the large amount of data generated.

Epigenetics is a relatively new field of research which refers to changes in gene expression due to mechanisms other than changes in the underlying DNA sequence. The molecular basis of epigenetics is complex; put simply, it refers to altered DNA structure. It involves modifications of the activation of certain genes, but not the basic DNA sequence. For example, DNA methylation refers to the addition of methyl groups to the DNA, which in turn affects transcriptional activity. Folate status can affect DNA methylation, which in turn can affect gene expression through mechanisms that are being actively researched.

Still, there is a considerable amount of research needed that investigates relationships between common genetic polymorphisms or epigenetics with disease because certain genetic variations may predispose an individual to a greater risk of developing a disease. The effect of genetic variation in response to dietary change is also of great interest because some polymorphisms/epigenetic states may determine an individual's response to dietary changes. Hence, genetic variation can determine the efficacy of nutritional approaches, which may in turn determine the outcome of certain disease states. The inter-relationship between diet, disease and genetic variation will be discussed in more detail in Section 2.5.

The *phenotype* is the observable biochemical, physiological or morphological characteristics of a cell or individual resulting from the expression of the cell's genotype, within the environment in which it is expressed. Allelic variation and expression can affect the phenotype of an organism. A *dominant allele* is the allele of a gene that contributes to the phenotype of a heterozygote. The non-expressing allele that makes no contribution to the phenotype is known as the *recessive allele*. The phenotype of the recessive allele is only demonstrated in homozygotes who carry both recessive alleles. *Co-dominant alleles* contribute equally to the phenotype. The ABO blood groups are an example of co-dominant alleles, where both alleles are expressed in an individual. In the case of *partial dominance*, a combination of alleles is expressed simultaneously and the phenotype of the heterozygote is intermediate between that of the two homozygotes. For example, in the case of the snapdragon, a cross between red and white alleles will generate heterozygotes with pink flowers. *Genetic heterogeneity* refers to the phenomenon whereby a single phenotype can be caused by different allelic variants.

DNA damage, genetic mutations, polymorphisms, and heritability (monogenic and polygenic disorders)

Human DNA sequence changes slowly under normal physiological conditions over time or with age, due to inevitable errors occurring during normal DNA synthesis associated with cellular renewal. In addition, many agents contribute to DNA changes, as they cause DNA damage, including ionising radiation, ultraviolet light, chemical mutagens, viruses, free radicals produced during many cellular processes and nutrition. A change in the nucleotide sequence is known as either a *mutation* or a *polymorphism*, depending on the frequency of the alteration. Polymorphisms are more frequent sequence variants or simply 'common mutation'.

A mutation may be defined as a permanent transmissible change in the nucleotide sequence of a chromosome, usually in a single gene, which may lead to loss or change of the normal function of the gene. A mutation can have a significant effect on protein production or function because it can alter the amino acid sequence of the protein that is coded by the DNA sequence in a gene. *Point mutations* include *insertions, deletions, transitions* and *transversions*. Two types of events can cause a point mutation: chemical modification of DNA, which directly changes one base into another, or a mistake during DNA replication that causes, for instance, the insertion of the wrong base into the polynucleotide during DNA synthesis. Transitions are the most common type of point mutations and result in the substitution of one pyrimidine (C–G) or one purine (A–T) by the other. Transversions are less common, where a purine is replaced by a pyrimidine or vice versa.

The functional outcome of mutations can vary significantly. Most of the DNA is non-coding and mutations may therefore not necessarily result in a change of our proteins. However, non-coding regions may be important in regulation, controlling transcription. On the other hand, a single-point mutation in a coding region can change the third nucleotide in a codon and not change the amino acid that is translated, or it may cause the incorporation of another amino acid into the protein – this is known as a *missense mutation*. The functional effect of a missense mutation varies greatly depending on the site of the mutation and the importance of the protein in relation to health. A missense mutation can have no apparent effect on health or it can result in a serious medical condition. For example, sickle cell anaemia is due to a missense mutation of the β-globin gene: a glutamine is changed to valine in the amino acid sequence of the protein. This has drastic effects on the structure and function of the β-globin protein, which causes aggregation of deoxygenated haemoglobin and deformation of the red blood cell. A nucleotide change can also result in the generation of a stop codon (*nonsense mutation*) and no functional protein will be produced. *Frameshift mutations* refer to small deletions or insertions of bases that alter the reading frame of the nucleotide sequence; hence, the different codon sequence will affect the expression of amino acids in the peptide sequence.

Heritability refers to how much a disease can be ascribed to genetic rather than environmental factors. It is expressed as a percentage, a high value indicating that the genetic component is important in the aetiology of the disease. Genetic disorders can simply be classified as *monogenic* (or single-gene disorders) or *polygenic diseases* (multifactorial diseases). In general, we have a far greater understanding of the *single-gene disorders* because they are due to one or more mutant alleles at a single locus and most follow simple *Mendelian inheritance*. Examples of such disorders are:

- autosomal dominant: familial hypercholesterolaemia, von Willebrand disease, achondroplasia
- autosomal recessive: cystic fibrosis, phenylketonuria, haemochromatosis, α-thalassaemia, β-thalassaemia
- X-linked dominant: vitamin D-resistant rickets
- X-linked recessive: Duchenne muscular dystrophy, haemophilia A, haemophilia B, glucose-6-phosphate dehydrogenase deficiency.

Some single-gene disorders show non-Mendelian patterns of inheritance, which are explained by different degrees of penetrance and variable gene expression. *Penetrance* means that a genetic lesion is expressed in some individuals but not in others. For example, people carrying a gene with high penetrance have a high probability of developing any associated disease. A low-penetrance gene will result in only a slight increase in disease risk. *Variable expression* occurs when a genetic mutation produces a range of phenotypes. *Anticipation* refers to the situation when a Mendelian trait manifests as a phenotype with decreasing age of onset and often with greater severity as it is inherited through subsequent generations (e.g. Huntington chorea, myotonic dystrophy). *Imprinting* refers to the differential expression of a chromosome or allele depending on whether the allele has been inherited from the male or female gamete. This is due to selective inactivation of genes according to the paternal or maternal origin of the chromosomes. Although there are only a few examples of diseases that arise as a result of imprinting (e.g. Prader–Willi and Angelman syndromes), it is thought that this form of gene inactivation may be more important than previously realised.

Polygenic (or multifactorial) diseases are those due to a number of genes (e.g. cancer, coronary heart disease, diabetes, obesity). Even though polygenic disorders are more common than monogenic disorders, we still do not understand the full genetic basis of many of these conditions. This reflects the fact that there is interaction between many candidate genes, that is, those genes that are thought to play an aetiological role in multifactorial conditions. Furthermore, in polygenic inheritance, a trait is in general determined by a combination of both the gene and the environment.

The Human Genome Project

Historically, the Human Genome Project (HGP) was an important source of genetic information to advance human genetics and molecular nutrition, especially in terms of providing the opportunity to understand the interaction between nutrition and the human genome. Interestingly, the HGP showed that the human genome was apparently less complex than was anticipated. For example, the size of the human genome is only 30 times greater than that of the fruit fly and 250 times larger than yeast. Also importantly, only 3% of the DNA in the human genome constitutes coding regions and codes for approximately 25 000 genes. Humans only have 2–3 times as many genes as a fruit fly. Compared with the fruit fly, the human genome has many more non-coding or intronic regions. The 2–3-fold difference between humans and fruit flies is largely accounted for by the greater number of control genes (e.g. transcription factors) in the human genome. Although these intronic regions do not code for genes as such, many have important functional roles, including promoter and/or repressor regions, contributing to isoform variations of genes, and interacting with regulatory RNA molecules, known as miRNA or long non-coding RNAs.

The real challenge is not sequencing the genetic code as such but understanding how the human genome interacts with the environment. The HGP showed that humans are very alike; it estimated that humans are 99.8% genetically similar. Nevertheless, it is very apparent that human phenotypes are very diverse. Thus, the implications of the HGP in relation to molecular nutrition are not yet fully known. The challenge is to identify the proportion of human genetic variation that is relevant to nutrition. The term 'gene mining' refers to the process that identifies new genes involved in nutrition, health and disease. At the most basic level, we already know that the human genome determines nutrient requirements; for example, gender determines iron requirements – the iron requirement of menstruating women is greater than that of men of the same age. In the case of folate, research would suggest that the methylene tetrahydrofolate reductase (MTHFR) polymorphism could determine an individual's folate requirements. Since there are fewer genes than anticipated, it has been proposed that different isoforms of the same gene with different functionality are important. Already there are examples of this; for example, the three isoforms of the apolipoprotein E (Apo E) gene determine the magnitude of postprandial triacylglycerol metabolism. Furthermore, it has been proposed that the interaction between the human genome and the environment is an important determinant of within- and between-individual variation.

Within the context of molecular nutrition, we will have to determine how nutrients alter gene expression and determine the functional consequences of genetic polymorphisms. With the information generated from the HGP, we will have a more complete understanding and information in relation to the relevance of genetic variation, and how alterations in nutrient intake or nutritional status affect gene expression in a way that is relevant to human health and disease processes. In essence, the challenge is to bridge the gap between the genome sequence and whole-organism biology, nutritional status and intervention.

2.3 Gene and protein expression: transcription and translation

Gene structure and function

Gene expression refers to the process whereby the information encoded in the DNA of a gene is converted into a final product, which confers the observable phenotype upon the cell. A *gene* may be defined as the nucleic acid sequence that is necessary for the synthesis of a functional transfer RNA (tRNA), ribosomal RNA (rRNA), a peptide or protein in a temporal and tissue-specific manner. In the case of a peptide or protein, a gene is not directly translated into a protein; it is expressed via a nucleic acid intermediary called *messenger RNA* (*mRNA*). The transcriptional unit of every gene is the sequence of DNA transcribed into a single mRNA molecule, starting at the promoter and ending at the terminator regions.

The essential features of a gene and mRNA are presented in Figure 2.3, including key elements such as the two non-coding (or untranslated) promoter and terminator regions at the beginning and end of the gene. Also, the coding regions are a mix of exons and introns. Each gene contains DNA sequences that code for the amino acid sequence of the protein, which are called *exons*. These exons are interrupted by non-coding DNA sequences, which are called *introns*. Whilst introns are non-coding regions, they may contain critical regulatory elements that may alter gene expression. The *promoter region* is located immediately upstream of the gene-coding region; it contains DNA sequences, known as the TATA and CAAT boxes, which define the DNA binding sites at which general transcription factors bind to start. The promoter region also contains enhancer and inhibitor elements where specific transcription factors can bind that enhance or inhibit the possibility of the general transcription factors to start gene expression. This promoter region can be sensitive to nutritional and hormonal cues, for example insulin response elements. The last exon ends with a *stop codon* (TAA, TAG or TGA), which represents the end of the gene-coding region and is followed by the terminator DNA sequence that defines the end of the gene-coding region.

Ribonucleic acid

Ribonucleic acid, like DNA, carries genetic information. The composition of RNA is very similar to DNA, and it plays a key role in all stages of gene expression. RNA is also a linear polynucleotide, but it differs from DNA in that it is mostly single-stranded and the sugar of the bases is polymers of ribose instead of deoxyribose. Furthermore, the pyrimidine base *uracil* (*U*) is used as a substitute for thymine (T) in DNA, and it is relatively unstable when compared to DNA. There are at least five different types of RNA in eukaryotic cells and all are involved in gene expression.

- Messenger RNA (mRNA) molecules are long, linear, single-stranded polynucleotides that are direct copies of DNA. mRNA is formed by transcription of DNA.
- Small nuclear RNA (snRNA) is a short, ± 150 nucleotide (nt) RNA molecule that forms, together with a protein, the small nuclear ribonucleoprotein (snRNP). Several of these

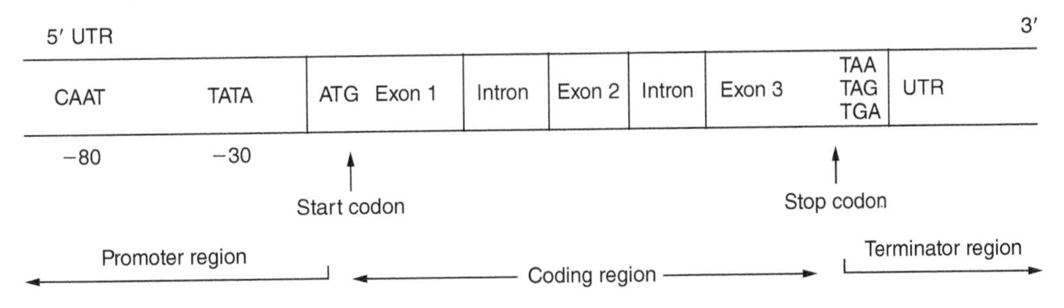

Figure 2.3 Essential features of the gene.

snRNPs form with other proteins the spliceosome that facilitates the removal of introns from the precursor mRNA.

- Ribosomal RNA (rRNA) is a structural and functional component of ribosomes. Ribosomes, which are present in the cytoplasm and on the rough endoplasmic reticulum (rER), are the machines that synthesise mRNA into amino acid polypeptides. The ribosomes are composed of rRNA and ribosomal proteins.
- Transfer RNA (tRNA) is a small RNA molecule that donates amino acids during translation or protein synthesis.
- A group of small RNAs such as microRNAs (miRNA), small interfering RNAs (siRNAs), Piwi-interacting RNAs (piRNAs) and repeat-associated siRNAs (rasiRNAs). These small RNA molecules have various functions at the transcriptional and/or post-transcriptional level, including the breakdown of mRNA, repression of transcription and chromatin remodelling.

RNA is about 10 times more abundant than DNA in eukaryotic cells; ~80% is rRNA, ~15% is tRNA and ~1–5% is mRNA. In the cell, mRNA is normally found associated with protein complexes called *messenger ribonucleoprotein (mRNP)*, which package mRNA and aid its transport into the cytoplasm, where it is decoded into a protein.

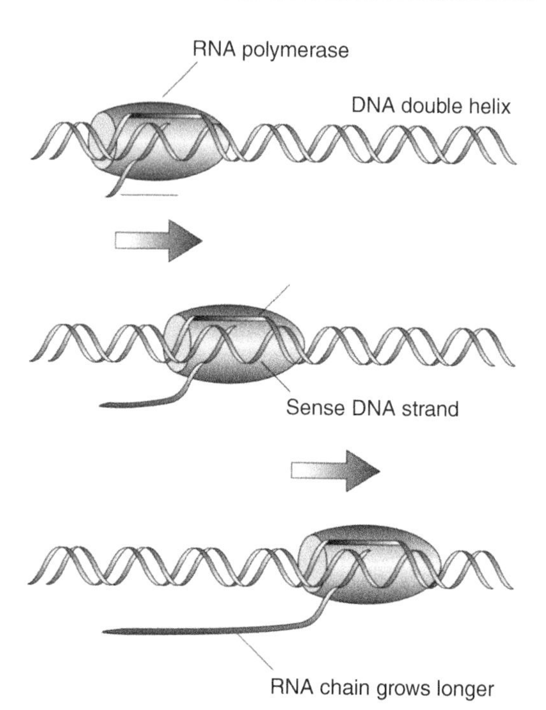

Figure 2.4 RNA polymerase II transcribes the information in DNA into RNA.

Transcription

RNA transcription is the process whereby the genetic information encoded in DNA is transferred into a *heterogeneous nuclear RNA (hnRNA)* because of its considerable variation in size. Instead of hnRNA, often the term *precursor mRNA (pre-mRNA)* is used to designate that it refers to a just produced, unprocessed 'raw' RNA molecule. Formation of the pre-mRNA is the first step in the process of gene expression and occurs in the nucleus of the cell. It can be divided into four stages: template recognition, initiation, elongation and termination.

Transcription is catalysed by DNA-dependent *RNA polymerase* enzymes (Figure 2.4). In eukaryotic cells, RNA polymerase II synthesises mRNA, while RNA polymerase I and RNA polymerase III synthesise tRNA and rRNA. The strand of DNA that directs synthesis of mRNA via complementary base pairing is called the template or *antisense strand*. The other DNA strand that bears the same nucleotide sequence is called the coding or *sense strand*. Therefore, RNA represents a copy of DNA.

Transcription is a multistep process. When the specific transcription factors switch on the promoter, transcription is initiated by the assembly of an initiation complex formed by the general transcription factors (TFIID, TFIIA, TFIIB, TFIIF, TFIIE, TFIIH) at the promoter. They then 'recruit' RNA polymerase II that binds to these general transcription factors. More specifically, the general transcription factor, TFIID, recognises the promoter and binds to the TATA box at the start of the gene. TFIIH unwinds double-stranded DNA to expose the unpaired DNA nucleotide and the DNA sequence is used as a template from which RNA is synthesised. TFIIE and TFIIH are required for promoter clearance, allowing RNA polymerase II to commence movement away from the promoter. Initiation describes the synthesis of the first nine nucleotides of the RNA transcript. Elongation describes the phase during which RNA polymerase moves along the DNA and extends the growing RNA molecule. The RNA molecule is synthesised by adding nucleotides to the free 3′-OH end of the growing RNA chain. As new nucleotides can only be attached at this free 3′-OH end, RNA synthesis always takes place in the 5′–3′ direction. Growing of the RNA chain is ended by a process called *termination*, which involves recognition of the *terminator sequence* that signals the dissociation of the polymerase complex.

Post-transcriptional processing of RNA

After transcription of DNA into RNA, the newly synthesised pre-mRNA is modified. This process is called post-transcriptional processing of RNA (Figure 2.5). The primary transcript is a pre-mRNA molecule that represents a full copy of the gene extending from the promoter to the terminator region of the gene and includes introns and exons. While still in the nucleus, the newly synthesised RNA is capped, polyadenylated and spliced. *Capping* refers to the addition of a modified guanine (G) nucleotide (7-methylguanosine) at the 5′ end of the mRNA. This 7-methylguanosine cap has several functions: it protects the synthesised RNA from enzymatic attack, it aids pre-mRNA splicing, helps to transport the mRNA out of the nucleus, and enhances translation of the mRNA. *Polyadenylation* involves the addition of a string of adenosine (A) residues (a *poly-A tail*) to the 3′ end of the pre-mRNA. Then the pre-mRNA is spliced, which is performed by the spliceosome with the help of small nuclear RNA (snRNA). This RNA–protein complex recognises the consensus sequences at each end of the intron (5′-GU and AG-3′) and excises the introns so that the remaining exons are spliced together to form the mature mRNA molecule. After capping, polyadenylation and splicing, the mRNA is transported from the nucleus to the cytoplasm for translation.

Alternative splicing describes the process whereby other, alternative splice recognition sites lead to the removal of certain exons or retention of (parts of) introns, thereby generating different mature mRNA molecules and ultimately different proteins. As illustrated in Figure 2.6, one primary RNA transcript can be spliced in several different ways, in this example leading to three mRNA isoforms, one with five exons or two with four exons. Upon translation, the different mRNA isoforms will give rise to different isoforms of the protein product of the gene. This is a relatively common phenomenon. Importantly, those variants can have different functionality.

Although the physiological or metabolic relevance of the different isoforms of many proteins is not fully understood, it surely will be relevant to molecular nutrition. For example, the peroxisome proliferator activator receptor-γ (PPARγ) gene can produce seven different isoforms of mRNA (PPAR$\gamma_{1\text{-}7}$), because of different promoters and alternative splicing. To illustrate, the PPARγ_2 mRNA isoform is responsive to nutritional states. PPARγ_2 mRNA expression is increased in the fed state; its expression is

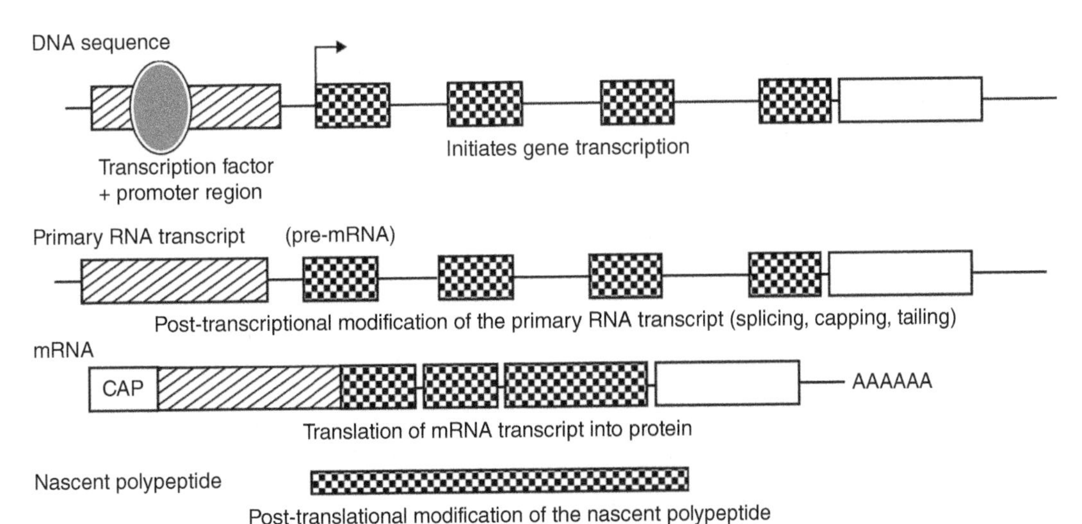

Figure 2.5 Transcription and processing of mRNA. Primary RNA transcript (pre-mRNA). Post-transcriptional modification (splicing, capping and tailing) of the primary RNA transcript (pre-mRNA).

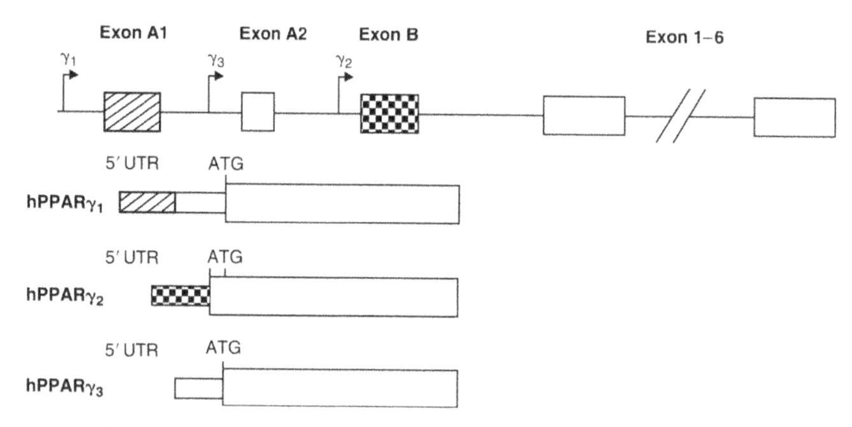

Figure 2.6 Alternative RNA splice variants of the peroxisome proliferator activator receptor-γ (PPARγ) gene. Note: This is a simplified representation of the full picture of all seven PPARγ isoforms.

positively related to adiposity and is reduced on weight loss. Therefore, PPARγ_2 mRNA may be the isoform that mediates nutrient regulation of gene expression. AMP-activated protein kinase (AMPK) is another important example, again with multiple variants that regulate a host of metabolic events relating to fatty acid, glucose and protein metabolism, often with tissue and cell specificity.

RNA editing is another way in which the primary RNA transcript can be modified. Editing involves the binding of proteins or short RNA templates to specific regions of the primary transcript and subsequent alteration or editing of the RNA sequence, either by insertion of one or more different nucleotides or by base changes. This mechanism allows the production of different proteins from a single gene under different physiological conditions. The two isoforms of apo B are examples of RNA editing. In this case, RNA editing allows for tissue-specific expression of the apo B gene, such that the full protein apo B100 is produced in the liver and secreted on very low density lipoprotein (VLDL). By contrast, in the intestine the apo B48 isoform is produced because a premature stop codon is inserted so that transcription of the protein is halted and the resultant apoprotein is only 48% of that of the full-length apo B100 protein.

Control of gene expression and transcription

All individual cells of an organism contain the complete genetic blueprint of the organism. Therefore, it is of critical importance that the right genes are expressed in the correct tissue, at the proper level and at the correct time. Temporal and tissue-specific gene expression in eukaryotic cells is mostly controlled at the level of transcription initiation. Two types of factors regulate gene expression: *cis-acting control elements* and *trans-acting factors*. *Cis*-acting elements do not encode proteins; they influence gene transcription by acting as binding sites for proteins that regulate transcription. These DNA sequences are usually organised in clusters, located in the promoter region of the gene, that influence the transcription of genes. The TATA and CAAT boxes are examples of *cis*-acting elements.

The *trans*-acting factors are known as *transcription factors* or DNA binding proteins. Transcription factors are proteins that are encoded for by other genes, and include steroid hormone receptor complexes, vitamin receptor proteins and mineral–protein complexes. These transcription factors bind to specific DNA sequences in the promoter region of the target gene and promote gene transcription. The precise mechanism(s) of how transcription factors influence gene transcription is not fully understood. However, transcription factors in general share some properties. They often have amino acid domains that contain one or more zinc ions, so-called *zinc fingers* that enable binding to DNA in a sequence-specific manner. Another type of domain contains predominantly the amino acid leucine, the leucine zipper, and has a function in binding ('zipping') to similar domains in other proteins based on the charge of the individual amino acids. Beside these domains, transcription factors often contain a nuclear localisation signal to direct the protein from the cytoplasm to the nucleus and specific amino acids that can be modified by, for instance, phosphorylation, ubiquitinylation or acetylation and thereby get activated or inactivated. These modifications enable fine-tuning of the action of transcription factors. In addition, they contain protein binding elements to recruit proteins that can modulate the histones that open up the DNA to allow the general transcription factors and the RNA polymerase to enter and attach to the DNA.

As mentioned, transcription factors play a key role in temporal and tissue-specific gene expression. On binding, they can influence the transcription of typically more than one gene. There is ample evidence that transcription factors can remodel the chromatin structure in such a way that certain parts of the genome become available for transcription, while other parts of the chromosomes become tightly packed and inaccessible for transcription. This process is called *chromatin remodelling*. The transcription factors recruit histone acetyl transferases (HAT) that add acetyl groups to the histone tails, thereby making them repel each other, which opens up the chromatin structure for transcription. On the other hand, transcription factors may recruit histone deacetylase (HDAC), which removes the acetyl groups from the histone tails. The nucleosomes are then tightly packed together and not accessible for transcription.

Beside modification of the histones by the action of the transcription factors, the degree of DNA methylation also plays an important role in the regulation of transcription. It is generally accepted that hypermethylated DNA (the cysteines are mainly prone to be methylated) inhibits gene expression while unmethylated DNA can be transcribed. Here also nutrition is known to be involved. A typical example is the agouti mouse. This also introduces the term *epigenetics*, the study of inheritable characteristics that change the expression of the genes in the offspring but do not alter the genetic code. If the agouti mother mouse is fed a low-folate (methyl donor) diet, the DNA cannot be methylated properly, including the Avy allele which is normally highly methylated. Among other phenotypical changes, this results in offspring that has a yellow coat colour while on a folate-rich diet the offspring has a brown coat.

Master regulatory proteins regulate the expression of many genes of a single metabolic pathway. For example, in the lipogenic pathway the fatty acid synthase complex codes for seven distinct genes that have to be co-ordinately expressed to form the enzymes required for fatty acid synthesis. This ensures that sufficient levels of all the enzymes of this metabolic pathway are available simultaneously.

Beside regulation by transcription factors, another mechanism using microRNAs (miRNA) regulates the final translation of an mRNA molecule (Figure 2.7). miRNA are small non-coding RNA molecules. After transcription from DNA, first primary miRNA molecules (pri-miRNA) are formed. They mostly arise from intronic regions during normal transcription though a few are exon derived or even produced independently of gene transcription, as they have their own promoter. There are two routes to produce a functional miRNA molecule: canonical and non-canonical. The non-canonical route differs in the start of production of the pri-miRNA. Here the canonical route is described as an example. Following processing of the pri-miRNA hairpin structure by Drosha and DGCR8, a precursor miRNA (pre-miRNA) protein complex is formed. The pre-miRNA is then recognised by exportin-5, that facilitates transport of the pre-miRNA from the nucleus to the cytoplasm where it is recognised by a dicer that cleaves the hairpin. Next, one of the two miRNA strands is transferred to Argonaut (AGO) protein and together they form the RNA induced silencing complex (RISC). Based on the miRNA sequence, an mRNA molecule that contains the complementary sequence can be silenced by this complex. Interestingly, the AGO coupled miRNAs can be excreted into the blood and may become effective in other cells of the body. These pathways include AGO/miRNAs excreted via vesicles, apoptotic bodies, exosomes, HDL particles or even directly into the blood.

First, intronic mRNA or a miRNA gene is transcribed and forms a pri-miRNA. The pri-miRNA is recognised by Drosha and DGCR8, trimmed and the pre-miRNA is exported to the cytoplasm by exportin. Then, the pre-miRNA is released of its hairpin by Diced following binding of Argonaut and RISC loading. Within the RISC complex, one of the two miRNA molecules is selected and used to recognise the RNA target which is thought to be degraded upon a full match and inhibited upon a partial match. miRNA* is the reverse complement of the miRNA.

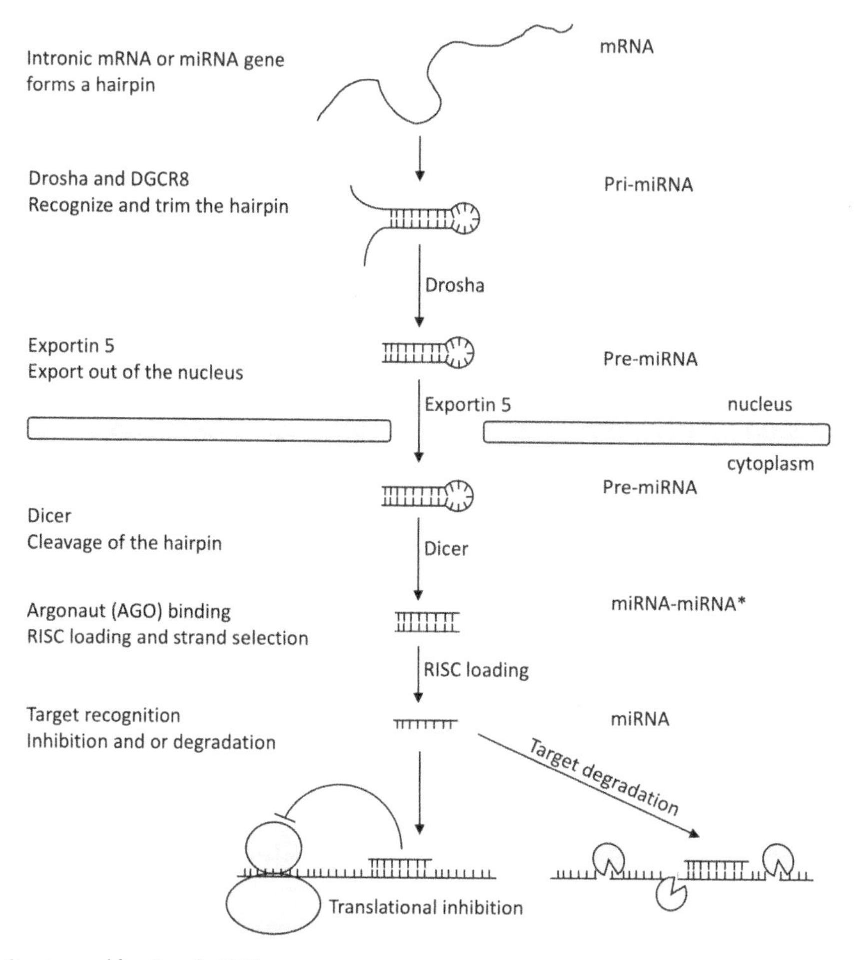

Figure 2.7 Structure and function of miRNA.

RNA translation: protein synthesis

Translation is the process by which the genetic information coded in mRNA is converted into an amino acid sequence (polypeptide) through reading the mRNA coding sequence as a continuous, non-overlapping, tri-letter code (Figure 2.8 and Table 2.1). This three-letter or trinucleotide code (triplets or codons) can thus be read in three possible *reading frames* within a single-strand mRNA molecule. To read this code, the *ribosomes* dock on the mRNA to start protein synthesis in the cytoplasm or on ribosomes attached to the cytoplasmic side of the rough endoplasmic reticulum membrane (called rough due to the ribosomes attached that make this look 'rough' under an electron microscope). In humans, ribosomes consist of two subunits, a 40S (small) and 60S (large), which combine to form the 80S particle.

Protein synthesis begins by the formation of a complex involving a 40S ribosomal subunit carrying a methionine tRNA, which base pairs with the initiation codon AUG on the mRNA molecule. Translation is largely regulated by controlling the formation of the initiation complex, which is the 40S ribosomal subunit, the mRNA and specific regulatory proteins. The structure of the 5′ UTR is critical for determining whether mRNA is translated or sequestered in the untranslated ribonucleoprotein complex. Initiation of translation is also dependent on the presence of the 5′ cap structure and secondary structure of the mRNA next to the initiation codon. Secondary structures such as stem loops in the 5′ end of the mRNA inhibit the initiation of translation. Some of the translation regulatory proteins are cell specific and mRNA specific, allowing precise post-transcriptional control over the synthesis of specific proteins. The poly(A) tail also modulates translation. Although it is not essential for translation to occur, mRNA lacking the poly(A) tail is less efficiently translated.

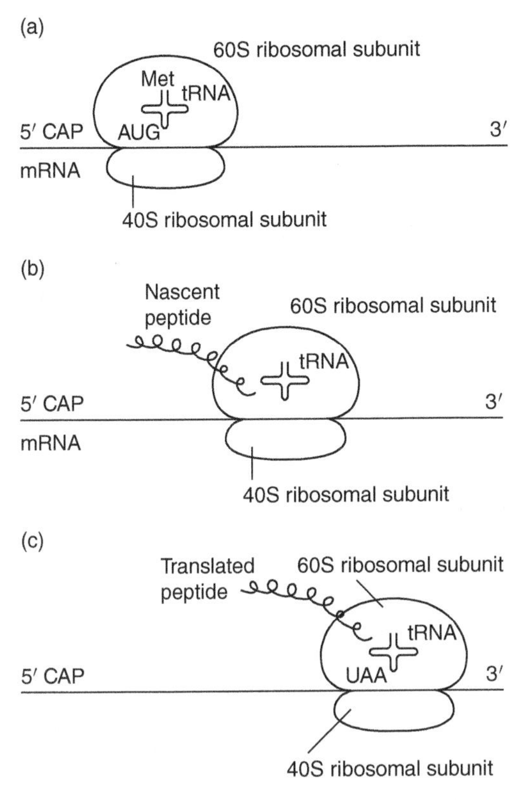

Figure 2.8 Three phases of translation. (a) Initiation: first codon methionine; (b) elongation: translation continues elongating the nascent peptide; (c) termination: translation ends on recognition of the stop codon.

Once the initiation complex has formed, synthesis of the polypeptide chain is driven by the interaction between *elongation factors* (elFs), the ribosome and tRNA along the length of the mRNA molecule. At each codon, the ribosome and elFs promote the interaction between mRNA and tRNA. tRNA donates an amino acid, which is added to the newly synthesised polypeptide that corresponds to the codon in the mRNA. The tRNA achieves this because it bears a triplet of bases (*anticodon*) that are complementary to the mRNA codon, and there is an amino acid attached to the acceptor arm of the tRNA that corresponds to the codon in the mRNA. When the anticodon of the tRNA matches the codon or trinucleotide sequence in the mRNA, the tRNA donates the amino acid to the newly synthesised polypeptide. For example, the first tRNA that carries methionine has the anticodon UAC; it recognises the methionine codon AUG on mRNA. Similarly, the anticodon AAC recognises the leucine codon TTG on the mRNA. This process continues until the ribosome reaches a termination codon (UAA, UAG, UGA) and then the completed polypeptide is released from the ribosomal units.

Post-translational modification

Post-translational modification refers to the process that converts the pre-mature polypeptide into the mature protein product of the gene. The nascent polypeptide may start to form the complex structure of the protein as it is being synthesised directly in the cytoplasm or at the membrane of the rough endoplasmic reticulum. When formed at the ribosomes attached to the rough endoplasmic membrane, the proteins may end up in/on the endoplasmic reticulum (ER) membrane or in the ER lumen. Within the ER, other proteins, the chaperones, are present that help proteins to form properly. Furthermore, they are glycosylated. When the pre-mature polypeptide chain is properly folded and glycosylated, it is likely to be transported to the Golgi body by COPII coated vesicles for further processing. In the Golgi, the proteins are further processed and sorted for transport to their final locations. For instance, when a protein in the Golgi needs to be transported back to the ER, it will be packed in vesicles coated by the COPI protein that directs 'the protein' back to the ER.

Besides modifications within the ER or Golgi, mature proteins are also modified which may trigger different effects. A well-known modification is phosphorylation, hydroxylation, ubiquitination or by proteolytic activities, which will confer functional characteristics to the protein. For example, the phosphorylation status of a protein may determine whether it is active or inactive, while extensive ubiquitination typically directs the protein to the proteasome to be degraded.

2.4 Research tools to investigate molecular aspects of nutrition

For many reasons, it is not always possible to examine the molecular aspects of nutrition in humans. In many cases, it is simply not possible to obtain the tissues or cells of interest. Also, such human studies may be too expensive. Another pragmatic problem is that responses to dietary interventions may vary widely between individuals owing to differences in genetic background and dietary diversity, such as variable intake and/or absorption affecting bioavailability. To overcome these problems, various approaches, such as animal and cell studies, are used to gain a better understanding of the effects of diet or specific dietary constituents at the cellular and molecular level. Although such approaches are valuable, it should always be kept in mind that each has its strengths and limitations. Extrapolation to the human situation will always be a problem and questions that can be tackled in humans should therefore be addressed in humans.

In vivo models – animal studies

In vivo studies have the advantage of determining the impact of a given dietary intervention on the whole organism, albeit in animals rather than humans. One advantage of such studies is that they can be done in a highly controlled way. Food intake can easily be monitored and manipulated, while factors such as temperature, humidity and the stage of the disease can be controlled for. In addition, tissues and cells can be obtained that cannot be readily sampled in humans. It should be appreciated that each animal model has specific advantages and disadvantages, therefore many factors need to be considered before a model is chosen, including animal size, the time needed for breeding, litter size, costs of animal housing and the feasibility of performing blood sampling and vivisection. Then, as with human studies, it is possible to use techniques from the fields of genetics and molecular biology.

An important advantage of animal studies is that the effect of a nutrient can be investigated in animals with a similar genetic background. In this way, variation in responses between animals due to genetic heterogeneity is excluded. Genetic homogeneity can be obtained by inbreeding. For this, genetically related animals, such as brothers and sisters, are mated for many generations. The overall result will be that heterozygosity between animals will reduce and that in the end, an inbred strain is obtained with a similar genetic, homozygous background. Genetic homogeneity, however, is not always an advantage. If, for example, the mode of action of a food component is not known or if the main purpose is to test the safety of nutrients or drugs, genetic heterogeneity may be preferable because it increases the extrapolation of results. In addition, differences in responses between various strains can be helpful for elucidating responses at the molecular level. Thus, it is evident that a clear research question is needed before the choice of animal model can be made.

Using molecular tools applied to the germ cells of animals, it is possible to insert an isolated (foreign) gene sequence into an animal's genomic material. The resultant animals that carry such a foreign gene are called transgenic animals and have traits coded by these genes. Another approach is to knock out a certain endogenous gene. With this technique, a specific gene is modified, which results in a loss of function of that specific gene. With these approaches, it is possible to test candidate genes for their effects on the parameters of interest. Transgenic and knockout animals are valuable models for nutrition research. For example, mice have been generated that express human Apo AI. In this way, it was possible to examine in detail the role of this apolipoprotein in atherogenesis. By inserting or knocking out genes, the importance of specific genes on the genesis of disease and on complex biological pathways in various tissues can be addressed. Finally, by interbreeding mouse strains with different genetic backgrounds, the interaction between genes and the relative contribution of genes and diet on phenotype can be determined.

Another tool involves the generation of transgenic mice with tissue-specific expression of a particular gene. This can be accomplished in several ways. It is, for example, possible to insert a tissue-specific promoter before the foreign gene of interest. This DNA construct can then be used to create a transgenic animal. It is possible to make genes whose expression is dependent on the presence or absence of dietary components, such as tetracycline. In this way, it is even possible to switch on the expression of a gene in a specific tissue at any stage of the disease just by taking away tetracycline from (or adding it to) the diet. Alternatively, the so-called Cre–Lox recombination system is widely used to control tissue-specific expression of knockout of genes. In this system, the enzyme Cre-recombinase is used to recognise specific DNA sequences, the so-called LoxP sites. When two LoxP sites flank a certain DNA code, this DNA code is removed by the Cre-recombinase. Using this system, tissue-specific expression, knockout or even replacement of the gene of interest can be achieved by crossing animals that contain a LoxP cassette with animals that express Cre-recombinase in a tissue-specific manner. Alternatively, the Cre-recombinase mRNA can be delivered by injection or using viruses to the animal.

Other ways to manipulate gene expression in the intact animal can also be manipulated by DNA electroporation or with adenovirus-mediated gene delivery. With these methods, DNA can also be inserted into a specific tissue. This results in increased levels of mRNA expression and subsequent protein synthesis from the gene of interest.

Another way to study the function of specific genes is through the use of tissue or cell transplantation. For example, apoE gene expression in macrophages, which are made by bone marrow, can be studied in vivo by bone marrow transplantation. For these experiments, two strains of inbred mice with similar genetic background are used: one strain has no functional apoE gene locus (apoE knockout mice) but the other strain has. After lethal irradiation of the bone marrow of the apoE knockout mice, a bone marrow cell suspension from the wild-type animal is injected. Within about four weeks, the knockout animals are completely reconstituted with the bone marrow from the wild-type animals. As a consequence, their macrophages do not contain apoE and the specific effects of macrophage apoE production can be studied.

In vitro models – tissue cultures

In vitro studies with intact tissues or isolated cells are an alternative to in vivo human and animal studies. If conditions are optimal, cells survive, multiply and may even differentiate after isolation. Tissue cultures can be divided into two categories: organ cultures and cell cultures. An organ culture is composed of a complete or a small part of an intact organ or tissue that is brought into the culture medium. The advantage is that, at least to some extent, the normal biochemical and morphological differentiation and communication routes between the various cell types of a tissue are maintained. Furthermore, it is possible to mimic closely the physiological environment. The survival time of organ cultures, however, is limited and generally not more than 24 h. In addition, it is not possible to culture the cells further and fresh material is needed for each set of experiments.

Cell suspensions are different from tissue cultures in that the extracellular matrix and intercellular junctions between cells are disrupted. For this, tissues are first treated with proteolytic enzymes and with components that bind calcium. The proteolytic enzymes cut the proteins that hold together the tissue in which the cells are embedded whereas chelating agents, like EDTA, capture di- and tricationic metal ions like calcium to prevent cadherin binding between cells. The tissue is then mechanically dispersed into single cells. By doing this, a suspension is obtained that is composed of various cell types. If necessary, it is possible to isolate one specific cell type that can be used for experiments or serve as starting material for a cell culture.

Cell cultures prepared directly from the organs or tissues of an organism are referred to as primary cultures. These cells may still have the potential to divide and in this way secondary and tertiary cell cultures are obtained. Depending on the type of cell and technique used, these cells can be cultured in a suspension or as a monolayer. As for tissue cultures, survival time is limited. However, cell lines can be used that have obtained the ability to replicate indefinitely. Such cells are frequently derived from cancerous tissue and are all derived from the same stem cell, and have therefore a similar genetic make-up. It should be emphasised, however, that after each cell passage these cells lose some of their biochemical and morphological characteristics: the older they are, the more differences can exist from the original stem cell. However, it is feasible to freeze cells after each passage. After thawing properly, these cells can be cultured further from the point where they were frozen, which makes it possible to go back to a previous passage.

Cell cultures are typically monolayers. However, in vivo cells normally are part of a tissue in a three-dimensional (3D) multicellular environment. To mimic tissues in a laboratory system, cells can be grown in a 3D structure using several cell types in so-called *organoids*. Organoids are now becoming very popular to investigate the impact of an intervention on the whole organ. This allows investigation of the interaction between the multiple heterogeneous cells that constitute the organ of interest. For example, hepatic organoids would contain more than just hepatocytes, but also reflect constitutive endothelial cells, vascular cells and immune cells, etc. that are present in the functioning liver.

Table 2.2 details some common tissue culture models. The effects of adding nutrients or

Table 2.2 Common tissue culture models.

Cell line	Organism	Tissue	Morphology
HUVEC	Human	Umbilical vein, vascular endothelium	Epithelial
HepG2	Human	Liver, hepatocellular carcinoma	Epithelial
Caco-2	Human	Colon, colorectal adenocarcinoma	Epithelial
CV-1	African green monkey	Kidney	Fibroblast
293	Human	Kidney	Epithelial
HeLa	Human	Cervical adenocarcinoma	Epithelial
THP-1	Human	Monocyte, acute monocytic leukaemia	Suspension culture

combinations of nutrients can be studied in detail. The responses of cells to these stimuli can be examined under the microscope or biochemically. It is also possible to examine the effects on mRNA or protein expression, and interactions between two or more different cell types. Finally, it is possible to insert isolated (foreign) gene sequences into the cells or to downregulate gene expression of specific genes using RNA interference. Studies with isolated tissue or cell models are very useful for mechanistic purposes. However, extrapolation of the results to the intact animal or to the human situation is always a problem.

Molecular cloning, overexpression, and downregulation

[fo]Molecular cloning refers to the process of making copies of a DNA segment, which is not necessarily the entire gene. This segment can be a piece of genomic DNA, but also complementary DNA (cDNA) derived from mRNA. After isolating the DNA fragment, it is inserted into a plasmid that serves as a vector. A plasmid is a circular double-stranded, relatively small (1–1000 kb) DNA molecule that can replicate independently of the bacterial chromosome. The vector with its piece of foreign DNA is subsequently inserted into bacterial cells. As a vector has the capability to replicate autonomously in its host and the host itself can also be grown indefinitely in the laboratory, huge amounts of the plasmids containing the DNA segment of interest can be obtained. Many vectors are available and the selection of the vector depends on, among other things, the size of the DNA fragment, whether it is genomic DNA or cDNA, and on the host to be used. The purpose of cloning is to generate sufficient amounts of a particular DNA sequence for further study – to provide direct mechanistic evidence relating to a DNA sequence of interest.

Molecular cloning has important applications in a number of areas of molecular nutrition. For example, this method can test the functionality of an SNP and determine whether or not the SNP is differentially affected by a nutrient. In this case, the impact of two DNA constructs, one with and the other without the SNP, can be determined. Each construct can be inserted into a vector that can be used to transfect a cell line of interest. After exposure of these cells to different nutritional interventions, it is, for instance, possible to determine whether dietary effects on mRNA or protein expression are affected by this particular SNP.

DNA libraries have been generated to facilitate faster gene research. These libraries, which are widely available nowadays, are collections of bacterial or fungal clones in which each individual clone contains a vector with a piece of foreign DNA inserted. Similarly, cDNA libraries have been constructed from the mRNA population present within a particular cell at the exact moment of mRNA isolation. A potential advantage of a cDNA library over a genomic library is that the cloned material does not contain the non-coding regions and the introns as present in the genomic DNA. Furthermore, it is possible to make cDNA libraries from specific tissues or cells. In this way, cDNA libraries can be constructed that are enriched with clones for genes expressed preferentially in a certain tissue, or during a specific disease or state of development.

Besides molecular cloning to overexpress a gene of interest, downregulation of genes is also a valuable tool to study the function of genes. This is often done using RNA interference (RNAi). This is similar to miRNA though now the short interfering RNA molecules are made by the scientist and delivered to the cells. RNAi molecules can be synthesised synthetically and delivered to the cells as short dsRNA molecules or via a plasmid using conventional transfection systems or by transduction (brought to the cells by a viral delivering system). In the case of plasmids, the cells use their own machinery to synthesise short hairpin RNA molecules (shRNA). These molecules are successively recognised by the naturally present system that is used by the miRNA. They follow the same route as the miRNAs to downregulate the mRNA expression.

Lastly, to study genetic variation, lab-made mutations can be used. In the laboratory, genes can be changed using the polymerase chain reaction (described in the next section) with modified primers. Also, mutagens can be used by inducing mutations in living organisms by exposure to radiation or chemicals. The biggest disadvantage of these methods is that the mutations are quite random, while it is laborious to find and select the changed genes. Currently, the Clustered Regularly Interspaced Short Palindromic Repeats (CRISPR)/CRISPR associated protein 9 (CRISPR/Cas9) system can be used to highly specifically mutate genes. This technique is based on a bacterial inheritable defence system that protects the bacteria from viral attacks that their ancestors encountered and survived. In laboratories, this bacterial immune defence system is adapted and used to specifically alter the genetic code.

Quantification of gene expression: single-gene mRNA expression

Single-gene expression is measured by quantification of the mRNA transcripts of a specific gene. It gives information on the effects of, for example, nutrients at the transcriptional level. mRNA quantification can also provide a good estimate of the level of protein present in a sample, at least when production of the protein is transcriptionally regulated. Since only very small amounts of mRNA are present and mRNA is relatively unstable, mRNA samples are first translated into cDNA with the reverse transcriptase (RT) enzyme. To make detection possible, cDNA molecules corresponding to the transcribed gene of interest need to be amplified by the polymerase chain reaction (PCR). To control for experimental variation in the RT and the PCR step, an internal control RNA can be used in the entire RT-PCR or a DNA competitor in the PCR only. The total procedure is then called competitive RT-PCR or, when a DNA competitor is used, RT-competitive-PCR.

If an internal control is used during RT-PCR, the original number of mRNA molecules can be calculated by comparing the intensity of the PCR product of the mRNA of interest with the intensity of the PCR product from the internal control. Since the initial amount of the internal standard is known, the unknown amount of mRNA can be calculated. If no internal control is used, expression of a gene is related to the expression of a constitutive gene such as β-actin or GAPDH. These housekeeping genes are always expressed to the same extent and can therefore be used as a measure for the amount of RNA used in the reaction as well as for cDNA synthesis.

Detection of the PCR products formed is carried out by different methods. A method frequently used in the past was separation of the multiplied DNA fragments by gel electrophoresis followed by visualising the fragments. Alternatively, it is also possible to use a labelled (radioactive, biotin or digoxigenin) probe, which is complementary to the amplified gene of interest. Probes hybridise with the PCR product, which can be visualised and quantified.

Currently, most methods make use of intercalating agents (like cyber green) of fluorescent-labelled primers during PCR. Further details on the relative benefits of RT-PCR versus more comprehensive transcriptomics approaches are given in Chapter 13 'Application of "Omics" Technologies' in the Nutrition Research Methodologies textbook.

Quantification of gene expression: multiple-gene mRNA expression

In general, the expression of large numbers of genes (i.e. an expression profile) changes when environmental conditions, such as nutrient exposure or nutritional status, are altered. For example, dietary interventions known to upregulate the expression of the low-density lipoprotein (LDL) receptor may also change the expression of genes coding for cholesterol-synthesising enzymes. Furthermore, expression profiles may vary between tissues, and it is also known that not all individuals respond in a similar way to changing conditions. This may be due to differences in genetic background and variables such as gender, age, amount of physical activity or state of disease. DNA microarrays are useful to understand better the interactions between a large number of genes and to examine events at a transcriptional level.

DNA microarrays or chips are used as a tool to detect differences in the expression level of a large number of genes between two or more samples in a single assay. More details on different transcriptomics are present in Chapter 13 'Application of "Omics" Technologies' in the Nutrition Research Methodologies textbook.

In summary, transcriptomics provides the tools to determine expression profiles of numerous known and unknown genes at the same time in different tissues at various stages of a disease. Such studies will give an enormous amount of data, which need to be analysed in a meaningful way. As with the other high-throughput 'omics' technologies, all rely on excellent bioinformatics tools, such as clustering analysis and principal component analysis, which integrate multiple gene expression profiles. Bioinformatic approaches that look at the effect of nutrients on metabolic/biochemical pathways and networks are also of use, as they allow the integration of genes that are functionally related to each other. These approaches facilitate characterisation of a 'transcriptomic signature' that describes the effect of a nutrient or metabolic state of a cell/tissue.

Quantification of proteins

Proteins fulfil a very diverse role, which varies from being structural and contractile components to being essential regulatory elements in many cellular processes in the form of enzymes or hormones. Although mRNA serves as a template for a protein, it is important to realise that the amount of mRNA is not necessarily correlated with the amount of protein. For example, mRNA can be degraded without being translated. In addition, proteins can be modified after translation in such a way that their half-life, functions or activities are altered. Finally, genetic variation may also result in different molecular forms of the protein. It is therefore important to obtain information not only on gene expression but also on protein synthesis, modification and activity.

Various techniques are used for protein analysis. The most sensitive techniques are a combination of one or two methods that make use of differences in molecular weight, electric charge, size, shape or interactions with specific antibodies. A frequently used technique is Western blotting in combination with specific protein detection using antibodies. It is also possible to detect and quantify amounts of protein using enzyme-linked immunosorbent assays (ELISAs) or radio-immuno-assays (RIAs). These techniques are mainly based on immunoreactivity and can detect many proteins at physiological concentrations. However, they do not make use of differences in molecular weight, as with Western blotting, and may therefore be less specific for determining post-translational modification of proteins. Furthermore, unknown proteins may be present on a Western blot.

The most direct approach to identify these unknown proteins of interest, that are differentially expressed after different nutritional experiments, is to determine the amino acid sequence. Typically, a purified protein is first cleaved into a number of large peptide fragments, which are isolated. The next step is to determine the amino acid sequence of each fragment. If a part of or the total amino sequence is known, this amino acid sequence is compared to a library of known protein stretches to identify the protein. Alternatively, the amino acid sequence can be reverse translated to the mRNA and the exon sequence of the corresponding gene can be generated, as information on the amino acid sequence provides information on the nucleotide sequence.

Cleavage of the protein into fragments and then into amino acids is also helpful to determine those parts of the protein that are transformed after translation. In this way, for example, it is possible to identify sites and the degree of phosphorylation or glycosylation. Comparable to DNA microarrays, arrays have also been developed for protein profiling. Instead of gene-specific polynucleotides, antibodies raised against proteins are spotted on a flat solid support.

It is estimated that the human proteome contains some 20 000–25 000 proteins, an estimate which might increase as we identify more proteins generated via RNA splicing and proteolysis, as well as post-translational modifications. Proteomics refers to the most comprehensive profiling of almost all proteins in a biological sample, utilising mass spectrometry, after different cell, tissue or fraction (e.g. phosphoproteomics) specific isolation procedures. To this end, proteomics provides the opportunity to better characterise the cell proteome, using a variety of well-established as well as emerging approaches. More details in relation to different proteomic approaches are presented in Chapter 13 'Application to "Omics" Technologies' in the Nutrition Research Methodology textbook.

Briefly, there are two broad methodological approaches. The first uses traditional separation techniques (e.g. two-dimensional [2D] gel electrophoresis) of proteins from tissues and/or biofluids, followed by mass spectrometry to identify the differentially expressed proteins. Alternatively, 'shotgun' proteomics involves direct proteolytic enzyme digestion, wherein the composition of the resultant peptide mixture is analysed by liquid chromatography coupled mass spectrometry (LC-MS). The beauty of shotgun proteomics is that it can be combined with stable isotope labelling to quantify proteomic expression profiles. This can be particularly useful if assessing the impact of a particular metabolite and/or nutrition intervention.

Stable isotopes: a tool to integrate cellular metabolism and whole-body physiology

With DNA arrays and protein arrays, it is possible to obtain information on the expression and formation of many genes and proteins, respectively. This provides valuable information for delineating how patterns are influenced by internal and external stimuli. It does not, however, provide information on enzyme activity or quantify metabolic events in vivo. For example, an increase in the synthesis of a certain protein for gluconeogenesis does not necessarily mean that glucose production is also increased. Glucose production is very complex and controlled at many levels. To address such issues, stable isotope technology is helpful, with which quantitative information can be obtained in humans on in vivo rates of synthesis, degradation, turnover, fluxes between cells and tissues, and so on.

Stable isotopes are molecules that differ slightly in weight owing to differences in the number of neutrons of one or more atoms. For example, ^{12}C and ^{13}C refer to carbon atoms with atomic masses of 12 and 13, respectively, which behave similarly metabolically. In nature, about 99% of carbon atoms are ^{12}C and only 1% are ^{13}C. As ^{13}C atoms are non-radioactive, they can be used safely in human experiments. In contrast, radioactive isotopes are frequently used in animal or cell studies.

These characteristics make stable isotopes useful tools to integrate cellular metabolism and whole-body physiology. With the appropriate analytical techniques, it is possible to separate atoms and molecules based on differences in mass. Experiments are based on the low natural abundance of one of the isotopes. To use a very simplistic example, if one is interested in the rate of appearance in breath of $^{13}CO_2$ from the oxidation of an oral dose of glucose, ^{13}C-labelled glucose (which is prepared on a commercial scale) can be given. The expired air will become enriched with $^{13}CO_2$, which can be measured. One now has the certainty that this $^{13}CO_2$ is derived from ^{13}C-labelled glucose. Ultimately, this gives information on the rate, proportion and amount of glucose oxidised. Such an approach is instrumental in increasing our understanding of the metabolic consequences of the effects found at the cellular and molecular level.

Metabolomics

Metabolomics represents the newest 'omics' technology being applied within molecular nutrition and nutrigenomics. It utilises analytical chemistry technologies such as nuclear magnetic resonance (NMR) spectroscopy and mass spectrometry (MS) to capture complete data on the low molecular weight metabolites, nutrients and other compounds in various human biofluids, cells and tissues. Lipidomics, which quantifies lipid-based metabolites, is a subsidiary of metabolomics, in which LC-MS-based techniques are used to quantify lipophilic molecules. This polar and non-polar chemical fingerprint/small molecule metabolite profile that reflects cellular processes is referred to as the *metabolome*. The metabolome is viewed as the functional outcome of all metabolites, that reflect prior modulation of gene and protein expression.

Metabolomics studies can take either targeted or non-targeted approaches. The former, being very hypothesis based, just focuses on specific metabolite targets, in an experiment. In contrast, untargeted metabolomics takes a hypothesis-free approach to holistically characterise the metabolome in different nutritional or metabolic states to generate biomarkers, using NMR, GC-MS and LC-MS techniques. The nutrition field is still challenged by the requirement to assess nutritional status and to this end, the metabolomics field assesses the extent to which metabolomics may provide the comprehensive biomarker of multiple metabolites to assess nutrient status, metabolic responses, disease predisposition, etc.

Although the metabolome can be readily defined, it is not possible to analyse the entire range of metabolites by a single analytical method. To this end, a number of international consortia such as the Human Metabolome Database have been established wherein the key metabolites, drug derivatives and food components that can found in the human body are documented.

The biological interpretation of metabolomic data is also a great challenge within the context of human nutrition – the data analytical tools are not yet as sophisticated, with well-characterised pathway analysis tools which are more readily available in the transcriptomic or proteomics field. Also, it is important to appreciate that it can be difficult to directly relate metabolites to specific biomarkers of nutritional status. Metabolomics data are not readily interoperable from one platform to another, which makes merging data from different cross-sectional and dietary intervention studies difficult. However, there are several initiatives under way to resolve these challenges.

2.5 Genetic variability: determinants of health and response to nutrients

Nutrigenetics is a term which describes how genetic variation determines an individual's risk of a diet-related disease, their nutrient requirement and metabolic response, as well as responsiveness to a bioactive dietary component or nutritional intervention. Genetic polymorphisms may affect metabolic responses to diet by influencing the production, composition and/or activity of proteins. A considerable amount of research has been conducted on the effects of common genetic polymorphisms on the response to nutrients, commonly referred to as *genetic association studies*. This is of great interest for two reasons. First, such studies provide information on specific molecular processes underlying dietary responsiveness. Second, if a common genetic polymorphism determines the dietary response, then people may be identified who will, or will not, benefit from specific dietary recommendations. In theory, this *personalised nutrition* approach may allow targeted gene-based dietary treatment for specific metabolic aberrations of certain disease states. This paradigm might work well in monogenic conditions but given the polygenic basis of most diet-related chronic diseases, this 'gene only' approach may be somewhat naive.

It is important to appreciate that several general criteria have been formulated to assess the impact of the results from *genetic association studies* which attempt to link genetic variation and dietary responses. First, the polymorphism should affect the metabolic response by influencing the production, composition and/or activity of the protein. Mutations/polymorphisms in the promoter region will not necessarily affect the composition of the protein but may affect its production. In contrast, mutations in an exon may not affect the production, but can change the composition and consequently the structure or activity of a protein. However, it is also possible that mutations in introns are associated with dietary responses. In such cases, it is very likely that this mutation is in *linkage disequilibrium* or associated with another functional genetic variation that is directly responsible for the effect observed. Essentially, there should be pronounced, biologically relevant effects of the polymorphism on dietary responses. In terms of the practical applicability of a gene–nutrient interaction, it is important that sufficient subjects in the population carry the variant, for example >10%. Finally, there should be a plausible biological mechanism or, ideally, well-characterised functional studies to explain the association and demonstrate the gene–nutrient interaction.

Most of the initial gene–nutrient interaction studies focused on single genetic variants (SNPs) in individual genes, to determine if those variants interacted with different dietary patterns or biomarkers of dietary status. MTHFR is a nice example, described below. This approach provided the concept of gene–nutrient interactions but it was somewhat limited. The challenge is to identify the other candidate genes involved in diet-related diseases, which often involve perturbations in multiple metabolic pathways, that are common to several conditions including obesity, type 2 diabetes (T2D) and cardiovascular disease (CVD). Also, within any pathway there might be several genes involved; for example, there are more than 50 involved in complex processes such as lipid metabolism, insulin signalling or inflammation and redox status.

In contrast, *genome-wide association studies (GWAS)* provide a more comprehensive approach that attempts to characterise most of the whole genome to identify new genetic variants associated with disease. This approach has identified many new candidate genes, such as TCF7L2, associated with greater risk of type 2 diabetes. There is no doubt that this approach will identify new candidates for diet-related diseases in the future. GWAS data are also a very useful resource for retrospective analysis of existing datasets. It is now possible to conduct reanalysis of GWAS datasets, with detailed clinical phenotype and dietary information, in order to investigate a new hypothesis. For example, potential interactions between habitual dietary saturated fat intake and genetic variants of the NLRP3 inflammasome/IL-1β activation pathway to modulate insulin resistance and diabetes risk were studied by retrospective meta-analysis of some 19 000 individuals. Whilst such data need to be validated, the availability of GWAS data provides an efficient means to explore and investigate pathway-related gene–nutrient interactions.

Example: 677C-T polymorphism of methylene tetrahydrofolate reductase

The 677C-T MTHFR polymorphism is a good example of a gene–nutrient interaction with functional consequences. Increased levels of plasma homocysteine, a sulfur-containing amino acid, are associated with an increased risk of neural tube defects, thrombotic disease and vascular disease. These associations do not prove a causal role for homocysteine in these diseases, but they do indicate that homocysteine plays an important role in many physiological processes. In homocysteine metabolism, a crucial role is fulfilled by the enzyme methylene tetrahydrofolate reductase (MTHFR). The MTHFR gene is located on chromosome 1. Some subjects with MTHFR deficiency have been identified and they all have increased homocysteine levels. To study the in vivo pathogenetic and metabolic consequences of MTHFR deficiency, an MTHFR knockout mouse model has been generated. Homocysteine levels are slightly elevated in heterozygous animals and are increased 10-fold in homozygous knockout mice. The homozygous animals show similar diseases to human MTHFR patients, supporting a causal role of hyperhomocystinaemia in these diseases. Nevertheless, extrapolation from animal studies to the human situation remains a problem.

A common DNA polymorphism in the MTHFR gene is a C-to-T substitution at nucleotide 677, which results in the replacement of alanine for valine. The allelic frequency of the 677C–T genotype, which can be identified with the restriction enzyme *HinfI*, is as high as 35% in some populations, but the allelic frequency of the 677C–T polymorphism varies widely between populations. Substitution of alanine for valine results in reduced enzyme activity and individuals homozygous for this polymorphism have significantly increased plasma homocysteine concentrations. Thus, the 677C–T mutation may have functional consequences.

It has been hypothesised that the lack of association between MTHFR polymorphisms and disease in certain populations may be due to differences in dietary status. Homocysteine metabolism requires the participation of folate and vitamin B_{12}. In the plasma, a negative relationship exists between homocysteine levels and levels of folate and vitamin B_{12}. Moreover, supplementation with folate, but also with vitamin B_{12}, reduced fasting homocysteine levels. In this respect, subjects carrying the 677C–T mutation were more responsive. Thus, a daily low-dose supplement of folic acid will reduce, and in many cases normalise, slightly increased homocysteine levels, which is explained by a possible effect on the stability of folate on MTHFR. Whether this will result in a reduced disease risk or whether certain population groups are more responsive to folate intervention needs to be established.

Methylmalonic aciduria cblB type: result from GWAS

In GWAS, large numbers of genetic markers such as SNPs are screened that cover (a large part of) the total genome of a species. The aim of this approach is to discover a group of genetic markers that are associated with a certain trait, for example the risk of developing type 2 diabetes, blood pressure, weight gain or abnormalities in serum lipid and lipoprotein concentrations. Groups of markers that associate with the trait can then be investigated further. Ultimately, results should be used to predict long-term outcomes early in life and to enable personal (nutritional) advice to prevent possible adverse effects.

GWAS demonstrated that SNPs within the gene methylmalonic aciduria (cobalamin deficiency) cblB type (MMAB) are associated with serum lipid and lipoprotein concentrations. This gene encodes a protein that catalyses the final step in the conversion of vitamin B_{12} to adenosylcobalamin. Adenosylcobalamin is derived from vitamin B_{12} and is a co-factor for the enzyme methylmalonyl-CoA mutase. Mutations in the MMAB gene could result in diminished adenosylcobalamin levels and cause accumulation of methylmalonic acid. Consequently, patients with a defect in the MMAB gene are prone to life-threatening acidosis. MMAB defects are sometimes treated by vitamin B_{12} supplementation, although this is successful in only a limited number of MMAB patients.

Why SNPs within the MMAB gene were related to abnormalities in serum lipid and lipoprotein concentrations is currently unknown. What is known is that these SNPs, together with several other SNPs, are associated with lower HDL cholesterol concentrations on a carbohydrate-rich diet even when corrected for differences in fat and protein intake. However, to determine whether these relationships are truly causal requires further study. For instance, in a GWAS using a Japanese genetic SNP database, no associations were found between MMAB and serum lipoprotein concentrations. This suggests that ethnic background may modulate the outcome of such studies. Another explanation is that, as GWAS only reveal associations, the observed relationships were due to chance. Thus, it is evident that associations found using GWAS always need to be verified using other approaches.

Summary

Gene–nutrient interaction studies investigating the effects of genetic polymorphisms on the response to dietary interventions are often inconsistent. Several explanations can be offered for this. First, many of the earlier, pioneering studies were carried out in retrospect. This means that genotyping was performed after the study had ended and that groups were not well balanced. For example, if a study is carried out with 150 subjects and the allele frequency of a certain mutation is 10%, then the expected number of subjects homozygous or even heterozygous for that mutation will be too low. As a consequence, groups will be small and it will be difficult to find any differences between the groups with respect to differences in responses to the diets, simply because of a lack of statistical power. Another explanation is that the effects of the polymorphism of interest depend on the make-up of many other genes (gene–gene interactions). Human gene–gene interaction studies are even more difficult to design. Suppose that the frequency of a certain polymorphism is 10% and that of another is 20%. Such values are not uncommon, but it means that only 2% of the subjects may have the combination of the two mutations of interest. Thus, one needs to screen many subjects to find appropriate numbers. Similarly, effects may depend on gender, age or other factors, such as body mass index, smoking or state of the disease.

Without doubt, studies on gene–gene or environment–gene interaction will expand our knowledge of the effects of the genetic code on the response to nutrients. Finally, it is possible that in a certain population the polymorphism is a marker for another, unknown genetic defect. Ideally, results should therefore always be confirmed in different, independent populations with different genetic backgrounds. Nevertheless, given greater availability of more comprehensive GWAS data, in very large cohorts, with good dietary intake data or biomarkers of nutrient status, some of the inconsistencies will be addressed. This will provide greater clarity with respect to valid and robust gene–nutrient interactions.

As already mentioned, one of the rationales for studies on genetic polymorphisms in relation to dietary responses was that people can be identified who will, or will not, benefit from specific dietary recommendations – often referred to as *personalised nutrition*. However, it has become clear that dietary responses are a complex combination of both genetic and environmental factors. In general, the known relationships between responses to diet with genetic polymorphisms are not very strong and different combinations of genetic and environmental factors may ultimately lead to a similar response. To address such issues, cell and animal studies are also relevant, to validate the functional effects of any association. Another approach is to make a complete inventory of differences in gene expression or protein synthesis in a particular organ or cell type after changing dietary intake. For this, microarray analyses can be useful, to characterise the impact of a genetic variant in conjunction with a dietary element.

Clearly there is still a long way to go, but studies on specific molecular processes underlying dietary responsiveness remain a major challenge.

2.6 Nutrient regulation of gene expression

While nutritional recommendations strive to promote good health through good nutrition, it is clear that such population-based strategies do not account for variations in an individual's nutritional requirements. These variations are due to our genetic background and environmental factors, including nutrition. The previous section showed how different genetic variations or polymorphisms can alter an individual's nutritional responses to diet. This section will examine how nutrients affect gene, protein and metabolite expression. The expression of a gene that results in an active protein can be regulated at any number of points between transcription and the synthesis of the final protein product. While the process of gene expression is well understood, as detailed in Section 2.3, relatively little is known about how nutrients affect gene expression at the level of mRNA, protein and/or metabolite. Nevertheless, there are a few examples of nutrients that affect gene expression to alter mRNA and/or active protein levels by interacting with transcriptional, post-transcriptional and post-translational events.

The effect of nutrients on gene expression is a very intensive area of research. In light of the technological advances in molecular biology, as detailed in Section 2.4, the number of examples of nutrient regulation of gene expression will expand rapidly over the next few years. Ultimately, it is important not only to understand the concept of nutrient regulation of gene expression (at the mRNA, protein and metabolite level), but also to know how changes at the level of the gene relate to whole-body metabolism and health.

Nutrient regulation of gene transcription

Theoretically, gene expression can be regulated at many points between the conversion of a gene sequence into mRNA and into protein. For most genes, the control of transcription is stronger than that of translation. As reviewed in Section 2.3, this is achieved by specific regulatory sequences, known as *cis*-acting control elements, in the promoter region of the gene, and *trans*-acting factors, known as transcription factors or DNA-binding proteins, that interact with the promoter region of genes and modulate gene expression. Nutrients can alter the transcription of target genes; some examples of this are detailed in Table 2.3.

Direct and indirect nutrient regulation of gene transcription

Certain dietary constituents can influence gene expression by direct interaction with regulatory elements in the genome, altering the transcription of a given gene. Examples include retinoic acid, vitamin D, fatty acids and zinc. However, nutrients can also have an indirect effect on gene transcription. It is often difficult to discern whether a gene–nutrient interaction is a direct effect of a particular nutrient or an indirect effect of a metabolite or a secondary mediator such as a hormone, eicosanoid or a secondary cell message that alters transcription. For example, many genes involved in fat and carbohydrate metabolism have genes that have an insulin response element in their promoter region. Hence, a particular fatty acid or carbohydrate could mediate its effect via insulin. The indirect route may be particularly important for the more complex dietary components because they have a number of bioactive

Table 2.3 Effect of nutrients on gene transcription.

Nutrient	Gene	Transcriptional effect
Glucose	Glucokinase	Increase
Retinoic acid	Retinoic acid receptor	Increase
Vitamin B_6	Steroid hormone receptor	Decrease
Zinc	Zinc-dependent enzymes	Increase
Vitamin C	Procollagen	Increase
Cholesterol	HMG CoA reductase	Decrease
Fatty acids	SREBP	Increase

HMG CoA, 3-hydroxy-3-methylglutaryl-coenzyme A; SREBP, sterol regulatory response element binding protein.

constituents. For example, part of dietary fibre is metabolised by the colonic flora to produce butyric acid. Butyric acid, in turn, may affect gene expression by selective effects on G-proteins (which are intracellular messengers) or by direct interactions with DNA regulatory sequences.

Nutrient regulation of transcription factors

Nutrients can also regulate transcription factor activity and thereby alter gene expression. The peroxisome proliferator activated receptors (PPARs) are a good example of *trans*-acting transcription factors that can be modulated by nutritional factors. PPARs are members of the nuclear hormone receptor superfamily. They regulate the expression of many genes involved in cellular differentiation, proliferation, apoptosis, fatty acid metabolism, lipoprotein metabolism and inflammation. PPARs are ligand-dependent transcription factors, which are activated by a number of compounds, including fatty acids. There are several members of the PPAR family – PPARα, PPARγ and PPARδ (β) – each with multiple isoforms. PPARα is primarily expressed in the liver, PPARγ in adipose tissue and PPARδ (β) is ubiquitously expressed. When activated (by fatty acids, eicosanoids and/or pharmacological agonists), the PPARs dimerise with the retinoid X receptor (RXR). This PPAR–RXR heterodimer binds to a PPAR response element (PPRE) in the promoter region of the target gene and induces transcription of the target gene (see Figure 2.9).

Many genes involved in lipid and glucose metabolism have PPAR response elements (PPRE), some of which are listed in Table 2.4. Therefore, the PPARs represent an example of how nutrients can regulate gene expression through transcription factors. Sterol regulatory response element binding proteins (SREBPs) are another group of transcription factors that mediate the effects of dietary fatty acids on gene expression. There are two forms of SREBP: SREBP-1 regulates fatty acid and triacylglycerol synthesis, whereas SREBP-2 regulates the genes involved in cholesterol metabolism. Therefore, the SREBPs modulate the expression of the genes involved in cholesterol and fatty acid metabolism, in response to different fatty acid treatments.

Table 2.4 PPAR-responsive genes and their metabolic effect.

PPAR target genes	Target cell	Metabolic effect
aP2	Adipocyte	Adipogenesis
FABP, ACS	Adipocyte	Fatty acid synthesis
Apo CIII, LPL	Hepatocyte	VLDL metabolism
Apo AI, Apo AII	Hepatocyte	HDL metabolism

PPAR, peroxisome proliferator activator receptor; VLDL, very low-density lipoprotein; HDL, high-density lipoprotein; FABP, fatty acid binding protein; ACS, acyl-CoA synthetase; LPL, lipoprotein lipase.

Nutrients and post-transcriptional control of gene expression

It is generally accepted that initiation of transcription is the primary mode of regulating gene expression, and there are good examples of different nutrients increasing and decreasing mRNA expression. However, there is increasing evidence that the response of gene expression to nutrients involves control of post-transcriptional events. Much of the evidence for post-transcriptional control comes from observed discrepancies between mRNA abundance and transcriptional rates (altered mRNA abundance associated with unchanged gene transcription implies altered mRNA stability). In addition, mRNA abundance is not necessarily correlated with protein concentration (altered protein concentration in the absence of any changes in mRNA abundance implies either altered

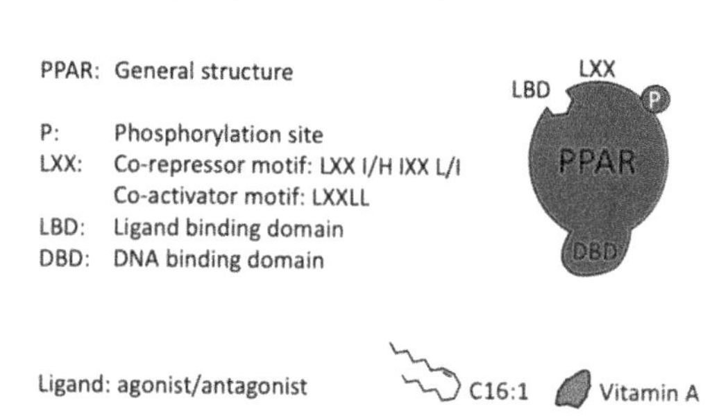

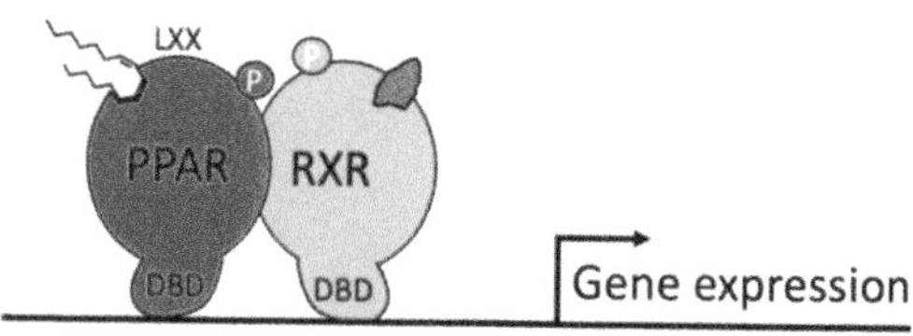

Figure 2.9 PPAR transcriptional activity by dimerisation with RXR and successive DNA attachment following ligand binding.

translation of the mRNA or changes in the proteolytic breakdown of the protein). Since nutrients can regulate mRNA translation and stability, mRNA abundance may not reflect amounts of protein or rates of synthesis. Therefore, it is incorrect to assume that if a nutrient alters the mRNA level, then there is a concomitant change in protein levels. To assess with confidence the effect of a nutrient on gene expression, mRNA analysis should be accompanied by measurements of the protein product.

Some examples of post-transcriptional control of gene–nutrient interactions are presented in Table 2.5. It is important to note that often the non-coding region of the gene may play a key role in the regulation of gene expression whereby nutrients interact with regulatory elements located in the 5′ and 3′ UTRs of a range of target genes, mediating the effect of nutrients on gene expression.

Iron is a classic example of how a nutrient regulates the expression of the genes involved in its metabolism (Figure 2.10). Transferrin and ferritin are key proteins involved in iron metabolism, and the expression of each is determined by the non-coding region of transferrin and ferritin mRNA. The transferrin receptor is required for the uptake of iron in cells. Transferrin receptor has five regulatory sequences, known as the iron response elements (IREs) in the 3′ UTR. In the absence of iron, *trans*-acting transcription repressor proteins, the iron regulatory proteins (IRPs), bind to the IRE and protect the mRNA from degradation. In the presence of iron, the IRPs are removed from the transferrin receptor mRNA, leading to mRNA destabilisation,

Table 2.5 Nutrient regulation of gene expression: post-transcriptional control.

Gene	Nutritional factor	Control point	Regulatory element
Ferritin	Iron	Translation	5′ UTR
Transferrin receptor	Iron	Stability	3′ UTR
Glutathione peroxidase	Selenium	Translation	3′ UTR
Glucose transporter-1	Fed/fasted state	Translation	5′ UTR
Lipoprotein lipase	Fatty acid supply	Translation	3′ UTR
Apolipoprotein CIII	Hyperlipidaemia	Unknown	3′ UTR

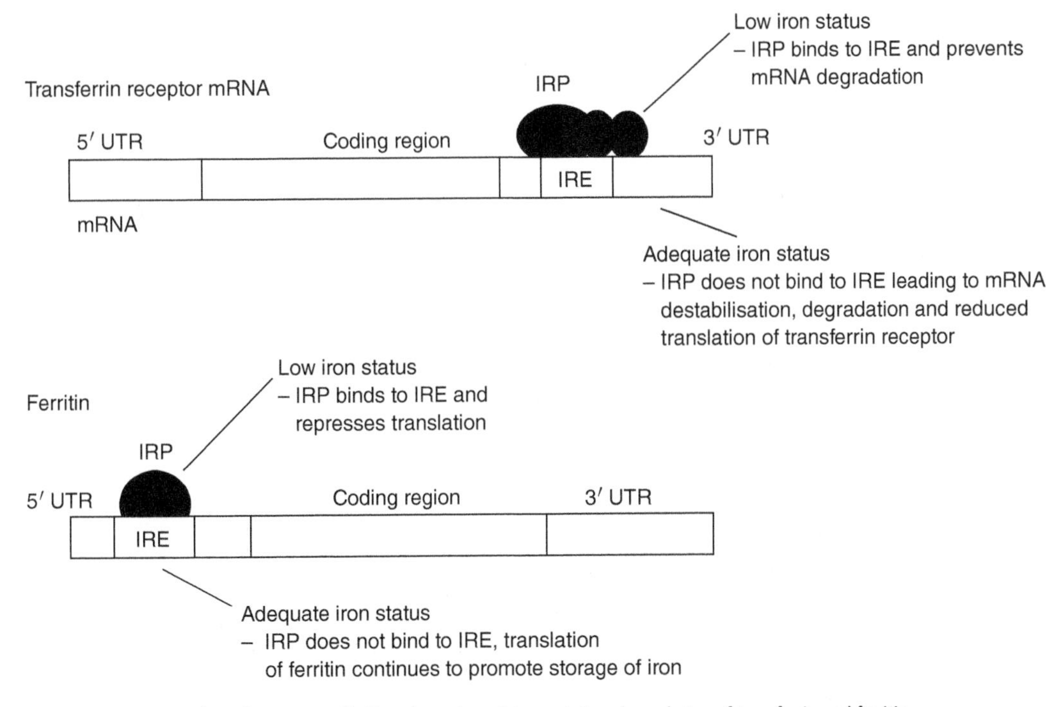

Figure 2.10 Iron regulates its own metabolism through post-transcriptional regulation of transferrin and ferritin.

decreased translation of mRNA and decreased transferrin receptor synthesis. This mechanism therefore prevents excessive amounts of iron being taken up by a cell.

Ferritin is required for the storage of iron. It sequesters cellular iron, which would otherwise be toxic. The expression of ferritin is also regulated post-transcriptionally. This controls the supply of free iron within the cell according to cellular iron levels. Ferritin has an IRE in the 5′ UTR that regulates transcription. When cellular iron levels are low, the IRP binds the IRE and represses translation of ferritin, thus supplying free iron as required for cell metabolism. In the presence of iron, the IRP does not bind the IRE and ferritin translation is increased to allow storage of iron. The presence of similar IREs in the transferrin receptor and ferritin mRNA is important because it allows co-ordinated regulation of the synthesis of the two proteins according to different cellular levels and requirements for protein.

Nutrient regulation of translation and post-translational protein modification

Hypothetically, a nutrient could regulate the translation of mRNA into protein. However, to date there are no examples of a direct effect of a nutrient on protein translation that is independent of alterations in mRNA. Similarly, there is little information on the effects of nutrients on post-translational protein modification.

Vitamin K is one of the very few examples of nutrient regulation of post-translational protein modification, and has this effect by regulating the activation of prothrombin. Prothrombin is an essential protein involved in the coagulation system. It is the proenzyme for thrombin, which is an inherent component of the clot. Prothrombin cannot function correctly unless its glutamic acid residues are carboxylated. Carboxylation of prothrombin allows it to bind to calcium, and prothrombin can only participate in the clotting process if it can bind to calcium. This post-translational modification means that the nascent prothrombin protein is dependent on the supply of vitamin K. Apart from the effect of overt vitamin K deficiency on coagulation, the extent to which an individual's vitamin K status affects clotting is unknown. Further research that specifically addresses whether and/or how nutrients modulate protein expression is required to gain a greater understanding of this aspect of molecular nutrition.

Metabolomic signatures of nutritional intervention

Nutritional metabolomics attempts to characterise the metabolome that reflects a certain nutritional status/sensitivity at the whole organism, tissue, cellular and biochemical process levels. It needs to be appreciated that an individual metabolomic profile represents highly complex regulation of many simultaneous biochemical pathways in different organs. Furthermore, the human metabolome will reflect the formation of endogenous metabolites, as well as exogenous compounds derived from the diet and/or the microbiome. Therefore, interindividual metabolomic variation is very high, and nutritional metabolomics needs to account for the multifactorial nature of an individual's metabolomic phenotype. For example, the gut microflora may have a huge impact on individual metabolomic profiles. Consequently, it is advised that within dietary intervention trials to characterise the nutritional metabolome, the subjects should consume a defined diet with/without the nutritional intervention to exclude the effect of other non-diet-related complex interactions.

There are increasing examples of studies that employed metabolomics to characterise the human metabolic phenotype. Often metabolomic studies present clusters of metabolites that represent risk versus non-risk or response versus non-response. When groups of metabolites are differentially expressed, it is important to understand the functional relevance of the differentially expressed metabolite(s) associated with health or disease within those metabolomic signatures. This will allow greater understanding of cause–effect relationships that are demonstrated in the initial study, which then should be validated in an independent model/cohort.

There is no doubt that there will be a wealth of data generated in the near future characterising the nutritional metabolome, which will help to advance our understanding of nutrition and metabolism.

2.7 Personalised or precision nutrition

Advanced understanding of molecular nutrition and the application of nutrigenomics in human nutrition have created the potential opportunity for more accurate understanding of metabolic responses to dietary intake. To this end, personalised and precision nutrition has become very topical. Like personalised medicine, the objective is to improve health in a more targeted and effective way.

Initially, personalised nutrition was based on the premise that we might be able to categorise risk and response to diet based on an individual's genetic signature. However, given the highly polygenic nature of most diet-related conditions, combined with the complexity of dietary patterns and their interaction with other lifestyle factors, it is now evident that this initial paradigm was naive. Nevertheless, it is possible that a multi-'omic' approach, that integrates key processes identified via transcriptomic, proteomic and/or metabolomic signatures, sometimes referred to as a 'metabotype' or metabolic phenotype, might more effectively reflect human nutrition and metabolism to predict risk and response. It is proposed that such a multi-omic approach may allow nutrition research that robustly reflects the dynamic interactions between food intake, nutrient status and the human genome.

A cautionary note needs to be sounded with respect to commercial personalised nutrition offerings – currently, the science is not at a stage to accurately and robustly provide personalised nutrition tools. Therefore, the full efficacy of such products should be challenged and not assumed.

2.8 Perspectives on the future

Molecular nutrition and nutrigenomics represent a new phase of nutrition research that will provide a much greater understanding of the interactions between nutrients and the human genome. This chapter has explored the technological opportunities in molecular biology that can be applied to nutrition research. These rapidly advancing technologies present tremendous opportunities for improving our understanding of nutritional science. The Human Genome Project has undoubtedly improved our understanding of genetic background and diversity. From the perspective of nutrition, we should be able to develop a greater understanding of how nutrients interact with genetic destiny, alter disease processes and promote health. Within the context of human nutrition, probably the greatest challenge is to develop experimental models that mimic the in vivo response and human whole-body metabolism. These models are needed in order to maximise the potential of the novel molecular technologies that then can investigate the molecular and cellular effects of nutrients, wherein the results can be extrapolated and applied to human health.

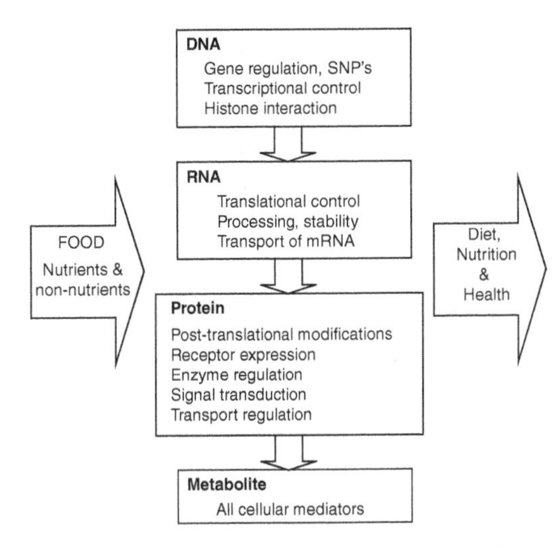

Figure 2.11 Molecular interactions between nutrients and non-nutrient food components with the genome, transcriptome, proteome and metabolome as predictors of nutrition and health. *Source:* Roche HM 2006/With permission of John Wiley & Sons.

There is no doubt that the next decade will be very exciting in terms of applying the novel molecular technologies in an intelligent and meaningful way to nutrition research. One of the greatest challenges for nutrigenomics will be to characterise the nutritional phenotype. This will require a systems biology and functional genomics approach to integrate genetic, transcriptomic, proteomic and metabolomic information to give a more complete picture of human nutrition and health (Figure 2.11). It is hoped that the next phase of molecular nutrition research will generate a more comprehensive understanding of the cellular and molecular effects of nutrients and improve our understanding of whether and how nutrients modulate disease processes. In the future, the personalised nutrition approach might provide the ability to define scientifically sound, evidence-based and effective nutritional strategies to promote health.

References

Cox, T.M. and Sinclair, J. (1997). Molecular biology in medicines. Blackwell Sciences, Oxford, 25(4), 247–248.

Roche, H.M. (2006). Nutrigenomics – new approaches for human nutrition research, Journal of the Science of Food and Agriculture, 86: 1156–63.

3
Integration of Metabolism 1: Energy

Jennifer L. Miles-Chan

Key messages

- Energy metabolism refers to the ways in which the body obtains and spends energy from food. In terms of energy transduction, part of nutrient energy is converted into chemical, mechanical, electrical or osmotic forms of energy.
- The chemical energy in nutrients (redox energy) is converted into adenosine triphosphate (ATP). ATP is the universal currency of cellular energy metabolism; it is formed from adenosine diphosphate (ADP), essentially by oxidative phosphorylation.
- At the cellular level, ATP can be synthesised aerobically or anaerobically, that is, in the presence or absence of oxygen. When considering the whole body, interorgan aerobic and anaerobic energy metabolism must be complementary and steady-state energy metabolism must be completely aerobic.
- Not all ingested energy is available for use as metabolic fuel, with losses occurring primarily during digestion and absorption, as well as in the incomplete oxidation of proteins.
- When not immediately required as fuel, energy may be stored, although the energy content, energy density and autoregulation of the different body energy stores differ considerably.
- Energy balance refers to the difference between metabolisable energy intake and total energy expenditure, with a steady state evidenced by long-term stability of body weight and body composition.
- Energy expenditure is primarily determined by body size, body composition and physical activity level.
- In accordance with the first and second laws of thermodynamics, energy expenditure can be measured directly as heat loss or indirectly as heat production (oxygen consumption and carbon dioxide production).
- Inter-relationships exist between the different components of energy balance (energy intake, energy stores, energy expenditure), with each component able to effect change in the others.

3.1 Introduction

Rien ne se perd, rien ne se crée, tout se transforme.
(Nothing is lost, nothing is created,
everything is transformed)
Antoine Lavoisier, 1789.

Conforming to the *first law of thermodynamics*, the human body is a metabolic machine, transforming energy yielded from the catabolism of nutrients contained within food into chemical, mechanical, electrical and osmotic forms of energy, with the heat dissipated during these conversions used to maintain body temperature. As such, to develop an understanding of human nutrition and metabolism, it is important to explore these energy transformations in terms of the biochemical and biophysical laws and processes involved, and our ability to measure, monitor and predict these energy transformations in a clinical (*in vivo*) context. Indeed, the ability of the human machine to operate efficiently under conditions of both nutritional abundance and paucity underpins our risk for

Nutrition and Metabolism, Third Edition. Edited on behalf of The Nutrition Society by Helen M. Roche, Ian A. Macdonald, Annemie M.W.J. Schols and Susan A. Lanham-New.

Companion Website: www.wiley.com/go/nutrition/metabolism3e

the entire gamut of cardiometabolic disease; from disorders of energy excess (e.g. obesity, diabetes, hyperlipidaemia, atherosclerosis) to disorders of energy deficit (e.g. anorexia, disease-related cachexia).

As introduced in Chapter 1, the human body may be defined as being in a dynamic state of *energy balance*, whereby the energy content of the body (and its composition) reflects its inter-relationships with energy intake and energy dissipation (expenditure). In healthy adults, this metabolic steady state can be defined as a peculiar condition where production of final metabolite(s) equals consumption of the initial precursor(s). In other words, for energy stored in the body to remain stable, energy expenditure must be equivalent to energy intake. Furthermore, as the body's energetic precursors derive from dietary macronutrients, this notion can be extrapolated to individual substrate balance (expressed as energy); that is to say, the balance of an individual macronutrient (carbohydrate, protein, lipid or alcohol) is equal to its absorbed, metabolisable energy minus its oxidation.

It is important to note, however, the dynamic nature of energy balance – with rates of turnover and overall flux altering in response to energy supply (i.e. nutrient quantity and composition) and demand (physiological or pathophysiological state), the human body operates across a large range of energy intake and energy expenditure.

In terms of human physiology, one might also express the inter-relationship between energy intake and expenditure within the context of the *second law of thermodynamics*; that is to say, when energy derived from food is used by the body, such as for muscle contraction or tissue synthesis, these processes must inevitably be accompanied by the generation of heat (*thermogenesis*) which is not available for performing work, thus increasing entropy. Indeed, *'la vie est donc une combustion'* ('life is therefore a process of combustion')[1], but albeit a rather inefficient one, with approximately 75% of food energy ultimately lost as heat during its transformation into usable metabolic fuel; somewhat comparable to the efficiency of a modern petrol or diesel car engine. As such, for core body temperature to maintain constant, heat production through the biochemical reactions associated with energy transformations must be equivalent to heat loss (dissipation) through radiation, evaporation, convection and conduction (Box 3.1).

Box 3.1 A note about units

The joule (J) is the SI unit of energy, and represents the energy required to move a mass of 1 kilogram (kg) a distance of 1 metre (m) by a force of 1 newton (N). In the context of nutrition, values are largely expressed as kilojoules (kJ, i.e. 10^3 J) or megajoules (MJ, i.e. 10^6 J or 10^3 kJ). In contrast, while a kilocalorie (kcal) is defined as the energy required to raise the temperature of 1 gram (g) of water by 1 degree Celsius (°C), and is often still used in the field of nutrition and metabolism, this definition depends on the starting temperature of the water and the atmospheric pressure. It is recommended that the *thermochemical calorie conversion factor* (1 kcal = 4.184 kJ, or 1 kJ = 0.239 kcal) be used when necessary.

3.2 Cellular and mitochondrial energetics

While we consider the body to be fuelled by the energy substrates consumed via our diets (exogenous macronutrients) or from liberation of the storage form of these energy substrates from our body tissues (endogenous sources), the energy that is harnessed from the oxidation of these substrates requires the conversion of the chemical energy contained within food and bodily stores into a more useable form (i.e. adenosine triphosphate [ATP]), as outlined in Figure 3.1. The first process of this conversion from exogenous energy sources (food) is *digestion*, during which the large polymeric molecules are enzymatically broken down into their monomer subunits before entry into the cytosol, whereupon two further steps are required. First, the chemical energy contained in the nutrient must be converted into a redox form and second, this redox potential must be transduced (via ATP synthesis) as a phosphate potential (ATP/(ADP · Pi), where ADP is adenosine diphosphate and Pi is inorganic phosphate).

While macronutrients are catabolised via different pathways (outlined in Chapters 4 and 5), all of them converge towards a common intermediate: acetyl-coenzyme A (acetyl-CoA). Complete oxidation of this acetyl-CoA is achieved within the mitochondrial matrix

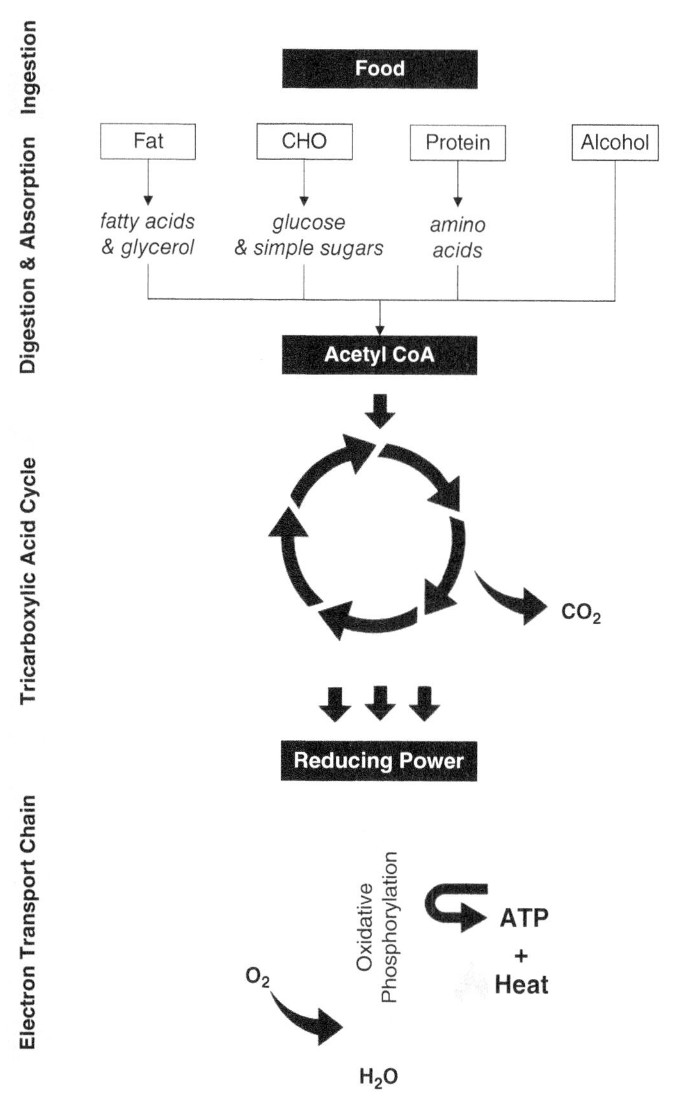

Figure 3.1 Energy transformations from food to biochemical energy. CHO, carbohydrate; CO_2, carbon dioxide, O_2, oxygen; ATP, adenosine triphosphate.

through the *tricarboxylic acid cycle*, also known as the Krebs cycle, resulting in the formation of reducing equivalents (three NADH, H^+ and one $FADH_2$) and carbon dioxide (CO_2). The high-energy electrons formed subsequently pass along an *electron transport chain* within the mitochondrial inner membrane, during which time proton pumping from the matrix to the intermembrane space establishes an electrochemical gradient, termed the *proton-motive force*. The final electron acceptor of the transport chain is oxygen, thus resulting in the formation of water (H_2O) and ultimately, the proton-motive force-driven *oxidative phosphorylation* of ADP to ATP, and dissipation of heat.

Although proton transport across the inner mitochondrial membrane is generally *coupled* to ATP synthesis (using ATP synthase as a proton channel), this process can be 'uncoupled', such that heat may be generated via the established proton gradient without the associated formation of ATP. This process is perhaps best observed within the mitochondrial-rich brown adipose tissue (BAT) where *uncoupling proteins* (UCPs), and in particular UCP1, provide an alternative proton channel to that of ATP synthase. Rather than a continual process, this proton gradient discharge and resultant liberation of heat can be activated via UCP1 by the sympathetic nervous system. The overall effect of such activation in BAT is an increase in both fatty acid release from stored triacylglycerol and tissue blood flow – the latter serving to increase oxygen supply to the tissue and dissipate the released heat to other areas of the body.

It is not surprising, therefore, that there has been considerable interest in increasing activation of uncoupling proteins for the treatment and prevention of obesity. This interest has led to the discovery of uncoupling proteins in other tissues, with homologues, such as UCP2 and UCP3, expressed in most tissues, including white adipose tissue, muscle, pancreas and brain, although their role in energy metabolism is less defined than that for UCP1 and somewhat controversial. For example, overexpression of UCP3 in skeletal muscle has been shown to reduce white adipose tissue mass, but adipose tissue mass is not altered in UCP2 and UCP3 knockout mice. Associations between UCP2 and 3 expression and diabetes risk have now been established, with UCP2 demonstrated to negatively regulate β-cell insulin secretion, while an upregulation of UCP3 in skeletal muscle may prevent the development of insulin resistance. However, despite several UCP polymorphisms having been identified in relation to lipid metabolism, obesity and diabetes, effects of these polymorphisms are largely dependent on individual and population characteristics such as ethnicity, diet, age, environment and lifestyle.

Interestingly, UCP1-containing adipocytes are also present within white adipose tissue depots, and owing to the considerable mass of white adipose tissue within the human body, the potential 'browning' of this tissue by increasing the number of these so-called beige or brite adipocytes offers an exciting opportunity for targeting thermogenesis in the context of weight management. Nevertheless, although adrenergic stimulation has been shown to lead to transformation of white adipocytes to brown in rodents, and a number of pharmacotherapies have been shown to stimulate BAT activity in humans, effective, safe and sustainable approaches are yet to be realised.

Energy may also be dissipated via a process known as *substrate cycling*, during which opposing metabolic pathways are in operation, with no net biochemical change beyond an increase in ATP utilisation (i.e. no net anabolic or catabolic transformation is performed). Examples of such cycling can be observed in pairs of kinase and phosphatase reactions commonly observed in human hepatocytes and adipocytes, for example the glucose-6-phosphate/glucose cycle. Often termed *futile cycling*, the purpose of many such cycles beyond the consumption of excess energy and the dissipation of heat is not entirely understood. However, it is recognised that substrate cycling allows for maintenance of a pool of intermediary metabolites and the active enzymes which produce them, thus ensuring a rapid cellular response to any event that may require their use.

The glycerolipid-free fatty acid cycle is another example of substrate cycling consuming ATP and dissipating heat, whereby synthesis of glycerolipid occurs via esterification of free fatty acids (also known as non-esterified fatty acids, NEFA) onto a glycerol backbone and is subsequently followed by its hydrolysis, and release of the free fatty acid for re-esterification. Each 'cycle' consumes seven molecules of ATP, with the anabolic and catabolic segments completed either within the same cell or between tissues (e.g. gluconeogenesis and de novo lipogenesis occurring in the liver, and lipolysis and re-esterification occurring in adipose tissue). Beyond its role in energy homeostasis and thermogenesis, the glycerolipid-free fatty acid cycle is considered important for the generation of metabolites involved in the control of multiple biological processes, including insulin secretion, cell proliferation and gene expression and protection from lipotoxicity. These established and hypothesised functions are supported by evidence of associations between disordered lipid cycling and several pathological states, such as insulin resistance, type 2 diabetes, non-alcoholic fatty liver disease and cancer.

The Cori cycle is also considered as a 'futile' interorgan cycle, with a net consumption of four ATP equivalents (two ATP molecules produced in the skeletal muscle and six consumed by the liver) and provides a mechanism to transfer aerobically synthesised ATP from lipid oxidation in the liver to anaerobic glycolytic ATP in peripheral cells. However, it is important to note that while anaerobic ATP production does occur at a cellular level, whole-body energy metabolism is always purely aerobic. This is best illustrated by considering lactate excretion: anaerobic metabolism (in human cells) implies that the endproduct of glycolysis must be excreted as lactate. This is indeed the case for isolated cells and tissues, but overall body lactate excretion is negligible, with lactate ultimately metabolised or oxidised. Under conditions of lactate accumulation, as occurs during a short bout of exercise, anaerobic metabolism does contribute to net ATP synthesis, but the accumulated lactate is subsequently metabolised without net excretion during the recovery period. Similarly, energy metabolism remains fully aerobic in a diseased state where blood lactate concentrations are altered but stable (e.g. sepsis, liver failure), regardless of the concentration itself.

3.3 Energy intake

Unable to harness energy from sunlight as plants do, animals, including humans, must derive energy from the catabolism of nutrients, requiring a complex pathway of energetic transformations (as described above). However, when considering energy metabolism at a whole-body level, it is important to consider the control of food intake (for example, patterns of consumption and food choice, discussed in Chapter 15), the energetic properties of the food itself and the physiological processes that occur well before its molecular components reach the body's cellular machinery.

Measurement of food energy

Food consists of a complex matrix of macro- and micronutrients, but fortunately its gross chemical energy can be measured fairly simply using a *bomb calorimeter*. In its original design, this device consists of an insulated chamber surrounded by a water jacket, into which a small sample of food is placed. The chamber is then pressurised with oxygen and its contents ignited. The heat released from the combustion of the food sample, and therefore its gross energy, is calculated by measuring the change in the temperature of the surrounding water. However, as discussed below, not all of this combustible energy is available to fuel the human body.

From ingestion to metabolism

While the total ingested energy (IE) of a consumed food is equal to its gross energy (GE), a cascade of energetic losses occurs between its ingestion and usage of the metabolic substrates it provides. The first such loss is the excretion of undigested, or indeed undigestible, food in the faeces. The magnitude of this faecal energy (FE) loss is dependent not only on the properties of the food that is ingested (e.g. the cooking method used, whether macronutrients are of animal or plant origin), but also the functionality of an individual's gastrointestinal system in terms of both physiological and pathophysiological variability (see Chapter 10).

In general, the digestibility of a typical Western diet is fairly high: approximately 97% for ingested carbohydrates, 95% for ingested fats and 92% for proteins. Unabsorbed carbohydrates within the colon may also undergo microbial fermentation, resulting in the production and loss of combustible gas energy (GaE). Furthermore, some short-chain (volatile) fatty acids may be formed, absorbed and utilised as an energy source either locally (e.g. butyrate usage by intestinal epithelial cells) or via the portal vein (e.g. propionate as a gluconeogenic precursor, or butyrate and acetate as lipid precursors). Therefore:

$$\text{Digestible energy(DE)} = \text{IE(or GE)} - \text{FE} - \text{GaE}$$

Following digestion, the incomplete oxidation of protein and other nitrogenous material leads to further energetic losses through the formation of nitrogenous waste compounds, predominantly urea, which are excreted in urine (urinary energy, UE). Notably, urea contains approximately a quarter of the gross energy of the original protein, i.e. ~5.2 kJ/g of protein oxidised (gross energy of protein ~22.9 kJ/g). In addition to nitrogenous waste compounds, energy may be lost through the urine in various disease states, for example in individuals with type 2 diabetes, which is characterised by hyperglycaemia and inability of the kidneys to reabsorb all the filtered glucose, resulting in glucose (and therefore energy) loss in urine. A very small quantity of DE is also lost from the body surface (surface energy, SE), with the remaining energy deemed 'metabolisable':

$$\text{Metabolisable energy(ME)} = \text{DE} - \text{UE} - \text{SE}$$

However, it should be noted that not all of this gross metabolisable energy is available for ATP production owing to obligatory energy losses associated with the digestive and absorptive processes and the intermediary metabolism of food, and the heat associated with microbial fermentation (Box 3.2).

Box 3.2 Atwater conversion factors

In order to simplify the calculation of metabolisable energy (ME), a number of conversion factor systems have been proposed, undoubtably the most well known and widely adopted being that of Wilbur Olin Atwater, proposed more than 120 years ago from experiments on a small number of young men, consuming an American diet typical at the time.

- Carbohydrate: $ME_{Atwater}$ = 17 kJ/g
- Fat: $ME_{Atwater}$ = 37 kJ/g
- Protein: $ME_{Atwater}$ = 17 kJ/g
- Alcohol: $ME_{Atwater}$ = 29 kJ/g

Whilst generally accepted to be a good approximation of the metabolisable energy of the macronutrients, these conversion factors are not without limitation and criticism. In particular, the Atwater conversion factors are based on averages, despite considerable variability in terms of the gross and digestible energy densities within macronutrient categories, and generally overlook dietary fibre content. However, despite acknowledgement by bodies such as FAO/WHO that a move toward a conversion system based on *net metabolisable energy* (the ATP-generating potential of a food) would be beneficial, a feasible, large-scale alternative to the Atwater factors is yet to be proposed.

Measurement of energy intake

Under conditions of uncontrolled energy intake, or whenever constant supervision is not possible/undertaken, the accurate measurement of energy intake is notoriously difficult. Current methods include retrospective assessment (for example, food frequency questionnaires, 24-hour recall, dietary history) which relies on the memory of the individual, estimation of portion size and/or objectivity of both the investigator and subject. On the other hand, prospective methods are labour-intensive and dependent on strict reporting or observation, and the exact composition of each meal and its components being known. As such, the assessment of energy intake is prone to a large degree of measurement error. This is particularly true when dietary fat is considered, where a small under- or overestimate of intake of the order of one tablespoon of fat (~20 g, ~740 kJ) could lead to an error in calculated intake of the order of 10%. Therefore, energy intake is often determined in the context of energy balance, where, providing body weight remains stable across the measurement period, energy intake can be assumed to equal energy expenditure. Hence, where energy intake is independently assessed in a free-living setting, validation against an objective measure of total energy expenditure (such as via the doubly labelled water technique) is recommended (see Section Measurement of energy expenditure).

3.4 Energy storage

Metabolisable energy that is not used directly may be stored by the body in order to provide metabolic fuel to tissues when direct supply from the gastrointestinal tract is limited, such as occurs during fasting (see Chapter 2). The balance between oxidation and storage of each fuel type is under the regulation of several feedback systems related to energetic supply and demand, and ultimately controlled by hormones – principally insulin, glucagon and thyroid hormones – and the sympathetic nervous system. In his 1970 seminal work 'Starvation in man', Cahill estimated a reference man of 70 kg, with a basal metabolic rate of 7531 kJ, to have approximately 473 kJ of circulating metabolic fuel, composed of ~20 g glucose, 3 g triglycerides and 0.3 g fatty acid. However, significant inter- and intraindividual variability in this circulating energy pool undoubtedly exists based on nutritional and disease status, for example. It should also be remembered that several other substrates within the circulation can act as metabolic fuels (Table 3.1).

Table 3.1 Circulating energy sources.

Fuel	Source
Glucose	Direct carbohydrate; glycogen stores; gluconeogenesis in liver and kidney from lactate, amino acids and glycerol
Free fatty acids (FFA)	Dietary fats; triacylglycerol stores (especially in adipose tissue); synthesised from carbohydrate in liver and adipose tissue, especially after feeding on low-fat diets
Amino acids	Dietary protein; tissue protein stores; synthesised from carbohydrates
Ketone bodies (acetoacetate, 3-hydroxybutyrate)	Produced from FFA and some amino acids in liver
Glycerol	Produced from triacylglycerol breakdown
Lactate	Anaerobic glycolysis
Acetate	Gut fermentation of carbohydrates; produced from FFA in liver and muscle, and from ethanol in liver
Ethanol	Dietary intake; gut fermentation
Fructose	Dietary sucrose
Galactose	Dietary intake, especially as milk lactose

Source: M. Elia 2000/With permission of Elsevier.

In general, with the exception of glycogen, tissue energy stores are less prone to short-term fluctuation. Furthermore, while alcohol is not stored by the body, but rather the ethanol appearing in the circulation following consumption is rapidly oxidised, the storage capacity for carbohydrate, fat and protein, and the degree to which each store is autoregulated, do vary considerably (Table 3.2). Alcohol also plays a role here, since it is oxidised in priority to other fuel substrates, thereby sparing fat from oxidation and hence disrupting nutrient balance.

The exact mechanisms underlying the overall control of *nutrient (energy) partitioning* towards storage as either fat or lean tissue remain elusive, although this energy-partitioning characteristic is highly variable between individuals and influenced by genetics, habitual body composition (in the event of weight gain or loss), endocrine factors such as circulating glucocorticoids, and diet. With regard to the latter, there is now evidence of a role for essential polyunsaturated fatty acid intake in enhancing the partitioning of nutrient energy towards lean mass versus fat mass. However, given the inter-relationships between body composition, energy intake and energy expenditure, determining causality remains problematic. Regardless, it should be noted that these energy stores play an important role in both the control of energy intake (as discussed in Chapter 15) and the determination of energy expenditure (see Section 3.4).

Measurement of stored energy

The assessment of stored energy requires body composition analysis at the molecular level, where the body is composed of water, lipid, proteins, glycogen and minerals. However, while such analysis is possible post mortem, whole-body measurements of energy storage *in vivo* are difficult, if not impossible, and estimation techniques require a number of assumptions to be made.

Glycogen storage

The size of the carbohydrate storage pool (as glycogen within muscle and liver) is limited, and as such this pool is under tight autoregulatory control – that is, it can regulate its own oxidation. However, it is considered an inefficient energy reserve as each gram of glycogen can only be stored with 3 g of water, thereby resulting in an energy reserve with a low energy density (Table 3.2).

Quantification of glycogen *in vivo* is relevant to a number of disease states (e.g. diabetes, glycogen storage disease), although current methodologies are largely impractical within a clinical setting. These methodologies include ultrasound, positron emission tomography (PET) and ^{13}C and ^{1}H magnetic resonance spectroscopy (MRS) but several important limitations exist. For example, whilst MRS techniques have a higher specificity than ultrasound, they require specialised equipment and the more sensitive of the techniques, ^{1}H MRS, may still underestimate

Table 3.2 Macronutrient stores.

Substrate	Storage form	Tissue	Pool size	% associated stored water	Energy density (kJ/g)	Autoregulation	Postprandial thermogenesis
Carbohydrate	Glycogen	Liver and muscle	Small (limited)	~75	~4	Accurate	Average
Fat	Triglyceride	Adipose tissue	Moderate – large (unlimited)	~5	~33	Poor	Low
Protein	Protein	Lean tissue	Moderate (limited)	~73	~4	Accurate	High
Alcohol	(none)						

Source: Adapted from Schutz Y, Garrow JS. Energy and substrate balance, and weight regulation. In: Garrow JS, James WPT, Ralph A (eds) *Human Nutrition and Dietetics*, 10th edn. Edinburgh: Churchill Livingstone, 2000, pp. 137–148.

in vivo glycogen levels. Similarly, PET using exogenous isotope labelling only detects the synthesis of labelled glycogen rather than the total glycogen pool. Recently, a new technique has been proposed, using the magnetic coupling between glycogen and water protons through the nuclear Overhauser enhancement (NOE), which may overcome these issues. However, this technique is still in its infancy.

Body protein

In common with glycogen, protein is stored with water within the body's fat-free mass (FFM). In adults, a hydration constant of ~73% is assumed for FFM although variability is known to occur with, for example, age, sex, race, weight loss and various disease states. Assuming that all body nitrogen is incorporated into protein, with a nitrogen-to-protein ratio of 0.16, measurement of total body nitrogen by *in vivo* neutron activation (IVNA) is considered the criterion method for the quantification of total body protein. IVNA allows measurement of the main elements of atomic body composition, including hydrogen, carbon, nitrogen, oxygen, phosphorus, calcium, chlorine, potassium and sodium. In brief, following exposure of either the total or partial body to a beam of neutrons, these elements can be quantified by delayed-γ neutron activation, prompt- γ neutron activation or inelastic neutron scattering, and subsequently molecular-level body composition (i.e. fat, protein, water and minerals) can be calculated. However, due to associated radiation exposure, high costs and the technical expertise required, usage of the technique is now rare.

Instead, non-IVNA models have been proposed, including the use of total body potassium (measured by quantifying γ rays emitted by a naturally radioactive isotope [^{40}K] contained within all potassium) as a proxy for total body protein, given potassium is only present in the FFM compartment and assuming a constant content. In brief, potassium is the major cation within intracellular water, at an approximate concentration of 140 mmol/l, compared to approximately 4 mmol/l within extracellular fluid (i.e. a 35:1 intracellular-to-extracellular ratio). Combined with a 2:1 ratio of intracellular-to-extracellular water, the vast majority of potassium (~98%) is therefore located within cells of the FFM. However, other assumptions need to be made regarding, for example, bone mineral content. Hence, multicompartment models have been proposed, such as that of Heymsfield and colleagues, which combines the assessment of total body potassium by whole-body ^{40}K counting, total body water by tritium dilution and bone mineral by dual-energy X-ray absorptiometry (DEXA).

However, inaccessibility of the criterion method (IVNA) means that scope to develop new methodologies and models for the *in vivo* assessment of body protein per se is restricted. Protein storage is therefore largely approximated by the assessment (or calculation) of FFM rather than directly assessed, for example by densitometry techniques, DEXA or bioimpedance analysis.

Body fat

The largest energy storage pool is body fat, with a theoretically limitless storage potential and poor autoregulation. However, as with protein storage, direct assessment of fat storage *in vivo* is currently impossible beyond cadaver dissections. However, a number of techniques are used within clinical and research settings to estimate either fat mass or adipose tissue mass, the latter comprising both fat and its cellular and intracellular supporting structures, with an average composition of 80% lipid, 14% water, 5% protein and <1% mineral, and a density of 0.92 g/cm^3 at body temperature. Importantly, there is known variability in the fat content of adipose tissue, with an increase in the lipid fraction observed with increasing adiposity.

As with protein mass assessment, the most accurate assessments of fat are made by combining multiple indirect techniques, with the *four-compartment model* widely regarded as a gold standard. This model is based on body weight being the sum of four distinct compartments – namely, fat, water, bone mineral and the residual (protein, non-bone mineral, glycogen) – each assumed to have a fixed density. By measuring body volume (for example, by underwater weighing or air displacement plethysmography), total body water (for example, by stable isotope dilution) and bone mineral content (for example, by DEXA), total body fat can be derived.

3.5 Energy expenditure

The total amount of energy expended daily (*total energy expenditure*, TEE) by the human body represents the energetic cost of the maintenance of vital functions, the energy cost associated with the digestion and absorption of nutrients and physical activity energy expenditure. There is considerable variability in TEE between individuals, some of the sources of which are discussed below. However, a large-scale 2021 study by the International Atomic Energy Association (IAEA) Doubly Labelled Water Consortium has revealed that TEE adjusted for FFM increases rapidly from birth to one year of age, after which it declines slowly until ~20 years, remaining stable from 20–60 years (even during pregnancy), and then further declining in older adults. Work remains to be done to characterise life course trajectories of the individual compartments of energy expenditure (see Section Measurement of stored energy) to determine the primary drivers of this pattern of TEE.

Compartments of energy expenditure

Energy expenditure across the day is often divided into different compartments, each of which can be individually determined. However, such divisions depend on the context in which energy expenditure is being investigated and assessed (Figure 3.2). For example, different compartmentalisation will be necessary if considering expenditure in a thermal context (i.e. heat loss) versus a context of physical activity.

In most people, in the absence of excessive physical activity, the largest compartment of energy expenditure is *basal metabolic rate* (BMR). Accounting for, on average, ~60–75% of total daily energy expenditure, BMR refers to the minimum energy cost of preserving the integrity of the body's vital functions, such as maintaining cardiorespiratory function, ionic equilibrium across cell membranes and cell and protein turnover. Intraindividual variability in BMR (i.e. day-to-day fluctuations) is small, but higher in premenopausal women (~2–5%) than men (~1–3%) owing to menstrual cycling. However, the main determinant of BMR is FFM, accounting for approximately two-thirds of interindividual variability.

Fat-free mass can be divided into two main components – skeletal muscle mass and mass of the organs and remaining tissues – and it should be noted that each component differs in terms of its contribution to BMR (Table 3.3), with the relative metabolic rate (MJ/kg) of the organs being ~15–40 times greater than that of skeletal muscle and ~50–100 that of adipose tissue. In agreement with this, data indicate that in individuals with obesity these specific metabolic

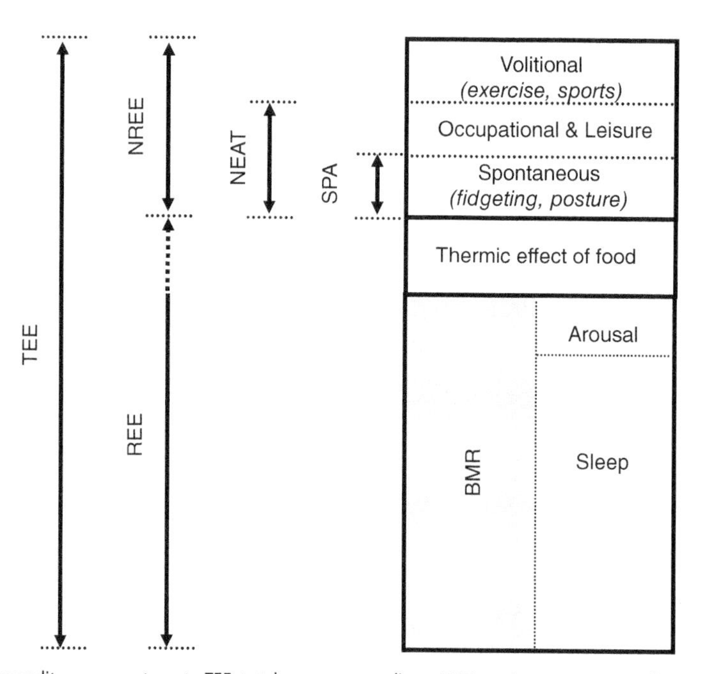

Figure 3.2 Energy expenditure compartments. TEE, total energy expenditure; REE, resting energy expenditure; NREE, non-resting energy expenditure; NEAT, non-exercise activity thermogenesis; SPA, spontaneous physical activity; BMR, basal metabolic rate. Note that, as stated in Box 3.3, best practice dictates measurement of REE be undertaken in the fasted state. *Source:* Adapted from Dulloo AG, Schutz Y. Adaptive thermogenesis in resistance to obesity therapies: issues in quantifying thrifty energy expenditure phenotypes in humans. Curr Obes Rep 2015;4(2):230–240.

Box 3.3 Basal metabolic rate (BMR) versus resting metabolic rate (RMR)

Despite the term often being used erroneously, particularly in the popular media, to describe any measure of metabolic rate at rest, BMR is a strictly defined physiological variable. Its measurement requires a strict set of standardised conditions to be upheld. Specifically, it should be measured in the morning, following a 10–12h overnight fast (i.e. the postabsorptive state) and with the subject in thermal comfort, complete rest (supine), free from emotional stress and familiar with the environment and experimental procedures, and fully awake. In contrast, beyond restriction of physical activity, RMR (also called resting energy expenditure, REE) measurement lacks such standardisation, although best practice dictates that such measurements be undertaken with subjects in a fasted, comfortable state and in a supine or seated position, following a defined period of rest. As such, RMR ≥ BMR.

rates for the various tissues and organs expressed per unit body weight may be slightly overestimated by ~2%. Nevertheless, even after accounting for age, gender, body weight and FFM, ~10% of variance in BMR remains unaccounted for but is probably primarily attributable to genetic factors and underlying measurement error.

The *thermic effect of food* (TEF) refers to the acute rise in metabolic rate that occurs following ingestion of food or a meal, and which can persist for several hours thereafter. About three-quarters of our day is spent in the postprandial state, and the TEF contributes to ~10% of daily energy expenditure. Although numerous factors (including gender, menstrual cycle, age, nutritional status and genetic factors) are known to influence the TEF, its magnitude and duration, to a large extent, are dependent on the energy content of the food (or meal) consumed and its macronutrient composition. This is in part due to the energy costs associated with digestion, absorption, processing and/or tissue synthesis (energy storage) of macronutrients from food.

The TEF of specific fuel substrates, expressed as a percentage of their energy content, is about 5–8% for carbohydrates, 2–3% for fats, 20–30% for proteins and 22% for alcohol. Furthermore, in addition to the *obligatory* energetic costs associated with the metabolic processing of nutrients, the TEF also encompasses a *facultative* component. That is, an increase in metabolic rate associated partially with the sensory aspects of the food consumed (taste, smell) and also partially due to the stimulation of the sympathetic nervous system, in particular in response to glucose and high carbohydrate-containing foods or meals. A number of foods and bioactive food ingredients (e.g. caffeine, capsaicin, certain polyphenols) have been demonstrated to enhance energy expenditure via sympathomimetic stimulation, perhaps the best characterised of which is green tea. Rich in catechin-polyphenols and also containing caffeine, acute green tea consumption increases daily energy expenditure by ~4%, through an increase in fat oxidation.

The *non-resting* compartment of daily energy expenditure is the most variable. As elaborated upon in Chapter 18, the increase in expenditure

Table 3.3 Relative contribution of organs and tissues to basal metabolic rate (BMR) in a 70 kg 'reference man'.

	Weight of tissue		Protein content	Organ/tissue metabolic rate (per day)			
	(kg)	(% body weight)	(kg)	(MJ/kg)	(MJ/kg protein)	(MJ/day)	(%BMR)[a]
Liver	1.8	2.6	0.32	0.84	4.7	1.51	21
Brain	1.4	2.0	0.11	1.00	12.8	1.41	20
Heart	0.33	0.5	0.055	1.84	11.0	0.61	9
Kidney	0.31	0.4	0.053	1.84	10.8	0.57	8
Muscle	28	40.0	4.80	0.054	0.32	1.52	22
Adipose tissue	15	21.4	0.75	0.019	0.37	0.28	4
Miscellaneous by difference (e.g. skin, intestine, bone, glands)	23.16	33.1	4.51	0.049	0.25	1.13	16
Whole body	70	100	10.6	0.10	0.66	7.03	100

[a] BMR ~7.03 MJ/day.
Source: M. Elia 2000/With permission of Elsevier.

resulting from physical activity is dependent on the type of activity undertaken, its intensity and duration; the energy cost of particular physical activities is often expressed as a multiple of BMR, termed a *metabolic equivalent* (MET). In highly active individuals (for example, manual labourers or competitive athletes), this compartment may account for up to 70% of TEE, but is likely to average only 10–15% in most people living in industrialised (relatively sedentary) societies. However, it should be noted that most physical activities comprise both dynamic and static (isometric) work, albeit in varying proportions. Dynamic work is considered as work performed on the environment and to be mostly inefficient in that the muscle contraction produces 3–4 times more heat than mechanical energy; the mechanical efficiency is thus about 25%. In contrast, isometric work refers to an increase in muscular tension (without muscle shortening), when no external work is performed on the environment and only heat is generated. Many of the activities associated with daily life comprise a large portion of isometric activities, and indeed it is a change in muscle tone that underlies differences in metabolic rate during the maintenance of different body postures (e.g. lying versus sitting versus standing).

Non-resting energy expenditure can be broadly divided into two categories: volitional structured physical activity (e.g. exercise, sports) and non-exercise physical activity. The latter, referred to as 'non-exercise activity thermogenesis' (NEAT), can be subdivided further into energy expenditure related to occupational/leisure physical activity and spontaneous physical activity (SPA), the latter being essentially involuntary and subconscious (see Figure 3.2). SPA is highly variable between individuals, with studies indicating an eight-fold range (100–800 kcal/day) in the contribution of SPA to TEE. It includes activities such as postural changes and fidgeting, and is probably primarily biologically driven although also heavily influenced by environmental factors. However, despite their importance to energy balance and body weight regulation, the energetic costs of such low-level physical activities (i.e. generally below 2 METs) are difficult to measure owing to their magnitude relative to intraindividual variability in RMR and potential measurement errors.

Measurement of energy expenditure

As discussed in sections 3.1 and 3.2, energy expended by the body is liberated as heat, and as such assessment of heat production or heat loss will provide a measure of metabolic rate, noting, of course, that in accordance with the first law of thermodynamics, energy cannot be produced (or destroyed) but rather the body is transforming energy to produce and dissipate heat.

Energy expended by the body by oxidative processes and anaerobic glycolysis is ultimately transformed into heat and external work. Therefore, energy expenditure can be directly measured by quantifying *heat loss*. The technique for measuring heat loss, termed *direct calorimetry*, was first pioneered in the eighteenth century by Lavoisier and Laplace, who constructed a triple-walled calorimetry chamber. Snow was packed into the outer shell, and ice within the inner shell, with a guinea pig placed within the inner chamber. As the heat of the animal melted the ice, the weight of the resulting water allowed for the calculation of heat loss and therefore metabolic rate. The same principles are applied by modern human direct calorimeters, which must measure all components of heat transfer – usually achieved via the combined measurement of radiant, convective and conductive heat transfer (i.e. the 'dry' heat component), and a separate measurement of evaporative heat transfer, including insensible water vapour loss through the skin, respiratory tract and sweat glands. However, such devices are rare owing to the high costs associated with building and maintaining the equipment, and several limitations, including the inability to measure short-term changes related to feeding or other situations where an increase in heat storage may occur (when body temperature transiently rises) – with heat loss only reflective of heat production when total body heat content (composition and mean body temperature) remains constant.

Owing to the difficulties described above, while direct calorimetry remains an important technique in the study of thermoregulation, the technique of *indirect calorimetry* now dominates in the assessment of energy expenditure. This involves the assessment of *heat production* by the measurement of oxygen consumption ($\dot{V}O_2$) and carbon dioxide production ($\dot{V}CO_2$), with measurements made over minutes to hours using a facemask or ventilated canopy, or hours to days using a whole-body respiration chamber, to determine the difference between inhaled and exhaled gas concentrations and respiratory volumes. As discussed earlier (Section 3.2), with the exception of anaerobic glycolysis, ATP synthesis is coupled to substrate oxidation, with the rate of ATP utilisation determining the rate of substrate oxidation (metabolic rate), and therefore oxygen consumption. That is:

$$\text{Substrate} + O_2 \rightarrow CO_2 + H_2O - \Delta\text{enthalpy}$$

Hence, if we can measure the volume of oxygen consumed, we are able to calculate the amount of heat produced, and consequently the energy expended. Note the convention of using a negative sign to indicate enthalpy change (heat production) when heat is released. This can be demonstrated by solving the stoichiometric equation for oxidation of one mole of glucose:

$$C_6H_{12}O_6 + 6O_2 \rightarrow 6CO_2 + 6H_2O - \text{heat}$$

$$\begin{aligned} &(180\text{g glucose}) + (6 \times 22.4\text{L } O_2) \\ &\rightarrow (6 \times 22.4\text{L } CO_2) + (6 \times 18\text{g } H_2O) \\ &-2.8\text{MJ} \end{aligned}$$

Energy yields from other nutrients can be similarly calculated, with values differing slightly within macronutrient class depending on the exact substrate being oxidised, as shown in Table 3.4. Various equations are available to convert measurements of $\dot{V}O_2$, either alone or in combination with $\dot{V}CO_2$ and urinary nitrogen excretion, into whole-body energy expenditure (EE). For example, the Weir equation:

$$\text{EE(kJ / day)} = 4.184 \times \left[5.68\dot{V}O_2 + 1.59\dot{V}CO_2 - 2.17N_u\right]$$

where $\dot{V}O_2$ and $\dot{V}CO_2$ are measured in ml/min, expressed at standard temperature (0°C) and pressure (760 mmHg) and dry (STPD), and N_u is urinary nitrogen excretion in g/day. Note the dot over the $\dot{V}$, indicating that this measure is a rate rather than a volume.

Besides its use in the assessment of whole-body energy expenditure, indirect calorimetry can also provide important information regarding the nature of the substrates being oxidised, by considering the ratio of the volume of carbon dioxide produced to the volume of oxygen consumed ($\dot{V}CO_2/\dot{V}O_2$). This ratio, termed the *respiratory quotient*, varies from ~1 when exclusively glucose is being oxidised to ~0.8 for protein, ~0.7 for fat and ~0.66 for alcohol (Table 3.4). By measuring urinary nitrogen excretion, and assuming oxidation of 1 g of protein requires the consumption of 0.97 ml of O_2, and produces 0.77 ml of CO_2, liberating 17 kJ of heat, one can improve estimates of the relative proportion of fat and carbohydrate being oxidised by subtracting these parameters to obtain *non-protein RQ*. However, RQ should be interpreted with some caution as there is large potential for error due to changes in ventilation rate (hyper- or hypoventilation) or acid–base balance (acidosis or alkalosis). Under such situations, RQ measurements are no longer representative of cellular level respiration, which is why some prefer the use of the term *respiratory exchange ratio* (RER). Similarly, RQ may also exceed 1 in situations where rates of fat synthesis from non-lipid sources (de novo lipogenesis) are high, or be less than 0.7 when rapid weight loss and net ketogenesis are occurring.

Although useful, the concomitant measurement of urinary nitrogen excretion during gas exchange assessments is not a requirement of indirect calorimetry, as an estimate of nitrogen

Table 3.4 Energy yields from substrate oxidation.

Substrate	O_2 consumed (l/g)	CO_2 produced (l/g)	Respiratory quotient (RQ)	Heat released (kJ/g)	Energy equivalent[a] VO_2 (kJ/l)	Energy equivalent[a] VCO_2 (kJ/l)
Starch	0.829	0.829	1.00	17.6	21.2	21.2
Saccharose	0.786	0.786	1.00	16.6	21.1	21.1
Glucose	0.746	0.746	1.00	15.6	21.0	21.0
Lipid	2.019	1.427	0.71	39.6	18.6	27.7
Protein	1.010	0.844	0.83	19.7	18.9	23.3
Lactic acid	0.746	0.746	1.00	15.1	20.3	20.3

[a] STPD.

Source: Adapted from Schutz Y, Jéquier E. Handbook of Obesity: Etiology and Pathophysiology, 2nd edn. New York: Marcel Dekker, 2004.

excretion can generally be used. However, provided that little or no alcohol is consumed during the measurement period, urinary nitrogen excretion measurements allow for the calculation of individual oxidation rates (g/min) of protein (PRO_{OX}), carbohydrate (CHO_{OX}) and fat (FAT_{OX}). Specifically, as ~1 g urinary nitrogen is excreted per 6.25 g protein:

$$PRO_{OX} = 6.25N$$

$$CHO_{OX} = 4.59\dot{V}CO_2 - 3.25\dot{V}O_2 - 3.68N$$

$$FAT_{OX} = 1.69\dot{V}O_2 - 1.69\dot{V}CO_2 - 1.72N$$

Caution is needed in the interpretation of these calculations under conditions where, for example, lipid synthesis is taking place (and hence values of CHO_{OX} are instead representative of both oxidation and conversion to fat, i.e. the net rate of carbohydrate utilisation), or if there is gluconeogenesis from amino acids, when CHO_{OX} reflects the net rate of carbohydrate disappearance. Under these conditions, FAT_{OX} will be underestimated by ~10% of the rate of glucose synthesis, and urinary nitrogen will originate from alanine deamination, leading to similar underestimation of PRO_{OX}.

Indirect calorimetry equipment is relatively accessible, and unlike direct calorimetry, it has a relatively quick response time, allowing assessment of various components of daily energy expenditure. However, in the assessment of daily energy expenditure by indirect calorimetry in a respiration chamber, the mobility of the subject is somewhat restricted, meaning that results may not necessarily be representative of free-living conditions and alternative methods may be required.

Despite not being able to provide information about the nature of the macronutrient being oxidised per se, and its relatively high cost, the technique of *doubly labelled water* (DLW) is considered the gold standard for measuring energy expenditure under free-living conditions. It provides an integrated measure of total energy expenditure over a time period of several days to ~2 weeks, depending on an individual's rate of water turnover (with children having faster turnover than adults and active individuals faster than sedentary). This DLW technique is centred on the knowledge that energy expenditure involves the production of water and CO_2, and the enrichment of body water with two stable (non-radioactive) isotopes: ^2H (deuterium) and ^{18}O. Following consumption of a heavy water drink containing both isotopes, equilibration of the isotopes within the body water occurs within a few hours. The rate of disappearance of the two isotopes is then determined from body fluid samples (for example, urine or saliva) collected over the subsequent days/weeks. $\dot{V}CO_2$ is calculated from the difference between these disappearance rates, based on the assumption that ^2H is only excreted as water, while ^{18}O will be excreted in both as water and CO_2. In order to then calculate TEE, VO_2 must be estimated using an alternative technique, most commonly by concurrently assessing food intake, with the assumption that under conditions of energy balance, substrate oxidation will reflect substrate supply and therefore RQ will be equal to the food quotient (the macronutrient ratio of the diet; FQ). Specifically:

$$\dot{V}O_2(L/day) = 0.966Protein_{Intake} + 2.019Fat_{Intake} + 0.829CHO_{Intake}$$

$$RQ = FQ = \dot{V}CO_2 / \dot{V}O_2$$

where the intake of each macronutrient is expressed as g/day. However, despite its accuracy in assessing free-living TEE, the DLW technique is unable to provide time-course data or differentiate between the different compartments of energy expenditure described above (Section Measurement of stored energy).

A number of lower cost methods have been employed to estimate energy expenditure under free-living conditions, with growing popularity within retail consumers; for example, heart rate monitoring and accelerometery. However, the utility of these techniques is hampered by a low level of accuracy. In particular, the relationship between heart rate and energy expenditure is only linear within the zone of moderate-to-high intensity physical activity (above ~40% of maximal aerobic capacity), making estimations of low-level physical activity within the sedentary zone erroneous, and further confounded by disproportionate changes in heart rate versus energy expenditure during common activities such as postural change and food ingestion. Similarly, accurate and precise algorithms to extrapolate measurements of acceleration to values of energy expenditure across the spectrum of daily body movements are currently lacking. Therefore, while these techniques may provide estimates of energy expenditure at the group or population level, their validity at the individual level, and particularly within the sedentary (low physical activity) zone, remains limited.

3.6 Energy requirements

In the previous sections, we have discussed the energy balance triad – energy intake, stored energy and energy expenditure – but it is also important to consider how much energy the human body actually requires, and how energy requirements can be calculated. Indeed, matching energy intake to requirements is key to the maintenance of body size and body composition (i.e. energy balance) and ultimately, long-term metabolic health. In children and pregnant or lactating women, this energy requirement includes energy needs associated with tissue accretion and milk production/secretion.

When body weight is known to be stable, and therefore implying that the individual is in energy balance, energy requirements can be determined by measuring TEE, or its individual components, using the methodologies described above (Section Measurement of energy expenditure). However, when such measurements are unavailable or unfeasible, a number of predictive equations are available, taking into account the primary determinants of BMR (in particular, age, gender, body size and/or composition) and physical activity level (PAL). The most widely used predictive equations for BMR, and still recommended by the FAO/WHO/UNU, are those derived by Schofield in 1985 (Table 3.5). However, it has also been widely acknowledged that as almost half the data from which these equations were derived were collected in the late 1930s and early 1940s from Italian men with relatively high BMRs, the appropriateness of their use in other geographic and ethnic population groups is questionable. Similarly, in healthy, normal-weight adults, the Harris–Benedict equations, originally derived in 1918, generally perform well in the prediction of RMR, but their accuracy is reduced in older adults or individuals with high or low body mass index.

To address such deficits, several population-specific predictive equations are now available (such as those of Henry and Cole), with continual developments occurring, particularly with increasing use of machine-learning techniques, enabling researchers and clinicians to select predictive equations based on the demographics of their particular study population where possible.

Where PAL cannot be measured, it is estimated as a multiple of BMR based on the type of activities habitually performed by an individual or population group, their intensity and duration. For example, a sedentary lifestyle may correspond to a PAL of ~1.4 compared to a vigorous lifestyle with an

Table 3.5 Schofield equations for estimating BMR from body weight.

	Age (years)	N	BMR equation (MJ/day)	SEE
Males				
	< 3	162	0.249 × BW – 0.127	0.292
	3–10	338	0.095 × BW + 2.110	0.280
	10–18	734	0.074 × BW + 2.754	0.441
	18–30	2879	0.063 × BW + 2.896	0.641
	30–60	646	0.048 × BW + 3.653	0.700
	≥60	50	0.049 × BW + 2.459	0.686
Females				
	< 3	137	0.244 × BW – 0.130	0.246
	3–10	413	0.085 × BW + 2.033	0.292
	10–18	575	0.056 × BW + 2.898	0.466
	18–30	829	0.062 × BW + 2.036	0.497
	30–60	372	0.034 × BW + 3.538	0.465
	≥60	38	0.038 × BW + 2.755	0.451

BW, body weight in kg; SEE, standard error of estimate.
Source: Adapted from Food and Agriculture Organization of the United Nations. Human energy requirements: report of a joint FAO/WHO/UNU Expert Consultation, Rome, 17–24 October 2001. Rome: FAO, 2003a.

estimated PAL of ~2.25. In other words, TEE of these individuals would equal the estimated BMR × 1.4, and estimated BMR × 2.25, respectively. Another estimate of PAL can be derived from the ratio of TEE (assessed by DLW) to BMR or RMR, the latter either measured by indirect calorimetry or obtained from predictive equations. The most recent population estimates of PAL for the UK population, published by the Scientific Advisory Committee on Nutrition (SACN 2011) and derived from DLW measures, recommend the use of PAL values of 1.49, 1.63 and 1.78, equating to the 25th, median and 75th centiles, respectively.

3.7 Perspectives on the future

Despite centuries of scientific investigation, significant knowledge gaps remain in our understanding of the regulation of human energy metabolism, in particular (i) unravelling the regulatory factors that underlie energetic pathways at the cellular level, and determining our ability (or otherwise) to manipulate them; and (ii) gaining better insight into the determinants of interindividual variability of energy expenditure. Their elucidation will open avenues for targeted and individualised nutritional intervention strategies to optimise energy balance, and consequently metabolic health. With ongoing advancements in 'omics' technologies and increasing interest in personalised nutrition, such knowledge will give us the best chance to overcome the limitations of current, largely ineffective approaches to prevent and treat energy imbalance disorders such as obesity and diabetes.

Acknowledgement

I would like to sincerely acknowledge the work of Xavier M. Leverve (1950–2010) who authored the previous versions of this chapter upon which the initial sections of the current chapter are based, and also valuable feedback provided by Abdul Dulloo.

Note

1 Attributed to French chemist Antoine Lavoisier (1743–1794).

Further reading

Dulloo, A.G. (2011). The search for compounds that stimulate thermogenesis in obesity management: from pharmaceuticals to functional food ingredients. Obesity Reviews 12: 866–883.

Elia, E. (2000). Fuel of the tissues. In: Human Nutrition and Dietetics, 10th edn (ed. J.S. Garrow, W.P.T. James, and A. Ralph), 37–59. Edinburgh: Churchill Livingstone.

Food and Agriculture Organization of the United Nations. (2003a). Human energy requirements: report of a joint FAO/WHO/UNU Expert Consultation, Rome, 17–24 October 2001. Rome: FAO.

Food and Agriculture Organization of the United Nations. (2003b). Food energy – methods of analysis and conversion factors. FAO food and nutrition paper 77. Report of a technical workshop. Rome, 3–6 December 2002. Rome: FAO.

Heymsfield, S.B., Wang, Z., Baumgartner, R.N., and Ross, R. (1997). Human body composition: advances in models and methods. Annual Review of Nutrition 17: 527–558.

Salway, J.G. (2012). Medical Biochemistry at a Glance, 3rd edn. Oxford: Wiley-Blackwell.

Scientific Advisory Committee on Nutrition. (2011). Dietary Reference Values for Energy. London: TSO.

4 Integration of Metabolism 2: Macronutrients

Abayomi O. Akanji and Keith N. Frayn

Key messages

- We take in carbohydrate, fat and protein; ultimately (if we are not growing) we oxidise them, liberating energy, but they may be directed to storage pools before this happens.
- The metabolism of each of these macronutrients is highly regulated, partly by direct metabolic interactions between them, but largely through the secretion of hormones.
- Integration of macronutrient metabolism between tissues and organs requires a flow of information. This is carried by nerves and hormones although we now recognise that circulating metabolites also carry signals.
- In particular, the metabolic fates of fat and carbohydrate are intimately related: when one is predominant, use of the other tends to be minimised. This is achieved both by hormonal effects (e.g. insulin suppresses fat mobilisation) and by metabolic interaction (e.g. fatty acids tend to inhibit glucose oxidation in muscle).
- The interplay of these various regulatory systems in different tissues enables humans, as intact organisms, to adapt to a wide variety of metabolic demands: starvation, overfeeding or a sudden increase in energy expenditure during exercise.

4.1 Introduction: fuel intake and fuel utilisation

The human body as a machine

The human body is like a machine. It takes in fuel (chemical energy in food) and converts this to useful forms of energy: heat, physical work and other forms of chemical energy, including biosynthesis and pumping of substances across membranes. The chemical energy is liberated from the fuels by oxidation.

The fuels that we take in are the macronutrients: carbohydrate, fat and protein. Each of these may be burned in a bomb calorimeter. The products are carbon dioxide, water and oxides of nitrogen from the nitrogen content of the protein. Their combustion also liberates heat. Similarly, after their oxidation in the body, waste products are excreted. These are essentially carbon dioxide, water and urea (which contains the nitrogen from the protein). Within the body, these macronutrients may be partially oxidised (e.g. glucose to pyruvic acid) or converted to other substances, but essentially in the end they are either oxidised completely in the body or stored: humans do not excrete significant amounts of lactate, ketone bodies, amino acids or other products of their metabolism. It is often useful to maintain this 'global' view of the body's metabolic activities (Figure 4.1).

The pattern of energy intake is sporadic; people usually take in regular, discrete meals, which are digested and absorbed into the circulation over discrete periods. Even though the pattern of food

Nutrition and Metabolism, Third Edition. Edited on behalf of The Nutrition Society by Helen M. Roche, Ian A. Macdonald, Annemie M.W.J. Schols and Susan A. Lanham-New.

Companion Website: www.wiley.com/go/nutrition/metabolism3e

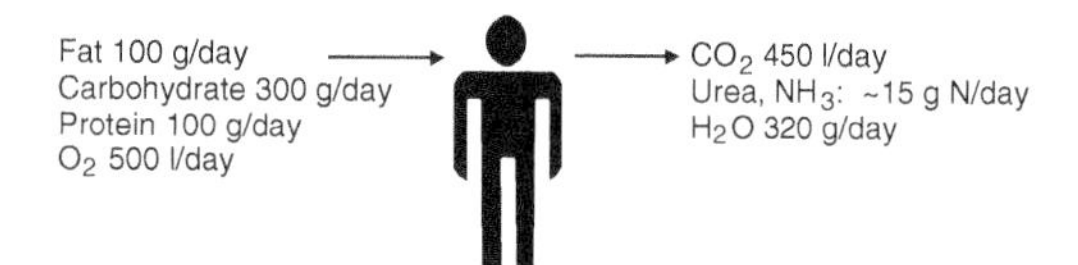

Figure 4.1 Global view of the body's macronutrient utilisation. Figures are approximate and refer to a 70 kg person. Note that, ultimately, combustion of the dietary macronutrients is complete except for the conversion of protein–nitrogen to urea and ammonia.

consumption in developed countries may be approaching one of continuous 'grazing', the pattern of energy intake is not geared in general to the pattern of energy expenditure. Therefore, the body must be able to take in fuels (macronutrients), store them as necessary and oxidise them when required. This clearly requires control mechanisms that are similar to the throttle that determines when gasoline (petrol) is used from the storage tank in a car. The situation is more complex than the flow of gasoline, however. In the car, there is just one engine demanding fuel. In the human body, there are multiple organs, each with its own requirements that vary with time on an individual basis. Furthermore, whereas we fill up our car tank with just one fuel, as humans we take in the three macronutrients. Each of the body's organs has its own particular requirements for these macronutrients, and so the flow of individual substrates into and out of storage pools must be regulated in a complex, highly co-ordinated way. This regulated flow of 'energy substrates', and the way in which it is achieved, is the theme of this chapter.

Macronutrient stores and the daily flow of fuel

The body's macronutrient stores are summarised in Table 4.1. Also summarised are the daily intakes (in very round figures) of the macronutrients, for comparison. It will immediately be obvious that the store of carbohydrate, glycogen, is very limited in relation to the daily turnover. In comparison, most people have vast stores of fat and protein.

The concept of energy balance was introduced in Chapters 1 and 3. Here, the concept of substrate balance is introduced (Figure 4.2). It is less clear-cut than that of energy balance because of potential interconversion of substrates. However, if someone is in a steady state of body weight, then the amount of each macronutrient completely disposed of each day must equal that ingested (on average – there will be fluctuations from day to day).

It is useful here to think of the body's 'strategy' in using its fuel stores. The term 'strategy' implies that these are the patterns that have evolved because they have survival benefits. Some important organs can, under normal circumstances, only use carbohydrate as a fuel source. The brain is the biggest example. The adult brain requires about 100 g glucose per day, close to the amount of glycogen stored in the liver. (As we will see, glycogen stored in skeletal muscles has a local role as a fuel source for the

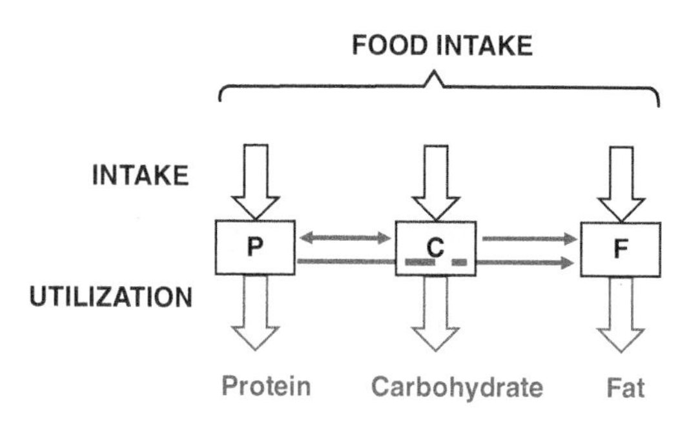

Figure 4.2 Concept of macronutrient balance. As with energy balance, what goes in must come out, with the exception of what is stored. In the case of macronutrients, however, some interconversions are possible. *Source:* Reproduced with permission from Frayn K.N. 1995/Springer Nature.

Table 4.1 The body's macronutrient stores in relation to daily intake.

Macronutrient	Total amount in body (kg)	Energy equivalent (MJ)	Days' supply if the only energy source	Daily intake (g)	Daily intake as % of store
Carbohydrate	0.5	8.5	<1	300	60
Fat	12–18	550	56	100	0.7
Protein	12	200	(20)	100	0.8

These are very much typical, round figures. Days' supply is the length of time this store would last if it were the only fuel for oxidation at an energy expenditure of 10 MJ/day: the figure for protein is given in parentheses since protein does not fulfil the role of energy store in this way.

Table 4.2 Fat and carbohydrate as fuel stores.

	Energy liberated on oxidation (kJ/g)	Water associated (g/g of fuel)	Energy stored (MJ)/kg carried
Carbohydrate	17	3	4.3
Fat	37	0.2	31

Carbohydrate (glycogen) is stored with about three times its own weight of water; fat with only a small amount of adipocyte cytoplasm.

muscles themselves.) Therefore, the glycogen store may be regarded more as a daily buffer than as a long-term fuel reserve. Although there appears to be plenty of protein, there is no specific storage form of protein. All proteins in the body have a defined role: structural, enzymic and so on. Therefore, to use protein as a fuel involves some loss of bodily function. Hence, it is not surprising that protein is relatively well protected and the body's protein is not, in general, used as a fuel for energy beyond an amount equivalent to the daily intake. In contrast, the body's fat stores are there primarily as a source of energy.

There is a very clear reason why fat (triacylglycerol, TAG) is the major energy store in mammals. Because TAG molecules are hydrophobic, they coalesce into lipid droplets that are stored, in adipocytes, with only a small amount of cytoplasm. The efficiency of energy storage in terms of kilojoules stored per gram is around eight times that for carbohydrate (Table 4.2). Protein is similar to carbohydrate, although some proteins may be less heavily hydrated. During starvation, the body's strategy is to minimise the use of carbohydrate and protein, and to obtain as much energy as possible from fat stores.

4.2 Regulatory mechanisms

The body needs mechanisms for regulating the flow of individual macronutrients in and out of storage pools. There are various ways in which this regulation is achieved.

It is useful to think of short-term and longer-term mechanisms. Short-term means minutes or hours and covers what might happen in between meals or during a bout of exercise. Longer term is taken to mean a period of several hours or days.

Short-term regulation of macronutrient flux

In the short term, some regulation is achieved simply by substrate availability affecting rates of reaction by 'mass action' effects. There is an old observation that if an unusual excess of protein is ingested, it will be oxidised over the next 24 h or so. Krebs investigated the means by which this is achieved and concluded that it reflects the kinetic properties of the initial enzymes of amino acid degradation, including the aminotransferases (transaminases), which have a high K_m (typically several mmol/l, but for alanine aminotransferase 34 mmol/l) (Krebs 1972). A high K_m means that the higher the concentration of substrate, the faster will be the reaction. Similarly, the first steps in glucose uptake and metabolism in the liver, transport across the cell membrane by the facilitated transporter GLUT2 and phosphorylation by hexokinase IV (glucokinase), are both characterised by a high K_m and high capacity. Therefore, glucose will be taken up and enter metabolic pathways in the liver according to its extracellular concentration. As glucose is absorbed from the small intestine and reaches the liver via the hepatic portal vein, so it will be taken out of the bloodstream (thus helping to minimise fluctuations in blood glucose concentration). Another example is that of ethanol: the first enzyme in the metabolism of ethanol has a low K_m but high capacity, and ethanol is oxidised at a constant rate when it enters the bloodstream.

Beyond that, many pathways are regulated by the effects of pathway products or intermediates on the enzymes of that pathway. Often this is achieved by binding of substrates to enzymes, causing allosteric effects that regulate enzyme activity. One example is that of phosphofructokinase in the pathway of glycolysis. The activity of this enzyme is regulated by a number of compounds, including activation by adenosine monophosphate (AMP) and inhibition by adenosine triphosphate (ATP) and citrate. It is also activated by the compound fructose-2,6-bisphosphate, a by-product of the pathway of glycolysis (note that fructose-1,6-bisphosphate is the product of phosphofructokinase), generated by a separate enzyme apparently purely for regulatory purposes.

These examples help to explain the regulation of pathways within cells. When considering the regulation of macronutrient-derived substrates in the body, one has to consider effects that involve more than one tissue. For instance, the use of fatty

Table 4.3 Some response elements in the promoter regions of genes regulating macronutrient metabolism that are affected by dietary macronutrients.

Response element	Examples of genes with this element	Notes
Carbohydrate response element	Pyruvate kinase (glycolysis) Pyruvate dehydrogenase E1 subunit Fatty acid synthase	The transcription factor is ChREBP. It is activated by a product of glucose metabolism, possibly xylulose 5-phosphate (pentose phosphate pathway)
Insulin response element	Hexokinase II (+ve) Acetyl-CoA carboxylase (+ve) Glucose-6-phosphatase (−ve) Carnitine palmitoyltransferase (−ve)	May be positive (stimulates transcription) or negative (suppresses transcription)
PPAR-response element	α: Enzymes of peroxisomal fatty acid oxidation in liver β or δ: Enzymes for fatty acid oxidation in muscle γ: Factors leading to adipocyte proliferation	There are three major isoforms of the transcription factor PPAR: α, β or δ, and γ. The ligand for PPAR is a fatty acid or a lipid derivative, perhaps an eicosanoid
Sterol regulatory element	LDL receptor HMG-CoA reductase (cholesterol synthesis)	The transcription factor is SREBP; it is activated by a low cellular cholesterol concentration

ChREBP, carbohydrate response element binding protein; PPAR, peroxisome proliferator-activated receptor; acetyl-CoA, acetyl-coenzyme A; LDL, low-density lipoprotein; HMG-CoA, 3-hydroxy-3-methylglutaryl-coenzyme A; SREBP, sterol regulatory element binding protein.

acids as a fuel by skeletal muscle during exercise requires that adipocytes increase the release of fatty acids as muscle uses them. Such intertissue co-ordination is brought about largely through the nervous and hormonal systems. Hormones are released from the endocrine glands in response to a signal, transmitted either through the blood (e.g. an increase in the blood glucose concentration stimulates insulin secretion from the pancreas) or through the nervous system (e.g. release of adrenaline [epinephrine] from the adrenal medulla is brought about by nerve stimulation). The hormone travels through the circulation to transmit a signal to another tissue, by binding to a specific receptor (a protein), which may be on the cell surface (this is the case for adrenaline, and for peptide hormones such as insulin and glucagon) or within the cell (steroid hormones, thyroid hormones). Binding of hormone to receptor brings about changes in signal transduction pathways, often involving reversible phosphorylation of proteins, and ultimately changes in enzyme activity. Short-term changes in enzyme activity are themselves often the result of reversible phosphorylation or dephosphorylation (e.g. adrenaline stimulates fatty acid release from adipocytes by phosphorylation of the enzyme hormone-sensitive lipase, causing its activity to increase many-fold).

Longer-term regulation of macronutrient flux

In the longer term, regulation is achieved in many cases by alteration of gene expression, usually increased or decreased transcription, but sometimes via alterations in the stability of messenger RNA (mRNA). This would be the case, for instance, if someone switches from a high-fat to a high-carbohydrate diet. Adaptation, through alterations of gene expression, will occur over a period of days. This form of regulation will also operate during a normal day – for instance, the expression of the lipoprotein lipase gene alters during the day in response to fasting overnight and feeding during the day – but usually the acute effects of hormones are more dominant over this period.

Alterations in gene expression can be brought about by both substrates and hormones. The genes for many enzymes concerned with energy metabolism have specific promoter sequences that recognise the availability of carbohydrate, fatty acids and related hormones (e.g. insulin-response elements). Some of these response elements are listed in Table 4.3. See also Tables 2.3 to 2.5 in Chapter 2 for more information.

4.3 Hormones and other substances that regulate macronutrient metabolism

Pancreatic hormones

The pancreas is mainly an exocrine organ, producing digestive juices that are discharged into the small intestine. Only 1–2% of the volume of the pancreas is occupied by endocrine (hormone-producing) cells, arranged in groups (the islets of Langerhans) surrounded by exocrine tissue. Nevertheless, the products of these endocrine cells are of enormous importance for the regulation of macronutrient metabolism according to nutritional state. Each islet is supplied with blood through a small arterial vessel, and drained by veins that lead to the hepatic portal vein. The islet cells can therefore respond to changing concentrations of substrates in the blood (e.g. glucose), and the hormones they release act first on the liver. The liver extracts a large proportion (40–50%) of insulin and glucagon, so other tissues are exposed to lower concentrations (Edgerton et al. 2021).

Insulin

Insulin is produced by the β-cells of the pancreatic islets. The insulin molecule is synthesised as one polypeptide chain, but during processing in the β-cell it is cleaved to produce two peptide chains linked by disulfide bonds. Although the β-cell responds to concentrations of various macronutrients in the blood (Table 4.4), the major factor regulating insulin secretion under most circumstances is the blood glucose concentration. Thus, insulin responds to nutritional state: in the fed state, since most meals contain carbohydrate, insulin secretion is stimulated, and during fasting, when glucose concentrations fall, insulin secretion is low. Insulin secretion after a meal that contains carbohydrate is potentiated by hormones released from the gut called 'incretins'. They are described in more detail below.

Insulin exerts its effects on other tissues by binding to specific receptors in the plasma membrane. Insulin receptors are composed of four protein subunits: two α- and two β-subunits. (The α- and β-subunits are synthesised initially as one polypeptide chain.) When insulin binds to the cytoplasmic face of the insulin receptor, the intracellular domains become activated and initiate phosphorylation of tyrosine residues, both in themselves and in other proteins. Among these other proteins are a family known as insulin receptor substrate (IRS) proteins, particularly IRS-1 and IRS-2. This initiates a chain of events, which for metabolic signals includes activation of the enzyme phosphatidylinositide-3-kinase. The signal passes via other steps to the enzyme to be regulated. Enzymes are regulated in the short term by insulin, usually by dephosphorylation. Insulin also brings about longer-term regulation by alteration of gene transcription. Some of the important effects of insulin on macronutrient metabolism are summarised in Table 4.5.

Overall, the metabolic effects of insulin may be summarised as anabolic. It brings about a net deposition of glycogen in liver and muscle, a net storage of fat in adipose tissue and a net synthesis of protein, especially in skeletal muscle. Note that these effects are brought about at least as much by inhibition of breakdown as by stimulation of synthesis; in the case of muscle protein, this is probably the major mechanism for the anabolic effect of insulin. Patients with untreated type 1 diabetes mellitus, who lack insulin, display marked wasting, which is reversed when insulin is given.

Glucagon

Glucagon is a single polypeptide chain of 29 amino acids, secreted from the α-cells of the islets. Its main metabolic effects are on the liver; in fact, it is debatable whether glucagon has metabolic effects outside the liver. Its major role is to maintain glucose output during fasting, and its secretion is stimulated when the plasma glucose concentration falls. In many respects hepatic glucose output is regulated by the ratio of insulin to glucagon (insulin/glucagon high, glucose output suppressed; insulin/glucagon low, glucose output increased). Glucagon secretion is also stimulated by amino acids (see Table 4.4). It has been suggested that this is important if a meal of pure protein is eaten (as might have been the case after a hunt, for our hunter–gatherer ancestors), as glucagon would then prevent hypoglycaemia caused by amino acids stimulating insulin secretion.

Table 4.4 Macronutrients in the circulation and their effects on insulin and glucagon secretion.

	Insulin	Glucagon	Comments
Glucose	Stimulates	Inhibits	
Amino acids	Stimulates	Stimulates	Some amino acids are more potent than others
Non-esterified fatty acids	Short term: potentiates glucose stimulation Long term: inhibits glucose stimulation	No effect	

Table 4.5 Major metabolic effects of insulin.

Tissue	Pathway/enzyme	Short or long term?	Key enzyme	Comments
Liver	Stimulation of glycogen synthesis/suppression of glycogen breakdown	Short	Glycogen synthase/glycogen phosphorylase	Regulates glucose storage in liver
	Stimulation of glycolysis/ suppression of gluconeogenesis	Short and long	Short term mainly via fructose 2,6-bisphosphate Long term via altered expression of a number of enzymes	Regulates hepatic glucose output
	Stimulation of de novo lipogenesis	Short and long	Acetyl-CoA carboxylase	De novo lipogenesis does not (under most circumstances tested) make a major contribution to triacylglycerol synthesis in liver, but this pathway is important for regulation of fatty acid oxidation
	Stimulation of triacylglycerol synthesis	Short and long	Phosphatidic acid phosphohydrolase, diacylglycerol acyltransferase (and others)	
	Stimulation of cholesterol synthesis	Short and long	3-Hydroxy-3-methyl-glutaryl-CoA reductase	
	Suppression of fatty acid oxidation/ketogenesis	Short	Carnitine palmitoyl transferase-1	Via malonyl-CoA (product of acetyl-CoA carboxylase)
Skeletal muscle	Stimulation of glucose uptake	Short	Glucose transporter GLUT4	Regulates glucose uptake by muscle (GLUT4 is translocated to the cell membrane upon stimulation by insulin)
	Stimulation of glycogen synthesis	Short	Glycogen synthase	
	Net protein anabolic effect	Short	Not clear	Insulin may suppress protein breakdown more than stimulation of protein synthesis
Adipose tissue	Stimulation of triacylglycerol removal from plasma	Short and medium	Lipoprotein lipase	'Medium term' is during periods between meals, by increased transcription plus altered intracellular processing
	Stimulation of triacylglycerol synthesis	Short and long	Phosphatidic acid phosphohydrolase, diacylglycerol acyltransferase (and others)	
	Suppression of fat mobilisation	Short	Hormone-sensitive lipase	Suppresses release of non-esterified fatty acids

Incretins

Incretin is a term used to describe a peptide hormone released from the gut that potentiates the secretion of insulin after a meal containing carbohydrate. It has long been recognised that if glucose is given by mouth, the insulin response is considerably greater than if the same amount of glucose is given intravenously. That led to the discovery of the incretins.

There are two major incretins in humans: glucagon-like peptide-1 (GLP-1) and gastric inhibitory polypeptide (GIP) (also known as glucose-dependent insulinotrophic polypeptide). GLP-1 and the closely related GLP-2 (which has a role in regulation of intestinal growth) are produced by the intestinal L cells, while the K cells produce GIP. GLP-1 and GLP-2 are derived from the breakdown of proglucagon by specific small intestinal prohormone convertases. (They get their name because they are similar in sequence to glucagon, also derived from proglucagon.) The GLP-1 receptor is a G-protein-coupled receptor (GPCR) (described below). These glucagon-like peptides are rapidly inactivated by the enzyme dipeptidyl peptidase IV (DPP-IV) which is transmembrane and also circulates in a soluble form in blood.

Incretins influence metabolism by stimulating glucose-dependent insulin release and inhibiting gastric secretion and motility. They are pleiotropic, with other described effects on the brain (appetite regulation); heart (increased cardiac output and left ventricular function); liver (reduced gluconeogenesis); skeletal muscle (increased glucose uptake and insulin sensitivity); pancreatic β-cells (increased proliferation, reduced apoptosis, reduced glucagon synthesis); bone (increased osteoblastic and reduced osteoclastic activity); and kidneys (reduced albumin excretion).

Incretin analogues and incretin mimetic drugs are now in established clinical use in the treatment of type 2 diabetes. These drugs – GLP-1 analogues (examples liraglutide, exenatide, semaglutide) and DPP-IV antagonists (examples sitagliptin, saxagliptin, linagliptin, alogliptin) – which increase the duration of action of the incretins, also have potential benefits in preventing and/or reducing risk of cardiovascular disease in diabetic patients (Meier 2012).

Catecholamines

The catecholamines that are relevant to macronutrient metabolism are adrenaline (epinephrine) and noradrenaline (norepinephrine). Adrenaline is a hormone released from the central part (medulla) of the adrenal glands, which sit over the kidneys. Its release is initiated by nervous signals that come from the hypothalamus, the integrating centre of the brain. Stimuli for adrenaline release include stress and anxiety, exercise, a fall in the blood glucose concentration and a loss of blood. Adrenaline acts on tissues through adrenergic receptors (sometimes called adrenoceptors) in cell membranes. These receptors are again proteins. The family of adrenergic receptors, which are all GPCRs, is summarised in Table 4.6. They are linked with metabolic processes through a signal chain.

The first step is the interaction of the receptor with a trimeric protein that can bind guanosine triphosphate, a G-protein. There are inhibitory and stimulatory G-proteins, named for their effects on the next step in the sequence, the enzyme adenylate cyclase. Adenylate cyclase produces cyclic 3′,5′-adenosine monophosphate (cAMP), which then acts on the cAMP-dependent protein kinase (protein kinase A) to bring about phosphorylation of key proteins, including glycogen phosphorylase and hormone-sensitive lipase. (In the case of glycogen phosphorylase, there is another step – protein kinase A phosphorylates phosphorylase kinase, which then acts on glycogen phosphorylase.) Therefore, adrenaline acting on β-receptors will cause mobilisation of stored fuels, glycogen and TAG, raising plasma concentrations of glucose and non-esterified fatty acids (NEFA). This was termed by the American physiologist Walter Cannon in 1915 the 'fight or flight' response, implying that adrenaline, released in response to stress or anxiety, produces fuels that may be used to run away or stand up to an aggressor.

The inhibitory effects of adrenaline, mediated by α_2-adrenergic receptors, may be seen as moderating the effects of overstimulation via β-receptors. For instance, adipocytes have both β- and α_2-adrenergic receptors, the latter opposing excessive lipolysis that might be brought about by high concentrations of (nor)adrenaline.

Noradrenaline is not strictly a hormone, at least under normal circumstances. It is a neurotransmitter. It is released at the ends of sympathetic nerves (nerve terminals) in tissues. It acts on adrenergic receptors, which are identical to those acted on by adrenaline and listed in Table 4.6. The stimuli for noradrenaline release are similar to those for adrenaline, and in many cases it is not clear which exerts the more

Table 4.6 Adrenergic receptors.

	Receptor subtype		
	β	α_1	α_2
Second messenger system	Adenylate cyclase/cAMP	Phospholipase C/intracellular [Ca^{2+}]	Inhibition of adenylate cyclase
Metabolic effects	Glycogen breakdown Fat mobilisation	Glycogen breakdown	Inhibition of lipolysis

There are at least three subtypes of β-adrenergic receptor, not distinguished here.

important effect. Most of the noradrenaline released from sympathetic nerve terminals is taken up again by the nerve ending for degradation or resecretion, but some always escapes or spills over and reaches the plasma. Plasma concentrations of noradrenaline are usually higher than those of adrenaline, and when noradrenaline is present at elevated concentrations (e.g. during strenuous exercise), it is believed to act as a hormone as well.

Cortisol

Cortisol is a steroid hormone, released from the outer layer (cortex) of the adrenal glands. It responds to stress in a similar way to adrenaline. Its secretion is stimulated by another hormone, adrenocorticotrophic hormone (ACTH), which in turn is released from the pituitary gland at the base of the brain. About 95% of circulating cortisol is bound to plasma proteins, especially cortisol binding globulin (CBG or transcortin). Thus, only a relatively small fraction (~5%) circulates free; however, it is this free fraction that is the physiologically active form. As will be discussed in further detail later, in relation to thyroid hormones, the highly significant protein binding of cortisol has important implications.

Cortisol acts on receptors, but these are not in the cell membrane; they are within the cell, and once cortisol is bound, they migrate and bind to the chromosomes where they regulate gene transcription. Therefore, the metabolic effects of cortisol are mainly long term, mediated by increased gene expression. (There is some evidence for short-term effects mediated via a GPCR.) Its metabolic effects are generally catabolic, including increased fat mobilisation, stimulation of gluconeogenesis and increased breakdown of muscle protein.

Growth hormone and insulin-like growth factors

Growth hormone is a peptide hormone released from the pituitary gland, and has some direct metabolic effects on tissues. These include increased fat mobilisation and stimulation of hepatic glucose output. Its secretion is stimulated by stress, including a fall in the plasma glucose concentration. However, the main effect of growth hormone, as its name suggests, is an anabolic one, promoting growth, especially through increased cartilage synthesis, an important aspect of longitudinal growth (lengthening of bones). This is not a direct effect; rather, growth hormone acts on the liver to stimulate production of further peptide hormones, the insulin-like growth factors (IGF-1 and IGF-2) that directly mediate these effects. IGF-1 and -2 have structural similarities to insulin, as their name suggests, and they act via similar (but specific) receptors. However, it has been proposed that when insulin is present at abnormally high concentrations, it can bind to and activate IGF receptors, and vice versa.

Thyroid hormones

The thyroid hormones, thyroxine (also known as T_4 since it contains four atoms of iodine per molecule) and tri-iodothyronine (T_3), are produced by the thyroid gland in the neck, responding in turn to the peptide hormone thyroid-stimulating hormone (TSH) released from the pituitary. Most (about 80%) of the circulating T_3 concentration is derived from deiodination of T_4 in peripheral tissues, especially the liver and the kidney. T_3 is a significantly more potent thyroid hormone than T_4; many would indeed consider T_4 as only a prohormone for T_3. Deiodination may also produce another thyroid hormone, called reverse tri-iodothyronine (rT_3); this hormone is essentially metabolically inactive. Its levels increase particularly during stressful situations, for example major surgery, starvation and severe sepsis. Increased production of rT_3 in these situations may be considered part of the adaptive energy-conserving response to stress.

More than 99% of the circulating thyroid hormones, T_3 and T_4, is bound to plasma proteins, especially thyroxine binding globulin (TBG), thyroxine binding prealbumin (TBPA) and albumin. Only a tiny fraction (<1%) circulates free; however, it is this free fraction, designated as free T_4 or free T_3, that is the physiologically active form that promotes thyroid hormone activity at the peripheral tissues. The highly significant protein binding of the thyroid hormones (as with cortisol) has two important implications: (i) variations in plasma protein concentrations will affect the plasma levels of these hormones, for example the hypoproteinaemia of severe protein malnutrition is associated with low total thyroid hormone levels, but free hormones are retained at normal levels by the action of pituitary TSH; and (ii) certain important drugs such as salicylates and phenytoin can displace thyroid hormones from binding sites on plasma proteins; this may also reduce the total but not free hormone levels.

The action of free thyroid hormones on other tissues is mediated via nuclear receptors, as described for cortisol, and therefore again thyroid hormones have long-term rather than short-term effects. Their effects are again mainly catabolic and include net breakdown of muscle protein. However, their most important metabolic effect is stimulation of energy expenditure. People with elevated thyroid hormone concentrations have an elevated metabolic rate and may become thin, whereas people deficient in thyroid hormone concentrations have a low metabolic rate and easily gain weight. It should be added here that alterations in thyroid function are not considered to be responsible for the vast majority of cases of human obesity, and treatment with thyroid hormones is not useful in weight reduction (unless there is a deficiency) as the system is highly regulated and administration of thyroid hormones simply leads the thyroid gland to produce less.

The metabolic explanation for the increased energy expenditure brought about by thyroid hormones is not entirely clear, but is usually considered to represent some effect on the efficiency of coupling respiration with ATP synthesis in mitochondria.

Leptin and other peptides secreted by adipose tissue

The peptide hormone leptin was only discovered at the end of 1994, although its existence had been postulated much earlier. It is produced almost exclusively by adipocytes in white adipose tissue. It is secreted in amounts that correspond to the degree of fat storage in the adipocyte: bigger fat cells secrete more leptin. It acts on receptors that are present in a number of tissues, although probably the most important are in the brain, particularly the hypothalamus. A short isoform of the leptin receptor is expressed in the choroid plexus, the region that governs the blood–brain barrier and is believed to transport leptin into the brain.

In the brain, leptin signals to decrease appetite and, in small animals, to increase energy expenditure. The latter does not seem to be true in humans. This constitutes an important system for regulation of energy stores. When fat stores are low, leptin levels are low and this alters signals in the hypothalamus with the net result of an increase in appetite. When fat stores are high, leptin levels are increased and the hypothalamic signals tend to lead to a net reduction in food intake. The system is summarised in Figure 4.3.

This system was discovered through genetic work on the *ob/ob* mouse, a mutant mouse with spontaneous high food intake, low energy expenditure and massive obesity that leads to diabetes. This mouse is homozygous for a mutation in the *ob* gene, now known to code for the protein leptin. Treatment of *ob/ob* mice with synthetic leptin leads to a reduction in body weight through decreased food intake and increased energy expenditure. The *db/db* (diabetic) mouse is another mutant with identical phenotype: it has a mutation in the leptin receptor and cannot be treated with synthetic leptin. A small number of people have been found with mutations in either the leptin gene or the leptin receptor. They all display massive obesity, associated with an intense drive to eat. Some people with mutations in the leptin gene have been treated with synthetic leptin and have shown reductions in weight for the first time in their lives. However, the vast majority of obese humans have a normal leptin gene and leptin secretion that appears to be operating normally, in that they have high levels of leptin in the blood. It has been postulated that these people suffer from 'leptin resistance', implying that there is some problem with access of leptin to the brain or with its function within the brain. In one sense this must be true, but a precise molecular explanation has not yet been found. It seems that in humans, low or absent leptin levels are a potent stimulus for appetite, whereas high levels have less effect: the system operates to protect against starvation rather than against overconsumption.

The leptin system also has an important role in reproduction. The *ob/ob* mouse is sterile, and people with mutations in leptin or its receptor have delayed sexual maturity. Leptin seems to be a signal from adipose tissue to the reproductive organs, relaying that there are sufficient energy reserves to begin the energy-demanding processes of reproduction and nurturing children (see Figure 4.3). Leptin also affects the immune system. Leptin-deficient humans treated with

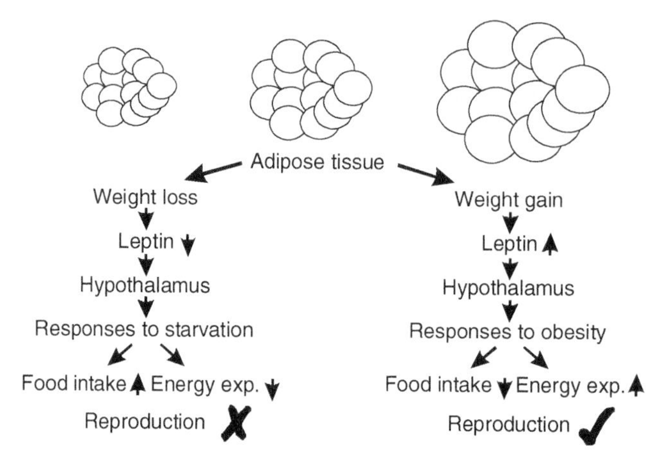

Figure 4.3 Leptin system and energy balance. Well-filled adipocytes secrete leptin, which signals to the hypothalamus to decrease food intake and increase energy expenditure (the latter only in rodents, not humans). They also signal to the reproductive system that energy reserves are sufficient. *Source:* Adapted from Friedman J.M. (1997).

leptin have shown improvements in reproductive and immune function.

It is now recognised that adipose tissue secretes a number of peptides other than leptin. These are known as 'adipokines'. Some are secreted by the adipocytes themselves, some by other cells in the tissue, for instance macrophages. Adiponectin is a protein secreted from adipocytes, which appears to confer protection against insulin resistance and cardiovascular disease. It is unusual in that, as adipocytes grow larger, they secrete less adiponectin. Therefore, in general, plasma adiponectin concentrations are higher in lean than in overweight people. Other adipokines have been suggested to be associated with, and perhaps causes of, insulin resistance and metabolic disturbances associated with adipose tissue accumulation. These include resistin, retinol binding protein 4 and some inflammation-related cytokines such as tumour necrosis factor (TNF)-α and interleukin (IL)-6.

Other hormones and the regulation of appetite

Appetite and food intake are regulated both centrally (in the brain) and peripherally (the vagus nerve and the gastrointestinal tract, GIT). These components must work in synergy to maintain energy homeostasis.

The brain control resides primarily in the hypothalamus – in the arcuate, paraventricular, solitary tract, vagal, lateral hypothalamic nuclei – with receptors for hormones and peptides that generate signals that may be anorexigenic (satiety) or orexigenic (hunger). The peripheral control, on the other hand, exists in two phases: short term (postprandial sensing of stretch from nutrients in the gut) and long term (sensing of status of body's energy stores; leptin, described above, is part of this system).

The initial signals for short-term peripheral control are generated from GIT epithelial cells exposed to ingested food. These cells in turn release mediators that stimulate vagal afferents and receptors in the hypothalamic arcuate nucleus. The processes involved include one or more of:

- stimulation of mechanoreceptors (by stretch) and chemoreceptors (by composition)
- orexigenic effect during fasting derived from increased ghrelin secretion by gastric neuroendocrine cells; this results in increased GIT activity
- anorexigenic effects postprandially, from increased intestinal production of hormones:
  - cholecystokinin, from I cells of the duodenal mucosa and causing stimulation of gall bladder contraction and bile release, release of pancreatic enzymes, inhibition of gastric and stimulation of intestinal motility
  - hormone PYY (peptide tyrosine tyrosine) from L cells in the ileum and colon
  - incretins (GIP, GLP-1), discussed above.

The peripheral long-term regulation of food intake is mediated by leptin (see above) and insulin secreted in relation to the quality and quantity of food intake and in proportion to the adipose tissue mass. In the hypothalamus, both hormones stimulate POMC (pro-opiomelanocortin) and CART (cocaine- and amphetamine-regulated transcript) neurons which secrete melanocyte stimulating hormone (α-MSH). This results in stimulation of the satiety centres. At the same time, the orexigenic pathways through the hypothalamic neuropeptide Y (NPY) and agouti-related peptide (AgRP) are inhibited. These neural pathways are described in greater detail in other sections.

It is important to understand the mechanisms involved in the regulation of food intake because of their role in the development of obesity and other feeding disorders, such as anorexia nervosa. Numerous attempts at pharmacological manipulation of these pathways are currently under way, to develop effective antiobesity drugs. Indeed, the success of bariatric surgery in the management of morbid obesity may be at least partly due to the postsurgical changes in the secretion of some of these hormones and signalling peptides.

Metabolites that act through receptors like hormones

Several of the hormones discussed above (including glucagon and the incretins) act through receptors belonging to the large group of GPCRs, described in Section Catecholamines. In recent years, it has become clear that metabolites may also act via GPCRs to regulate intracellular pathways. In general, such metabolites are present at concentrations many-fold higher than the true hormones. This can be seen as adding yet another layer of regulation to daily metabolism. A few of these metabolites are listed in Table 4.7.

4.4 Macronutrient metabolism in the major organs and tissues

As mentioned in the introduction, different tissues have their own characteristic requirements, or preferences, for metabolic fuels, and their demand for fuel may vary from time to time. Some of the major consumers of metabolic energy will be discussed in this section. Many of these tissues or organs also play roles in energy metabolism other than simply consuming fuel. Often they need energy derived from oxidative metabolism to support these activities.

Brain

The brain is a large organ (1.5 kg in an adult) and has a high requirement for oxidative metabolism to support its continuous electrical activity. This is usually met almost entirely by glucose. Fatty acids cannot cross the blood–brain barrier in significant amounts for use as an energy substrate, although the brain also has a large need for fatty acids for structural purposes, especially during development. Brain glucose consumption has been estimated by drawing blood from the carotid artery (supplying the brain) and the jugular vein (draining the brain). It is around 100–120 g/day. In the overnight fasted state, the liver produces about 2 mg glucose/kg body weight per minute which, for a 70 kg person, is equivalent to about 200 g/24 h. Thus, the brain would consume about half of the liver's glucose output after an overnight fast.

The brain can use other water-soluble fuels, notably the ketone bodies, 3-hydroxybutyrate and acetoacetate. When their concentration rises during starvation, they displace glucose as a fuel and can sustain about two-thirds of the brain's oxidative fuel requirement during prolonged starvation.

Within the hypothalamus, the blood–brain barrier operates differently, and signals from blood may enter and play regulatory roles. For instance, although insulin does not regulate brain glucose utilisation to any significant extent, there are insulin receptors in the hypothalamus which regulate appetite and energy metabolism. There are also sites where fatty acids are oxidised, again leading to regulation of energy metabolism in the whole-body.

Liver

The adult human liver weighs about 1.5 kg and is a highly active organ in the regulation of carbohydrate, fat and amino acid metabolism. To support its metabolic activities, it has a large requirement for oxidative fuel metabolism (its oxygen consumption is about 20% of the whole body's at rest, the largest of any single organ or tissue, excluding skeletal muscle during exercise). The fuels used by the liver are amino acids, fatty acids and glucose, usually in that order of importance.

However, the liver's importance in energy metabolism is more as a regulatory organ, controlling the uptake and release of compounds

Table 4.7 G-protein-coupled receptors (GPCRs) activated by metabolites and related molecules.

GPCR number	Other names	Gene name	Ligand	Tissue expression (major tissues)	Physiological role and comments
GPR40	FFA1, FFAR1 (free fatty acid receptor 1)	*FFAR1*	FAs with chain lengths in the range 12C–16C	Pancreatic β-cells	Potentiates glucose-stimulated insulin secretion
GPR41	FFA3, FFAR3	*FFAR3*	Short-chain FAs (see Section 4.4)	Adipose tissue, GI tract (enteroendocrine cells)	Stimulation of leptin production; stimulation of gut hormone secretion
GPR43	FFA2, FFAR2	*FFAR2*	Short-chain FAs (see Section 4.4)	Adipose tissue, GI tract (enteroendocrine cells)	Adipogenesis, reduction of lipolysis; stimulation of gut hormone secretion
GPR81		*HCAR1* (hydroxycarboxylic acid receptor 1)	Lactate	Adipocytes	Suppresses lipolysis. Appears to respond to lactate produced by the adipocytes rather than systemic lactate concentration; acts as part of the mechanism whereby insulin suppresses lipolysis
GPR109A	HM74A, NIACR1	*HCAR2*	3-Hydroxybutyrate	Adipocytes	Identified initially as the receptor for nicotinic acid (a component of the B-vitamin niacin), used in large doses to treat high triacylglycerol concentrations (see Section Intestinal secretions). Suppresses adipocyte lipolysis. Since 3-hydroxybutyrate is a product of hepatic fatty acid oxidation, this provides a feedback loop
GPR119	Oleoylethanolamide receptor	*GPR119*	Oleoylethanolamide and other lipids containing oleic acid, e.g. 2-oleoyl glycerol	Pancreatic β-cells, GI tract	Oleoylethanolamide has appetite-suppressing activity (although not entirely via GPR119). It is related to the endogenous cannabinoids. 2-monoacylglcyerol stimulation in the GI tract may enhance GLP-1 secretion (together with GPR40)
GPR120		*FFAR4*	*n*-3 polyunsaturated fatty acids (PUFAs)	Macrophages, GI tract, adipose tissue, brain (hypothalamus)	Has been suggested to modulate anti-inflammatory effects of *n*-3 PUFAs. Human genetic variation associated with obesity and insulin resistance
GPR131	GPBAR1 (G-protein-coupled bile acid receptor 1), TGR5	*GPBAR1*	Bile acids	Liver, adipose tissue, intestine, gall bladder	Regulates gall bladder filling with bile, gut motility and secretion of GI tract hormones
	CB_1, CB_2	*CNR1*, *CNR2*	Endogenous cannabinoids (signalling molecules related to the drug cannabis)	CB_1: brain, neurons. CB_2: immune cells. Also, both: adipose tissue and muscle	Respond to endogenous cannabinoids (related to the drug cannabis) released from nearby cells (acting in a paracrine or autocrine fashion). Activation of CB_1 was shown to reduce obesity but had unwanted side-effects
GPR26 (LPA_1), GPR23 (LPA_4), GPR92 (LPA_5)	$LPAR_{1-5}$	*LPAR1–LPAR6*	Lysophosphatidic acid (LPA)	Widespread, including blood and immune cells, cells of blood vessel walls, fibroblasts	There may be nine LPARs in total. They respond to LPA produced locally from lysophosphatidylcholine. Biological effects include immune activation and many others
	$S1P_{1-6}$	*S1PR1–S1PR5*	Sphingosine 1-phosphate	Widespread, including vascular and immune cells	S1PRs respond to sphingosine 1-phosphate produced locally (intracellularly) by the phosphorylation of sphingosine, derived from the deacylation of ceramide. Widespread biological effects

FA, fatty acid; GI, gastrointestinal; LPA, lysophosphatidic acid. When each GPCR was first discovered, it was given a sequential number. However, as their ligands have been identified, GPCRs have mostly been given other names relating to their function. Some have never been given numbers. Note that most receptors for eicosanoids are also GPCRs (not listed here). *Source:* Reproduced from Frayn and Evans, (2019).

to maintain homeostasis. The best example is that of blood glucose. When the glucose concentration in the blood is high, the liver takes up glucose and phosphorylates it to glucose-6-phosphate. The fate of that glucose-6-phosphate is determined by hormones, particularly the balance of insulin and glucagon, which stimulate glycogen synthesis and degradation, respectively. When the glucose concentration is low, the liver will release glucose, formed from glycogen breakdown and from gluconeogenesis. The liver is the major organ releasing glucose into the blood, although the kidney plays an increasing role during starvation. Thus, the liver plays a major role in keeping blood glucose concentrations relatively constant throughout the day. Major control points for glucose metabolism in the liver are illustrated in Figure 4.4.

The liver is a major site for fatty acid oxidation. It derives fatty acids from plasma NEFA (released from adipose tissue) and from the uptake of lipoprotein particles that carry TAG and cholesteryl esters. Fatty acid oxidation leads directly to production of acetyl-coenzyme A (acetyl-CoA), but in the liver this may be converted to ketone bodies, 3-hydroxybutyrate and acetoacetate. The liver is the only organ producing ketone bodies, which, as mentioned above, are an important fuel for the brain during starvation. The alternative fate for fatty acids in the liver is esterification, especially to form TAG. The balance between fatty acid oxidation and esterification is regulated by the mechanism shown in Figure 4.5. This mechanism, involving malonyl-CoA, is central to the integration of carbohydrate and fat metabolism in liver and skeletal muscle.

The liver is the only site of urea production. Amino acids derived from dietary protein, and from protein breakdown in peripheral tissues, are oxidised and their nitrogen is transferred to the urea cycle. As well as providing a route for disposal of excess amino acids and their nitrogen content, this provides the major source of oxidative fuel for the liver under most circumstances. The rate of amino acid oxidation is largely determined by amino acid availability, as mentioned earlier. See Chapter 5 for more information on amino acid metabolism in the liver.

Kidneys

Each kidney, weighing about 150 g in an adult, is composed of many cell types. However, a broad distinction can be made between the outer layer, or cortex, and the inner part, or medulla. Most

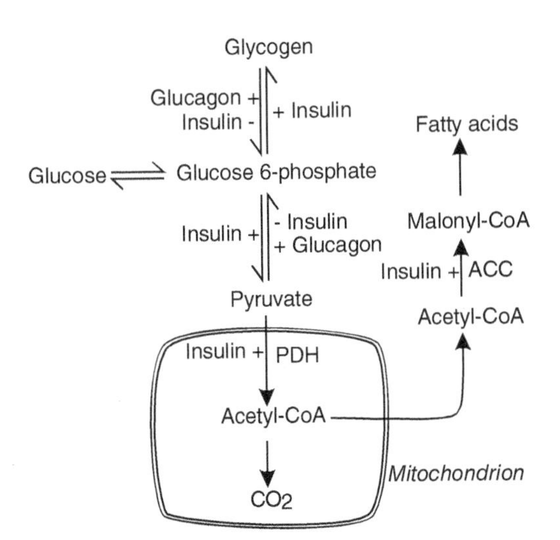

Figure 4.4 Regulation of glucose metabolism by the liver. When blood glucose concentrations are elevated, the liver extracts glucose from the circulation and phosphorylates it to glucose 6-phosphate. The metabolic fate of glucose 6-phosphate is determined by the balance of insulin and glucagon. Conversely, when blood glucose levels fall, the liver releases glucose from stored glycogen or from gluconeogenesis. PDH, pyruvate dehydrogenase; ACC, acetyl-coenzyme A carboxylase.

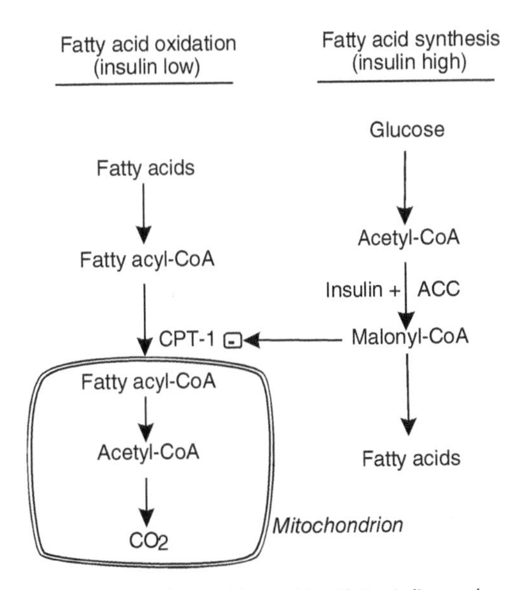

Figure 4.5 Regulation of fatty acid oxidation in liver and other tissues by malonyl-coenzyme A (CoA). Insulin stimulates synthesis of malonyl-CoA (see Figure 4.4) and this inhibits fatty acid entry into the mitochondrion for oxidation. In tissues other than liver (e.g. skeletal muscle), this system operates to regulate fatty acid oxidation, although the later part of the pathway of fatty acid synthesis (beyond malonyl-CoA) is absent. ACC, acetyl-CoA carboxylase; CPT-1, carnitine-palmitoyl transferase-1 (sometimes called carnitine-acyl transferase-1).

of the metabolic activity involved in pumping substances back from the tubules after filtration goes on in the cortex. The cortex, therefore, has a larger requirement for oxidative fuel, and appears able to oxidise most fuels (fatty acids, ketone bodies, glucose). It has a correspondingly high blood flow and oxygen consumption; the oxygen consumption of the two kidneys is about 10% of the whole body's at rest. The major role of the renal cortex in energy metabolism in the body is as a relatively large consumer. The medulla, in contrast, has a rather poor blood supply and an anaerobic pattern of metabolism.

The kidneys also play a specific role in amino acid metabolism. Glutamine is used by the kidney, especially during starvation, and ammonia is liberated during its metabolism and excreted into the renal tubules (see Chapter 5, Figure 5.1). This can be an important means of reducing an acid load in the circulation which, as shown later, is a potential problem during starvation.

Blood is filtered in the renal glomeruli. Many substances will pass through into the renal tubules. It is reabsorption of these substances, usually against a concentration gradient, that accounts for the large energy requirement of the kidneys. The renal tubules are responsible for reabsorption of virtually all of the filtered glucose in healthy persons. This is achieved with the aid of sodium-glucose co-transporters, especially the high-capacity, low-affinity type 2 (SGLT2), located in the early segments of the renal proximal convoluted tubules. The type 1 co-transporter (SGLT1) is mainly present in the small intestine and contributes significantly to intestinal glucose absorption but is also seen in the latter part of the renal tubules.

Glycosuria (i.e. the excretion of glucose through the kidneys) only occurs if the maximal capacity of these glucose transporter proteins (350 mg glucose/min) is exceeded. This is classically seen with diabetes mellitus. The kidneys try to mitigate the physiological tendency to hyperglycaemia by increasing glucose excretion. There is thus the practical possibility that compounds which inhibit renal glucose transporters and thereby promote glucose excretion will reduce a tendency to prolonged hyperglycaemia. This is the basis for the recent and increasing clinical utility of the class of drugs, called SGLT2 inhibitors (examples canagliflozin, dapagliflozin, empagliflozin), in patients with type 2 diabetes. They act to promote glycosuria and thereby modestly lower elevated blood glucose levels. An advantage is that their mechanism of action is dependent only on renal function and blood glucose levels and independent of insulin action. Thus, there is minimal potential for hypoglycaemia, and there is no potential risk of pancreatic β-cell failure and worsening of the diabetic state. SGLT2 inhibitors have also been demonstrated to have additional long-term benefits for renal and cardiovascular health (Verma and McMurray 2018).

Adipose tissue

There are two types of adipose tissue, white and brown. The metabolic function of brown adipose tissue is to generate heat, largely by oxidation of fatty acids. It is important in small animals, especially the newborn, and in animals that hibernate. Until recently, brown adipose tissue was considered unimportant in adult humans but this view has recently been challenged. 'Hot spots' of metabolism in the neck region in some people, visualised using positron emission tomography (PET) which detects regions with a high rate of glucose utilisation, have been shown to represent brown fat depots. They become activated especially when the person is exposed to cold. Their significance for energy balance is, however, not yet clear.

This section will concentrate on white adipose tissue, the major site for storage of excess dietary energy in the form of TAG. The amount of TAG stored within each adipocyte is large in comparison to daily turnover, so the half-life for turning over adipocyte TAG stores is around one year (Arner et al. 2011). This, coupled with the fact that the requirement of white adipose tissue for oxidative fuel consumption is very low, led to the view that white adipose tissue is rather inert metabolically. It has been recognised in recent years, however, that white adipose tissue has a highly active pattern of metabolism. Adipose tissue is the only site of release into the circulation of NEFA, a major metabolic fuel for many tissues, and it controls the flow of NEFA on a minute-by-minute basis. To match this, it is also responsible for a large proportion of the uptake of dietary fatty acids via the enzyme lipoprotein lipase (LPL), situated in adipose tissue capillaries, and the pathway of TAG synthesis within adipocytes. All this is achieved with a very small consumption of fuel, which in white adipose tissue appears to be mainly glucose. Regulation of the major pathways of fat mobilisation and storage is shown in Figure 4.6.

Skeletal muscle

Skeletal muscle constitutes typically 40% of body weight. Resting muscle has a rather low blood flow and metabolic activity, but because of its bulk it makes a significant contribution to whole-body fluxes of the macronutrients. During exercise, however, the metabolic activity of skeletal muscle can increase 1000-fold, and it may dominate the body's metabolic activities.

Skeletal muscle is composed of fibres, or multinucleate cells. There are different types of fibre, adapted for either short-duration, rapid contractions, using fuels present within the fibre, or slower, rhythmic contractions that can be continued for long periods, largely using fuels and oxygen supplied from the blood. The differences between these fibre types are summarised in Table 4.8.

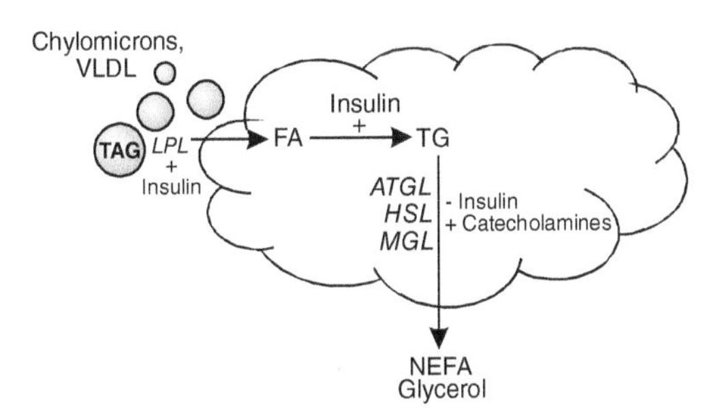

Figure 4.6 Regulation of the major pathways of fat mobilisation and storage in white adipose tissue. + denotes stimulation. Of the three enzymes that sequentially bring about complete hydrolysis of stored TAG, HSL is subject to short-term regulation as shown. The activity of ATGL is also regulated, although mechanisms are not yet clear. Insulin and catecholamines also affect the phosphorylation of proteins, including perilipin, that coat the lipid droplet in the cell, and this may regulate the access of the lipases. ATGL, adipose triglyceride lipase; FA, fatty acids; HSL, hormone-sensitive lipase; LPL, lipoprotein lipase; MGL, monoacylglycerol lipase; NEFA, non-esterified fatty acids; TAG, triacylglycerol; VLDL, very low-density lipoprotein particles.

The oxidative fuels used by skeletal muscle are mainly fatty acids and glucose. Amino acid oxidation typically accounts for around 15% of muscle fuel oxidation, similar to the whole body, and this figure does not increase appreciably with exercise. Skeletal muscle plays important roles in whole-body macronutrient metabolism, both as a consumer (glucose, fatty acids, amino acids) and as a supplier (lactate, particular amino acids, especially glutamine and alanine).

The main factors regulating muscle fuel utilisation are nutritional state (feeding and fasting) and exercise. In the fed state, insulin stimulates glucose uptake and utilisation, by recruitment of the insulin-regulated glucose transporter GLUT4 to the cell membrane and activation of glycolysis and glycogen synthesis. When fatty acids are available at high concentrations (e.g. during fasting, when insulin concentrations will also be low), they will be used as the preferred fuel. The reciprocal utilisation of glucose and fatty acids by muscle is therefore determined partly by extracellular factors (substrate availability, insulin), but also by mechanisms within the cell. Oxidation of fatty acids, when available, will suppress glucose oxidation, but a high rate of glucose utilisation (and high insulin) will suppress fatty acid oxidation via the malonyl-CoA/CPT-1 system, which operates in both muscle and liver (see the section on the liver, above).

The store of glycogen in muscle is large, typically 300–500 g in the whole body, compared with 100 g or so in the liver. However, because muscle lacks the enzyme glucose-6-phosphatase, this cannot be delivered as glucose into the blood. It cannot therefore be used as a fuel by the brain, except by conversion to lactate and export to the

Table 4.8 Metabolic characteristics of different muscle fibre types.

	Type I red	Type IIa white	Type IIb white
Other names	Slow-twitch oxidative (SO)	Fast-twitch oxidative/ glycolytic (FOG)	Fast-twitch glycolytic (FG)
Speed of contraction	Slow	Fast	Fast
Myoglobin content	High	Low	Low
Capillary density	High	Low	Low
Mitochondrial (oxidative) enzyme activity	High	Low	Low
Triacylglycerol content	High	Low	Low
Myofibrillar ATPase activity	Low	High	High
Glycogenolytic enzyme activity	Low	High	High

liver, where lactate can be converted to glucose. It seems to be present primarily as a fuel for local utilisation, especially in fast-twitch (type II) fibres.

There are different patterns of macronutrient utilisation in skeletal muscle during brief, intense exercise (anaerobic exercise, involving mainly type II fibres) and during sustained exercise (aerobic exercise, involving mainly type I fibres). In either case, the primary requirement of metabolism is to generate ATP, which fuels the sliding of myosin along actin filaments, which underlies muscle contraction.

Aerobic metabolism (e.g. oxidation of glucose or fatty acids) is efficient: about 30 ATP molecules are produced per molecule of glucose completely oxidised. In contrast, anaerobic metabolism is inefficient (three molecules of ATP per molecule of glucose from glycogen). One might imagine that skeletal muscle would only use the former, but aerobic metabolism requires the diffusion of substrates and oxygen from the blood into the muscle fibres, and the diffusion of carbon dioxide back to the blood. The rates of these diffusion processes are low in comparison with the need to generate ATP during intense exercise. Therefore, during brief, intense (anaerobic) exercise (e.g. weight lifting, high jumping, sprinting 100 m), ATP is generated by anaerobic metabolism of glucose 6-phosphate derived from intracellular glycogen.

The stimulus for glycogen breakdown is not initially hormonal (that would also require time for movement through the blood); instead, the processes of muscle contraction and glycogen breakdown are co-ordinated. The mechanism is release of Ca^{2+} from the sarcoplasmic reticulum, which initiates muscle contraction and also activates glycogen breakdown. At the same time, allosteric mechanisms activate glycolysis and the net flux through this pathway may increase 1000-fold within a few seconds.

Despite the rapid utilisation of ATP during intense exercise, the concentration of ATP in muscle only falls slightly. This is because of the existence of a reservoir of energy in the form of creatine phosphate (also called phosphocreatine). There is about four times as much creatine phosphate as ATP in skeletal muscle. As ATP is hydrolysed, so it is re-formed from creatine phosphate (with the formation of creatine). The activity of the enzyme concerned, creatine kinase, is high and it operates close to equilibrium. At rest, creatine phosphate is re-formed from creatine and ATP. Gradual non-enzymic breakdown of creatine forms creatinine, which is excreted in the urine at a remarkably constant rate, proportional to muscle mass.

During sustained exercise, blood-borne fuels and oxygen are used to regenerate ATP through aerobic metabolism. This requires co-ordinated adjustments in other tissues (e.g. adipose tissue must increase fatty acid release; the heart must deliver more blood) and this is brought about by the hormonal and nervous systems. Fatty acids predominate as the oxidative fuel in low- or moderate-intensity sustained exercise, but carbohydrate is the predominant fuel for high-intensity exercise (e.g. elite long-distance running). Although most of the fuel is blood-borne, still the intramuscular glycogen store seems essential for maximal energy output, and when this is depleted the athlete feels a sensation of sudden intense fatigue or 'hitting the wall'. Dietary preparation to maximise muscle glycogen stores before an event is now common practice.

Gut

The primary role of the intestinal tract in macronutrient metabolism is to ensure the uptake of dietary nutrients into the body. However, the intestinal tract has its own requirements for energy. There is a high rate of cell turnover, especially in the small intestine, and this requires a supply of amino acids to act as substrates for protein synthesis and for purine and pyrimidine synthesis (to make DNA and RNA). In addition, there are active transport mechanisms (e.g. for glucose absorption) that require energy. A major metabolic fuel for the small intestine appears to be the amino acid glutamine. This can be partially oxidised to produce ATP, and at the same time acts as a precursor for purine and pyrimidine synthesis. In fact, glutamine appears to be a major fuel for most tissues that have the capacity for rapid rates of cell division (e.g. lymphocytes and other cells of the immune system).

In the colon, the situation is somewhat different. Bacterial fermentation of non-starch polysaccharides and resistant starch in the colon produces the short-chain fatty acids, acetic, propionic and butyric acids. Acetic and propionic acids are absorbed and used by tissues in the body (propionic mainly in the liver), but a large proportion of the butyric acid is used as an oxidative fuel by the colonocytes. The supply of butyric acid to the colonocytes appears to protect them against neoplastic change.

4.5 Substrate fluxes in the overnight fasting state

The nutritional state of the human body typically cycles through feeding and fasting over each 24 h period. The macronutrients of the three or more meals eaten during the day enter the system over a period of several hours, and are either oxidised or sent into stores. By about 8 h after a meal (variable, depending on the size and the nature of the meal), the macronutrients have been fully absorbed from the gastrointestinal tract and the body enters the postabsorptive state. This lasts until a further meal is eaten or, if no further food is forthcoming, the body gradually enters a state of early starvation. Many studies of nutritional physiology are conducted after an overnight fast (typically around 10 h after last eating), when there is a relatively steady metabolic state.

The regulation of macronutrient flux through the circulation during these different periods of the 24-h cycle is brought about by a number of mechanisms, as outlined in Section 4.2. After an overnight fast, the concentration of insulin will be relatively low, and the glucagon/insulin ratio reaching the liver will be high.

Carbohydrate metabolism after an overnight fast

After an overnight fast, no new dietary glucose is entering the circulation and yet the glucose in the blood is turning over at a rate of about 2 mg/min per kilogram of body weight (equivalent to around 200 g/24 h; see Section 4.4). The concentration will be steady at about 5 mmol/l. New glucose is coming almost entirely from the liver, partly from glycogen breakdown (stimulated by the high glucagon/insulin ratio reaching the liver) and partly from gluconeogenesis. Substrates for the latter will include lactate and pyruvate coming from blood cells and from peripheral tissues (muscle, adipose tissue), the amino acid alanine released from muscle and adipose tissue, and glycerol, released from adipose tissue as a product of lipolysis. A major consumer of glucose at this time will be the brain, with skeletal muscle, renal medulla and blood cells also using significant amounts. (Skeletal muscle glucose utilisation after an overnight fast, at rest, is low per gram of muscle, but because of its large mass this becomes significant.) The pattern of glucose metabolism is illustrated in Figure 4.7.

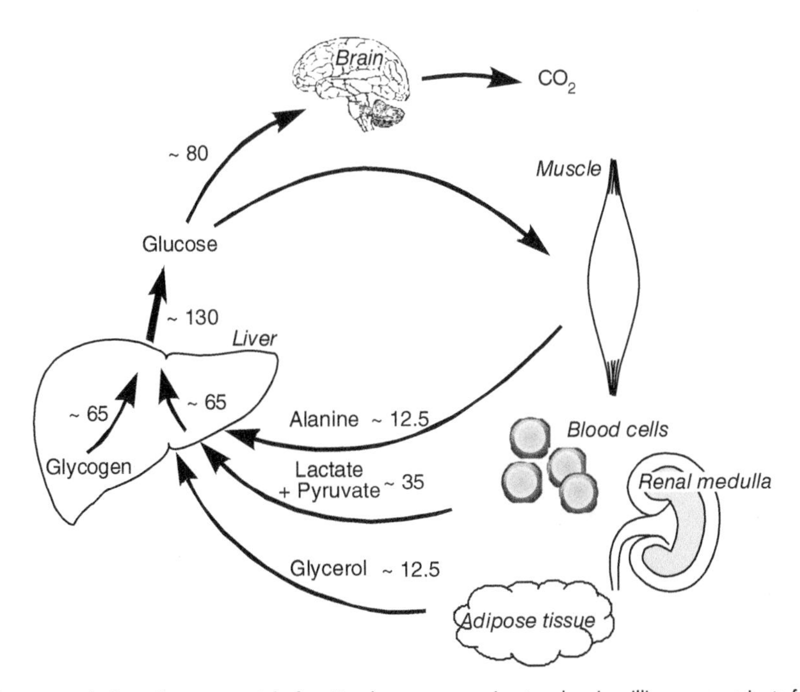

Figure 4.7 Glucose metabolism after an overnight fast. Numbers are approximate values in milligrams per minute for a typical person. *Source:* From Frayn and Evans, (2019).

Note that, of the tissues using glucose, only those carrying out complete oxidation lead to irreversible loss of glucose from the body. In other tissues, a proportion (muscle, kidney) or all (red blood cells) of the glucose is released again as three-carbon compounds (mainly lactate) and can be taken up by the liver for gluconeogenesis. This gives the enzyme pyruvate dehydrogenase (PDH) a key role in regulating loss of glucose from the body. Not surprisingly, perhaps, PDH is controlled by many factors reflecting the nutritional state of the body, including insulin, which activates it (Figure 4.8).

There is a potential metabolic cycle between peripheral tissues and the liver; muscle, for instance, releases lactate, the liver converts it to glucose, muscle can take this up and produce lactate. This is known as the Cori cycle, after its discoverer. It will be discussed again and illustrated below.

Fat metabolism after an overnight fast

Fatty acids in the circulation are present in a number of forms: NEFAs, TAG, phospholipids and cholesteryl esters. In all cases, these molecules are hydrophobic or at most somewhat amphipathic (having both hydrophobic and hydrophilic qualities), so they cannot circulate in solution in the blood plasma.

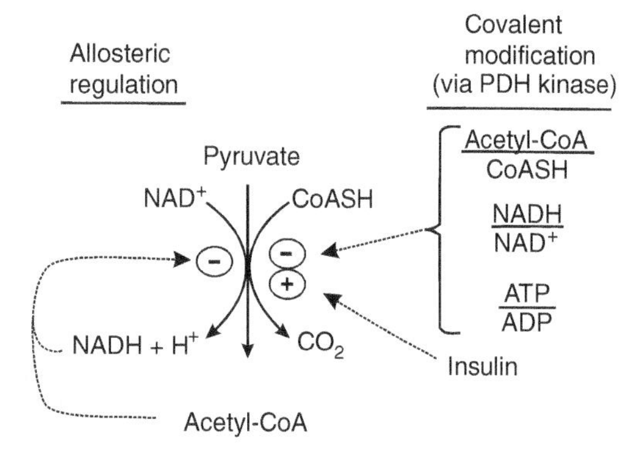

Figure 4.8 Summary of regulation of pyruvate dehydrogenase (PDH). PDH is a multienzyme complex associated with the inner mitochondrial membrane. The dehydrogenase subunit is subject to both allosteric and covalent regulation in accordance with nutritional state. Covalent modification is brought about by reversible phosphorylation, catalysed by the enzyme PDH kinase. PDH is inactivated by phosphorylation by PDH kinase, which occurs when cellular energy status is high (e.g. during rapid oxidation of fatty acids). PDH is activated by dephosphorylation, brought about by a specific PDH phosphatase, stimulated by insulin.

NEFAs are carried bound to albumin. Each molecule of albumin has around three high-affinity binding sites for fatty acids. Albumin, in a partially delipidated state, arrives in the capillaries of adipose tissue and picks up NEFAs released from adipocytes. It arrives in the capillaries of another tissue (e.g. skeletal muscle or liver) and, because of a concentration gradient between plasma and tissue, fatty acids tend to leave it and diffuse into the tissue. A typical plasma albumin concentration is 40 g/l; with a relative molecular mass of 66 kDa, that is about 0.6 mmol/l. There is therefore an upper limit under many physiological conditions of about 2 mmol/l for NEFAs. After an overnight fast, their concentration is typically 0.5–1.0 mmol/l. They enter the circulation only from adipose tissue, where fat mobilisation is stimulated after overnight fast mainly, it seems, by the fall in insulin concentration compared with the fed state. Catecholamines and other hormones, such as growth hormone and cortisol, may exert short- or longer-term stimulatory effects.

NEFAs are removed from the circulation by tissues that can use them as an energy source, such as liver and skeletal muscle. Fatty acid utilisation in these tissues is stimulated after an overnight fast (compared with the fed state) by the increased NEFA concentration in plasma. Within the tissues, fatty acid oxidation is stimulated by the low insulin concentration and low glucose utilisation, leading to a low intracellular malonyl-CoA concentration, so that fatty acids enter mitochondria for oxidation (see Figure 4.5).

The other forms of fatty acids in the circulation, TAG, phospholipids and cholesteryl esters, are transported in specialised macromolecular aggregates, known as 'lipoprotein particles'. Phospholipids are also carried as components of blood cell membranes. The lipoprotein particles are like droplets in an emulsion. They consist of a core of hydrophobic lipid (TAG and cholesteryl esters) stabilised by an outer shell, which is a monolayer of phospholipid molecules with their polar head groups facing outwards into the aqueous plasma and their tails pointing into the hydrophobic core. There are various classes of lipoprotein particle. A detailed description is outside the scope of this chapter but important characteristics are given in Table 4.9. Phospholipids and cholesteryl esters will not be

Table 4.9 Major lipoprotein fractions in plasma.

	Density range (g/ml)	Major lipids	Function, comments
Chylomicrons	<0.950	Dietary TAG	Transport dietary TAG from small intestine to peripheral tissues
Very low-density lipoproteins (VLDL)	0.950–1.006	Endogenous TAG	Transport hepatic TAG to peripheral tissues
Low-density lipoproteins (LDL)	1.019–1.063	Cholesterol/cholesteryl esters	Particles remaining after removal of TAG from VLDL: main carriers of cholesterol in the circulation; elevated levels are a risk factor for atherosclerosis
High-density lipoproteins (HDL)	1.063–1.210	Cholesteryl esters/ phospholipids	Transport of excess cholesterol from peripheral tissues back to liver for excretion (protective against atherosclerosis)

TAG, triacylglycerol.

discussed further because they are not directly relevant to macronutrient metabolism, although cholesteryl esters are highly relevant to cardiovascular disease.

In the overnight fasted state, most TAG is carried in the very low-density lipoprotein (VLDL) particles secreted from the liver. A typical concentration would be 0.5–1.5 mmol/l in plasma, but this is very variable from person to person. Since each TAG molecule contains three fatty acids, this is potentially a much greater source of energy than the plasma NEFA pool (TAG-fatty acid concentration 1.5–4.5 mmol/l). However, the turnover is slower, and in practice plasma NEFA is the predominant substrate for fatty acid oxidation in tissues.

VLDL-TAG derives within the liver from a number of sources: plasma NEFAs taken up by the hepatocytes, plasma cholesteryl esters and TAG taken up by the hepatocytes, and TAG stored within the cells. In the fasting state, plasma NEFAs are the major source, although they are processed through a pool of TAG within the hepatocyte.

VLDL particles give up their TAG-fatty acids to tissues through the action of the enzyme LPL. LPL is expressed in many extrahepatic tissues, especially muscle, adipose tissue and mammary gland, where it is increased enormously during lactation. Within these tissues, it is bound to the luminal aspect of the capillary endothelium. Here it can interact with the VLDL particles as they pass through the capillaries, hydrolysing their TAG and releasing fatty acids that diffuse into the tissues down a concentration gradient. This concentration gradient is generated by the binding of fatty acids to intracellular fatty acid binding proteins, and then their subsequent metabolism. After hydrolysis of some of its core TAG, the VLDL particle may recirculate a number of times through capillary beds, losing TAG until it has a core composed almost entirely of cholesteryl esters. Then it is called a low-density lipoprotein (LDL) particle.

The distribution of TAG-fatty acids to tissues is regulated by the tissue-specific regulation of LPL expression in the capillaries. In turn, this is achieved through effects on gene transcription (mRNA abundance) and on intracellular, post-translational processing of the immature enzyme. In adipocytes in particular, a significant proportion of LPL molecules is degraded without reaching the capillary endothelium, and nutritional regulation of adipose tissue LPL expression largely involves switching between these pathways. LPL is relatively inactive in adipose tissue after an overnight fast, which makes good sense because in that state adipose tissue is a net exporter of NEFAs and has no need to take up additional fatty acids. Its activity in adipose tissue is increased after a meal, in a process involving insulin. In skeletal muscle, in contrast, LPL activity is regulated more by the level of physical activity.

Amino acid metabolism after an overnight fast

There is a complex pattern of flow of amino acids into and out of the circulation, covered in detail in Chapter 5. Some general points are relevant here. As shown in Figure 5.2, glutamine and alanine predominate amongst the amino acids released from skeletal muscle after an overnight fast, much more than would be predicted from their abundance in muscle protein. This shows that other amino acids 'donate' their amino groups to form glutamine and alanine, which are exported from the muscle. Similarly, the uptake of amino acids across the abdominal tissues (liver and gut) shows almost the exact counterpart: the greatest removal is of glutamine and alanine. Hence, muscles and probably other tissue, including adipose tissue, are using glutamine and alanine as 'export vehicles' for their amino acid nitrogen, sending it to the liver where urea can be formed.

The pathway whereby alanine comes to play such a prominent role will be outlined here, since it is relatively well understood. Alanine is formed by transamination of pyruvate (Figure 4.9). Hence, other amino acids can donate their amino group to pyruvate, a substantial proportion of which is formed from the pathway of glycolysis. The remaining carbon skeleton can be oxidised in the muscle. The branched-chain amino acids (leucine, isoleucine, valine) play a predominant role in skeletal muscle amino acid metabolism. Their corresponding carbon skeletons, the branched-chain 2-oxo acids, are oxidised by an enzyme, the branched-chain 2-oxo acid dehydrogenase, which is similar in many ways to PDH (which is also a 2-oxo acid dehydrogenase). Thus, they contribute to oxidative fuel metabolism in muscle.

The production of alanine by transamination of pyruvate gives rise to a metabolic cycle that has been termed the glucose–alanine cycle. It operates in parallel with the Cori cycle, as illustrated in Figure 4.10.

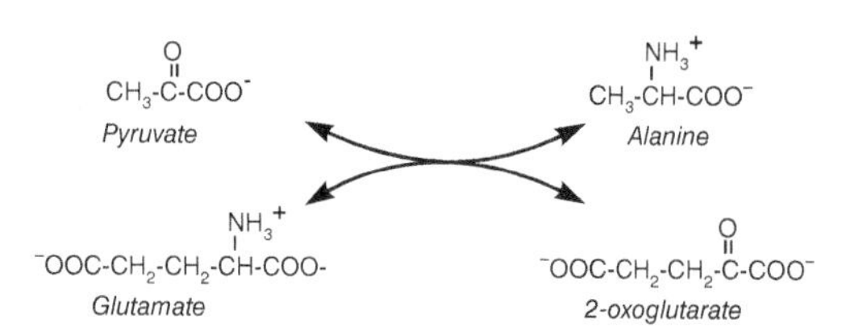

Figure 4.9 Transamination (aminotransferase) reaction involving pyruvate/alanine and 2-oxoglutarate/glutamate. All amino acids can participate in transamination reactions, which are usually the first step in their degradation.

4.6 Postprandial substrate disposal

This section examines how the relatively steady metabolic state after an overnight fast is disturbed when macronutrients are ingested and enter the circulation. This will demonstrate clearly some of the ways in which the metabolism of the different macronutrients is co-ordinated. It is unusual to eat a meal that contains only fat, and since both glucose and amino acids will stimulate the secretion of insulin, this is an important signal of the transition from postabsorptive to fed, or postprandial, state.

Impact on endogenous metabolism

Among the most rapid changes detectable in macronutrient metabolism following ingestion of a meal is the suppression of mobilisation of endogenous fuels. The production of glucose by the liver is switched off, as is the release of NEFAs from adipose tissue. Therefore, the body preserves its endogenous macronutrient stores and switches to using incoming macronutrients and storing any excess.

Glucose

Glucose enters the circulation through the hepatic portal vein. Hence, it reaches the liver in high concentrations; concentrations of almost 10 mmol/l have been measured in the portal vein when the systemic concentration is still only 4–5 mmol/l. Glucose will enter hepatocytes and be phosphorylated, as described in Section 4.4. Nevertheless, despite the liver's high capacity for soaking up glucose, much will still pass though into the systemic circulation, otherwise there would be no rise in glucose concentration and no stimulation of insulin secretion. Within the liver, the rise in insulin/glucagon ratio switches off gluconeogenesis and glycogenolysis, and stimulates glycogen synthesis. Until recently, these were considered as 'on-off' switches: either gluconeogenesis or glycolysis operated; glycogen synthesis and glycogen breakdown operated at different times. Now it is recognised that the system is more fluid; there appears always to be some glycogen synthesis and breakdown, so there is cycling between glycogen and glucose-6-phosphate. The nutritional state determines which pathway predominates. Similarly, it is now

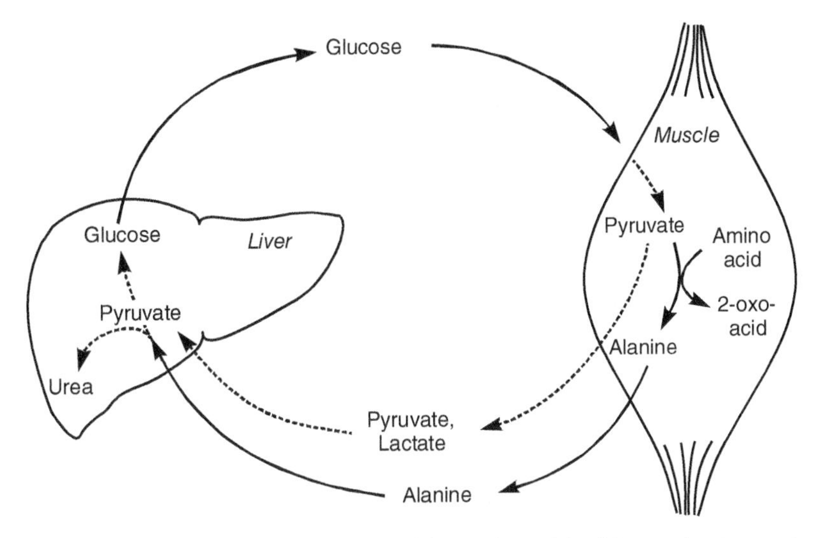

Figure 4.10 The glucose–alanine and Cori cycles operate in parallel. Note that peripheral tissues other than muscle may be involved (e.g. adipose tissue). *Source:* Adapted from Frayn and Evans, (2019).

recognised that gluconeogenesis must continue in the postprandial period. Studies with isotopic tracers show that a significant proportion of liver glycogen laid down in the period following a meal is not synthesised directly from blood glucose. Instead, the glucose is first converted to three-carbon compounds (lactate, pyruvate, alanine) and then, via pyruvate, converted by the gluconeogenic pathway to glycogen. The latter is known as the indirect pathway for glycogen synthesis and, depending on the experimental design, accounts for around 25–40% of liver glycogen synthesis.

Skeletal muscle will also take up glucose, insulin activating the glucose transporter GLUT4 (recruiting it to the plasma membrane) and, within the cell, insulin stimulating both glycogen synthesis and glucose oxidation. This is aided by the fall in plasma NEFA concentration (see below), so removing competition for oxidative disposal. There is no indirect glycogen synthesis in muscle since the gluconeogenic pathway does not operate. Other insulin-responsive tissues such as adipose tissue also increase their glucose uptake; although this does not make a major contribution to glucose disposal from the plasma, it has important effects within the adipocytes (see Section 4.3). The brain and the red blood cells continue to use glucose at the same rate as before, since their glucose uptake is not regulated by insulin.

Lipids

NEFA release from adipose tissue is suppressed very effectively by insulin. Thus, the body's fat stores are 'spared' at a time when there are plenty of dietary nutrients available. This suppression is brought about by the mechanisms shown in Figure 4.6. Because the flux of NEFAs to muscle is reduced, competition for glucose uptake is removed. The reduced delivery of NEFAs to the liver will tend to suppress the secretion of VLDL-TAG. In addition, insulin seems to suppress this directly. However, the short-term regulation of VLDL-TAG secretion is not clearly understood *in vivo*; it has been studied mainly in isolated hepatocytes.

Dietary TAG is absorbed into the enterocytes and packaged into large, TAG-rich lipoprotein particles, the chylomicrons. These enter the circulation relatively slowly (peak concentrations after a meal are typically reached at 3–4 h) and give the plasma a turbid appearance (postprandial lipaemia). It seems beneficial for the body to be able to remove chylomicron-TAG from the circulation quickly; there is considerable evidence that delayed removal of chylomicrons is associated with an increased risk of developing coronary heart disease. Rapid removal is achieved by insulin activation of adipose tissue LPL. Chylomicrons and VLDL compete for clearance by LPL, but chylomicrons are the preferred substrate, perhaps because their larger size enables them to interact with a larger number of LPL molecules at once. Nevertheless, it would make good physiological sense if VLDL-TAG secretion were suppressed after a meal, as discussed above, to minimise competition and allow rapid clearance of chylomicron-TAG. Because of the tissue-specific activation of LPL by insulin, adipose tissue plays an important role in clearance of chylomicron-TAG, although muscle, because of its sheer mass, is also important and in a physically trained person, muscle LPL will itself be upregulated (compared with a sedentary person), so making a greater contribution to minimising postprandial lipaemia.

Within adipose tissue, the fatty acids released from chylomicron-TAG by LPL in the capillaries diffuse into the adipocytes; this is aided by the suppression of intracellular lipolysis, reducing intracellular NEFA concentrations. Insulin also acts to stimulate the pathway of TAG synthesis from fatty acids and glycerol phosphate. The precise steps at which insulin acts are not entirely clear, but it may be that a number of enzymes of fatty acid esterification are activated. In addition, increased glucose uptake will produce more glycerol 3-phosphate (from dihydroxyacetone phosphate, an intermediate in glycolysis), and this itself will stimulate TAG synthesis. Thus, dietary fatty acids can be stored as adipose tissue TAG by a short and energy-efficient pathway (Figure 4.11).

Amino acids

Again, the pattern of amino acid metabolism is complex (with 20 different amino acids, each having its own pathways), but some generalisations can be drawn. Further detail is given in Chapter 5. The small intestine itself may remove some amino acids such as glutamine for use as a metabolic fuel. A further selection of amino acids is removed by the liver, and the mixture of amino acids entering the systemic circulation is depleted of glutamine and enriched in the branched-chain amino acids. These will be taken up largely by muscle, where they play a special role in oxidative metabolism. Amino acids are secretagogues for (i.e. they stimulate secretion of) both insulin and glucagon (see Table 4.4).

The rate of protein synthesis in muscle, the largest single reservoir of protein in the body, is regulated by many factors, including anabolic hormones (androgens, growth hormone), physical activity of the muscle and catabolic hormones (e.g. thyroid hormones, cortisol). In the short term, insulin also has a net anabolic effect. Measurements made using isotopic tracers suggest that this reflects not so much a stimulation of muscle protein synthesis as an inhibition of muscle protein breakdown. Nevertheless, the net effect is an increased sequestration of amino acids in muscle in the fed state.

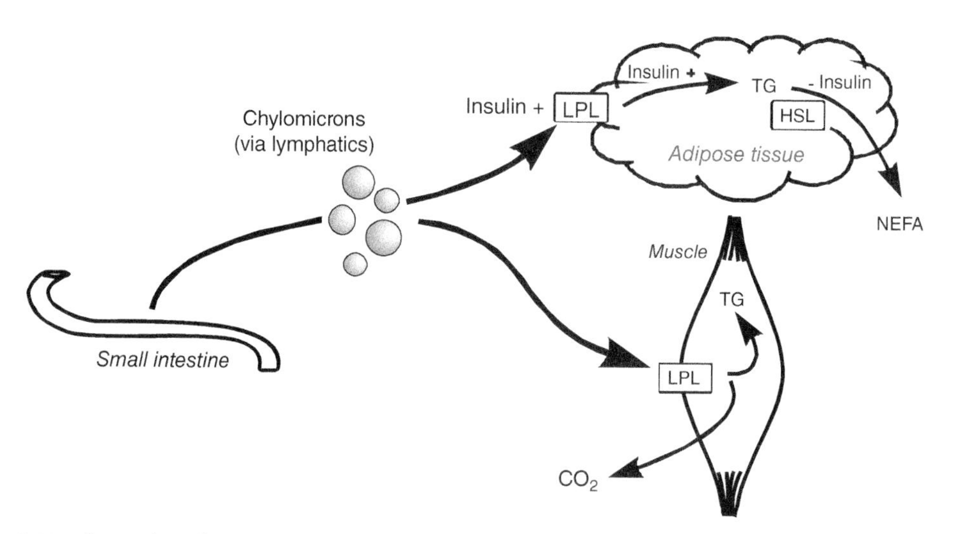

Figure 4.11 A direct pathway for storage of dietary fat as adipose tissue triacylglycerol (TAG). Skeletal muscle is also shown as this will be the destination of some of the dietary fat. HSL, hormone-sensitive lipase; LPL, lipoprotein lipase; NEFA, non-esterified fatty acids.

Cardiovascular changes

The regulation of the metabolic disposal of the macronutrients in the period following a meal was discussed above. It is important to realise that eating a meal leads to a series of co-ordinated changes. As well as the purely metabolic responses, there are changes in the cardiovascular system. Cardiac output will rise (slightly, a relatively small effect compared with physical exercise) and the distribution of blood flow to different tissues will change. There is an increase in blood flow to the abdominal viscera to help in the process of absorption and transport of nutrients into the circulation. The blood flow through certain tissues also increases: skeletal muscle blood flow is stimulated by insulin (there is some debate about the physiological relevance of this, although it is certainly seen when insulin is infused at high concentrations) and adipose tissue blood flow rises after a meal. These changes may help to deliver substrates to the tissues where they will be used.

4.7 Short-term and longer-term starvation

Starvation: general aspects

Earlier, the postabsorptive state was defined as that in which the absorption of nutrients from the gastrointestinal tract is essentially complete and the body is in a relatively stable metabolic state (typically, after an overnight fast in humans). In addition, the events were described that occur when this state is interrupted, as it usually is, by ingestion of a meal. Another possible outcome is that no meal is forthcoming, perhaps because of food shortage or for therapeutic reasons (to achieve rapid weight loss or in preparation for surgery), or perhaps because the individual chooses not to eat, for instance for religious reasons. In that case, the body gradually enters a state of early starvation, progressing eventually to a relatively steady metabolic state of complete starvation, which can last for a matter of several weeks. Some initially obese people, starved under close medical supervision and with supplementation with vitamins and minerals, have survived several months of starvation. The ability of the human body to cope with long periods of starvation illustrates perfectly the interaction and co-ordination of metabolism in different organs and tissues.

Starvation may be total or partial (when energy intake is not sufficient to maintain a steady body mass). Our understanding of partial starvation was increased enormously by studies carried out in Minnesota, USA, during World War II, by the celebrated nutritionist Ancel Keys (1904–2004). The volunteers were conscientious objectors. They were fed for a period of 24 weeks on about 40% of their estimated energy requirements and were carefully monitored during this time. Noticeable features of starvation were lethargy and depression. These were rapidly reversed when full feeding was resumed. The responses to partial starvation seem essentially to be similar to those to total starvation, albeit somewhat less marked.

Our metabolic understanding of complete starvation largely comes from studies carried out in the 1960s on obese women fasting under medical supervision to lose weight. (The responses observed may not be typical because the excess initial fat stores of the subjects may have produced a different set of responses from those of initially slimmer people.) There are some observations of normal-weight people starved for various reasons: during periods of famine in Europe in World War II, during periods of famine elsewhere in the world today and people who have chosen to starve themselves, for instance for political reasons. Tragic though these cases are, the observations that have been made help us to understand the adaptations of the body to deprivation and may in the future help those whose access to food, or ability to consume it, is limited.

If no carbohydrate is entering the body from the gut, then the body's limited carbohydrate reserves (see Table 4.1) become very precious, and much of the metabolic pattern in starvation can be understood in terms of preservation of carbohydrate. Some tissues have a continual need for glucose. If the glucose is oxidised, specifically if its carbon atoms in the form of pyruvic acid go through the PDH reaction, then it is irreversibly lost to the body. Glucose can then be generated *de novo* only from non-carbohydrate precursors: glycerol from lipolysis and the carbon skeletons of some amino acids. While TAG stores may be used without detriment to the organism, all the body's protein, as discussed earlier, has some specific function other than as a fuel for oxidation. As discussed later, the pattern of metabolism adapts to starvation in such a way that protein oxidation is minimised, so far as possible, and as much energy as possible derived from fat.

Short-term starvation

A rapid response to starvation is loss of the liver glycogen store; this is virtually completely depleted within 24 h. At this stage, there is still a need for generation of glucose at a high rate for oxidation in the brain. Net protein breakdown may be relatively rapid. This is sometimes called the gluconeogenic phase. Because of the lack of incoming glucose, plasma glucose concentrations will fall slightly, insulin concentrations will fall and glucagon secretion will increase. These changes reduce glucose utilisation by tissues such as skeletal muscle, which can use fatty acids instead, and in particular suppression of PDH activity in tissues that are responsive to insulin will reduce irreversible disposal of glucose. There will be stimulation of lipolysis in adipose tissue, and of hepatic glucose production through gluconeogenesis. At this stage, gluconeogenesis is largely proceeding at the expense of muscle protein. This phase may last for 3–4 days.

Longer-term starvation

At around one week, short-term starvation merges into a period when glucose and insulin concentrations fall further, lipolysis increases still further and a sparing of protein breakdown is observed. This sparing of protein oxidation is reflected in a gradual decrease in the excretion of nitrogen in the urine. It is understandable in terms of the body's need to protect its protein, but the mechanism is not fully clear. Some features of this phase are clear. Increased lipolysis leads to increased delivery of NEFAs and glycerol from adipose tissue. Since glycerol is a gluconeogenic precursor, the need for amino acids is reduced. Furthermore, the oxidation of fatty acids within the liver is increased by the high glucagon/insulin ratio, and ketone bodies are produced in high concentrations in the circulation. Although the brain cannot use NEFAs, it can use the water-soluble ketone bodies that are derived from them. In the phase of adapted starvation, say from 2–3 weeks of starvation onwards, ketone bodies reach a relatively steady concentration around 7–9 mmol/l (compared with less than 0.2 mmol/l typically after an overnight fast) and largely replace glucose as the oxidative fuel for the brain; they have been shown to meet about two-thirds of brain oxidative fuel requirements. The need for protein breakdown to feed gluconeogenesis is therefore again reduced. Because of the inactivation of pyruvate dehydrogenase (by the low insulin concentration), much of the glucose that is used by tissues outside the brain is only partially broken down, to pyruvate and lactate, which can then be recycled in the liver through gluconeogenesis. Thus, red blood cells, for instance, which have an obligatory requirement for glucose, are not depleting the body of glucose.

Although this shows how the need for protein oxidation is reduced, the cellular mechanism by which this is achieved also needs to be understood. There are several suggestions. Energy expenditure (metabolic rate) falls during starvation, sparing the body's fuel stores for as long as possible. The main mechanism is a fall in concentration of the active thyroid hormone, T_3 (with increased production of the inactive form, rT_3; see Section 4.3). The reduction in protein breakdown has been suggested simply to reflect this overall slowing of metabolism. Since thyroid hormones have a catabolic effect on muscle protein, the fall in their concentration may in itself spare muscle protein. In addition, it has been suggested that the high concentrations of ketone bodies exert a protein-sparing effect, possibly through suppression of the activity of the branched-chain 2-oxo acid dehydrogenase in skeletal muscle that is responsible for irreversible breakdown of the branched-chain amino acids.

The high concentrations of NEFAs and ketone bodies (note that both 3-hydroxybutyric acid and acetoacetic acid are acids) might cause a metabolic acidosis (low blood pH), but this is corrected in part by the mechanism described earlier, whereby glutamine is metabolised in the kidney, releasing ammonia that is excreted along with H^+ ions. Urinary nitrogen excretion is normally largely in the form of urea, but during starvation the relative amount of ammonia increases, as does the contribution of the kidney to gluconeogenesis (the carbon skeleton of glutamine will contribute to this).

Ultimately, the body's fat store limits the length of survival. It appears that once the fat store is essentially depleted, then there is a phase of rapid protein breakdown that leads quickly to death. Thus, fatness is a useful adaptation when food is plentiful, if there are likely to be long periods of famine. In present-day industrialised societies, the latter is not a feature of life, but it may explain why there is such a strong tendency for people to become obese.

4.8 Perspectives on the future

The human genome encodes about 25 000 genes, each of which may lead to several different protein products by differential splicing and post-translational modifications. The function of most of these proteins is unknown. A major challenge for the next few decades will be to bridge the gap between molecular biology and whole-body function. As this understanding develops, 'genomic medicine' will offer the possibility of personalisation or individualisation of metabolic regulation, and 'genomic (or personalised) nutrition' will offer the hope of matching an individual's diet to their metabolic make-up. Gene editing using CRISPR-Cas9 methodology also offers hope for the management of genetic disorders of metabolic regulation.

New techniques such as transcriptomics, exomics, metabolomics, proteomics and lipidomics should allow a better understanding of hormones, enzymes and intermediary metabolism by assessing 'snap-shot' levels of different intermediary metabolites under different metabolic stresses.

A better understanding of the role of the microbiome in colonic heath and the development of metabolic diseases such as obesity, type 2 diabetes could change the way we view the human body. This is a rapidly evolving field. Similarly, the potential role of usefulness of nutraceuticals, natural products and phytochemicals in metabolic regulation, antioxidant defence and immune function will all be further explored over the coming years.

Nutrition is a prime example of a science in which an integrated approach is necessary. We can study the nutritional needs of a single cell, but most cells require a constant supply of nutrients to keep them alive in cell culture systems. They will die within a day or two if nutrients are not provided. A human being, in contrast, can survive for perhaps two months without food because of the co-ordination that occurs between different cells. This chapter has described metabolism at the level of the cell, the tissues and organs, and the whole body. It is hoped that this approach will help the reader to see the grander picture of life: there is more to it than molecules!

References

Arner, P., Bernard, S., Salehpour, M. et al. (2011). Dynamics of human adipose lipid turnover in health and metabolic disease. Nature 478: 110–113.

Edgerton, D.S., Moore, M.C., Gregory, J.M., Kraft, G., and Cherrington, A.D. (2021). Importance of the route of insulin delivery to its control of glucose metabolism. American Journal of Physiology, Endocrinology and Metabolism 320: E891–E897.

Frayn, K.N. (1995). Physiological regulation of macronutrient balance. 19 (Suppl 5): S4–S10.

Frayn, K.N. and Evans, R.D. (2019). Human Metabolism: A Regulatory Perspective, 4e. Oxford: Wiley-Blackwell.

Friedman, J.M. (1997). The alphabet of weight control. Nature 385: 119–120.

Krebs, H.A. (1972). Some aspects of the regulation of fuel supply in omnivorous animals. Advances in Enzyme Regulation 10: 397–420.

Meier, J.J. (2012). GLP-1 receptor agonists for individualized treatment of type 2 diabetes mellitus. Nature Reviews Endocrinology 8: 728–742.

Verma, S. and McMurray, J.J.V. (2018). SGLT2 inhibitors and mechanisms of cardiovascular benefit: a state-of-the-art review. Diabetologia 61: 2108–2117.

Further reading

Ahima, R.S. and Antwi, D.A. (2008). Brain regulation of appetite and satiety. Endocrinology and Metabolism Clinics of North America 37: 811–823.

Frayn, K.N. and Evans, R.D. (2019). Human Metabolism: A Regulatory Perspective, 4e. Oxford: Wiley-Blackwell.

Friedman, J.M. and Halaas, J.L. (1998). Leptin and the regulation of body weight in mammals. Nature 395: 763–770.

Gurr, M.I., Harwood, J.L., Frayn, K.N., Murphy, D.J., and Michell, R.H. (2016). Lipids: Biochemistry, Biotechnoloigy and Health, 6th edn. Oxford: Wiley-Blackwell.

Rodwell, V.W., Bender, D.A., Botham, K.M., Kennelly, P.J., and Weil, P.A. (2018). Harper's Illustrated Biochemistry, 31st edn. New York: Lange Medical Books/McGraw-Hill.

Salway, J.G. (2017). Metabolism at a Glance, 4th edn. Oxford: Wiley-Blackwell.

5 Integration of Metabolism 3: Protein and Amino Acids

Emily J. Ferguson, Nicholas A. Burd, Tyler A. Churchward-Venne, and Chris McGlory

Key messages

- Protein and amino acid metabolism are tightly regulated processes that involve the integration of numerous organ systems and support a variety of biological pathways.
- Dietary protein and indispensable amino acid ingestion promote a state of positive net protein balance, i.e. rates of protein synthesis > rates of protein breakdown, to support protein accretion.
- Dietary protein quality is measured according to the amino acid composition and digestibility of the protein source. Intact protein digestibility impacts the subsequent bioavailability of its constituent amino acids for protein synthesis and other biological processes in splanchnic and peripheral tissues.
- The measurement of protein turnover was initially limited to crude measurements of whole-body protein balance; however, advancements such as tissue biopsies and application of stable isotopic tracers enable measurement of tissue-specific protein turnover rates.
- Resistance exercise promotes skeletal muscle protein turnover and potentiates the anabolic response to dietary protein and indispensable amino acid ingestion.
- Dietary protein feeding and resistance exercise enhance the activity of various signalling pathways upstream of cytosolic mRNA translation resulting in elevated rates of skeletal muscle protein synthesis.

5.1 Introduction

Dietary protein is a key macronutrient required for optimal health and bodily function. Protein ingestion is crucial for the synthesis of new tissue and maintenance of whole-body energy homeostasis with roles in membrane transport, enzymatic function, cellular signalling and structure.

Proteins are polypeptide chains composed of amino acids. Amino acids are precursors for protein synthesis and, interestingly, some amino acids serve as substrates for the production of neurotransmitters, catecholamines and nucleotides. Net protein mass (i.e. the amount of protein we have) is determined by the algebraic difference between rates of protein synthesis and rates of protein breakdown that can be evaluated at whole-body or tissue-specific levels. In the rested, fasted state, we exist in a state of negative whole-body net protein balance where rates of protein breakdown are greater than rates of protein synthesis. Enhanced rates of protein breakdown result in the liberation of amino acids that are either recycled for the formation of new proteins or metabolised to support energy production. In contrast, the ingestion of dietary protein, especially sources rich in indispensable/essential amino acids, has been shown to be most effective to elevate rates of protein synthesis,

Nutrition and Metabolism, Third Edition. Edited on behalf of The Nutrition Society by Helen M. Roche, Ian A. Macdonald, Annemie M.W.J. Schols and Susan A. Lanham-New.

Companion Website: www.wiley.com/go/nutrition/metabolism3e

inducing a shift towards a positive state of net protein balance. However, most of this work has been based on the ingestion of isolated protein sources or free amino acids with less information available on the influence of whole-food sources of protein and their associated food matrices on modulating the muscle protein synthesis (MPS) response. The continuous formation of new proteins and subsequent clearance of damaged or dysfunctional proteins, referred to as protein turnover, is critical to maintain the structure and function of all tissues in the body.

In this chapter, we will examine the integrated metabolism of amino acids across multiple organs and discuss the cellular and molecular mechanisms regulating protein turnover with a specific emphasis on protein synthesis. We begin by discussing dietary-derived protein and amino acid requirements and the assessment of dietary protein quality. The mechanisms of dietary protein digestion, amino acid absorption and the metabolism of amino acids across multiple organ systems will also be reviewed. Then, we discuss current methodologies to measure whole-body and tissue-specific protein turnover and the impact of diet and lifestyle factors on protein turnover. Finally, we examine the molecular signalling mechanisms underpinning the stimulation of protein synthesis in response to protein intake and resistance exercise. By the end of this chapter, you will have a more in-depth understanding of how protein and amino acid metabolism is regulated and be able to critically discuss factors that affect these processes.

5.2 Protein quality and amino acids in the diet

Dietary protein requirements are often presented as the estimated average requirement (EAR) and/or recommended dietary allowance (RDA). The EAR is the average daily protein intake estimated to meet the needs of half the population aged 19 years and older. In contrast, the RDA refers to the average daily protein intake sufficient to meet the protein requirements of 97–98% of the population aged 19 years and older. The relative EAR and RDA for dietary protein for males and females aged 19 years and older are 0.66 and 0.80 g/kg body weight/day, respectively (Institute of Medicine 2005). One important consideration is that the RDA and its European equivalent, the population reference intake (PRI), are designed to offset protein malnutrition, but not optimise health, something that we discuss in later sections (Burd et al. 2019).

To date, there is insufficient evidence to establish a tolerable upper intake level (UL) for dietary protein, i.e. 'the highest average daily intake level likely to pose no risk of adverse health effects to almost all individuals in the general population'. The inability to set a UL for dietary protein consumption may be driven, in part, by numerous studies demonstrating that protein intake above the RDA is tolerable and can be safely maintained in healthy individuals without kidney disease (Devries et al. 2018).

Whilst the RDA for total dietary protein is 0.80 g/kg body weight/day, not all dietary protein sources have similar amino acid composition. Individual amino acids themselves have varying daily intake requirements. Of the 20 amino acids commonly found in proteins, 11 are characterised as dispensable (non-essential) because they can be synthesised by the body (Table 5.1). The remaining nine amino acids are deemed indispensable (essential) amino acids because they cannot be synthesised by the body in quantities sufficient to meet requirements and so must be obtained through dietary sources (see Table 5.1). Thus, many dietary protein recommendations emphasise the ingestion of mixed sources of high-quality proteins to meet daily protein and amino acid requirements.

Protein quality is defined as the ability of a given dietary protein to fulfil human amino acid requirements based on its amino acid composition (the indispensable amino acid profile in particular) and digestibility. The Digestible Indispensable Amino

Table 5.1 Indispensable and dispensable amino acids.

Indispensable amino acids	Dispensable amino acids
Histidine	Alanine
Isoleucine[a]	Arginine
Leucine[a]	Asparagine
Lysine	Aspartate
Methionine	Cysteine
Phenylalanine	Glutamate
Threonine	Glutamine
Tryptophan	Glycine
Valine[a]	Proline
	Serine
	Tyrosine

[a] denotes branched-chain amino acids.

Acid Score (DIAAS) is the current measure of protein quality recommended by the Food and Agriculture Organization of the United Nations. The shift away from the use of the protein-digestibility-corrected amino acid score (PDCAAS) in favour of the DIAAS was prompted by, among other limitations, inaccuracies associated with the use of truncated scores and faecal digestibility to measure protein quality. Ultimately, the use of the DIAAS is believed to provide greater accuracy when determining protein quality within the diet while also allowing for the ability to differentiate between high-quality proteins. This protein quality scoring metric has mainly been applied in the growing pig model due to the challenges involved in applying it in vivo in humans. It is relevant to keep in mind that this scoring system provides no downstream correlate related to where the amino acids are being disposed (accretion, oxidation, etc.) and/or whether physical activity or exercise patterns affect the subsequent determined DIAAS. This is relevant as increased physical activity and/or exercise are both important aspects relevant to supporting human health.

The DIAAS characterises intact dietary protein quality based on indispensable amino acid content and ileal digestibility. Ileal digestibility is calculated as the difference between dietary amino acid intake and the appearance of unabsorbed dietary amino acids at the terminal ileum. When using the DIAAS, amino acids are viewed as individual nutrients which is congruent with their role both as precursors for protein synthesis and as functional molecules with varying metabolic fates in the body. Consideration for differences in the digestibility of individual indispensable amino acids is evident in the calculations used to derive the DIAAS for a dietary protein source. The DIAAS of a dietary protein source is determined according to the most limiting indispensable amino acid (lowest digestible indispensable amino acid reference ratio) in the dietary protein source and can be calculated as:

DIAAS % = [(mg of digestible dietary indispensable amino acid in 1 g of the dietary protein)/(mg of the same dietary indispensable amino acid in 1 g of the reference protein)] × 100

where [(mg of digestible dietary indispensable amino acid in 1 g of the dietary protein)/(mg of the same dietary indispensable amino acid in 1 g of the reference protein)] can also be expressed as the digestible indispensable amino acid reference ratio (Wolfe et al. 2016). DIAAS values of <75 are classified as low-quality proteins, while scores of 75–99 and ≥100 are considered to be of good and high quality, respectively. The protein quality scores of common dietary protein sources and their respective limiting amino acids can be found in Table 5.2.

Dietary protein quality is assessed, in part, according to its amino acid composition such that proteins that contain the full complement of indispensable amino acids are deemed to be high-quality proteins whereas protein sources that are deficient in at least one indispensable amino acid are classified as lower-quality proteins. Dietary protein quality is also affected by the digestibility and subsequent bioavailability of its constituent amino acids. Amino acid digestibility refers to the amount of ingested amino acids that are absorbed by the digestive tract, whereas amino acid

Table 5.2 Limiting amino acids (AAs) and protein quality scores of common foods.

Food source	Limiting AAs	PDCAAS	DIAAS
Chicken breast	Tryptophan	1	1.08
Whole milk	Methionine + cysteine	1	1.14
Egg (hard-boiled)	Histidine	1	1.13
Chickpeas	Methionine + cysteine	0.74	0.83
Kidney beans (cooked)	Methionine + cysteine	0.65	0.59
Cooked rice	Lysine	0.62	0.59
Milk protein concentrate	Methionine + cysteine	1 (1.25)	1.41
Pea protein concentrate	Methionine + cysteine	0.89	0.82
Soy protein isolate A	Methionine + cysteine	0.95	0.90
Soy protein isolate B	Methionine + cysteine	1 (0.97)	0.90
Whey protein isolate	Histidine	1 (1.12)	1.25

DIAAS, digestible indispensable amino acid score; PDCAAS, protein-digestibility-corrected amino acid score.
Source: Data were compiled from Phillips 2017 and Nunes et al. 2021 where original data sources are given.

bioavailability describes the amount of dietary-derived amino acids that are 'absorbed in a form that can be utilised for body protein synthesis' and other metabolic pathways. Lower-quality proteins need to be consumed in greater amounts or in combination with other protein sources (complementary proteins) to meet dietary indispensable amino acid requirements. However, simply consuming more of a single dietary protein source is often not appropriate or feasible, thus the ingestion of a variety of protein sources is a more favourable approach to ensure adequate amino acid intake.

Animal-based proteins such as meat and dairy products are typically classified as high-quality protein sources, whereas plant-based protein sources, with the notable exception of isolated soy protein, are more frequently classified as lower-quality protein sources (Phillips et al. 2015). Studies examining protein digestion and amino acid absorption kinetics using isolated whey, casein and soy protein ingestion provide unique insight into the impact of high-quality protein ingestion. The digestibility and amino acid profiles of these protein sources produce observable differences in the duration and magnitude of postprandial aminoacidaemia which is often used as a proxy of amino acid availability for protein synthesis in peripheral tissues. Whey and soy protein are rapidly digested while casein is slowly digested as it clots upon interaction with stomach acid, slowing the rate of gastric emptying. When consumed as a drink, the ingestion of whey hydrolysate results in a large yet relatively short-lived rise in aminoacidaemia while micellar casein ingestion results in a lower but protracted rise in plasma amino acid concentrations (Tang et al. 2009). Relative to soy protein, whey protein has a greater proportion of constituent amino acids that can escape first-pass splanchnic extraction, e.g. branched-chain amino acids (see Section 5.3), resulting in the appearance of a large proportion of amino acids in the systemic circulation capable of promoting protein synthesis in peripheral tissues (Tang et al. 2009).

Thus, although numerous dietary protein sources may be classified as higher-quality protein sources, differences between these proteins with respect to amino acid composition and digestibility produce notable differences in postprandial aminoacidaemia.

5.3 Dietary protein and amino acid metabolism

Now that we have discussed protein quality and sources of amino acids in our diet, in this section we will expand on this knowledge by examining the mechanisms by which intact proteins and amino acids are digested, absorbed and metabolised across multiple organ systems.

The period of time following the ingestion of a protein-rich meal is divided into the postprandial period (~3–5h) and the postabsorptive period which correspond to the *fed* and *fasted* states, respectively. The postprandial period is characterised by a positive net protein balance in favour of protein synthesis over protein breakdown in response to the digestion, absorption and incorporation of amino acids into tissue proteins. However, during the postabsorptive period, rates of protein breakdown exceed rates of protein synthesis until ingestion of the next protein-rich meal.

Postprandial indispensable amino acid bioavailability for protein synthesis in peripheral tissues is dependent on various factors such as protein dose and digestibility, amino acid content and absorption, and splanchnic extraction (Gorissen et al. 2020). In a study by Groen and colleagues (2015), it was observed that ~55% of dietary protein-derived phenylalanine becomes available in the systemic circulation after ingestion of 20 g of casein (equivalent to the amount of protein consumed in a single meal). Further, ~20% of circulating phenylalanine was absorbed into skeletal muscle, with ~11% of dietary-derived phenylalanine being used for de novo muscle protein synthesis (Groen et al. 2015). That only ~55% of exogenous phenylalanine appears as free amino acids in the circulation suggests that some amino acids are absorbed by splanchnic tissues, i.e. the gastrointestinal tract and liver (first-pass splanchnic extraction), to support protein synthesis in these tissues. The seemingly inverse relationship between splanchnic amino acid extraction and circulating amino acid concentrations highlights the integrated nature of postprandial protein synthesis across multiple organ systems.

The following section will discuss the events that underpin protein and amino acid metabolism during the postprandial and postabsorptive periods with a specific emphasis on branched-chain amino acids.

Gastrointestinal tract

Stomach

Whilst food is mechanically digested in the mouth, chemical digestion of proteins does not occur until proteins arrive at the stomach. The initial stages of chemical digestion for dietary proteins are in contrast to carbohydrate and fat digestion due to a lack of protein-digesting enzymes in saliva. Protein breakdown into polypeptides, which are further broken down into free amino acids, dipeptides and tripeptides, is accomplished via hydrolysis of peptide bonds (Jeukendrup and Gleeson 2018). This reaction is dependent on the acidity of the stomach and the presence of proteases from both the stomach and exocrine pancreas. In order to prevent the degradation of the cells that store proteases, proteases are stored and secreted as their inactive precursors.

The initial stages of protein digestion in the stomach are accomplished by the protease pepsin (Figure 5.1). Hydrochloric acid and the acidic condition of the stomach also denature dietary proteins. Pepsinogen, the inactive form of pepsin, is activated by hydrochloric acid which is secreted in response to a protein-rich meal. Pepsin converts proteins into smaller polypeptides via cleavage of peptide bonds. Chemical digestion continues in the small intestine, but because the pH is higher in the duodenum than in the stomach, pepsin is inactivated and the remainder of digestion is accomplished by pancreatic proteases.

Intestines

To summarise, the initial stages of protein digestion in the stomach result in the degradation of proteins into polypeptides. In the small intestine, these polypeptides are further broken down into tri- and dipeptides as well as free amino acids by pancreatic proteases (trypsin, chymotrypsin, carboxypeptidases and elastase) and brush border enzymes

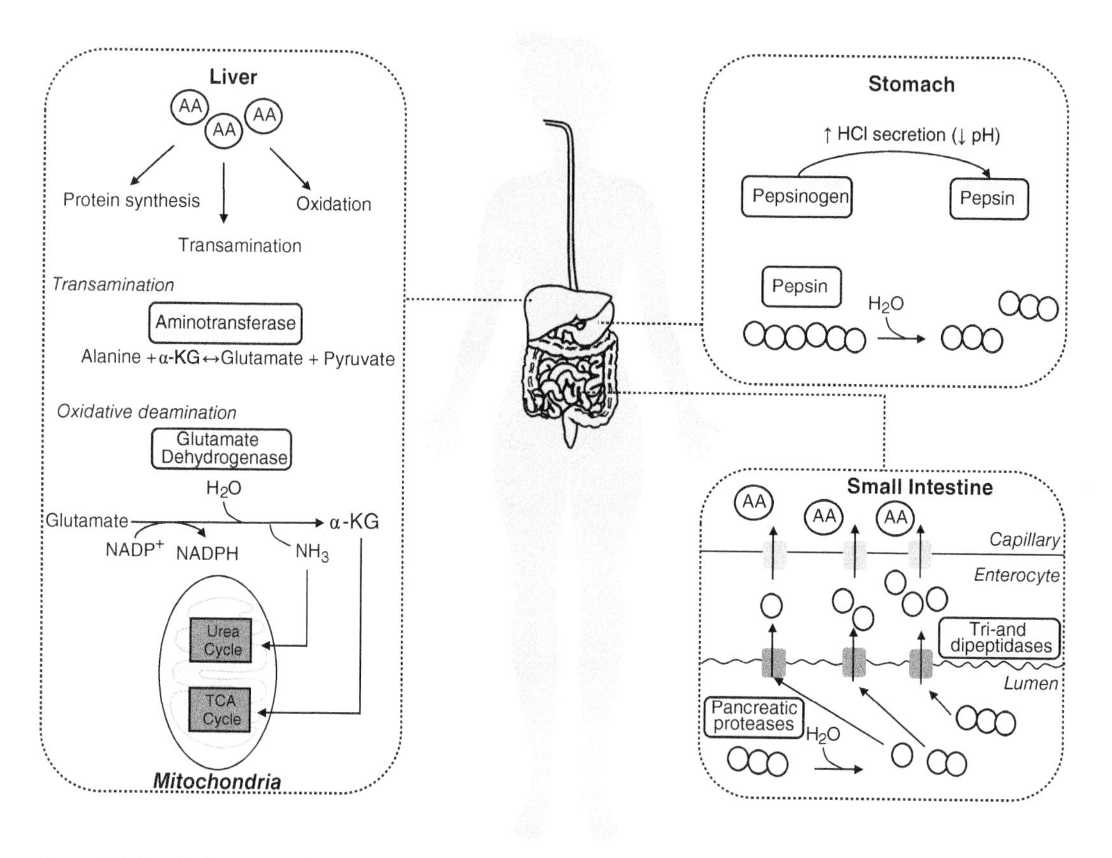

Figure 5.1 Protein digestion and amino acid absorption. The stomach and small intestine serve as sites of chemical digestion for amino acids. Following amino acid absorption across the enterocyte and into the bloodstream, the amino acids are transported to the liver or continue on to peripheral tissues. AA, amino acid; α-KG, α-ketoglutarate; HCl, hydrochloric acid; NADP+, nicotinamide adenine dinucleotide phosphate; NH_3, ammonia.

(Trommelen et al. 2021). Once activated, pancreatic proteases cleave peptide bonds, releasing individual amino acids, dipeptides and tripeptides from polypeptides. The remaining tripeptides, dipeptides and free amino acids are then transported across the apical membrane into the enterocyte via specific transporters utilising mechanisms such as electrochemical gradients, proton co-transport and amino acid exchange to fuel flux into the enterocyte (Broer and Broer 2017). Within the enterocyte, amino acids are liberated from tri- and dipeptides by intracellular tri- and dipeptidases, respectively.

Amino acid transport across the basolateral membrane is mediated by a set of uniporters and antiporters that facilitate free amino acid efflux from the enterocyte into capillaries that feed the hepatic portal vein. Remnant undigested and unabsorbed components of dietary proteins travel to the large intestine where they are metabolised by microbiota or digested by remaining proteases and peptidases. Whilst the small intestine is the site of the majority of amino acid absorption, the large intestine is also proposed to contribute to whole-body protein and amino acid metabolism via protein and amino acid fermentation and de novo amino acid synthesis (van der Wielen et al. 2017).

Liver

Upon amino acid absorption across the enterocyte, amino acids are transported to the liver via the hepatic portal vein. Once absorbed by the liver, amino acids contribute to hepatic protein synthesis and other processes such as gluconeogenesis via transamination and deamination reactions or are released into the circulation for transport to the periphery (e.g. skeletal muscle). Transamination reactions are reversible and involve the transfer of the amino group from one amino acid to a keto acid such as α-ketoglutarate via specific aminotransferase enzymes.

The products of transamination reactions vary in nature from amino acids like glutamate, a common endproduct of these reactions, to substrates for energy production and gluconeogenesis such as pyruvate. Amino acids can also undergo deamination whereby the amino group is removed, producing ammonia. For example, glutamate undergoes oxidative deamination, producing α-ketoglutarate and ammonia via glutamate dehydrogenase in the liver (Cynober 2018). Ammonia is protonated to form ammonium prior to the urea cycle which leads to the formation of urea. Finally, urea is transported to the kidneys for excretion as urine. α-Ketoglutarate itself is a tricarboxylic acid cycle intermediate and therefore even removal of excess amino acids provides another avenue by which substrates for energy-producing pathways can be synthesised.

It is important to note that excess amino acids can enter the circulation after protein catabolism from various tissues and travel to the liver for further breakdown. Nitrogenous by-products from amino acid degradation in peripheral tissues can also travel to the liver for processing via the urea cycle. This is significant because amino acids are not stored in the body to the same extent as glycogen and triglycerides. Therefore, the events in the liver provide a clear example of the elegant mechanism by which interorgan transport of amino acids supports numerous synthetic and energetic pathways as well as removal of dangerous by-products such as ammonia.

First-pass splanchnic extraction

It is important to remember that not all dietary-derived amino acids are released into the circulation following absorption in the splanchnic tissues. The uptake and utilisation of amino acids by the gastrointestinal tract and liver to support protein turnover and metabolic processes is referred to as first-pass splanchnic extraction (Figure 5.2). Splanchnic tissues exhibit high rates of protein turnover at a rate of ~50% per day. This high turnover rate is probably caused by the metabolic activity and environmental conditions of these tissues (e.g. high acidity in the stomach). High rates of splanchnic extraction limit the amount of free amino acids in the circulation that can contribute to protein synthesis in peripheral tissues. Though splanchnic protein synthetic pathways are beyond the scope of this chapter it is important to consider the contributions of splanchnic amino acid uptake to whole-body protein turnover and amino acid metabolism.

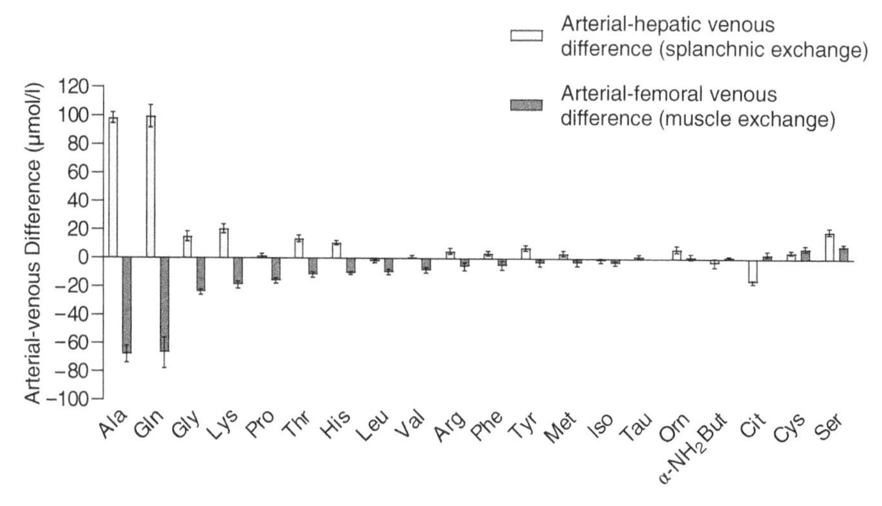

Figure 5.2 Amino acid release from skeletal muscle (dark grey) versus amino acid uptake by splanchnic tissues (light grey) after an overnight fast. *Source:* Printed with permission, from the Annual Review of Biochemistry, Vol.44 © 1975 by Annual Reviews www.annualreviews.org.

Skeletal muscle branched-chain amino acid metabolism

Branched-chain amino acid metabolism primarily occurs in the peripheral tissues such as skeletal muscle. Human skeletal muscle has high transamination capacity and a large proportion of ingested branched-chain amino acids are said to escape sequestration by the gastrointestinal tract and liver. Specifically, unlike other amino acids, ingested branched-chain amino acids largely bypass metabolism in the liver due to low hepatic expression of the enzyme branched-chain amino acid aminotransferase (Suryawan et al. 1998). Thus, a large proportion of branched-chain amino acid catabolism is initiated in skeletal muscle tissue.

Circulating amino acids are transported into skeletal muscle through carrier proteins such as sodium-coupled neutral amino acid transporter (SNAT-2) and, in the case of branched-chain amino acids, L-type amino acid transporter (LAT-1). Given the concentration gradient that exists for amino acids between plasma and skeletal muscle (Table 5.3), it is clear that the regulation of amino acid transport into the skeletal muscle cell is an important factor in the muscle response to amino acid ingestion (see Section 5.6). Branched-chain aminotransferase catalyses the transamination of leucine, isoleucine and valine. This transamination reaction results in the formation of glutamate and branched-chain α-keto acids via the transfer of an amino group to α-ketoglutarate (Holececk 2018). Each branched-chain amino acid has a corresponding α-keto acid (Figure 5.3). Branched-chain amino acid catabolism continues with the irreversible oxidative decarboxylation of branched-chain α-keto acids by the branched-chain α-keto acid dehydrogenase complex.

Table 5.3 Postabsorptive free amino acid concentrations in human plasma and muscle.

Amino acids	Plasma (mM)	Intracellular muscle (mM)
Alanine	0.33	2.34
Arginine	0.08	0.51
Asparagine	0.05	0.47
Citrulline	0.03	0.04
Cysteine	0.11	0.18
Glutamate	0.06	4.38
Glutamine	0.57	19.45
Glycine	0.21	1.33
Histidine	0.08	0.37
Isoleucine[a]	0.06	0.11
Leucine[a]	0.12	0.15
Lysine	0.18	1.15
Methionine	0.02	0.11
Ornithine	0.06	0.30
Phenylalanine	0.05	0.07
Proline	0.17	0.83
Serine	0.12	0.9
Taurine	0.07	15.44
Threonine	0.15	1.03
Tyrosine	0.05	0.10
Valine[a]	0.22	0.26

[a] denotes branched-chain amino acids.

Source: Data compiled from Bergström J, Fürst P, Norée L, Vinnars E. Intracellular free amino acid concentration in human muscle tissue. J Appl Physiol 1974;36:693.

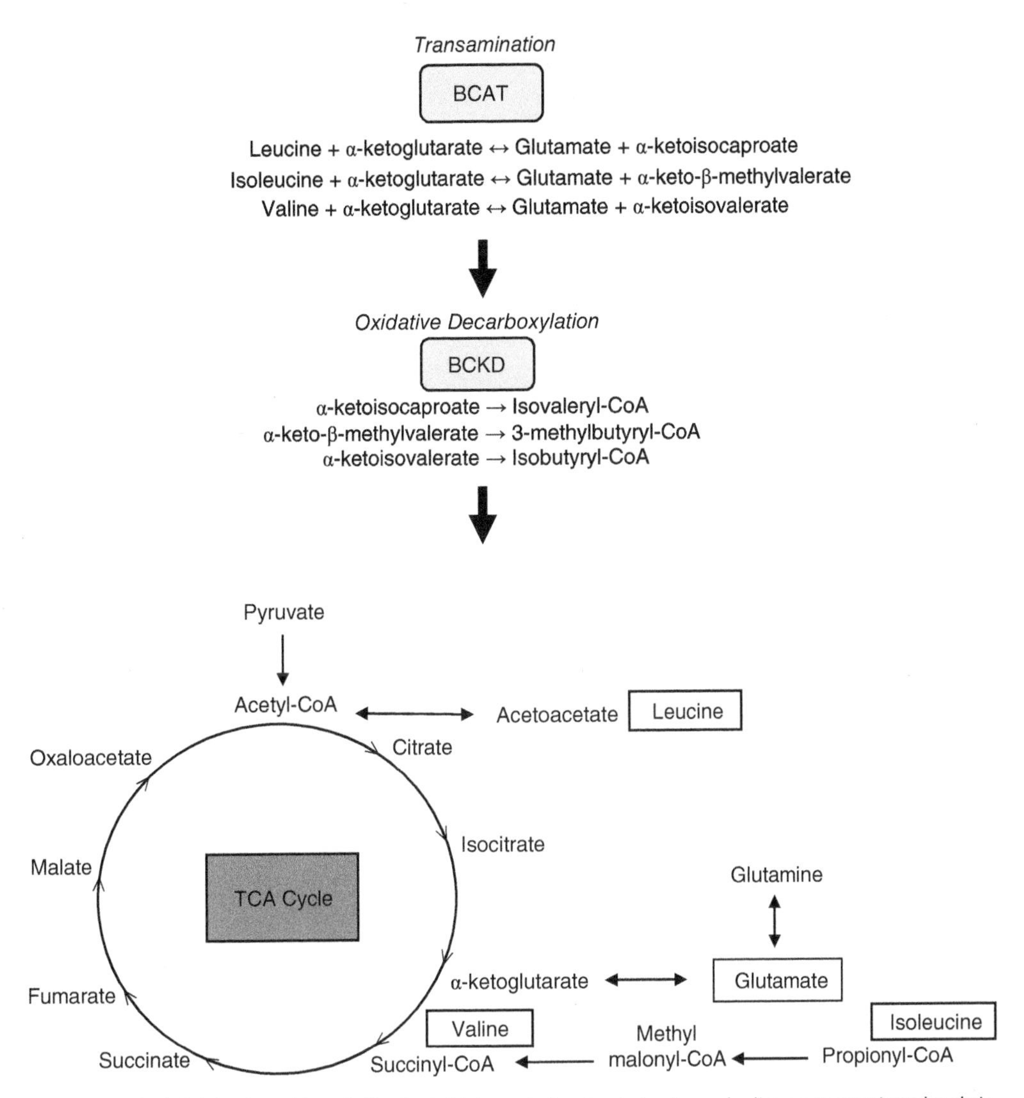

Figure 5.3 Branched-chain amino acid catabolism in skeletal muscle. Leucine, isoleucine and valine are transaminated to their corresponding branched-chain α-keto acid via branched-chain aminotransferase. The resulting α-keto acid undergoes oxidative decarboxylation and further catabolism yields glucogenic and ketogenic endproducts that contribute to energy production via the tricarboxylic acid cycle. BCAT, branched-chain aminotransferase; BCKD, branched-chain α-keto acid dehydrogenase; TCA, tricarboxylic acid cycle. *Source:* Adapted from Jeukendrup and Gleeson (2018).

Leucine catabolism yields acetyl-CoA and acetoacetate while isoleucine and valine catabolism yield acetyl-CoA and propionyl-CoA, and succinyl-CoA, respectively.

Branched-chain amino acids can be classified as glucogenic or ketogenic according to the metabolic fates of their endproducts. For example, leucine is considered ketogenic because its endproducts acetyl-CoA and acetoacetate are involved in fatty acid synthesis (Institute of Medicine 2005).

It is also important to note that enzymes responsible for branched-chain amino acid metabolism can be found in other tissues such as the brain, liver and kidney. The expression of enzymes responsible for branched-chain amino acid metabolism in multiple tissues and the various endproducts associated with branched-chain amino acid catabolism highlight the integrated nature of amino acid metabolism across multiple tissues and its capacity to support numerous processes in the body (Figure 5.4).

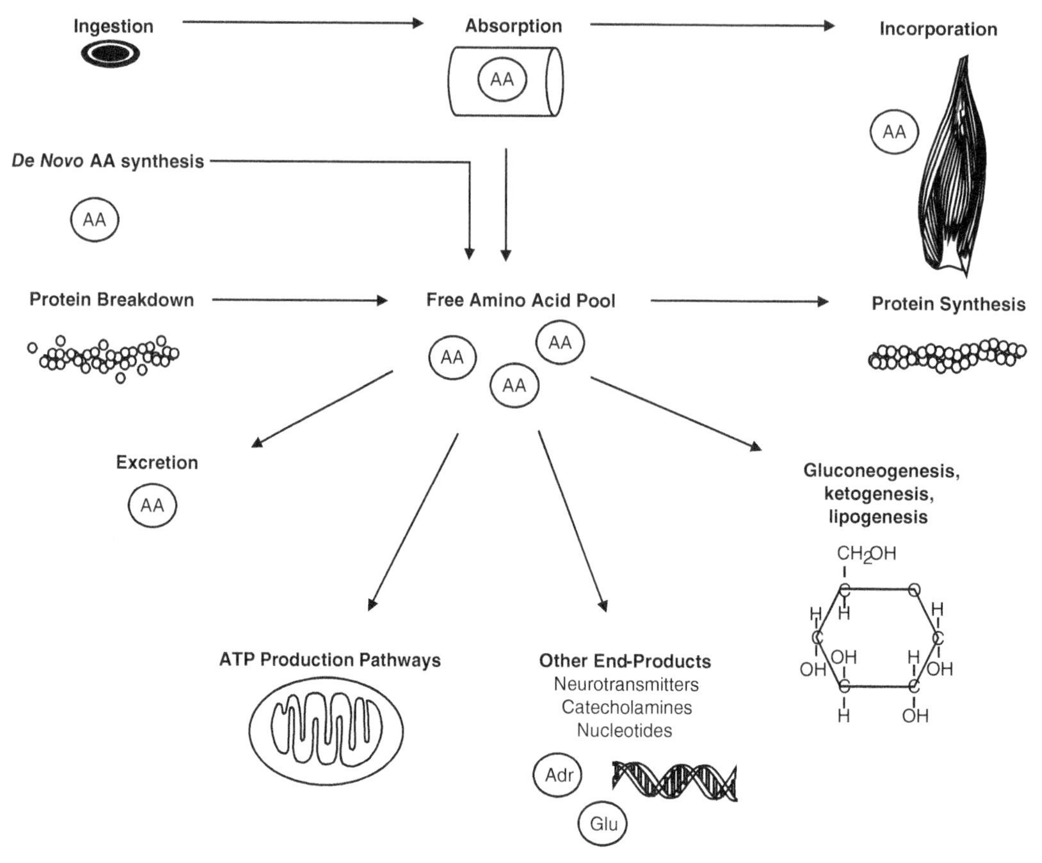

Figure 5.4 Schematic of protein and amino acid metabolism in the human body. Intact proteins are ingested and digested. Absorption of amino acids leads to their direct incorporation into tissue proteins or entry into a free amino acid pool prior to uptake for protein synthesis, or involvement in other metabolic pathways. De novo amino acid synthesis and liberation of amino acids due to protein breakdown also contribute to the free amino acid pool. AA, amino acid; Adr, adrenaline; Glu, glutamate.

5.4 Measurement of protein turnover

Initial work examining protein turnover in humans was limited to the analysis of whole-body protein turnover using the nitrogen balance method that was used to inform dietary reference intakes for protein. The nitrogen balance method measures nitrogen balance as the algebraic difference between nitrogen intake via protein ingestion and nitrogen loss through urinary and faecal excretion as well as other processes. Whilst whole-body measures of protein turnover are beneficial, they do not account for the contribution of varying tissue-specific rates of protein turnover, such as splanchnic versus skeletal muscle, to aggregate rates of whole-body protein turnover. However, the re-emergence of stable isotope tracer analysis and advent of skeletal muscle biopsies promoted the development of methodologies to measure tissue-specific rates of protein turnover. Further, these advances allowed for the more accurate analysis of mechanisms of protein turnover in order to improve upon existing crude measurements of whole-body protein turnover in humans.

While the use of stable isotopes to trace protein and amino acid metabolism is the sole focus of this section, it is important to note that stable isotope tracer analysis can be applied to a broad range of substrates, including aspects of carbohydrate and fat metabolism.

Stable isotopes are species of an element that contain the same number of protons but differ in mass because of the addition of one or more neutrons to their nucleus. Despite this difference in mass, these species, such as ^{13}C, possess the same chemical and functional properties as their base elements (Wilkinson et al. 2021). This difference in mass allows the identification of specific species of the same element in various tissues via mass spectrometry (Wilkinson et al. 2021).

The basic premise of stable isotope tracer analysis of amino acid metabolism is that the isotopically labelled amino acid (tracer) is 'tracing' the path of the unlabeled amino acid (tracee) within the body (Wolfe et al. 2019). Stable isotope tracers can be infused or orally consumed. Stable isotope tracers can be substrate specific, whereby labelled amino acids such as L-[1-^{13}C] leucine are used to 'trace' the path of specific amino acids throughout the body over a matter of hours. In contrast, studies utilising deuterium oxide, a non-specific stable isotope tracer, are not constrained in their analysis by the stringent conditions associated with stable isotope infusion studies, such as shorter study durations and laboratory settings. Indeed, there are many benefits to using deuterium oxide as a stable isotope tracer, with the ability to measure tracer enrichment in free-living conditions over more prolonged periods (days to weeks) being chief among them.

Although stable isotope tracer techniques capable of measuring rates of both protein synthesis and breakdown at the whole-body and tissue-specific levels are available, the measurement of rates of protein breakdown is challenging and, as a result, analysis of fractional breakdown rates remains limited in the current literature. Therefore, this section will discuss the basis of stable isotope tracer infusions and oral deuterium oxide consumption in contemporary studies with a particular emphasis on the measurement of protein synthesis.

Application of stable isotopes in human research

Primed constant infusion

The priming dose technique is used to reduce the time required to reach an isotopic steady state between precursor and product pools. Amino acid pools are 'primed' through the provision of a bolus of amino acid tracer allowing amino acid pools to reach isotopic steady state within ~1–2h such that infusion-based measurements can be completed in a shorter period of time. The stable isotope tracer is continuously infused following the administration of the priming dose.

Though calculations to determine the metabolic flux of amino acids throughout the body in non-steady-state conditions exist, many stable isotope tracer approaches require an isotopic steady state in the precursor pool. *Tracer enrichment* refers to the ratio of tracer to tracee in a particular compartment and is usually expressed as tracer-to-tracee ratio (TTR) or mole percent excess (MPE). The precursor pool is the amino acid pool from which the amino acids used for de novo protein synthesis are derived. Whilst aminoacyl-transfer RNA (tRNA) is considered the true precursor pool for protein synthesis (see Section 5.6), it is technically challenging to measure, so surrogate measures such as the enrichment of plasma or muscle free amino acid pools are used in lieu of aminoacyl-tRNA.

Arterial venous balance approach

The arterial venous balance approach is used to examine tissue-specific protein turnover and amino acid metabolism. The flux of substrates, such as amino acids, across a tissue bed is determined by sampling arterial blood entering the tissue as well as the venous blood leaving the tissue. The arterial venous balance approach can be coupled with stable isotope tracer administration in two- and three-pool models (Figure 5.5).

The two-pool model measures the rate of appearance of an isotopically labelled amino acid in venous blood draining a specific tissue, such as skeletal muscle, as a proxy of protein breakdown, whereas the rate of disappearance from arterial blood is indicative of protein synthesis. The three-pool model is an expansion of the two-pool model that includes skeletal muscle biopsies to measure the labelled amino acid enrichment of the muscle intracellular compartment, thus providing a more accurate measure of intramuscular protein turnover.

The arterial venous balance approach is an excellent example of a method where the choice of stable isotope tracer employed affects the accuracy of conclusions drawn from the uptake and release of amino acids into the systemic circulation. Specifically, intramuscular metabolism or recycling of the amino acid for protein synthesis would affect the measurement of rates of protein turnover. In addition, the potential confounding effect of alterations in blood flow on rates of appearance and disappearance of a substrate must be accounted for via measurement of blood flow to the tissue being examined.

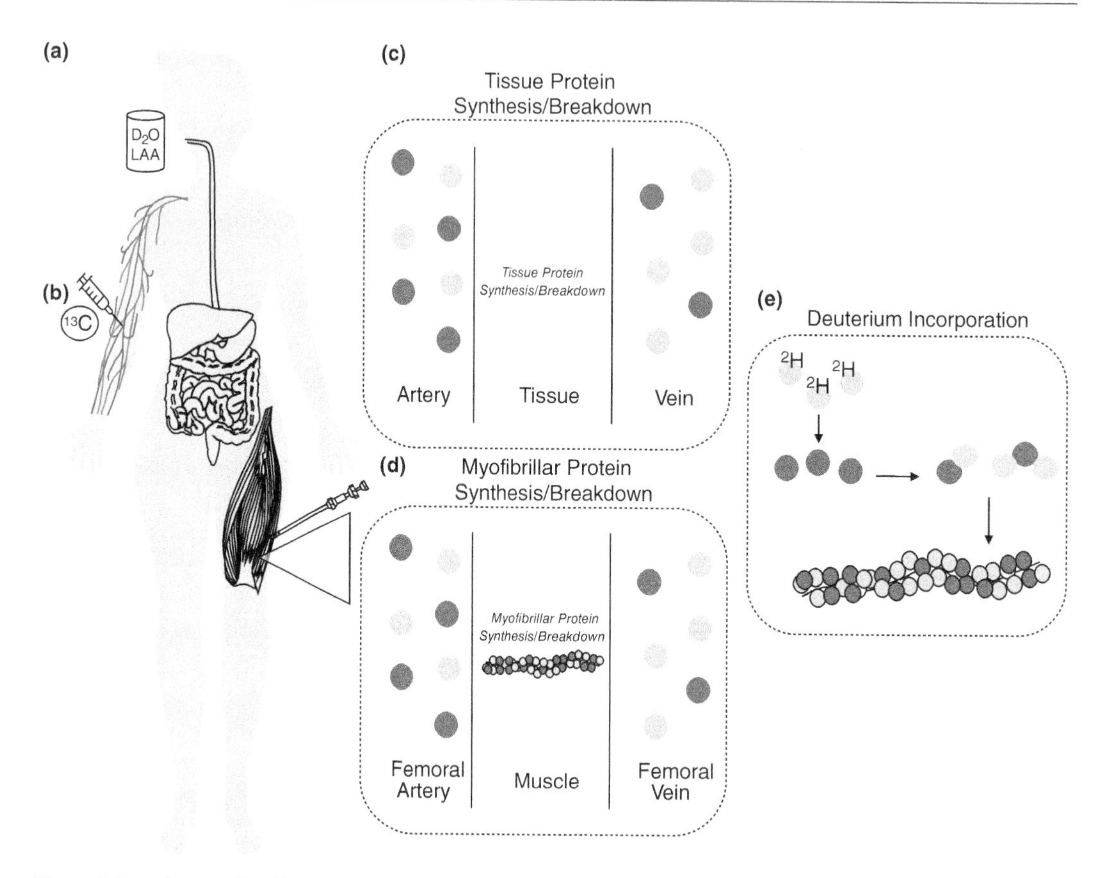

Figure 5.5 Application of stable isotope tracers to measure skeletal muscle protein turnover. (a) Oral consumption of deuterium oxide (D_2O) or intrinsically labelled proteins (LAA). (b) Primed continuous infusion of substrate-specific stable isotope tracers to trace the path of the unlabeled 'tracee' (light grey). (c) The two-pool model measures the rate of disappearance of the labelled amino acid tracer (dark grey) from arterial blood entering the tissue and the rate of appearance of the tracer in venous blood draining the tissue. However, if you incorporate direct muscle biopsies into (c) you can use the two-pool model to determine fraction-specific protein turnover. (d) The three-pool model includes tissue biopsies to determine intramuscular tracer incorporation in addition to tracer flux across the tissue bed. (e) Incorporation of deuterium (^2H) into amino acids followed by labelled amino acid incorporation into peptides and proteins permitting the measurement of the synthetic rates of tissues and individual proteins.

Finally, one noteworthy application of the arterial venous balance approach is that it can be coupled with the provision of intrinsically labelled dietary protein sources to examine dietary protein digestion and amino acid absorption kinetics across splanchnic and skeletal muscle tissues.

Deuterium oxide

Deuterium is a stable isotope of hydrogen that can be orally consumed in the form of deuterium oxide (^2H_2O), known as 'heavy water', to measure rates of protein synthesis at both tissue-specific and individual protein levels (see Figure 5.5). Deuterium equilibrates with the body water pool and its subsequent incorporation into amino acids, such as alanine, results in the enrichment of newly synthesised proteins with deuterated amino acids. Rates of tissue-specific protein synthesis can be calculated using the 'gold standard', precursor-product approach based on the average enrichment of the body water pool (precursor) and enrichment of newly synthesised proteins in a tissue at two separate biopsy time points (product). This equation can be adjusted to calculate the synthetic rates of individual proteins as well.

In studies using deuterium oxide, typically participants orally consume a loading dose of deuterium prior to study initiation followed by smaller doses at regular intervals to maintain the enrichment of deuterium oxide in the body water pool. The enrichment of deuterium oxide in the body water pool can be obtained via saliva samples and used as a surrogate for plasma amino acid enrichments. Therefore, with the notable exception of tissue biopsies, deuterium oxide tracer analysis provides a less invasive method to examine protein synthesis in the body.

Finally, the non-specific nature and ability of deuterium oxide to be incorporated into many tissues and substrates makes it a clear choice to pair with 'omic' analysis. For example, dynamic proteomics is an emerging area of study in human research whereby the effect of various conditions, such as exercise and ageing, on the turnover of numerous individual proteins can be investigated. To date, stable isotope tracer analysis has been integral to our understanding of protein and amino acid metabolism in the body and will undoubtedly provide the foundation for future research in this field of study.

5.5 The effect of amino acid feeding and resistance exercise on skeletal muscle

Skeletal muscle supports vital bodily functions such as locomotion and respiration and is a key contributor to whole-body protein and amino acid metabolism. Skeletal muscle mass homeostasis is regulated by changes in skeletal muscle protein synthesis and breakdown with alterations in muscle mass primarily driven by changes in protein synthesis. Ingestion of a meal rich in indispensable amino acids stimulates muscle protein synthesis and suppresses rates of muscle protein breakdown, thus promoting a state of positive net protein balance.

The stimulation of muscle protein synthesis via high-quality protein ingestion is a saturable process. Specifically, high-quality protein ingestion up to ~0.24 and 0.40 g/kg body mass/meal appears to maximally stimulate myofibrillar protein synthesis, i.e. the contractile elements of skeletal muscle, in young and older men, respectively (Moore et al. 2015). Amino acid feeding-induced increases in muscle protein synthesis rates are transient in nature and remain elevated above fasting levels for only a few hours after amino acid provision. Thus, repeated stimulation of muscle protein synthesis via high-quality protein ingestion, consistent with diurnal feeding patterns, facilitates periods of positive net protein balance throughout the course of the day.

In addition to high-quality protein ingestion, rates of skeletal muscle protein synthesis are stimulated by resistance exercise. An acute bout of resistance exercise significantly elevates skeletal muscle fractional synthetic rates above basal levels for up to 48h post exercise (Figure 5.6). However, the muscle damage incurred during the resistance exercise session also enhances rates of muscle protein breakdown above resting levels for 24h post exercise (see Figure 5.6). Despite the fact that rates of muscle protein synthesis increase post exercise, net muscle protein balance remains negative as a result of exercise-induced increases in rates of muscle protein breakdown. However, when combined, amino acid ingestion and resistance exercise training have an additive effect on muscle protein synthesis, inducing a state of positive net protein balance. Specifically, resistance exercise 'sensitises' skeletal muscle to the anabolic effects of amino acids, producing a protracted increase in muscle protein synthesis above rates achieved following amino acid

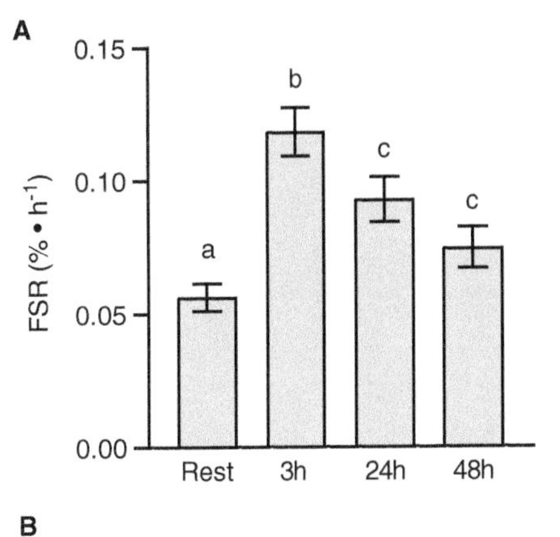

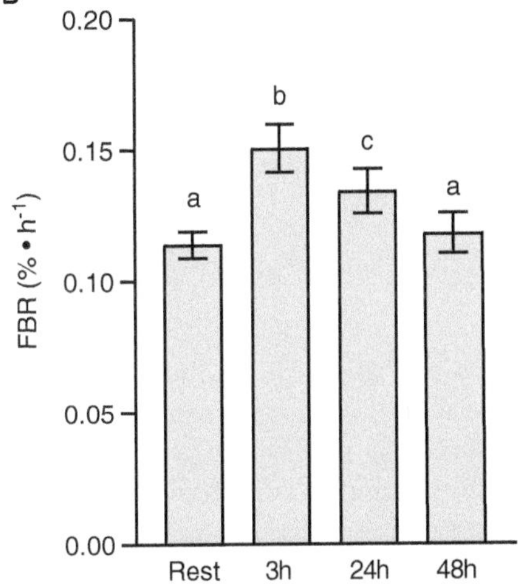

Figure 5.6 Mixed muscle protein fractional synthetic (FSR; A) and breakdown (FBR; B) rates at rest and following an acute bout of resistance exercise (3h, 24h and 48h). Means with different letters are statistically different. Data presented as mean ± SEM. *Source:* Data compiled from Phillips et al. (1997).

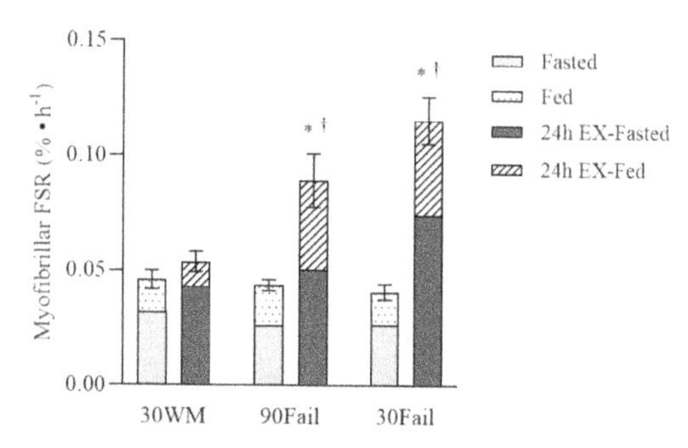

Figure 5.7 Myofibrillar fractional synthetic rate (FSR) after whey protein (15 g) ingestion at rest (Fed) and 24h after unilateral lower body resistance exercise (24 h EX-Fed). 30WM: 30% work-matched to 90Fail; 90Fail: 4 sets at 90% of 1 repetition maximum (RM) until volitional failure; 30Fail: 4 sets at 30% of 1RM until volitional failure. * Significantly different from Fed ($P < 0.05$). † Significantly different from 30WM ($P < 0.05$). Data are presented as mean ± SEM. *Source:* Data compiled from Churchward-Venne et al. (2012) where original data sources are given.

feeding alone (Figure 5.7). Similar to the synthetic response when amino acids are provided in isolation, the stimulation of muscle protein synthesis via amino acid provision following resistance exercise is also represented by a saturable dose-dependent relationship. Ingestion of up to 20 g of high-quality intact protein maximises skeletal muscle protein synthesis following an acute bout of resistance exercise, with intakes of 40 g providing no additional benefit in young adult males (Moore et al. 2009). Further, the enhanced sensitivity of skeletal muscle to amino acid provision following resistance exercise training results in potentiated rates of myofibrillar protein synthesis that are present up to 24h post exercise and may persist up to 48h.

Increased daily protein intake up to ~1.6 g/kg body mass, achieved via protein supplementation, has been speculated to maximise gains in fat-free mass (a proxy of skeletal muscle mass) during prolonged resistance exercise training (Morton et al. 2018). As such, dietary protein intake above the current RDA is probably required to support resistance exercise-induced gains in fat-free mass. The RDA, however, was never meant to maximise the resistance training-mediated skeletal muscle adaptive response. Instead, it was developed to prevent a protein deficiency.

5.6 Molecular regulation of protein turnover

Mechanistic basis of skeletal muscle protein synthesis

The molecular events responsible for the synthesis of a protein from strands of DNA are transcription and translation. Transcription yields strands of messenger ribonucleic acid (mRNA), corresponding to specific genes, as a result of RNA polymerases transcribing strands of DNA in the nucleus. Following transcription, strands of mRNA are exported from the nucleus and undergo translation in the cytosol. The end result of cytosolic translation is a polypeptide which undergoes stages of folding prior to becoming a fully functional protein.

The process of translation can be divided into stages including initiation, elongation, termination and ribosome recycling. Translation initiation is mediated by eukaryotic initiation factors (eIFs) and results in the association of the 40S and 60S ribosomal subunits (80S initiation complex) permitting translation of the mRNA strand. Briefly, eIF3 and eIF1A bind to the 40S ribosomal subunit followed by eIF2 and guanosine triphosphate (GTP) bound methionyl-transfer RNAi (tRNAi) to form a 43S preinitiation complex. Next, the hydrolysis of ATP and binding of eIF4 proteins facilitates the assembly of the 43S preinitiation complex on the 5′ end of the mRNA strand. The preinitiation complex scans the mRNA until it reaches the start codon (AUG) where a 48S initiation complex is formed. eIFs that were bound to the initiation complex are subsequently released, promoting the joining of the 60S ribosomal subunit and formation of the 80S initiation complex with Met-tRNAi remaining in the ribosomal peptidyl (P) site. Translation elongation is facilitated by elongation factors and begins as aminoacyl-tRNAs bound to specific amino acids are recruited to the aminoacyl (A) site of the ribosome. Peptide bond formation between the carboxyl terminal of one amino acid and the amino terminal of the next creates a nascent chain of amino acids that grows as the steps of elongation repeat. The movement of tRNA from the ribosomal A site to the P site simultaneously occurs alongside the translocation of mRNA as the ribosome moves along the strand of mRNA until the stop codon is reached. The high energetic demand of this process is met by the hydrolysis of GTP.

Once a stop codon (UAA, UAG or UGA) has been reached, translation is terminated. Translation termination is facilitated by various release factors and results in the release of the polypeptide chain from the ribosome. Finally, the ribosome can then be 'recycled' whereby the 60S ribosomal subunit, deacetylated tRNA and mRNA dissociate from the 40S ribosomal subunit.

Translation initiation is generally considered to be the main rate-limiting step of the translational process. It is therefore unsurprising that the regulation of this stage in response to nutritional and exercise interventions has received much experimental attention. Indeed, such investigations highlight the impact of amino acid feeding and resistance exercise on rates of skeletal muscle protein synthesis as a result of alterations in the activity of targets upstream of mRNA translation initiation and elongation.

One common target is mechanistic target of rapamycin complex 1 (mTORC1). mTORC1 is composed of the serine/threonine kinase mTOR, regulatory-associated protein of mTOR (RAPTOR), DEP domain-containing mTOR-interacting protein (DEPTOR), proline-rich AKT substrate of 40 kDa (PRAS40), and mammalian lethal with SEC13 protein 8/G protein β subunit-like (mLST8/GβL). mTORC1 is a key regulator of cell growth and becomes activated when associated with lysosomes. Lysosomal membranes are rich in amino acids, Rheb and the glycerophospholipid phosphatidic acid, all of which are capable of directly or indirectly activating mTORC1. Two well-known downstream targets of mTORC1 that affect translation initiation are ribosomal protein S6 kinase 1 (S6K1) and eukaryotic translation initiation factor 4E-binding protein 1 (4EBP1).

Phosphorylation by mTORC1 activates S6K1, allowing it to phosphorylate ribosomal protein subunit 6 (rpS6) as well as eukaryotic initiation factor 4B (eIF4B). In contrast, phosphorylation of eukaryotic elongation factor 2 (eEF2) kinase (eEF2K) by S6K1 inhibits eEF2K, resulting in elevated rates of translation elongation. Phosphorylation of 4EBP1 by mTORC1 enables its dissociation from eukaryotic initiation factor 4E (eIF4E). eIF4E forms part of the eIF4F complex which facilitates the assembly of the 43S preinitiation complex on the 5′ end of the mRNA strand to promote translation initiation. mTORC1 also phosphorylates eukaryotic initiation factor 4G (eIF4G) and protein phosphatase 2A (PP2A). PP2A is inhibited by mTORC1-dependent phosphorylation, resulting in reduced rates of phosphate group removal from mTORC1's downstream targets.

Interestingly, amino acid provision and resistance exercise seem to activate mTORC1 via different mechanisms, which may help to explain why, when combined, the two anabolic stimuli appear to have an additive effect on rates of muscle protein synthesis. However, targets upstream of mRNA translation, such as the eIF4 family, can be activated by mechanisms independent of mTORC1 in response to skeletal muscle contraction as well.

Regulation of protein synthesis by amino acids

Collectively, it appears that amino acids enhance mTORC1 activity via the translocation of mTORC1 towards Rheb positive membranes, e.g. lysosomal membranes (Figure 5.8a). Specifically, increased amino acid availability promotes the binding of Rag proteins and RAPTOR, allowing mTORC1 recruitment to the lysosomal membrane. Rag proteins (RagA, B, C and D) are G-proteins that exist as heterodimers with RagA or B bound to one RagC or D (Marcotte et al. 2015). During conditions when amino acid concentrations are high, particularly leucine, methionine and arginine, GATOR2 inhibits GATOR1, therefore removing its GTPase activating protein (GAP) activity towards RagA/B. The Ragulator complex, which exists adjacent to Rheb at the lysosomal membrane, acts as a guanine exchange factor (GEF) and activates RagA/B by facilitating GTP binding. Once activated, i.e. RagA/B is bound to GTP and RagC/D is bound to GDP, Rag proteins bind to RAPTOR, resulting in the recruitment of mTORC1 to the lysosome and subsequent activation of mTORC1 by GTP-bound Rheb. Thus, the binding of Rag proteins to RAPTOR is mediated by the Ragulator complex at the site of the lysosomal membrane during conditions of high amino acid availability.

Whilst it has been established that amino acids stimulate mTORC1-dependent pathways via events that localise mTORC1 to lysosomal

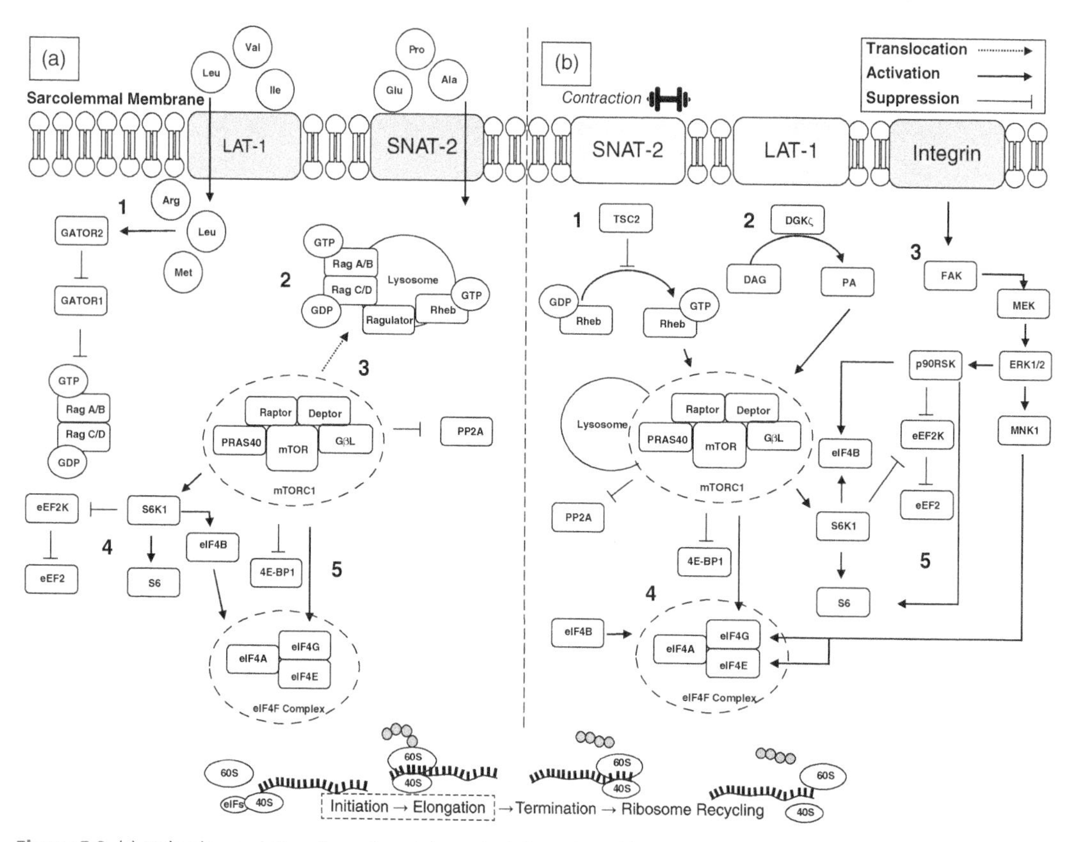

Figure 5.8 (a) Molecular regulation of muscle protein synthesis by amino acids. 1 Activation of GATOR2 by amino acids inhibits GATOR1; 2 the Ragulator complex activates Rag proteins promoting binding to RAPTOR; 3 recruitment of mTORC1 to the lysosome and subsequent activation of mTORC1 by GTP-bound Rheb; 4 mTORC1-dependent activation of S6K1 activates S6 and eIF4B and inhibits eEF2K activity; 5 phosphorylation of 4EBP1 and eIF4G promotes preinitiation complex assembly on the 5′ end of the mRNA strand. (b) Molecular regulation of muscle protein synthesis in response to resistance exercise training. 1 Skeletal muscle contraction phosphorylates TSC2, leading to its dissociation from Rheb and the subsequent activation of Rheb; 2 diacylglycerol kinase ζ activity is elevated in response to muscle contraction, increasing phosphatidic acid production; 3 mechanical stimuli activate the MAPK/ERK1/2 pathway; 4 phosphorylation of 4EBP1 and eIF4G by mTORC1; 5 activation of p90RSK results in downstream signalling that mirrors mTORC1-dependent activation of S6K1 (activation of S6 and eIF4B and inhibition of eEF2K). *Source:* Adapted from Hodson et al. (2020).

membranes, the exact mechanisms by which amino acids are sensed upstream of mTORC1 have yet to be fully elucidated.

Regulation of protein synthesis by resistance exercise

Resistance exercise enhances translational activity through increased mTORC1 activity via Rheb and phosphatidic acid as well as by mTORC1-independent mechanisms (Figure 5.8b). Rheb is a direct activator of mTORC1, but during basal conditions tuberous sclerosis complex 2 (TSC2) inhibits Rheb and renders it inactive by maintaining its GDP-bound state. During resistance exercise, skeletal muscle contraction leads to the phosphorylation of TSC2 and its subsequent dissociation from Rheb such that Rheb can bind to GTP and activate mTORC1. However, to date, the exact mechanism underpinning signal transduction upstream of TSC2 to enhance mTORC1 activity during mechanical loading has not been fully elucidated.

Skeletal muscle contraction also enhances the production of another direct activator of mTORC1, phosphatidic acid, via elevations in the activity of diacylglycerol kinase ζ. Diacylglycerol kinase ζ produces phosphatidic acid through the phosphorylation of diacylglycerol.

One additional pathway that appears to be involved in contraction-induced increases in skeletal muscle protein synthesis is the mitogen activated protein kinase/extracellular signal-regulated kinases 1/2 (MAPK/ERK1/2) pathway. Ligand–receptor binding or mechanical stimuli-, i.e. muscle contraction, induced activation of focal adhesion kinase or integrins at the cellular membrane initiate the signalling events involved in this pathway. ERK1/2 is the main downstream target of this pathway that, when activated via phosphorylation, phosphorylates its own downstream targets, namely p90 ribosomal protein S6 kinase (p90RSK) and MAP kinase-interacting kinase 1 (MNK1). p90RSK enhances rates of translation initiation and elongation by phosphorylating its shared targets with S6K1, rpS6, eIF4B and eEF2K, whereas MNK1 exerts its stimulatory effect on protein synthesis via phosphorylation of eIF4G followed by eIF4E.

Whilst muscle protein synthetic signalling pathways are commonly presented as if they occur in isolation, it is important to note that these pathways are often interconnected. For example, inhibition of 4EBP1 by mTORC1 may be required to maximise the effect of MNK1 on eIF4E. Thus, when examining alterations in rates of skeletal muscle protein synthesis in response to various stimuli, it may be pertinent to consider the roles of multiple signalling pathways and the potential for cross-talk between them.

5.7 Conclusion

Dietary protein and amino acid metabolism are vital processes that support the structure and function of all bodily tissues. Whilst amino acids have numerous metabolic fates, perhaps their most important role is to serve as substrates for protein synthesis. As discussed in this chapter, whole-body net protein mass is dictated by the algebraic difference between rates of whole-body protein synthesis and protein breakdown. However, methodological advancements, such as skeletal muscle biopsies and stable isotope tracer analysis, permitting the direct determination of tissue-specific protein turnover, have highlighted that rates of protein turnover vary in a tissue-specific manner. Moreover, tissue-specific rates of protein synthesis, such as skeletal muscle, are highly responsive to various anabolic stimuli such as indispensable amino acid feeding and resistance exercise training.

To date, research in the field of protein and amino acid metabolism has examined protein turnover and amino acid kinetics in response to protein and amino acid ingestion from a variety of dietary sources and exercise. Recent advances in the field of 'omic' analysis, such as dynamic proteomics, provide a promising avenue by which future research will be able to elucidate the effect of stimuli including, but not limited to, indispensable amino acid ingestion and resistance exercise, on the synthesis of individual proteins from various cellular fractions.

Protein and amino acid metabolism is an important field of research that will continue to provide the foundation for the development and optimisation of nutritional and nutraceutical interventions as we seek to find ways to improve human health throughout the lifespan.

References

Broer, S. and Broer, A. (2017). Amino acid homeostasis and signalling in mammalian cells and organisms. Biochemical Journal 474: 1935–1963.

Burd, N.A., McKenna, C.F., Salvador, A.F., Paulussen, K.J.M., and Moore, D.R. (2019). Dietary protein quantity, quality, and exercise are key to healthy living: a muscle-centric perspective across the lifespan. Frontiers in Nutrition 6: 83.

Churchward-Venne, T.A., Burd, N.A., and Phillips, S.M. (2012). Nutritional regulation of muscle protein synthesis with resistance exercise: strategies to enhance anabolism. Nutrition and Metabolism 9: 40.

Cynober, L. (2018). Metabolism of dietary glutamate in adults. Annals of Nutrition and Metabolism 73 (Suppl 5): 5–14.

Devries, M.C., Sithamparapillai, A., Brimble, K.S., Banfield, L., Morton, R.W., and Phillips, S.M. (2018). Changes in kidney function do not differ between healthy adults consuming higher- compared with lower- or normal-protein diets: a systematic review and meta-analysis. Journal of Nutrition 148: 1760–1775.

Gorissen, S.H.M., Trommelen, J., Kouw, I.W.K.et al. (2020). Protein type, protein dose, and age modulate dietary protein digestion and phenylalanine absorption kinetics and plasma phenylalanine availability in humans. Journal of Nutrition 150: 2041–2050.

Groen, B.B., Horstman, A.M., Hamer, H.M.et al. (2015). Post-prandial protein handling: you are what you just ate. PLoS One 10: e0141582.

Hodson, N., Moore, D.R., and McGlory, C. (2020). Resistance exercise training and the regulation of muscle protein synthesis. In: The Routledge Handbook on Biochemistry of Exercise. (ed. P.M. Tiidus, R MacPherson, P. LeBlanc, and A.Josse). Abingdon: Routledge

Holececk, M. (2018). Branched-chain amino acids in health and disease: metabolism, alterations in blood plasma, and as supplements. Nutrition and Metabolism 15: 33.

Institute of Medicine. (2005). Dietary Reference Intakes for Energy, Carbohydrate, Fiber, Fat, Fatty Acids, Cholesterol, Protein, and Amino Acids. Washington, DC: National Academies Press.

Jeukendrup, A. and Gleeson, M. (2018). Sport Nutrition. Champaign, IL: Human Kinetics.

Marcotte, G.R., West, D.W.D. and Baar, K. (2015). The molecular basis for load-induced skeletal muscle hypertrophy. Calcified Tissue International 96(3):196–210.

Moore, D.R., Churchward-Venne, T.A., Witard, O.et al. (2015). Protein ingestion to stimulate myofibrillar protein synthesis requires greater relative protein intakes in healthy older versus younger men. Journals of Gerontology A: Biological Sciences and Medical Sciences 70: 57–62.

Moore, D.R., Robinson, M.J., Fry, J.L.et al. (2009). Ingested protein dose response of muscle and albumin protein synthesis after resistance exercise in young men. American Journal of Clinical Nutrition 89: 161–168.

Morton, R.W., Murphy, K.T., McKellar, S.R.et al. (2018). A systematic review, meta-analysis and meta-regression of the effect of protein supplementation on resistance training-induced gains in muscle mass and strength in healthy adults. British Journal of Sports Medicine 52: 376–384.

Nunes, E.A., Currier, B.S., Lim, C., and Phillips, S.M. (2021). Nutrient-dense protein as a primary dietary strategy in healthy ageing: please sir may we have more? Proceedings of the Nutrition Society 80: 264–277.

Phillips, S.M. (2017). Current concepts and unresolved questions in dietary protein requirements and supplements in adults. Frontiers in Nutrition 4: 13.

Phillips, S.M., Fulgoni, V.L., 3rd, Heaney, R.P., Nicklas, T.A., Slavin, J.L., and Weaver, C.M. (2015). Commonly consumed protein foods contribute to nutrient intake, diet quality, and nutrient adequacy. American Journal of Clinical Nutrition 101: 1346S–1352S.

Phillips, S.M., Tipton, K.D., Aarsland, A., Wolf, S.E., and Wolfe, R.R. (1997). Mixed muscle protein synthesis and breakdown. American Journal of Physiology 273: E99–E107.

Suryawan, A., Hawes, J.W., Harris, R.A., Shimomura, Y., Jenkins, A.E., and Hutson, S.M. (1998). A molecular model of human branched-chain amino acid metabolism. American Journal of Clinical Nutrition. 68: 72–81.

Tang, J.E., Moore, D.R., Kujbida, G.W., Tarnopolsky, M.A., and Phillips, S.M. (2009). Ingestion of whey hydrolysate, casein, or soy protein isolate: effects on mixed muscle protein synthesis at rest and following resistance exercise in young men. Journal of Applied Physiology 107: 987–992.

Van der Wielen, N., Moughan, P.J., and Mensink, M. (2017). Amino acid absorption in the large intestine of humans and porcine models. Journal of Nutrition 147: 1493–1498.

Wilkinson D.J., Brook M.S., Smith K. (2021). Principles of stable isotope research – with special reference to protein metabolism. Clinical Nutrition Open Science, 36: 111–125.

Wolfe, R.R., Rutherford, S.M., Kim, I.Y., and Moughan, P.J. (2016). Protein quality as determined by the digestible indispensable amino acid score: evaluation of factors underlying the calculation. Nutrition Reviews 74: 584–599.

Wolfe, R.R., Park, S., Kim, I.Y.et al. (2019). Quantifying the contribution of dietary protein to whole body protein kinetics: examination of the intrinsically labeled proteins method. American Journal of Physiology Endocrinology and Metabolism 317: E74–E84.

Further reading

Dickinson, J.M. and Rasmussen, B.B. (2013). Amino acid transporters in the regulation of human skeletal muscle protein metabolism. Current Opinion in Clinical Nutrition and Metabolic Care 16 (6): 638–644.

Gorissen, S.H.M., Trommelen, J., Kouw, I.W.K. et al. (2020). Protein type, protein dose, and age modulate dietary protein digestion and phenylalanine absorption kinetics and plasma phenylalanine availability in humans. Journal of Nutrition 150 (8): 2041–2050.

Groen, B.B., Horstman, A.M., Hamer, H.M. et al. (2015). Post-prandial protein handling: you are what you just ate. PLoS One 10 (11): e0141582.

Joanisse, S., Lim, C., McKendry, J., Mcleod, J.C., Stokes, T., and Phillips, S.M. (2021). Understanding the effects of nutrition and post-exercise nutrition on skeletal muscle protein turnover: insights from stable isotope studies. Clinical Nutrition Open Science 36: 56–77.

Miller, B.F., Reid, J.J., Price, J.C., Lin, H.L., Atherton, P.J., and Smith, K. (2020). CORP: the use of deuterated water for the measurement of protein synthesis. Journal of Applied Physiology 1985 128 (5): 1163–1176.

Trommelen, J., Tomé, D., and van Loon, L.J.C. (2021). Gut amino acid absorption in humans: concepts and relevance for postprandial metabolism. Clinical Nutrition Open Science 36: 43–55.

White, J.P. (2021). Amino acid trafficking and skeletal muscle protein synthesis: a case of supply and demand. Frontiers in Cell and Developmental Biology 9: 656604.

Wolfe, R.R., Rutherfurd, S.M., Kim, I.Y., and Moughan, P.J. (2016). Protein quality as determined by the digestible indispensable amino acid score: evaluation of factors underlying the calculation. Nutrition Reviews 74 (9): 584–599.

6
Pregnancy and Lactation

Sharleen L. O'Reilly and Michelle K. McGuire

Key messages

Pregnancy

- Critical periods exist in pregnancy during which adequate levels of specific nutrients, such as iodine, DHA, protein, iron, choline and folate, are required for optimal fetal development. Suboptimal development during critical periods cannot be corrected in pregnancy or postpartum.
- Common genetic variants can influence nutrient requirements in pregnancy. For example, single nucleotide polymorphisms (SNPs) influence pathways for the biosynthesis of DHA, folate and choline. The influence of different genetic variants on a range of nutritional pathways needs greater characterisation.
- The approach to estimating nutritional needs in pregnancy is mainly focused on the amount and composition of pregnancy weight gain, increased maternal metabolism and altered absorption rates.
- The energy costs of pregnancy require about 15% more food intake, compared with pre-pregnancy, primarily in the third trimester.
- Tailoring of dietary recommendations for pregnant individuals is needed due to the potential of a variety of options to support the energy and nutrient needs during pregnancy.
- Recommendations generally include increasing dietary intake, decreasing physical activity and/or limiting maternal adipose tissue accretion.

Lactation

- Human milk represents a biological system composed of nutrients, complex oligosaccharides, cells (both maternal and microbial), hormones and growth factors, immune factors and myriad additional biologically active substances that collectively nourish, protect and communicate important information to the infant.
- Milk production and milk composition are controlled by a complex web of anatomy, maternal hormones (e.g. prolactin and oxytocin), infant inputs (e.g. suckling intensity) and environmental factors (e.g. nutrient availability).
- Homeorhetic shifts in maternal metabolism, blood flow and nutrient absorption are important physiological responses that support lactogenesis and milk production.
- Human milk composition varies greatly among women and across populations. Composition is affected by factors such as maternal genetics, time of day, time since last feed, time during nursing bout, maternal diet and health status, stress and time postpartum.
- Some constituents in milk (e.g. thiamine, riboflavin, vitamin C, fatty acids) are affected by variation in maternal diet whereas others (e.g. folate, calcium, iron, copper) are not.
- Breastfeeding confers myriad benefits to both mothers and infants, and exclusive breastfeeding is generally recommended for the first six months with continued breastfeeding up to two years or beyond.

6.1 Introduction

Maternal nutrition and nutritional status during pregnancy and lactation have substantial impacts on the health of both the woman and her offspring, and the potential impact of nutrition on health and well-being is probably greater during these times than during any other life stage.

There are three main phases within a woman's reproductive life stage: preconception, pregnancy and lactation. During the preconception phase, low percentage body fat can negatively influence ovulation and menses and challenge

Nutrition and Metabolism, Third Edition. Edited on behalf of The Nutrition Society by Helen M. Roche, Ian A. Macdonald, Annemie M.W.J. Schols and Susan A. Lanham-New.

Companion Website: www.wiley.com/go/nutrition/metabolism3e

the beginning of pregnancy. Equally, excessive fat stores may also reduce fertility by affecting ovulation because of insensitivity to insulin, an excess of male reproductive hormones and the overproduction of leptin.

Achieving adequate nutritional status prior to pregnancy is important as this will lay the foundation for healthy pregnancy and lactation. Entering pregnancy either underweight or with nutritional deficiencies is linked with poor pregnancy outcomes, further depleted maternal stores, increased risk for lactation insufficiency and suboptimal milk composition. The impact of undernutrition also influences infant birth outcomes such as low birth weight, cognitive and physical disability, and stillbirth. Longer-term infant health impacts include increased prevalence of coronary heart disease, raised lipids, obesity and decreased glucose tolerance in adult life. Entering pregnancy with overweight or obesity is linked with increased risks of gestational hypertension, gestational diabetes, stillbirth, caesarean delivery, macrosomia and certain birth defects. Similarly, if the in utero weight or fat mass increase is too high, there may be short- and long-term adverse effects.

The effects of maternal malnutrition across the reproductive period of a woman's life mostly depend on the demands being placed on nutrient stores. Other external factors need to be considered as well, such as the woman's age, genetics, exposure to stressors or toxins including alcohol and nicotine, illicit drug use and the presence of disease or other illnesses. This chapter, however, will concentrate on the nutritional aspects.

6.2 Pregnancy

Physiological stages of pregnancy

Pregnancy can be divided into three main physiological stages: implantation, organogenesis and growth.

Implantation

The implantation stage includes the first two weeks of gestation, when the fertilised ovum becomes embedded in the wall of the uterus. The nutrients provided by the secretions of the uterine glands pass directly into the fertilised ovum and developing embryo. Specific nutritional needs are undoubtedly required at this stage but, quantitatively speaking, these demands are negligible.

Organogenesis

The next six weeks of pregnancy are known as the period of organogenesis or embryogenesis. During this stage, the cells of the embryo begin to differentiate into distinct tissues and functional units that later become organs, such as heart, lungs and liver. The neural tube knits together and closes to form the brain, spinal cord and spinal column. The development of the skeleton, including dentition buds, also begins. During organogenesis, the fetus obtains nourishment from the intervillous space and yolk sac (spaces around the developing fetus and amniotic cavity) because fetal and maternal circulation to the placenta are not fully established until the end of week 8 and the first trimester respectively. When organogenesis is complete, the fetus weighs approximately 6 g and is less than 3 cm long.

Evidence indicates that this specific stage of pregnancy is critical. It determines embryonic viability and has far-reaching consequences for development and growth postnatally, as originally proposed in Barker's developmental origins hypothesis (now referred to as the developmental origins of health and disease hypothesis). This window is typically before a pregnancy is confirmed and, as a result, underlines the importance of achieving optimal nutritional status prior to pregnancy.

There are specific nutrients of note in this stage, namely folate, vitamin D, vitamin A, vitamin B_{12}, riboflavin, niacin, pyridoxine and manganese. Low folate status is inversely related to the risk of neural tube defects (NTDs) which include anencephaly and spina bifida. Folic acid supplementation (400 μg/day) is recommended to effectively reduce this risk; women with known risk factors for NTDs such as obesity, previous NTD birth and diabetes are recommended to consume 5 mg/day folic acid to achieve risk reduction (WHO 2007). Dietary folate alone is insufficient to achieve the circulating levels of folate required to reduce risk, therefore supplementation either through folic acid food fortification or supplements is necessary and this should ideally be at least six weeks prior to attempting pregnancy to ensure all folate stores are replete.

Riboflavin deficiency is associated with poor skeletal formation, pyridoxine and manganese deficiencies with neuromotor problems, vitamin D with enamel hypoplasia, and vitamin B_{12}, vitamin A and niacin deficiencies with central nervous system problems.

Growth

The remaining seven months of pregnancy are known as the growth period and the nutrients are delivered through the placenta. Blood flow is not directly from the woman's circulation to the fetus; the placenta provides about 10–11 m^2 of surface area for the exchange of nutrients and this is influenced by uteroplacental blood flow. The placental transport mechanisms vary from simple diffusion (not requiring ATP or a transport protein), facilitated diffusion (diffusion assisted by the presence of a protein embedded in the membrane) and active transport (ATP-requiring transport against a concentration gradient powered), to endocytosis. Diffusion is the main transport mechanism for free fatty acids, long-chain polyunsaturated fatty acids, steroids, electrolytes, fat-soluble vitamins, caffeine and gases.

Transporter-mediated transfers fall into two main categories: facilitated transport and active transport. Glucose is the main source of fetal energy, crossing the placenta via facilitated diffusion, while amino acids, calcium, iron, iodine, phosphate and water-soluble vitamins are transferred by active transport. Endo/exocytosis is used to transfer immunoglobulins and harmful substances (drugs, viruses, toxins).

During the growth period, there are three main phases during which the tissues and organs formed during organogenesis continue to grow and mature. The first phase, known as hyperplasia, is characterised by a rapid increase in cell number through cellular division; this requires sufficient supplies of folate and vitamin B_{12}. The second growth phase is characterised by increased cell size and replication continuing with the hypertrophy, which requires sufficient supplies of amino acids and vitamin B_6. The final phase is characterised by hypertrophy wherein cellular growth predominates and cellular division ceases.

Inadequate nutrition during the growth phase can lead to intrauterine growth restriction (IuGR) and low birth weight, but because the main organs and tissues have formed, the more serious abnormalities associated with deficiencies at earlier times are not seen as readily. IuGR is generally defined as less than 10% of predicted fetal weight for gestational age, whereas low birth weight is defined as < 2500 g body weight at birth. IuGR and low birth weight are associated with higher perinatal morbidity and mortality as well as growth failure and cognitive developmental delay during infancy and childhood. The fetus is, however, considered resilient to short periods of suboptimal nutrition and can compensate by compensatory growth when nutrition returns to improved levels, which means that nutritional interventions are worthwhile throughout all pregnancy stages.

In summary, pregnancy is clearly a life stage during which optimal nutrition is particularly important, and different phases require adaptations to ensure nutritional needs are met. The impact of nutritional inadequacy is evident in terms of pregnancy outcomes, but less is known about the exact consequences of individual nutrients for each phase.

Estimating nutritional requirements during pregnancy

The ability to perform standard nutritional requirement studies in pregnancy is limited by the nature of the risks associated with more invasive methods. As a result, there are few direct human data available, and determining nutritional requirements takes a 'factorial approach'. This factorial approach uses the known variables in non-pregnant and non-lactating women and applies additional consideration to achieving optimal maternal body composition changes occurring in pregnancy to ensure healthy fetal growth and adequate stores for lactation. For this purpose, a predefined desirable pregnancy outcome of a live, full-term birth weight between 3 and 4 kg is typically used.

We will now examine the energy content of the pregnancy weight gain and the increased energy metabolism in pregnancy.

Energy requirements

Average weight gain during pregnancy is 12 kg, which will deliver an average infant birth weight of 3.4 kg. This is based on well-nourished women in developed countries, who will typically have a pre-pregnant body weight of 60–65 kg. The velocity of weight gain aligns with the stages of growth in pregnancy and as a result, the spread is calculated to be 650, 3350, 4500 and 4000 g, respectively, over the four quarters of pregnancy.

Pregnancy weight gain can be divided into two categories: (i) the products of conception, that is the fetus, amniotic fluid and placenta, and (ii) the increased amounts of maternal tissues, that is extracellular fluid, uterus, breasts, blood and adipose stores. Distribution of the total weight gain during pregnancy that can be attributed to each component is shown in Table 6.1. Hytten and Leitch were the first to describe the theoretical estimates of pregnancy-driven nutritional requirements in the 1960s. These estimates are regularly used by international and national bodies to define a 'reference pregnancy' in their efforts to develop recommended intakes in pregnancy.

The nutritional cost of developing these additional body composition changes can be estimated by understanding their distribution. The energy equivalent of the reference pregnancy weight gain is about 200 MJ, which is composed of two components. The first is the energy deposited in adipose stores, which is about 150 MJ (154 MJ based on 3345 g adipose, at an energy cost of 46 kJ/g), and the second is the energy deposited in other tissues such as the fetus, placenta, uterus, breasts, amniotic fluid and blood, which is about 50 MJ (47 MJ based on 440 g lipid and 925 g protein, at an energy cost of 46 kJ/g and 29 kJ/g, respectively).

Table 6.1 The components of weight gain over 40 weeks of pregnancy.

	Weight gain (g)	Percentage of gained weight (%)
Products of conception		
Fetus	3400	27.2
Amniotic fluid	800	6.4
Placenta	650	5.2
Maternal tissues		
Extracellular fluid	1680	13.4
Uterus and breasts	1375	11.1
Blood	1250	10.0
Adipose stores	3345	26.8
Total weight gain	12 500	100.0

Source: Adapted from Hytten F, Leitch I. The Physiology of Human Pregnancy, 2nd edn. Oxford: Blackwell Publishing, 1971.

Metabolism and nutrients

An increase in the maintenance of metabolism during pregnancy is approached by studying basal metabolic rate (BMR) or resting energy expenditure (REE) throughout pregnancy. Measurements are made under standardised conditions (rest, postabsorptive state, thermoneutrality). Any increase in the maintenance of metabolism is calculated as the cumulative area under the curve, represented by the rise in a woman's BMR/REE above the pre-pregnancy rate.

Wide variation exists when looking at the cumulative increases in metabolic maintenance during pregnancy, for example +210 MJ in Swedish women to −45 MJ in non-supplemented Gambian women. When results from all available studies are combined, a good correlation appears between pregnancy weight gain and the cumulative increase of metabolic maintenance during pregnancy.

Energy metabolism

The effect of pregnancy on energy expenditure varies over the course of a pregnancy and among individuals. Pregnancy increases REE because of the energy needed to maintain increased tissue mass, the heart and lungs increasing their work as the pregnancy progresses alongside the metabolic load of the uterus and fetus. For the reference weight gain of 12 kg, the increase in REE is estimated to be 147.8 MJ, which is very close to the estimate based on theoretical calculations by Hytten and Leitch (150 MJ) (see Table 6.1), so for a reference pregnancy a total energy cost has been projected to be approximately 320 MJ (200 MJ energy content of weight gain plus 120 MJ for increased metabolism). The corresponding cumulative increases in REE are estimated to be 5%, 10% and 25% across the trimesters, which results in extra energy expenditure of 0.35, 1.2 and 2.0 MJ/day, respectively. However, the energy required will depend on the woman's pre-pregnancy nutritional status, and women with BMIs higher or lower than the normal range will require adjustments.

Macronutrient metabolism

Changes in maternal hormone production shift the utilisation of macronutrients in pregnancy, an overarching concept referred to as *homeorhesis*. Homeorhetic shifts in what are considered to be the physiological 'normal' are also important to support lactation (described in the subsequent section). Amino acids and glucose are continually required by the developing foetus. Human chorionic somatotropin (also referred to as human placental lactogen and human chorionic somatomammotropin) is produced by the placenta and is thought to drive greater use of lipids to make glucose and amino acids available for energy uptake by the fetus. Other hormones support conservation of maternal lean tissue.

Shifts in protein metabolism are complex and change gradually throughout pregnancy to ensure nitrogen conservation for fetal growth, especially during the third trimester. Nitrogen balance studies in pregnant and non-pregnant women show no evidence to suggest that nitrogen deposited in early pregnancy will be mobilised later. However, nitrogen retention is increased in late pregnancy and is due to a reduction in urinary nitrogen excretion. A 30% decrease in urea synthesis seems to account for this reduction in urinary urea nitrogen and suggests that amino acids are conserved for tissue synthesis.

While it is likely that the increased requirements of late pregnancy might, at least in part, be met by physiological (homeorhetic) adjustments that enhance dietary protein utilisation, nevertheless protein requirements rise substantially in the third trimester. A linear relationship exists for nitrogen intake and deposition for this trimester that is statistically significant. The slope of this equation gives a very low nitrogen efficiency (21–47%), and this is used to calculate the additional protein requirements for pregnancy.

Micronutrient metabolism

More than 90% of fetal growth occurs in the latter half of pregnancy. This in turn drives the increased demand for nutrients during this time. The increase in a woman's plasma volume also affects metabolism of certain nutrients, and the increase in red blood cell mass (200–250 ml) requires additional iron. In pregnancy, plasma concentrations of fat-soluble vitamins, lipids and some carrier proteins typically rise; concentrations of albumin, most minerals, water-soluble vitamins and amino acids drop. The lower levels in circulation generally reflect homeorhetic shifts in metabolism associated with pregnancy and potentially increased urinary excretion rather than altered nutritional status.

Adaptive responses are key for effective nutrient metabolism. For many nutrients, absorption rates increase in pregnancy although the underlying mechanisms are not clear. These absorption rates need to be taken into consideration when suggesting nutrient intake recommendations. Homeorhetic responses in pregnancy include upregulation of iron and calcium absorption. Total plasma calcium levels fall very early in pregnancy, mediated by a fall in plasma albumin, as part of the process of haemodilution. However, ionised calcium and phosphate levels remain at levels consistent with non-pregnant levels throughout. Conversely, plasma levels of 1,25-dihydroxy-D_3 are elevated early in pregnancy and remain elevated throughout. Intestinal calcium absorption doubles in pregnancy, probably owing to the changes in vitamin D status. The earliest study of calcium absorption was conducted at 12 weeks and even at this early stage the rise in calcium absorption was noted. The observed elevation in 1,25-dihydroxy-D_3 is associated with an observed rise in calcium binding protein $_{9K}$-D.

Given that fetal need for calcium does not increase until later in pregnancy, it is possible that the increased calcium flow from the intestine leads to storage in the maternal skeletal pool. Indeed, increased bone density has been observed in animal models of pregnancy. Zinc and copper absorption may also be upregulated, while riboflavin and taurine excretion in urine is reduced. Aldosterone production is increased particularly towards the end of pregnancy, which raises blood pressure and increases sodium reabsorption from the renal tubules, thereby increasing fluid retention.

Haemodilution is a natural consequence of pregnancy. It is important that the fall in serum nutrient concentration caused by haemodilution should not be interpreted as a sign of nutritional deficiency but rather a homeorhetic shift in 'normal' within pregnancy. It is therefore essential that, to achieve a valid assessment of a pregnant woman's micronutrient status, appropriate pregnancy standards are used.

Behavioural adaptations to meet nutrient requirements during pregnancy

Several behavioural options exist to adapt the energy and nutrient needs of pregnancy, including consuming more and/or a more varied dietary pattern, consumption of dietary supplements and reduced physical activity.

Altered diet

Traditionally, women were told to 'eat for two' but this mantra is not actually evidence based. To cover the energy costs of pregnancy solely by increasing food intake, women should consume about 10–15% more than pre-pregnancy. However, most studies in well-nourished women revealed relatively minor increases in the amount of energy and nutrients consumed in pregnancy; where an increased intake was found, the observed level of increment only partly covered the energy cost of pregnancy. The practical application is advising two to five 0.5 MJ snacks daily in the second and third trimesters to cover the additional energy requirements.

Dietary quality

Dietary quality is also an important aspect to consider in pregnancy. A diverse diet containing core food groups in line with dietary guidelines should ensure nutritional adequacy. Cultural practices, food cravings (i.e. strong desires to eat particular foods, such as dairy and sweet foods) and food aversions (i.e. strong desires to avoid particular foods, such as fried foods, eggs and tea) may influence dietary quality in pregnant women. Cravings and aversions that arise during pregnancy are most likely due to hormone-induced changes in sensitivity to taste and smell, and do not seem to reflect altered physiological needs.

Dietary supplements

Pregnant women are encouraged to obtain their nutrients from a well-balanced, varied diet, rather than from vitamin or mineral supplements, with the noted exception of folic acid. When this cannot be achieved and for specific risk groups or populations, dietary supplementation is an option.

Physical activity

The energy costs for a person's daily physical activity depend on (i) the time–activity pattern (amount of time spent on various activities), (ii) the pace or intensity of performing the various activities, and (iii) the person's body weight.

Reviews of physical activity levels during pregnancy across a variety of countries suggest that women do not change their activity levels during pregnancy. However, the reviews do not provide data on the level of effort exerted to undertake normal activities, and this may change as pregnancy progresses. The energy expended through physical activity (EEPA) for the third trimester ranges from -22 to + 17%. And when expressed in per kilo body weight, there is a tendency towards lower EEPA/kg/day. However, the energy expenditure is not significantly different from non-pregnant women, so it is assumed that EEPA is not altered in pregnancy. This assumption is underpinned by women's ability to adapt their daily activities or to change the pace or intensity of the work performed. This assumption may not hold for low-income women in developing countries who often have to continue their strenuous activity pattern until delivery, and the option to save energy by reducing physical activity is not available. Equally, if a woman has very low levels of physical activity and further reduces her activity levels during pregnancy, the potential for excess energy storage is high. Furthermore, even if women voluntarily decrease their pace and consequently the energy expenditure per minute, the energy cost to complete a task may be unchanged or even increased, since the performance of the task may take more time and/or require more energy due to increased maternal body weight.

Women with high levels of pre-pregnant EEPA who maintain their activity levels in pregnancy will tend to have less gestational weight gain and deliver smaller babies. Systematic review supports an inverted U-shaped curve for physical activity levels in pregnancy and birth weight. Observational studies support the benefits of EEPA in maternal outcomes such as fewer birth interventions and lower risks for gestational diabetes and depressive disorders.

Dietary recommendations for pregnancy

The nutritional demands of pregnancy can be met through a variety of ways: increasing and/or altering dietary intake, decreasing physical activity and limiting body storage (e.g. adipose accretion). The approach selected should be tailored to the individual to ensure it can be achieved. Factors to consider in these options are the availability of food; current physical activity level and physical abilities; social and cultural acceptability; and pre-pregnancy nutritional status, including adiposity.

Table 6.2 shows the Average Requirement (AR) dietary reference values for energy, macronutrients, vitamins and minerals for pregnant as well as non-pregnant/non-lactating (NPNL) women.

Energy

The reference pregnancy, based on an increase in maternal adipose stores of 3741 g and a protein deposition of 597 g, takes an extra 158.9 MJ approximately over the whole pregnancy. The ARs are +0.29, +1.1 and +2.1 MJ/day in the first, second and third trimesters respectively. In other words, on average, pregnant women must consume 0.29, 1.1, and 2.1 MJ more than NPNL women on a daily basis during the first, second and third trimesters. It is worth noting that these are average requirements and that some women will need more while others will need less.

Protein

The recommendations are based on the deposition of protein mainly in the second (20%) and third (80%) trimesters. Accordingly, the daily recommended increases in energy from protein (RI levels) are 0, 30 and 121 kJ for the first, second and third trimesters, respectively. The population reference intake (PRI) for NPNL women is 0.83 g/kg/day and increases for pregnant women at rates of +1, +9 and +28 g/day across the trimesters.

Table 6.2 Nutrient reference values of vitamins and minerals for pregnant women.

	Women, ≥18 years			
	Non-pregnant, non-lactating		Pregnant	
	AI	PRI	AI	PRI
Water-soluble vitamins				
Thiamin (mg/MJ)	0.072[a]	0.1	0.072	0.1
Riboflavin (mg/day)	1.3[a]	1.6	1.5	1.9
Niacin (mg NE/MJ)	1.3[a]	1.6	1.3	1.6
Vitamin B_6 (mg/day)	1.3[a]	1.6	1.5	1.8
Vitamin B_{12} (μg/day)	4		4.5	
Folate (μg DFE/day)	250	330	600	
Pantothenic acid (mg/day)	5		5	
Biotin (μg/day)	40		40	
Vitamin C (mg/day)	80	95		105
Fat-soluble vitamins				
Vitamin A (μg RE/day)	490[a]	650	540	700
Vitamin D (μg/day)	15		15	
Vitamin E (mg/day)	11		11	
Vitamin K (μg/day)	70		70	
Choline (mg/day)	400		480	
Minerals				
Calcium (mg/day)	860/750[a]	1000/950	860/750[b]	1000/950[b]
Phosphorus (mg/day)	550		550	
Zinc (mg for 300 mg LPI/day)	6.2[a]	7.5	(+)1.3	(+)1.6
Iron (mg/day)	7[a]	16	7	16
Magnesium (mg/day)	300		300	
Iodine (μg/day)	150		200	
Selenium (μg/day)	70		70	
Molybdenum (μg/day)	65		65	
Copper (mg/day)	1.3		1.5	
Manganese (mg/day)	3		3	
Fluoride (mg/day)	2.9		2.9	
Potassium (mg/day)	3500		3500	

[a] AR instead of AI.
[b] 18–24 years/25 years and older, + added to non-pregnant, non-lactating values.
PRI, population reference intake; AI, adequate intake; AR, average requirement; LPI, level phytate intake; RE, retinol equivalents; DFE, dietary folate equivalents; NE, niacin equivalents.
Source: Adapted from European Food Safety Authority (2019).

Essential fatty acids

Transplacental transfer of essential fatty acids is of crucial importance for fetal development, particularly membrane formation and cellular development in the brain. Demand for n-6 and n-3 polyunsaturated fatty acids (PUFAs) for placental and fetal tissue must be met from maternal stores or by increased dietary intake. The AIs for pregnancy are based on those for NPNL women, with additional amounts where required. α-Linolenic acid (ALA) is 0.5% energy, docosahexaenoic acid (DHA) is an additional 100–200 mg/day, linoleic acid is 4% energy. DHA is the major long-chain n-3 PUFA needed to build fetal brain tissue and is therefore important in pregnancy. To make more DHA available for brain development, the maternal diet should include fish, especially oily fish (1–2 portions per week).

For pregnant women, there are no other specific recommendations with respect to dietary fat other than to consume levels needed during the NPNL period. This is also the case for *carbohydrates* and dietary *fibre*. Guidelines recommend abstaining from alcohol and reducing caffeine intake to <200 mg/day, the amount contained in approximately two cups of coffee.

Fat-soluble vitamins

While adaptive mechanisms can improve the body's use of some minerals during pregnancy, there are some that do not adapt (e.g. copper and iodine). There is less evidence of similar mechanisms for adapting to vitamin requirements. Dietary reference values for pregnancy include a small increase for vitamin A and no increments for vitamins D, E and K.

The newborn infant appears to require 100 μg/day retinol for growth, and this requirement is assumed to hold for the third trimester. Alternatively, the amount of retinol accumulated in the fetus is assumed to be 3600 μg with an assumed efficacy of 50% storage, so an additional 50 μg retinol equivalents (RE) are required for pregnancy. It is important to note that high intakes of vitamin A (3000 μg RE/day is the upper limit, UL) will damage the developing fetus and should be avoided.

Active forms of vitamin D readily cross the placenta to play an active role in the metabolism of calcium in the fetus. However, the amounts required are too small to affect the woman's vitamin D requirement, particularly since serum calcitriol is increased and there is an improvement in calcium resorption during late pregnancy. Guidelines support supplementation with vitamin D (10–15 μg/day) where minimal cutaneous vitamin D synthesis is occurring; where greater sunlight exposure exists, the requirements are lower or may even be zero.

Water-soluble vitamins

The requirements for most B vitamins are increased by 10–50% in pregnancy, based on maternal and fetal growth needs and increased energy use. Folate requirements increase substantially in pregnancy (by 60–70%). This does not include the additional needs to prevent neural tube defects, as the neural tube is formed before most women are aware of their pregnancy. Average requirement for vitamin C also increases slightly to account for fetal needs.

Choline

Choline also belongs to the nutrients for which there is limited capacity for endogenous biosynthesis and therefore it is considered by many to be conditionally essential. Choline is required for membrane synthesis, methylation reactions and neurotransmitter synthesis, and is metabolically inter-related to folate. The AI for pregnancy is based on the fetal and placental accumulation of choline, plus turnover in the woman. As such, the AI for choline increases during pregnancy.

Minerals

Mineral requirements in pregnancy are estimated from the amounts transferred to the fetus. These amounts can come from the woman's stores, increased maternal consumption or increased bioavailability (% absorption). The extent to which these options contribute depends on maternal pre-pregnancy nutritional status and access to food during pregnancy. Calcium, iron, zinc and iodine are the main minerals of interest during pregnancy.

The woman transfers about 30 g of calcium to her infant before birth, most of it during the last trimester of gestation. A well-nourished woman has more than 1000 g of stored calcium to draw from, but metabolic adaptations mean that more efficient uptake and utilisation generally cover fetal skeletal development needs. The adaptations include placental synthesis of active vitamin D_3 which improves calcium absorption, oestrogen, lactogen and prolactin hormones stimulating increased absorption, greater calcium reabsorption in the kidney tubules, and bone density decreasing during the first trimester to provide an internal reservoir that is refilled by the third trimester. For populations with low dietary calcium intakes, 1.5–2.0 g/day calcium is recommended in pregnancy to reduce pre-eclampsia (WHO 2020).

Infants are born with a supply of iron stored in the liver, sufficient to last for 3–6 months. To achieve this storage, the woman must transfer about 200–400 mg of iron to the fetus during gestation. In addition, the pregnant woman requires additional iron for the formation of the placenta, blood volume expansion (haemoglobin formation), and to compensate for blood loss during delivery. This brings the total iron requirement to 835 mg for pregnancy, so approximately 3 mg/day must be supplied from diet or maternal stores. Menstrual iron losses are about 0.5–1 mg/day with noted variability among women, so their cessation during pregnancy contributes between one-sixth and one-third of the iron requirements. Non-haem iron absorption increases across pregnancy from 7% in the first trimester to 36% in the second and 66% in the third, plus 25% absorption from haem iron throughout pregnancy. These metabolic adaptations mean that no additional dietary iron is required where a woman has an adequate diet.

Additional zinc is required for the formation of new tissue (placenta, uterus, breast, amniotic fluid, maternal blood) and accumulation of sufficient zinc in the fetus. The PRI for zinc is an additional 1.6 mg/day throughout pregnancy.

In pregnancy, thyroxine (T_4) production is increased, and this creates an additional 25 μg/day iodine demand. The development of the fetus, placenta and amniotic fluid create a net transfer of 1 μg/day and there is an additional fetal need for thyroid hormone synthesis. These add up to an extra 50 μg/day iodine required, in addition to the AI for NPNL women (150 μg/day).

Perspectives on the future

Growing evidence exists that the fetus and infant are not protected from maternal dietary inadequacy, and that critical windows exist during pregnancy where future health, and even future generational health, is influenced. Understanding where specific nutrients affect fetal programming and plasticity and the extent to which they can be addressed are areas of active research. Researchers are also investigating how common genetic variants influence nutrient requirements during these periods. For example, single nucleotide polymorphisms (SNPs) have been shown to exist in pathways for the biosynthesis of DHA, folate and choline. Many more SNPs must influence metabolism and as our knowledge of SNPs increases, we will be able to identify women who appear to be consuming enough of a nutrient, but who need to consume more because of a metabolic inefficiency, thereby delivering more personalised nutrition.

Dietary reference values (DRVs) can be used for designing appropriate intervention strategies in pregnancy. Further studies should be performed to establish whether such strategies indeed result in more optimal health outcomes in the woman and neonate. In general, pregnant women are encouraged to obtain their nutrients from a well-balanced, varied diet, rather than from vitamin and mineral supplements. Dietary supplements are required for folic acid and are potentially required in other situations such as iron deficiency anaemia. If supplements are used, we need to know more about the risk associated with high levels of intake, interactions with other nutrients, food matrix components, and even the gastrointestinal microbiome. The same applies to fortified foods.

Greater understanding of how energy requirements and body composition affect outcomes in the woman and child are needed, especially in relation to adipose and iron stores and their retention after pregnancy. Excessive gestational weight gain is an additional area that would be important for future study as it appears to have important impacts on nutrient metabolism, gastrointestinal microbiome and potentially fetal health.

Many physiological (homeorhetic) adaptations occur during pregnancy to supply the growing fetus with its increasing demand for macro- and micronutrients. This increase in nutrient bioavailability, manifested by an increase in gastrointestinal absorption, has not been fully elucidated and work is required in this area.

6.3 Lactation

Introduction and mammary development

Lactation, the complex process whereby mammalian females produce milk, is an integral part of the reproductive cycle. Indeed, the presence of mammary glands (breasts) is the fundamental characteristic of all mammals. Milk production, therefore, has been evolutionarily optimised over the millennia, and the inability to produce adequate amounts of high-quality milk has (until recently) been associated with suboptimal reproductive fitness. Consequently, exclusive breastfeeding is generally recommended for the first six months with extended breastfeeding recommended for up to two years and beyond (WHO/UNICEF 2003).

Although mammary gland development begins during fetal life, there is very little additional growth or development until puberty when increased circulating oestrogen along with insulin-like growth factor-1 (IGF-1) and growth hormone (hGH) stimulate additional maturational events. Once menarche occurs, progesterone further stimulates mammary development. Additional development and growth occur during pregnancy, when human chorionic gonadotropin (hCG) released by the placenta stimulates increased oestrogen and progesterone release by the ovaries. Eventually, the placenta produces these hormones in addition to human placental lactogen (hPL) which is key to mammary development, referred to as the process of mammogenesis, during this time. Following delivery of the infant and placenta, a sharp fall in circulating oestrogen and progesterone occurs, and prolactin (PRL) is released from the anterior pituitary gland. This combination of events stimulates the initiation of milk synthesis, referred to as secretory activation.

Maintenance of lactation is complex but relies mainly on the removal of milk via suckling, hand expression or pumping. Thus, it is essential that women breastfeed, hand express or pump frequently to stimulate and maintain milk production. Because the most common causes of poor lactational performance are the infant's lack of access to the breast or inappropriate suckling behaviour, seeking lactation support during the early postpartum period is critical. In early lactation, milk production is stimulated by PRL along with myriad other hormones such as insulin, cortisol and thyroid hormone. The hormone oxytocin, produced in the hypothalamus, is also essential because it causes myoepithelial cells surrounding the milk-containing alveoli to contract, resulting in the milk letdown reflex and the movement of milk into the ducts and eventually the nipple. Milk is then removed from the breast when the infant suckles. Oxytocin also interacts with receptors in the maternal brain to facilitate bonding.

As in pregnancy, multifaceted homeorhetic adjustments occur to support successful lactation. These shifts govern the regulation of nutrient use to maintain maternal health while supporting milk production and are generally mediated by changes in circulating hormones, hormone membrane receptors, and alterations in intracellular signalling pathways. Most of what is known about homeorhetic shifts supporting lactation comes from research conducted with cows, but it is likely that most if not all of them also occur in women. For example, to ensure sufficient glucose to support lactation (including production of lactose), the liver increases gluconeogenesis and glycogen mobilisation, and skeletal muscles decrease their use of glucose. These shifts are driven by a blunting of pancreatic insulin release in response to increases in circulating glucose coupled with reduced muscle sensitivity and responsiveness to insulin. These and other homeorhetic adjustments help repartition nutrients away from other maternal tissues to the mammary gland for milk synthesis.

Colostrum, transitional milk, and mature milk

Colostrum, which begins to be produced during pregnancy, is the first secretion consumed by the breastfed infant during the first few days after birth. Colostrum is a thick, sticky and often yellowish fluid that contains substantial levels of nutrients, immune factors (including cells), complex carbohydrates, and a range of microbes and biologically active compounds thought to collectively nourish and protect the infant. The amount of colostrum produced varies greatly among women and is highly dependent on breastfeeding frequency, but it averages about 50–400 ml/day from one to three days postpartum. Although colostrum contains relatively little water, lactose and lipids, it contains a greater percentage of protein, minerals and fat-soluble vitamins than milk produced later. Moreover, colostrum is rich in constituents that enhance the neonate's immune system and protect the infant during the first few months of life.

Colostrum gradually becomes mature milk; during the 'transition period' (typically defined as days 5–14 postpartum), concentrations of lipid and lactose increase, while those of protein and minerals decrease. After this time, milk production by a woman who is exclusively breastfeeding averages about 780 ml/day.

Composition of human milk

The nutrient composition of human milk varies greatly among women. It also changes over lactation, within a day, and even within a feed. Because of this, researchers must be extremely careful to use appropriate milk collection procedures when studying human milk composition and, even when they do, it is often difficult to interpret results because what is considered 'normal' milk composition in one context might not be typical in another context. Nonetheless, average values for the concentrations of nutrients are used to estimate infant nutrient requirements, establish levels of nutrients contained in infant formulas, and inform nutrient requirements for lactating women. Typical milk composition values are provided in Table 6.3.

Table 6.3 Estimates of the mean concentrations of nutrients in mature human milk. Values represent means ± standard deviations.

Nutrient	Concentration
Energy-yielding macronutrients	
Lactose	72.0 ± 2.5 g/l
Protein	10.5 ± 2.0 g/l
Lipids	39.0 ± 4.0 g/l
Minerals	
Calcium	280 ± 26 mg/l
Phosphorus	140 ± 22 mg/l
Magnesium	35 ± 2 mg/l
Iron	0.3 ± 0.1 mg/l
Zinc	1.2 ± 0.2 mg/l
Copper	0.25 ± 0.03 mg/l
Vitamins	
Vitamin A	670 ± 200 RE (2230 IU)
Vitamin E	2.3 ± 1.0 mg/l
Vitamin D	0.55 ± 0.10 μg/l
Vitamin K	2.1 ± 0.1 μg/l
Vitamin C	40 ± 10 mg/l
Thiamin	0.210 ± 0.035 mg/l
Riboflavin	0.350 ± 0.025 mg/l
Niacin	1.5 ± 0.2 mg/l
Vitamin B_6	93 ± 8 mg/l
Folate	85 ± 37 μg/l
Vitamin B_{12}	0.97 μg/l

Source: Lawrence R.A. et al. 2005/With permission of Elsevier.

Carbohydrates

The disaccharide lactose is the most abundant nutrient in milk, and its concentration is relatively stable – not seemingly impacted by maternal diet. Nonetheless, lactose concentrations have been shown to vary around the globe, and the source of this variation is not understood. In addition to providing energy to the infant, lactose regulates milk volume. Milk also contains a wide array of complex carbohydrates that are not digested by human enzymes but can be metabolised by microbes. These substances, collectively referred to as human milk oligosaccharides (HMO), are thought to have a variety of health implications for the infant, including nourishment of the gastrointestinal microbiome and providing immune modulation and training in early life. HMO concentrations and profiles vary among women and populations. This variation is largely driven by maternal genetics and probably nutritional factors.

Protein and non-protein nitrogen

Compared with the milk of other mammals, human milk contains very low concentrations of protein, and levels decline over the first six months postpartum. Human milk proteins consist mainly of whey proteins (60%, mostly α-lactalbumin) and caseins (40%). There are hundreds of proteins in milk having multiple functions, including the following.

- Supplying all essential amino acids in needed ratios.
- Protecting against infection (e.g. immunoglobulins, lactoferrin and lysosyme).
- Participating in lactose synthesis within the mammary gland.
- Mineral transport (e.g. calcium, zinc and magnesium).
- Acting as hormones, growth factors and enzymes (e.g. leptin, insulin, insulin-like growth factor, bile salt-stimulated lipase).

Milk also contains myriad nitrogen-containing compounds that are not proteins. These are collectively referred to as 'non-protein nitrogen' and include substances such as urea, free amino acids, small peptides, choline and nucleic acids.

The exact functions of non-protein nitrogen compounds in human milk are largely unknown, although they may provide the infant with metabolic and protective functions, at least in an indirect manner. For example, urea (which makes up ~50% of the non-protein nitrogen in human milk) may contribute to nitrogen homeostasis of the breastfed infant's gastrointestinal microbes.

Lipids

On a percentage basis, lipids have the second highest concentration of the nutrients in milk, and provide most of the energy to the infant. Fatty acids (in the form of triglycerides) comprise most lipids in milk, but others (e.g. lipid-soluble vitamins, cholesterol, phospholipids) are also present. Milk lipid content varies greatly within and among women and is dependent on a variety of factors such as maternal body fat, time postpartum, time of day and time since last feed. Importantly, maternal fatty acid intake greatly influences the fatty acid profile of milk. For example, women who consume fatty fish have higher levels of DHA than women who do not consume fish. Both DHA (an important LC n-3 PUFA) and arachidonic acid (an important n-6 PUFA) are considered to be 'conditionally essential' fatty acids during the neonatal period, because synthetic rates can be limited. Therefore, it is important that lactating women consume sufficient amounts. Good sources of arachidonic acid include canola, soybean, corn and sunflower oils, whereas cold-water fish (e.g. salmon) are rich in DHA.

Vitamins and minerals

At least when produced by well-nourished individuals, milk contains all the essential water-soluble vitamins, fat-soluble vitamins and minerals needed by infants. A variety of factors affect variation in the concentrations of micronutrients, with maternal diet and nutritional status being particularly important. A useful categorisation of milk micronutrients is into one of two groups: (i) those for which the milk concentrations are affected by maternal nutrient status and diet (including supplementation); and (ii) those for which the milk concentrations are independent of maternal nutrient status, leading to depletion of maternal body stores when dietary intake is inadequate. For example, maternal consumption of thiamin, riboflavin, vitamin C, all the fat-soluble vitamins, iodine and selenium affects their concentrations in milk. Conversely, maternal diet has little to no impact on milk folate, calcium, iron and sodium contents.

Protective components in human milk

In addition to nutrients, milk contains a complex system of biologically active factors that augment the infant's systems and provide protection against infection. These components include the following.

- Antibodies, such as secretory immunoglobulin A (sIgA) and immunoglobulin G (IgG), which prevent binding and proliferation of pathogens in the infant's gastrointestinal tract and may actively prime the newborn's immune system.
- White blood cells, such as neutrophils and macrophages, which can kill pathogenic micro-organisms.
- Biologically active proteins, for example lactoferrin (an iron-binding protein that inhibits proliferation of iron-requiring bacteria) and lysosyme (an enzyme that attacks microbial pathogens).
- Oligosaccharides, complex indigestible carbohydrates, which inhibit binding of certain bacterial pathogens to gastrointestinal epithelial cells; promote growth of beneficial bacteria; and create a relatively low pH environment of pH 5, which discourages growth of potential pathogens.
- Antioxidants, which are important for the integrity of epithelial surfaces.
- Epidermal growth factor, which stimulates maturation of the lining of the infant's intestine to receive the nutrients in milk.
- Anti-inflammatory and antimicrobial agents, such as certain fatty acids.
- Inherent (commensal) bacteria, viruses and fungi thought to help maintain and establish a healthy gastrointestinal microbiome and prevent dysbiosis (an imbalance in the microbiome composition) and establishment of pathogenic microbes.

Such protection is most valuable during the first year, when the infant's immune system is not fully prepared to mount a response to infection. The protective role of other milk components, such as hormones, growth promoting factors, cytokines and prostaglandins, is not fully understood, but it may go beyond infection to protection against non-communicable diseases of later life.

Microbes

Although once thought sterile, milk is now recognised as being a rich source of microbes, including bacteria, viruses and fungi. Major bacteria found in milk include *Staphylococcus* and *Streptococcus*, although which bacteria are present varies greatly among women and populations. Fungi include *Candida*, *Alternaria* and *Rhodotorula*. For most women, the presence of microbes in their milk appears to represent a healthy state, although an imbalance may lead to problems such as breast inflammation and mastitis. Although research on this topic is relatively new and little is known about the functions of the microbes in milk, it is likely that they act to protect the mammary gland and also colonise the recipient infant's gastrointestinal tract. It is also likely that they help educate the infant as to which microbes should be tolerated and which should be eliminated.

It is noteworthy that there are situations during which the presence of certain microbes in milk has negative implications for the infant. For instance, some pathogens (e.g. HIV and *Mycobacterium tuberculosis*) can be vertically transmitted from mother to infant and cause illness. Fortunately, researchers learned early in the global COVID-19 pandemic that SARS-CoV-2 could not be transmitted in this way.

Recommended energy and nutrient intakes during lactation

The total amounts of energy and nutrients needed to support lactation and milk production are related to a variety of factors such as maternal activity level (for energy), chronic nutritional status, milk composition, and the extent and duration of lactation. As such, estimating nutrient requirements during this period of the lifespan is particularly difficult. Nonetheless, dietary recommendations have been formulated utilising average values, and these are provided in Table 6.4 and discussed briefly next.

Energy

The factorial approach was used to determine energy requirements during lactation, and values were based on estimates of resting energy expenditure plus the energy needed for various levels of physical activity. As such, energy requirements are directly related to physical activity levels, with more energy needed as activity increases. On average, it is estimated that the typical woman requires 2.1 MJ/day more energy than she needed during the non-pregnant/non-lactating period.

Water

To provide what is needed for milk production, recommended water consumption increases to 2.7 l/day during lactation. This includes water from beverages of all kinds, including drinking and mineral water, and from moisture in food.

Energy-yielding nutrients

Experts recommend that lactating women consume 20–35% of their energy from lipids. Saturated and trans fatty acids should be limited; linoleic and α-linolenic acid should contribute 4% and 0.5% of energy respectively; and consumption of DHA should be increased by 100–200 mg/day. Consumption of protein should increase by 15 and 10 g/day (above that recommended in the NPNL period) from 0–6 and >6 months postpartum, respectively. There are no specific recommendations for total carbohydrate intake during lactation.

Vitamins

Recommended intakes of all water-soluble vitamins except for thiamin and niacin are higher in lactation than during the NPNL period, although it should be noted that there are insufficient data to establish population reference intakes (PRIs) for biotin, choline, vitamin B_{12} and pantothenic acid. As such, only adequate intake (AI) levels are available. AI/PRI values for vitamin A and choline are higher in lactation than during the NPNL period, whereas recommended intakes of vitamins D, E and K are the same.

Minerals

Milk is a rich source of minerals. As such, dietary requirements increase during lactation for many of these micronutrients such as zinc, iodine, selenium, copper and potassium. You may be surprised that dietary requirements for calcium do not increase during lactation. This is because homeorhetic adaptive changes (e.g.

Table 6.4 Summary of nutrient intake recommendations for lactating women as compared to non-pregnant, non-lactating women. Values preceded by '+' indicate that this amount should be consumed in addition to that recommended for NPNL women.[a]

	Non-pregnant, non-lactating women		Lactating women	
	AI/AR[b]	PRI	AI/AR[b]	PRI
Macronutrients				
Protein				
0–6 months postpartum	0.66 g/kg/day[b]	0.83 g/kg/day	+15 g/day	+19 g/day
>6 months postpartum			+10 g/day	+13 g/day
Linoleic acid (% energy)	4	NA	4	NA
α-Linolenic acid (% energy)	0.5	NA	0.5	NA
EPA + DHA (mg/day)	250	NA	250	NA
DHA (mg/day)	100	NA	+100–200[c]	NA
Water (l/day)	2.0	NA	2.7	NA
Water-soluble vitamins				
Thiamin (mg/MJ)	0.072[b]	0.1	0.072[b]	0.1
Riboflavin (mg/day)	1.3[b]	1.6	1.7	2.0
Niacin (mg NE/MJ)	1.3[b]	1.6	1.3	1.6
Vitamin B_6 (mg/day)	1.3[b]	1.6	1.4	1.7
Vitamin B_{12} (μg/day)	4.0	NA	5.0	NA
Folate (μg DFE/day)	250[b]	330	380[b]	500
Pantothenic acid (mg/day)	5	NA	7	NA
Biotin (μg/day)	40	NA	45	NA
Vitamin C (mg/day)	80[b]	95	145[b]	155
Fat-soluble vitamins				
Vitamin A (μg RE/day)	490[b]	650	1020[b]	1300
Vitamin D[d] (μg/day)	15	NA	15	NA
Vitamin E (mg/day)	11	NA	11	NA
Vitamin K (μg/day)	70	NA	70	NA
Choline (mg/day)	400	NA	520	NA
Minerals				
Calcium (mg/day)	750[b]	1000	860	1000
18–24 years		950	750	950
≥25 years				
Phosphorus (mg/day)	550	NA	550	
Zinc (mg for 300 mg LPI/day)	6.2–10.2	7.5–12.7	+2.4	+2.9
Iron (mg/day)	7[b]	16	7[b]	16
Magnesium (mg/day)	300	NA	300	NA
Iodine (μg/day)	150	NA	200	NA
Selenium (μg/day)	70	NA	85	NA
Molybdenum (μg/day)	65	NA	65	NA
Copper (mg/day)	1.3	NA	1.5	NA
Manganese (mg/day)	3.0	NA	3.0	NA
Fluoride (mg/day)	2.9	NA	2.9	NA
Potassium (mg/day)	3500	NA	4000	NA

[a] *Source:* Adapted from EFSA 2019.
[b] AI is listed instead of AR.
[c] In addition to combined intakes of EPA and DHA of 250 mg/day0.
[d] Under conditions of assumed minimal cutaneous vitamin D synthesis. In the presence of endogenous cutaneous vitamin D synthesis, the requirement for dietary vitamin D is lower or may even be zero.
AI, adequate intake; AR, average requirement; PRI, population reference intake; LPI, level phytate intake; RE, retinol equivalents; DFE, dietary folate equivalents; NE, niacin equivalents; EPA, eicosapentaenoic acid; DHA, docosahexaenoic acid.

increased calcium absorption) result in greater efficiency and use of dietary calcium during this time. Similar types of homeorhetic adjustments are experienced for copper, magnesium and phosphorus. The 'safe and adequate' level for both sodium and potassium is 3.1 g/day, which is the same for NPNL women.

Special nutritional considerations during lactation

In general, lactating women are considered at risk for nutrient inadequacies due to the increased nutritional requirements during this phase. This is particularly true for women living in low-income regions of the world where food insecurity is high and access to a varied diet can be limited and/or seasonal.

Conversely, many women (particularly in middle- and high-income countries) retain excess weight during the postpartum period, and this is related to increased risk of overweight and obesity. For these women, consuming a nutrient-dense diet that does not contribute excess calories while breastfeeding is particularly important. In fact, healthy weight loss diets that provide a balance of nutrient-dense foods should be considered to return to a healthy postpartum weight. In addition, participating in exercise and other forms of physical activity is important for weight loss. There is no evidence that normal levels of exercise affect lactation success, so women should not be concerned about negative impacts of exercise on their milk production or milk composition.

Vegetarian and vegan eating patterns

Although vegetarian and vegan dietary patterns can easily provide all the essential macronutrients and micronutrients, extra care must be taken so that enough iron, iodine, choline, DHA, zinc and vitamin B_{12} are consumed. Many foods (e.g. ready-to-eat cereals) are fortified with iron, and non-haem iron absorption can be enhanced by consuming this mineral with vitamin C. Iodine is typically consumed from dairy products, eggs, seafood and iodised table salt. As such, vegans are at especially high risk of not consuming enough iodine, particularly if they use salt that is not iodised. It is worth noting that many 'specialty' salts (e.g. pink Himalayan salt) are not iodised and are not good sources of this mineral. Vitamin B_{12} also is of concern because it is only found in animal-source foods. Similarly, choline, DHA and zinc are typically obtained from animal-source foods.

Women following a vegetarian or vegan dietary pattern should consult with a healthcare provider to determine whether supplementation of iron, vitamin B_{12} and/or other nutrients such as choline, zinc, iodine or DHA is necessary and if so, the appropriate levels to meet their unique needs.

Alcohol consumption

When breastfeeding women consume alcohol, some (<2%) of the alcohol becomes incorporated into their milk. Despite popular belief in some cultures that alcohol consumption increases milk production, research does not provide any evidence of benefits to lactation. Rather, alcohol consumption can delay the let-down reflex and reduce the quantity of milk produced. Nonetheless, moderate consumption of alcohol (up to one standard drink in a day) is not known to be harmful to the infant, especially if the woman waits at least 2h after a single drink before nursing or expressing milk. Clearly, though, not drinking alcohol is the safest option for women who are lactating.

Perspectives on the future

There is substantial need for additional research on human milk and lactation, particularly related to nutrition. For example, although there exists a rich literature regarding human milk composition, very little is known about the many factors (e.g. maternal diet, genetics) that influence its variation. This lack of knowledge hinders the ability of scientists and public health experts to understand what constitutes optimal nutrition for both the breastfeeding woman and her infant. In addition, although the dairy industry has been studying the physiology and genetics of milk production for decades, much less is known about these processes in women.

A more complete understanding of milk production and factors influencing its onset and maintenance is critical to helping women meet their breastfeeding goals and for society to realise national and global targets for rates of breastfeeding initiation, exclusivity and duration.

Acknowledgements

We would like to acknowledge Joop M.A. van Raaij and Lisette C.P.G.M. de Groot as the original authors of the chapter, which we updated for this edition.

References

European Food Safety Authority (2019). Dietary reference values for nutrients. Summary report. https://efsa.onlinelibrary.wiley.com/doi/epdf/10.2903/sp.efsa.2017.e15121.

WHO (2007). Standards for Maternal and Neonatal Care: Prevention of Neural Tube Defects. Geneva: World Health Organization.

WHO (2020). WHO recommendation on calcium supplementation before pregnancy for the prevention of pre-eclampsia and its complications. https://apps.who.int/iris/bitstream/handle/10665/331787/9789240003118-eng.pdf

World Health Organization, United Nations Children's Fund (2003). Global Strategy for Infant and Young Child Feeding. Geneva: World Health Organization.

Further reading

Allen, L.H. (2012).B vitamins in breast milk: relative importance of maternal status and intake, and effects on infant status and function. *Advances in Nutrition* 3 (3): 362–369.

Anderson, P.O. (2018). Alcohol use during breastfeeding. *Breastfeeding Medicine*13 (5): 315–317.

Australian Government(2020). Nutrient reference values for Australia and New Zealand including recommended dietary intakes. www.nrv.gov.au/nutrients

Bauman, D.E. and Currie, W.B. (1980). Partitioning of nutrients during pregnancy and lactation: a review of mechanisms involving homeostasis and homeorhesis. *Journal of Dairy Science* 63 (9): 1514–1529.

Baumgard, L.H., Collier, R.J. and Bauman, D.E. (2017). A 100-Year Review: regulation of nutrient partitioning to support lactation. *Journal of Dairy Science* 100 (12): 10353–10366.

Bell, A.W. and Bauman, D.E. (1997). Adaptations of glucose metabolism during pregnancy and lactation. *Journal of Mammary Gland Biology and Neoplasia* 2(3):265–278.

Carlson, S.E. (2001). Docosahexaenoic acid and arachidonic acid in infant development. *Seminars in Neonatology* 6 (5): 437–449.

Dror, D.K. and Allen, L.H. (2018). Overview of nutrients in human milk. *Advances in Nutrition* 9 (suppl_1): 278S–294S.

Food and Nutrition Board: Institute of Medicine (2002). Dietary Reference Intakes for Energy, Carbohydrates, Fiber, Fat, Fatty Acids, Cholesterol, Protein and Amino Acids. Washington, DC: National Academy Press.

German Nutrition Society (2000). Austrian Nutrition Society, Swiss Society for Nutrition Research, Swiss Nutrition Association: Reference Values for Nutrient Intake (D-A-CH). Frankfurt am Main: Umschau/Braus.

Grote, V., Verduci, E., Scaglioni, S. et al.(2016). European childhood obesity project. Breast milk composition and infant nutrient intakes during the first 12 months of life. *European Journal of Clinical Nutrition* ;70 (2): 250–256.

Haastrup, M.B., Pottegård, A. and Damkier, P. (2014). Alcohol and breastfeeding. *Basic and Clinical Pharmacology and Toxicology* 114(2):168–173.

Joint FAO/WHO Expert Consultation on Human Vitamin and Mineral Requirements (2004). Vitamin and Mineral Requirements in Human Nutrition, 2nd edn. Rome: World Health Organization and Food and Agriculture Organization of the United Nations.

Keikha, M., Bahreynian, M., Saleki, M. and Kelishadi, R. (2017). Macro- and micronutrients of human milk composition: are they related to maternal diet? A comprehensive systematic review. *Breastfeeding Medicine* 12 (9): 517–527.

Lackey, K.A., Williams, J.E., Meehan, C.L. et al.(2019). What's normal? Microbiomes in human milk and infant feces are related to each other but vary geographically: the INSPIRE study. *Frontiers in Nutrition* 6:45.

McGuire, M.K. and O'Connor, D.L. (2021). Human Milk. Sampling and Measurement of Energy-Yielding Nutrients and Other Macromolecules. Cambridge, MA: Elsevier.

Mennella, J. (2001). Alcohol's effect on lactation. *Alcohol Research and Health* 25 (3): 230–234.

Moberg, K.U., Handlin, L. and Petersson, M. (2020). Neuroendocrine mechanisms involved in the physiological effects caused by skin-to-skin contact – with a particular focus on the oxytocinergic system. *Infant Behavior and Development* 61: 101482.

Moossavi, S., Fehr, K., Derakhshani, H. et al.(2020). Human milk fungi: environmental determinants and inter-kingdom associations with milk bacteria in the CHILD Cohort Study. *BMC Microbiology* 20 (1): 146.

Motil, K.J., Thotathuchery, M., Bahar, A. and Montandon, C.M. (1995). Marginal dietary protein restriction reduced nonprotein nitrogen, but not protein nitrogen, components of human milk. |*Journal of the American College of Nutrition* 14 (2): 184–191.

Pace, R.M., Williams, J.E., Järvinen, K.M. et al.(2021). Characterization of SARS-CoV-2 RNA, antibodies, and neutralizing capacity in milk produced by women with COVID-19. *mBio* 12 (1): e03192–e03120.

Pace, R.M., Williams, J.E., Robertson, B. et al.(2021). Variation in human milk composition is related to differences in milk and infant fecal microbial communities. *Microorganisms* 9 (6): 1153.

Rudolph, M.C., Russell, T.D., Webb, P., Neville, M.C. and Anderson, S.M. (2011). Prolactin-mediated regulation of lipid biosynthesis genes *in vivo* in the lactating mammary epithelial cell. *American Journal of Physiology Endocrinology and Metabolism* 300 (6): E1059–1068.

SACN.https://assets.publishing.service.gov.uk/government/uploads/system/uploads/attachment_data/file/339317/SACN_Dietary_Reference_Values_for_Energy.pdf

Saini, R.K. and Keum, Y.S. (2018). Omega-3 and omega-6 polyunsaturated fatty acids: dietary sources, metabolism, and significance – a review. *Life Sciences* 203:255–267.

Saint, L., Smith, M. and Hartmann, P.E. (1984). The yield and nutrient content of colostrum and milk of women from giving birth to 1 month post-partum. *British Journal of Nutrition* 52(1):87–95.

Section on Breastfeeding (2012). Breastfeeding and the use of human milk. *Pediatrics* 129 (3): e827–41.

Seferovic, M.D., Mohammad, M., Pace, R.M. et al.(2020). Maternal diet alters human milk oligosaccharide composition with implications for the milk metagenome. *Science Reports* 10 (1): 22092.

Sela, D.A., Chapman, J., Adeuya, A. et al.(2008). The genome sequence of Bifidobacterium longum subsp.

infantis reveals adaptations for milk utilization within the infant microbiome. *Proceedings of the National Academy of Science USA* 105 (48): 18964–18969.
Williams, J.E., McGuire, M.K., Meehan, C.L. et al.(2021). Key genetic variants associated with variation of milk oligosaccharides from diverse human populations. *Genomics* 113 (4): 1867–1875.
World Health Organization. (2021). Infant and young child feeding. www.who.int/news-room/fact-sheets/detail/infant-and-young-child-feeding.
World Health Organization, Pan American Health Organization (2003). Guiding Principles for Complementary Feeding of the Breastfed Child. Washington, DC: World Health Organization.
World Health Organization, United Nations Children's Fund (2007). Reaching Optimal Iodine Nutrition in Pregnant and Lactating Women and Young Children. Joint Statement by the World Health Organization and the United Nations Children's Fund. Geneva: World Health Organization.

7
Growth and Ageing

Mark L. Wahlqvist

7A Early Life

Key messages

- Growth provides an indication of nutritional status in pre-adult years.
- Changes in body composition and anthropometry reflect changes in growth and thus nutritional status.
- Nutritional needs change in accordance with the demands of growth throughout the different stages of life.
- The interplay of genetic and environmental factors determines growth outcomes and disease risk.
- Inadequate nutrition during the early years of life can impair growth and affect stature and health outcomes in later life.
- 'Catch-up growth' is a phenomenon that compensates for deviations in growth from the genetic trajectory.
- Maximum height may not be equivalent to 'optimal' height with respect to positive health outcomes.
- Obesity and overweight in early life, and especially during adolescence, increase the likelihood of obesity and associated risk factors in adult life.

7A.1 Introduction

Nutrition plays an important role in human growth and development throughout life (Wahlqvist and Gallegos 2020). Infancy and childhood are important times for nutrition and growth, as they strongly predict health outcomes later in life. Nutrition once again plays an important role in later life when prevention of chronic disease and system degeneration becomes a major priority (Wahlqvist et al. 2005). All people require the same nutrients to maintain health and well-being, but these are required in differing amounts according to their stage of life. Optimal growth and healthy ageing will occur if nutritional requirements are met and environmental influences are conducive to health throughout life. The human life cycle is one of socioecological beings intimately part of nature, dependent on it and societal connectedness; failure of any of these characteristics may lead to food and health insecurity and disorder, and threaten planetary habitability and survival. The nutritional biology which underpins this life cycle has several fields, each with environmental connectedness: homeostatic, energetic, genetic, microbiomic, gut physiologic, sensory and immunoinflammatory (Figure 7A.1).

7A.2 Growth and development

'Growth' may be defined as the acquisition of tissue with a concomitant increase in body size. 'Development' refers to changes in the body's capacity to function both physically and intellectually through increased tissue and organ complexity. Different individuals experience these processes at different rates.

Nutrition and Metabolism, Third Edition. Edited on behalf of The Nutrition Society by Helen M. Roche, Ian A. Macdonald, Annemie M.W.J. Schols and Susan A. Lanham-New.

Companion Website: www.wiley.com/go/nutrition/metabolism3e

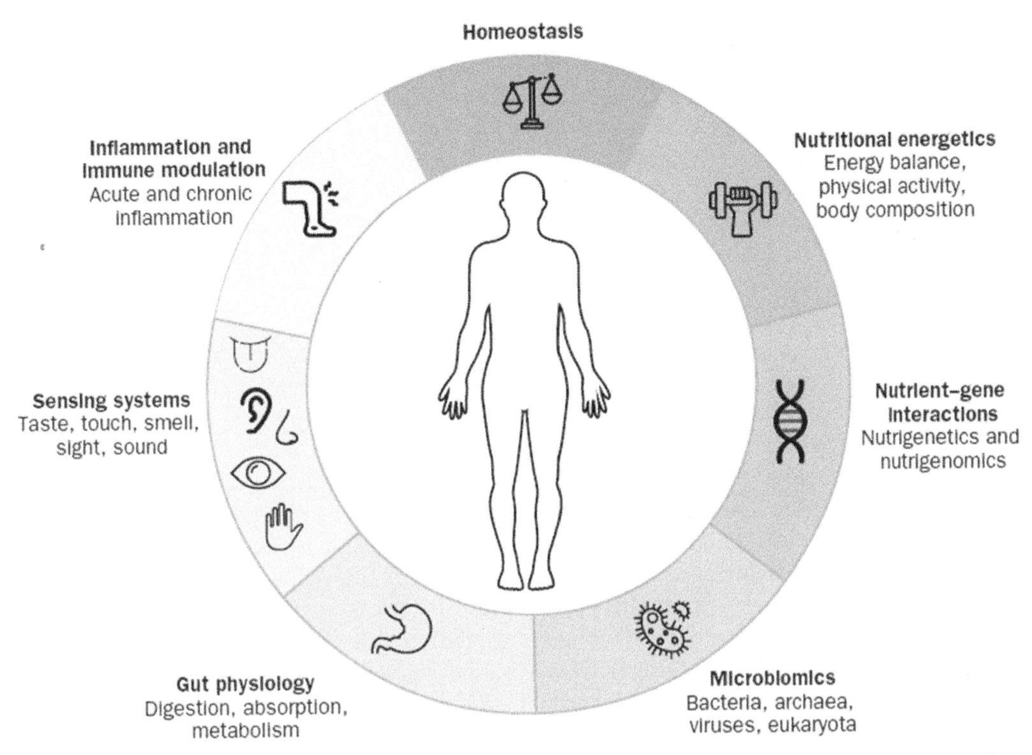

Figure 7A.1 The fields of nutritional biology, each of which has environmental connectedness. *Source:* Wahlqvist ML and Gallegos D 2020/Taylor & Francis.

There are five stages under which major growth and developmental changes occur in humans:

- infancy
- childhood
- adolescence
- adulthood
- late adulthood.

These stages can be distinguished by changes in growth velocity and distinct biological and behavioural characteristics. Nutritional needs change in response to the demands that these stages of growth place on the body. If nutritional needs are met and adverse social circumstances or diseases are not encountered, optimal growth will occur.

7A.3 Cellular aspects of growth and death

Cell division

Cells are subject to wear and tear, as well as to accidents and death, so we must create new cells at a rate as fast as that at which our cells die. As a result, cell division is central to the life of all organisms.

Cell division or the M phase (M = mitotic) consists of two sequential processes: nuclear division (mitosis) and cytoplasmic division (cytokinesis). Before a cell can divide, it must double its mass and duplicate all its contents to ensure that the new cell contains all the components required to begin its own cycle of cell growth, followed by division. Preparation for division goes on invisibly during the growth phase of the cell cycle and is called the interphase. During interphase, cell components are continually being made. Interphase can last for up to 16–24h, whereas the M phase lasts for only 1–2h. Interphase starts with the G_1 phase (G = gap) in which the cells, whose biosynthetic activities have been slowed during the M phase, resume a high rate of biosynthesis. The S phase begins when DNA synthesis starts and ends when the DNA content of the nucleus has doubled and the chromosomes have replicated. When DNA synthesis is complete, the cell enters the G_2 phase, which ends when mitosis starts. Terminally differentiated and other non-replicating cells represent a quiescent stage, often referred to as the G_0 phase (Figure 7A.2).

In multicellular animals, like humans, the survival of the organism is paramount rather than the survival of its individual cells. As a result, the 10^{13} cells of the human body divide at very different rates depending on their location and are programmed and co-ordinated with their neighbours. Something in the order of 10^{16} cell divisions take place in a human body during a lifetime. Different cell types sacrifice their potential for rapid division, so that their numbers are monitored at a level that is optimal for the organism as a whole. Some cells, such as red blood cells, do not divide again once they are mature. Other cells, such as epithelial cells, divide continually. The observed cell cycle times, also called generation times, range from 8h to 100 days or more. Cells that do not actively proliferate have a reduced rate of protein synthesis and are arrested in the G_1 phase. A cell that has become committed to division by passing through a special restriction point (R) in its G_1 cycle will then go on to make DNA in the S phase and proceed through to the following stages. Growth stops for those cells arrested at R, although biosynthesis continues. This specific restriction point of the growth control mechanism may have evolved partly due to the need of a safe resting state (at R) for those cells whose growth conditions or interactions with other cells demand that they stop dividing. Cells that have been arrested at this stable resting state are said to have entered a G_0 phase of the cell cycle.

Whether or not a cell will grow and divide is determined by a variety of feedback control mechanisms. These include the availability of space in which a cell can flatten (contact inhibition of cell division) and the secretion of specific stimulatory and inhibitory factors (peptides, steroids, hormones, short-range local chemical mediators and others still to be identified) by cells in the immediate environment.

Cancer cells have escaped from or respond abnormally to many of the control mechanisms that regulate cell division. Cancer cells require fewer protein growth factors than do normal cells in order to survive and divide in culture; in some cases, this may be because they produce their own growth factors. A second fundamental difference between normal and cancer cells is that the latter can go on dividing indefinitely. For example, cells taken from older animals will

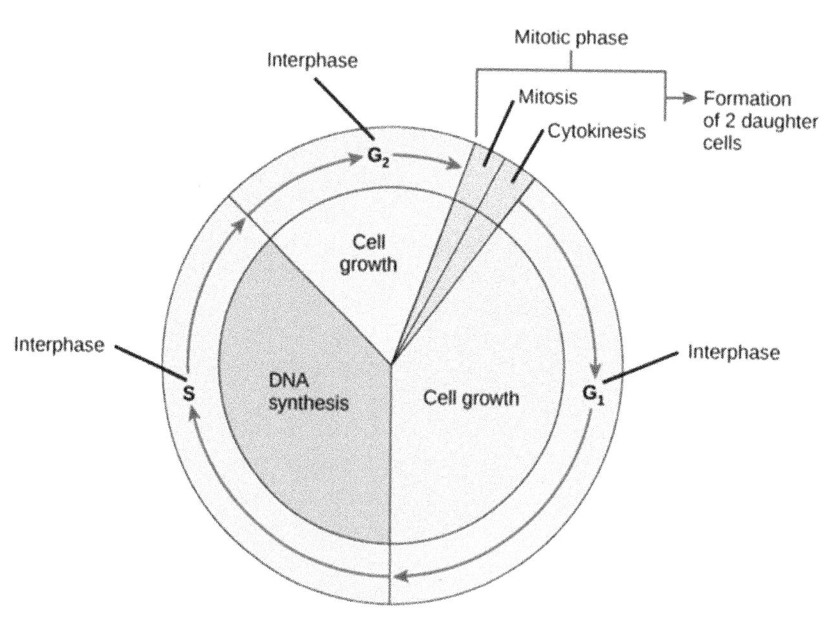

Figure 7A.2 Cell growth and division: the successive phases of a typical cell cycle.

divide fewer times in culture than the same cells taken from young animals, suggesting that older cells have used up many of their allotted divisions whilst in the animal.

As cells differentiate, they become programmed to die after a certain number of divisions. This programmed cell death is an additional safeguard against the unbridled growth of one particular cell. However, cancer involves something more than just abnormalities of proliferation and programmed senescence; it requires the coincidental occurrence of several specific mutations in a single cell, enabling it to proliferate in disregard to the usual constraints and to invade regions of the body from which it would normally be excluded.

Cell death

Apoptosis is an active process in which cells undergo genetically programmed death. Apoptosis occurs when a calcium-dependent enzyme (endonuclease) fragments the genome of the cell into approximately 180 base pairs. The dead cells are then removed by phagocytosis. It is still unclear how many mechanisms are involved in causing cell death, but it is known that alterations in intracellular calcium levels can trigger apoptosis.

Cell death appears to be activated by special genes in dying cells. For cell death to occur, genes identified as *ced-3* and *ced-4* must be expressed in a dying cell. A third gene, *ced-9*, is a major control factor, which negatively regulates the *ced-3* and *ced-4* genes. Animal studies have shown that mutations to *ced-9* inactivate the gene, causing the death of cells that were otherwise intended to survive and thereby killing the animal under study.

Cell senescence occurs when the cell does not divide or proliferate and DNA synthesis is blocked. Cells are programmed to carry out a finite number of divisions, which is called the Hayflick limit. If cells do not reach senescence, they continue to divide and in effect become immortal. Immortal cells eventually form tumours and this is therefore an area where research into ageing and cancer is inter-related. It is thought that senescence may have evolved exactly because its action protects against cancer. Cells exhibiting evidence of DNA damage and oxidative stress may recover from these stresses provided they are equipped with an adequate DNA repair system and an adequate level of stress response proteins and/or antioxidant defences. Most theories that aim to explain the ageing process suggest that senescence results from the accumulation of unrepaired damage.

Telomeres (the sections of DNA at the end of chromosomes) shorten with each division of the cell, resulting in a set lifetime for each normal cell. Inhibition of telomerase is thought to reduce cellular ageing. Although genes play an important role and can be an indicator of disease risk in later life, environmental factors also strongly influence the development of disease. The general degenerative properties of ageing are also a factor in disease development. Greater understanding of the molecular events regulating cell progression and apoptosis is rapidly emerging. The non-nutritional and nutritional factors affecting cell division and growth are outlined below.

7A.4 General factors affecting growth and development

Growth and development have intrinsic and interactive psychobiological determinants, including genetics and endocrine control, and are influenced by environmental, societal, cultural and economic factors. These factors do not affect growth in isolation but act collectively.

Genetic factors

Genetic factors arguably play the most important role in determining the outcome of *growth and development*. What has been inherited from previous generations clearly influences how we advance through life. The contribution that *epigenetics* makes to the intergenerational passage of acquired environmental and behavioural information is, however, substantial, and more or less *paternal or maternal* dependent on its X or Y chromosome location and on how much is *mitochondrial* and, therefore, maternal. An example of how the paternal diet may affect intergenerational behavioural transfer is that of taste preference, which can affect *spermatogenesis* (Ding et al. 2020; Luo et al. 2020; Mu et al. 2016). Twin studies show that patterns of growth, age of menarche, body shape, composition and size and deposition of fat are genetically linked. Growth seems dependent on '*imprinted maternal or paternal genes'*, especially in utero on placental function, but also postnatally (Reik et al. 2001). For as long as the mother determines the nutrient supply, her genes will endeavour to match the supply to survival and overall growth and development, while those of the father will presumably begin to shape a growth pattern for the child's future. Both parents will have *epigenetic information* to do with their external world, including its food system, to pass on their progeny and the environment it is expected to encounter (Reik et al. 2001). The *growth co-ordinating system* which develops is one with a genetic, epigenetic maternal and increasingly socioenvironmental complexity mediated by growth factors operative in neuroendocrine pathways (Russo et al. 2005). It must also take account of *microbiomic transfers* from the maternal reproductive tract, nasopharynx, gut, integument and breast, and whether the offspring was born by caesarean section or vaginally (Mirpuri 2021).

The ultimate achievement of *nutritional homeostasis* or its failure, as with energy balance and body composition, may only become apparent with a major shift in the socioeconomic and behavioural background of the genetically susceptible. This has been evident with disorders like obesity, rare until the latter twentieth century. Its endemic manifestation depended partly on whether there was intrauterine growth retardation (IUGR) and short stature attained ahead of a plentiful, nutritionally poor food supply. Height itself has fluctuated greatly historically in the same populations with famine, conflict, migration, livelihood and food pattern, albeit with genetic predisposition. But it was not necessarily associated with obesity. The propensity to obesity has depended on a series of nutritional and non-nutritional intergenerational exposures, and the resilience or adaptability available to counter them. Adjustment 'success' is a judgement about the prioritisation between various socioecological and biological outcomes in the short to medium and long term.

Genetic control can be quite *explicit and limited* in terms of its action, particularly in early life. Dental maturation, for example, appears to be independent of skeletal maturation; genes regulating limb growth may also be independent by limb section.

Evolutionary, intergenerational and historical determinants of gene expression are modulated by socioeconomic and educational advantage and disadvantage, together with living conditions, including food, a scenario which tends to be underestimated. One reason for this is that attributability of genetic effects may not be population or internationally based. The complexity of inheritance begins before conception with maternal and paternal gametes, continuing in intrauterine life, at birth, during infancy and beyond with child rearing, all important qualifiers of what we regard as genetic.

Gender differences for the majority are predominantly explained by genetic factors, such as differences in both the timing of the adolescent growth spurt and the *sex-specific changes* that occur during puberty. At the age of 10 years, males and females are of a similar height, weight and body fatness. However, after puberty, major differences are evident. Although the timing of the growth spurt is largely genetically controlled, hormonal factors are thought to control its intensity and duration, while environmental factors also play an important role. Females typically achieve skeletal maturation at an earlier age than males. Genes on the Y chromosome, which are found only in males, are thought to retard skeletal maturation relative to females.

A difference in the expression of a *single gene, or group of genes, or SNPs* (single nucleotide polymorphisms) can have widespread and even drastic effects, leading to compromised growth and development. Furthermore, epigenetic effects on gene expression during intrauterine life in particular can have important consequences in later life in terms of growth and development, disorder and disease in the short, medium and long term. Inborn errors of metabolism such as phenylketonuria, galactosaemia and hereditary fructose intolerance are genetic conditions which require early recognition and life-long dietary management to avoid nutrient deficiencies and tissue damage.

Hormones as growth factors

Hormones are responsible for co-ordinating much of the appropriate timing and rates of growth. There are many hormones that have recognised effects on growth. Generally, there are three distinct endocrine phases of linear growth: infancy (including fetal growth), childhood and puberty. Each of these phases is regulated by different endocrine growth-promoting systems. Thyroid and parathyroid hormones, through thyroxine and tri-iodothyronine, stimulate general metabolism. Other hormones and growth factors include epidermal growth factor (important for actions in the epidermis), platelet-derived growth factor (involved in blood clots and possibly cell division) and melatonin (plays a role in regulating puberty and possibly growth velocity). Hormones are largely responsible for many changes in body composition and the development of secondary sexual characteristics, which occur as individuals grow older. The effects of hormones are often influenced and regulated by other hormones and growth factors.

One of the most important hormones that regulates growth is the human growth hormone (GH). GH exerts a powerful effect on growth, at least in part, through somatomedins or insulin-like growth factors (IGF-1, IGF-2). IGF acts by stimulating muscle cell differentiation and is therefore important in fetal and postnatal growth and development. GH is important in regulating protein synthesis and cellular division and is thus thought to be the major regulator of the rate of human growth and development from the latter few months of the first year of life. Synthetic forms of GH are used in a clinical setting to encourage growth in children of a small stature. It is thought that sex steroids, which are active around puberty, may trigger significant secretions of GH and thus IGF. Hormones such as thyroxine and cortisol may also influence IGF plasma levels.

Disturbances to genetic programming in utero during critical phases of fetal development may result in the impairment of the endocrine factors, responsible for growth in later life. There are associations between growth retardation and abnormalities in levels of hormones, such as GH, and hormone-mediated factors, such as IGF (Russo et al. 2005).

Neural control

It could be argued that the *hypothalamus* has connections and functions which would allow it to be considered a possible growth centre, closely associated with the anterior pituitary, a hormonal gland integrally involved in growth. Any deviation of growth from its predetermined genetic pathway would be identified by the hypothalamus and compensated for by the subsequent release of growth-specific hormones from the anterior pituitary gland, thereby inhibiting the potential for damage to any growth or development. Growth hormone-releasing hormone is produced in the hypothalamus. It stimulates the pituitary gland to produce and release growth hormone. This promotes growth in body tissues and stimulates production of IGF-1 in the liver and elsewhere to control growth. Somatostatin is also produced by the hypothalamus and prevents growth hormone release.

The peripheral nervous system may also play an important role in growth. With denervation and loss of neurochemical secretion, muscle atrophy occurs.

Local biochemical control

To attain optimal growth, areas of localised growth rely on complex regional influences, whether they are mechanical or chemical stimuli. These can be paracrine factors, such as neighbouring proteins, which act as local signalling and growth factors. Alternatively, they may be autocrine factors, such as hormones or growth factors that influence the growth of quite localised areas, despite being secreted from a more distant position. These mechanisms appear to be responsible for the growth of specific tissue. Furthermore, the age of groups of cells, or a specific tissue, can determine the amount of cell replication and division that is possible. This is the case where cells undergo a finite number of mitotic divisions.

Social and cultural factors and practices

The major social factors affecting growth are culture, age, family, gender and socioeconomic status (McLoyd 1998). Emotional disorders have also been related to growth abnormalities. Different countries or regions may exhibit their own cultures and cuisines, which reflect a mixture of geographical, agricultural, historical, religious and economic factors, among others. These factors may either directly or indirectly affect growth through their influence on a number of factors, such as health, nutrition, food selection and cooking methods. Infants and children who emigrate to other areas may experience changes to growth patterns. Generational differences in stature, for example, often reflect changing environments. Many adolescents eagerly seek independence and factors associated with this have an effect on nutritional status. Levels of physical activity, peer pressure, self-esteem and distorted body image, chronic dieting and disordered eating, together with substance abuse, are all factors that can affect nutritional status and, consequently, growth. Chronic dieting, in particular, can also place an individual at risk of many micronutrient deficiencies, infertility, long-term degenerative diseases, such as osteoporosis and a compromised immune system, which can negatively affect growth and health outcomes.

Family factors that have been associated with compromised growth and development include single-parent families, family conflict and disturbance to the family unit, such as divorce and separation. Of these, *parental education* (mother and father in somewhat different ways) and maternal autonomy are critically important (Balaj et al. 2021; Chilinda et al. 2021; Huebener 2019; Huang et al. 2021). Growth retardation in conflict situations may partly be a result of stress, which is thought to affect GH levels. The amount of care and attention that an infant receives has an impact on growth. This can be related to the number of children and birth position in the family. In some cultures, *gender bias* is in evidence, favouring boys, who receive more care and nourishment. In both developing and industrialised countries, *socioeconomic status* is associated with many behavioural, nutritional and health outcomes, which can influence growth. Usually, children who are socioeconomically advantaged are taller, display faster growth rates and become taller adults. Conversely, disadvantaged children may be smaller at birth, shorter and be prone to body compositional disorders with excess body fat and its maldistribution (Chiang et al. 2017).

However, socioeconomic status, locality and food system are also predictors of various nutrient and other bioactive food component deficiencies which may or may not be associated with *stature*, and need to be addressed in their own right. Shortness associated with nutritional problems is often referred to as *stunting* but may have other pathogeneses, notably socioeconomic, domestic and comorbidities such as infectious disease. Care should be taken that prevention and intervention does not resort inappropriately to nutrient interventions. If in doubt, a dietary pattern strategy will be safer than one which is nutrient specific (Cannon 2010).

Educational attainment and income are positively associated with growth and development, through the alleviation of poverty and acquisition of the facility of self-determination. Girls, often denied education, demonstrably benefit from schooling in their own biological, mental and general health status. In part, this follows from greater control over their environment. It is also evident in their reproductive life, pregnancy and its outcomes. Typically, as income increases, dietary quality, judged by its biodiversity and inclusion of nutrient-dense foods such as animal products like meat, eggs and dairy, can improve and be of some prestige and pleasure. Preconceptional dietary quality, sustained intrapartum, reduces the risk of intrauterine growth retardation (IUGR) and low birth weight (LBW), of relevance to offspring childhood and long-term health. Underweight children are more likely to have been of lower birth weight. Learning ability and social competence of girls, more than boys, depend on birth weight and diet; those with a lower birth weight can offset its risk by a diverse diet (Lee et al. 2012).

Infant feeding practices, such as breastfeeding and weaning, may follow culturally or socially acceptable patterns but are amenable to internationally recommended nutritional optima, through systematised maternal and childcare programs.

Oral health and dental caries are an example of a nutritional problem with relevance to growth and development. They are problematic in both developing and industrialised countries, but contingent on environmental and health system governance as well as household awareness, priorities and practice. Dental hygiene is often least satisfactory in areas where water fluoridation is absent. Under such conditions, dietary determinants of dental caries, such as the regular consumption of foods high in sucrose, particularly in sticky sweet form, are paramount. It is now known that disease patterns in later life, such as cardiometabolic disorders, cognitive impairment and life expectancy, depend on oral health.

7A.5 Nutritional factors affecting growth

For growth to proceed at its predetermined genetic rate, adequate nutrition is essential. Food supplies the individual with the required energy, nutrients and food components to influence growth. The strong link between food intake and growth is supported from studies of food intake patterns during famines and was encountered at the time of the two world wars. Children of Dutch and German families living in these areas exhibited impaired growth, while in Japan a reduction in the mean height of adults was observed between 1945 and 1949. It remains a global problem with population displacement, food deprivation and safety. Nutrition affects all body systems and factors influencing growth. For example, a compromised nutrient intake can influence gene replication and expression, hormonal and neural control. As stated earlier, other environmental factors are also important for growth, the nexus between ecosystem dysfunction, limited societal support, food insecurity and transmissible disease is often inseparably causative in growth failure.

Malnutrition can be expressed as undernutrition, overnutrition, dysnutrition or permutations among these. Undernutrition is rarely only a matter of inadequate energy intake, being also a lack or imbalance of specific food components and nutrients. Chronic energy deficiency, commonly referred to as protein–energy malnutrition (PEM), occurs in both developing and industrialised countries but is more prevalent in the former. Characteristic features of PEM include stunted growth, delayed maturation, reduced muscle mass and decreased physical working capacity.

The food environment at home (Balaj et al. 2021), in and around school and recreationally matters for growth and development. Within the family, it may be bidirectional between children and elders (Wahlqvist et al. 2014). The trend with time in food system exposure can be profound and lasting as reported with so-called 'ultraprocessed' foods.

In addition to sufficient energy, adequate supplies of macronutrients, micronutrients and phytonutrients are required to promote optimum growth. The proportions and amounts of these food components may change according to the various stages of growth. For example, protein is a prerequisite for optimal growth at all life stages, while fat may arguably have its most important role during infancy and childhood, as a major supplier of energy and long-chain polyunsaturated (n-3) fatty acids, which are important in neural development and the inflammatory responses to infection. Components of certain foods, called growth factors, may affect growth. For example, some of the proteins present in milk are thought to promote growth. In dairy-consuming populations, tallness tends to be more common.

Nutritional status affects hormonal status. GH, for example, will not stimulate linear growth if nutritional status is compromised. Plasma IGF responds to acute directional changes in the body's nitrogen balance. Thus, the ingestion of several dietary factors, such as essential amino acids, adequate energy and an optimal nitrogen balance may be critical for optimal hormonal control. When this is not the case, there may be negative effects on growth and development. Food and its components, such as metabolites of vitamins A and D, fatty acids, together with some sterols and zinc, can directly influence gene expression and thus growth. Components of dietary fibre are thought to influence gene expression indirectly by a number of pathways, including altering hormonal signalling, mechanical stimulation and through metabolites, which are produced by the intestinal microflora.

Micronutrients which may be of functional consequence in growth are shown in Table 7A.1. Although some may assume specific relevance by locality or food availability, as with iodide, iron or zinc deficiency, it is a policy error to use nutrient-specific interventions to manage growth and development unless the particular diagnosis is made and with knowledge of the background diet. For example, children with growth impairment due to energy dysnutrition or iodine deficiency would not be given zinc to resolve the problem.

Iodide

Iodide deficiency at different life stages produces differing health outcomes. Its intake is of most critical importance in utero and during the first

Table 7A.1 Micronutrients affecting growth.

Children	Adolescents
Iodide	Calcium
Zinc	Folate
Iron	
Vitamin A	
Vitamin K	
Riboflavin	
Vitamin C	
Vitamin D	

two years of life, when neural cells in the brain undergo major cellular division. Adequate iodide intakes during these times are therefore essential for mental and cognitive growth and development. The extent of mental dysfunction may be lessened if sufficient dietary iodide is consumed or administered in the early years of life. Complications associated with iodide deficiency in childhood and adolescence may appear as goitre, hypothyroidism, mental dysfunction, retarded mental and physical growth and reduced school performance (iodine deficiency disorders, IDD) (Hetzel 2012).

Vitamin A

Severe deficiency in vitamin A is associated with impaired vision, retarded growth and development, poor bone health, compromised immune functioning and complications with reproductive health and outcomes. In industrialised countries, vitamin A deficiency is rare. Too much vitamin A in the diet can also slow growth. Subclinical vitamin A deficiency increases the risk of morbidity and mortality in vulnerable population groups. Reductions in mortality rates of around 20–25% can be achieved by improving the vitamin A status in young children in populations where deficiency has been identified.

Zinc

Zinc is of crucial importance in over 200 enzyme reactions. It is of structural and functional importance in biomembranes, DNA, RNA and ribosomal structures. Zinc deficiency has been linked with disturbed gene expression, protein synthesis, immunity, skeletal growth and maturation, gonad development, pregnancy outcomes, behaviour, skin integrity, eyesight, appetite and taste perception. Zinc deficiency can cause intrauterine growth retardation (IUGR) if the maternal diet provides inadequate sources of zinc. It is of importance for linear growth, as well as the development of lean body mass.

Iron

In industrialised countries, iron represents the major micronutrient deficiency, but in developing countries iron deficiency occurs on a much larger scale. Although iron is important at all life stages, iron deficiency commonly affects preschool and school-aged children who, as a consequence, face compromised growth if dietary intake is inadequate. Iron is especially important in pregnant women, as low intakes can have wide implications for the newborn infant with limited iron stores. The effects of iron deficiency are varied, but a major effect is its impairment of cognitive development in children. Other consequences of iron deficiency include reduced work capacity and decreased resistance to fatigue.

Other micronutrients

Other nutrients of importance to growth include vitamin B_2, which affects general growth, vitamin C, which is important in bone structure, and vitamin D, which is involved in calcium absorption from the intestines and its tissue turnover, together with cell differentiation and immune function. Chronically low dietary intakes of these vitamins impair growth and bone health. Calcium and folate appear to be micronutrients of importance for growth during adolescence. Vitamin K is a factor in bone, vascular, kidney and brain health and growth through dependent proteins like osteocalcin.

Phytonutrients

Certain phytochemicals are advantageously bioactive with wide-ranging potential functions and reinforce the nutritional value of plant-derived foods. They can be referred to as phytonutrients. Examples are the isoflavones genistein and daidzein (found in legumes, especially soy) which may help to inhibit tumour formation by regulating cell cycle progression and by promoting cell differentiation and apoptosis (cell death). Formation of new vasculature is required for a cancer to grow and metastasise; isoflavones have also been identified as antiangiogenic agents that inhibit the formation of new vasculature and thus the development and dissemination of tumours.

7A.6 Nutrition and the life cycle

Energy, nutrient and food needs differ according to *life stage* so that food intake must reflect these changing demands. Inadequate nutrition exerts a detrimental impact on fetal, infant, childhood, prepubertal and pubertal growth, although this is somewhat methodologically difficult to study with the dynamism and sensitivities involved. Food-based supplementation and other interventions that are provided before and at this time will inevitably have consequential growth and development outcomes, but clarity about these relationships continues to evolve.

The list begins before conception, with the nutritional needs and behaviours of the father in spermatogenesis, the mother in her nutritional reserves as with folacin, zinc and iron, then during gestation of variable lengths and with maternal exposures of concern like alcohol and substance abuse, vaginal or caesarean delivery and acquisition of microbiomic type, breastfeeding success, infancy affected by the newborn's nutrient reserves, growth rate, activities and the elements of care when wholly dependent.

Early *weaning* (four months or earlier) has been associated with negative health outcomes, such as the formation of allergies, diarrhoea and even death. Chronic or episodic diarrhoea may affect the absorption of nutrients which, if not addressed, may lead to growth impairment. Common reasons for introducing complementary feeding with formula milk or solids include a perceived inferior quality of breast milk, poor weight gain, difficulties or pain with feeding, mother's employment, refusal by the infant to feed and lack of mother's confidence. Many people perceive formula to be of a higher quality than breast milk, particularly because of aggressive marketing strategies from infant formula manufacturers. This may cause the mother to abandon exclusive breastfeeding from an early age.

The World Health Organization (WHO) has systematically reviewed the optimal duration of exclusive breastfeeding. This indicates that *breastfeeding should be exclusive for the first six months of life* and that healthcare systems should adhere to and promote the consequential guidelines. Insofar as the value of breastfeeding as a protectant against infective diarrhoea is concerned, hygienic environments may allow weaning to be acceptable from around four months, a practice frequently adopted among those of socioeconomic advantage, even if not ideal. However, the baby's gut is not sufficiently mature for solids until after 17 weeks, and the early introduction of solids increases the risk of childhood obesity.

The adequacy of *infant formulae*, should they be used upon weaning, continues to be under review, especially in regard to breast milk substitution (Fewtrell et al. 2017; O'Reilly et al. 2021; Wang and Brand-Miller 2003). Human breast milk has vesicles or nanoparticles capable of transferring much nutritional information from mother to child beyond conventional genomics. How vital this is for child development is likely to be an unfolding story for some time to come. The essential long-chain fatty acids (LCPUFA), such as the n-3 fatty acids docosahexaenoic acid (DHA) and arachidonic acid (AA), are structurally important in cell membranes, particularly in the central nervous system. Most infant formulae contain only the precursor essential fatty acids, α-linolenic acid (ALA, the n-3 precursor) and linoleic acid (LA, the n-6 precursor), from which infants must assemble their own DHA and AA, respectively. Infant formulae may not be effective in fully meeting the essential fatty acid requirements of most infants. Another example of a conditionally essential nutrient might be sialic acid, a monosaccharide involved in synaptogenesis and neural transmission, and present in human milk oligosaccharides in high concentration. Reduced cognitive, motor and visual acuity outcomes have been reported in formula-fed infants, compared with their breastfed counterparts.

From six to 24 months of infant life, breastfeeding alone cannot provide all the nutrients and energy needed to promote and sustain adequate growth, so complementary feeding is necessary, whether with family foods and meals or commercial products. If complementary feeding is introduced too late, there is a risk of impaired growth, macronutrient deficiency, impaired cognitive and physical development and nutritionally compromised stature, referred to as stunting (shortness can also be healthy).

Individuals may experience *erratic growth* during childhood, largely reflecting changes in appetite and food intake, or even an underlying illness. Nutrient needs increase throughout childhood, reflecting the continuing growth of all body systems and recovery from intercurrent illness. Most children progressively assume an omnivorous diet, although traditionally, and increasingly, in health-conscious households it is plant based. Children can also grow and thrive on most lacto-ovo, vegetarian and *vegan diets*, provided they are well planned and have sufficient biodiversity. Essential nutrients which may be marginal with veganism are vitamin B_{12} and vitamin D, although the latter is found in sun-dried fungi and mushrooms where ergosterol has been converted to ergocalciferol. Sunlight exposure is critical in all children in its own right, but especially in vegan households for the conversion of cholesterol into cholecalciferol. Growth delays are reported in children fed severely restricted diets (such as macrobiotic, Rastafarian or fruitarian).

By school age, the growth of vegetarian and non-vegetarian children is similar; few differences are found in the timing of puberty or completed adult growth; nor is the intelligence quotient (IQ) different with an unrestricted vegetarian diet. Typically, girls tend to consume less than boys at all ages but during adolescence males begin to increase their intake to levels well above those of most females. Adolescence is a time that requires the greatest total energy intake of all the life stages, as a result of the body being in a highly metabolically active state. Inadequate intakes of nutrients and energy during this time can potentially impede growth and delay sexual maturation.

Adolescent pregnancy has many increased risks for both the mother and the child, as the fetus and the mother must compete for nutrients to maintain and promote their respective growths. This is of even greater concern if the adolescent is malnourished. Calcium is illustrative of this concern as it is needed for continuing bone development in the mother while also being required in large quantities by the developing fetus. Birth and maternal complications during adolescent pregnancy are greater than those for older women of similar nutritional status. Adolescent mothers face increased risks of infant and maternal mortality, preterm delivery and in giving birth to babies prematurely and of a low birth weight.

Complications of pregnancy with possible nutritional contributors include *pre-eclamptic toxaemia* (PET) and *perinatal depression* (PND) (Leung and Kaplan 2009). In the former (PET), Eben Hipsley (Hipsley 1953), for the first time, attributed a disease, PET, to a deficiency of what he called dietary 'fibre', especially from green leafy and root vegetables, fruit, nuts and unrefined bread or brown rice. His report was based on observations in Australia, New Guinea and Fiji in 1953, and from other sources back to before World War II. In New Guinea, where there was little or no PET, not only was there a high intake of various unrefined plant foods but also fish consumption was recorded at a relatively high 120 g/day. Nutrition counselling, especially in gestational diabetes, reduces the risk of PET, but the foods or nutrients involved are unresolved (Allen et al. 2014). In the latter (PND), the focus has been on possible nutrient deficiencies, notably n-3 fatty acids, rather than dietary patterns, although it has been found that 'nutritious foods' decrease risk and support the case for a biodiverse food intake.

The *dietary guidelines* for children and adolescents can be accommodated in and consistent with those for all ages, adjusted for energy needs and sociocultural factors, in the manner of FBDG implementation (Buttriss 2016; DeSalvo et al. 2016; Wahlqvist et al. 1999; World Health Organization 2021) (Figure 7A.3). They should recognise the need for dietary patterns to take account of food security and climate change (Willett et al. 2019).

7A.7 Effects of undernutrition

The causes of undernourishment are not confined to inadequate dietary intake. Social, cultural, genetic, hormonal, economic and political factors may be involved. Underlying health problems and inadequate care and hygiene may also contribute. Ultimate outcomes reflect severity and duration. Consequences of undernutrition include death, disability and compromised mental and physical development. Poor nutrition often commences in utero and extends into adolescence and adult life (Gudmundsson et al. 2005; Nohr et al. 2009). There is a causal relationship between fetal undernutrition and increased risks of impaired growth, development and adult disease.

Much disease originates in fetal life. IUGR or LBW and wasting are among the earliest signs of undernutrition (Brodsky and Christou 2004; Pallotto and Kilbride 2006; World Health Organization 2021). Wasting can be detected by reduced measures of weight for age and skinfold thickness, reflecting a loss of weight or a failure to gain weight. In severe cases, individuals may exhibit the clinical signs of hair loss, skin discolouration or pigmentation (in marasmus) and oedema (in kwashiorkor), and evidence of specific nutrient deficiencies. Stunting or pathological shortness reflects chronic undernutrition manifested in impaired linear growth.

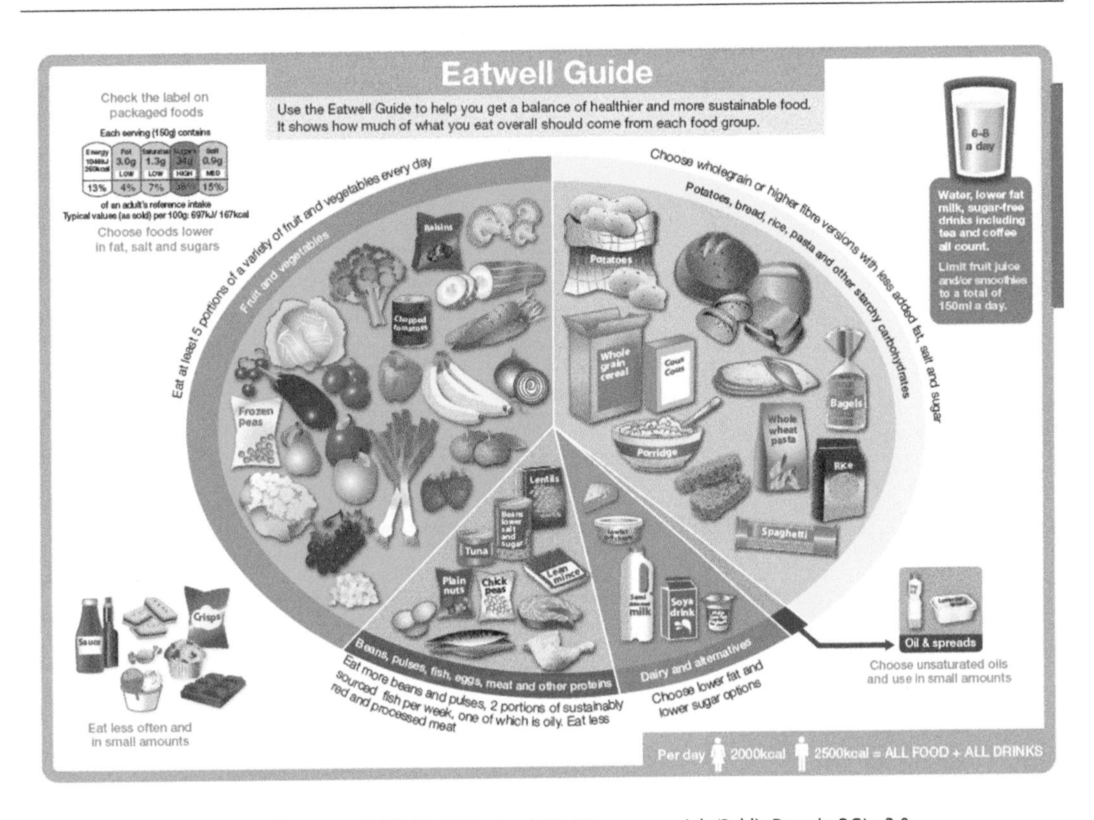

Figure 7A.3a UK Eatwell Guide. *Source:* Public Health England 2016/Crown copyright/Public Domain OGL v3.0.

However, where stunting and wasting are both present, growth charts may not detect abnormal weights (or heights) for length owing to proportional growth retardation of both weight and height (or length). A stunted infant is likely to remain stunted throughout childhood and adolescence and become a stunted adult, particularly if the individual continues to live in the same environment that instigated the stunting. In girls and women, life-long poor nutrition has wide family community and intergenerational adverse consequences. Adult stunting and underweight affect not only a woman's health, well-being, autonomy and role, but also the risks of pregnancy complications, such as gestational diabetes, peripartum depression and the likelihood that her offspring will have low birth weight (LBW).The first 1000 days of life from conception are a crucial nutritional window for life-long health (Schwarzenberg and Georgieff 2018).

Most growth impairment, of which underweight and stunting are outcomes, occurs within a relatively short period, from before birth until about two years of age (the so-called *first 1000 days* from conception) (Georgiadis and Penny 2017; Linnér and Almgren 2020). Severe undernutrition during infancy can be particularly damaging to brain growth. This can result in major retardation of cognitive growth and functioning. Delayed intellectual development is a risk factor for absenteeism from school and poor school performance. Infants born with LBW and who have suffered IUGR are born undernourished and face a greatly increased risk of mortality in the neonatal period or later infancy. Suboptimal intakes of energy, protein, vitamins A and D, zinc and iron, among all other nutrients, during the early years of life may exacerbate the effects of fetal growth retardation. There is a cumulative negative impact on the growth and development of an LBW infant if undernutrition continues during childhood, adolescence and pregnancy. The LBW infant is thus more likely to be underweight or stunted in early life. Undernourished girls tend to have delayed menarche and grow at lower velocities, but for longer periods, compared with their better nourished counterparts. They may attain similar heights to better nourished

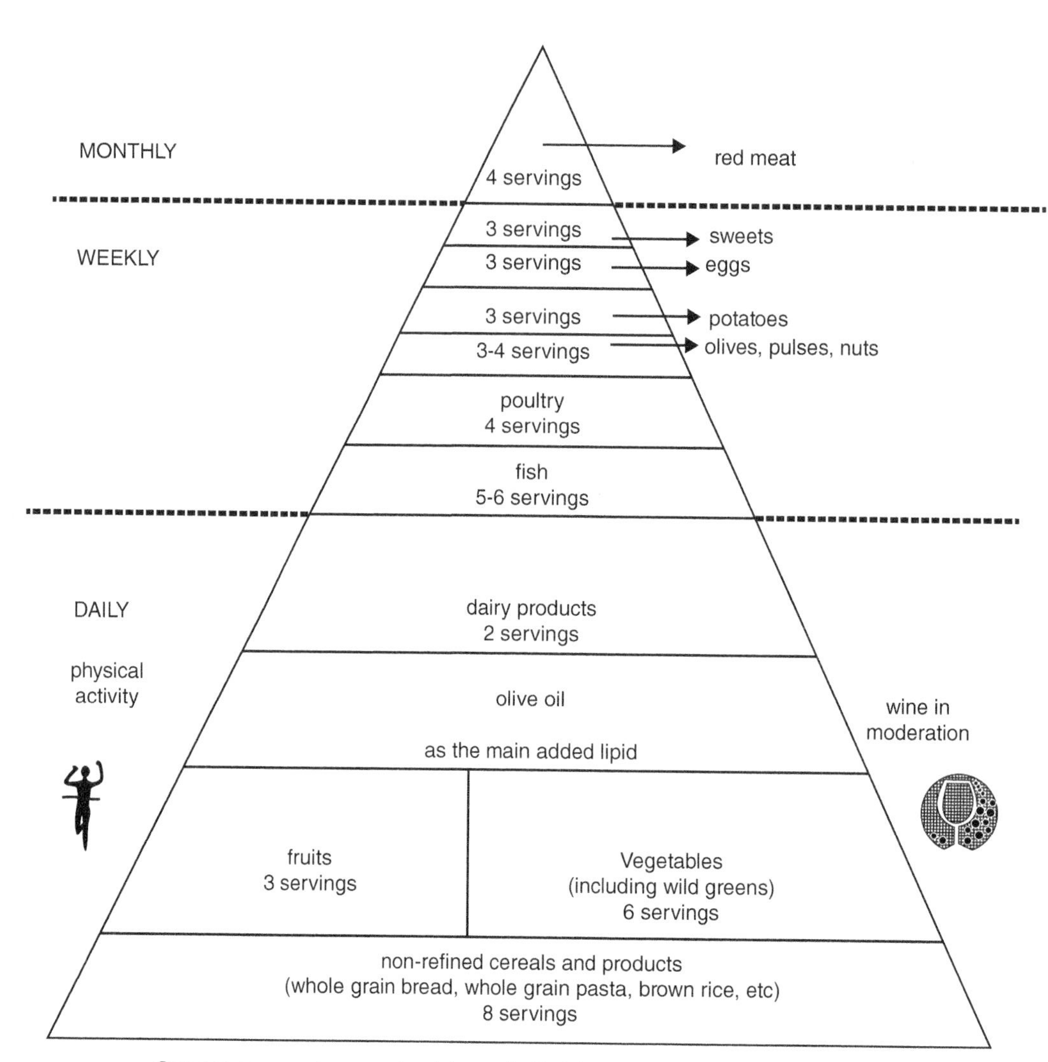

Figure 7A.3b Mediterranean diet guide. *Source:* Supreme Scientific Health Council, 1999. www.mednet.gr/archives/1999-5/pdf/516.pdf.

girls if undernutrition is limited to adolescence. However, if childhood stunting was also experienced, undernourished adolescent females are unlikely to reach similar heights to well-nourished girls.

Optimal development during adolescence is reliant on both present and past nutritional intake. Malnourishment and impaired growth during infancy and early childhood can affect an individual's attainment of height and its related biology. Stature itself may not be the pathway that matters most in ultimate physical and mental health outcomes since these associations can disappear with nutritious biodiverse diets, socioeconomically supportive households and access to education. Shortness does not constitute an inevitable barrier to favourable health and societal outcomes.

Catch-up growth

'Catch-up' or 'catch-down' growth is a phenomenon that appears to compensate for retarded or accelerated intrauterine growth, whereby children return to their genetic trajectory. Many studies have suggested that the potential for catch-up growth and reversal of cognitive impairment among children who have suffered growth retardation during infancy and/or early childhood is thought to be limited after the age of two years, particularly when children remain in poor environments. However, other studies have shown that undernourished children in poor environments can display spontaneous catch-up, even without environmental change. As adolescence is a time of rapid growth, this provides an opportunity for further catch-up growth. It is thought, however, that the potential for significant catch-up during this time is limited, perhaps when the damage of undernutrition is less reversible.

Reversed stunting in women may reduce the risk of adverse pregnancy outcomes that may be associated with women of small stature; dystocia in pregnancy in short women occurs principally with fetal macrosomia. Females appear to display greater catch-up than males. In a Taiwanese study, shorter girls with poorer school performance, mainly dependent on parental education and household socioeconomics rather than diet, were less compromised as their grades progressed, which suggests that education is key to catch-up. Nevertheless, where nutrient deficiencies like zinc are endemic, growth retardation is an outcome. At all ages, zinc requirements are higher for males than females and therefore a zinc limitation may explain restricted growth in males to a greater extent than in females, and clarify why males are less able to catch up to the same extent as females. Catch-up growth may still be associated with adverse outcomes in later life.

It is not known why catch-up growth may be detrimental, but one theory suggests that IUGR restricts cell numbers, with catch-up growth achieved by the inappropriate overgrowth of a restricted cell mass. Notwithstanding the concerns that problematic early life nutrition may have lasting and long-term adverse effects on health, extensive studies of adoption reveal remarkable plasticity and resilience in growth and development given opportunity.

Short stature and plant food environments

Many populations consume diets that are predominantly plant based. They may be associated with micronutrient deficiencies, chronic energy deficiency and short stature. Plant-based diets are of relatively low energy density and high protective phytonutrient density may limit the amount of 'metabolic dysregulation' and the development of degenerative diseases. Plant antinutrients, on the other hand, may limit growth for better or worse. If the food pattern is one of biodiversity and little refinement, micronutrient deficiencies will be less likely. For example, vitamin C-replete fruits will increase the bioavailability of iron from food seeds and grains. Phytases will release zinc from phytic acid. Thus, short stature need not be an index of poor nutrition, but of food patterns conducive to stature appropriate and even optimal for the environment in question. Tallness is not necessarily better than shortness.

Slow growth rates modulated by restricted zinc intake may represent a survival advantage. Indeed, in a Taiwanese study of schoolchildren's food environment, girls responded to fast food outlet exposure with increased height velocity, as opposed to boys who were fatter. This was evident in a society where breast cancer incidence is increasing and, possibly, a reflection of more rapid linear growth.

Maximal versus optimal height

As undernutrition and stunting during infancy and early childhood have been found to be associated with adverse short- and long-term health, the health policy focus has been to encourage tallness and favour quicker growth and bigger size. Secular increases in height have been apparent in many cultures in association with socioeconomic development in apparent confirmation of such policy. However, in several developed regions, such as North America, Western Europe and Australasia, this linear growth trend is slowing and reaching a plateau. The presumption has been that an individual should achieve maximal height potential in order to achieve the best health outcomes. This is now questioned and replaced by a quest for 'optimal' rather than 'maximal' stature. Shorter stature may actually confer health benefits in some environments (like high altitude or where foodstuffs are less growth promoting but no less conducive to successful development). Shortness may represent an adaptive response to an individual's environment or be in anticipation of food shortage (likely with climate change). Short stature may carry an increased risk of avoidable abdominal obesity and heart disease, while tall stature, which usually reflects increased growth velocity, has a greater risk of neoplastic and degenerative disease (Chiang et al. 2011). That shortness may be associated with adverse health outcomes does not mean that it causes them. Moreover, shortness may occur for non-nutritional reasons and may or may not be amenable to or desirably addressed by nutritional means.

It is, however, noteworthy that health disadvantage in general, associated with being indigenous, and on account of female gender, may be offset, at least in part, by education and a biodiverse diet (Huang et al. 2021; Lee et al. 2012).

7A.8 Overnutrition and childhood obesity

Obesity is of global epidemic proportions, affecting children, adolescents and adults in growing numbers. In some countries, over half of the adult population is affected, thus leading to increasing death rates from heart disease, hypertension, stroke and diabetes. The growing prevalence of obesity in younger age groups has raised alarm.

Obesity is a body compositional, multifactorial disorder and disease. It is a problem of energy intake, the state of basal energy expenditure and self-paced activity, insufficient physical activity (with direct and indirect effects on energy metabolism) and the efficiency of energy utilisation. It may reflect an inappropriate loss of beige or brown fatty tissue with growth and development and a dysfunctional dietary pattern, which would otherwise maintain a level of 'inefficient energy expenditure' for thermogenesis (e.g. instead of having to heat living environments to achieve a similar or, preferably, lesser outcome) (Symonds 2013). For younger people, the decline in physical activity levels with increased screen viewing times, more sedentariness and a host of labour-saving devices has made energy dysregulation the norm. Added to that are less access to the natural environment with its ecobiological role in homeostasis, an environmental food system polluted with endocrine disruptors (especially from plastic waste) and stressors that disrupt the neuroendocrine pathways which connect brain, gut microbiome and abdominal fat. The assault on our and our children's nutrition biology is potentially overwhelming. It is no surprise that effective programmes to stem the tide of childhood obesity have so far only been possible with a whole-community involvement, as in the Community-Based Childhood Obesity Intervention (CBCOI), based on the EPODE approach (Slot-Heijs et al. 2020).

Effects of overweight and obesity

Being overweight or obese as a child and adolescent has many associated health, social and psychological implications. Overweight and obese children may suffer from impaired social interaction and self-esteem. They may be taller than their non-overweight peers and viewed as more mature. This inappropriate expectation may result in adverse effects on their socialisation. Effects on academic performance and socialisation may contribute to lower job prospects in later life. Mobility and physical fitness suffer and exacerbate the problem. Serious physical complications of child obesity depend on severity and may be uncommon, but include cardiomyopathy, pancreatitis, orthopaedic disorders and respiratory disorders such as upper airway obstruction and chest wall restriction. By adolescence, cardiovascular risks, abnormal glucose tolerance and even type 2 diabetes, hypertension and lipid disorders may manifest, especially as fat distribution becomes abdominal and muscle mass less well developed. Abdominal fatness in children of both genders is associated with early maturity.

The limited recoverability and long-term effects of childhood obesity on adult health outcomes have become an increasing health challenge in more and more countries, irrespective of economic development, threatening the gains made in disability-free life expectancy.

Age of onset, gender, equity, and obesity in later life

The prevalence of obesity differs by gender and with socioeconomic inequity. Obesity is also familial but not necessarily genetic. Less than

10% of obese children have both parents of a normal weight, with 50% and 80% of obese children having one and two obese parents, respectively. Obese children are now more likely to remain obese as adolescents and as adults, but this was less so in the past where overweight children often normalised during puberty and adolescence. The older the obese child, the more probable it is that he or she will become an obese adult. The rate of spontaneous weight reduction decreases with age. The proportion of overweight or obese adults, who had been overweight or obese in adolescence, ranges from 20% to 45%, while from 25% to 50% of overweight or obese adolescents had been overweight or obese in early childhood. Obesity that begins during, or close to, adolescence is often abdominal in type of fat distribution, which has been considered 'android' or masculine. Women have a greater proportion of body mass as fat in any case for purposes of reproduction. Interestingly, while the same BMI (Body Mass Index, weight (kg)/height (m^2)) is associated with relatively more body fat in women, its health predictive significance is similar to men, with abdominal fat distribution also of similar relevance.

Predictors of body size and fatness in adolescence

Genetic and environmental factors interact to determine human body fatness. Predictors in early life of body size and fatness in adolescence include birth weight, weight and height velocities during the first year, body composition at 12, 50 and 80 months, major illness and parental socioeconomic status in early life. Tienboon et al. found that the historically common practice of consumption of fish liver oil (high in vitamins A and D and in eicosapentaenoic acid [EPA] and DHA) in early life reduced the risk of developing obesity, particularly abdominal obesity (Tienboon et al. 1992). Perhaps this programmed adipocyte development was a consequence of this practice. Self-reported exercise, appetite and food item consumption were more closely linked to contemporary adolescent body size and fatness.

Factors affecting weight status in children and adolescents

Many factors have been reported to affect the development of obesity in children and adolescents, including season, geographical region, population density, ethnicity, socioeconomic status, family size, gender, parental education, levels of physical activity, maternal age and maternal preference for a chubby baby. These associations underscore the role of the household, community and their socioecology in the realisation of healthy growth, development and related body composition. The self-correcting value of linear growth, especially with the pubertal growth spurt, liberal physical activity in natural precincts and nutritious food cannot be overstated, in contrast to restrictive and developmentally adverse dietary restriction.

Rate of weight gain and mode of feeding in early life

Early feeding experience is related to the development of excess weight in infancy. Both breast-feeding and the delayed introduction of solid foods appear to exert a protective effect against adiposity up to two years of age, and probably in later life. However, not all studies have shown this effect. The ages where most obesity arises before adulthood are between 0 and four years, seven and 11 years and during adolescence. During infancy, rapid increases in weight have been associated with obesity at adolescence in boys. Early catch-up growth, between the ages of 0 and two years, has been reported as a predictor of childhood obesity, particularly for central or abdominal obesity.

Heritability and familial aggregation of weight status

The heritability of BMI as an index of obesity is generally considered to be about 40–70%, but this almost certainly overstates what may be genetic and encourages an understatement of what is environmental (Barness et al. 2007; Locke et al. 2015; Silventoinen et al. 2010; Yang et al. 2007). The obesity epidemic began to emerge in the 1960s and 1970s at a time when the food system lost satiety signals and the needs for energy expenditure decreased. There were other environmental changes which have not been understood for what they might have meant for health, particularly an abundance of petrochemical-based products and plastics like textiles, furnishings and packaging. These have radically increased our exposure to endocrine-disrupting particles which now contaminate the water we

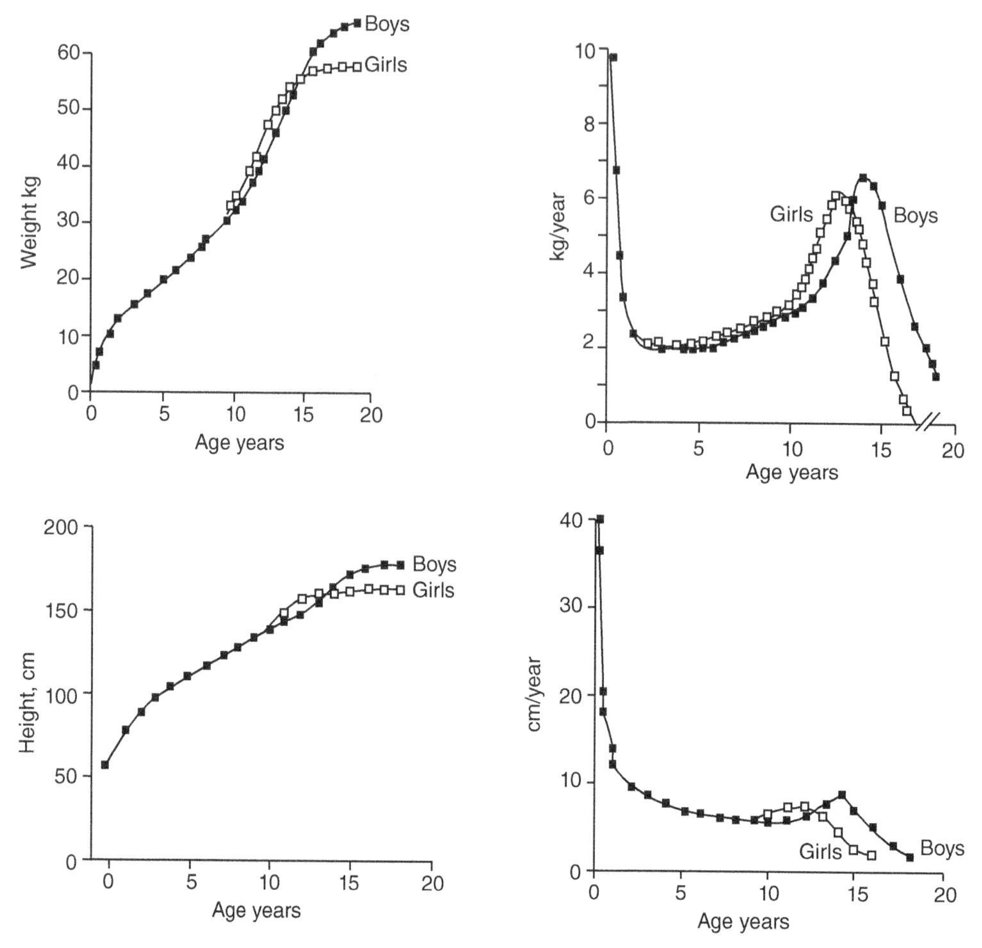

Figure 7A.4 Growth and growth velocities throughout childhood and adolescence for girls and boys as 50th centiles for weight and its velocity and for height and its velocity. *Source:* Reproduced with permission from World Health Organization (2021) and Wahlqvist ML, Vobecky JS, Eds. The Medical Practice of Preventive Nutrition. London: Smith-Gordon & Co. Ltd., 1994, pp223–243.

drink, the food we eat, the air we breathe, the places we live and the clothes we wear.

The estimates of genetic contributions to excessive body fatness and its maldistribution come principally from twin and adoptive studies, mainly in people of European ancestry. More than 50 or so genes with single nucleotide polymorphisms (SNPs) could be involved in altering energy throughput and balance; these could alter a range of pathways such as appetite, food choice, movement, bioenergetics or their combination. Single gene determination of obesity is rare. More likely is epigenetic modulation of gene expression, itself environmentally determined but potentially transmissible from generation to generation. Adoptive studies allow an assessment of whether and the extent to which early life rearing may explain what, at first sight, may appear genetic, but itself might reveal projection of risk from early exposure into later life and subsequent generations. Our gut microbiomics, an environmental interface, may also be involved in the risk of obesity – a huge prokaryotic genome, multiples of our eukaryotic genome.

The reality is that, for most of the human experience, obesity has been uncommon, notwithstanding our genomics. Moreover, the prevalence of obesity (BMI >30 kg/m^2) by country remains wide, from <5% in north-east Asia, as in Japan, to >50% in some Pacific Islands and the Middle East, irrespective of differentials in industrialisation and urbanisation. These differences point to the importance of social, cultural, familial, behavioural and environmental factors in the pathogenesis of overfatness. This variability is not taken into sufficient account in studies by particular location, leading to excessive assignment of aetiology to genetics. That overweight or obese individuals generally have at least one overweight or obese parent where overall prevalence is high is an example of this misinterpretation.

7A.9 Growth during childhood and adolescence

The most rapid periods of growth take place during the first few months of life and during adolescence. Growth velocity slows significantly after the first year of life but accelerates again as an individual enters puberty in adolescence over a period of 1–3 years. After the peak velocity of growth in puberty has been attained, the growth rate slows considerably until growth in height ceases at around 16 years of age in girls and 18 years in boys (Figure 7A.4).

During the first year of life, a well-nourished infant will ideally increase in length by 50% and in weight by 300%. Rapid and essentially linear growth occurs in well-nourished infants, with rates of weight and height increases keeping pace with each other. Infants have a high surface area to body weight, which means that their energy and nutritional needs are much higher than adults on a per kilogram of body weight basis. As a result of this, infants are prone to heat stress and dehydration, but the ratio of surface area to body weight decreases with age.

The period of infancy ends when the child is weaned from the breast (or bottle), which in pre-industrialised societies occurs at a median age of 36 months and less in more developed countries. Alternative definitions of infancy are that it extends to the second birthday or is the first 1000 days from conception during which epigenetic programming is most possible (Linnér and Almgren 2020). During early childhood, height and weight increase in an essentially linear fashion. Steady growth necessitates a gradual increase in the intake of most nutrients to support growth and development. It is thought that the age of 6–7 years is a critical period for determining future weight and height status. During childhood, an individual is still largely dependent on the caregiver for providing nourishment but this begins to change during late childhood, when a child begins to develop increased control over their food intake and relies less on caregivers.

During adolescence, the body undergoes many changes as a result of puberty. Once puberty is reached, an individual is capable of sexual reproduction. The onset of puberty is characteristically earlier in females than in males (10.5–11 years and 12.5–13 years, respectively). As males reach puberty later than females, they experience on average two more years of prepubertal growth than females, resulting in typically taller statures at the onset of puberty for males.

Before puberty, there are no significant height differences between girls and boys. It is during the adolescent growth spurt that major skeletal differences between males and females become apparent. In males, there is a widening of the shoulders with respect to the pelvis, while the opposite is seen in females. Typically, height gains of approximately 20 and 15 cm are realised in males and females, respectively.

In both genders, a sequential pattern of growth is observed: the feet and hands, calves and forearms, hips and chest, and shoulders, followed by the trunk. Skeletal growth ceases once the epiphyses, the active areas of bone at the end of long bones, have closed. Once this occurs, bones cannot become any longer and are largely unresponsive to exogenous GH administration. In males, puberty is marked by growth of the sexual organs, followed by changes in the larynx, skin and hair distribution. The growth of the ovaries signifies the beginning of puberty in females. Fat is deposited in the breasts and around the hips, dramatically changing the female body shape. Menarche coincides with the peak growth spurt of adolescence, after which growth decelerates.

Adolescence is a period of dietary vulnerability.

Body composition changes during childhood and adolescence

Growth and development in children and adolescents are associated with changes in body composition, which affect body fatness and leanness. Body composition is therefore one measure of growth. It also provides an indication of both nutritional status and physical fitness. For the young person, the transitions in body size, composition, related function and appearance are enmeshed in the sense of identity and contributory to interpersonal relationships. The nutritionist observer has a duty to respect the wider meaning of the measurements.

Changes in fat-free mass and body water content

Lean body mass (LBM) includes all non-fat tissue. In adolescence, LBM increases to a much greater extent in males than in females, with muscle and bone representing the largest gains in growth. Fat-free mass (FFM) is similar to LBM, but it excludes all essential fats and phospholipids. FFM density (effectively lean mass) increases from the first year of life through to 10 years of age, and then again dramatically during puberty. Both males and females show a linear increase in FFM mineral content from the age of 8–15 years. The body's water content, measured as FFM–water content, decreases from 81% to 72% during the period of birth to adulthood. This drop in water content coincides with the abrupt rise in mean FFM density seen between these ages.

Changes in body fat

Full-term infants have 10–15% body fat. In infancy, the amount of fat is correlated with body weight. Body fat shows a remarkable increase during the first year of life. By the end of this year, it is estimated that body fat represents 20–25% of total body weight. Thereafter, there is a decline in the percentage of body fat to its lowest level in the mid-childhood years, and then an increase in adolescence. Girls experience a much larger increase in body fat than boys during adolescence. Adolescent boys typically have significantly less superficial fat, less total body fat and less percentage body fat than adolescent girls. Furthermore, adolescent boys have a significantly higher abdominal or central fat distribution than girls, as shown by the waist–hip ratio and the ratio of subscapular–triceps skinfold thickness.

Fat patterning in children and adolescents

The waist-to-hip circumference ratio (WHR) measures the predominance of fat storage in the abdominal region, relative to the gluteal region. A high WHR is indicative of excess abdominal fat; that is, a central fat distribution. In adults, the WHR has been related to a number of metabolic diseases and is a strong predictor of mortality, but the implications of a high WHR in children are not clear. WHR is significantly influenced by age and gender. Boys generally have higher measures of WHR than girls. The WHR decreases with age from approximately 1.1 in the youngest children to approximately 0.8 in pubertal children. From puberty onwards, the WHR approaches the values reported for adults. In children, WHR is independent of the total body fatness. In adults, however, the WHR is positively correlated with body fatness, as measured by BMI.

In general, subcutaneous adipose tissue is distributed peripherally for most children up until puberty. For most boys but only a few girls, fat begins to be stored more centrally. The rate of change towards a more central fat distribution decreases in girls after about 13–14 years of age but continues in boys. Parental and sociodemographic factors together with BMI in childhood predispose to increased WHR in adolescence, with implications for later life. Physical activity in girls and, in boys, the use of fish liver oil in past generations have been found protective against increased WHR in adolescence (Tienboon et al. 1992). In adults, a central fat pattern is common in men but not in women. However, a central fat pattern is more prevalent in elderly women than in younger, adult women.

7A.10 Assessment of growth

Growth charts are routinely used to assess growth and are used in the diagnosis and management of diseases, and in monitoring the efficacy of therapy. Chronic undernutrition or overnutrition is typically reflected in growth rates and therefore the monitoring of growth is used as an integral part of nutritional assessment. Common growth charts include weight for age, length for age (for children below two years of age) and height for age (for children two years and over). Charts to 60 months of age, based on breast-fed infants in various cultural settings, are provided by the WHO (World Health Organization 2021).

Although growth charts that show attained growth may indicate a steady progressive increase in anthropometric body measures with age, growth velocity charts reveal the changing rates of growth more visibly (see Figure 7A.4). The growth charts may suggest that an infant or child who is above the 97th centile, or below the third centile, is unusually large or small.

However, such a conclusion is not always appropriate. A well-nourished individual will follow a typical growth curve and it is therefore possible that a misinterpretation may occur for an individual whose growth is considered to be small or large from their chart but is simply a reflection of individual variation. Cause for concern is when an individual crosses a centile, particularly if this happens over a short period. This may indicate inappropriate weight gain or acute growth failure or retardation, such as wasting or stunting, because of chronic undernutrition due to either chronic malnourishment or an underlying disease. Wasting is detectable on weight-for-age growth charts and by reduced skinfold measures, while stunting is seen as impairment of linear growth, as detected by length- or height-for-age growth charts. It should be noted, however, that there are large variations between individuals in rates and patterns of growth, and these variations must be taken into consideration when determining whether a child's growth is abnormal.

Appropriate measures must be used for specific population groups. Growth charts based on Caucasian bottle-fed babies are unlikely to be appropriate for use in developing countries, as such charts may misclassify an individual or population group as underweight. International growth references have been developed by the WHO based on pooled data from seven countries, combining measurements from over 13 000 breastfed healthy infants and children. They are available to 60 months of age. This provides a more biologically valid tool for global use to monitor growth and nutritional status. Other anthropometric measures that are routinely used to monitor growth and nutritional status include head circumference, which is used in young infants, mid-upper arm circumference, which is widely used in developing countries as a measure of muscle and fat in both children and adults, and skinfold thicknesses, which measure subcutaneous fat (World Health Organization 2021). BMI centiles (Table 7A.2), using Quetelet's index or BMI (weight in kilograms divided by the square of height in metres), have been developed to detect overweight and obesity in children (Cole et al. 2000).

Table 7A.2 Child and teen healthy weight ranges.

Weight status category	Percentile range
Underweight	Less than the 5th percentile
Healthy weight	5th percentile to less than the 85th percentile
Overweight	85th to less than the 95th percentile
Obesity	95th percentile or greater

Source: Centers for Disease Control, 2021/US Department of Health & Human Services/Public Domain.

Tienboon (2003, personal communication) has tabulated cut-off levels for BMI for children, both boys and girls. They have utility in the evaluation of service delivery in ambulatory care departments or in field work in developing countries. They are for apparently healthy children (Table 7A.3) and consistent with WHO and Centers for Disease Control publications (www.cdc.gov/obesity/childhood/defining.html; de Onis et al. 2009).

Both GH and IGFs can be measured in the plasma and reflect changes in growth and cellular activity. Plasma IGF is considerably more stable and does not display the large fluctuations evident with GH. This greater stability of IGF provides a sensitive measure of growth. Growth is a multifactorial process with genetic and environmental contributions to growth and development. Both overnutrition and undernutrition exert negative and often irreversible impacts on growth at all life stages which need to be addressed early in life.

Table 7A.3 International cut-off points for Body Mass Index (BMI) for overweight and obesity by gender between the ages of 2 and 18 years.

	Overweight		Obese	
Age (years)	Males	Females	Males	Females
2	18.41	18.02	20.09	19.81
2.5	18.13	17.76	19.80	19.55
3	17.89	17.56	19.57	19.36
3.5	17.69	17.40	19.39	19.23

(Continued)

Table 7A.3 (Continued)

	Overweight		Obese	
Age (years)	Males	Females	Males	Females
4	17.55	17.28	19.29	19.15
4.5	17.47	17.19	19.26	19.12
5	17.42	17.15	19.30	19.17
5.5	17.45	17.20	19.47	19.34
6	17.55	17.34	19.78	19.65
6.5	17.71	17.53	20.23	20.08
7	17.92	17.75	20.63	20.51
7.5	18.16	18.03	21.09	21.01
8	18.44	18.35	21.60	21.57
8.5	18.76	18.69	22.17	22.18
9	19.10	19.07	22.77	22.81
9.5	19.46	19.45	23.39	23.46
10	19.84	19.86	24.00	24.11
10.5	20.20	20.29	24.57	24.77
11	20.55	20.74	25.10	25.42
11.5	20.89	21.20	25.58	26.05
12	21.22	21.68	26.02	26.67
12.5	21.56	22.14	26.43	27.24
13	21.91	22.58	26.84	27.76
13.5	22.27	22.98	27.25	28.20
14	22.62	23.34	27.63	28.57
14.5	22.96	23.66	27.98	28.87
15	23.29	23.94	28.30	29.11
15.5	23.60	24.17	28.60	29.29
16	23.90	24.37	28.88	29.43
16.5	24.19	24.54	29.14	29.56
17	24.46	24.70	29.41	29.69
17.5	24.73	24.85	29.70	29.84
18	25	25	30	30

References

Allen, R., Rogozinska, E., Sivarajasingam, P., Khan, K.S., and Thangaratinam, S. (2014). Effect of diet- and lifestyle-based metabolic risk-modifying interventions on preeclampsia: a meta-analysis. Acta Obstetrica et Gynecologica Scandinavica 93: 973–985.

Balaj, M., York, H.W., Sripada, K. et al. (2021). Parental education and inequalities in child mortality: a global systematic review and meta-analysis. Lancet 398: 608–620.

Barness, L.A., Opitz, J.M., and Gilbert-Barness, E. (2007). Obesity: genetic, molecular, and environmental aspects. American Journal of Medical Genetics 143a: 3016–3034.

Brodsky, D. and Christou, H. (2004). Current concepts in intrauterine growth restriction. Journal of Intensive Care Medicine 19: 307–319.

Buttriss, J.L. The Eatwell guide refreshed. 2016. https://assets.publishing.service.gov.uk/government/uploads/system/uploads/attachment_data/file/528193/Eatwell_guide_colour.pdf.

Cannon, G. (2010). Are short people 'stunted'? World Nutrition 1.

Chiang, P.H., Huang, L.Y., Lee, M.S., Tsou, H.C., and Wahlqvist, M.L. (2017). Fitness and food environments around junior high schools in Taiwan and their association with body composition: gender differences for recreational, reading, food and beverage exposures. PLoS One 12: e0182517.

Chiang, P.H., Wahlqvist, M.L., Lee, M.S., Huang, L.Y., Chen, H.H., and Huang, S.T. (2011). Fast-food outlets and walkability in school neighbourhoods predict fatness in boys and height in girls: a Taiwanese population study. Public Health Nutrition 14: 1601–1609.

Chilinda, Z.B., Wahlqvist, M.L., Lee, M.S., and Huang, Y.C. (2021). Higher maternal autonomy is associated with reduced child stunting in Malawi. Science Reports 11: 3882.

Cole, T.J., Bellizzi, M.C., Flegal, K.M., and Dietz, W.H. (2000). Establishing a standard definition for child overweight and obesity worldwide: international survey. BMJ 320: 1240–1243.

de Onis, M., Garza, C., Onyango, A.W., and Rolland-Cachera, M.F. (2009). WHO growth standards for infants and young children. Archives of Pediatrics 16: 47–53.

DeSalvo, K.B., Olson, R., and Casavale, K.O. (2016). Dietary guidelines for Americans. *JAMA* 315: 457–458.

Ding, N., Zhang, X., Zhang, X.D. et al. (2020). Impairment of spermatogenesis and sperm motility by the high-fat diet-induced dysbiosis of gut microbes. Gut 69: 1608–1619.

Fewtrell, M., Bronsky, J., Campoy, C. et al. (2017). Complementary feeding: a position paper by the European Society for Paediatric Gastroenterology, Hepatology, and Nutrition (ESPGHAN) Committee on Nutrition. Journal of Pediatric Gastroenterology and Nutrition 64: 119–132.

Georgiadis, A. and Penny, M.E. (2017). Child undernutrition: opportunities beyond the first 1000 days. Lancet Public Health 2: e399.

Gudmundsson, S., Henningsson, A.C., and Lindqvist, P. (2005). Correlation of birth injury with maternal height and birthweight. British Journal of Obstetrics and Gynaecology 112: 764–767.

Hetzel, B.S. (2012). The development of a global program for the elimination of brain damage due to iodine deficiency. Asia Pacific Journal of Clinical Nutrition 21: 164–170.

Hipsley, E.H. (1953). Dietary 'fibre' and pregnancy toxaemia. BMJ 2: 420–422.

Huang, L.Y., Lee, M.S., Chiang, P.H., Huang, Y.C., and Wahlqvist, M.L. (2021). Household and schooling rather than diet offset the adverse associations of height with school competence and emotional disturbance among Taiwanese girls. Public Health Nutrition 24: 2238–2247.

Huebener, M. (2019). Life expectancy and parental education. Social Science and Medicine 232: 351–365.

Lee, M.S., Huang, L.Y., Chang, Y.H., Huang, S.T., Yu, H.L., and Wahlqvist, M.L. (2012). Lower birth weight and diet in Taiwanese girls more than boys predicts learning impediments. Research in Developmental Disabilities 33: 2203–2212.

Leung, B.M. and Kaplan, B.J. (2009). Perinatal depression: prevalence, risks, and the nutrition link–a review of the literature. Journal of the American Dietetic Association 109: 1566–1575.

Linnér, A. and Almgren, M. (2020). Epigenetic programming – the important first 1000 days. Acta Paediatrica 109: 443–452.

Locke, A.E., Kahali, B., Berndt, S.I. et al. (2015). Genetic studies of body mass index yield new insights for obesity biology. Nature 518: 197–206.

Luo, D., Zhang, M., Su, X. et al. (2020). High fat diet impairs spermatogenesis by regulating glucose and lipid metabolism in Sertoli cells. Life Science 257: 118028.

McLoyd, V.C. (1998). Socioeconomic disadvantage and child development. American Psychologist 53: 185–204.

Mirpuri, J. (2021). Evidence for maternal diet-mediated effects on the offspring microbiome and immunity: implications for public health initiatives. Pediatrics Research 89: 301–306.

Mu, Y., Yan, W.J., Yin, T.L., and Yang, J. (2016). Curcumin ameliorates high-fat diet-induced spermatogenesis dysfunction. Molecular Medicine Reports 14: 3588–3594.

Nohr, E.A., Vaeth, M., Baker, J.L., Sørensen, T.I., Olsen, J., and Rasmussen, K.M. (2009). Pregnancy outcomes related to gestational weight gain in women defined by their body mass index, parity, height, and smoking status. American Journal of Clinical Nutrition 90: 1288–1294.

O'Reilly, D., Dorodnykh, D., Avdeenko, N.V. et al. (2021). Perspective: the role of human breast-milk extracellular vesicles in child health and disease. Advances in Nutrition 12: 59–70.

Pallotto, E.K. and Kilbride, H.W. (2006). Perinatal outcome and later implications of intrauterine growth restriction. Clinical Obstetrics and Gynecology 49: 257–269.

Reik, W., Davies, K., Dean, W., Kelsey, G., and Constância, M. (2001). Imprinted genes and the coordination of fetal and postnatal growth in mammals. Novartis Foundation Symposium 237: 19–31. discussion 31–42.

Russo, V.C., Gluckman, P.D., Feldman, E.L., and Werther, G.A. (2005). The insulin-like growth factor system and its pleiotropic functions in brain. Endocrine Reviews 26: 916–943.

Schwarzenberg, S.J. and Georgieff, M.K. (2018). Advocacy for improving nutrition in the first 1000 days to support childhood development and adult health. Pediatrics 141.

Silventoinen, K., Rokholm, B., Kaprio, J., and Sørensen, T.I. (2010). The genetic and environmental influences on childhood obesity: a systematic review of twin and adoption studies. International Journal of Obesity 34: 29–40.

Slot-Heijs, J.J., Collard, D.C.M., Pettigrew, S. et al. (2020). The training and support needs of 22 programme directors of community-based childhood obesity interventions based on the EPODE approach: an online survey across programmes in 18 countries. BMC Health Service Research 20: 870.

Symonds, M.E. (2013). Brown adipose tissue growth and development. Scientifica 2013: 305763.

Tienboon, P., Wahlqvist, M.L., and Rutishauser, I.H. (1992). Early life factors affecting body mass index and waist-hip ratio in adolescence. Asia Pacific Journal of Clinical Nutrition 1: 21–27.

Wahlqvist, M.L., Darmadi-Blackberry, I., Kouris-Blazos, A. et al. (2005). Does diet matter for survival in long-lived cultures? Asia Pacific Journal of Clinical Nutrition 14: 2–6.

Wahlqvist, M.L. and Gallegos, D. (2020). Nutrition across the life course. In: Food and Nutrition: Sustainable Food and Health Systems (eds M.L. Wahlqvist and D. Gallegos). London: Routledge.

Wahlqvist, M.L., Huang, L.Y., Lee, M.S., Chiang, P.H., Chang, Y.H., and Tsao, A.P. (2014). Dietary quality of elders and children is interdependent in Taiwanese communities: a NAHSIT mapping study. Ecology of Food and Nutrition 53: 81–97.

Wahlqvist, M.L., Worsley, A., Harvey, P., Crotty, P., and Kouris-Blazos, A. (1999.). Food-based dietary guidelines for the Western Pacific: the shift from nutrients and food groups to food availability, traditional cuisines and modern foods in relation to emergent chronic non-communicable diseases. Manila, Philippines: WHO.

Wang, B. and Brand-Miller, J. (2003). The role and potential of sialic acid in human nutrition. European Journal of Clinical Nutrition 57: 1351–1369.

Willett, W., Rockström, J., Loken, B. et al. (2019). Food in the Anthropocene: the EAT-Lancet commission on healthy diets from sustainable food systems. Lancet 393: 447–492.

World Health Organization Physical status: the use of and interpretation of anthropometry. Report of a WHO Expert Committee. https://apps.who.int/iris/bitstream/handle/10665/37003/WHO_TRS_854.pdf?se

World Health Organization Development of food-based dietary guidelines for the Western Pacific region: the shift from nutrients and food groups to food availability, traditional cuisine and modern foods in relation to emerging chronic noncommunicable diseases. https://iris.wpro.who.int/bitstream/handle/10665.1/5470/1.pdf?sequence=1

World Health Organization The WHO child growth standards, 2006. On-line scientific background papers for each guideline. www.who.int/tools/child-growth-standards

Yang, W., Kelly, T., and He, J. (2007). Genetic epidemiology of obesity. Epidemiology Reviews 29: 49–61.

Further reading

Chiang, P.H., Huang, L.Y., Lee, M.S., Tsou, H.C., and Wahlqvist, M.L. (2017). Fitness and food environments around junior high schools in Taiwan and their association with body composition: gender differences

for recreational, reading, food and beverage exposures. PLoS One 12: e0182517.
Gracey, M., Hetzel BS, R., Strauss, B.J., Tasman-Jones, C., and Wahlqvist, M. (1989). Responsibility for Nutritional Diagnosis: A Report by the Nutrition Working Party of the Social Issues Committee of the Royal Australasian College of Physicians. London: Smith-Gordon.
Lo, Y.T., Wahlqvist, M.L., Chang, Y.H., and Lee, M.S. (2016). Combined effects of chewing ability and dietary diversity on medical service use and expenditures. Journal of the American Geriatric Society 64: 1187–1194.
Longo-Silva, G., Silveira, J.A.C., Menezes, R.C.E., and Toloni, M.H.A. (2017). Age at introduction of ultra-processed food among preschool children attending day-care centers. Journal of Pediatrics 93: 508–516.
Liu, C.K., Huang, Y.C., Lo, Y.C., Wahlqvist, M.L., and Lee, M.S. (2019). Dietary diversity offsets the adverse mortality risk among older indigenous Taiwanese. Asia Pacific Journal of Clinical Nutrition 28: 593–600.
Li, K., Liu, C., Wahlqvist, M.L., and Li, D. (2020). Econutrition, brown and beige fat tissue and obesity. Asia Pacific Journal of Clinical Nutrition 29: 668–680.
Loos, R.J. (2012). Genetic determinants of common obesity and their value in prediction. Best Practice Research in Clinical Endocrinology and Metabolism 26: 211–226.
Meltzer, H.M., Brantsæter, A.L., Trolle, E. et al. (2019). Environmental sustainability perspectives of the Nordic diet. Nutrients 11.
Rauber, F., Chang, K., Vamos, E.P. et al. (2021). Ultra-processed food consumption and risk of obesity: a prospective cohort study of UK Biobank. European Journal of Nutrition 60: 2169–2180.
Stinson, L.F., Payne, M.S., and Keelan, J.A. (2018). A critical review of the bacterial baptism hypothesis and the impact of cesarean delivery on the infant microbiome. Frontiers in Medicine 5: 135.
Swinburn, B.A., Sacks, G., Hall, K.D. et al. (2011). The global obesity pandemic: shaped by global drivers and local environments. Lancet 378: 804–814.
Wahlqvist, M.L. (2009). Connected Community and Household Food-Based Strategy (CCH-FBS): its importance for health, food safety, sustainability and security in diverse localities. Ecology of Food and Nutrition 48: 457–481.
Yang, Y.X., Wang, X.L., Leong, P.M. et al. (2018). New Chinese dietary guidelines: healthy eating patterns and food-based dietary recommendations. Asia Pacific Journal of Clinical Nutrition 27: 908–913.

7B Later Life

Key messages

- Chronological and biological age do not necessarily correlate.
- Energy requirements generally decrease with age, but nutrient needs remain relatively high. Animal studies suggest that energy restriction promotes longevity, but human studies suggest that 'eating better, not less' is desirable.
- Physical activity can improve health and well-being, and reduce morbidity risk at any stage of the lifespan.
- Many health problems commonly associated with older age are not necessarily products of 'ageing'; instead, they can be prevented or delayed by consuming a nutritious biodiverse diet and engaging in regular physical activity.
- The human life cycle is one of socioecological beings intimately part of nature and dependent on it and societal connectedness; failure of any of these characteristics may lead to food and health insecurity and disorder, and threaten planetary habitability and survival

7B.1 Ageing

Ageing is not a disease, nor are the so-called diseases of ageing – cancer, heart disease, arthritis and senility – inevitable consequences of advancing years. If we live long enough, changes in body composition, physical function and performance will occur in all of us and affect our well-being (Wahlqvist et al. 2001). Many of these changes, as well as health problems which become more common in old age, have long been attributed to the 'normal ageing process' or 'senescence'. Ageing is a construct which may be chronological, biological or societal. Much more than early life, it has great variability and can be subgrouped in various ways as, for example, chronologically 'young-old' (65–74), 'old' (74–84), and 'old-old' (85 +). This section will highlight that many age-related health problems can be delayed to the last few years of life (i.e. there can be a compression of morbidity into a narrower time-frame). Food systems have the capacity to contribute to this compression.

7B.2 Theories of ageing

The theory of *programmed ageing* suggests that the body has a built-in clock that begins ticking at birth (or maybe conception) (Jansen et al. 2021). This theory is supported by the discovery that normal cells have a limited capacity to divide because telomeres (sections of DNA at the end of chromosomes) shorten at each division, resulting in a fixed lifespan for each normal cell.

Nutrition and Metabolism, Third Edition. Edited on behalf of The Nutrition Society by Helen M. Roche, Ian A. Macdonald, Annemie M.W.J. Schols and Susan A. Lanham-New.

Companion Website: www.wiley.com/go/nutrition/metabolism3e

The ageing process accelerates so rapidly in some individuals that they become biologically 'old' in their teens.

The *error theory* attributes ageing to increasing damage to DNA and progressive decline in the function of specialised enzymes that repair DNA. It is thought that diseases such as cancer, heart disease, osteoporosis and diabetes may be the result of an accumulation of errors.

The *free radical theory* proposes that free radicals (highly reactive oxygen molecules) are produced by oxygen-consuming biological reactions in the body. Free radicals damage cells and have been implicated in the development of cancer and heart disease. There is no evidence that taking antioxidants will improve longevity, but antioxidants consumed as food components may have a more reliable and appropriate ability to reduce the damage produced by free radicals. A free radical is a molecule with an unpaired, highly reactive electron, which is often associated with the development of cancer, arteriosclerosis, autoimmune diseases and ageing. Antioxidant activity is provided by various phytonutrients (e.g. flavonoids) and nutrients (vitamins C and E and β-carotene), endogenous uric acid and bodily enzymes such as superoxide dismutase (SOD), catalase and glutathione peroxidase. These together and in concert prevent most, but not all, oxidative damage. Bit by bit, the damage mounts and contributes, so the theory goes, to the deterioration of tissues and organs. Antioxidants may reduce the risk of cancer by protecting cellular DNA from free radical damage. However, oxidation is an important part of the body's physiology and defence system; it makes no sense to take antioxidant supplements which are disconnected from each other, and which damage this function. Indeed, the body has an elaborate antioxidant system of its own, in which uric acid is the major vehicle in addition to the enzymic systems.

Protracted and inappropriate inflammation will also hasten ageing. Among the several *biological clocks* which have been identified, and which could determine the tenure of life, there may be one which is expressed by multimorbid inflammation as incurred by the vascular system and other vital tissues with chemokine CXCL9, so-called *iAge*. Dietary patterns can be more or less anti-inflammatory and conceivably alter the expression of such a clock (Sayed et al. 2021; Wahlqvist et al. 2001). Regular physical activity in its various forms can also mitigate inflammation, including that which contributes to atherosclerosis or arterial disease.

One popularised view, derived mainly from animal experiments, has argued that *energy restriction* may decrease tissue damage and the risk of cancer, so increasing longevity. However, extrapolation for the whole of life to humans would create an unethical risk of increased risk of illness and mortality in early life and, in later life, potentially inappropriate weight loss, a predictor of premature mortality. Several longitudinal studies, like that in Zutphen, The Netherlands, show that increased energy intakes, without increased body fatness, are associated with increased life expectancy. The relatively greater energy intakes need to be of *nutritious foods in a plant-based pattern, preferably with a regular fish intake as well.*

Yet, with advancing years, physical activity often declines, which leads to a loss of muscle and, as a direct consequence, basal metabolic rate (BMR) falls. In addition to less physical activity, *sedentariness tends to increase* and together these increase the burden of *chronic cardiometabolic disease and certain cancers* (e.g. colorectal cancer); sedentariness matters irrespective of physical activity. Similarly, greater physical activity, requiring energy utilisation, and energy intakes are associated with improved survival. That said, higher BMIs in later rather than earlier life are compatible with longer life expectancy. To some extent, this is dependent on recourse to *healthcare expenditure* to manage health problems as shown by Pan et al. (2011) (Gracey et al. 1989; Jansen et al. 2021; Sayed et al. 2021; Wahlqvist et al. 2001; Wolf and Ley 2019).

7B.3 Sociodemography

Humans are living longer than ever before, with several population life expectancies at birth now exceeding 80 years. Since the early 1970s, life expectancies have increased globally by 1–2 years every five years, particularly in socioeconomically advantaged societies. The proportion of centenarians is increasing (upwards of 1 in 1000 of the population in some countries). But individuals do not appear to exceed a maximal lifespan of about 120 years. Maximal lifespan may yet increase as biotechnology, socioecology and healthcare develop in favour of greater longevity, unless we are overtaken by the increasing threats of inequity and climate change.

Presently, many adults are reaching older age in better health and the majority are able to live independently. Life expectancy is increasing for men and women although, where maternal mortality is low, women still generally live longer than men and make up a greater proportion of elders. The proportion of older people in the population has increased, especially for those aged over 70. Our ability to live longer is, in part, attributable to better nutrition and other changes in personal behaviour (e.g. reduced substance abuse, greater recreational opportunities), improved healthcare (e.g. reduced infant, child and maternal mortality, earlier diagnosis and management of cancers and heart disease), educational and economic improvements, and better housing (especially less crowding) and social support systems. But as we live longer, our nutritional needs may change, either with 'healthy' ageing or because of the advent of disease.

Keeping an elderly population well is of great importance for the individuals themselves and the well-being of society in general. The availability and transfer of knowledge and skills to younger people, especially descendants, a reduced burden on others, and access to aged care resources are among the multiple reasons for healthy ageing (Wahlqvist et al. 2014). The numbers of elderly people in developing countries will exceed those in developed countries, so that the problem is global.

7B.4 Biological and chronological age

As indicated above, ageing may be defined as chronological (a person's age in years since birth) or biological (the heterogeneous decline in function that occurs in every human being with time). Some look and function as though they were older and others as though they were younger, even at the same 'chronological age'. Prospective studies, where some assessment of biological age has been made during the twentieth century, as in Sweden, indicate that people are less biologically old at the same chronological age than they used to be, and that this difference may be as much as 10 years of biological age. This is a rather remarkable change and some of it is likely to be attributable to improved lifelong nutrition. Indeed, it may well be that much of what we currently regard as ageing is preventable by nutritional means. In other words, even though genes, some of them 'clock genes', affect biological age, environmental and behavioural factors are modulatory. You may remain biologically younger if you take care in your younger years.

The question is, what aspects of ageing are biologically inevitable, having to do, for example, with the programmed death of cells (apoptosis), and how much is age related? While the clock cannot be turned back in terms of chronological age, the search for prolonged youth continues to invoke much interest and research. The older we are, the more dissimilar we become from others of the same chronological age. Poor eating habits, being sedentary, disordered sleep and smoking are among the most evident hasteners of a decline in biological age, a loss of organ and system functionality and well-being and the development of disease. The Baltimore Longitudinal Study of Aging was among the earliest studies to draw attention to this phenomenon, particularly in regard to, but irrespective of, cardiovascular health. However, after careful exclusion of those with heart disease, no consistent declines in function with age remained. The accumulating effects of years of poor eating habits can increase the risk of many health conditions as one grows older, but it is never too late to change!

7B.5 Compression of morbidity

The decline in LBM and increases in body fat, which tend to occur as people grow older, cannot be entirely attributed to the ageing process per se. If we endeavour to keep linked to the natural world of which we are biologically part, be socially connected and supportive, and engage in regular exercise and healthy eating, body compositional changes and functional decline within the limits set by genetics can be slowed. This may mean taking a recreational walk for 30–40 minutes, cycling or gardening most days, the very least a living person might be expected to do during waking hours! This will compress morbidity into the last few years of life (i.e. increase health span potential), even once we enter later life. It is an advantage if this is achieved, insofar as exercise is concerned, by our mid-50s as judged by measures of disability.

Major contributors to age-related changes are sedentariness and loss of appetite for a nutritious diet, both rectifiable by physical activity. In turn, this contributes to the loss of bone and muscle, being major components of LBM; they not only comprise our musculoskeletal system, but they are also regulatory endocrine organs in their own right, involved in immunoprotection, and are storage organs for vital nutrients like amino acids, the building blocks of proteins, and of the elements calcium, magnesium and zinc.

By taking socioecological care of ourselves, which is who we are, we can compress morbidity towards the end of life, and maintain or even increase our physiological and nutritional reserves.

7B.6 Physiological reserves

Many bodily functions remain relatively unaffected until about 75 years of age when, on average, they start to decrease more noticeably. Nutritionally related health problems are often compounded in later life by reduced physiological reserves of many organs and functions. This applies to both reduced metabolic tissues (e.g. insulin resistance or reduced insulin response to a meal load, or a greater glycaemic response to the same food) and organ tissues (e.g. reduced cardiac reserve means that an added salt load may tip someone into heart failure, whereas otherwise it would not). While a younger person will be able to consume an inadequate diet with no foreseeable consequences, an elderly person is more likely to experience problems because of diminished physiological function. Many studies have shown significant reductions in different body functions with age. These may not be inevitable, however. For example, what used to be regarded as a decline in brain function at about the age of 70 may not be seen until much later, raising the possibility that biological age in some body functions may be occurring at a later and later chronological age.

Measures of physiological and nutritional reserves may be important indicators of health in older adults. Prevention of associated health problems may be possible, if physiological and nutritional reserve levels are known and developed. Being active and engaged on several fronts throughout life's stages has arguably the most potential according to the evidence (Savige et al. 2001).

7B.7 Frailty

Avoidance of frailty is one of the major challenges facing older people and their carers. Frailty among older people has been defined as 'a condition or syndrome which results from a multi-system reduction in reserve capacity, to the extent that a number of physiological systems are close to, or past, the threshold of symptomatic clinical failure (Rockwood et al. 2005). Consequently, the frail person is at an increased risk of disability and death from minor external stresses'. As the number of chronic conditions increases with age, they contribute to disability and frailty, which in turn reduce a person's level of independence, sometimes resulting in institutionalisation.

Falls, incontinence and confusion are regarded as clinical consequences of frailty, and several risk factors are associated with each of these conditions. The risk of falling is increased as muscle strength and flexibility decline, and if balance and reaction time are impaired. *Urinary incontinence* is also a risk factor for falls among elderly people. Dehydration and PEM are two nutritional factors that can contribute to the *confusion* often experienced by elderly adults. Urinary incontinence often results in elderly people restricting their fluid intake in an effort to control their incontinence or reduce their frequency of urination. In Taiwan, those who have a *dietary pattern* with more phytonutrient-rich plant foods, tea, omega-3-rich deep-sea fish and other protein-rich foods such as shellfish and milk are less frail. Frailty prevention, management and reversibility remain clinical and public health challenges.

7B.8 Prevention strategies

The major prevention strategies that elderly individuals can take to increase their physiological and nutritional reserves include:

- Consuming a wide variety of foods
- Engaging in physical activity, as this maintains lean muscle and bone mass, thus increasing nutritional and physiological reserves to prevent major health problems
- Engaging in social activity
- Cooking

- Avoiding substance abuse (including alcohol, tobacco, excessive caffeine intake and unnecessary intake of medications).

By focusing on the complete way of life rather than on just one component, such as nutrition, elderly people can enjoy life without experiencing major consequences of nutritional error.

Food biodiversity and patterns

Food variety has an important role to play in preventing the onset of diseases such as diabetes, cancer and cardiovascular disease. A varied diet, ideally containing 20–30 biologically distinct foods a week, is associated with reduced morbidity and mortality. Increased food variety is associated with less atherosclerotic vascular disease in apparent health and diabetes (Wahlqvist et al. 1989). The Zutphen studies of Kromhout et al. demonstrate the relatively important role of a plant-based diet including apples, onions and tea, in better all-cause and disease-specific mortalities. As shown in the IUNS (International Union of Nutritional Sciences) studies of Food Habits in Later Life (FHILL), a greater dietary diversity is predictive of greater longevity in people aged over 70 years and reduces the risk of death by more than 50%. This mortality advantage was observed in Greece and several other countries with higher intakes of vegetables (>300 g/day), legumes (>50 g/day), fruits (>200 g/day) and cereals (>250 g/day), together with moderate intakes of dairy products (<300 g of milk/day or equivalent in cheese/yogurt), meat and meat products (<100 g/day), alcohol (<10 g/day), being high in monounsaturated fat (mainly from olive oil) and low in saturated fat (i.e. high monounsaturated:saturated fat ratio). This pattern was consistent with food patterns prevalent in Greece in the 1960s, when Greeks enjoyed one of the longest life expectancies in the world. Greater mortality advantage obtained if they followed the entire food pattern with a high dietary variety score (Figure 7B.1).

There is likely to be synergy between food groups, which signals the need to follow dietary recommendations as a whole, rather than focusing on just one food group or nutrient. For this and other reasons of food culture and security, *Food-Based Dietary Guidelines* (FBDGs) were recommended in the FAO-WHO Cyprus declaration of 1995. The evidence for optimal dietary patterns being ones of biodiversity is internationally consistent (Miyamoto et al. 2019) irrespective of food culture. It can also allow indigenous peoples to offset some of the health disadvantage they suffer (Liu et al. 2019).

Given that, across cultures and despite socioeconomic gradients, *hypertension* is the most avoidable risk factor for morbidity and mortality, it is noteworthy that the furtherance of food biodiversity with fruit generally lowers blood pressure and reduces the risk of stroke, a major cause of premature death and disability. This is probably in part because, as a source of potassium, it decreases the sodium (Na):potassium (K) molar ratio (Chang et al. 2006).

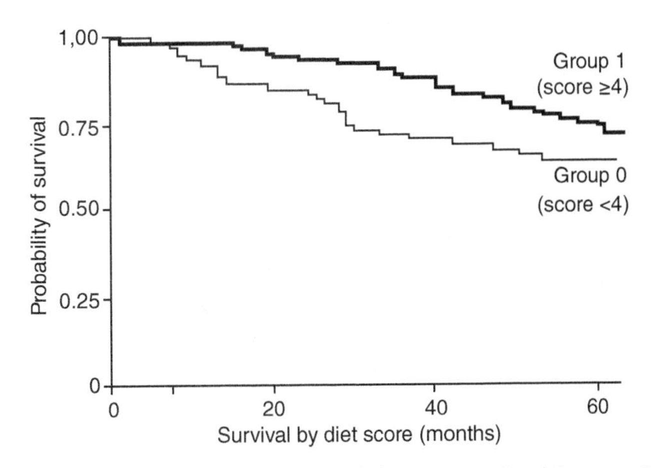

Figure 7B.1 Kaplan–Meier survival curves for individual subjects with diet score up to 3 and 4 or more. *Source:* Reproduced from Diet and overall survival in elderly people, Trichopoulou A, Kouris-Blazos A, Wahlqvist M, 311: 1457–1460 (1995) © with permission from BMJ Publishing Group Ltd.

Physical activity

Ageing in modern society is, in many ways, an exercise deficiency syndrome, implying that we may have more control over the rate and extent of the ageing process than we previously thought (Frontera et al. 2000).

Some of the most impressive changes that we see with age are changes in body composition. A decline in muscle mass and increases in body fat tend to occur as people grow older. What is often not appreciated, however, is that these cannot be blamed on the ageing process per se. A major contributor to these changes is the increasingly sedentary nature of people's lifestyles as they grow older. Reduced physical activity leads to loss of muscle and, as a direct consequence, BMR falls. A lower metabolic rate means that we need to eat less in order to maintain the same body weight. If one does indeed eat less to avoid weight gain, rather than remaining (or becoming) active, it becomes increasingly difficult to meet the needs for essential nutrients. Without doubt, it is preferable to keep physically active, maintain muscle mass and continue to enjoy eating.

Physical activity is associated with greater energy intakes and, hopefully, essential nutrient intakes, so advancing the quality of life as we age. This higher plane of energy nutrition preference with age runs counter to the advocacy that energy restriction prolongs life. Such views are usually based on animal studies where physical activity and overall energy balance have not been considered in ways that are applicable to humans. To suggest that elderly people restrict their food intake to prolong life is inappropriate when it may contribute to frailty and loss of lean mass. Any extra energy intake must, however, be from nutrient (and phytochemical)-dense foods, without excessive abdominal fatness.

Several studies have shown that energy intake tends to decline with age, making a nutritionally adequate diet more difficult to achieve. Older men consume about 800 kcal less than younger men, and older women consume about 400 kcal less than younger women. A reduction in BMR is partly responsible for this decline in energy intake, but physical inactivity appears to be the major cause. Increased energy intakes above those usually prevailing in sedentary societies, of the order of 300–500 kcal/day, balanced with increased physical activity to avoid fat gain, confer either decreased cardiovascular or total mortality and improve life expectancy. Physical activity also protects against osteoporosis and fractures, diabetes, and breast and colon cancers, improves mental health and cognitive function, reduces symptoms of anxiety and depression and enhances feelings of well-being in older people. Exercise intervention in mid-life compresses morbidity (measured as disability) towards the end of life. Even in the presence of morbidity, minimal physical activity improves survival (Wen et al. 2011). The combination of endurance and strength training continued into later life can compress morbidity nearer to death at a later time.

Cooking

Active involvement in the food system provides for independence and a measure of control over this part of one's life. In itself, it is a health behaviour. Cooking several times a week is associated with increased longevity (Erlich et al. 2012).

Social activity

Social activity is one of the most important determinants of longevity. Participation in fewer social activities outside the home and limited social networks have been linked with higher mortality in old age. The impact of social activity on longevity could be through its impact on psychological well-being and nutrition. For example, elderly people who are socially isolated, lonely, institutionalised, recently bereaved and socially inactive have been found to have inadequate food intakes. Glass et al. (1999) found social and productive activities to be as effective as fitness activities in lowering the risk of death. Further studies of this kind indicate the importance of social activity to the health and mortality of older people (Huang et al. 2017).

Substance abuse

A major reason for improvements in life expectancy in recent history is the success of quit *smoking* campaigns. The reductions in cardiovascular disease and death have occurred within days and months of cessation, while those from neoplastic disease like lung cancer and chronic obstructive lung disease have taken longer, even years. Less recognised have been gains in bone health and reduced skin wrinkling.

Effects of ageing on physiological function

Physiological changes that occur with ageing contribute to the body's declining function which, in turn, influences nutritional status, just as growth and development do in the earlier stages of the life cycle. Physiological changes include the following.

- Hormonal and metabolic activity which alters *body composition, energy metabolism* and *insulin resistance.*
- Changes in the *immune system* which raise the risk of infections and some chronic diseases.
- Atrophic gastritis which interferes with *nutrient digestion and absorption.*
- *Oral health* (e.g. gingivitis) and *chewing ability* which are associated with nutritional status, cognition, well-being and survival.
- *Sensorineural factors* with blunted sensory inputs including taste and smell, sound, sight, touch.
- *Mood and affective disorders* such as depression which can adversely influence food choice and be affected by it.
- 'Eating better, but not less' which is possible with more physical activity, providing a higher plane of energy throughput and the ability to eat more food without it being in excess.
- *Cell differentiation*, turnover and death which accounts for organ and system function, failure and excessive proliferation (neoplasia).

Body composition

A decline in muscle and bone mass and increases in body fat tend to occur as people grow older, and subcutaneous fat is redistributed from limbs to the trunk. Some of these changes occur because some hormonal activities that regulate metabolism and body composition decrease with age (e.g. insulin, GH, androgens), and the activity profile of others becomes adverse (e.g. prolactin). Being physically active and avoiding sedentariness while consuming a nutritious biodiverse diet are ways to counter these effects. The incidence of underweight increases with age. A lower body weight is more strongly linked with morbidity in the elderly than mild to moderate excess weight, and the problem is often insidious. Survival rates in elderly Finnish people (85 years and over) during a five-year period showed the highest mortality to be in those with a BMI less than 20 kg/m^2 and the lowest mortality to be in the group with a BMI of 30 or more. Other studies have shown that elderly people with BMIs below 27 kg/m^2 lived shorter lives than those with higher BMIs. Weight changes, and especially weight loss, are of greater concern in the elderly than too much fat. In developed countries, 30–50% of older adults have been reported to be at high risk of developing health problems because of an inadequate food and nutrient intake.

Immune system

Both physical stressors (e.g. alcohol abuse, other drug abuse, smoking, pain, heat and illness) and psychological stressors (e.g. divorce, exams, migration and loss of a loved one) elicit the body's stress response: the classic fight-or-flight response. Stress that is prolonged or severe can increase vulnerability to illness. As people age, adaptation to both external and internal stressors, especially via the immune system, is diminished. The immune system can also be compromised by suboptimal dietary patterns with essential nutrient and phytonutrient deficiencies and gut dysbiosis (see below). A combination of age and subclinical nutrient deficiencies increases the vulnerability of older adults to infectious, autoimmune and inflammatory disease, with proneness to arthritic and neoplastic illness (Dorshkind et al. 2009).

Innate immunity may wane with age, but be of increasing importance in resilience to outbreaks of infectious disease and to food-borne illness. In the COVID-19 pandemic of 2019–2021, older persons, often dependent on institutional care, were among the most vulnerable, and their immune responses were faulty. Risk minimisation depended on hygiene, distancing, mask wearing and innate immunity.

Gastrointestinal tract

The intestine loses strength and elasticity with age; this slows motility and increases the risk of developing *constipation* (which is 4–8 times more common in the elderly than in younger adults). *Atrophic gastritis* is also more common among older adults; about 30% of adults aged over 60 have this condition. It is characterised by chronic inflammation of the stomach, accompanied by a diminished size and functioning of the mucosa and glands, resulting in less hydrochloric acid being secreted and increased levels of bacteria. These changes in the stomach can impair digestion and the absorption of nutrients, especially vitamin B_{12}, biotin, calcium and iron.

Oral health and chewing ability

Chewing can be painful or difficult in old age as a result of tooth loss, gum disease and ill-fitting dentures. This can result in a reduced variety of foods consumed and an increased risk of developing nutrient deficiencies. Chewing difficulty is a risk factor for premature mortality, cognitive impairment and dementia. Attention to the oral microbiome (Dominy et al. 2019; Tzeng et al. 2016), gingival health and relevant diets are ways in which general health and survival can be improved and health expenditure reduced (Lo et al. 2016).

Sensory loss

Taste and smell are often impaired or dysfunctional in later life, to some extent associated with lifelong cigarette smoking, poor dental hygiene and disease. This makes eating less enjoyable and partly explains why older people tend to increase salt intake and welcome caffeinated beverages (caffeine may briefly suppress, but then increase appetite). Many of the receptors located in the nose and tongue are also represented in organ tissues and influence function, making them of general health relevance. They provide an interface with the natural environment and its place in our ecobiology and health. The physical and mental health-promoting potential of public open space (POS) throughout life, especially in later life, has become clear, as has its dependency on the sensory inputs into our nutritional biology. This is underscored with the consequences of ecosystem loss and disorder, and the advent of viral pandemics affecting and damaging our sensory apparatus (see Figure 7A.1).

Ageing is associated with a decrease in the opioid (dynorphin) *feeding drive* and an increase in the *satiety effect* of cholecystokinin. Early satiety in older people may be caused by a nitric oxide deficiency, which decreases the adaptive relaxation of the fundus of the stomach in response to food.

Psychological changes and mental health

Depression is common among older adults but is not an inevitable component of ageing. It is frequently accompanied by loss of appetite and the motivation to cook. Loss of appetite itself is a risk factor for survival, as is recent unintended weight loss. Cooking is a healthy behaviour and up to five times per week is associated with more favourable life expectancy.

As with many aspects of health, energy dysregulation with overfatness and diabetes in turn compromises mood and increases the risk of neurodegeneration (Hsu et al. 2011; Wahlqvist et al. 2012a; Wahlqvist et al. 2012b).

Neoplastic disease and cancer

The World Cancer Research Fund (WCRF) regularly reviews the general recommendations for diet and physical activity to reduce the global burden of cancer in its entirety, and on account of particular cancers, as well as policies to give effect to its recommendations (www.wcrf.org/wp-content/uploads/2021/02/Summary-of-Third-Expert-Report-2018.pdf). These are relevant to older people and to those who have cancer, and are consistent with more age-specific evidence.

Control of cellular differentiation is a well-developed feature of human biology, so that errors may need to be multiple for neoplastic disease to manifest. Cumulative DNA damage and failed cellular differentiation are age related (Hsu et al. 2011; Wahlqvist et al. 2012a; Wahlqvist et al. 2012b). Their minimisation is provided by a biodiverse diet replete with phytonutrients (e.g. their presence in skin reduces actinic damage and skin cancer), by avoidance of cured meats and by the limited use of alcoholic beverages.

Those nutritional factors which increase the risk of *body compositional and metabolic disorders* like obesity and diabetes in turn increase the risk of certain cancers, particularly hepatic, pancreatic, oesophageal and colorectal. Nevertheless, the *nutritional pathway to cancer* may be complex. In the example of breast cancer, rapid growth velocity may be an early life risk factor, then inadequate sunlight exposure itself accompanied by limited formation of antineoplastic vitamin D in the skin, a diet with variable phytoestrogenicity (see phytonutrients), energy nutrition and body composition, excessive alcohol intake, an inadequately protective reproductive history and the secular trends in menarche and menopausal status.

7B.9 Nutritionally vulnerable older adults

There are several *instruments for the detection and evaluation of nutritional status and risk of elderly people* such as the Mini Nutritional Assessment (MNA) (Vellas et al. 1999) used in communities or institutions, Malnutrition Universal Screening Tool (MUST) (Stratton et al. 2004) in clinical and hospital care, and Health and Nutrition Index (HANI) (Huang et al. 2018) a *relatively non-invasive* tool for prognostication.

However, *high alert at its most simple* is provided by questions to do with (Huang et al. 2014; Huang et al. 2018):

- Recent unplanned weight loss
- Loss of appetite
- Chewing difficulty
- Food biodiversity.

These can be readily and efficiently incorporated into regular clinical practice (Wahlqvist 2000).

Adult *energy requirements* decline by an estimated 5% per decade. And although energy intakes fall with advancing age in socioeconomically developed countries from about 2800 to 2000 kcal for men and from 1900 to 1500 kcal for women, average macro- and micronutrient intakes are relatively well maintained by the selection of nutrient-dense foods and meals and as long as social and physical activities are maintained. Susceptibilities emerge with the loss of traditional and familiar constraints and interest, preference and ease of consumption of highly refined and salty foods, along with alcoholic beverages. Essential nutrient marginality may be of concern where the cost and availability of wholegrains, nuts and seeds, fruits, vegetables, seafood, dairy and culinary herbs and spices are prohibitive. These are the items which round out the biodiversity of food patterns now known to be associated with healthy ageing whatever the basic food culture, European, Asian, African, Latin American or Indigenous, and within already long-lived cultures. They take care of the breadth and required bioavailability of human diets.

One of the present and increasing difficulties with this appreciation of dietary optimisation is that climate change, lack of potable and agricultural water, compromised food systems, population displacement, conflict, inequity, pollution and contamination will make deliverability, accessibility, affordability and sustainability of this way of eating a privilege and a risk of imponderable ethical difficulty – unless urgent global action is taken.

In the meantime, the health problems attributable to inadequate and inappropriate food and food nutrient intake include cognitive impairment, poor wound healing, anaemia, bruising, an increased propensity for developing infections, neurological disorders, stroke and some cancers.

Some groups of older people are more likely to have inadequate diets, especially if cooked meals are irregular, they live alone or are in organisational care. These risk settings can be made more food intake secure if there is regular visitation with nutritious pre-prepared meals, as obtains with so-called Meals-on-Wheels programmes, although the impact of social isolation, itself a risk factor for poor nutrition, is not fully met. Encouragement to eat with family or friends has been shown to increase food intake.

An elderly person may eat less food for several reasons. Nutritionally vulnerable 'at-risk' groups within older populations, who are more likely to be consuming inadequate diets (e.g. less regular consumption of cooked meals) and to be at risk of protein–energy malnutrition, include those who are:

- Institutionalised
- Older men living alone
- From low socioeconomic status groups
- Socially isolated and lonely
- Recently bereaved
- Depressed or cognitively impaired
- Physically and socially inactive

and those with:

- Physical handicaps, impaired motor performance and mobility
- Presence of chronic diseases (e.g. arthritis, diabetes, hypertension, heart disease, cancer)

- Polypharmacy (unnecessary intake of medications, drug–nutrient interactions; some drugs affect appetite/mood and cause nausea)
- Sensory impairment: taste/smell (reduction in taste), eyesight (cataracts)
- Reduced sense of thirst (hypodypsia)
- Problems with chewing (loss of teeth and poorly fitting dentures)
- Limited food storage, shopping difficulties and inadequate cooking skills
- Erroneous beliefs and food faddism, food preferences.

Medications, depression, dementia, chronic illness, disability, loneliness and diminished senses of smell and taste may decrease the pleasure of eating. Food beliefs in relation to health can be strongly held among elderly people and lead to both food fads and undesirable food avoidance. There may be a significant association between food beliefs and food habits, as evidenced in studies of various elderly communities around the world.

Low intakes of some nutrients have implications for bone health (calcium), wound healing (zinc, protein, energy), impaired immune response (zinc, vitamin B_6, protein, energy) and vascular disease via elevated homocysteine levels (folate, vitamin B_6).

7B.10 Risky food patterns

When older people are physically active, marginal food patterns are less likely to lead to problems of the aged, such as:

- Frailty
- Protein energy dysnutrition
- Micronutrient and phytochemical deficiency, because greater amounts of nutritious food can be eaten without positive energy balance
- Chronic metabolic disease (non-insulin-dependent diabetes mellitus [NIDDM], cardiovascular disease, osteoporosis) and certain cancers (breast, colonic, prostate)
- Depression (there is growing evidence that n-3 fatty acid deficiency can contribute to depression in some individuals, and that exercise can alleviate it)
- Cognitive impairment with the apoE4 genotype; excess dietary saturated fat is likely to increase the risk of Alzheimer disease and some antioxidants such as vitamin E and glutathione may reduce the risk.

Specific risky food patterns in later life include:

- Large rather than smaller, frequent meals or snacks because of the inability of insulin reserve to match the carbohydrate load in those proven to have impaired glucose tolerance (IGT) or in those with NIDDM, or because, where appetite is impaired, nutritious snacks can help to avoid chronic energy undernutrition
- Alcohol excess, no alcohol-free days and/or alcohol without food (since food reduces the impact of alcohol ingestion on blood alcohol concentration and its consequences)
- Eating alone most of the time (since social activity encourages interest in food and, usually, healthy food preferences)
- Use of salt or salty food rather than intrinsic food flavour (especially as taste and smell tend to decline with age), as excess sodium contributes to hypertension through an increased Na:K molar ratio, and to salt and water retention in cardiac decompensation, and promotes the loss of urinary calcium
- Problems with availability of bioactive food components of nutritional value.

7B.11 Nutrients at risk of inadequate intake

Protein–energy dysnutrition

Protein–energy malnutrition (PEM) in the aged, as in earlier life, is not only a body compositional disorder but one of general malnutritional complexity. In the aged, there is a decrease in lean mass (comprising water and protein-dominant tissues such as muscle, organs such as liver and also bone) and an increase in abdominal fat. It might better be described as protein–energy dysnutrition (PED). Illness or inadequate food intake may result in PED, a condition more common among elderly adults, especially in institutionalised care. It is associated with impaired immune responses, infections, poor wound healing, osteoporosis/hip fracture and decreased muscle strength (frailty) and is a risk factor for falls in the elderly.

About 16% of elderly people living in the community consume <1000 kcal/day, an amount that cannot maintain adequate nutrition, and up to some 60% in long-term care institutions are malnourished. Being underweight in middle age and beyond places a person at greater risk of death than does overweight. *Marasmus* is a condition of borderline nutritional compensation in which there is marked depletion of muscle mass and fat stores but normal visceral protein and organ function. As there is a depletion of nutritional reserves, any additional metabolic stress (e.g. surgery, infection, burns) may rapidly lead to kwashiorkor (hypoalbuminaemic PEM). However, the presence of oedema in the face of malnutrition in elders is often complicated by other disease and its management. Characteristically, elderly people undergo nutritional deterioration more rapidly than do younger people, so that even minor stress may lead to being less biologically resilient. Recent tissue loss in older people may be under-recognised because of body fatness and fluid retention in the presence of sarcopenia. This is of added concern since it may represent further loss of lean mass and dehydration. It exemplifies the value of clinical nutrition astuteness in complex care.

The protein requirements of older people seem to be similar to, or higher than, those of younger people. The dietary recommendation of protein for older people is slightly higher, at 0.91 g/kg (recommended to be even higher in the USA), than for younger adults, at 0.75–0.8 g/kg. This may not be achievable in the presence of renal impairment or the need to preserve renal function with age. In elderly people with PEM, the interpretation and utility of serum albumin measurements may be fraught without comprehensive patient management and other indices, especially of the commonly associated problems of immunodeficiency and inflammation. These can include anergy (failure to respond to common antigens), autoimmunity and dysregulated immune responses in malnourished elderly persons. The COVID-19 pandemic posed these difficult questions about how best to optimise innate immunity and vaccinate older persons, especially those in dependent situations.

Folate

Case study

Awareness, nutritional literacy, poverty, inequity and food system failure have combined to allow folate deficiency to develop in many settings. An example in institutionalised elderly care in Melbourne, Australia, was the unsolicited switch by a contracted food caterer from fresh citrus fruit juice to a vitamin C orange-flavoured beverage. Within a few weeks (folates stores in the liver can last three months if replete at baseline), there was an unexplained outbreak of folate deficiency disorders. The introduced drink had no folate and the background diet was marginal in folate-rich vegetables. This was salutary on several fronts for the nutritional susceptibility of older dependent people.

Mandatory folate fortification of cereals was introduced in countries like Australia in order to prevent neural tube defects (e.g. spina bifida) among neonates, but without a knowledge of the broader risks in the community at large, including the elderly, or even its association with later life problems. It is considered a success story but one, it might be argued, that was achieved with an ill-understood benefit/risk/cost ratio.

It is the case that many older adults do not consume enough folate. This is compounded by the fact that folate absorption is compromised by atrophic gastritis and elevated plasma homocysteine is consequential for this and other reasons. It is linked with an increased risk of heart disease and strokes. Folate metabolism may also be altered by medication–nutrient interactions, but not as often or as importantly as for other nutrients. The fortification exercise was to reduce NTDs. As it turned out, the companion nutrition education programme to fortification probably achieved as much if not more than the fortification amounts safe for others if not over 1000 mcg.

Could folate fortification have adverse effects on elderly people? Vitamin B_{12} deficiency is quite common in older adults owing to inadequate absorption (see below). Vitamin B_{12} is needed to convert folate to its active form, therefore one of the most obvious vitamin B_{12} deficiency symptoms is the anaemia of folate deficiency. Vitamin B_{12}, but not folate, is also needed to maintain the myelin that surrounds and protects nerve fibres. If folate is consumed via fortified foods without B_{12}, the outcome may be permanent nerve damage and paralysis because folate can mask a vitamin B_{12} deficiency. This underscores safety issues with fortification of the food supply.

Vitamin B_6

Plasma vitamin B_6 falls by approximately 3.6 μmol/l per decade. Age-related changes occur in both the absorption and metabolism of this vitamin and, consequently, aged adults may have a higher requirement. Vitamin B_6 deficiency can result in decreased immune response. Vitamin B_6 deficiencies (as well as vitamin B_{12} and folate) also result in higher concentrations of homocysteine associated with cardiovascular and neurological disease.

Vitamin B_{12}

The prevalence of pernicious anaemia increases with age, as does atrophic gastritis; the absorption of vitamin B_{12} is reduced in individuals with either condition. The prevalence of *Helicobacter pylori* also increases with age and has been shown to be associated with vitamin B_{12} malabsorption, possibly because it contributes to gastric atrophy. Among older adults, this increases the risk of irreversible neurological damage and may contribute to macrocytic and megaloblastic anaemias and to homocysteinaemia associated with vascular disease (see sections on B_6 and folate).

Vitamin D

Older adults are at risk of vitamin D deficiency and at least its adverse effects on bone health (osteopenia and osteoporosis). Risk factors for vitamin D deficiency include:

- Lack of exposure to sunlight (may be due to less physical activity or sunscreen use)
- Decline in renal function
- Impaired skin synthesis (may be due to ageing skin)
- Low fish intake (especially fatty fish)
- Low intake of egg yolks, butter, vitamin D-fortified margarine and cheese.

The diet becomes an important source of vitamin D in people who are not exposed to enough sunlight, and who make less in the skin in later life. The diets of elderly people are often deficient in vitamin D-rich foods such as oily fish, and so fat-soluble vitamin absorption may be impaired. This contributes to vitamin D deficiency in older people. In the USA and the UK, some 30–40% of older patients with hip fractures are vitamin D deficient. However, even in old age, improving vitamin D status can be beneficial for bone health (Zhao et al. 2017). In a Finnish study of outpatients over the age of 85 years and municipal home residents aged 75–84 years, those randomly assigned to receive an annual vitamin D injection had significantly fewer fractures over a five-year follow-up period. In a nursing-home population of 3270 women with an average age of 84 years, randomised to vitamin D (20 μg/day) and calcium (1200 mg/day), those receiving the supplement experienced 43% fewer hip fractures and 32% fewer non-vertebral fractures over an 18-month period (Chapuy et al. 1992).

Vitamin D resistance is also relatively common with impaired renal function in later life, indicated by elevated parathyroid hormone (PTH) concentration in blood, referred to as secondary hyperparathyroidism. Vitamin D is important not only for bone but also for immune function muscle strength, and as a cell differentiator to reduce the risk of neoplastic disease.

Zinc

Zinc plays an important role in wound healing, taste acuity, immune function and albumin turnover in older adults. It is the element in numerous metalloenzymes. Food intake is usually from meat, with limited intake from plant foods, in which it is bound to phytic acid, oxalate and dietary fibre. It is more bioavailable in cereals that are leavened because of the presence of phytase in yeast, which breaks down phytic acid. Older adults may absorb zinc less effectively than younger people and so a diet including zinc-rich foods is important in later life. Zinc deficiency in older people is likely to compromise immune function with a greater risk of infection, particularly respiratory, such as pneumonia. Some of the symptoms of zinc deficiency are similar to the symptoms associated with normal ageing, such as diminished taste and dermatitis; the difficulty is in determining whether to attribute these symptoms to zinc deficiency or simply to the ageing process.

Calcium

Ageing is associated with a decrease in calcium absorption and turnover, partly attributable to alterations in vitamin D physiology. This awareness increases as osteoporosis becomes a practical problem in later life, although a consequence of life-long nutritional and other factors. Postmenopausal women may have higher calcium needs and benefit from intakes as much as 1000 mg/day or more, although these are not realistically achievable from diet alone, and their utility is vitamin D dependent. It is also likely to be food culturally and ethnically dependent. For example, where prevailing food habits do not include dairy and where lactose intolerance is found, the food habits of consequence may be fruits and vegetables with bone health protective properties (e.g. carotenoids and vitamin K) and sodium which, in excess, promotes urinary calcium excretion and loss; and, of course, sunlight exposure will also matter insofar as risk of fracture is concerned.

Despite wide variations in dietary patterns and fracture rates in Europe and elsewhere, there is no clear-cut overall association with calcium intake or benefit of supplement usage.

Magnesium

Magnesium is a major intracellular divalent cation and an essential nutrient involved in calcium and potassium absorption. Its status is difficult to determine and not sufficiently represented by its extracellular presence in blood and urine because it is mainly intracellular. Measurements of turnover demonstrate how sensitive and critical it is to homeostasis judged by blood pressure, nerve conduction, muscle function and insulin resistance which are affected when it is deficient. It has protective properties in inflammatory disease, coronary heart disease, diabetes and stroke. Commonly used medications like diuretics and gastric acid suppressants can cause magnesium deficiency and might therefore increase the risk of osteoporosis, sarcopenia and frailty.

Important food sources include natural chlorophyll (as a commercial food colourant, the magnesium is replaced with copper) in green vegetables, together with seeds, nuts, beans and grains. However, the effective bioavailability of magnesium, reflected in blood measurements, depends on the background diet being biodiverse. This, in turn, affects the extent to which magnesium intake can favourably affect life expectancy (Huang et al. 2015).

Phytonutrients

Phytochemicals (from the Greek *phyton*, meaning plant) are unlike vitamins and minerals in that their nutritional value has been recognised more recently and may not reside exclusively in any one phytochemical. Phytochemicals are numerous and often chemically and physicochemically complex, and not necessarily essential to life in the short to medium term, but conducive to longer term health and well-being. For those that are bioactive and health favourable, the term 'phytonutrient' is preferable (Wilcox et al. 1990).

Phytochemicals are naturally occurring plant secondary metabolites which plants produce to protect themselves against bacteria, viruses and fungi. Many function as antioxidants, which protect cells from the effects of oxidation and free radicals within the body but are, more importantly, multifunctional. Thus, they may offer protection from conditions such as heart disease, diabetes, some cancers, arthritis, osteoporosis and ageing. They are present in frequently consumed foods, especially fruits, vegetables, grains, legumes and seeds, culinary herbs and spices, with some more culturally specific such as soy and its products like tofu and tempeh, in beverages like tea with infusions of *Camellia sinensis*, whether black, green or other, and in chocolate and liquorice.

Phytonutrients may play a protective role in cardiovascular disease and certain hormonally sensitive cancers (breast and prostate). They are likely to lend protection against some disorders associated with ageing. A diet rich in phytoestrogens (isoflavones, lignans) may lessen the symptoms and impact of the menopause by improving vaginal health, reducing the incidence of hot flushes and improving bone mineral content (BMC). Food sources of phytoestrogens include soy, chickpeas, sesame seeds, flax seed (linseed) and olives.

Some phytonutrients of consequence, like the carotenoid lutein (and the related zeaxanthin) (Johnson 2014), are important for retinal health (avoidance of macular degeneration) and probably cognition. They are obtainable from food sources like greens, coloured fruits, yellow corn, marigold flowers used in Mediterranean salads or on which chickens are fed and the yellow colour of their eggs may depend. Gou qi (wolfberries) are also a source, which are both food and medicine in Chinese culture. Culinary herbs and spices which are candidates for reducing cognitive impairment include cinnamon and turmeric (Lee et al. 2014).

Water

Total body water declines with age. As a result, an adequate intake of fluids, especially water, becomes increasingly important in later life, as thirst regulation is impaired and renal function declines. Dehydration is a particular risk for those who may not notice or pay attention to thirst, or who may find it hard to get up at night to make a drink or reach the bathroom. Older people who have decreased bladder control may particularly be at risk because of bathroom access and avoidance of intake.

Dehydrated elderly people appear to be more susceptible to urinary tract infections, pressure ulcers, pneumonia and confusion. Recommended intakes for the elderly are approximately 6–8 glasses of fluid a day, preferably water (Jéquier and Constant 2010; Lukaski et al. 2019).

7B.12 Nutrition-related health problems in the aged

There is growing awareness that major health problems in the aged, and even mortality, have nutritional contributors and can be (in part) prevented by food intake. These health problems do not necessarily need to occur with ageing and death can be delayed. As the number of chronic conditions increases with age, they contribute to disability and frailty, which in turn reduce a person's level of independence, sometimes resulting in institutionalisation. The primary nutritional problems affecting the elderly are:

- Physical inactivity (sedentariness, endurance/aerobic, strength, balance, stretch/dexterity)
- Inflammatory disorders and immunodeficiency
- Body compositional with sarcopenia, osteopenia and fat maldistribution (especially abdominal fatness)
- Cardiometabolic with energy dysregulation, insulin resistance and hypertension
- Neurodegenerative.

Each is amenable to some extent to prevention and management by socioecological attention to the food system. Yet all can contribute to the development of chronic conditions seen with ageing. Some common nutrition-related problems in the aged are outlined below.

Sarcopenia

The condition or state of sarcopenia is the specific involuntary loss of muscle that occurs with age and is more marked in women. Reduced muscle mass and body cell mass are associated with loss of muscle strength, and impaired immune and pulmonary function. The decline in muscle strength is responsible for much of the disability observed in older adults and in the advanced elderly, as muscle strength is a crucial component of walking ability. If levels of body cell mass fall below 60% of the normal levels of young adults, life cannot be sustained. As muscle mass declines, survival prospects also decline (Bachettini et al. 2020).

The prevalence and incidence of sarcopenia are increasing globally with age and sedentariness (10% in men and 13% in women in 2017) and are determinants of health system costs (Lo et al. 2017) and reduced survival (Chuang et al. 2016), but the pathogenesis is poorly understood. Decreasing physical activity and GH levels are likely contributing factors to the development of sarcopenia, along with poor general nutrition, disease and the ageing process. It is difficult to arrest, but physical activity with strength training and dietary pattern are inevitably important. Frequent nutrition counselling where sarcopenia is problematic in diabetes has been demonstrated to interrupt the decline (Chan et al. 2021). Such interventions can be expected to reduce overall health system utilisation and expenditure attributable to sarcopenia (Lo et al. 2017).

Obesity

Overweight and obesity are common in the aged, not because they are an inevitable part of growing older but because of the associated sedentariness. Although a less serious problem in older people than PEM, obesity can impair *functional status*, increase the risk of pulmonary embolus and pressure sores, and aggravate chronic diseases such as diabetes mellitus and hypertension. Greater body fatness, especially if centrally distributed, increases the risk of insulin resistance, hypertension and dyslipidaemia in the aged. In contrast, heavier women have a lower risk of *hip fracture*. This is partly due to 'padding' and better muscle development, but may also be due to maintenance of higher oestrogen levels from the conversion of precursor steroids to oestrogen in adipose tissue. Abdominal obesity is defined as an abdominal circumference of greater than 102 cm for men and 88 cm for women. However, abdominal obesity can be reduced in old age by engaging in daily physical activity. An appropriate body weight is a protective factor in older people with advancing age. Older people tolerate and exhibit *survival advantage with higher BMIs* than do their younger counterparts, but this is partly dependent on *health service availability and expenditure* for associated morbidity (Pan et al. 2012). Body weight maintenance, at a suitable level, is desirable to maintain physical strength and activity, resistance to infection and skin breakdown, and quality of life.

However, in overall health outcomes, cardiometabolic factors are more consequential than body mass.

Immune function

Infections are a common cause of illness and death among the aged. Ageing adults are more susceptible to *infection* and less responsive to *vaccination*. This is probably due, in part, to the age-associated decline in immune function, but this decline may be preventable with good nutrition and physical activity (Dorshkind et al. 2009). A decline in immunity and the development of autoimmune disease may also increase the risk of cancer and arthritis, respectively. The observed decline in immune function with ageing may be prevented with greater attention to dietary patterns or, on occasion, nutrient intakes greater than those currently recommended for 'normal' health. Nutrients important in immune function include protein, zinc, vitamin A, vitamin C, pyridoxine, riboflavin and tocopherols, along with various phytonutrients. Food components not considered to be essential for health in earlier life may become more important with age.

The non-essential amino acid *glutamine* has an important role in DNA and RNA synthesis. It is stored primarily in skeletal muscle and is utilised by intestinal cells, lymphocytes and macrophages. As the contribution of skeletal muscle to whole-body protein metabolism declines with age, the rate of glutamine formation and availability may be impaired. As such, it may compromise immune function, resulting in a suboptimal response to infection or trauma. Glutamine can be synthesised from glutamic acid, which is found in wheat, soybeans, lean meat and eggs. Glutathione (a tripeptide) and phytochemicals, such as flavonoids and carotenoids, from a variety of plant-derived foods play a role in immune function. Meat is a good source of glutathione, with moderate amounts being found in fruits and vegetables. Whey proteins, although low in glutathione, are capable of stimulating endogenous glutathione production.

The coronavirus pandemic of 2019–2021 made it strikingly clear that food and health security, climate, *innate immune function*, proneness to infection and vaccinatability are inseparable. The world's most common and threatening health problems are still the nexus between nutrition and infection. They require socioecological approaches to minimise the risk (Watanabe and Wahlqvist 2020) and ways in which we can personally and collectively identify and address the risks, as much as possible by enhancing our innate immunity by nutritional and other means.

Osteoporosis and fractures

Older age is associated with decreased bone mass, and osteoporosis is one of the most prevalent diseases of ageing. Amongst the ageing population, there is an increase in the incidence of osteoporosis, with women most affected. About 25% of the female population over 60 years is affected by osteoporosis, and 70% of the fractures that occur can be attributed to osteoporosis. *Hip fractures* result in both mortality and morbidity.

Two types of osteoporosis have been identified. Type I involves the loss of trabecular bone (calcium-containing crystals that fill the interior of the bone). Women are more affected by this type of osteoporosis, which is associated with oestrogen deficiency. Type II osteoporosis progresses more slowly than type I and involves the loss of both the cortical (exterior shell of the bone) and the trabecular bone. As the person ages, the disease becomes evident with compressed vertebrae forming wedge shapes with kyphosis, sometimes referred to as a 'dowager's hump'. *Vertebral fractures* and the associated pain and disability are an under-rated cause of suffering among elderly women. Women are more prone to osteoporosis than men for two reasons: bone loss is accelerated after the menopause and women have a lower bone mineral density than men. In Australia, after the age of 60 years, some 60% of women and 30% of men sustain an osteoporotic fracture.

Calcium intake and vitamin D status together prevent or reduce bone loss in postmenopausal women., but the effect on fracture rate is controversial (Chapuy et al. 1992). Other potentially protective factors include fish oil as a source of n-3 fatty acids, vitamins C and K, protein, boron, copper, possibly phytoestrogens and other phytonutrients (Sacco et al. 2013). Soy consumption, a source of phytoestrogens, may provide benefit to bone health. Curiously, high-dose vitamin D supplementation does not reduce non-vertebral fracture rates. Despite the differences in calcium intakes within Europe, this is not an apparent determinant of osteoporosis, which is negatively associated with the dietary diversity which characterises the so-called Mediterranean diet. While nutrition and physical activity can maximise peak bone mass during growth, other factors such as excess sodium, caffeine, smoking and alcohol can accelerate bone loss in later life.

What may also be important is season, since in a temperate climate like Australia, vitamin D status is less good in winter, fracture rates higher and reducible by ergocalciferol supplementation. Proneness to falling is greater in darker wet weather and skin synthesis of vitamin D is compromised. The problem is *lack of sunlight* itself not correctable by diet or nutrient supplements.

Cardiovascular disease

Cardiovascular disease is the most common cause of death and disability in the developed world. Dietary habits may contribute to, or provide protection against, risk factors associated with cardiovascular disease. In a longitudinal health survey of elderly people living in The Netherlands, an inverse relationship was found between *fish consumption* and coronary heart disease mortality. Elevated serum *homocysteine* concentrations have been identified as an independent risk factor for cardiovascular disease. In the Framingham study, elderly adults with better folate status had lower homocysteine concentrations. Inadequate intake of folate, together with vitamins B_6 and B_{12}, can lead to homocysteinaemia and then to vascular damage and proneness to thrombosis.

The *dietary sodium:potassium* molar ratio is not only a determinant of blood pressure but a risk factor for stroke. In north-east Asia where these problems account for much of the disability and mortality, and where there is often resistance to reduction in dietary sodium intakes, increased potassium intakes, especially from fruits, can go some way towards mitigation of this health risk. In so doing, the adverse effects of excess sodium on bone density and risk of gastric cancer are also addressed.

The problem of the dose-related increased risk of stroke with *alcohol* is too frequently ignored, especially as it is a generally greater risk than any benefit there might be for protection against ischaemic heart disease (Chang et al. 2006; Hillbom et al. 2011; Larsson et al. 2011).

Cancer

Dietary patterns that protect against cancer may differ by cancer type, but overall reduction in cancer incidence by diet and physical activity has been possible to distill into guidelines by the World Cancer Research Foundation (WCRF) (www.wcrf.org/wp-content/uploads/2021/02/Summary-of-Third-Expert-Report-2018.pdf) (Table 7B.1). They are quite synchronous with guidelines to prevent the major causes of chronic diseases, as we presently understand them. These would be mainly cardiometabolic, neurodegenerative and musculoskeletal. Indeed, the risk factors for more and more cancers are similar to, say, atherosclerotic vascular disease or diabetes. For example, diabetes is now a major risk factor for neurodegenerative diseases like dementia (Hsu et al. 2011), Parkinson disease (Wahlqvist et al. 2012b) and affective disorders like depression (Wahlqvist et al. 2012a). Diabetes is also a risk factor for several gastrointestinal cancers – colorectal, liver and pancreatic. Physical inactivity is a risk factor for colorectal cancer. Alcohol is a risk factor for breast cancer; so is vitamin D or sunlight deficiency. However, certain food groups are associated with a reduced risk of cancer, for instance a high intake of fruit and vegetables appears to be associated with a reduced risk of cancer at many sites.

Fruit and vegetables are excellent sources of antioxidants, phytochemicals and dietary fibre. Particular foods that may protect against prostate cancer include soy products, tomatoes and pumpkin seeds. Foods high in resistant starch

Table 7B.1 WCRF/AICR recommendations for cancer prevention.

Recommendations	Details	Goals
1. Be a healthy weight	Keep your weight within the healthy range and avoid weight gain in adult life	• Ensure that body weight during childhood and adolescence projects towards the lower end of the healthy adult BMI range • Keep your weight as low as you can within the healthy range throughout life • Avoid weight gain (measured as body weight or waist circumference) throughout adulthood
2. Be physically active	Be physically active as part of everyday life – walk more and sit less	• Be at least moderately physically active, and follow or exceed national guidelines • Limit sedentary habits
3. Eat a diet rich in wholegrains, vegetables, fruit and beans	Make wholegrains, vegetables, fruit and pulses (legumes) such as beans and lentils a major part of your usual diet	• Consume a diet that provides at least 30 g/day of fiber from food sources • Include in more meals foods containing wholegrains, non-starchy vegetables, fruit and pulses (legumes) such as beans and lentils • Eat a diet high in all types of plant foods, including at least five portions or servings (at least 400 g or 15 oz in total) of a variety of non-starchy vegetables and fruit every day. If you eat starchy roots and tubers as staple foods, eat non-starchy vegetables, fruit and pulses (legumes) regularly too if possible
4. Limit consumption of 'fast foods' and other processed foods high in fat, starches or sugars	Limiting these foods helps control calorie intake and maintain a healthy weight	Limit consumption of processed foods high in fat, starches or sugars – including 'fast foods', many prepared dishes, snacks, bakery foods and desserts and confectionery (candy)
5. Limit consumption of red and processed meat	Eat no more than moderate amounts of red meat, such as beef, pork and lamb. Eat little, if any, processed meat.	If you eat red meat, limit consumption to no more than about three portions per week. Three portions are equivalent to about 350–500 g (about 12–18 oz) cooked weight of red meat. Consume very little, if any, processed meat
6. Limit consumption of sugar-sweetened drinks	Drink mostly water and unsweetened drinks	Do not consume sugar-sweetened drinks
7. Limit alcohol consumption	For cancer prevention, it's best not to drink alcohol	For cancer prevention, it's best not to drink alcohol
8. Do not use supplements for cancer prevention	Aim to meet nutritional needs through diet alone	High-dose dietary supplements are not recommended for cancer prevention – aim to meet nutritional needs through diet alone
9. For mothers: breastfeed your baby, if you can	Breastfeeding is good for both mother and baby	This recommendation aligns with the WHO advice which recommends infants are exclusively breastfed for six months, and then up to two years of age or beyond alongside appropriate complementary foods
10. After a cancer diagnosis, follow our recommendations if you can	Check with your health professional what is right for you	• All cancer survivors should receive nutritional care and guidance on physical activity from trained professionals • Unless otherwise advised, and if you can, all cancer survivors are advised to follow the cancer prevention recommendations as far as possible after the acute stage of treatment

[a]AICR, American Institute for Cancer Research; WCRF, World Cancer Research Fund.
Source: Shams-White et al. 2019/MDPI/Licensed under CC BY 4.0.

and dietary fibre may protect against colorectal cancer. Foods that appear to increase the risk of cancer at specific sites include salt, smoked/cured foods (stomach cancer) and alcohol (oesophageal cancer). Factors that occur early in life may affect the risk of breast cancer in later life. For instance, rapid early growth, greater adult height and starting menstruation at a younger age are associated with an increased risk of breast cancer. Although it is unlikely that appropriate interventions could be undertaken to avoid these, other nutritional and behavioural factors are amenable to change and may reduce the risk of breast cancer. These include consuming diets high in vegetables and fruits, avoiding alcohol, maintaining a healthy body weight and remaining physically active throughout life. There is some evidence that phytoestrogens (compounds found in plants that possess mild oestrogenic properties) may reduce the risk of breast cancer. Soy and linseed are two sources of phytoestrogens and Australian food manufacturers have added them to a variety of breads and cereals. Increased prevalence of nutritionally related immunodeficiency with ageing is likely to contribute to the development of neoplastic disease.

The WCRF guidelines for cancer prevention by diet and physical activity have assumed a pattern approach (Table 7B.1) (www.wcrf.org/wp-content/uploads/2021/02/Summary-of-Third-Expert-Report-2018.pdf).

Even for nasopharyngeal carcinoma (NPC) which has a generally understood multifactorial pathogenesis, presumptively genetic (descendants of Pearl River inhabitants, Guandong province, China), viral (EBV) and dietary (small salty fish), its nutritionally protective profile in Taiwan is a dietary pattern. This is phytonutrient rich and plant based (fruits and vegetables), includes milk, fresh fish and eggs, and has tea as a beverage.

Diabetes

Ageing is associated with increased insulin resistance, and more prevalent prediabetes and type 2 diabetes mellitus (T2DM). Risk factors include obesity and physical inactivity. Even in older people, modest weight reduction can improve glycaemic control. In turn, this reduces the likelihood and severity of stroke, cardiovascular disease, visual impairment, nephropathy, infections and cognitive dysfunction (Hsu et al. 2011; Wahlqvist et al. 2012a; Wahlqvist et al. 2012b). In addition, a relatively small reduction in salt and/or improved dietary sodium:potassium molar ratio can lower blood pressure and reduce the risk of stroke with its disability and mortality. Similar benefits accrue from reduced alcohol intake. And as indicated elsewhere, diabetes is a risk factor for several gastrointestinal (GI) tract cancers which would also be expected to have a reduced incidence.

Endocrine function

Several hormonal systems decline with age. A decline in human GH may play a role in the ageing process, at least in some individuals. It does play a role in body composition and bone strength. Oestrogen also drops with age, and this is associated with bone thinning, frailty and disability. Low testosterone levels may weaken muscles and promote frailty and disability. Melatonin responds to light and seems to regulate seasonal changes in the body; as it declines with age, other changes in the endocrine system may be triggered. A decline in dehydroepiandrosterone (DHEA) may affect the immune system and its potential to prevent disease.

Cognitive function

Prevention of cognitive loss, or dementia, poses a particular challenge in older people and for the community. Multifactorial *neurodegeneration and atherosclerotic disease*, with risk factors like hypertension, dyslipidaemia and insulin resistance, are principally involved. Thus preventive approaches through environmental and societal measures and with diet and physical activity are uppermost. Genetic propensities like the apo4 lipoprotein type are recognised and may direct dietary approaches, especially where more prevalent. But walking and immersion in nature, higher educational status and not living alone illustrate how environmental, social and mental stimulation may be neuroprotective, probably in part through activating neurotrophic factors like brain-derived neurotrophic factor (BDNF) (O'Mara 2019), brain metabolism and dampening neuroinflammation.

Thus, it may be possible to delay the onset of poor cognitive function in old age by a number of accessible and affordable measures. Most promising are the several reports that *dietary patterns* characterised by biodiversity and minimal use of refined, fatty and salty foods are protective against cognitive decline with age (Kendig and Morris 2019). These include populations in Europe (Mediterranean diets), North America (MIND and DASH diets), Australia (Mediterranean) and Asia (China and Taiwan) (Cherian et al. 2019; Chuang et al. 2019; Johnson 2014; Lee et al. 2014).

Long-term moderate (subclinical) *nutrient deficiencies* can produce memory impairments. Certain nutrients or toxic substances may directly affect brain development. Water-soluble vitamins and fat-soluble vitamin E and the carotenoid lutein, along with essential n-3 fatty acids, are candidate nutrient deficiencies for attention.

Several of these, if deficient, may elevate homocysteine which may have toxic metabolites. Deficiencies of folacin and B_{12} are especially problematic for the central nervous system (CNS) if not recognised with veganism, atrophic gastritis or pernicious anaemia because giving folacin can exacerbate B_{12} deficiency and precipitate subacute combined degeneration of the spinal cord. Some nutrients like iron, whose deficiency can impair brain function, might also be damaging in excess. Vitamin K may also be protective against cognitive decline and Alzheimer dementia, as certain proteins are dependent on it.

Depression and affective disorders and Parkinson disease occur with varying frequency in the elderly. An increasingly common association with them is prediabetes or diabetes which, if prevented or treated, can be minimised. There is a growing body of evidence to suggest that n-3 polyunsaturated fatty acids may play an important role in the aetiology of depression. Caffeine ingested as either tea or coffee may improve mood. Tea contains theanine which can be a calming agent. Flavonoids in these beverages may also be neuroprotective.

Alcohol is a neurotoxin and associated with neurodegeneration. It also increases the risk of hypertension and stroke. In excess, it causes vitamin B_1 (thiamin) deficiency and the specific brain damage known as Wernicke–Korsakoff psychosis. For this reason, fortification of beer with thiamin has been considered.

Cognitive impairment and dementia can themselves exacerbate nutritional problems. Examples would be forgetting to eat, indifference to food, failure to see the need to eat and behavioural abnormalities, such as holding food in the mouth. Changes in smell and taste with age may also disrupt food choice and patterns, even resulting in anorexia. Weight loss is common in older adults with dementia.

7B.13 Nutritional assessment of the aged

One of the greatest difficulties in making any assessment of the aged is *biological heterogeneity* ('biological age'). There are clearly many health problems seen in the aged in some communities that are not seen in others, making them more age related than ageing; nutritional assessment of the aged needs to pay attention to several sociodemographic variables and the food culture in which the elderly person has lived. Another challenge for nutritional assessment in the aged is the question of *'timing'*; that is, when nutritional factors will have operated during the lifespan and the resulting consequences for health in later life. With these considerations, the areas of nutritional assessment required are historical (if possible) and current for:

- Food and nutrient intake
- Anthropometry and body composition
- Laboratory investigations by way of biochemistry, haematology, and immunology
- Nutritionally related risk factors for various health problems in the aged.

The *Mini Nutritional Assessment (MNA)* is used in community and institutional settings and for comparative studies (Vellas et al. 1999). It comprises:

- Anthropometric measurements (weight, height and weight loss)
- Global assessment (six questions related to lifestyle, medication and mobility)
- Dietary questionnaire (eight questions, related to number of meals, food and fluid intake, and autonomy of feeding)
- Subjective assessment (self-perception of health and nutrition).

The *Malnutrition Universal Screening Tool (MUST)* has application in clinical settings (Stratton et al. 2004).

In Australia, a tool has been developed that identifies older adults at risk of poor nutritional health by giving warning signs with context provided by assessment of general health.

Food and nutrient intake

Assessment of food and nutrient intake is an important tool in health assessment of the aged. To take account of decline in memory, instruments used for food intake assessment should be as simple and practical as possible, and should involve corroboration from other observers, such as family or friends. Knowledge of appetite, the special senses for smell and taste and the overall food patterns facilitates an understanding of the various factors that may affect food intake (Morley 2001; Schiffman and Warwick 1993).

Food and nutrient intake can alert the healthcare worker to *possible nutritionally related disease*. For example, risk of *osteoporosis* could be assessed by asking 'What do you have in the way of dairy products, fish, sesame-based foods [sources of calcium]?, How much salt or salty items [sodium] do you use or eat [sodium increases calcium loss in urine]? What is your sunlight exposure?'

A systematic enquiry about food intake usually requires asking about each episode of eating during the day – main meals and snacks. The notation generated can indicate, for example, the frequency with which a given food or beverage item is consumed on average per week. This, in turn, provides *an estimate of food biodiversity on a weekly basis*. It can be obtained at interview, in consultation or from a food diary. In this way, a time- and cost-effective approach to ascertaining relevant *food intake information in clinical settings* is possible (Wahlqvist 2000).

Anthropometry and body composition

Anthropometry is a simple, non-invasive, quick and reliable form of obtaining objective information about a person's nutritional status (World Health Organization 2021).

Weight

Digital scales are now widely available, but may not be recalibrated regularly. For accuracy, ambulatory elderly persons may be weighed on an upright balance beam scale or microprocessor-controlled digital scale. A movable wheelchair balance beam scale can be used for those elderly who can only sit. A bed scale should be available in geriatric hospitals for measuring the weight of bed-ridden elderly patients. Weights less than 20% of the ideal body weight indicate a significant loss of total body protein, requiring immediate investigation and action. They are associated with reduced tolerance to trauma and an increased risk of morbidity, infection and mortality. Low body weight and/or unintended weight loss are significant risk factors as the ageing process progresses and require careful intervention and monitoring. General guidelines requiring action would be:

- A 2% decrease in body weight in one week
- A 5% decrease in body weight in one month (3.5 kg in a 70 kg man)
- A 7% decrease in body weight in three months
- A 10% decrease in body weight in six months.

Interpretation of the weight of elderly people should be done with circumspection. Increases in body weight may indicate overweight/obesity or oedema. Decreases in body weight can signify the correction of oedema, development of dehydration or emergence of a nutritional disorder.

Height

For the elderly who are agile and without stooped posture, height should be measured in an upright position. When this cannot be done, knee height (using a knee-height calliper) in a recumbent position can be used to estimate stature. The following formulae are used to compute stature from knee height:

$$\text{stature for men} = (2.02 \times \text{knee height}) - (0.04 \times \text{age}) + 64.19$$

$$\text{stature for women} = (1.83 \times \text{knee height}) - (0.24 \times \text{age}) + 84.88$$

The knee-height measurement in these equations is in centimetres, and the age is rounded to the nearest whole year. The estimated stature derived from the equation is in centimetres. These equations are derived from observations which presume that elderly people will have lost some height, an inevitability that may not always continue as healthcare improves.

Arm span

Arm span is another substitute for height and happens to be the same as maximal height achieved. It is sometimes necessary to ask for maximum adult height to be recalled by the subject or by a carer. Gradual reduction in height may be an indicator of vertebral crush fractures due to osteoporosis or it may be due to loss of vertebral disc space.

Mid-arm circumference

Combined with triceps skinfold (TSF), mid-arm circumference (MAC; taken at the midpoint between the acromion and olecranon) can be used to calculate mid-arm muscle area (MAMA), which is an index of total body protein mass. The equation to estimate MAMA is:

$$\text{MAMA} = \left[\text{MAC} - (3.14 \times \text{TSF}/10)\right]^2 / 12.56$$

The MAC measurement in this equation is in centimetres and the TSF is in millimetres. The calculated MAMA derived from the equation is in centimetres squared. MAMA of less than 44 cm^2 for men and less than 30 cm^2 for women may indicate protein malnutrition.

Calf circumference

Calf circumference (taken at the largest circumference using non-elastic flexible measuring tape) in the absence of lower limb oedema can be used to calculate weight in a bed-ridden patient. Several anthropometric measurements, apart from calf circumference itself (calf C), are required to compute weight. They are knee height (knee H), MAC and subscapular skinfold thickness (subsc SF) (taken at posterior, in a line from the inferior angle of the left scapula to the left elbow). There are separate equations for men and women:

$$\begin{aligned} &\text{body weight for men} \\ &= (0.98 \times \text{calf C}) + (1.16 \times \text{knee H}) \\ &\quad + (1.73 \times \text{MAC}) + (0.36 \times \text{subsc SF}) \\ &\text{body weight for women} \\ &= (1.27 \times \text{calf C}) + (0.87 \times \text{knee H}) \\ &\quad + (0.98 \times \text{MAC}) + (0.4 \times \text{subsc SF}) \\ &\quad - 62.35 \end{aligned}$$

All measurements should be in centimetres and the resulting computed weight is in kilograms. Calf circumference is an assessment of lean mass. It can also be used as a measure of physical activity in the aged.

Anthropometric indices

Body Mass Index has been used widely to estimate total body fatness. BMI can be obtained by using the formula BMI = weight (kg)/height (m^2). It can be calculated to help classify whether or not the subject is in the reference range. Interobserver errors are possible. Height and weight having coefficient of variations in the order of less than 1% may be altered by kyphosis in the aged and make interpretation of BMI invalid. Circumferences ratio (AHR) measured at the abdomen (taken at the midpoint between lower ribcage and iliac crest) and hip (taken at the maximal gluteal protrusion) is another anthropometric index used to estimate fat distribution and the one now recommended by the WHO. It is fat distribution reflected in abdominal fatness, which may account for a number of chronic non-communicable diseases in the elderly if the ratio is above 0.9 for men and 0.8 for women.

Several studies are now showing that umbilical measurements alone can be used to safely decide whether weight loss is necessary to reduce the risk from diseases such as heart disease and diabetes. Statistical analyses of umbilical circumferences of Caucasian men and women aged 25–74 years indicated that the ideal circumference for men is less than 102 cm and for women less than 88 cm. These conclusions are drawn from Caucasian subjects and thus may not apply in ethnic groups where the build is slight, such as in many Asian countries, and where a lesser degree of abdominal fatness may still put the person at risk of developing chronic diseases.

Laboratory investigations: biochemistry, haematology and immunology

Biochemical, haematological and immunological assessments are useful to confirm nutritional disorders and identify specific complications that accompany them in the elderly (see also the chapter on metabolic and nutritional assessment in *Clinical Nutrition* in this series).

7B.14 Guidelines for healthy ageing

Sometimes the assumption is made that, after reaching the age of 65 or 70 years, personal behavioural changes no longer confer useful health benefits. Are the remaining years sufficient to reap the benefits of modifications to food choice or exercise patterns? Several recent intervention and survival studies reveal that improvements in nutrition and regular exercise can benefit health, even in advanced old age.

Biodiverse dietary patterns are associated with less cognitive impairment as a precursor of dementia. Older muscles are responsive to strength-training exercises as are young muscles. Nonagenarians show increases in muscle mass, muscle strength and walking speed with weight-training programmes. Chronological age is clearly no justification for deciding whether it is worthwhile to pursue potentially healthful changes in personal habits and environs. Mentally, socially, domestically and physically active lives, and food system engagement from garden to kitchen, provide physiological reserves with resilience and margin for error, making nutritional errors and dysfunction of less consequence.

Being not only physically active but also active in other ways, socially, mentally and occupationally, confers long-term health and survival advantage. In most active settings, food and beverage consumption is involved, and inseparable in the analysis. One of the prime functions of food is as a social facilitator (Glass et al. 1999; Savige et al. 2001; Welin et al. 1992).

Physical activity

The type of physical activity undertaken can play an important role in the health of older people. The two principal forms of physical activity or exercise that are important in promoting health and well-being are endurance/aerobic exercise and strength training. Endurance activities improve heart and lung fitness and psychological functioning, while strength training enhances muscle size and strength, thus limiting muscle atrophy.

The level of physical activity required for older adults to achieve optimal health benefits is guided by practicalities, opportunities, the need for daily activities and any medical restraints. Nevertheless, even as little as 15 minutes a day of moderate physical activity makes a survival difference, while 30–40 minutes realises most benefit (Wen et al. 2011). Separately, being less sedentary is also health advantageous. Resistance (or strength) training reduces lean muscle atrophy more effectively than aerobic activity. Both may improve and maintain psychological functioning in older people, particularly if in like-minded groups. Strength training in older people has the potential to reduce or prevent the decline in muscle mass found with ageing. It can improve walking ability and balance and the associated risk for falls. Strength training also contributes to improved tendon and ligament strength, bone health and improvements in glycaemic status.

The benefits of physical activity, such as strength training, should make activities of daily living easier for older people. Such activities might include climbing stairs, getting out of a chair, pushing a vacuum cleaner, carrying groceries and crossing a road with sufficient speed. Daily endurance activities (e.g. a walk of 30–40 min duration, if necessary in bouts of 8–10 min), along with some strength training are to be encouraged. Added to them can usefully be those of balance, posture, flexibility and stretching (Savige et al. 2001).

Protective foods and food variety

The foremost consideration about nutrition and health is whether there is enough food for basic energy and nutritional needs. Then, food variety serves as an indicator of nutritional adequacy and the nature of health outcomes (Lee et al. 2011; Miyamoto et al. 2019; Wahlqvist et al. 1989). It is also an environmental imperative, and in this way basic to who and what we are – socioecological beings (Wahlqvist 2016). Food variety is invariably linked to food availability, but is not guaranteed by it. Consuming a wide variety of biodiverse foods (especially plant foods) is associated with longevity and a reduced burden of disease. A way to increase variety in the diet is to choose foods from across all major food groups, aiming for a wide selection overall.

Energy and total food intakes tend to decline with age, making a nutritionally adequate diet more difficult to achieve. Older men consume about 800 kcal less than younger men and older women about 400 kcal less than younger women. A reduction in BMR is partly responsible for this decline, especially as muscle mass declines, but physical inactivity is generally the major contributor to reduced food intake. However, men without a partner fare less well nutritionally than men with partners for complex and poorly understood reasons, including limited food system knowledge and skills, and psychosocial dependency.

Compared with younger adults, older adults need at least to achieve similar levels of nutrient intakes, and in some cases higher levels. If these need to be obtained from substantially lower food intakes, a nutrient- and phytochemical-dense diet becomes a high priority in later life. The corollary is that, given the tendency for activity levels and total food intakes to decline with advancing years, there is less dietary room for energy-dense foods (e.g. indulgences, treats) which supply fewer essential nutrients. Therefore, older adults need to be selective about what they eat, in order to avoid excessive fat tissue gain, and to develop a preference for foods that are nutrient dense (e.g. nuts, lean red meat, low-fat dairy products, legumes, seeds). This principle also applies to younger adults who are sedentary.

A traditional food cultural context may, with its familiarity and known benefit–risk–cost profile, provide a measure of food security for the aged. This is one of the arguments for an emphasis on food-based dietary guidelines (FBDGs) in later life. The International Union of Nutritional Sciences (IUNS) Food Habits in Later Life (FHILL) initiative showed that a range of traditional food cultures could be associated with comparable advantageous health outcomes. Food pattern scores have been developed on this basis for European (Mediterranean) and Asian (Taiwanese-TEA) diets which are associated with better within-population survivals and health outcomes like cardiovascular disease and cognitive impairment (Chuang et al. 2019; Chuang et al. 2021; Trichopoulou et al. 1995).

While a biodiverse diet makes essential nutrient intake coverage more likely, the consumption of nutrient-dense foods further reduces the risk of essential nutrient deficiencies. These foods and beverages include:

- Eggs (little, if any, effect on serum cholesterol if not eaten with saturated fat)
- Liver
- Lean meat
- Meat alternatives such as legumes (especially traditional soy products, e.g. tofu and tempeh) and nuts
- Fish
- Low-fat milk and dairy products
- Fruits, vegetables and plant shoots
- Wholegrain cereals
- Wheat germ
- Yeast
- Unrefined fat from whole foods (nuts, seeds, beans, olives, avocado, fish)
- Refined fat from liquid oils (cold pressed, from a variety of sources, predominant in n-3 and/or n-9 fatty acids)
- Tea.

The protective nutritional value of fruits and vegetables is derived particularly from their content of phytochemicals, which are multifunctional compounds, usually of benefit to health, (i.e. antioxidants, antimutagenics, antiangiogenics, immunomodulators, phytoestrogens). Other relevant nutritional advice for older people includes the adequate intake of water and other fluids, but limited alcohol intake, avoidance of sugary beverages and preference for low-salt foods.

The FBDGs published in conjunction with the WHO and FAO encourage the use of traditional foods, dishes and, most importantly, cuisine, making such guidelines more practical and user friendly on an individual level. These principles with a livelihood orientation can be addressed by countries around the world when developing their own country-/culture-specific FBDGs. These considerations can be documented non-invasively and brought together in an index like the HANI (Healthy Aging Nutritional Index) (Huang et al. 2018).

To summarise, the nutritional factors involved in healthy ageing include food variety, and nutrient and phytochemical density. A 'Mediterranean' or food culturally comparable food pattern may also reduce the risk of death in older adults. In the frail elderly, there should be more emphasis on the need for support and increased nourishment, together with the prevention of malnutrition. Early nutritional warning signs of health deterioration in later life include unintentional weight loss and poor appetite. The main message for an older person at home is to be well nourished, to be as active as possible without overdoing it, to eat better, not less, to keep a proportionate weight and to drink plenty of fluids every day.

7B.15 Livelihood and household

As the life cycle progresses from conception to later life, our *social and environmental dependencies* change and evolve. Survival, health, the quality of life and well-being are assigned priorities. The understanding, acceptability, affordability and availability of livelihood allows *a humanistic and sustainable approach* to this prioritisation. *Livelihoods* require the basics of food, potable water, shelter, clothing and textiles, hygiene, healthcare, fuel (energy), education and information, walkability, public open natural space and personal safety. One's *household* is an indicator of how functional, co-operative and

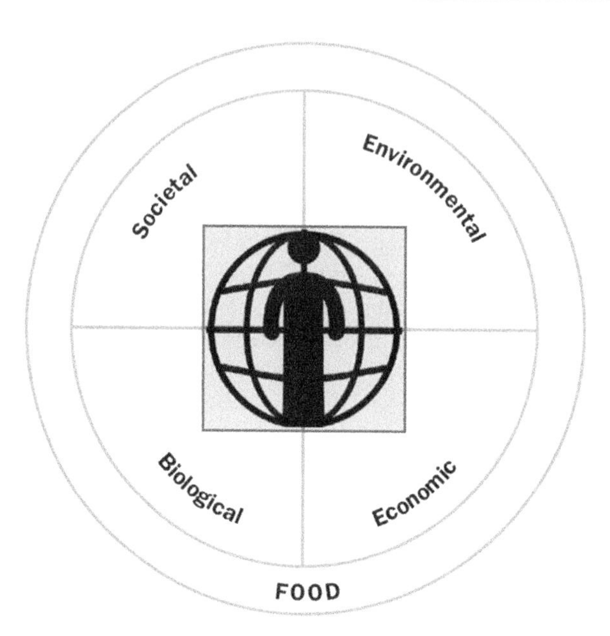

Figure 7B.2 Dimensions of nutrition science. *Source:* Wahlqvist ML and Gallegos D 2020/Taylor & Francis.

secure a livelihood might be; by definition, it has a hearth, a place for eating together (Wahlqvist 2009). Together, livelihood and the household illustrate the multidimensionality and continuum of nutrition as biomedical, environmental, societal and economic, as expounded by the IUNS in its Giessen Declaration (Figure 7B.2). Its underlying tenet is that what we need takes precedence over what we want. Enough is enough in this and if there are to be ensuing generations (Ostrom 2000).

7B.16 Future perspectives

While *maximal lifespan* is mainly genetically determined, the probability of reaching it in good health is determined by environmental, livelihood and personal behavioural factors. Thus, in order to increase our lifespan and associated quality of life, we need to make alterations in the way we live our day-to-day lives, *being as mentally, socially and physically active as practical*. Aside from avoiding harmful pursuits, activities could include continued learning, cooking, socialising, consuming more fruit and vegetables, daily walks and gardening.

With the potential decrease in energy requirements of older age, diet plays an integral part in maintaining health and vitality. The quality of the diet is critical in ensuring that nutritional needs are met. It should comprise nutrient-rich, low-energy-dense natural foods that are more concentrated in dietary fibre, essential fats and other healthful bioactive phytonutrients. The difficulty is that there may be not much room for indulgences in a sedentary elderly person's diet. Even though energy requirements are lower, nutrient needs are the same as, or higher than, those in younger adults. To reduce abdominal obesity and the development of subclinical nutrient deficiencies, older people must choose foods wisely and maintain physical activity as best they can. At the same time, the social role of food should not be neglected.

Of increasing importance is the *cost-effectiveness of healthcare*, especially for older people. Much of this cost arises through disability, mental health disorders and dementia, which are somewhat amenable to improved nutrition, and the use of multiple and costly pharmaceuticals may be reduced by nutritious diets. It is known that dietary diversity in later life is associated with reduced health system costs (Lo et al. 2012a; Lo et al. 2012b; Lo et al. 2016a; Lo et al. 2016b; Lo et al. 2017). A more socioecological approach to nutritionally related health problems across the life cycle is required (Wahlqvist 2014).

The pace of *climate change* is quickening and its mitigation across the life span is a priority of the highest order. The greater and more pervasive risk of socioecological disruption and loss, along with *food, water and health insecurity*, are challenges for nutritional science, food system management and medical biotechnology. Attention to the inevitable *ethical and equity nutritional dilemmas* will also be required.

References

Bachettini, N.P., Bielemann, R.M., Barbosa-Silva, T.G., Menezes, A.M.B., Tomasi, E., and Gonzalez, M.C. (2020). Sarcopenia as a mortality predictor in community-dwelling older adults: a comparison of the diagnostic criteria of the European working group on Sarcopenia in older people. European Journal of Clinical Nutrition 74: 573–580.

Chan, L.C., Yang, Y.C., Lin, H.C., Wahlqvist, M.L., Hung, Y.J., and Lee, M.S. (2021). Nutrition counseling is associated with less sarcopenia in diabetes: a cross-sectional and retrospective cohort study. Nutrition 91–92: 111269.

Chang, H.Y., Hu, Y.W., Yue, C.S. et al. (2006). Effect of potassium-enriched salt on cardiovascular mortality and medical expenses of elderly men. American Journal of Clinical Nutrition 83: 1289–1296.

Chapuy, M.C., Arlot, M.E., Duboeuf, F. et al. (1992). Vitamin D3 and calcium to prevent hip fractures in elderly women. New England Journal of Medicine 327: 1637–1642.

Cherian, L., Wang, Y., Fakuda, K., Leurgans, S., Aggarwal, N., and Morris, M. (2019). Mediterranean-Dash Intervention for Neurodegenerative Delay (MIND) diet slows cognitive decline after stroke. Journal of Prevention of Alzheimer's Disease 6: 267–273.

Chuang, S.Y., Chang, H.Y., Fang, H.L. et al. (2021). The healthy Taiwanese eating approach is inversely associated with all-cause and cause-specific mortality: a prospective study on the Nutrition and Health Survey in Taiwan, 1993–1996. PLoS One 16: e0251189.

Chuang, S.Y., Hsu, Y.Y., Chen, R.C., Liu, W.L., and Pan, W.H. (2016). Abdominal obesity and low skeletal muscle mass jointly predict total mortality and cardiovascular mortality in an ederly Asian population. Journal of Gerontology A 71: 1049–1055.

Chuang, S.Y., Lo, Y.L., Wu, S.Y., Wang, P.N., and Pan, W.H. (2019). Dietary patterns and foods associated with cognitive function in Taiwanese older adults: the cross-sectional and longitudinal studies. Journal of the American Medical Directors Association 20: 544–550. e544.

Dominy, S.S., Lynch, C., Ermini, F. et al. (2019). Porphyromonas gingivalis in Alzheimer's disease brains: evidence for disease causation and treatment with small-molecule inhibitors. Science Advances 5: eaau3333.

Dorshkind, K., Montecino-Rodriguez, E., and Signer, R.A. (2009). The ageing immune system: is it ever too old to become young again? Nature Reviews Immunology 9: 57–62.

Erlich, R., Yngve, A., and Wahlqvist, M.L. (2012). Cooking as a healthy behaviour. Public Health Nutrition 15: 1139–1140.

Frontera, W.R., Hughes, V.A., Fielding, R.A., Fiatarone, M.A., Evans, W.J., and Roubenoff, R. (2000). Aging of skeletal muscle: a 12-yr longitudinal study. Journal of Applied Physiology 88: 1321–1326.

Glass, T.A., de Leon, C.M., Marottoli, R.A., and Berkman, L.F. (1999). Population based study of social and productive activities as predictors of survival among elderly Americans. BMJ 319: 478–483.

Gracey, M., Hetzel BS, R., Strauss, B.J., Tasman-Jones, C., and Wahlqvist, M. (1989). Responsibility for Nutritional Diagnosis: A Report by the Nutrition Working Party of the Social Issues Committee of the Royal Australasian College of Physicians. London: Smith-Gordon.

Hillbom, M., Saloheimo, P., and Juvela, S. (2011). Alcohol consumption, blood pressure, and the risk of stroke. Current Hypertension Reports 13: 208–213.

Hsu, C.C., Wahlqvist, M.L., Lee, M.S., and Tsai, H.N. (2011). Incidence of dementia is increased in type 2 diabetes and reduced by the use of sulfonylureas and metformin. Journal of Alzheimer's Disease 24: 485–493.

Huang, Y.C., Cheng, H.L., Wahlqvist, M.L., Lo, Y.C., and Lee, M.S. (2017). Gender differences in longevity in free-living older adults who eat-with-others: a prospective study in Taiwan. BMJ Open 7: e016575.

Huang, Y.C., Wahlqvist, M.L., Kao, M.D., Wang, J.L., and Lee, M.S. (2015). Optimal dietary and plasma magnesium statuses depend on dietary quality for a reduction in the risk of all-cause mortality in older adults. Nutrients 7: 5664–5683.

Huang, Y.C., Wahlqvist, M.L., and Lee, M.S. (2014). Appetite predicts mortality in free-living older adults in association with dietary diversity. A NAHSIT cohort study. Appetite 83: 89–96.

Huang, Y.C., Wahlqvist, M.L., Lo, Y.C., Lin, C., Chang, H.Y., and Lee, M.S. (2018). A non-invasive modifiable Healthy Ageing Nutrition Index (HANI) predicts longevity in free-living older Taiwanese. Scientific Reports 8: 7113.

Jansen, R., Han, L.K., Verhoeven, J.E. et al.(2021). An integrative study of five biological clocks in somatic and mental health. Elife 10.

Jéquier, E. and Constant, F. (2010). Water as an essential nutrient: the physiological basis of hydration. European Journal of Clinical Nutrition 64: 115–123.

Johnson, E.J. (2014). Role of lutein and zeaxanthin in visual and cognitive function throughout the lifespan. Nutrition Reviews 72: 605–612.

Kendig, M.D. and Morris, M.J. (2019). Reviewing the effects of dietary salt on cognition: mechanisms and future directions. Asia Pacific Journal of Clinical Nutrition 28: 6–14.

Larsson, S.C., Orsini, N., and Wolk, A. (2011). Dietary potassium intake and risk of stroke: a dose-response meta-analysis of prospective studies. Stroke 42: 2746–2750.

Lee, M.S., Huang, Y.C., Su, H.H., Lee, M.Z., and Wahlqvist, M.L. (2011). A simple food quality index predicts mortality in elderly Taiwanese. Journal of Nutrition, Health and Aging 15: 815–821.

Lee, M.S., Wahlqvist, M.L., Chou, Y.C. et al. (2014). Turmeric improves post-prandial working memory in pre-diabetes independent of insulin. Asia Pacific Journal of Clinical Nutrition 23: 581–591.

Liu, C.K., Huang, Y.C., Lo, Y.C., Wahlqvist, M.L., and Lee, M.S. (2019). Dietary diversity offsets the adverse mortality risk among older indigenous Taiwanese. Asia Pacific Journal of Clinical Nutrition 28: 593–600.

Lo, Y.C., Wahlqvist, M.L., Huang, Y.C., Chuang, S.Y., Wang, C.F., and Lee, M.S. (2017). Medical costs of a low skeletal muscle mass are modulated by dietary diversity and physical activity in community-dwelling older Taiwanese: a longitudinal study. International Journal of Behavioral Nutrition and Physical Activity 14: 31.

Lo, Y.T., Chang, Y.H., Lee, M.S., and Wahlqvist, M.L. (2012a). Dietary diversity and food expenditure as indicators of food security in older Taiwanese. Appetite 58: 180–187.

Lo, Y.T., Chang, Y.H., Wahlqvist, M.L., Huang, H.B., and Lee, M.S. (2012b). Spending on vegetable and fruit consumption could reduce all-cause mortality among older adults. Nutrition Journal 11: 113.

Lo, Y.T., Wahlqvist, M.L., Chang, Y.H., and Lee, M.S. (2016a). Combined effects of chewing ability and dietary diversity on medical service Use and expenditures. Journal of the American Geriatric Society 64: 1187–1194.

Lo, Y.T., Wahlqvist, M.L., Huang, Y.C., and Lee, M.S. (2016b). Elderly Taiwanese who spend more on fruits and vegetables and less on animal-derived foods use less medical services and incur lower medical costs. British Journal of Nutrition 115: 823–833.

Lukaski, H.C., Vega Diaz, N., Talluri, A., and Nescolarde, L. (2019). Classification of hydration in clinical conditions: indirect and direct approaches using bioimpedance. Nutrients 11.

Miyamoto, K., Kawase, F., Imai, T., Sezaki, A., and Shimokata, H. (2019). Dietary diversity and healthy life expectancy-an international comparative study. European Journal of Clinical Nutrition 73: 395–400.

Morley, J.E. (2001). Decreased food intake with aging. Journal of Gerontology A 56 (Spec No 2): 81–88.

O'Mara, S. (2019). In Praise of Walking: The New Science of How We Walk and Why It's Good for Us. Random House.

Ostrom, E. (2000). Reformulating the commons. Swiss Political Science Review 6: 29–25.

Pan, W.H., Yeh, W.T., Chen, H.J. et al. (2012). The U-shaped relationship between BMI and all-cause mortality contrasts with a progressive increase in medical expenditure: a prospective cohort study. Asia Pacific Journal of Clinical Nutrition 21: 577–587.

Rockwood, K., Song, X., MacKnight, C. et al. (2005). A global clinical measure of fitness and frailty in elderly people. Canadian Medical Association Journal 173: 489–495.

Sacco, S.M., Horcajada, M.N., and Offord, E. (2013). Phytonutrients for bone health during ageing. British Journal of Clinical Pharmacology 75: 697–707.

Savige, G.W., Wahlqvist, M.L., Lee, D., and Snelson, B. (2001). Agefit: Fitness and Nutrition for an Independent Future. Pan Macmillan.

Sayed, N., Huang, Y., Nguyen, K. et al. (2021). An inflammatory aging clock (iAge) based on deep learning tracks multimorbidity, immunosenescence, frailty and cardiovascular aging. Nature Aging 1: 598–615.

Schiffman, S.S. and Warwick, Z.S. (1993). Effect of flavor enhancement of foods for the elderly on nutritional status: food intake, biochemical indices, and anthropometric measures. Physiology and Behavior 53: 395–402.

Stratton, R.J., Hackston, A., Longmore, D. et al. (2004). Malnutrition in hospital outpatients and inpatients: prevalence, concurrent validity and ease of use of the 'malnutrition universal screening tool' ('MUST') for adults. British Journal of Nutrition 92: 799–808.

Trichopoulou, A., Kouris-Blazos, A., Wahlqvist, M.L. et al. (1995). Diet and overall survival in elderly people. BMJ 311: 1457–1460.

Tzeng, N.S., Chung, C.H., Yeh, C.B. et al. (2016). Are chronic periodontitis and gingivitis associated with dementia? A nationwide, retrospective, matched-cohort study in Taiwan. Neuroepidemiology 47: 82–93.

Vellas, B., Guigoz, Y., Garry, P.J. et al. (1999). The Mini Nutritional Assessment (MNA) and its use in grading the nutritional state of elderly patients. Nutrition 15: 116–122.

Wahlqvist, M.L. (2000). Clinicians changing individual food habits. Asia Pacific Journal of Clinical Nutrition 9 (Suppl 1): S55–59.

Wahlqvist, M.L. (2009). Connected Community and Household Food-Based Strategy (CCH-FBS): its importance for health, food safety, sustainability and security in diverse localities. Ecology of Food and Nutrition 48: 457–481.

Wahlqvist, M.L. (2014). Ecosystem Health Disorders – changing perspectives in clinical medicine and nutrition. Asia Pacific Journal of Clinical Nutrition 23: 1–15.

Wahlqvist, M.L. (2016). Ecosystem dependence of healthy localities, food and people. Annals of Nutrition and Metabolism 69: 75–78.

Wahlqvist, M.L., Lee, M.S., Chuang, S.Y. et al. (2012a). Increased risk of affective disorders in type 2 diabetes is minimized by sulfonylurea and metformin combination: a population-based cohort study. BMC Medicine 10: 150.

Wahlqvist, M.L., Lee, M.S., Hsu, C.C., Chuang, S.Y., Lee, J.T., and Tsai, H.N. (2012b). Metformin-inclusive sulfonylurea therapy reduces the risk of Parkinson's disease occurring with Type 2 diabetes in a Taiwanese population cohort. Parkinsonism and Related Disorders 18: 753–758.

Wahlqvist, M.L., Huang, L.Y., Lee, M.S., Chiang, P.H., Chang, Y.H., and Tsao, A.P. (2014). Dietary quality of elders and children is interdependent in Taiwanese communities: a NAHSIT mapping study. Ecology of Food and Nutrition 53: 81–97.

Wahlqvist, M.L., Lo, C.S., and Myers, K.A. (1989). Food variety is associated with less macrovascular disease in those with type II diabetes and their healthy controls. Journal of the American College of Nutrition 8: 515–523.

Wahlqvist, M.L., Setter, T.L., Savige, G.S., and Kouris-Blazos, A. (2001). Nutritional well-being in the elderly. World Review of Nutrition and Diet 90: 102–109.

Watanabe, S. and Wahlqvist, M.L. (2020). Covid-19 and dietary socioecology: risk minimisation. Asia Pacific Journal of Clinical Nutrition 29: 207–219.

Welin, L., Larsson, B., Svärdsudd, K., Tibblin, B., and Tibblin, G. (1992). Social network and activities in relation to mortality from cardiovascular diseases, cancer and other causes: a 12 year follow up of the study of men born in 1913 and 1923. Journal of Epidemiology and Community Health 46: 127–132.

Wen, C.P., Wai, J.P., Tsai, M.K. et al. (2011). Minimum amount of physical activity for reduced mortality and extended life expectancy: a prospective cohort study. Lancet 378: 1244–1253.

Wilcox, G., Wahlqvist, M.L., Burger, H.G., and Medley, G. (1990). Oestrogenic effects of plant foods in postmenopausal women. BMJ 301: 905–906.

Wolf, D. and Ley, K. (2019). Immunity and inflammation in atherosclerosis. Circulation Research 124: 315–327.

World Health Organization. Physical status: the use of and interpretation of anthropometry. Report of a WHO Expert Committee. www.who.int/childgrowth/standards/technical_report/en/index.html

Zhao, J.G., Zeng, X.T., Wang, J., and Liu, L. (2017). Association between calcium or vitamin D supplementation and fracture incidence in community-dwelling older adults: a systematic review and meta-analysis. JAMA 318: 2466–2482.

Further reading

Angeloni, C., Businaro, R., and Vauzour, D. (2020). The role of diet in preventing and reducing cognitive decline. Current Opinion in Psychiatry 33: 432–438.

Beauman, C., Cannon, G., Elmadfa, I. et al. (2005). The principles, definition and dimensions of the new nutrition science. Public Health Nutrition 8: 695–698.

Benetou, V., Orfanos, P., Pettersson-Kymmer, U. et al.

(2013). Mediterranean diet and incidence of hip fractures in a European cohort. Osteoporosis International 24: 1587–1598.
Bijnen, F.C., Caspersen, C.J., Feskens, E.J., Saris, W.H., Mosterd, W.L., and Kromhout, D. (1998). Physical activity and 10-year mortality from cardiovascular diseases and all causes: the Zutphen Elderly Study. Archives of Internal Medicine 158: 1499–1505.
Brownie, S., Myers, S.P., and Stevens, J. (2007). The value of the Australian nutrition screening initiative for older Australians – results from a national survey. Journal of Nutrition, Health and Aging 11: 20–25.
Campbell, A.J. and Buchner, D.M. (1997). Unstable disability and the fluctuations of frailty. Age and Ageing 26: 315–318.
Chen, R.C., Lee, M.S., Chang, Y.H., and Wahlqvist, M.L. (2012a). Cooking frequency may enhance survival in Taiwanese elderly. Public Health Nutrition 15: 1142–1149.
Chen, X., Huang, Y., and Cheng, H.G. (2012b). Lower intake of vegetables and legumes associated with cognitive decline among illiterate elderly Chinese: a 3-year cohort study. Journal of Nutrition, Health and Aging 16: 549–552.
Chen, Y.C., Huang, Y.C., Lo, Y.C., Wu, H.J., Wahlqvist, M.L., and Lee, M.S. (2018). Secular trend towards ultra-processed food consumption and expenditure compromises dietary quality among Taiwanese adolescents. Food and Nutrition Research 62.
Dalais, F.S., Meliala, A., Wattanapenpaiboon, N. et al. (2004). Effects of a diet rich in phytoestrogens on prostate-specific antigen and sex hormones in men diagnosed with prostate cancer. Urology 64: 510–515.
Dalais, F.S., Rice, G.E., Wahlqvist, M.L. et al. (1998). Effects of dietary phytoestrogens in postmenopausal women. Climacteric 1: 124–129.
Darnton-Hill, I.C., Wahlqvist, E.T., and Ratnaike, M.L. (2001). Assessment of Nutritional Status. A Practical Guide to Geriatric Practice. Sydney: McGraw-Hill.
De Stefani, F.D.C., Pietraroia, P.S., Fernandes-Silva, M.M., Faria-Neto, J., and Baena, C.P. (2018). Observational evidence for unintentional weight loss in all-cause mortality and major cardiovascular events: a systematic review and meta-analysis. Scientific Reports 8: 15447.
Erdman, J.W., Jr., Smith, J.W., Kuchan, M.J. et al. (2015). Lutein and brain function. Foods 4: 547–564.
Fried, L.P. (2016). Interventions for human frailty: physical activity as a model. Cold Spring Harbor Perspectives in Medicine 6.
Frontera, W.R., Hughes, V.A., Fielding, R.A., Fiatarone, M.A., Evans, W.J., and Roubenoff, R. (2000). Aging of skeletal muscle: a 12-yr longitudinal study. Journal of Applied Physiology 88: 1321–1326.
Fu, R., Sun, Y., Zhai, J. et al. (2021). Dietary patterns and sarcopenia in a Chinese population. *Asia Pacific Journal of Clinical Nutrition* 30: 245–252.
Gibson, R.S., Perlas, L., and Hotz, C. (2006). Improving the bioavailability of nutrients in plant foods at the household level. Proceedings of the Nutrition Society 65: 160–168.
Hanna, K.L., O'Neill, S., and Lyons-Wall, P.M. (2010). Intake of isoflavone and lignan phytoestrogens and associated demographic and lifestyle factors in older Australian women. Asia Pacific Journal of Clinical Nutrition 19: 540–549.
Hertog, M.G., Feskens, E.J., Hollman, P.C., Katan, M.B., and Kromhout, D. (1993). Dietary antioxidant flavonoids and risk of coronary heart disease: the Zutphen Elderly Study. Lancet 342: 1007–1011.
Hodgson, J.M., Hsu-Hage, B.H., and Wahlqvist, M.L. (1994). Food variety as a quantitative descriptor of food intake. Ecology of Food and Nutrition 32: 137–148.
Hoffman, R. and Gerber, M. (2012). Virgin olive oil as a source of phytoestrogens. European Journal of Clinical Nutrition 66: 1180.
Hotz, C. and Gibson, R.S. (2007). Traditional food-processing and preparation practices to enhance the bioavailability of micronutrients in plant-based diets. Journal of Nutrition 137: 1097–1100.
Hsu, C.C., Wahlqvist, M.L., Wu, I.C. et al. (2018). Cardiometabolic disorder reduces survival prospects more than suboptimal body mass index irrespective of age or gender: a longitudinal study of 377,929 adults in Taiwan. BMC Public Health 18: 142.
Jernerén, F., Cederholm, T., Refsum, H. et al. (2019). Homocysteine status modifies the treatment effect of omega-3 fatty acids on cognition in a randomized clinical trial in mild to moderate Alzheimer's disease: the OmegAD Study. Journal of Alzheimer's Disease 69: 189–197.
Khaw, K.T. (1997). Healthy aging. BMJ 315: 1090–1096.
Khaw, K.T., Stewart, A.W., Waayer, D. et al. (2017). Effect of monthly high-dose vitamin D supplementation on falls and non-vertebral fractures: secondary and post-hoc outcomes from the randomised, double-blind, placebo-controlled ViDA trial. Lancet Diabetes and Endocrinology 5: 438–447.
Kouris-Blazos, A., Gnardellis, C., Wahlqvist, M.L., Trichopoulos, D., Lukito, W., and Trichopoulou, A. (1999). Are the advantages of the Mediterranean diet transferable to other populations? A cohort study in Melbourne, Australia. British Journal of Nutrition 82: 57–61.
Kromhout, D., Bosschieter, E.B., and de Lezenne Coulander, C. (1982). Dietary fibre and 10-year mortality from coronary heart disease, cancer, and all causes. The Zutphen study. Lancet 2: 518–522.
Lai, Y.H., Leu, H.B., Yeh, W.T., Chang, H.Y., and Pan, W.H. (2015). Low-normal serum potassium is associated with an increased risk of cardiovascular and all-cause death in community-based elderly. Journal of the Formosa Medical Association 114: 517–525.
Lanham-New, S.A. (2008). Importance of calcium, vitamin D and vitamin K for osteoporosis prevention and treatment. Proceedings of the Nutrition Society 67: 163–176.
Lee, M.S., Hsu, C.C., Wahlqvist, M.L., Tsai, H.N., Chang, Y.H., and Huang, Y.C. (2011). Type 2 diabetes increases and metformin reduces total, colorectal, liver and pancreatic cancer incidences in Taiwanese: a representative population prospective cohort study of 800,000 individuals. BMC Cancer 11: 20.
Lee, M.S., Huang, Y.C., and Wahlqvist, M.L. (2010). Chewing ability in conjunction with food intake and energy status in later life affects survival in Taiwanese with the metabolic syndrome. Journal of the American Geriatric Society 58: 1072–1080.
Lee, M.S. and Wahlqvist, M.L. (2005). Population-based studies of nutrition and health in Asia Pacific elderly. Asia Pacific Journal of Clinical Nutrition 14: 294–297.
Lennartsson, C. and Silverstein, M. (2001). Does engagement with life enhance survival of elderly people in Sweden? The role of social and leisure activities. Journal of Gerontology B 56: S335–342.
Lo, Y.L., Hsieh, Y.T. et al. (2017). Dietary pattern associated with frailty: results from Nutrition and Health Survey in Taiwan. Journal of the American Geriatric Society 65: 2009–2015.
Mei, Z., Chen, G.C., Hu, J. et al. (2021). Habitual use of fish oil supplements, genetic predisposition, and risk of

fractures: a large population-based study. American Journal of Clinical Nutrition 114: 945–954.
Murkies, A., Dalais, F.S., Briganti, E.M. et al. (2000). Phytoestrogens and breast cancer in postmenopausal women: a case control study. Menopause 7: 289–296.
National Research Council (1992). Applications of Biotechnology in Traditional Fermented Foods. Washington, DC: National Academies Press.
Nöthlings, U., Schulze, M.B., Weikert, C. et al. (2008). Intake of vegetables, legumes, and fruit, and risk for all-cause, cardiovascular, and cancer mortality in a European diabetic population. Journal of Nutrition 138: 775–781.
Nouchi, R., Suiko, T., Kimura, E. et al. (2020). Effects of lutein and astaxanthin intake on the improvement of cognitive functions among healthy adults: a systematic review of randomized controlled trials. Nutrients 12.
Nurk, E., Refsum, H., Drevon, C.A. et al. (2009). Intake of flavonoid-rich wine, tea, and chocolate by elderly men and women is associated with better cognitive test performance. Journal of Nutrition 139: 120–127.
Olsen, I. (2021). Can porphyromonas gingivalis contribute to Alzheimer's disease already at the stage of gingivitis?. Journal of Alzheimer's Disease Reports 5: 237–241.
Pan, W.H., Wu, H.J., Yeh, C.J. et al.(2011). Diet and health trends in Taiwan: comparison of two nutrition and health surveys from 1993–1996 and 2005–2008. Asia Pacific Journal of Clinical Nutrition 20: 238–250.
Pastorino, G., Cornara, L., Soares, S., Rodrigues, F., and Oliveira, M. (2018). Liquorice (Glycyrrhiza glabra): a phytochemical and pharmacological review. Phytotherapy Research 32: 2323–2339.
Patel, A.V., Friedenreich, C.M., Moore, S.C. et al. (2019). American College of Sports Medicine Roundtable Report on Physical Activity, Sedentary Behavior, and Cancer Prevention and Control. Medicine and Science in Sports and Exercise 51: 2391–2402.
Perissinotto, E., Pisent, C., Sergi, G., and Grigoletto, F. (2002). Anthropometric measurements in the elderly: age and gender differences. British Journal of Nutrition 87: 177–186.
Prince, R.L., Austin, N., Devine, A., Dick, I.M., Bruce, D., and Zhu, K. (2008). Effects of ergocalciferol added to calcium on the risk of falls in elderly high-risk women. Archives of Internal Medicine 168: 103–108.
Psaltopoulou, T., Kyrozis, A., Stathopoulos, P., Trichopoulos, D., Vassilopoulos, D., and Trichopoulou, A. (2008). Diet, physical activity and cognitive impairment among elders: the EPIC-Greece cohort (European Prospective Investigation into Cancer and Nutrition). Public Health Nutrition 11: 1054–1062.
Shafiee, G., Keshtkar, A., Soltani, A., Ahadi, Z., Larijani, B., and Heshmat, R. (2017). Prevalence of sarcopenia in the world: a systematic review and meta- analysis of general population studies. Journal of Diabetes and Metabolic Disorders 16: 21.
Shams-White, M.M., Brockton, N.T., Mitrou, P. et al. (2019). Operationalizing the 2018 World Cancer Research Fund/ American Institute for Cancer Research (WCRF/AICR) Cancer Prevention Recommendations: a standardized scoring system. Nutrients 11.
Shin, H.R., Boniol, M., Joubert, C. et al. (2010). Secular trends in breast cancer mortality in five East Asian populations: Hong Kong, Japan, Korea, Singapore and Taiwan. Cancer Science 101: 1241–1246.
Skidelsky, E.S. and Skidelsky, R. (2012). How Much is Enough? Money and the Good Life. London: Penguin.
Trichopoulou, A., Kouris-Blazos, A., Vassilakou, T. et al. (1995). Diet and survival of elderly Greeks: a link to the past. American Journal of Clinical Nutrition 61: 1346s–1350s.
Wahlqvist, M.L. (2016). Future food. Asia Pacific Journal of Clinical Nutrition 25: 706–715.
Wahlqvist, M.L. (2020a). Self-monitoring networks for personal and societal health: dietary patterns, activities, blood pressure and Covid-19. Asia Pacific Journal of Clinical Nutrition 29: 446–449.
Wahlqvist, M.L. (2020b). Benefit risk and cost ratios in sustainable food and health policy: changing and challenging trajectories. Asia Pacific Journal of Clinical Nutrition 29: 1–8.
Wahlqvist, M.L. (2021). Imagining a habitable planet through food and health. European Journal of Clinical Nutrition 75: 219–229.
Wahlqvist, M.L. and Dalais, F.S. (1997). Phytoestrogens: emerging multifaceted plant compounds. Medical Journal of Australia 167: 119–120.
Wahlqvist, M.L., Day, A.J., and Tume, R.K. (1969). Incorporation of oleic acid into lipid by foam cells in human atherosclerotic lesions. Circulation Research 24: 123–130.
Wahlqvist, M.L., Keatinge, J.D., Butler, C.D. et al. (2009). A Food in Health Security (FIHS) platform in the Asia-Pacific Region: the way forward. Asia Pacific Journal of Clinical Nutrition 18: 688–702.
Wahlqvist, M.L., Lee, M.S., Lee, J.T. et al. (2016). Cinnamon users with prediabetes have a better fasting working memory: a cross-sectional function study. Nutrition Research 36: 305–310.
Wahlqvist, M.L. and Specht, R.L. (1998). Food variety and biodiversity: econutrition. Asia Pacific Journal of Clinical Nutrition 7: 314–319.
Wattanapenpaiboon, N., Lukito, W., Wahlqvist, M.L., and Strauss, B.J. (2003). Dietary carotenoid intake as a predictor of bone mineral density. Asia Pacific Journal of Clinical Nutrition 12: 467–473.
Wilhelmsen, L., Svärdsudd, K., Eriksson, H. et al. (2011). Factors associated with reaching 90 years of age: a study of men born in 1913 in Gothenburg, Sweden. Journal of Internal Medicine 269: 441–451.
Woo, J., Ong, S., Chan, R. et al. (2019). Nutrition, sarcopenia and frailty: an Asian perspective. Translational Medicine of Aging 3: 125–131.
Worsley, A., Wahlqvist, M.L., Dalais, F.S., and Savige, G.S. (2002). Characteristics of soy bread users and their beliefs about soy products. Asia Pacific Journal of Clinical Nutrition 11: 51–55.
Yeh, T.S., Yuan, C., Ascherio, A., Rosner, B., Willett, W., and Blacker, D. (2021). Long-term dietary flavonoid intake and subjective cognitive decline in US men and women. Neurology 97: e1041–e1056.

8

Nutrition, the Brain and the Gut Microbiota

Harriët Schellekens, Mara Loana Iesanu, and Siobhain M. O'Mahony

Key messages

- Brain chemistry and function are influenced by short-term and long-term changes in the diet.
- Glucose is the principal energy substrate of the brain, although ketone bodies become major energy substrates during starvation.
- Dietary intake of carbohydrates does not have a major impact on the availability of glucose to the brain, as blood glucose concentrations and cerebral blood flow are tightly regulated, to ensure the required energy supply to the brain.
- The dietary intake of protein and amino acids (AAs) can influence the uptake of certain AAs into the brain, depending on the properties of AA transporters located at the blood–brain barrier.
- In the brain, AAs are incorporated into proteins and functionally important small molecules, such as neurotransmitters. The production of these AA-derived small molecules is often surprisingly sensitive to the supply of AA precursors from the circulation and thus the diet.
- The gut microbiota, consisting of trillions of microorganisms that inhabit the digestive tract, have been shown to play an essential role in the synthesis and availability of nutrients that can reach the brain, including glucose, glutamate and tryptophan metabolism.
- During early life, optimal nutrition and a healthy microbiota play crucial roles in shaping neurodevelopmental processes and laying the foundation for lifelong cognitive, emotional and behavioural outcomes.
- The gut microbiota needs to be included as a key factor in future studies of the impact of nutrition on the brain.

8.1 Introduction

Nutrition plays a critical role in maintaining overall health, and the brain is no exception. Until the 1960s, it was believed that the brain was protected from nutritional changes, but this idea of a metabolically invulnerable brain has now been replaced by the concept that nutrition has a direct and profound impact on brain function. In recent years, there has been mounting interest in the relationship between nutrition and the brain. Initial studies in the 1970s demonstrated that fairly modest changes in the composition of the diet could directly influence brain function, via the alteration of neurochemistry and the production of key signalling molecules in the brain (the neurotransmitters) (Fernstrom 1977). The diet–gut–brain concept is now receiving increased attention in the nascent disciplines of nutritional neuroscience and nutritional psychiatry.

The impact of nutrition on brain chemistry has also been attributed to the blood–brain barrier (BBB), a network of capillary endothelial cells and basement membrane, neuroglial membrane and glial podocytes (astrocyte projections), which can

Nutrition and Metabolism, Third Edition. Edited on behalf of The Nutrition Society by Helen M. Roche, Ian A. Macdonald, Annemie M.W.J. Schols and Susan A. Lanham-New.

Companion Website: www.wiley.com/go/nutrition/metabolism3e

modulate the flow of circulating molecules into and out of the brain. The BBB contains different types of transport carriers, promoting or restricting the entry of a substantial variety of peripherally derived molecules into the brain, as well as their removal. Indeed, the brain cannot generate energy without a unique, rapid manner of acquiring glucose from the blood. The same scenario arises for other carbohydrates, amino acids (AAs), ketones, vitamins and minerals. They are needed in the brain in amounts that could not be acquired rapidly enough if diffusion governed their rate of uptake into the brain. The solution is that capillary endothelial cells have transport carriers for each of these (and other) nutrients that accelerate their uptake into the brain (relative to diffusion). Furthermore, in many cases, transport carriers have been found to be 'plastic', that is, the properties of specific carriers can change in relation to current demands for the nutrients they transport. A contemporary appreciation of how nutrition influences the brain thus requires an understanding of the nutrient transporters on capillary cells.

Data show that nutrition is essential for optimal human physiology and body composition and play a key role in different brain processes and functions, including cognitive performance, memory retention, mood regulation and mental well-being. Conversely, a poor diet and malnutrition can negatively influence and exacerbate mood disorders as well as stress-related conditions and neuropsychiatric disorders. This realisation is driving more researchers to investigate the potential of nutritional interventions for mental health, neurological and neuropsychiatric disorders. Data from epidemiological studies and population-based observations are beginning to emerge and demonstrate the impact of nutrition on brain health across the life span, from early in neurodevelopment until old age. Current studies, however, do not provide information about causality or underlying mechanisms, and future investigations should include large, randomised control trials and preclinical investigations to elucidate mechanisms.

Importantly, the gut microbiota, which is made up of trillions of microorganisms living in the gastrointestinal system, are well-recognised mediators of the impact of diet and nutrition on gut–brain axis signalling (Cryan et al. 2019). For example, the gut microbiota are involved in the production of the neurotransmitter serotonin from dietary tryptophan, which plays important roles in regulating gut health, including motility. Moreover, microbiota-mediated conversion of tryptophan in the gut is also poised to have an impact on central levels of serotonin, which impact mood, behaviour and cognitive function. In this chapter, we will explore the connection between nutrition and the brain, discussing how key nutrients (e.g. glucose, amino acids, vitamins and minerals) in the diet enter the brain. We will also discuss the importance of these nutrients for neurobiology, the role of nutrition in brain development, and how nutritional deficiencies may impact brain health. We will discuss the emerging role of the gut microbiota in synthesis and availability of nutrients to the brain. By the end of this chapter, readers will have a deeper understanding of how nutrition can reach and impact the brain to optimise brain function and the key role of the gut microbiota in nutrient availability to the brain.

8.2 The gut microbiota in the nutrition–brain connection

Humans and other animals share a mutualistic relationship with resident microorganisms that inhabit the gastrointestinal tract. The community has extensive metabolic capabilities which complement the activity of the liver and the gut mucosa and include functions essential for host digestion, yet we know that the role of the gut microbiota goes beyond digestion alone. Indeed, the human microbiota is now firmly established as a key player in the regulation of metabolism, glucose homeostasis, energy harvesting from nutrition, energy storage and expenditure (Schellekens et al. 2022). One of the primary factors that influences the composition of gut microbiota throughout life is diet. Given this strong connection, and the profound impact of the gut microbiota on digestion and metabolic processes, the diet–gut microbiota interaction also needs to be fully appreciated in the context of the effects of nutrition and diet on the brain.

Interestingly, the gut microbiota has gained recognition as a key factor that can influence brain processes, impacting behaviour. However, only recently has attention been given to how diet affects central nervous system function via microbiota-mediated processes. Various mechanisms for gut-to-brain communication have been identified, including microbial metabolites and immune, neuronal and metabolic pathways, which are potentially modulated by diet. Advancements in understanding the role of the gut microbiota in mediating the effects of diet in this bidirectional communication have been made through animal studies investigating the potential of nutritional interventions on the microbiota–gut–brain axis. With this increased understanding, the concept of 'psychobiotics' has recently emerged, referring to external factors that affect the microbiota (e.g. probiotics, prebiotics, diet) that have positive effects on mental health through bacterial mediation (Anderson et al. 2017). The consumption of Western-style diets – highly processed, high-fat and sugar-rich foods but low in plant foods (high in fibre and beneficial compounds including polyphenols) – has been shown to reduce microbial diversity (e.g. the range of different kinds of unicellular organisms, bacteria, archaea, protists and fungi) and function, leading to the reduction, and sometimes the extinction, of certain beneficial microbes and hence the expansion of opportunistic pathogens and less beneficial microbes. This has significant consequences for human health, and healthy diets can positively modulate gut–brain communication, providing possibilities for the prevention and treatment of common mental disorders. Although there are studies on the impact of supplementation with single food items high in prebiotic fibres, such as fruits and vegetables, the focus on individual foods may not capture the synergistic effects of different dietary components on microbiota diversity and composition, as well as on overall health. Therefore, investigating whole-dietary approaches is a more complete way to develop new dietary interventions that could inform national healthy-eating guidelines and policies. Likewise, the precise interaction of baseline gut microbiota with individual dietary components and microbiota-derived metabolites produced following dietary conversion is only beginning to be elucidated.

Although extensive evidence is accumulating and demonstrating that the gut microbiota plays a key mediating or modulating role in the effect of diet and nutrition on brain function, the precise molecular mechanisms remain unclear. The next 10 years will, no doubt, see a vast expansion in the field of gut microbiota and nutritional neuroscience. In the paragraphs to follow we will discuss examples of nutritional factors that are modulated by, and those that modulate, the gut microbiota, which are important for brain function. We will discuss how they reach the brain to ensure brain health and the potential role of the microbiota herein.

8.3 The brain's energy substrates: a role for the gut microbiota?

The human brain represents approximately 2% of the total body mass but consumes ~20% of glucose-derived energy, making it the main consumer of glucose. As a hydrophilic (water-soluble) molecule, glucose – the primary substrate of the brain – does not easily penetrate lipid membranes and requires the glucose 1 transporter (GLUT1) located in the BBB (Bentsen et al. 2019). The GLUT1 is non-energy-requiring, saturable and competitive, non-concentrative and not insulin-responsive. Using simple Michaelis–Menten kinetics, the K_m (i.e. the substrate concentration at which the reaction rate is 50% of the V_{max}.) for glucose at this transporter has been determined to be about 10 mM, which is moderately higher than normal blood glucose levels (4–6 mM). This means that the carrier is about half-saturated at normal blood glucose concentrations. The V_{max} of the transporter for glucose is 1.4 μmol/min per g of brain, or about 1200 g/day for the entire brain (a human brain weighs about 1400 g).

The human brain consumes 15–20% of the body's oxygen consumption, allowing an estimation of glucose utilisation by the brain to be about 100 g/day. The carrier thus has a capacity for transporting glucose that is well in excess of that demanded daily by the brain. This fact reflects the importance of glucose as the brain's primary fuel. Once glucose is transported across the BBB into the extracellular fluid, it is rapidly taken up into neurons mainly by GLUT3, a glucose transporter on the neuronal cell membrane. This transporter has a glucose affinity several times that of the BBB transporter (GLUT1). Within the cell, the glucose can then enter the glycolytic pathway. The initial enzyme, hexokinase, also has a very high affinity for glucose and is therefore fully saturated at normal brain glucose concentrations (3–4 mM).

The transport of glucose from the circulation to brain neurones is finely tuned to maximise glucose supply for energy production, and each step is optimised to ensure adequate glucose supply. It only fails when the blood glucose supply is abruptly curtailed (e.g. the rapid hypoglycaemia that follows the accidental administration of too high a dose of insulin to a diabetic patient). Blood glucose levels are normally regulated within bounds, except under unusual circumstances, such as starvation and diabetes. Hence, variations in the intake of carbohydrates are not thought to influence the uptake of glucose into the brain or cerebral function. Consuming more or less sugar on any given day does not influence brain glucose uptake. The brain regulates glucose homeostasis, by sensing circulating nutrients (glucose, free fatty acids, amino acids, etc.) and gut hormones (insulin, leptin, ghrelin). These peripheral metabolic signals and nutrients converge on the hypothalamus and brainstem, after they cross the BBB. The brain monitors blood glucose concentrations to ensure adequacy of supply and employs food intake behaviour as one mechanism for regulating blood glucose levels. Hunger ratings are elevated with lower blood glucose levels.

It is interesting to hypothesise that a role for the gut microbiota in glucose sensing and the regulation of eating behaviour may also involve the fine-tuning of glucose homeostasis in the brain. Indeed, two recent proof-of-concept animal studies, using faecal microbiota transfer from obese donor mice, demonstrated that the gut microbiota can alter food preference and affect the motivational components of food intake (de Wouters d'Oplinter et al. 2023). Food intake also has a direct impact on the availability of other nutrients to the brain.

In normal conditions, blood glucose levels do not fall below required levels and this is only seen under extreme environmental circumstances such as starvation. In starvation mode, the brain switches to utilise an additional energy source, ketone bodies. These are synthesised by the liver and are by-products of the breakdown of stored fat (triacylglycerols), providing an extended supply of energy without food-derived energy. The brain will use ketone bodies whenever provided with them (i.e. whenever blood ketone body levels rise to a sufficient level).

The brain participates in the mobilisation of the body's fat stores by increasing the firing rate of the sympathetic nerves that innervate adipose tissue. This causes the fat stores to break down and release non-esterified (or free) fatty acids, which is the initial event in a cascade leading to increased blood ketone body concentrations. The BBB transporter for ketone bodies, required because ketone bodies are hydrophilic compounds that do not readily cross cell membranes, is induced during starvation, further promoting the flow of ketone bodies into brain. This transporter, like the glucose transporter, has a K_m that exceeds the concentrations of circulating ketone bodies that occur during starvation, and a V_{max} well in excess of energy demands. Ketone body delivery to the brain will therefore never be limited by this transporter. During prolonged starvation, more than half of the energy used by the brain is determined by ketone bodies. Continued use of some glucose is mandatory and is supplied via hepatic gluconeogenesis. Moreover, the ingestion of certain foods, such as very-high-fat diets, elevates blood ketone body concentrations, promoting their use by brain as an energy substrate. However, extremely high levels of fat intake are required to produce this effect and such diets are found by most to be unpalatable. Hence, in practice, ketogenic diets are not sought after and are hard to follow for extended periods. Diet is therefore not normally thought to influence cerebral energy production via dietary fat manipulation of ketone body supply to the brain (although it can).

High-fat diets are occasionally used clinically with success to treat intractable seizures, and the effect is linked to elevations in circulating ketone bodies. Although the mechanism of this effect is not understood, one hypothesis is that such diets somehow modify neuronal excitability to reduce seizure occurrence through the provision of this energy source to brain neurons.

Blood glucose concentrations can also fall abruptly following an accidental overdose of insulin administered to a diabetic individual. The importance of maintaining blood glucose levels for the benefit of brain energy production quickly becomes evident, as confusion, delirium, seizures, coma and death occur as blood glucose concentrations drop below 50 mg/100 ml (2.5–3.0 mmol/l), about half normal. This can effectively be reversed by the infusion of glucose, indicating that there is no other compound that can readily substitute glucose as the primary energy source for the brain. When gradual, long-term declines in blood glucose are produced experimentally, the brain adapts by increasing the functioning of BBB transport carriers for glucose.

The BBB hexose transporter is thus somewhat 'plastic'. Indeed, faced with chronic hyperglycaemia, this hexose carrier can also reduce its ability to transport glucose into the brain.

The high energy demands of the brain are due, in part, to the need to generate and maintain neuronal electrochemical gradients across cell membranes. Neuronal depolarisation and repolarisation are high-energy processes with action and resting potentials utilising 21% and 20% of the energy, respectively. When neuronal activity increases, local glucose consumption needs to rise. This increase in glucose is provided by increased blood flow to areas of enhanced neuronal activity. This increase in blood flow is initiated by the release of molecules such as lactate, adenosine, potassium, nitric oxide and prostaglandins into the local brain extracellular fluid by neurons and adjacent glial cells. These molecules cause the smooth muscle cells surrounding the capillaries to relax, dilating local capillaries and increasing blood flow. Within the cerebral cortex, the bulk of signalling energy (50%) is directed towards glutamatergic neurons. Glutamate is the major excitatory neurotransmitter in the brain, and elevated glutamate levels indicate increased excitability. When glutamate is released into synapses, in addition to depolarising adjacent neurons, glutamate interacts with specific receptors located on the membranes of surrounding glial cells (astrocytes), causing them to release many of the same molecules (such as prostaglandins). In the cerebellar cortex, excitatory neurons consume 75% of the signalling energy, while inhibitory neurons consume the remaining 25%. Interestingly, the majority of signalling energy is devoted to the information processing undertaken by non-principal neurons. Purkinje cells, which are principal neurons, utilise only 15% of the signalling energy. Within the cerebellar cortex, the maintenance of resting potentials represents the largest expenditure of signalling energy, accounting for 54% of total usage. Postsynaptic receptors are the next largest consumer of energy, utilising 22%. Presynaptic transmitter release and transmitter recycling account for 5% and 4%, respectively. By contrast, action potentials only use up 17% of the signalling energy within the cerebellar cortex (Bentsen et al. 2019).

Increased blood flow leads to greater glucose extraction to supply the energy needs of both neurons and glia. Glucose can be transported directly from the capillary to the neuron via GLUT1 into the extracellular space and then via GLUT3 transporter into the neuron (Figure 8.1). In addition, glucose is taken up by astrocytes (via membrane GLUT1), metabolised into lactate, and released for immediate use by adjacent glutamatergic neurons. Interestingly, this mechanism seeks to couple glutamate neurotransmission directly to the supply of an energy substrate needed to drive neurotransmission (lactate) and astrocytic glutamate recycling to the neuron. When glutamate is released by neurons, it interacts with receptors on adjacent neurons and is then rapidly cleared, so that the synapse can be reset for the next depolarisation event (see Figure 8.2). Glutamate is removed from the synapse by surrounding astrocytes, which then convert it to glutamine with no activity as a neurotransmitter. The glutamine is then released by the astrocyte and is taken up by the nerve terminal, reconverted to glutamate and reused as a transmitter. The transport of one glutamate molecule into the astrocyte stimulates the uptake of one glucose molecule from local capillaries (the same astrocyte can both surround a portion of an adjacent capillary and enclose a glutamate synapse). The glucose is rapidly converted to lactate, generating for each glucose molecule two ATP molecules to drive (i) the synthesis of one glutamine molecule (requiring 1 ATP), and (ii) the transport out of the cell of the sodium ions that accompanied the uptake of the glutamate molecule (requiring 1 ATP; see Figure 8.2). The lactate is released by the astrocytes and taken up by adjacent glutamate neurons via a membrane transporter specific for monocarboxylic acids (lactate, ketone bodies), where it is converted to pyruvate and used to generate energy in the Krebs cycle. The lactate generated by astrocytes is taken up by adjacent glutamate neurons and used as an energy substrate. Although this model is supported by experimental evidence, it cannot explain how non-glutamate synapses receive an increased flow of energy substrates.

As a further note, although neurons can certainly take up and use lactate as an important source of energy, lactate derives from other cells within the brain (e.g. astrocytes). While lactate can cross the BBB, and a transporter for it (a monocarboxylic acid transporter) is found on capillary endothelial cells, the capacity of the uptake process is not sufficient to supply biologically meaningful

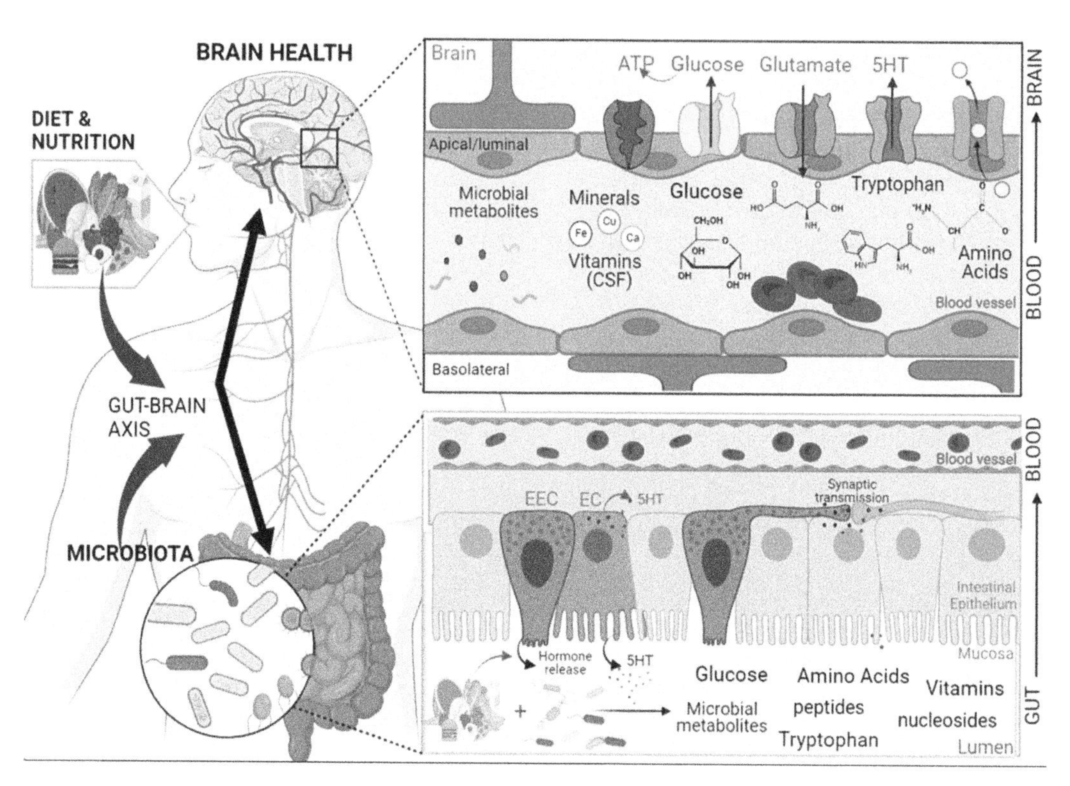

Figure 8.1 Diet and the microbiota, and their interaction influence brain chemistry and function. Diet is metabolized in the gut into key nutrients (e.g., glucose, amino acids, vitamins and minerals) which can enter the blood stream and can be transported into the brain via specific transporters. The gut microbiota plays an essential role in synthesis and availability of nutrients, including glucose, glutamate and tryptophan metabolism. Glucose is the principal energy substrate of the brain and enters the brain from the blood via the glucose 1 transporter (GLUT1). Blood glucose concentrations and cerebral blood flow are tightly regulated, to ensure the required energy supply to the brain. A role for the gut microbiota has been shown in the hepatic glucose production and glucose transporter expression in the intestine. The gut microbiota plays a role in the production of the neurotransmitter serotonin from dietary tryptophan, which plays important roles in regulating gut health including motility. The gut microbiota can also influence 5HT concentrations in the brain through via the microbiota-mediated conversion of tryptophan in the gut. The dietary intake of protein and amino acids (AAs) can influence the uptake of certain AAs into the brain, depending on the properties of AA transporters located at the blood–brain barrier (BBB). Dietary glutamate does not enter the brain, but high levels of glutamate are produced directly in the brain. High levels of glutamate are transported out of the brain. Water-soluble vitamins are transported from blood to brain across the BBB and, in some cases, from CSF to brain, across the blood–CSF barrier. The gut microbiota needs to be included as a key factor in future studies of the impact of nutrition on the brain. CSF, cerebral spinal fluid; EC, enterochromaffin; EEC, enteroendocrine cell; 5HT, serotonin.

amounts to brain neurons from the circulation. Simply put, this means that changes in blood lactate levels are thought to have minimal effects on brain lactate levels or on the use of lactate by brain neurons. Hence, any changes in blood lactate that might be produced by food ingestion or gut microbiota would not be expected to impact on the brain (at least in relation to energy supplies).

Ethanol, another energy substrate, provides approximately 26.8 kJ/g when metabolised by the body, mainly in the liver and intestines. However, it is not clear if the brain metabolises ethanol and if it contributes to its energy supply. Although ethanol is lipid-soluble and can therefore cross the BBB easily, the brain has low levels of alcohol dehydrogenase, which is needed for ethanol metabolism. Therefore, the brain does not significantly derive energy from ethanol metabolism. However, as the by-product of ethanol metabolism – acetaldehyde – is toxic to the brain and is not taken up from the circulation, this may be a protective mechanism. Moreover, ethanol consumption can decrease glucose transport and phosphorylation in the brain, potentially reducing the brain's energy supply. This effect may contribute to the behavioural effects of ethanol, such as central nervous system depression.

Interestingly, the gut microbiota has been implicated in energy and glucose homeostasis. Studies highlight that the gut microbiota can

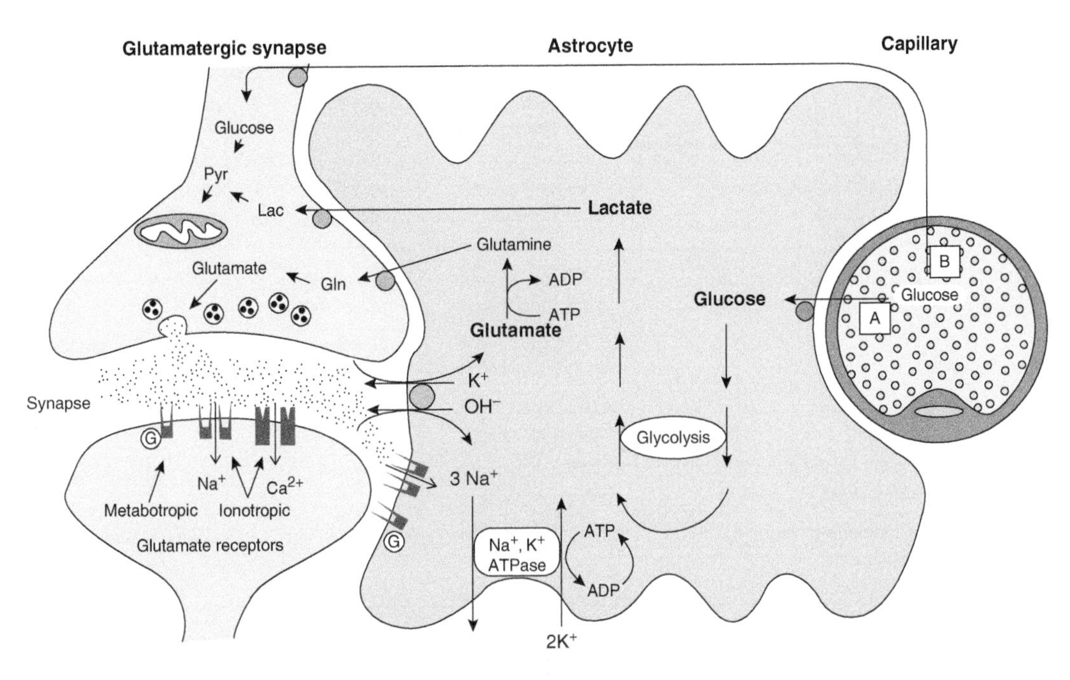

Figure 8.2 Model of the mechanism for glutamate-induced glycolysis in astrocytes during physiological activation. Glutamate release into synapses depolarises postsynaptic neurons by acting at specific receptor subtypes (ionotropic and metabotropic glutamate receptors). The action of glutamate is terminated by an efficient uptake system located primarily in astrocytes. Glutamate is co-transported with sodium ions, resulting in an increase in the sodium concentration in the astrocytes, which activates Na^+, K^+-ATPase. Activation of this enzyme stimulates glycolysis (glucose utilisation and lactate production). The stoichiometry of this process is such that, for one glutamate molecule taken up with three Na^+ ions, one glucose molecule enters the astrocytes, two ATP molecules are produced by glycolysis and two lactate molecules are released. Within the astrocytes, one ATP molecule fuels the extrusion of the Na^+ ions taken up with the glutamate molecule and the other provides the energy required to convert glutamate to glutamine by glutamine synthase. Lactate, once released by astrocytes, can be taken up by neurons and serve as an energy substrate. This model, which summarises in vitro experimental evidence indicating glutamate-induced glycolysis, is taken to show cellular and molecular events occurring during activation of a given cortical area (arrow labelled A, activation). Direct glucose uptake into neurons under basal conditions is also shown (arrow labelled B, basal conditions). Pyr, pyruvate; lac, lactate; Gln, glutamine; G, G-protein. *Source:* Reproduced with permission from Pellerin L, Magistretti J, Glutamate uptake into astrocytes stimulates aerobic glycolysis: A mechanism coupling neuronal activity to glucose utilisation. Copyright © (1994) National Academy of Sciences, USA.

regulate hepatic glucose production via nutrient-sensing mechanisms in the small intestine. The gut microbiota has also been shown to modulate glucose transporters (SGLT1 and GLUT2), as antibiotic-mediated depletion of gut microbiota was shown to upregulate the mRNA levels of these transporters in the small intestine, leading to a subsequent increase in fasting blood glucose (Ota et al. 2022) (Figure 8.1). Future studies will need to investigate the impact of gut microbiota on glucose homeostasis in the brain.

8.4 Transporters, amino acids and the brain

Amino acid availability and transport across the BBB are critically important for human cellular physiology. Neuronal and glial cells in the brain use AAs to produce proteins and use certain AAs for neurotransmission (i.e. glutamate is the major excitatory neurotransmitter in the brain). In addition, transmitter products of AAs such as serotonin (5HT), dopamine and norepinephrine (noradrenaline) are synthesised in the body from precursor AAs obtained from the diet (Figure 8.1). Importantly, diet intake influences the flow of the AAs into the brain. The first step from diet to brain involves absorption by the gastrointestinal tract, followed by insertion into the circulation, and finally extraction by the brain, which involves the BBB and specialised AA transporters. The brain capillary endothelial cells of the BBB contain many different transporters on their cell surfaces, which are responsible for the traffic and balance of AAs in and out of the brain (Table 8.1). These transporters are expressed in a polarised distribution between the luminal (facing

Table 8.1 Blood–brain barrier amino acid transport carriers.

System	Name	Hugo name	Polarity	Substrates
Antiporter, facilitative, Na^+-independent transporters				
System L	LAT1 (amino acid transporter light chain)	SLC7A5 (solute carrier family 7 member 5)	Luminal and abluminal	Large neutral AAs (Phe, Trp, Leu, Ile, Met, His, Tyr, Val, Thr)
System CAT or y^+	CAT1 (cationic amino acid transporter)	SLC7A1 (solute carrier family 7 member 1)	Luminal and abluminal	Cationic AA (Lys, Arg, Orn)
	CAT2	SLC7A2 (solute carrier family 7 member 2)	Luminal and abluminal	Cationic AA (Lys, Arg, Orn)
	CAT3	SLC7A3 (solute carrier family 7 member 3)	Luminal and abluminal	Cationic AA (Lys, Arg, Orn)
System BAT (y^+ LAT2)	y^+ LAT2 (amino acid transporter light chain, y^+ L system)	SLC7A6 (solute carrier family 7 member 6)	Unknown	Arg/Gln exchange
System Xc-	xCT (anionic amino acid transporter light chain, xc- system)	SLC7A11 (solute carrier family 7 member 11)	Luminal	Cys2/Glu exchange
Symporter, active, Na^+-dependent transporters				
System BAT ($ATB^{0,+}$)	$ATB^{0,+}$	SCLC6A14 (solute carrier family 6 member14)	Luminal	Neutral and basic AA
System A	ATA2, SAT2, SNAT2	SLC38A2 (solute carrier family 38 member 2)	Abluminal	Small neutral AA (Ala, Pro, His, Ser, Asn)
System ASC	ASCT1	SLC1A4 (solute carrier family 1 member 4)	Abluminal (Embryonic development)	Ala, Ser, Cys
	AAAT; ASCT2	SLC1A5 (solute carrier family 1 member 5)	Abluminal	Ala, Ser, Cys
System N	SN1; SNAT3	SLC38A3 (solute carrier family 38 member 3)	Luminal & Abluminal	Nitrogen-rich AA: Gln, His, Asn
	SN2; SNAT5	SLC38A5 (solute carrier family 38 member 5)	Unknown	Nitrogen-rich AA: Gln, His, Asn
EAATs	EAAT1; GLAST solute carrier family 1 (glial high-affinity glutamate transporter), member 3	SLC1A3 (solute carrier family 1, member 3)	Luminal	Glu, Asp
	EAAT2; GLAST solute carrier family 1 (glial high-affinity glutamate transporter), member 2	SLC1A3 (solute carrier family 1, member 2)	Luminal	Glu, Asp
	EAAT3; GLAST solute carrier family 1 (glial high-affinity glutamate transporter), member 1	SLC1A3 (solute carrier family 1, member 2)	Luminal	Glu, Asp

Nomenclature adopted from the HUGO Gene Nomenclature Committee (Human Genome Organisation). *Source:* Reproduced and adapted from (Zaragozá 2020) under the Creative Commons CC–BY license.

into the capillary) and abluminal (facing into the brain extracellular fluid) cell membranes, and AAs must cross both membranes to enter or leave the brain. The properties of these transporters dictate how much of each AA enters and exits the brain, and the choice of carrier relates to the size and charge of the AA sidechain. There is significant substrate overlap between the distinct types and classes of transporters, which ensures that there is always more than one transporter facilitating movement of one specific AA between the extracellular fluid and plasma. The polarised distribution and substrate overlap together control the proper balance of AA in the brain, which is key for its proper function (Errasti-Murugarren et al. 2019).

There are several AA transporters involved in transporting AAs into and out of the brain and the majority are found in the luminal membrane, some of which are described in the following (and also in Table 8.1). Facilitative Na^+-independent transporters belong to the SLC7 family, which has 15 members in total, two of which are pseudogenes. The 13 encoded proteins are grouped into the cationic AA transporters (CATs, members SLC7A1–SLC7A4 and SLC7A14), which function as monomers and are N-glycosylated, and the L-type AA transporters (LATs, SLC7A5–11 and SLC7A13), which are non-glycosylated light chain transporters that form heterodimeric AA transporters.

The system L and system y^+ transporters are located in both luminal and abluminal membranes and maintain a bidirectional transport across the endothelial cells. The main transporter in the L system is the LAT1 (SLC7A5) transporter and is primarily responsible for the transport of essential AAs into the brain. LAT1 facilitates the exchange of the large neutral AAs, such as phenylalanine, tryptophan and tyrosine, the AAs that are precursors for neurotransmitters. It is expressed on both the luminal (blood-facing) and abluminal (brain-facing) sides of the BBB. LAT1 functions as a heteromeric AA transporters (HAT), where the light chain (LAT1) interacts with the glycosylated heavy chain transporters from the SLC3 family (4F2hc/SLC3A2 or rBAT/SLC3A1). This large neutral AA transporter is a non-energy-requiring, bidirectional transporter that is saturable and stereospecific. Because the carrier is almost saturated at normal plasma concentrations of these AAs, it is competitive. Hence, changes in the plasma concentration of any one large, neutral AA will affect not only that AA's transport into the brain, but also that of each of its transport competitors. Because glutamine is present in the brain in extremely large concentrations (it is produced there for a variety of uses), this large neutral AA is thought to drive the uptake into the brain of the other large neutral AAs by serving as the principal AA that is counter-transported from brain to blood each time a large neutral AA is taken up into the brain.

The light chain facilitates substrate selectivity and AA transport across the membrane, while the heavy chain is responsible for transporter stability and plays a regulatory role (Fotiadis et al. 2013). The transport process involves the binding of the AA substrate to the transporter, followed by conformational changes that facilitate the translocation of the AA across the membrane.

The system y^+ transporters are also non-energy-requiring, bidirectional, saturable and stereospecific. They are responsible for the competitive transport of arginine, lysine and ornithine.

The symporter transport systems are active transporters and transport AAs against their electrochemical concentration gradient across the brain in a Na^+-dependent fashion. The majority of these active transporters are present exclusively at the abluminal membrane and are responsible for AA efflux from the brain into the endothelial cells. These transporters transport the AA together with a sodium ion and, hence, are co-expressed on the abluminal membrane with the sodium pump Na^+/K^+-ATPase.

The main Na^+-dependent transporters are the system A and system ASC transporters. The system A transporters (SLC38A1, SLC38A2, SLC38A3) are sodium-dependent, and include the sodium-coupled neutral AA transporter 1 (SNAT1). They facilitate the uptake of small neutral AAs, including alanine and serine. They are primarily expressed on the abluminal side of the BBB, allowing the transport of AAs from the bloodstream into the brain.

The Na^+-dependent system ASC transporters, ASCT1 (SLC1A4) and ASCT2 (SLC1A5), facilitate the transport of the small AAs, alanine, serine and cysteine. They are primarily expressed on the abluminal side of the BBB, allowing the uptake of AAs into the brain.

It is important to note that these transporters work in concert to regulate the transport of AAs into and out of the brain. The specific AA transported by each transporter depends on their affinity and expression levels in different regions of the BBB. Together they play an essential role in regulating the balance of dietary AAs and AA-derived neurotransmitters in the brain and maintain overall brain function.

8.5 A balancing act: the gut microbiota and amino acid homeostasis in the brain

The brain utilises AAs to synthesise proteins by mechanisms common to all cells in the body. In adult animals, variations in dietary protein intake have no effects on brain protein synthesis.

This includes chronic ingestion of very low levels of dietary protein and indicates that brain cells are quite efficient in reusing amino acids liberated during intracellular protein breakdown. Low levels of protein intake by neonatal and infant animals, however, are associated with below-normal rates of protein synthesis in the brain. To date, one putative mechanism of this association, namely reduced uptake of essential AAs into the brain and abnormally low brain concentrations of these AAs, has not been proven.

Nevertheless, the availability of precursor AAs in the diet can significantly affect the production and function of the neurotransmitters in the brain. For example, tryptophan is an essential AA that is used to synthesise serotonin, and a lack of tryptophan in the diet can result in decreased serotonin levels and associated neurological and psychological disorders (Figure 8.1). Similarly, tyrosine is necessary for producing dopamine and norepinephrine, and inadequate intake of this AA can lead to reduced levels of these neurotransmitters and their associated symptoms.

The carriers and transporters orientated to move AAs into the brain are those that transport mostly essential AAs (the large, neutral and basic AAs), while those orientated to move AAs out of brain are those transporting non-essential AAs (the acidic and small neutral AAs). A small net influx into the brain of the essential AAs, notably tryptophan, tyrosine, phenylalanine, histidine and arginine, is required for the synthesis of the neurotransmitters that are derived from them (and are ultimately metabolised). The net efflux of the non-essential AAs, notably aspartate, glutamate, glycine and cysteine, may serve as a means of removing AAs that act directly as excitatory transmitters or co-transmitters in the brain. The brain carefully compartmentalises these AAs metabolically, because they can excite (i.e. depolarise) neurons. It is clear why glucose supply to the brain needs to occur without interruption (cutting off the glucose supply has an immediate, drastically negative impact on the brain) and why so many nutrient pools should be buffered from metabolic and dietary vagaries. It is thus somewhat surprising that the transmitter products of some AAs are so vulnerable to diet-induced variations. This rapid diet–brain link is required to monitor protein intake. Therefore, a balanced diet that provides adequate amounts of these precursor AAs is essential for the optimal function of neurotransmitter systems in the body.

Recent evidence shows that the commensal microbiota of the gastrointestinal tract also plays a key role in AA homeostasis. Several bacterial species present in the gut have been shown to utilise dietary AAs as building blocks for microbiota-derived protein synthesis. For example, microbiota in the small intestines have been shown to digest over 50% of threonine and valine. In addition, microbial-produced hydrogen sulphide is primarily produced by the cysteine degradation pathway (Blachier et al. 2019). In addition, gut microbial species secrete extracellular proteases and peptidases, which catabolise dietary protein into specific AAs. This bacterial digestion results in the synthesis of essential AAs and uses fermentation to drive nutrient metabolism (Lin et al. 2017). For example, the AAs glutamine, glutamic acid, lysine, tryptophan and leucine have been reported to be produced by the colonic microbiota phyla of Firmicutes (Matsumoto et al. 2018). Specific AAs have been shown to affect levels of the gastric-derived hormone ghrelin in vivo, with L-glutamine, L-glutamic acid, L-lysine, L-threonine, and L-valine increasing, and L-cysteine, L-tryptophan and L-leucine reducing, ghrelin plasma levels. Preclinical evidence has also demonstrated associations between gut microbiota changes and AA transporter expression. Differences in gut microbiota in a mouse model of hyperuricemia (HUA), a metabolic disorder characterised by abnormal uric acid (UA, a purine) metabolism, were correlated with dysregulation of AA transportation (Song et al. 2022). Moreover, a negative correlation was recently found in human adolescent individuals between the level of circulating branched-chain AAs (BCAA) and bacterial BCAA inward transporter genes, mainly encoded by *Faecalibacterium prausnitzii* (Moran-Ramos et al. 2021). The systemic changes in AAs mediated by the gut microbiota can therefore also influence AA availability in the brain and subsequent alterations in AA-derived neurotransmitters. Specifically, it was shown that gut microbiota can restore the increased excitation/inhibition (E/I) ratio in a mouse model of autism, which was characterised by elevated levels of glutamate in the brain due to higher levels of intestinal AA transport (Yu et al. 2022).

Future studies investigating the interaction of the gut microbiota with diet need to consider how the gut microbiota interacts with diet and utilises dietary AAs. For example, AA supplementation has been shown to modulate the gut microbiota composition and its metabolic activity and may thus also be explored in the context of brain health. A better understanding of the fundamentally unique way in which the brain uses transport carriers to handle the essential AAs that are neurotransmitter precursors, and the non-essential AAs that are neurotransmitters, can be gained by studying an example of each. These examples also reveal important distinctions regarding the impact of diet and microbiota on AA transport into the brain. Examples are tryptophan, a large neutral AA, and glutamate, an acidic AA.

8.6 The impact of diet and the gut microbiota on glutamate neurotransmission

Glutamate is a non-essential AA, but it is also the most abundant AA in the brain. Glutamate is used as the main neurotransmitter by more than half of all nerve terminals in the brain. It is not known why there is such extensive use in neuronal transmission of a molecule of such great metabolic ubiquity. Glutamate is an excitatory neurotransmitter, which means that when it is applied to neurons with appropriate glutamate receptors on their surface, the neurons depolarise and become electrically active. Within nerve terminals, the acidic glutamate is packaged in small vesicles, where it resides until it is released. Once released into the synapse, it stimulates glutamate receptors on adjacent neurons and is then quickly cleared from the synapse (see Figure 8.2). Glutamate receptors fall into two broad categories, ionotropic and metabotropic. Ionotropic receptors are linked to ion channels and, when stimulated, directly alter transmembrane ion fluxes, thereby initiating depolarisation. Metabotropic receptors are linked to intracellular second messengers and, when stimulated, modify signalling pathways (such as those initiated by adenylate cyclase or protein kinases) that precipitate changes in neuronal function.

Dietary glutamate does not enter the brain, except in certain areas like the circumventricular organs, but high levels of glutamate are produced directly in the brain (physiological range is 0.5–2 mM in the brain's extracellular fluid). Nevertheless, when brain levels of glutamate get too high, this can cause neuronal overexcitation and neurotoxicity. Extremely large increases in plasma glutamate concentrations are required before signs of excitotoxicity are evident in the brain (10- to 15-fold greater than would ever be produced by food ingestion), increases that can only be produced by administering extremely large doses of free glutamate. The brain concentration of glutamate is tightly regulated via both facilitative transporters, present on the luminal side, and via sodium-dependent active co-transporters expressed on the abluminal side (Table 8.1). These findings led to the conclusion that an active transporter on the abluminal side releases Glu from cerebrospinal fluid (CSF) into endothelial cells, while facilitative carriers on the luminal side mediate Glu efflux from endothelial cells to plasma (Hawkins and Viña 2016).

Indeed, the luminal system XC- transporter is expressed in border areas of the brain, including endothelial cells, choroid plexus and ependymal cells. It consists of a catalytic subunit xCT (SLC7A11) that heterodimerises with a heavy chain 4F2hc (SLC3A2) and functions as an L-glutamate/L-cystine (Cys2) exchanger for removal of glutamate from endothelial cells to plasma (Burdo et al. 2006). The transporter generates a Cys2 influx in exchange for internal glutamate release. A second mechanism that protects brain neurons from excessive exposure to glutamate once released by other neurons involves the role of glial cells in intracellular glutamate trafficking in the brain (Figure 8.2). In order for information to flow rapidly and accurately in neuronal circuits, neurotransmitters, once released by nerve terminals, must quickly interact with receptors on adjacent neurons and then, just as rapidly, be removed from the synapse (to reset it for the next depolarisation). For glutamate synapses, this process of neurotransmitter removal is accomplished primarily by local glial cells, which have energy-requiring, high-affinity excitatory amino acid transporters (EAATs) on the abluminal membrane, encoded by genes from an SLC1 family. Three EAAT members (EAAT1-3), which are all Na^+-dependent, have been described using in vitro and in vivo models as the most effective transporters for glutamate efflux from the brain (Zaragozá 2020). Once inside the glial cell, the glutamate is quickly converted to glutamine (by glutamine synthetase) and released into the intercellular spaces, from which it is taken back up into neurons, reconverted to glutamate (by glutaminase) and stored for reuse (Figure 8.2). While the efficient removal of synaptic glutamate by glial cells functions primarily in glutamate–glutamine recycling between neurons and glia, it can also serve to remove from the brain extracellular fluid glutamate molecules that may penetrate the BBB. Hence, overall, the BBB transporter for glutamate (and brain glial cell glutamate transporters as a backup system) prevents the glutamate ingested in food from entering the brain or influencing brain glutamate neurotransmission.

It has been suggested that some gut bacteria can produce glutamate via their metabolic pathways while others can break down glutamate into other metabolites. This interaction between gut bacteria and glutamate metabolism can have a significant impact on the overall glutamate levels in the body and brain and, consequently, affect several physiological processes, including digestion, but also cognition and behaviour. Additionally, the gut microbiota influences glutamate receptor expression, which can impact neurological functions such as memory and learning (Dicks 2022). Researchers have proposed that manipulating the gut microbiota through specific dietary interventions or probiotic supplementation could help to regulate glutamate metabolism and improve overall neurological health (Bin-Khattaf et al. 2022). However, further research is needed to understand the mechanisms underlying the gut microbiota's role in glutamate metabolism fully. Finally, glutamate serves as a precursor for the neurotransmitter γ-amino butyric acid (GABA), synthesised in the cytoplasm of the presynaptic neuron following enzymatic conversion by glutamate decarboxylase, utilising vitamin B_6 (pyridoxine) as a co-factor. Interestingly, the gut microbiota has been shown to be capable of secreting not only AAs but also neuroactive amines, including GABA (Mazzoli and Pessione 2016).

8.7 A helping hand: microbiota boosts essential tryptophan

Tryptophan is an essential amino acid that needs to be obtained through dietary sources. Typically, it is found in proteins, and after being absorbed from the intestines and entering the bloodstream, tryptophan is present in both a free form and bound to albumin (Figure 8.1) and is the precursor for the neurotransmitter serotonin (5-hydroxytryptamine, 5-HT). Serotonin is a critical neurotransmitter that profoundly influences behaviour and mood regulation. Its impact extends to emotional well-being, appetite and satiety, sleep patterns, cognitive functions and social behaviour. Imbalances or deficiencies in serotonin function have been implicated in various psychiatric and neurological disorders. Adequate serotonin levels are associated with positive affect and emotional stability, while imbalances or deficiencies have been linked to mood disorders such as depression and anxiety. Serotonin acts on specific receptors in the brain, particularly the serotonin-1A (5-HT_{1A}) receptor, to modulate mood states, emotions and stress responses to promote feelings of well-being. Furthermore, serotonin is involved in the regulation of appetite and satiety. It plays a role in controlling food intake and influencing eating behaviour. Compounds targeting the 5-HT system, including d-fenfluramine, selective serotonin reuptake inhibitors (SSRIs) and 5-HT(2C) receptor agonists, have demonstrated notable reductions in weight gain in rodent models. This effect is closely linked to a significant decrease in food intake, indicating pronounced hypophagia (Halford et al. 2005). Dysregulation of serotonin signalling has been implicated in eating disorders, such as bulimia nervosa and binge-eating disorder.

The neurotransmitter serotonin is also intricately involved in the regulation of sleep. It contributes to the sleep–wake cycle, and disturbances in serotonin function have been associated with sleep disorders, including insomnia (Iesanu et al. 2022). Cognitive functions, including memory, learning and decision-making, are influenced by serotonin. Its receptors, such as the 5-HT_{2A} receptor, are abundantly expressed in brain regions involved in cognitive processing. Altered serotonin levels have been linked to cognitive impairments and disorders such as Alzheimer's disease and schizophrenia. This neurohormone also plays a role in social behaviour and impulse control. There is research evidence that it is involved in regulating aggression, impulsivity and social interactions. Deficiencies have been associated with impulsive and aggressive behaviour, as well as psychiatric disorders characterised by social dysfunction. Serotonin cannot cross the BBB, but tryptophan can cross into the brain by using the LAT1 transporter for amino acids, allowing it to contribute to serotonin synthesis in the central nervous system and exert considerable influence over behaviour and mood regulation in humans. Dietary tryptophan is converted to 5-hydroxy-L-tryptophan by tryptophan hydroxylase 2 (TpH2) in the brain, and by AA decarboxylase into serotonin (Höglund et al. 2019). Serotonin neurons have their cell bodies in small regions (termed raphe nuclei) of the brainstem and mid-brain (at the level labelled 'medulla'), and project their axons extensively throughout the spinal cord and brain. This relatively small population of neurons (less than 1% of the total brain population) project to a wide variety of brain regions and neuronal circuitries.

Interest in this neurotransmitter in the present context derives from the fact that the concentration of tryptophan in the brain rapidly influences the rate of serotonin synthesis in and release from neurons. This relationship holds because the enzyme that catalyses the initial and rate-limiting step in serotonin synthesis (tryptophan hydroxylase) is relatively unsaturated with substrate at normal brain tryptophan concentrations. Hence, a rise in tryptophan concentration leads to an increase in serotonin synthesis as tryptophan hydroxylase becomes more saturated with substrate (causing the rate-limiting reaction to proceed more quickly). A fall in tryptophan concentration causes a reduction in serotonin synthesis as enzyme saturation declines (and the reaction proceeds more slowly). Tryptophan concentrations in the brain are directly influenced by tryptophan concentrations in blood (serum). Hence, a change in serum tryptophan concentration can directly influence serotonin synthesis in brain neurons. However, because tryptophan is transported into the brain by the competitive carrier for large neutral AAs, tryptophan uptake and concentrations in the brain, and thus serotonin synthesis, can also be modified by changing the serum concentrations of *any* of its transport competitors. The serum concentrations of most AAs, including the large neutral AAs, are readily influenced by food intake, thereby forming a key link between diet and serotonin synthesis in the brain.

It is the competitive nature of the large neutral AA transporter (LAT1), however, that explains a seemingly confusing feature of how food intake modifies brain tryptophan concentrations and serotonin synthesis. When fasting animals ingest a meal of carbohydrates, serum tryptophan increases within 1–2 h. This effect is paralleled by a similar increase in brain tryptophan concentrations and a rise in serotonin synthesis (and neuronal release). When animals ingest a meal of carbohydrates to which protein has been added, however, although serum tryptophan rises even more than when carbohydrates are consumed alone, no rise in brain tryptophan concentration or serotonin synthesis occurs. The explanation of this effect involves tryptophan's competitors for transport at the BBB, that is, when carbohydrates alone are consumed, not only does serum tryptophan rise, but the serum levels of most of its transport competitors fall, giving tryptophan a great advantage in competing for available transport sites into the brain. Brain tryptophan uptake and concentration therefore increase. However, when a protein-containing meal is consumed, even though serum tryptophan rises considerably, the serum concentrations of its competitors not only fail to decline but rise, and by a proportionally similar amount to tryptophan (because they are present in the dietary protein). Consequently, tryptophan experiences no change in its competitive standing for access to available transporters into the brain, and brain tryptophan concentrations do not change. These effects of food on the serum concentrations of tryptophan and the other large neutral AAs, in relation to competitive tryptophan transport at the BBB, can be summarised with a single expression of this competition, the serum tryptophan ratio (the serum concentration of tryptophan divided by the summed concentrations of its transport competitors). Following carbohydrate ingestion, the serum tryptophan ratio rises; after a protein meal, it does not. A key assumption of this simple model is that it applies to all dietary proteins. Evidence has shown that the protein ingested with carbohydrates has a tremendous effect on the post-meal changes in the serum levels of tryptophan and its transport competitors, and thus on post-meal brain tryptophan levels and serotonin synthesis. Indeed, the impact of the meal on brain tryptophan and serotonin depends on how it changes the serum concentrations of tryptophan relative to those of its competitors. It is worth noting that since the amount of serotonin released by neurons tracks the rate of serotonin synthesis closely, the findings suggest that the ingestion of different proteins in a meal may produce different functional effects in the brain.

Altogether, such findings indicate that dietary proteins influence brain tryptophan and serotonin more than dietary carbohydrates, with some proteins causing increases that exceed those seen with carbohydrates. Indeed, carbohydrate effects seem modest in comparison. What might be the utility to the brain of having the function of a neurotransmitter so readily manipulated by dietary protein in a meal? Changes of the magnitude observed are not seen with any other transmitter (as far as is presently known). One possibility is that the effect provides a signal to the brain regarding dietary protein quality, which might be used to help the brain direct behaviour to ensure the intake of sufficient high-quality protein by an animal to maximise growth rate. It is likely that this signal is not the only cue informing the brain about the quality or quantity of ingested protein. For example, glutamate is the dominant AA in almost all dietary proteins (about 10% by weight). When proteins are hydrolysed during digestion, glutamate receptors in the gut, which are linked to sensory nerves, are stimulated. These nerves run into the central nervous system and may provide the brain with an assessment of the total amount of protein ingested (independent of its quality). If this is the case, the tryptophan and serotonin increases produced by a carbohydrate meal would not fool the brain into thinking that a high-quality protein had been ingested: the absence of a glutamate signal would indicate that no protein had been consumed. Such ideas suggest interesting hypotheses for future experimentation. The broader question is: does the brain assess – and if so, how – the nutritional adequacy of ingested matter to optimise growth and maintain optimal body function?

This example considers the rapid effects of food on tryptophan and serotonin. Longer-term changes are also observed. Generally, growth rate follows the diet's effects on brain tryptophan and serotonin. Given this fact, one might imagine that the brain of a rat containing low tryptophan and serotonin due to the ingestion of a low-quality protein would influence food-seeking behaviour to identify and consume foods containing better quality proteins. In addition, when animals consume diets low in protein (any protein) for extended periods, the plasma concentrations of all essential AAs decline, including those of tryptophan and the other large neutral AAs. In this case, even though the decline in plasma tryptophan may be proportionally similar to that of its transport competitors (i.e. competition for the brain transporters does not change), brain tryptophan and serotonin decline. The most likely explanation for this effect is that the transport carriers, which are about 95% saturated when normal amounts of protein are being consumed, become much less saturated at low levels of protein intake because of the decline in the plasma concentrations of all large neutral AAs. At low carrier saturation, competition ceases and simple changes in plasma tryptophan concentrations suffice to predict brain tryptophan uptake and concentrations. In such cases, tryptophan–serotonin signalling would not accurately communicate the quality of the dietary protein being ingested (if, indeed, that is what it does) and other signals would indicate that protein intake is low (e.g. if the glutamate signal considered earlier is active) and direct protein-seeking behaviour.

Other large neutral AAs serve as neurotransmitter precursors in substrate-driven pathways in the brain. Phenylalanine and tyrosine are substrates for the synthesis of the catecholamines (dopamine, norepinephrine, epinephrine), and histidine is the precursor of histamine. Like tryptophan, the brain concentrations of these AAs are directly influenced by their uptakes from the circulation, which in turn reflect the plasma concentrations of all the large neutral AAs. For these transmitter precursors as well, the influence of diet on their brain concentrations reflects the impact the diet has on the plasma concentration of each AA in relation to those of its competitors. Presently, too little experimental information is available for these transmitters to conjecture how they might be linked to dietary protein or carbohydrate intake or the possible sensing of the intake of these macronutrients.

Another mechanism that regulates brain serotonin levels involves selective serotonin reuptake inhibitors (SSRIs), which increase serotonin availability in the neuronal synaptic space. These are commonly used as antidepressant medications and enhance serotonin neurotransmission by blocking the reuptake of serotonin, thereby maintaining higher levels of the neurotransmitter in the brain (Stahl 1998). This pharmacological approach supports the notion that serotonin imbalance contributes to mood disorders.

It is noteworthy that most of the tryptophan to serotonin conversion occurs in the enterochromaffin cells of the gastrointestinal tract. Enterochromaffin cell activation induces voltage-gated

5-HT release, activating 5-HT receptors and activating sensory afferent neurons. leading to direct modulation of sensory neurons (Bellono et al. 2017). Here, the gut microbiota represents the final link between diet and serotonin availability in the brain (Legan et al. 2022). For example, specific microbial-derived metabolites (SCFAs, secondary bile acids, α-tocopherol, tyramine and p-aminobenzoate) can promote transcription of tryptophan hydroxylase 1 (Tph1), the rate-limiting serotonin biosynthetic enzyme, impacting downstream serotonin synthesis (Legan et al. 2022). Indeed, as the gut microbiota utilises dietary tryptophan, this influences its availability across the gut–brain axis. While the impact of diet on brain tryptophan and serotonin levels has been significantly investigated, the involvement of the gut microbiota in tryptophan and serotonin metabolism is only beginning to be understood.

8.8 Other nutritional components and the brain: a role for gut microbiota?

Fatty acids

Unlike other tissues in the body, the brain does not appear to use fatty acids directly as energy substrates. However, fatty acids are required by the brain on a continuing basis for the synthesis of the complex fat molecules that are used to construct neuronal and glial cell membranes, and of cellular signalling molecules (e.g. prostaglandins). Membrane construction is particularly active in infant and growing animals, although it is also active in adults (due to membrane turnover). The brain is thought to synthesise some fatty acids from smaller molecules, but uptake from the circulation is also considered to be an important, possibly the principal, source of most fatty acids found in the brain. (Indeed, the circulation is the ultimate source of all essential fatty acids in the brain.) In the blood, fatty acids circulate either as components of fat molecules or as non-esterified fatty acids. Some evidence suggests that lysophatidylcholine can be taken up into the brain, but non-esterified fatty acids are presumably the primary form in which fatty acids gain access to the brain. Indeed, uptake into the brain of both essential and non-essential fatty acids has repeatedly been demonstrated. The details of the uptake process are not presently understood, however, and fatty acid uptake presents certain conceptual problems. For example, the fact that fatty acids bind tightly to serum albumin suggests they might not be readily available for exchange with a transport carrier located at the BBB, and diffusion across capillary endothelial cell membranes may be limited, as fatty acids are almost completely ionised at physiological pH. At present, there is no generally accepted model of fatty acid transport to explain the available data for uptake into the brain and retina. Current hypotheses include multiple variants of the possibilities that fatty acids or albumin–fatty acid complexes bind to receptor/transporter sites on endothelial cells and are moved across the membrane into the cell interior, either as the free fatty acid or as the complex, or that fatty acids diffuse in the non-ionised form across cell membranes.

Whatever the ultimate mechanism of transport proves to be, from the nutritional perspective, diet does influence essential fatty acid availability to brain, with potentially important functional consequences. In almost all mammals (cats are an exception), there are two long-chain fatty acids that cannot be synthesised and which are thus essential in the diet. These are the polyunsaturated fatty acids linoleic acid and α-linolenic acid. In the body (including the nervous system), linoleic acid is converted into arachidonic acid, a key precursor in the synthesis of prostaglandins and leukotrienes, families of important second messenger molecules. α-Linolenic acid is converted into docosahexaenoic acid, a molecule found in very large amounts in the outer segments of rods and cones in retina, and in nerve terminal membranes in the brain. Docosahexaenoic acid is thought to be a key component of phototransduction cells and thus may be important in vision. The brain takes up both linoleic acid and α-linolenic acid, as well as arachidonic acid and docosahexaenoic acid. The calculation has been made that 3–5% of the arachadonic acid and 2–8% of the docosahexaenoic acid in adult (rat) brain turns over daily, which must be resupplied from the blood. As both the essential fatty acids (linoleic acid and α-linolenic acid) and their products (arachidonic acid and docosahexaenoic acid) readily enter the brain from the blood, all may contribute to the resupply effort. However, the relative contribution of each is not presently known, although studies in the developing brain suggest that, from the dietary perspective, dietary docosahexaenoic acid may be more important as a source for brain docosahexaenoic acid pools than dietary α-linolenic acid (see later), particularly when linoleic acid intake is high enough to retard the conversion of α-linolenic acid to docosahexaenoic acid.

Inside the nervous system, linoleic and α-linolenic acid are incorporated into phospholipid molecules and inserted into cellular membranes. Typical phospholipids, which contain two fatty acid molecules, contain one saturated and one unsaturated fatty acid. In retinal rods and cones, where the docosahexaenoic acid content is extremely high, both fatty acids in phospholipids can be unsaturated. Because they contain double bonds, essential fatty acids influence membrane fluidity and membrane-associated functions (e.g. the functionality of receptors, transporters and other membrane-embedded molecules). Arachidonic acid is a constituent of lipids in all cells, including those in the nervous system. It is released into the cytoplasm by phospholipase molecules that are activated by the occupancy of membrane receptors (of a variety of types). Once released, arachidonic acid is converted into prostaglandins and leukotrienes. These molecules influence cellular responses to second messenger molecules, such as cyclic adenosine monophosphate (cAMP), and are also released into the extracellular fluid, where they influence the functions of other cells. A dietary modification in essential fatty acid intake might therefore be expected to influence membrane functions in the brain (and elsewhere), leading to alterations in brain function.

Thus far, the most unambiguous studies linking the ingestion of an essential fatty acid to specific CNS functions have been those involving α-linolenic acid and docosahexaenoic acid, with impacts on brain myelination and vision. Unlike dietary amino acids, which can produce rapid changes in nervous system function (i.e. in hours), changes in the intake of α-linolenic acid require weeks to produce clearly observable functional effects. In addition, such effects are produced by restricting intake (producing a deficiency) or by restoring the essential fatty acid to the diet of deficient animals (i.e. correcting a deficiency; no effects are seen by increasing intake in normal animals). The functional effects of restricting α-linolenic acid intake are typically seen when applied to animals during the late gestational and early postnatal periods, times when neurons are developing axonal and dendritic extensions, and forming nerve terminals and synapses, and glial cells are proliferating and developing their variety of membrane structures (Gould et al. 2013). This also corresponds to the period when docosahexaenoic acid is most actively accumulating in the CNS. While it may ultimately be shown that vagaries in essential fatty acid intake influence brain functions at other times of life, current evidence most strongly links essential fatty acid deficiency to the developmental period of the life cycle.

Dietary fat also has a modulating effect on the composition of intestinal microbiota, which has been linked to obesity-associated low-grade inflammation. In addition, the gut microbiota metabolises dietary polyunsaturated fatty acids (Miyamoto et al. 2019), thereby affecting availability to the brain. The gut microbiota is also the main producer of short-chain fatty acids with a plethora of functional effects on gut–brain axis signalling and with significant impact on physiology as well as brain function and behaviour.

Choline

Choline is a water-soluble nutrient essential for human life and occurs in the body as: (i)a constituent of lipid molecules (phosphatidylcholine, lysophophatidylcholine, sphingomyelin, choline plasmalogen) that are building blocks of cell membranes; (ii) a source of methyl groups; and (iii) a precursor for the neurotransmitter acetylcholine. Choline is not considered to be an essential nutrient, at least in humans, as the body can synthesise it. Moreover, deficiencies are rarely seen, as it is ubiquitous in the diet. However, in recent decades, dietary choline has been a focus of interest in relation to brain function because of the possibility that changes in choline intake could influence neuronal acetylcholine synthesis. Acetylcholine is a neurotransmitter in neurons in the brain and in the periphery. Indeed, ample evidence indicates that acetylcholine synthesis in and release by brain cholinergic neurons is influenced by neuronal choline availability, and that neuronal choline concentrations can be altered by dietary choline intake, in the form of either free choline or phosphatidylcholine. In addition, oral choline and phosphatidylcholine have found some application in human disease and brain functions thought to involve cholinergic neurons.

For example, they have been used successfully to treat movement disorders such as tardive dyskinesia, a drug-induced muscular disorder seen in patients with schizophrenia. Cholinesterase inhibitors, which raise acetylcholine levels by inhibiting the transmitter's metabolism, have been used with success to reduce the uncontrollable muscle movements associated with antipsychotic treatment in these patients, and choline ingestion has also been found to produce some beneficial effect. Choline and phosphatidylcholine, however, have proved to be of little value in controlling the muscle movements associated with Huntington's disease. Dietary choline and phosphatidylcholine supplements have also been studied as potential memory enhancers, based on the notion that acetylcholine neurons in the hippocampus have an important role in memory and that enhancing acetylcholine levels might improve memory. (Memory is sometimes improved when acetylcholine levels are raised by administering a cholinesterase inhibitor.) Patients with Alzheimer's disease have been most studied and, in general, the disappointing outcome has been that neither choline nor phosphatidylcholine has offered much improvement in memory.

The production of choline deficiency in experimental animals (dietary choline deficiency in humans is rare) has been reported sometimes to reduce the brain content of choline, choline-containing lipids and acetylcholine. As this has not been a uniform finding, it is not possible at present to state whether or not occasional occurrences of choline deficiency in humans would be expected to diminish the brain's content of choline-containing and/or choline-derived molecules.

The metabolism of dietary choline by the intestinal microbiota produces trimethylamine (TMA), which is converted to trimethylamine-N-oxide (TMAO) in the liver. Excess TMAO has been shown to exacerbate atherosclerosis in mice and correlates with the disease severity in humans (Romano et al. 2015), which suggests that the gut microbiota needs to be considered in daily choline intake recommendations.

Vitamins

Vitamins are capable of modulating the microbiota via several mechanisms. They represent a large number of different functional molecules in the body and may be categorised into fat-soluble and water-soluble vitamins (Pham et al. 2021). Studies have shown that some vitamins (vitamins A, B_2, D, E and beta-carotene), have positive effects on the gut microbiota by increasing the abundance of beneficial microbes as well as increasing or maintaining microbial diversity (vitamins A, B_2, B_3, C, K) and richness (vitamin D) (Pham et al. 2021). This can occur as some vitamins are co-factors in energy-generation reactions and readily participate in energy metabolism in bacteria, thus directly supporting specific microbes. Furthermore, the production of short-chain fatty acid was noted to be increased by vitamin C, B_2, E, while those bacteria associated with its production were increased in abundance by the vitamins B_2 and E. Vitamins are also capable of indirectly affecting gut health and the microbiota through immune reactions of impact on the gut barrier, these include vitamins A and D (Pham et al. 2021).

The microbiota are also capable of producing vitamins, thus contributing to sufficient micronutrient availability and the balance of microorganisms within the gut. Therefore, vitamins have several effects on the gut microbiota, in a bidirectional, direct or indirect manner, without being used as an energy source.

Neurons and glia have the same functional demands for vitamins as other cells in the body. Their access to the brain is thus an important consideration, particularly given the existence of the BBB (Figure 8.1). *Water-soluble vitamins* appear to be transported across the BBB and, in some cases, the blood–CSF barrier, usually (though not always) by non-energy-requiring carriers. After they are taken up into neurons and glial cells, most are rapidly converted into biologically active derivatives, namely co-factors in enzyme-mediated reactions. As co-factors are recycled, dietary deficiencies in one or another vitamin do not lead immediately to brain dysfunction, because co-factor pools may take extended periods to become functionally compromised.

Folic acid is transported into brain as methylenetetrahydrofolic acid, the major form of folic acid in the circulation. The transport site is in the choroid plexus, at the blood–CSF barrier, and is about half-saturated at normal plasma concentrations of methylenetetrahydrofolate; little transport appears to occur at the BBB. CSF concentrations of methylenetetrahydrofolic acid are maintained at several times the circulating concentration, suggesting that transport is active (energy-requiring). Once inside cells, folates are polyglutamated.

Considerable interest has been generated in the past decade regarding the consequences of folate deficiency with regard to CNS development. The incidence of neural tube defects (NTDs, e.g. spina bifida, an abnormality in the formation of the spinal cord) has been found to be increased above the population mean in the children of women who are folate-deficient during pregnancy. Moreover, the occurrence of this abnormality can be reduced by folic acid supplementation before conception and during pregnancy. Initiating supplementation before conception is essential as the basic design of the CNS is laid down during the first trimester.

Folate is important in single-carbon metabolism, contributing carbon atoms to purines, thymidine and amino acids. Methylation reactions involving folate may also be important in the formation and maintenance of neuronal and glial membrane lipids. A folate deficiency, by impeding DNA, protein and/or lipid synthesis, could also conceivably influence neuronal and glial growth during critical points in neural tube development, leading to effects severe enough to induce NTDs. However, it is presently unknown which, if any, of these biochemical actions of folate might be involved in the production of NTDs.

Folate deficiency may also be linked to depression in adults. The clearest data supporting this connection come from findings in patients with megaloblastic anaemia (i.e. non-psychiatric patients). Patients having a clear folate deficiency in the absence of vitamin B_{12} deficiency showed a very high incidence of depression. Other findings then indicated that depressed patients (who do not have anaemia) have low plasma and red blood cell folate concentrations. Moreover, folate supplementation was found to improve mood in depressed patients. The mechanism(s) by which folate modifies mood may be related to the role of methylene tetrahydrofolate in methionine synthesis from homocysteine. It may thus help to maintain adequate methionine pools for *S*-adenosylmethionine synthesis. *S*-adenosylmethionine is a co-factor in methylation reactions in catecholamine synthesis and metabolism: catecholamines [dopamine, norepinephrine (or noradrenaline), epinephrine (or adrenaline)] are known to be important in maintaining mood, and *S*-adenosylmethionine ingestion is reputed to elevate mood. Folate has also been linked to the maintenance of adequate brain levels of tetrahydropterin, a co-factor in the synthesis of serotonin and catecholamines. It should be noted that, although the evidence described suggests that a connection may exist between folate deficiency and abnormal mood, the connection is still debated.

Interestingly, a role for the intestinal microbiota is becoming evident in the production of luminal folate and modulation of folate transporter. However, microbial folate production may not translate to an effect on human blood concentrations of folate (Malinowska et al. 2022). *Ascorbic acid* (*vitamin C*) is actively transported into the brain via the choroid plexus and the blood–CSF barrier, maintaining a concentration in the CSF that is four times the circulating concentrations. Transport through the BBB does not occur; the absence of such transporters suggests that the BBB might retard the efflux of ascorbate from brain. The choroid plexus transporter is about half-saturated at normal blood ascorbate concentrations, such that increases in blood ascorbate do not produce large increments in brain concentrations. Moreover, the active transport mechanism ensures that when plasma ascorbate falls, brain levels do not decline appreciably. Consequently, brain ascorbate concentrations show minimal fluctuations over a wide range of plasma ascorbate concentrations (indeed, brain ascorbate concentrations have been shown to vary only two-fold when plasma ascorbate concentrations were varied over a 100-fold range). Inside the CNS, ascorbate is actively transported into cells, although it is not known whether this uptake process is confined to neurons or glial cells, or occurs in both. Ascorbate is lost into the circulation at a rate of about 2% of the brain pool each hour, a loss thought to be caused by diffusion into the circulation and by bulk flow of CSF returning to the circulation. The active transport mechanism readily compensates for this loss. Presumably because ascorbate is actively transported by the choroid plexus, and brain concentrations decline minimally at very low plasma concentrations, CNS signs are not notable in ascorbic acid deficiency states. To date, the only defined biochemical function of ascorbic acid in the brain is to serve as a co-factor for dopamine β-hydroxylase, the enzyme that converts dopamine to norepinephrine (although ascorbate is thought by some to function as an antioxidant).

Thiamin (vitamin B_1) is taken up into the brain by a transporter located at the BBB. It is non-energy-requiring and is about half-saturated at normal plasma thiamin concentrations. Small amounts of thiamin also gain entry via transport through the choroid plexus (blood–CSF barrier). Thiamin is then transported into neurons and glia, and phosphorylation effectively traps the molecule within the cell. In nervous tissue, thiamin functions as a co-factor for the pyruvate dehydrogenase complex, the α-ketoglutarate dehydrogenase complex, α-ketoacid decarboxylase and transketolase. It may also participate in nerve conduction. Severe thiamin deficiency in animals reduces thiamin pyrophosphate levels and the activities of thiamin-dependent reactions. It also produces difficulties in the coordinated control of muscle movement, suggestive of a compromise of vestibular function (e.g. unsteady gait, poor postural control and equilibrium). However, despite these biochemical–functional associations, the exact biochemical mechanism precipitating the functional deficits is unclear. The functional deficits are rapidly corrected with thiamin treatment, suggesting that neurons have not been damaged or destroyed. Thiamin deficiency in humans (Wernicke's disease) produces similar deficits in the control of complex muscle movements and also mental confusion. Korsakoff's syndrome, which occurs in almost all patients with Wernicke's disease, involves a loss of short-term memory as well as mental confusion. Severe thiamin deficiency in humans does appear to produce neuronal degeneration in certain brain regions. While the motor abnormalities can be corrected with thiamin treatment, memory dysfunction is not improved. Interestingly, some bacteria also require vitamin B_1, and vitamin B_1 is involved in the production of the SCFAs, butyrate and acetate.

Riboflavin (vitamin B_2) enters the brain via a saturable transport carrier, by a mechanism that has not been well described. It is probably located on capillary endothelial cells, although some transport appears to occur at the choroid plexus (blood–CSF barrier) and is about half-saturated at normal plasma riboflavin concentrations. The vitamin is readily transported into neurons and glia from the extracellular fluid and trapped intracellularly by phosphorylation and subsequent conversion to flavin adenine dinucleotide and covalent linkage to functionally active proteins (flavoproteins). In nervous tissue, these functional forms of riboflavin participate in numerous oxidation–reduction reactions that are key components of metabolic pathways for carbohydrates, amino acids and fats. A key enzyme in neurotransmitter metabolism, monoamine oxidase, is a flavoprotein. The brain content of riboflavin and its derivatives is not notably altered in states of riboflavin deficiency or excess. Riboflavin has been shown to promote microbial butyrate production.

Pantothenic acid is transported into the brain by a saturable transport carrier located at the BBB. The carrier is almost completely unsaturated at normal plasma pantothenate concentrations, hence increases in plasma levels would never be likely to saturate the transporter completely and thereby limit transport. Neurons and glial cells take up pantothenic acid slowly by a mechanism of facilitated diffusion. Inside the cell, the vitamin becomes a component of coenzyme A, the coenzyme of acyl group transfer reactions. Relative to other tissues, the brain contains a high concentration of pantothenate, mostly in the form of coenzyme A. Brain coenzyme A concentrations do not become depleted in pantothenate deficiency states, indicating that the supply to the brain of this vitamin is not the key factor governing coenzyme A concentrations in this organ.

Niacin (vitamin B_3) is transported into the brain as niacinamide, primarily via the BBB (capillary endothelial cells), but possibly also to a small extent via the choroid plexus (blood–CSF barrier). Most niacin in the brain is derived from the circulation, although the brain may be able to synthesise small amounts. Niacin is taken up into neurons and glia and is rapidly converted to nicotinamide adenine dinucleotide (NAD). The half-life of NAD in the brain is considerably longer than in other tissues (e.g. seven to nine times that of liver). NAD and nicotinamide adenine dinucleotide phosphate (NADP) are involved in numerous oxidation–reduction reactions. Dietary niacin deficiency in the presence of a low intake of tryptophan causes pellagra in humans, a deficiency disease that includes changes in brain function such as mental depression and dementia, loss of motor coordination and tremor. The mechanisms for these effects have not been identified.

Pyridoxine (*vitamin B_6*) is taken up into the brain via a transport carrier that appears to be saturable; the details of the mechanism have not been well described. The carrier is probably located on capillary endothelial cells, although some transport appears to occur at the blood–CSF barrier. The vitamin can be transported in any of its non-phosphorylated forms (pyridoxine, pyridoxal, pyridoxamine). Once within the brain extracellular fluid compartment, the vitamin is readily transported into neurons and glia, and phosphorylated (primarily to pyridoxal phosphate or pyridoxine phosphate). In neurons and glia, pyridoxal phosphate serves as a co-factor in a variety of reactions, including decarboxylation reactions, such as those mediated by aromatic L-amino acid decarboxylase (which converts dihydroxyphenylalanine to dopamine, and 5-hydroxytryptophan to serotonin) and glutamic acid decarboxylase [which converts glutamate to GABA), and transamination reactions, such as that mediated by GABA transaminase (which catabolises GABA to glutamate and succinic semialdehyde). Dopamine, serotonin and GABA are neurotransmitters. In humans, pyridoxine deficiency is rare because of its widespread occurrence in foodstuffs. However, where it has been identified, it has been associated with increased seizure activity, an effect that is dissipated with pyridoxine treatment. Production of pyridoxine deficiency in laboratory animals leads to reductions in pyridoxal phosphate levels in the brain and in GABA concentrations. Reductions in GABA, an inhibitory transmitter, are known to be associated with increased seizure susceptibility, and electroencephalographic abnormalities have been recorded in deficient animals. Pyridoxine deficiency in rats also reduces serotonin concentrations in the brain, almost certainly the result of diminished 5-hydroxytryptophan decarboxylation (a pyridoxal phosphate-dependent reaction).

Biotin is transported into the brain by a saturable, sodium-dependent (energy-requiring), carrier-mediated mechanism located at the BBB. Essentially no transport occurs at the blood–CSF barrier. The affinity of the transporter for biotin is such that the carrier would never become saturated. The biotin carrier shows sufficient transport activity to compensate for normal daily turnover of the vitamin. Biotin is a coenzyme for a variety of carboxylation reactions (e.g. pyruvate carboxylase).

Cobalamin (vitamin B_{12}) is thought to be transported into the brain by a carrier-mediated mechanism. Relatively little is known about this process or about the function of vitamin B_{12} in the nervous system. However, vitamin B_{12} deficiency has been known for over a century to be associated with neurological abnormalities (Markun et al. 2021). The neurological deficits are presumed to derive from the demyelination of axons in spinal cord and brain that are seen in advanced deficiency cases. Left untreated, axonal degeneration eventually occurs. These effects can be reversed if vitamin B_{12} treatment is provided early enough. Vitamin B_{12} has been found to protect neurons from neurotoxin-induced damage, suggesting that it may be important in neuronal repair mechanisms, which may become compromised in deficiency states. Nervous system damage associated with vitamin B_{12} deficiency can occur at any age.

Fat-soluble vitamins are hydrophobic (lipophilic), suggesting that their uptake into the CNS may involve simple diffusion across barrier endothelial cell membranes. The reality, however, is not so simple. The transport into, and functions within, the brain of micronutrients have not yet been studied to the same extent as has been the case for the macronutrients. However, available data do begin to provide some definition of these processes.

Of the fat-soluble vitamins, *vitamin A* (retinol) has been the most studied in relation to CNS availability and function. The others have been much less well investigated, although vitamin E is currently a compound of some interest because of its function as an antioxidant. Vitamins D and K are not typically thought of as having the brain as a major focus of action and thus little information is available regarding their involvement in brain function. Vitamin A therefore presently provides the best candidate to illustrate some of the complexities of fat-soluble vitamin flux and function in the CNS. Some of vitamin E's putative brain actions will also be mentioned.

Vitamin E is an antioxidant and free radical scavenger that (among other functions) protects fatty acids in cellular membranes. It is transported in blood associated with lipoproteins. Dietary vitamin E deficiency is extremely rare in humans. It occurs in association with certain abnormalities of vitamin E transport and fat absorption, and sometimes in individuals with protein–calorie malnutrition. The neurological manifestations are peripheral nerve degeneration, spinocerebellar ataxia and retinopathy.

Vitamin E has been proposed to play a role in a number of diseases of the CNS that have been linked to oxidative damage. One example is Parkinson's disease, a disorder of movement control induced by the degeneration of certain populations of brain neurons. Evidence of oxidative damage is present in the brains of Parkinsonian patients, although controlled clinical trials of vitamin E supplementation have proved to be of no benefit in retarding the disease's progression, which brings into the question the likelihood of a vitamin E link to the degenerative changes. A second example is Alzheimer's disease (a form of senile dementia), which is associated with a progressive, generalised, ultimately catastrophic degeneration of the brain. Several types of oxidative damage have been found in the brains of patients with Alzheimer's disease. Unfortunately, clinical trials of the ability of vitamin E to slow the development and progression of Alzheimer's disease have produced conflicting results. Moreover, the prolonged administration of high doses of vitamin E, such as are typical of such trials, has been associated with an increased risk of mortality.

Minerals

All of the essential minerals are important for cellular functions in the brain, as they are elsewhere in the body. These are sodium, potassium, calcium, magnesium, iron, copper, zinc, manganese, cobalt and molybdenum. While most function as co-factors in enzymic reactions, sodium and potassium are key ions in electrical conduction in neuronal membranes, calcium functions as a second messenger within neurons, and magnesium functions at certain ligand-gated ion channels on neurons [e.g. the *N*-methyl-D-aspartate (NMDA) glutamate receptor] (Figure 8.1). The diet normally provides more than adequate amounts of almost all minerals, except possibly for calcium, iron, magnesium and zinc.

The gut microbiota can affect host mineral status via synthesis of enzymes that produce minerals from the diet, and also via modulating the absorption rate in the gastrointestinal tract (Barone et al. 2022). For example, bacterial phytases hydrolyse plant-based phytic acid, which release bioavailable forms of calcium, magnesium, iron and phosphorous.

The permeability of the BBB to most metals is quite low, generally much lower than in other organs. Indeed, as a point of reference, the brain extracts 20–30% of the glucose in blood in a single capillary transit, but less than 0.3% of any metal is normally extracted. The mechanisms of transport into the brain for most metals are unknown. However, some details regarding the transport and functions of iron, calcium and copper are available.

Iron (*ferric*) circulates bound to a protein, transferrin. More than 90% of plasma iron is bound to transferrin. Iron uptake into brain occurs primarily at the BBB. The process involves the uptake of the transferrin–iron complex by capillary endothelial cells, which express transferrin receptors on the luminal membrane. Circulating transferrin–iron complexes bind to transferrin receptors and are internalised in endosomes. For most cells, the typical mechanism of iron transport and secretion into the extracellular fluid involves the release of iron (in the ferrous form) from transferrin in the endosomes and its transport out into the cytoplasm by a divalent metal transporter, located on the endosomal membrane. Cytoplasmic ferrous iron is then transported out of the cell by a cellular membrane protein, transportin, and then converted back to ferric iron by a membrane ferroxidase. However, curiously, brain capillary endothelial cells do not appear to express the divalent metal transporter or transportin. Consequently, the current hypothesis for iron movement across the capillary endothelial cell posits that the endosomes carry the iron–transferrin complex to the abluminal membrane and release ferric iron (but not transferrin) into brain extracellular fluid. The endosome then recycles to the luminal membrane and releases apotransferrin back into the circulation.

Transferrin is found in brain extracellular fluid and the ferric iron released from capillary endothelial cells associates with it. Transferrin-bound iron is the main form in which iron is found in brain extracellular fluid. Neurons express both the transferrin receptor and the divalent metal transporter. Hence, iron enters neurons by the endosomal process, as described earlier, and is released from the endosome into the cytoplasm. The importance of the uptake process to neuronal iron homeostasis is suggested by the fact that transferrin receptor expression by neurons (but not other brain cells) is markedly increased in iron-deficient animals. Neurons also express ferroportin and thus have the capacity to release iron into the extracellular fluid. Ferroportin is found on all parts of the neuron (dendrites, soma, axons), and the neuron is thought to use it to maintain intraneuronal iron levels at values appropriate for metabolic needs. However, it should be noted that certain neuron groups in the brain store large amounts of iron intracellularly as ferritin, for reasons that are presently unknown.

Curiously, *astrocytes* lack transferrin receptors. A current hypothesis therefore posits that astrocytes absorb iron from the extracellular fluid complexed with ATP or citrate. The process of iron release from astrocytes is not presently understood, although astrocyte cell membranes express a ferroxidase (ceruloplasmin), which is presumed to be involved. *Oligodendrocytes*, the glial cells that insulate axons, also lack transferrin receptors, and thus are also thought to take up iron directly as an ATP or citrate complex. These cells, however, express ferroportin and thus can actively export iron.

Numerous enzymes in brain are iron-requiring, including several hydroxylases in neurons that mediate rate-limiting reactions in the production of neurotransmitters (tryptophan hydroxylase, phenylalanine hydroxylase, tyrosine hydroxylase) and monoamine oxidase, a key enzyme in the metabolism of the monoamine neurotransmitters (serotonin, dopamine, norepinephrine, epinephrine).

Iron deficiency can cause impairments in attention and cognition in children. Similar effects are seen in animals. In rats made iron-deficient, despite increases in the efficiency of iron uptake by the brain, brain iron concentrations decline, with newborn and infant animals showing more rapid declines than older animals. Iron repletion in the brain occurs (slowly) in infant and adult rats with iron supplementation, but not in animals depleted at birth. While outside the brain the activities of many iron-dependent enzymes are depressed by iron deficiency, inside the brain their activities are unaffected. The activities of the hydroxylase enzymes are not reduced by severe iron restriction, nor are the rates of production of the neurotransmitters derived from them (serotonin from tryptophan hydroxylase; dopamine, norepinephrine and epinephrine from phenylalanine and tyrosine hydroxylases). However, a reduction in dopamine receptors (D_2 subtype) does occur and coincides with aberrations in dopamine-dependent behaviours in rats. The inability of brain iron stores to recover in rats made iron-deficient as newborns coincides with a persistence of dopamine receptor-linked behavioural deficits, despite normal repletion of iron stores elsewhere in the body. Restoration of normal behaviour with iron supplementation, along with brain iron stores, is seen in animals made iron-deficient at other ages.

Iron deficiency also interferes with myelination. As marked glial proliferation and myelin formation occur early in infancy, iron deficiency during this period could compromise the optimal development of neuronal communications (e.g. glial cells provide insulation for axons and synapses). This effect could account for some of the behavioural deficits associated with neonatal iron deficiency. If brain function is permanently affected by the occurrence of iron deficiency shortly after birth, this vulnerability should be an issue of great concern in those parts of the world where iron deficiency in infancy is an endemic problem.

Calcium is transported into the CNS via a saturable, probably active, transport mechanism located in the choroid plexus (the blood–CSF barrier). At normal plasma calcium concentrations (calcium circulates mostly in the free form, i.e. unbound to a blood protein), it operates at near saturation, presumably enabling brain calcium levels to remain relatively constant in the face of increases in plasma calcium. Unlike intestinal calcium transport, transport into the brain is not vitamin D-sensitive. Some calcium is taken up into the brain via the brain capillary system (i.e. across the BBB), but this fraction is less than 40% of total calcium uptake each day. This transport mechanism has not been defined. The active transport of calcium at the blood–CSF barrier would presumably enable brain calcium levels to be maintained at, or near, normal levels in the face of dietary calcium deficits. As calcium concentrations in the circulation are also regulated, under most circumstances, this process would also be expected to help maintain brain calcium uptake and levels in the face of vagaries in calcium intake. Deficiencies in brain calcium should thus be a relatively rare occurrence.

Copper functions as a co-factor for numerous enzymes, including dopamine β-hydroxylase, which mediates the reaction converting dopamine to norepinephrine. Dietary copper deficiency in humans is fairly rare, but when experimentally induced in animals it lowers copper concentrations throughout the body, including the brain, and blocks the conversion of dopamine to norepinephrine in brain neurons. This enzyme block also occurs in peripheral sympathetic nerves and the adrenal gland, which synthesise norepinephrine and epinephrine. The mechanism of copper uptake into the brain is unknown, but there is recent evidence that while copper circulates bound to serum proteins (notably ceruloplasmin), free (non-protein-bound) copper is the form taken up into the brain, primarily across capillary endothelial cells (the BBB). One hypothesis is that copper transported into the endothelial cells is then directly transferred into astrocytes (through the 'end-feet' that surround brain capillaries) and is subsequently provided to neurons through astrocytic processes that impinge on neurons. In this manner, copper availability to neurons could be metered, as excess copper is neurotoxic. Copper deficiency occurs as an X-linked genetic disease of copper transport in Menkes disease, in which tissue and brain copper levels become extremely low and produce neurodegeneration. Children with Menkes disease die at a very young age.

8.9 The role of nutrition in brain development

Optimal nutrition during the prenatal and postnatal stages is crucial for appropriate brain development and long-term cognitive function. The developing brain exhibits a greater susceptibility to nutrient deficiencies compared with the adult brain, while also displaying remarkable plasticity during this period.

Maternal diet during pregnancy

During pregnancy, the foetus relies on the mother's nutrition for its growth and development, and inadequate nutrition can have a negative impact on the developing brain. For example, a lack of essential nutrients, such as iron, folate and omega-3 fatty acids during pregnancy can lead to impaired brain development and cognitive function later in life (Cusick and Georgieff 2022).

During pregnancy, the intrauterine environment serves as a critical interface between the mother and the developing baby, and it is a target for researchers looking into foetal programming mechanisms. Neurulation is a crucial process in early embryonic development that begins around day 18 after conception and continues until day 28 where the neural tube eventually gives rise to the brain and spinal cord. Interruption during neurulation may result in severe congenital malformations, including spina bifida and anencephaly. Additionally, during this period, a series of cellular events takes place, including neurogenesis from day 42 to mid-gestation, followed by neuronal migration, differentiation, synapse formation and apoptosis, which together form the basis of early brain development. The foetal brain is particularly vulnerable to stimuli and insults during this period, and proper brain development relies significantly on the maternal environment.

Nutrients and other bioactive compounds from the maternal diet are transported to the developing foetus via the placenta, which acts as an interface facilitating the transfer of substrates and regulating the maternal immune response to prevent immunological rejection of the foetus. These nutrients have a substantial impact on the foetal brain transcriptome and play critical roles in regulating the neurodevelopmental processes (Edlow et al. 2019).

For example, omega-3 fatty acids, such as docosahexaenoic acid, are important structural components of neuronal membranes and are essential for optimal brain development. Maternal consumption of omega-3 fatty acids during pregnancy has been shown to enhance foetal brain development, increase neural connectivity, and improve cognitive and visual function in the offspring. Additionally, folate supplementation during pregnancy has been widely recognised for its effectiveness in reducing the incidence of neural tube defects (MRC Vitamin Study Research Group 1991). Likewise, choline is a nutrient that plays a crucial role in brain development, as it is a precursor for the synthesis of the neurotransmitter acetylcholine and plays a key role in cell signalling and gene expression. Maternal choline intake during pregnancy has been linked to improved memory and cognitive function in the offspring.

Conversely, maternal consumption of diets high in saturated fats, trans fats or added sugars can lead to increased oxidative stress, inflammation and insulin resistance, all of which can negatively affect foetal brain development. Maternal high-fat diet and subsequent obesity are linked to sex-specific variations in foetal size and gene expression in the foetal brain, with males being more sensitive to environmental influences, with social deficits being noted that are indicative of behaviour similar to that observed in autism (Buffington et al. 2016).

C-reactive protein and leptin, which play a role in inflammation and energy homeostasis, also function as signalling molecules between the immune and nervous systems. Pregnant women with pre-existing obesity show an exaggerated systemic response to pregnancy, including elevated levels of interleukin-6, C-reactive protein and leptin. For example, in a mouse model, exposure to systemic maternal inflammation induced by lipopolysaccharide during pregnancy combined with neonatal hyperoxic exposure leads to a reduction in the number of oligodendrocytes in the cerebral cortex and hippocampus during adulthood. The offspring of rats, macaques and mice exposed to high-fat diets during pregnancy show increased reactivity of astrocytes and microglia, upregulated pro-inflammatory cytokines and impaired spatial learning. In a rat model, maternal fat-enriched diet was found to lead to an anxiogenic phenotype in adult offspring, along with significantly higher levels of hippocampal pro-inflammatory cytokine tumour necrosis factor-alpha (TNF-α) and monocyte-chemoattractant protein-1 (MCP-1), both of which were correlated with the degree of behavioural change (Winther et al. 2018).

Conversely, maternal malnutrition can lead to nutrient deficiencies and impaired foetal brain growth and development. Offspring born from malnourished mothers express hyperactivity and learning deficits related to the hippocampus into adult life. This can occur through epigenetic mechanisms associated with genes involved in neuronal development, leading to impaired cognitive and behavioural outcomes in the offspring.

Additionally, experiences during the perinatal period have a significant impact on how offspring respond to stressors later in life. In their studies, Isac et al. (2020) discuss the potential impact of maternal diet on the vulnerability of the offspring's brain to perinatal asphyxia. This condition is characterised by insufficient blood flow or gas exchange and is associated with a high incidence of mortality and morbidity among children. Maternal high-fat diet was shown to make the neonate more vulnerable, while other nutrients, such as trans resveratrol and a citicoline-enriched diet, provide neuroprotection against birth asphyxia in offspring (Isac et al. 2020).

Maternal gut microbiota during pregnancy

Inadequate ingestion of micro- and macronutrients during pregnancy has been associated with an altered maternal microbiota and also with a poor neurocognitive outcome in offspring. Interestingly, alterations in the enteric microbial communities occur concurrently with neurodevelopmental plasticity during pregnancy, suggesting a possible interaction between them. Additionally, it is now widely acknowledged that maternal gut bacteria have the ability to impact the development of the infant's brain and also that of the infant gut bacteria.

Therefore, early life provides a window of opportunity during which the composition and function of the microbiota can be altered. A newborn's colonisation has been suggested to begin before delivery, during the intrauterine period, as evidenced by the discovery of bacteria in the placenta. However, the presence of an intrauterine microbiota during a healthy pregnancy is a highly debated and controversial subject (Kennedy et al. 2023).

Maternal gut microbiota are dynamically remodelled during pregnancy, particularly in the third trimester, sustaining the increased requirements in foetal growth, including brain development. Altered maternal gut microbiota, due to infection, changes in diet and stress during pregnancy, has been associated with abnormalities in brain function and behaviour of the offspring (Vuong et al. 2020). Perturbations in the microbiota have been associated with impaired host social and communicative behaviour, stressor-induced behaviour, and performance in learning and memory tasks (Vuong et al. 2020). Furthermore, expression of genes associated with axonogenesis was impaired in mouse embryos from antibiotic-treated and germ-free dams, leading also to altered neurodevelopment and behaviour which was dependent on specific maternal microbial metabolites (Vuong et al. 2020).

Overall, these findings highlight the critical role of maternal nutrition and hence microbiota in shaping foetal brain development and the potential long-term consequences of poor maternal dietary habits. Optimal maternal nutrition during pregnancy can enhance foetal and infant neurodevelopment, reduce the risk of neurodevelopmental disorders, and promote lifelong health and well-being.

Diet during early life – first two years of life

It is important to note that the effects of maternal diet on offspring neurodevelopment are not solely limited to the prenatal period. For example, maternal diet during lactation can also influence the composition of breast milk, which can affect the offspring's nutritional status and neurodevelopmental outcomes.

The term 'nutritional programming' refers to the process and mechanisms through which nutrition, dietary behaviours and the environment during the first 1000 days of life (including pregnancy and early infancy) can have a significant impact on an individual's future health and disease risk. This concept sheds light on some of the observed effects of early-life nutrition on long-term health outcomes.

The first two years of life are characterised by rapid maturation of all systems, including the brain. Thus, any perturbance in nutrition at this stage can potentially impact the developing brain. Additionally, it is a critical period for the establishment of the individual's gut microbiota and immunological response. Recent research has focused on the crucial role of the infant microbiota in determining future health outcomes and risks. Many factors, including mode of delivery, nutritional provision in early life, drugs, antibiotic therapy and early-life stress, can influence the bacterial communities (Tamburini et al. 2016). As mentioned, the first 1000 days from conception until two years of age are crucial for the proper development of the host, and the adult-like complexity of the gut microbiota is attained by the end of this period. Afterwards, the microbiota undergoes continuous change, its diversity and composition being modified by numerous factors such as diet and lifestyle, drugs, stress, infections and ageing.

Breastfeeding has emerged as a significant factor in promoting a healthy microbiota, contributing to several of the acute benefits associated with breastfeeding. Human breast milk is a rich source of prebiotics in the form of human milk oligosaccharides and a diverse variety of probiotic bacteria. Breastfed infants are also exposed to the maternal skin microbiota, which reside on the nipple and areola, due to frequent nursing, while formula-fed infants do not receive this exposure. Breast milk contains antimicrobial peptides, such as defensins and lactoferrin, and human milk oligosaccharides that prevent competitive and pathogenic strains of bacteria from colonising the infant gut. Additionally, secretory IgA present in breast milk contributes to a distinct microbial profile, while hormones such as insulin and leptin found in breast milk are linked to the taxonomic and metagenomics profile of the breastfed infants' gut bacteria. All these factors collectively contribute to the stark differences between the microbial communities of breast- and formula-fed infants. Researchers found a significant association between prolonged breastfeeding and cognitive ability. Additionally, a correlation was found between extended breastfeeding and reduced scores for symptoms of attention deficit hyperactivity disorder, while consumption of no breast milk was identified as a risk factor for developing autism.

Malnutrition during early childhood can also have long-term effects on cognitive function. For example children who experience malnutrition, particularly protein-energy malnutrition, in the first years of life have poorer cognitive function and lower IQ scores compared with well-nourished children. Therefore, postnatal nutrition is also a critical determinant for the growth and development of the brain. Infants and young children require a balanced diet that provides essential nutrients, including protein, vitamins and minerals, to support the development of the brain and nervous system. In conclusion, proper nutrition during the prenatal and postnatal stages of brain development is critical for optimal brain growth and function.

8.10 Future perspectives

Good nutrition and a healthy diet can have beneficial effects on both the gut and brain due to the close interaction between the two systems via the gut–brain axis (Figure 8.1). Overall, suboptimal nutrition and nutritional deficiencies have a significant impact on brain health and cognitive function. Inadequate nutrition during critical periods of development can have long-lasting effects on cognitive and behavioural outcomes and increase the risk of various neurological disorders. Moreover, the ageing process can affect the brain's ability to absorb and utilise nutrients, which emphasises the need for a balanced and nutrient-dense diet throughout life. It is essential to maintain a balanced and nutrient-rich diet to ensure optimal brain development and function, particularly during critical periods of growth and development as well as maintenance during the life span. Through the gut–brain axis pathways, the gut microbiota plays a significant role in the development and functioning of the brain throughout all stages of life. Nutrients from the diet, such as fibre, prebiotics and probiotics, are crucial for maintaining a healthy gut microbiota and promoting optimal brain health. Therefore, promoting healthy dietary habits that include a variety of nutrient-dense foods is essential for maintaining optimal brain health and function throughout the life span.

Although this information is available, it is not always adhered to, due to lack of available nutrition in developing countries or the methods of sharing this information or the accessibility of the information. Education of healthcare professionals in maternity services and in healthcare settings for children and adults to distribute dietary advice is urgently needed. To have these data disseminated in an acceptable manner for diverse socioeconomic populations, as well as physical and mental abilities, would be key to achieving change. Given the potential of nutrition and the microbiota for brain health, Europe-wide commitments to research in enabling these changes is needed. Substantial action from policymakers to enable knowledge-building on diets supporting brain health is of paramount importance.

References

Anderson, S.C., Cryan, J.F., and Dinan, T. (2017). *The Psychobiotic Revolution: Mood, Food, and the New Science of the Gut-Brain Connection*. National Geographic Books.

Barone, M., D'Amico, F., Brigidi, P., and Turroni, S. (2022). Gut microbiome–micronutrient interaction: the key to controlling the bioavailability of minerals and vitamins? *Biofactors* 48 (2): 307–314. doi: 10.1002/biof.1835.

Bellono, N.W., Bayrer, J.R., Leitch, D.B., Castro, J., Zhang, C., O'Donnell, T.A., Brierley, S.M., Ingraham, H.A., and Julius, D. (2017). Enterochromaffin cells are gut chemosensors that couple to sensory neural pathways. *Cell* 170 (1): 185–198.e16. doi: 10.1016/j.cell.2017.05.034.

Bentsen, M.A., Mirzadeh, Z., and Schwartz, M.W. (2019). Revisiting how the brain senses glucose—and why. *Cell Metab* 29 (1): 11–17. doi: 10.1016/j.cmet.2018.11.001.

Bin-Khattaf, R.M., Alonazi, M.A., Al-Dbass, A.M., Almnaizel, A.T., Aloudah, H.S., Soliman, D.A., and El-Ansary, A.K. (2022). Probiotic ameliorating effects of altered GABA/glutamate signaling in a rodent model of autism. *Metabolites* 12 (8): 720. doi: 10.3390/metabo12080720.

Blachier, F., Beaumont, M., and Kim, E. (2019). Cysteine-derived hydrogen sulfide and gut health: a matter of endogenous or bacterial origin. *Curr Opin Clin Nutr Metab Care* 22 (1): 68–75. doi: 10.1097/MCO.0000000000000526.

Buffington, S.A., Viana Di Prisco, G., Auchtung, T.A., Ajami, N.J., Petrosino, J.F., and Costa-Mattioli, M. (2016). Microbial reconstitution reverses maternal diet-induced social and synaptic deficits in offspring. *Cell* 165 (7): 1762–1775. doi: 10.1016/j.cell.2016.06.001.

Burdo, J., Dargusch, R., and Schubert, D. (2006). Distribution of the cystine/glutamate antiporter system Xc- in the brain, kidney, and duodenum. *J Histochem Cytochem: Off J Histochem Soc* 54 (5): 549–557. doi: 10.1369/jhc.5A6840.2006.

Cryan, J.F., O'Riordan, K.J., Cowan, C.S.M., Sandhu, K.V., Bastiaanssen, T.F.S., Boehme, M., Codagnone, M.G. et al. (2019). The microbiota-gut-brain axis. *Physiol Rev* 99 (4): 1877–2013. doi: 10.1152/physrev.00018.2018.

Cusick, S.E. and Georgieff, M.K. (2022). Chapter 6 - early-life nutrition and neurodevelopment. In: *Early Nutrition and Long-Term Health*, 2e (ed. J.M. Saavedra and A.M. Dattilo), 127–151. Woodhead Publishing. doi: 10.1016/B978-0-12-824389-3.00007-6.

de Wouters d'Oplinter, W., de, A., Verce, M., Huwart, S.J.P., Lessard-Lord, J., Depommier, C., Van Hul, M., Desjardins, Y., Cani, P.D., and Everard, A. (2023). Obese-associated gut microbes and derived phenolic metabolite as mediators of excessive motivation for food reward. *Microbiome* 11 (1): 94. doi: 10.1186/s40168-023-01526-w.

Dicks, L.M.T. (2022). Gut bacteria and neurotransmitters. *Microorganisms* 10 (9): 1838. doi: 10.3390/microorganisms10091838.

Edlow, A.G., Guedj, F., Sverdlov, D., Pennings, J.L.A., and Bianchi, D.W. (2019). Significant effects of maternal diet during pregnancy on the murine fetal brain transcriptome and offspring behavior. *Front Neurosci* 13 (December): 1335. doi: 10.3389/fnins.2019.01335.

Errasti-Murugarren, E., Fort, J., Bartoccioni, P., Díaz, L., Pardon, E., Carpena, X., Espino-Guarch, M. et al. (2019). L Amino acid transporter structure and molecular bases for the asymmetry of substrate interaction. *Nat Commun* 10 (1): 1807. doi: 10.1038/s41467-019-09837-z.

Fernstrom, J.D. (1977). Effects of the diet on brain neurotransmitters. *Metabolism* 26 (2): 207–223. doi:10.1016/0026-0495(77)90057-9.

Fotiadis, D., Kanai, Y., and Palacín, M. (2013). The SLC3 and SLC7 families of amino acid transporters. *Mol Aspects Med* 34 (2–3): 139–158. doi: 10.1016/j.mam.2012.10.007.

Gould, J.F., Smithers, L.G., and Makrides, M. (2013). The effect of maternal omega-3 (n-3) LCPUFA supplementation during pregnancy on early childhood cognitive and visual development: a systematic review and meta-analysis of randomized controlled trials. *Am J Clin Nutr* 97 (3): 531–544. doi: 10.3945/ajcn.112.045781.

Halford, J.C.G., Harrold, J.A., Lawton, C.L., and Blundell, J.E. (2005). Serotonin (5-HT) drugs: effects on appetite expression and use for the treatment of obesity. *Curr Drug Targets* 6 (2): 201–213. doi: 10.2174/1389450053174550.

Hawkins, R.A. and Viña, J.R. (2016). How glutamate is managed by the blood-brain barrier. *Biology* 5 (4): 37. doi: 10.3390/biology5040037.

Höglund, E., Øverli, Ø., and Winberg, S. (2019). Tryptophan metabolic pathways and brain serotonergic activity: a comparative review. *Front Endocrinol* 10. https://www.frontiersin.org/articles/10.3389/fendo.2019.00158.

Iesanu, M.I., Denise Mihaela Zahiu, C., Dogaru, I.-A., Maria Chitimus, D., Gradisteanu Pircalabioru, G., Elena Voiculescu, S., Isac, S. et al. (2022). Melatonin-microbiome two-sided interaction in dysbiosis-associated conditions. *Antioxidants (Basel)* 11 (11): 2244. doi: 10.3390/antiox11112244.

Isac, S., Maria Panaitescu, A., Ioana Iesanu, M., Zeca, V., Cucu, N., Zagrean, L., Peltecu, G., and Zagrean, A.-M. (2020). Maternal citicoline-supplemented diet improves the response of the immature hippocampus to perinatal asphyxia in rats. *Neonatology* 117 (6): 729–735. doi: 10.1159/000512145.

Kennedy, K.M., de Goffau, M.C., Elisa Perez-Muñoz, M., Arrieta, M.-C., Bäckhed, F., Bork, P., Braun, T. et al. (2023). Questioning the fetal microbiome illustrates pitfalls of low-biomass microbial studies. *Nature* 613 (7945): 639–649. doi: 10.1038/s41586-022-05546-8.

Legan, T.B., Lavoie, B., and Mawe, G.M. (2022). Direct and indirect mechanisms by which the gut microbiota influence host serotonin systems. *Neurogastroenterol Motil* 34 (10): e14346. doi: 10.1111/nmo.14346.

Lin, R., Liu, W., Piao, M., and Zhu, H. (2017). A review of the relationship between the gut microbiota and amino acid metabolism. *Amino Acids* 49 (12): 2083–2090. doi: 10.1007/s00726-017-2493-3.

Malinowska, A.M., Schmidt, M., Kok, D.E., and Chmurzynska, A. (2022). Ex vivo folate production by fecal bacteria does not predict human blood folate status: associations between dietary patterns, gut microbiota, and folate metabolism. *Food Res Intl* 156 (June): 111290. doi: 10.1016/j.foodres.2022.111290.

Matsumoto, M., Kunisawa, A., Hattori, T., Kawana, S., Kitada, Y., Tamada, H., Kawano, S., Hayakawa, Y., Iida, J., and Fukusaki, E. (2018). Free D-Amino acids produced by commensal bacteria in the colonic lumen. *Sci Rep* 8 (1): 17915. doi:10.1038/s41598-018-36244-z.

Mazzoli, R. and Pessione, E. (2016). The neuro-endocrinological role of microbial glutamate and GABA signaling. *Front Microbiol* 7: https://www.frontiersin.org/articles/10.3389/fmicb.2016.01934.

Miyamoto, J., Igarashi, M., Watanabe, K., Karaki, S.-I., Mukouyama, H., Kishino, S., Li, X. et al. (2019). Gut microbiota confers host resistance to obesity by metabolizing dietary polyunsaturated fatty acids. *Nat Commun* 10 (1): 4007. doi: 10.1038/s41467-019-11978-0.

Moran-Ramos, S., Macias-Kauffer, L., López-Contreras, B.E., Villamil-Ramírez, H., Ocampo-Medina, E., León-Mimila, P., Del Rio-navarro, B.E. et al. (2021). A higher bacterial inward BCAA transport driven by faecalibacterium prausnitzii is associated with lower serum levels of BCAA in early adolescents. *Mol Med* 27 (1): 108. doi: 10.1186/s10020-021-00371-7.

Ota, T., Ishikawa, T., Sakakida, T., Endo, Y., Matsumura, S., Yoshida, J., Hirai, Y. et al. (2022). Treatment with broad-spectrum antibiotics upregulates Sglt1 and induces small intestinal villous hyperplasia in mice. *J Clin Biochem Nutr* 70 (1): 21–27. doi: 10.3164/jcbn.21-42.

Pham, V.T., Dold, S., Rehman, A., Bird, J.K., and Steinert, R.E. (2021). Vitamins, the gut microbiome and gastrointestinal health in humans. *Nutr Res* 95 (November): 35–53. doi: 10.1016/j.nutres.2021.09.001.

MRC Vitamin Study Research Group. (1991). Prevention of neural tube defects: results of the medical research council vitamin study. *Lancet* 338 (8760): 131–137.

Romano, K.A., Vivas, E.I., Amador-Noguez, D., and Rey, F.E. (2015). Intestinal microbiota composition modulates choline bioavailability from diet and accumulation of the proatherogenic metabolite trimethylamine-N-Oxide. *MBio* 6 (2): e02481. doi: 10.1128/mBio.02481-14.

Schellekens, H., Ribeiro, G., Cuesta-Marti, C., and Cryan, J.F. (2022). The microbiome-gut-brain axis in nutritional neuroscience. *Nutr Neurosci* (October): 1–13. doi: 10.1080/1028415X.2022.2128007.

Song, S., Lou, Y., Mao, Y., Wen, X., Fan, M., He, Z., Shen, Y., Wen, C., and Shao, T. (2022). Alteration of gut microbiome and correlated amino acid metabolism contribute to hyperuricemia and Th17-driven inflammation in Uox-KO mice. *Front Immunol* 13: https://www.frontiersin.org/articles/10.3389/fimmu.2022.804306.

Stahl, S.M. (1998). Mechanism of action of serotonin selective reuptake inhibitors. serotonin receptors and pathways mediate therapeutic effects and side effects. *J Aff Disord* 51 (3): 215–235. doi: 10.1016/s0165-0327(98)00221-3.

Markun, S., Gravestock, I., Jäger, L., Rosemann, T., Pichierri, G., and Burgstaller, J.M. (2021). Effects of vitamin B12 supplementation on cognitive function, depressive symptoms, and fatigue: a systematic review, meta-analysis, and meta-regression. *Nutrients* 13 (3): 923. doi: 10.3390/nu13030923.

Tamburini, S., Shen, N., Wu, H.C., and Clemente, J.C. (2016). The microbiome in early life: implications for health outcomes. *Nat Med* 22 (7): 713–722. doi: 10.1038/nm.4142.

Vuong, H.E., Pronovost, G.N., Williams, D.W., Coley, E.J.L., Siegler, E.L., Qiu, A., Kazantsev, M., Wilson, C.J., Rendon, T., and Hsiao, E.Y. (2020). The maternal microbiome modulates fetal neurodevelopment in mice. *Nature* 586 (7828): 281–286. doi: 10.1038/s41586-020-2745-3.

Winther, G., Elfving, B., Kaastrup Müller, H., Lund, S., and Wegener, G. (2018). Maternal high-fat diet programs offspring emotional behavior in adulthood. *Neuroscience* 388 (September): 87–101. doi: 10.1016/j.neuroscience.2018.07.014.

Yu, Y., Zhang, B., Ji, P., Zuo, Z., Huang, Y., Wang, N., Liu, C., Liu, S.-J., and Zhao, F. (2022). Changes to gut amino acid transporters and microbiome associated with increased E/I ratio in Chd8+/− mouse model of ASD-like behavior. *Nat Commun* 13 (1): 1151. doi: 10.1038/s41467-022-28746-2.

Zaragozá, R. (2020). Transport of amino acids across the blood-brain barrier. *Front Physiol* 11: https://www.frontiersin.org/articles/10.3389/fphys.2020.00973.

9 The Sensory Systems and Food Palatability

Conor M. Delahunty and Remco C. Havermans

Key messages

- Taste, smell, chemesthesis, somesthesis and vision are the human sensory systems that contribute to food perception and food palatability.
- Vision guides food selection, providing predictive signals for the other senses. Somesthesis evaluates texture during oral processing, preparing the food bolus for swallow and digestion. The chemical senses (taste, smell and chemesthesis) detect the flavour of foods, which is indicative of nutritional value, spoilage and potential toxicity.
- Each sense has distinct physiology and anatomy, but the psychophysical principles by which the senses detect and respond to stimuli are common to all.
- Cross-modal sensory interactions are observed when two or more perceptible components of a food system are studied together. The halo effect refers to how learning places greater reliance on one sensory modality over another. Flavour perception is influenced by the dominant visual sense and by changing food composition.
- Palatability can be defined as the hedonic reward provided by foods that are agreeable to the palate and is based on the integrated response to stimulation of all the human food senses. The performance of senses alters significantly across the lifespan; this in turn may change perceptions and palatability and influence nutritional status.

9.1 Introduction to the sensory systems

The human senses play a very important role in food choice and intake. Sight, hearing, smell, taste and touch collectively function as 'gatekeepers', distinguish the foods that are acceptable for consumption from those that should be rejected and provide an early cue to cease consumption or to switch the type of food consumed.

Visual food cues, such as overall appearance and colour of food, help us identify the food and estimate how it will taste. Touch contributes to mouthfeel, enabling us to detect and evaluate the texture of foods and changes during mastication preparing the food (bolus) for digestion. Mouthfeel includes the perception of food temperature and chemesthesis – chemically induced irritant sensations (e.g. heat, cooling, tingling) by specific food compounds (e.g. capsaicin in chili pepper or menthol). The chemical senses of taste, smell and chemesthesis (the common chemical sense) are in effect the flavour of food and are fundamental to food intake through detecting, identifying and evaluating food compounds. Importantly, experience dictates what flavours we learn to recognise indicating either nutritional value or potential harm.

The chemical senses have evolved to aid food selection. We select what is beneficial to eat and reject that which will cause ill health. From birth, we enjoy sweet taste (indicating sugars and hence energy) and reject bitter taste (associated with various toxins). In our modern world, the protective role of sensation is primarily to detect spoiled or tainted foods, as few now roam

Nutrition and Metabolism, Third Edition. Edited on behalf of The Nutrition Society by Helen M. Roche, Ian A. Macdonald, Annemie M.W.J. Schols and Susan A. Lanham-New.

Companion Website: www.wiley.com/go/nutrition/metabolism3e

the countryside hunting, gathering and sampling unusual 'food'. However, we retain the psychological foundations upon which a 'safe and nutritious' diet from initially unknown foods is built, where we observe others, are encouraged to taste by information, sample by tasting and experience postingestive consequences. This learning, often by association, results in a dietary habit of consuming foods with familiar sensory properties that tend to be rich in sugars, salt and/or fats. These dietary habits and food preferences are dynamic. Habits change as experiences with food accumulate and evolve slowly as we age. Individual differences in sensory acuity are genetically determined and inform these experiences with foods, thus affecting the acquisition of specific food preferences. Within the range of palatable foods in the marketplace, subtle differences in taste, smell or texture can therefore have a dramatic influence on individual preferences, ultimately determining quantity consumed, repeated intake and potential nutrition.

The chemical senses have independent physiology and will be considered separately in this chapter. However, during consumption, flavour is experienced as an integrated multisensory percept. Food palatability is thus determined by a combination of all sensations relevant to food acceptability, including taste, smell, temperature, touch, texture, pain, visual sensations and even auditory input. In this chapter, we will therefore consider the perception of the senses together, particularly regarding the understanding of food preferences.

9.2 The taste system

Unlike vision or olfaction, taste is a proximal sense, meaning that the primary stimulus for taste must make contact with the taste receptors, so the process of tasting begins in the oral cavity, primarily on the tongue. Taste receptors are stimulated by contact with liquid compounds, creating perceptions that ascribe distinctive taste qualities such as sweet, salty, sour, bitter and umami to these compounds.

Anatomy and physiology of the taste system

Taste receptors, located primarily on the tongue and soft palate, are termed taste buds (Figure 9.1). Each taste bud is a cluster of between 30 and 50 specialised epithelial cells. Taste bud cells end in hair-like cilia called microvilli, which make contact with the fluid environment in the mouth through a taste pore at the top of the taste bud. Taste receptors are primarily found on the microvilli, and stimulating compounds are believed to bind to these microvilli, initiating taste transduction.

Taste buds, of which there are about 6000 in total, are located within small but visible structures known as papillae. There are four types: fungiform papillae (located at the tip and sides of the tongue), foliate papillae (appear as a series of folds along the sides of the tongue), circumvallate papillae (found at the back of the tongue) and filiform papillae (cover the entire tongue surface and give the tongue a rough appearance). All papillae, apart from the filiform type, contain

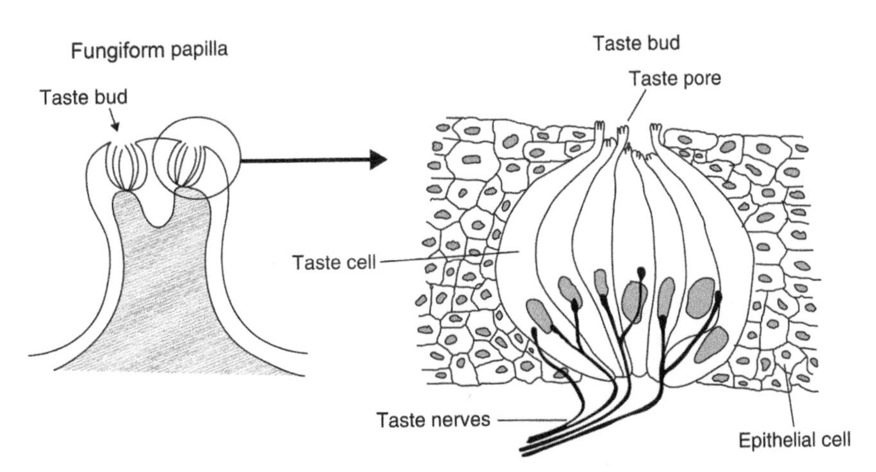

Figure 9.1 (Left) Cross-section of a fungiform papilla showing the location of taste buds. (Right) Cross-section of a taste bud, showing the taste pore where the taste enters, the taste cells and the taste nerve fibres.

taste buds. Filiform papillae grip food on the tongue and play a role in mouthfeel. Stimulation of the back or sides of the tongue produces a broad range of taste sensations. The soft palate, root of the tongue and upper part of the throat are also sensitive to tastes.

Taste papillae are supplied by a number of cranial nerves. The chorda tympani nerve conducts signals from the front and sides of the tongue, and from the fungiform papillae. The glossopharyngeal nerve conducts signals from the back of the tongue. The vagus nerve conducts signals from taste receptors in the mouth and larynx. The sense of taste is robust by comparison with other sensory modalities, and this may be partly explained by the fact that different nerves innervate the tongue. It would therefore be difficult to knock out all taste areas through trauma.

As a large range of different compound types can stimulate taste, it is not surprising that taste transduction involves a variety of mechanisms, dependent on the chemical stimulus type. In general, sweet, bitter and umami tastants signal via G-protein-coupled receptors that initiate a cascade of events leading to changes in calcium deposits and membrane permeability to various ions such as K^+ and Ca^{2+}. On the other hand, salt and sour taste transduction involves the movement of the taste active Na^+ or H^+ through ion channels. The movement of ions into taste cells produces depolarising potentials that lead to transmission of the nerve signal at synaptic terminals via release packets of chemical neurotransmitters, which stimulate the next cell in the sequence.

Each taste cell has a lifespan of about 10 days. New cells differentiate from the surrounding epithelium and during their lifetime migrate towards the centre of the bud, where they eventually die. Differences in cell type within the taste bud represent differences in age and development. As a person ages, taste cell replacement may slow, and this may have adverse effects on taste function.

Taste coding

The taste neurons branch before entering the taste papillae, and branch again inside the taste bud; therefore, a single nerve fibre may innervate more than one papilla and a large number of taste buds.

Two theories have been proposed to explain taste coding or how we can distinguish different taste qualities. The first is across-fibre pattern coding. The principle of this is that a taste compound stimulates many nerve fibres (groups of neurons), but not all to the same extent, and the pattern of stimulation across fibres is used to recognise, or code, the stimulus quality. The alternative theory is that of specificity coding, which assumes that some neurons respond best to specific compounds. However, results show that both mechanisms operate together. In general, individual taste cells respond to several types of chemical stimuli. However, taste cells can display selectivity, in particular with low threshold concentration compounds. The firing of specific neurons may identify, for example, compound group or type, and then more subtle qualities of a compound can be determined by the pattern of firing across large groups of neurons.

Taste quality

It is widely recognised that there are five basic taste qualities, referred to as sweet, salty, sour, bitter and umami. Umami is a recognised 'fifth taste' that has been accepted more recently, particularly in Japan and other cultures where it is most familiar. There is also evidence of additional 'tastes', including metallic taste, ability to detect calcium, starchiness and a 'fat taste'.

There have been attempts to define, or classify, the stimuli for specific taste qualities. Such a definition might explain an evolutionary basis for quality discrimination that is linked to the nutritional value of foods. Sweet taste is typical of sugars and other carbohydrates, although many artificial sweeteners are not carbohydrate. Salt taste is typical of salts of alkali metals and halogens, although again not without exception. Sour taste is very closely linked to the pH of a substance. Most bitter substances are organic compounds having biological activity, and many of them are poisonous. Thus, bitter taste seems to be a poison detector. Umami is used to describe the taste of monosodium glutamate (MSG) and ribosides, such as salts of 5′-ionosine monophosphate (IMP) and 5′-guanine monophosphate (GMP). Europeans or Americans generally describe the umami taste as 'brothy', 'savoury' or 'meaty'. Umami taste guides the intake of peptides and proteins. A 'fat taste', which is thought to detect free fatty acids, may have evolved to select high-energy foods, foods containing fat-soluble vitamins and foods containing essential fatty acids.

Some compounds have a predominant taste: sodium chloride (NaCl) is predominantly salty, hydrochloric acid (HCl) is predominantly sour, sucrose is predominantly sweet and quinine predominantly bitter. However, most compounds have more than one taste quality. For example, potassium chloride (KCl) has substantial salty and bitter tastes, and sodium nitrate ($NaNO_3$) tastes of salty, sour and bitter.

Table 9.1 Absolute taste thresholds for common taste compounds.

Compounds	Threshold[a] (molar concentration)[b]
Bitter	
Urea	0.015
Caffeine	0.0005
Quinine HCl	0.000014
Sour	
HCl	0.00016
Acetic acid	0.00011
Citric acid	0.00007
Salt	
NaCl	0.001
KCl	0.0064
Sweet	
Glucose	0.0073
Sucrose	0.00065
Aspartame	0.00002
Umami	
Monosodium glutamate	0.0005

[a]Thresholds are based on a compound's major taste quality, although several have additional tastes.
[b]Molar concentration represents the number of grams of solute divided by its molecular weight, per litre of solution.

The role of saliva in taste

Saliva plays an important role in taste function. It contains a weak solution of NaCl, and ionic constituents of chlorides, phosphates, sulfates and carbonates, as well as other organic components, proteins, enzymes and CO_2. It contains substances that modulate taste response, but saliva is mostly water acting as a carrier for sapid compounds to the taste buds. Saliva is not necessary for taste response as extensive rinsing of the tongue with deionised water does not inhibit the taste response. Such rinsing actually sharpens taste due to release from adaptation to some of saliva's components. For example, the absolute threshold for NaCl is decreased significantly.

Taste thresholds

Two types of taste threshold measurements can be considered. Absolute thresholds refer to the minimum concentration of a substance or compound that can be detected. Table 9.1 presents absolute thresholds for common compounds. Differential thresholds refer to the just noticeable difference between two levels of concentration above the absolute threshold (i.e. suprathreshold). At absolute threshold, a taste sensation may not have any discernible quality, but as its concentration in solution increases, quality becomes more defined. The taste quality of a compound may also change as concentration continues to increase. For example, NaCl tastes sweet at very low concentrations, and salt solutions below the concentration of saliva may give rise to bitter tastes. In general, intensity of perception increases with increases in the physical concentration of a compound. Psychometric taste functions relating concentration to response are sigmoidally shaped, with an initial flat portion where response at levels below threshold hover around a baseline or background noise level, an accelerating function as threshold is surpassed, and a decelerating function that eventually becomes flat as receptor sites are filled, or as the maximum number of sensitive nerves respond at or near their maximum frequency (Figure 9.2).

The full range of tastes can be detected on all parts of the tongue. This is contrary to the 'tongue map' myth, wherein it was proposed that different taste sensations are located exclusively on certain areas of the tongue. However, not all regions of the tongue are equally sensitive to all chemical stimuli: for sweet taste, threshold is lowest at the front; for sour taste, the rear sides are most sensitive; for salt taste, the front and sides are most sensitive; and for bitter taste, the front of the tongue and the soft palate are most sensitive. This relative localisation of taste sensitivity may aid in identification of food, manipulation of the food bolus or in removal of selected portions of the bolus that have an unacceptable taste quality. However, this spatial ability also creates an obstacle for the development of substitute sweeteners and salts, as these generally have more than one taste quality and differ in the part of the tongue that they stimulate from the pure tastes of sucrose and NaCl, which they seek to replace.

Widespread individual differences in taste sensitivity are observed for most compounds, especially for bitter compounds, intensive sweeteners

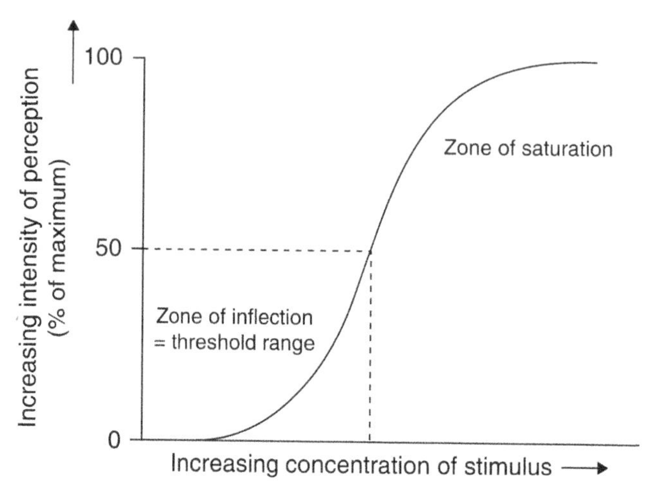

Figure 9.2 Psychometric function of increasing intensity of perception of a taste stimulus with increasing concentration of the stimulus in solution.

and multiquality salts. For some compounds, the range in sensitivity across population can be several orders of magnitude. There is a genetically inherited difference in sensitivity to the bitter compounds 6-n-propylthiouracil (PROP) and phenylthiouracil (PTC). Individuals may be classified into one of three PROP/PTC sensitivity statuses: non-tasters, medium tasters and super tasters. Anatomical differences have also been observed between groups with different PROP taster status, as papillae and taste bud numbers are correlated with taster status. PROP status has also been related to overall taste sensitivity, and to chemical irritant and somesthesis sensitivity. However, these relationships are tentative and a high threshold for one compound does not predict high thresholds for all compounds.

Taste mixtures

The taste qualities of mixed compounds are perceived separately and no taste quality results from a mixture that is not present in the individual constituents of the mixture. Sapid compounds in the mixture cannot mutually suppress one another to produce a tasteless mixture. Consumers often assume that sweet and sour are opposite in taste quality. In fact, when acetic acid and sucrose are mixed, the resultant solution tastes both sweet and sour.

Mutual suppression can raise the threshold of each mixed compound and if one compound is stronger than the other, it will dominate and may mask the weaker taste entirely. A solution of quinine and sucrose is less sweet than an equimolar concentration of sucrose tasted alone. Similarly, the mixture is less bitter than equimolar quinine. In general, this inhibitory criterion applies to mixtures of all four basic taste qualities and is referred to as mixture suppression. Since most foods contain more than one taste compound, mixture suppression is important for determining a balanced sensation and providing overall appeal. For example, in fruit juices the unpleasant sourness of acids can be partially masked by the pleasant sweetness from sugars. Salt may be added in processed foods to suppress bitterness. However, it is difficult to characterise taste mixture interactions without also considering the sequential effects of adaptation, a concept that will be dealt with later in this chapter. PROP taster status can also manifest itself in taste mixture sensitivity, as people who are not sensitive to PROP will not show some mixture suppression effects on other flavours, particularly when bitterness is present in the mixture. The relative intensity of other flavours may therefore be enhanced.

The sense of taste also demonstrates integrative, or additive, effects across the taste qualities. Solutions of taste compounds, regardless of quality, cannot be perceived when diluted below threshold. However, when two solutions of different taste quality that are diluted to 50% below absolute threshold are mixed, the mixture can be perceived. This additive effect has been demonstrated for up to 24 solutions of varied quality diluted to 1/24th of their absolute threshold.

9.3 The olfactory system

Like taste, the primary stimuli for smell must also interact with the olfactory receptors. However, the stimuli for smell are airborne compounds of volatile substances. The odour-stimulating compounds create perceptions that are endowed with distinctive smells. The olfactory system responds to odour (sensed orthonasally) and aroma (sensed retronasally from volatile compounds released in the oral cavity during consumption). Saliva plays an important role in retronasal olfaction. As stated above, saliva is a solvent that is mixed with food by mastication. The structural deformation and mixing of food have a significant influence on the release and availability of flavour-active compounds, both sapid and volatile.

The largest contribution to the diversity of food flavours comes from volatile compounds sensed by the olfactory system, and much of what we commonly refer to as 'taste' is incorrectly localised smell detection. The significant contribution of smell to flavour can easily be demonstrated if you pinch your nose shut whilst eating, effectively blocking air circulation through the nasal passages. Familiar foods will not be recognised and it is even possible to confuse apples with onions if tasting blind.

Anatomy and physiology of the olfactory system

The receptors for odour are olfactory cells (Figure 9.3). These cells are long, narrow, column-shaped cells, each less than 1 micron in diameter. Olfactory

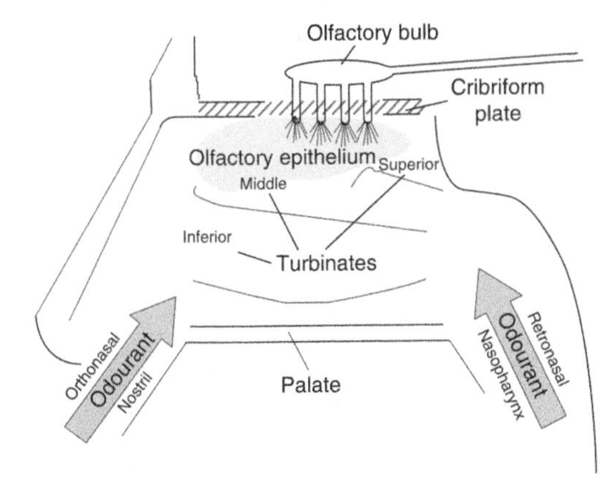

Figure 9.3 The human nasal cavity, illustrating the olfactory receptors, the olfactory epithelium and the different means of entry for volatile compounds.

cells are true nerve cells. There are between 6 and 10 million olfactory cells in the human nose, located within the olfactory epithelium, a region of tissue of around 5 cm^2 in area, located high in the upper part of each of two nasal cavities. A layer of mucus coats the olfactory epithelium. This mucus provides protection for the sensitive olfactory cells and regulates transport of olfactory stimuli to the receptors, aiding in smell quality determination. The olfactory receptors have very fine fibres (cilia), which project into the mucous layer, where they contact the stimulus. The cilia serve to enlarge the surface area of the cell enormously so that it is about the same size as the total skin area for humans. Extending from the other end of the olfactory cells are nerve fibres, which pass through the cribriform plate of the skull into the olfactory bulb and connect at a junction called the glomerulus, then further connect with other parts of the brain by olfactory tracts. Therefore, olfactory receptors both receive and conduct stimulus.

During inspiration, or when air refluxes retronasally from the mouth carrying aroma during eating, air is directed into the nasal cavity below the level of the olfactory epithelium. At the interface of the air with the mucus layer covering the olfactory receptors, there is an opportunity for chemically selective processes to take place. Compounds first partition from the air into the mucous medium and therefore for any compound to have smell potential, it must be soluble in the olfactory mucosa. These compounds may then bind to other compounds, called olfactory receptor proteins, of which there are many hundreds of different types. Compounds eventually stimulate the cilia and, by some poorly understood biochemical process, provoke the olfactory neural activity. The more vigorous the inhalation, the more of the olfactory epithelium is bathed by the odorant and the greater the stimulation. Olfactory receptors transmit signals directly to the olfactory bulb in the brain, where signals are then processed before being sent to the olfactory cortex and to the orbitofrontal cortex. In addition, olfactory nerves project to many different sites in the brain, some of them closely associated with emotion and memory.

Smell coding

Olfactory receptors respond to many hundreds of different odour-active compounds, and some scholars estimate that the human nose can discern trillions of odours. This may be an overestimation but the important point is that contrary to common belief, the human sense of smell is quite advanced, also in comparison to other species. We tend to underestimate the significant role our sense of smell plays in our daily lives.

To account for our incredible discriminative ability, it is likely that smell is coded by a pattern of responses across many receptors. The ability of the system to form a pattern is facilitated by differential movement of potential odorants through the olfactory mucosa. Some volatile compounds are strongly attracted to the mucosa, so they flow slowly through it and are potentially deposited close to its surface. Others are more weakly attracted to the mucosa and so flow more rapidly and stimulate receptors more uniformly throughout the mucosa. Each compound creates a different pattern of receptor stimulation because of differences in how each is deposited on the receptor surface, termed a regional sensitivity effect. In addition, the stimulus causes widespread and diffuse activity in the olfactory bulb.

Areas of the nervous system, higher than the olfactory bulb, may be important in interpreting the complex neural activity, and the orbitofrontal cortex has been implicated. The amount of neural activity increases with increases in the concentration of a given odorant and it is also possible that different odorants, owing to specific differences in molecular properties, produce specific and distinctive spatial and temporal patterns of activity.

Smell quality

The sense of smell can distinguish and recognise a wide range of qualities. However, researchers have yet to identify a relationship between a compound's smell and a physical property that causes that smell. Compounds that are structurally very different from one another can smell the same, whereas compounds almost identical in structure can have very different smell qualities. In addition, many smells that we perceive as unitary are in fact complex mixtures of many volatile compounds. For example, the smell of coffee contains over 800 volatile compounds, although only 20 or 30 of these are above absolute threshold. Adding to the complexity, different individuals may describe the same sensation in different ways, although fixed terminologies and objective descriptive analysis techniques are now used to overcome this problem. However, terminology systems work better on a product-by-product basis (e.g. there are agreed terminologies for wine, beer and other product categories) than across product categories.

There have been many attempts to classify odour quality and researchers have searched for primary odours, believing that their discovery will elucidate odour quality as did the discovery of the primary colours (see section on Vision below). The stereochemical theory proposes seven primary odours (camphoraceous, musky, floral, minty, ethereal, pungent, putrid). This theory proposes that all compounds with similar odour quality have similar geometric structures that fit specific olfactory receptor sites. However, this theory is not widely accepted as there is no evidence of clear and distinct receptor sites and some compounds of similar quality have different structures. Individual odorants can affect a range of olfactory receptors, allowing for a pattern of neural activation, inhibition and antagonism that changes over time. This pattern is a complex odorant code that is not the direct translation of the structure of the volatile. It explains why we can reliably distinguish an enormous number of odorants, even when odorants differ in merely a single molecule.

Smell thresholds

There are several million olfactory receptor cells on each side of the nose, and these are highly ciliated, with 6–12 cilia per cell. The cilia greatly increase the surface area of exposure. In the olfactory bulb, a relatively large number of olfactory cells converge on a single glomerular cell, which in turn proceeds directly to the higher cortical regions. There is therefore neural convergence from the cilia to the olfactory cells, and again from the olfactory cells to the olfactory bulb. This general funnelling of olfactory information makes the olfactory system extremely sensitive.

As for taste, absolute and differential thresholds can be determined for smell. Olfaction has far lower absolute thresholds to concentration than does taste and may be 10 000 times more sensitive. On the other hand, the sense of smell has a poor ability to discriminate intensity levels and differential thresholds are high relative to other sensory modalities (in the region of 15–25% concentration change is required before a difference can be detected). Odour thresholds are not related only to the concentration of compound. Odour activity is also related to the physical and chemical properties described earlier. The absolute odour thresholds for some common odour compounds are presented in Table 9.2. Methyl mercaptan, which has the objectionable smell of rotten cabbage, can be detected at extremely low concentrations. Methyl salicylate, the smell of wintergreen, has a relatively high threshold whereas water, which is also volatile, is odourless.

Odour thresholds can vary significantly, both within and between people, and some people with an otherwise normal sense of smell are unable to detect families of similar-smelling compounds. This condition is termed specific anosmia and can be defined as a smell threshold more than two standard deviations above the population mean. Specific anosmias identified include that to androstenone, a component of boar taint, cineole, a terpene found in many herbs, diacetyl, which is a butter-like smell important for dairy flavour, and trimethyl amine, a fish spoilage taint. Thresholds for odour may also be affected by an interaction of gender and hormonal variation in an individual. Olfactory sensitivity changes during puberty. For example, male sensitivity for various compounds has been reported to decrease with puberty, whereas female sensitivity remains constant. Sensitivity for food odours seems largely exempt of changes during puberty but that requires more research. It has been suggested that pregnant women show heightened smell sensitivity, especially in the first trimester, but there is no definitive evidence for such an effect.

The olfactory sense is fragile. Mild head trauma may cause complete loss of smell and this anosmia is often chronic. Most people experience temporary olfactory dysfunction due to nasal congestion caused by an upper respiratory tract infection (e.g. common cold). Anosmia in the absence of nasal congestion was a notable early symptom of initial variants of COVID-19 that was presumably due to virus-induced olfactory neuron dysfunction. The sense of smell does generally recover but olfactory dysfunction is one of the most persistent symptoms of COVID-19.

Smell mixtures

Like taste, the sense of smell also shows mixture interactions. In fact, it is smell mixtures and no single odorants that are encountered in almost every instance of everyday life. This may be why the sense of smell is limited in identifying individual odours even in the simplest of mixtures and in deciding whether a stimulus is a single odorant or a mixture.

There are several basic principles that hold in general: odours of different quality tend to mask or suppress one another, or may remain distinct, whereas odours of similar quality tend to blend to produce a third unitary odour quality. The extent to which these interactions occur depends on the compounds that are mixed and can be influenced by chemical interaction between compounds, or by interaction or filtering at the olfactory receptor sites. The perception of the mixture is also determined by the odour thresholds of the compounds that are mixed, their concentration response functions and their potential to cause adaptation (adaptation will be discussed in more detail later in this chapter).

In general, odour mixtures bear a strong resemblance in character to the quality characteristics of the individual components and obtaining

Table 9.2 Absolute odour thresholds for common odour compounds.

Compound	Smell quality	Threshold[a]
Methyl salicylate	Wintergreen	0.100
Amyl acetate	Banana	0.039
Butyric acid	Rancid butter	0.009
Pyridine	Burnt, smoky	0.0074
Safrol	Sassafras (woody-floral)	0.005
Ethyl acetate	Pineapple	0.0036
Benzaldehyde	Almond	0.003
Hydrogen sulfide	Rotten eggs	0.00018
Courmarin	Hay-like, nut-like	0.00002
Citral	Lemon	0.000003
Methyl mercaptan	Rotten cabbage	0.0000002

[a]Milligrams per litre of air.

a wholly new odour as a function of mixing is rare. However, in very complex mixtures, such as those that are typical of cheese or wine, often odours of dissimilar quality can be found. In cheese, for example Cheddar, subtle fruity character is provided by fatty acid esters. However, this character complements the dairy characteristics of the cheese because of the inherent complexity of a blend of up to 30 compounds that are above threshold. In fact, many natural odours are mixtures of many chemical components and none of the individual components of the mixture can produce the complexity of mixed odour on its own. It is this complexity and fine balance of natural odours that has made it difficult for flavour companies to reproduce accurately the character of natural flavours in manufactured foods.

9.4 Somesthesis and mouthfeel

Texture is a property not perceived by a single sense (or receptor type), as are taste or odour. As texture is a less specific construct of many variables, it requires multireceptor integration for perception, including exteroception (e.g. the touch of food structures) and interoception (e.g. the proprioception of muscle movement). In broad terms, the somesthetic sense is responsible for the detection of food texture during oral processing. Somesthetic receptors are found in all regions of the oral cavity, including the lips, tongue, teeth and mucosa. Table 9.3 presents the sensory basis of texture perception in mouth.

Sensations of the food attributes are gathered during each phase of the mastication process. As referred to earlier, the somesthetic sense is innervated by the trigeminal nerve (fifth cranial nerve). Therefore, somesthetic sense receptors perceive chemical stimuli by chemesthesis and structure, rheology, pain and temperature stimuli. The auditory sense also plays a role in texture perception as sounds perceived during oral processing are related to texture attributes such as crispness and crunchiness.

Chemesthesis

Chemesthesis is the term used to describe the sensory system responsible for detecting chemical irritants. Detection is more general than that of taste and smell, and takes place primarily in the eyes, nose and mouth. The perception is closely related to the somatosensory characteristics of texture, pain and temperature change, and provokes a strong behavioural response. The primary function of chemesthesis is to protect the body from noxious chemical stimuli, but this high-influence sense has also been exploited to great effect commercially. The fizzy tingle of carbon dioxide (CO_2) so desired in soft drinks, the cooling sensation of menthol in mint sweets or toothpastes and the burning sensation of chilli in curry or salsa are perhaps the best examples of how chemical irritation can provide additional character that is very much desired.

Anatomy and physiology of chemesthesis

The general chemical responsiveness to irritation is mediated by nerve fibres in the trigeminal nerve, which innervate the mucosae and skin of the mouth, nose and eyes, as illustrated in Figure 9.4. The trigeminal nerve has three main branches: the ophthalmic, the maxillary and the mandibular. The chorda tympani and glossopharyngeal nerves, which serve taste reception, may also convey information about chemical irritation. All the areas served by the trigeminal nerve are sensitive to chemical irritants. Tears caused when cutting onions readily demonstrate sensitivity in the eye, whereas the pungency of mustard readily demonstrates

Table 9.3 Somesthetic sense receptors and role in texture perception.

Mechanoreceptors	Detect mechanical pressure or distortion and therefore perceive tactile stimuli or the physical structure of food.
Proprioceptors	Sense the static position and movements of the tongue and jaws during mastication, and the resistance to jaw movements during biting and chewing. Proprioception and kinesthesis are sometimes used interchangeably.
Periodontal receptors	Detect forces applied to the teeth.
Nociceptors	Detect potentially damaging stimuli, resulting in the perception of pain.
Thermoreceptors	Detect absolute and relative changes in temperature. There are cold receptors and warmth receptors that work together.

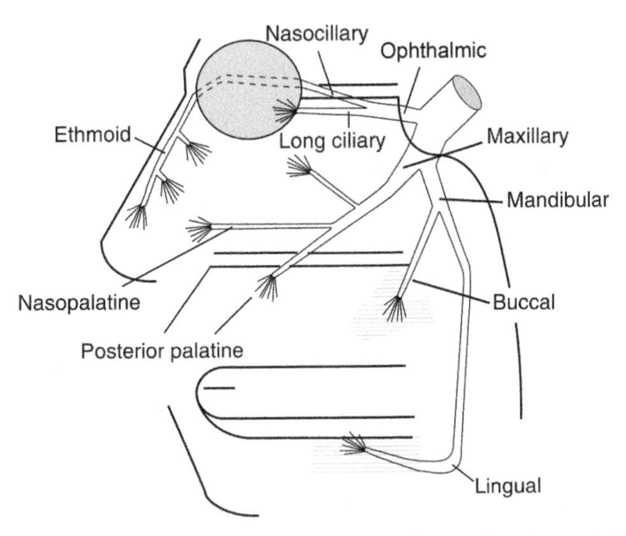

Figure 9.4 The branches of the trigeminal nerves that innervate the nasal and oral cavities. A person is bilaterally symmetric with two of each of the branches illustrated.

sensitivity in the nose. Chemical irritant sensitivity is by no means of less importance than that of taste or odour, and as already mentioned can have a strong influence on food acceptability.

Chemesthesis quality

The anatomical requirement to enable perception of different irritant qualities appears to be present, and there appears to be variety in the quality of irritative experiences. However, these differences could arise from intensity differences or additional taste or odour qualities. Examples of sensations arising from chemesthesis include the fizzy tingle of CO_2, the burn of hot peppers, the cooling of menthol, the nasal pungency of mustard and horseradish, the oral pungency of spices such as ginger, cumin and black pepper, and the bite of raw onions and garlic. However, each of these substances has additional qualities that stimulate other parts of the chemesthesis anatomy or stimulate olfaction and taste. Mustard, onions and garlic also have lachrymatory effects (tearing), not to mention strong odour. Menthol has several sensory properties, including cooling, warming, odour and other sensory effects depending on isomeric conformation, concentration and temporal parameters of exposure.

On the other hand, most compounds, including those that stimulate common odours and tastes, have some degree of irritation. For example, completely anosmic individuals can detect many odour compounds, presumably from the ability of odorants to stimulate the trigeminal nerve branches in the nasal cavity. The trigeminal nerve also signals somesthetic sensations (i.e. tactile, thermal and pain) and therefore the distinction between a true chemical sense and a somesthetic sense becomes blurred.

Chemesthesis thresholds

Psychophysical principles (absolute thresholds, differential thresholds, sigmoidal shaped concentration response functions) can be applied to understanding the relationship between concentration and perception of chemical irritant stimuli. However, such relationships are often difficult to study as irritant stimuli can cause desensitisation following stimulation, which can be long lasting. On first stimulation with capsaicin, the irritant compound of chilli pepper, a warm, burning, painful sensation is perceived. This is termed sensitisation. However, repeated stimulation with the same concentration of capsaicin can be without effect as the receptors become desensitised. The quantity of irritant compound consumed is also important. Irritation from capsaicin continues to build to higher levels if stimulation proceeds in rapid sequences without allowing sufficient time for the senses to recover. The absolute threshold of capsaicin has been measured at below 1 ppm. In pure form, capsaicin causes a warm or burning sensation, with little or no apparent taste or smell. Piperine, black pepper, is around 100 times less potent.

The rate of increase in perceived intensity as a function of concentration for common chemical irritants is often high relative to those for tastants and odorants. The relationship between the concentration of CO_2, which provides the fizziness in soft drinks, and perceived fizziness is greater than 1:2. This response to stimulus is much greater than that typical of other sensory modalities, apart from electric shock. The long-lasting nature of irritant stimulus is an important aspect of both quality and threshold of perception. Stimulation with capsaicin or piperine with concentrations above threshold may last 10 minutes or longer.

In addition, individual thresholds may change depending on regularity of exposure. People who regularly consume chilli peppers, or spices derived from chilli pepper, show chronic desensitisation. Interestingly, often reported 'addiction' to spicy food may have a physiological basis. Not only does frequent consumption of chili peppers desensitise a person to the burning sensation of capsaicin, but the chili itself triggers the release of endogenous opioids that may reinforce a liking for spicy foods.

Texture

The International Organisation for Standardisation defines the texture of a food product as all the rheological and structural (geometric and surface) attributes of the product perceptible by means of mechanical, tactile and, where appropriate, visual and auditory receptors. In this section, only the texture perception in-mouth will be considered, as visual perception will be described later.

It is most likely that the primary role for texture perception is to determine progress in oral processing towards food bolus preparation, ensuring that the bolus adequately passes to the stomach for further digestion. However, texture perception also contributes significantly to the palatability of foods (think soggy biscuits or hard bread). In picky eaters, food texture has been found to be the main determinant for food rejection. And texture contrast, both within and across foods (e.g. during a meal), is considered important for palatability of the food and of the meal. For most people, a meal would not be very appetising if all its foods were blended into a homogeneous blob.

Texture is important for food recognition. When texture is removed by blending, it is very difficult to recognise the food. Texture perception can also serve to determine if the food is homogeneous in texture or has the texture that is expected based on what is familiar.

Texture quality

Texture perception is dynamic in that it is continuously changing (dramatically when compared with flavour perception) as the food is manipulated and deformed, and by its very nature often requires movement of the food over the somesthetic sense receptors. Sensations of the food attributes are gathered during each phase of the mastication process. Sounds perceived during oral processing are related to crispness and crunchiness, and may give cues to differences in hardness, denseness and fracturability. In addition, the duration of sound may indicate differences in freshness (e.g. crisp apple) or toughness.

The most widely accepted classification of texture qualities contains three categories of terms: geometrical characteristics, mechanical characteristics and moisture-/fat-related characteristics. Geometrical characteristics relate to the size, shape and orientation of food particles, and are perceived mainly by the tongue and hard palate. Common quality descriptions include gritty, grainy, coarse, fibrous and crystalline. Mechanical characteristics relate to the reaction of the food to stress and include qualities such as hardness, cohesiveness, viscosity, elasticity, adhesiveness, brittleness, chewiness and gumminess. Moisture- and fat-related characteristics include dryness/wetness, oiliness and greasiness. To evaluate texture accurately, one should pay careful attention to the time during oral processing when an attribute can be perceived and include this in the quality description. Some researchers now evaluate the 'trajectory' of texture perception qualities during oral processing.

Texture thresholds

The oral cavity is one of the most sensitive regions of the body to touch, with the tip of the tongue and the hard palate having the lowest detection thresholds. Very little research has considered absolute thresholds and differential thresholds for specific food texture stimuli, or differences between individuals in this regard. However, there are large interindividual differences in the oral physiology parameters (i.e. chewing behaviour, number of chews and rate, tongue movements) that assist oral processing, and there are differences in sensitivity of the somesthetic receptors. It is therefore likely that individual food–subject interactions will lead to differences between individuals in texture perceptions, and this warrants further research. It has been demonstrated that older people are less sensitive to particle size than younger persons, with potential impact on food palatability.

Saliva and mouthfeel

When eating, the structural deformation and mixing of food with saliva to form a bolus gives rise to differences in texture perception. With regard to chemesthesis, there is strong correlation between increased oral chemical irritation and increased salivary flow rate. Saliva keeps the mouth lubricated, with salivary proteins and mucopolysaccharides giving it its slippery and coating properties. The importance of saliva lubrication is most noticeable when it is lost following strong astringent sensation. Astringency is a sensation that does not result from stimulation of the chemical senses per se, although it is caused by chemical stimuli. Astringency is not a taste sensation but a type of mouthfeel and is experienced as a dry and rough mouth, and a drawing, puckering or tightening sensation in the muscles of the face. It is believed that tannins or acids (as in wine, tea or coffee) binding to salivary proteins and inhibiting their lubricating function cause the dominant astringent sensations. It has also been demonstrated that astringent sensations can result without loss of saliva lubrication for some types of compounds, leading to the conclusion that astringency can result from only the tactile sensation of a precipitate in the mouth.

9.5 The visual system

The visual system is our main portal on the outside world and as a result tends to dominate other sensory systems. We use vision to gain information about food (colour and overall appearance), and about the immediate environment of the food (from where and from what packaging it comes) before choosing it for consumption. Visual assessment provides predictive signals for the other senses based on associations that have been learned by experience. For example, anyone familiar with bananas will expect a green banana to have unripe flavours and a firm texture, a blackened banana to be very sweet and soft, whereas a yellow banana will be most suitable for consumption and will probably be chosen as such. This visual dominance, and the expectation that it creates, is easily demonstrated when familiar colour–flavour combinations are confused in a food product, for example when a banana-flavoured food is coloured red or a clear beverage is intensely flavoured.

The effect of visual cues on flavour expectations is not limited to the colour of a food or packaging. An attractive presentation of a meal raises the expectation of exceptional care being invested in that meal. Beautiful plating may enhance the palatability of a meal.

Anatomy and physiology of the visual system

The visual system consists of the outer eye, or anterior cavity, which includes the cornea, aqueous humour and iris, and the inner eye, which includes the lens, vitreous body and retina, which is connected to the optic nerve. The iris, which is mounted between two ciliary muscles, regulates the amount of light entering the eye (the size of the pupil is determined by the iris) and the lens flexes to focus the image on the retina, accommodating vision of close or far objects. The fundus of the eye includes the retina, macula, fovea, optic disc and retinal vessels. The retina consists of two types of photoreceptor cells (which are the visual receptors): approximately 120 million rods, which are sensitive to low-intensity black and white vision (scotopic), and approximately 6 million cones, which operate at higher light intensities and are sensitive to colour (photopic vision). The optic disc, also called the blind spot, is the part of the retina where there are no photoceptors; it is where all the axons of the ganglion cells exit the retina to form the optic nerve. The macula is an oval area lateral to the disc used for central vision. The fovea is a small depression in the macula composed almost entirely of cones. The optic nerve leads to the lateral geniculate bodies and then to the cerebral cortex, where visual images are processed.

Visual quality

Colour perception is the brain's response to stimulus of the retina that results from the detection of light after it has interacted with an object. Wavelengths in the visual portion of the electromagnetic spectrum (~380–760 nm) may be refracted, reflected or absorbed by an object that is viewed. It is the reflected light that is seen by the eye. The perceived colour of an object is affected by three entities: the physical and chemical composition of the object, the spectral composition of the light source illuminating the object, and the spectral sensitivity of the viewer's eye. Changing any one of these entities can change the perceived colour of the object. Humans can perceive three primary colours (trichromacy): red, green and blue. This is because cones contain three colour-sensitive pigments, responding to red (two polymorphic variants at 552 nm and 557 nm), green (at 530 nm) or blue light (at 426 nm) most sensitively. Unlike taste and smell, colour can be relatively easily classified, as all colours consist of a combination of these three primaries.

The colour of an object can also vary in three dimensions. The hue describes the basic 'colour', such as red or green, and depends on differences in absorption by the object of various wavelengths. The value describes the relative lightness or darkness, indicating the relationship between reflected and absorbed light without regard for wavelength. The chroma describes the saturation or proportion of chromatic content. Chroma determines the purity or vividness of the colour (consider pure green compared with grey green).

The colour of food is due to compounds (termed pigments) that are contained in the food, such as chlorophylls (green), carotenoids (yellow/orange/red), anthocyanins (dark blue/purple) or myoglobin (a protein in meat giving it colour anywhere between pink and brown depending on the level of myoglobin oxidation). In processed foods, such compounds can be natural (either present naturally in the food or nature-identical but added) or artificial.

The visual sense has several characteristics useful for a consistent evaluation of foods in an environment that changes regularly. Object recognition is generally driven by the spatial, temporal and light–dark properties of the object, rather than by its chromatic properties. We can also discount the influences of illumination and recognise a prototypical colour in familiar objects. Objects also have a colour constancy that is served by colour memory and chromatic adaptation.

In addition to colour, we detect appearance attributes that help us to judge food quality. The appearance of an object, its gloss, size, shape and viscosity are determined by the optical properties associated with the object. Light can be distributed over the surface if the object is opaque or within the object if it is translucent. Gloss, transparency, haziness and turbidity are properties attributable to the geometric manner in which light is reflected. Uneven reflection of light from a surface can make an object appear dull or matte. In addition, irregular, patterned or particulate objects reflect light diffusely, whereas smooth objects reflect in a directional manner. Gloss or sheen results where reflection is stronger at a specific angle or in a beam.

As mentioned earlier, humans allow their visual sense to dominate their sensory judgement. This is based on learned experience, as colour is very often related to other sensory attributes, in particular flavour. Red is typical of strawberry, orange is typical of orange citrus, whereas green is typical of lime. Texture contrast, such as lumps, perceived by appearance, will also be perceived in the mouth. The colour of fruit may indicate how ripe it is, and hence its flavour intensity and firmness. The viscosity of a fluid can be assessed visually by pouring the fluid from a container or by tilting the container. However, if the association does not hold in a food due to manufacturing choices, the dominant visual sense can be manifest as a 'halo effect'. Examples of foods where colour gives the impression that is often not reflected in the sensory properties that are tasted (although most consumers think that it is) include white and yellow margarine or butter, white and red Cheddar cheese, brown and white shelled eggs, or bread made from bleached or unbleached flour. On the other hand, can one imagine eating a meal where foods have been unusually coloured, for example green steak accompanied by brown peas and red French fries? This can be relatively easily achieved using flavourless food colours, but it is certain to reduce the palatability of the familiar meal.

Importantly, there is no colour or appearance that makes any food unacceptable a priori. Experience dictates the relevance and meaning of visual food cues, which explains that certain dishes in a cuisine may look disgusting to anyone unfamiliar with that cuisine. The inability to form associations between visual cues and flavour sensations in cases of congenital blindness has been linked to increased taste thresholds and reduced taste identification ability.

9.6 Adaptation

Adaptation can be defined as the decrement in intensity or sensitivity to a stimulus under constant stimulation by this stimulus. Adaptation results in a higher threshold or a reduced perception of intensity. It is an important operating characteristic of the sensory systems that enables perception of change: the status quo is rarely of interest. We use adaptation constantly to adjust to our changing environment by becoming unresponsive to stimuli that are stable in space and time. An obvious example of adaptation in everyday life is as follows.

Consider entering an active kitchen, filled with the odour of food. After a short time there, you will no longer notice the odour which on entering you could perceive strongly. If you leave the kitchen for a time and return, the odour will be perceived as strongly as before. What is experienced in this example is a feature of all adaptation. Threshold will increase following periods of constant stimulation and ratings of perceived intensity will decrease over time and may even approach a judgement of 'no sensation'. If the stimulus is removed, then sensitivity returns. The fact that we cannot taste our own saliva, although it contains NaCl and other potentially sapid compounds, is another everyday example of necessary adaptation. Similarly, our odour world is continually filled, so that usually we are at least partially odour adapted.

In terms of visual adaptation, consider walking from a light to a dark room or cinema. The adjustment of vision to enable perception is termed dark adaptation. When one returns to the light, one adapts again; this is termed light adaptation. Chromatic adaptation also occurs where the same object is observed under various sources of illumination. This can be demonstrated by examination of white paper under these conditions, where the paper approximately retains its white colour.

Apart from visual texture adaptation, there is little reported regarding adaptation to somesthesis when food is in the mouth. However, it is expected that there will be adaptation, just as we stop feeling our clothes soon after putting them on. Differences have been observed for somesthesis that was direct or indirect; no adaptation was found for roughness perceived by direct touch whereas adaptation was observed for roughness perceived when using a probe.

Adaptation also has a role to play in the perception of mixtures. For example, the sweetness of sucrose and the bitterness of quinine are each partially suppressed when presented in a mixture. However, following adaptation to sucrose, the bitterness of the quinine sucrose mixture will be perceived at a higher intensity, comparable to that of an equimolar unmixed quinine solution. The opposite occurs following adaptation to quinine. Wine has dominant sweet and sour tastes. If wine is consumed with a salad dressed with a vinegar base (which is essentially sour), it will appear sweet to taste. If the same wine is consumed with dessert (which is essentially sweet), it will appear sour to taste. Similar principles hold for odour mixtures. Adaptation to one component of an odour mixture can make another stand out. Returning to wine, differential adaptation may explain why wine character appears to change with time of exposure.

Compounds of the same quality tend to cross-adapt. For example, after adapting to NaCl, the salty taste of other salts is also perceived to be less intense and other taste qualities are unaffected. Adaptation to sucrose reduces sensitivity to other sweet compounds and, similarly, adaptation to one odorant may affect the threshold of another odorant.

9.7 Cross-modal sensory interactions

Cross-modal sensory interactions are often observed when two or more perceptible components of a food system are studied together. The representations from each modality are brought together in multimodal regions of the brain, such as the orbitofrontal cortex, where the signals are integrated to form a complete picture, which is the perception. It is important to identify the nature of these interactions as they may influence consumers' perceptions and preferences in unexpected ways. The factors that cause apparent cross-modal sensory interaction are not always the same. Sensory differences can be caused by interactions between the components of the food prior to introduction to the senses. For example, differences in temperature can change the vapour pressure and partition coefficients of volatile compounds and therefore their release from a solute. Differences in microstructure or viscosity (perceived as texture) can also influence partition and availability of compounds for perception. Changing fat content or salt content also influences the physical chemistry of a food matrix dramatically, changing its flavour significantly.

Sensory interactions are also determined by a halo effect, which is due to learning to place greater reliance on one sensory modality over another to make behavioural decisions. This effect is most obvious by the dominance, or bias, of the visual sense over the taste or olfactory sense when familiar colour and flavour combinations are confused, or when the colour intensity is varied beyond expectation (e.g. imagine a dark beverage with no taste). This type of bias can be modified by directing attention to specific characteristics of the product or those of most interest. A true cross-modal sensory interaction is one where the function of one sense (e.g. threshold measures, concentration–response functions) is changed by stimulation of another sense. Some examples of all three interactions that influence sensory perception are given here.

Interactions between the chemical senses of taste and smell have received much attention. Consumers often do not make any distinction between taste and smell, and most often refer to the combined sensation as 'taste'. Incorrect localisation is therefore often misinterpreted as representing interaction. There is in fact very little specific interaction between taste and odour. However, when taste compounds and odour compounds are presented together, the perception of the combination can be more intense than when either is presented alone. In fact, it has been shown that a below-threshold tastant can be combined with a below-threshold odorant to provide a flavour stimulus above threshold. In addition, sucrose is perceived as sweeter when presented with fruit character odour compounds, but not when presented with savoury character odours. These interactions are learned by experience and the congruency develops from repeatedly experienced pairs, such as for fruity odour and sweet taste.

Interactions between chemesthesis and both odour and taste have been demonstrated. These interactions are truly cross-modal, as they are manifest as suppression effects that can be measured using psychophysical functions. When odour molecules and irritant molecules are mixed, they produce mutual suppression, much like when odours alone are mixed. As most odorants have irritant effect in addition to their odour quality, suppression may be a common occurrence. If a person has reduced sensitivity to irritation, then the balance of aromatic flavour they perceive may be changed. Capsaicin desensitisation has partial inhibitory effects on sweetness, sourness and bitterness perception, but not on saltiness. However, when capsaicin is presented mixed with these tastants, inhibitory effects are not found. This is most likely because the capsaicin has not had sufficient time to cause desensitisation before the tastants are perceived. There is less evidence to show how tastes can influence the burn of capsaicin, although studies suggest that sweet, sour and salt will reduce the intensity of burn, with sweet working best. Fat, or fat content, may also reduce burn as capsaicin is highly lipophilic, and fat may remove excess capsaicin that is available for perception. Irritation of various types can also improve preferences significantly. Soft drinks, or sodas, are not particularly pleasant without the fizziness provided by CO_2 and are in fact sweeter to taste.

Interactions between somesthesis, the visual senses and the chemical senses are also common. An intense sweet taste increases the perception of viscosity, although this is observed as a halo effect. The change in perception caused by sweetness intensity may be due to prior association between high levels of sugars and high viscosity. On the other hand, thresholds for the basic tastes, and for most odorants, are higher in solid foods, foams and gels than in water. However, this interaction is most influenced by physicochemical interactions occurring in the food matrix, rather than effects on the sensory receptors. This influence is particularly so for odour.

Colour has a strong influence on flavour perception. It influences absolute threshold measures, differential threshold measures, discrimination and even the ability of beverages to quench thirst. Darker coloured foods and beverages are rated as having more intense flavour. Uncoloured foods receive lower odour quality ratings, and incorrect identification is common when familiar flavour and colour combinations are confused. These interactions also demonstrate halo effects.

Temperature changes influence intensity of perception for all the chemical sense modalities. Thresholds for taste compounds are lowest between 22 °C and 32 °C. The effect of temperature can be represented by a U-shaped curve. In practical terms, salted foods taste saltier when they are heated or cooled to the 22–32°C range. Similarly, sweet-tasting foods will be most sweet in the 22–32°C range. A good cook knows to adjust seasoning to the temperature at which the food will be served. Temperature also influences chemical irritation perception. The most obvious example is that of the temporary, but significant, relief from chilli burn given by a cold drink.

9.8 Food preferences

Palatability is a term used to describe the hedonic reward provided by the sensory properties of a food or beverage. Evidence shows that a food or beverage that is highly palatable will be consumed in a larger quantity, although palatability also varies depending on the physiological need of the individual who tastes. Higher liking for a food will promote its consumption but especially food dislikes predict choice and intake. Undoubtedly, you will occasionally have refused to eat a food you really like, but it is unlikely that you often choose to eat a highly disliked food.

The palatability of a food is determined by the integrated response to stimulation of all the senses in combination. It is most logical to refer to palatability when attempting to relate combined chemical sense, somesthetic and visual sense perceptions to preferences. Indeed, consumers, whose preferences are of most interest, do not normally distinguish between the varied sensory stimuli when they respond to food during consumption.

In developed society, the significance of food has changed from providing basic nutritional needs to providing protocols that define you, your cultural identity and social standing. In fact, now palatability could be called the social sense because we tend to eat with other people when possible, sharing experiences as we eat. The colloquial reference to taste at the dinner table does not distinguish between taste and smell, often conflating sensory (or analytical) and hedonic (or evaluative) responses. Remarks over meals such as 'this soup is too salty' or 'I really like the taste of this sauce' are typical.

Of course, sensory perception and hedonic response to food are not fully independent. As the sensory intensity of a given stimulus increases, for example the sweetness of sugar, its hedonic tone becomes increasingly pleasant, reaches a maximum (ideal point), and then decreases in pleasantness to the point of neutral hedonic tone. Further increases in concentration become unpleasant (Figure 9.5). This inverted U-shaped curve is typical of the hedonic response to most chemical stimuli. However, there are many factors that influence an individual person's perception of a food's sensory properties. These include how much has already been eaten, what has already been eaten and how long ago, past experiences with different foods, individual genomics, nutritional status and, with reference to the preceding paragraph, the culture and personal company that have most influence on a person.

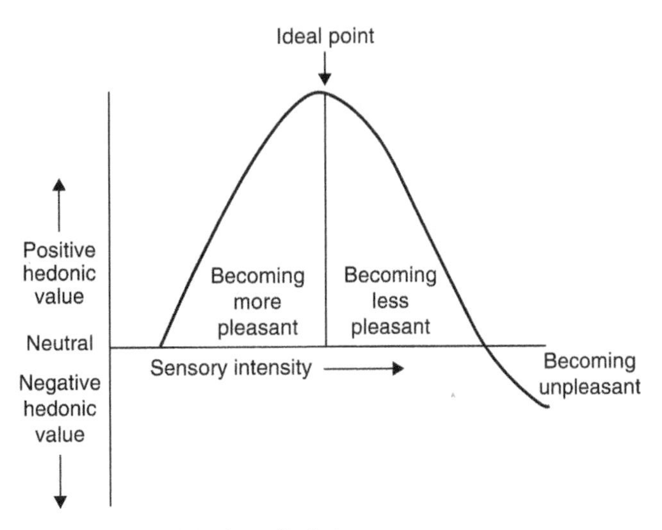

Figure 9.5 The relationship between sensory intensity and hedonic response.

Such factors will not negate the relationship between sensory perception and hedonic response described above, but they may alter the shape of psychophysical concentration–response curves, hedonic response curves or both.

Our reaction to sensory stimuli may be positive when we first try a food, but this positive response may become negative after we have eaten it for a while. This change in response can be due to changes in internal state, or satiety, signalling the body to stop eating. However, other factors may also be involved.

Sensory-specific satiation

Consider ordering a dark chocolate pudding, with intense chocolate flavour, at your local café. At first, you may feel that the piece you have received is rather small, but you will be doing well to finish it all off, even if you are quite hungry. This latter response is due to sensory-specific satiation – a decrease in the palatability (or hedonic decline) of a food with its consumption relative to other, unconsumed foods. It seems to depend mainly on orosensory exposure with no single sensory modality clearly dominating the hedonic decline. Furthermore, postingestive consequences do not affect sensory-specific satiation. Sham feeding, for example, reliably induces sensory-specific satiation.

The hedonic decline that characterises sensory satiation is thought to have two main functions: to prevent overeating and to promote variety seeking. A varied meal offers less opportunity for sensory-specific satiation for any one of its components and thus stimulates intake. Further, as sensory-specific satiation relies on orosensory stimulation in terms of degree, frequency and duration of intake, it is affected by eating rate. Eating tough, dry foods requires more effort and time to chew and swallow and they are hence more satiating. Conversely, beverages exert little sensory-specific satiation, which is why we can consume very large amounts of soda before feeling satiated.

Reduced liking for a meal can also occur across meal occasions. Unlike sensory-specific satiation, this monotony effect is sensitive to flavour intensity. Bland-tasting foods are less susceptible to the monotony effect, partly explaining why staple foods in our diet (i.e. typically starchy foods like potatoes, rice, bread, pasta and noodles) have little taste.

Food familiarity

People in certain cultures develop preferences for certain foods that are typically associated with that culture; consider the differences between Asian and European cuisines. However, there is little evidence to suggest that peoples of different cultures have different sensitivity to sensory properties. In addition, children, who often avoid unfamiliar foods because of their flavour or texture, come to prefer these foods if repeatedly exposed to them. Mere exposure to a certain food makes its sensory attributes more familiar and this promotes acceptance. People who perform blind preference tests among a range of products within type usually choose the product that they are most familiar with or the one that they themselves consume.

Evidence suggests that even subtle sensory preferences are learned and are strongly related to past experiences with food. However, sensory preferences are not fixed when one reaches a particular point in life. They are continuously changing as experience increases. There is evidence to suggest that the amount of repeated exposure to novel foods required to promote its acceptance is reduced as your palate develops.

Acquired tastes

Food preferences also demonstrate evolutionary design, instinctive response and adaptive value. At birth, infants prefer sweet taste, which indicates the presence of sugar, an important source of energy. They also tend to dislike bitter taste, which indicates substances that might cause sickness. These responses remain strong throughout life and only change through experience, as referred to above. Salt is liked at low concentrations but disliked at high concentrations, which makes sense as high intake might disrupt the body's careful osmotic balance. However, rats deprived of salt, a necessary component of their diets, have been shown to compensate and develop a specific hunger for sodium.

Food preferences can be acquired by pairings of a food with a positive outcome, such as an already preferred sweet taste or considerable energy repletion. This explains the near universal liking for the western diet, a diet rich in sugar, salt and fat. Food aversions can also be acquired by a single pairing of a food with gastrointestinal distress. This response is termed conditioned taste aversion. There is good evolutionary justification for this extreme response, as sickness was most likely caused by poisoning and conditioned flavour aversion prevents us from making the same mistake again.

Learning plays a dominant role in shaping idiosyncratic food preferences. We do not have genes for liking a specific dish but genes certainly influence the shaping of food preferences through experience. There are well-documented differences between individuals in ability to taste PROP, and this difference has a genetic cause. The ability to taste PROP is most strongly associated with overall bitter sensitivity. Significant individual differences in thresholds to other sapid and odour compounds have also been shown. There are also documented differences in saliva composition and flow rate, mastication behaviour and the quantities of sapid and volatile compounds that individuals can release from their food during eating. With these differences in mind, it may be that some foods taste differently to different people, and that these differences can be genetic. In line with this reasoning, a diversity of genes has been found to contribute to food preferences, notably a dislike for vegetables.

9.9 Changing function of the senses across the lifespan

When examining changing function of the senses across the lifespan, it is difficult to distinguish sensitivity from hedonic response. As discussed in the previous section of this chapter, it is palatability, preferences or hedonic response to stimuli, rather than the stimuli themselves, which determine behaviour towards food and relate best to food intake. To recap: the performance of the senses is responsible for determining initial preferences and may determine how easily new preferences can be learned during life; secondly, the senses constantly maintain their 'gate-keeping' function, screening stimuli to check that they meet preferred criteria, sourcing those that provide positive feelings and rejecting those associated with unhappy experiences or ill health.

There are several stages of life, from early infancy, through childhood, adolescence, adulthood, middle age and older people to the oldest old. Food preferences and food intake change across the lifespan. Nutritional requirements also change. Malnutrition, manifest in undereating, overeating or insufficient nutrient intake is widespread among almost all age groups or life stages. When one considers the contribution of the human senses to food choice and intake, one should approach the subject from two perspectives. The first is to seek knowledge of how sensitivity and hedonic response change across the lifespan, and ultimately to determine relationships between these factors and eating behaviour that can be exploited in age-appropriate new product development. The other perspective is to understand how a lack of knowledge on changing sensory function (and its significance in dietary intake) is currently contributing to poor dietary habits and incorrect dietary guidance.

New-born infants can discriminate between basic tastes. They demonstrate this by responding with positive facial expressions to sweet taste and with negative facial expressions to sour and bitter tastes. They are indifferent to salty taste, probably because they are relatively insensitive to it. However, a preference for salt emerges at around four months of age due to taste system development. Innate taste preferences remain strong throughout the lifetime, but may be modified by experience, and it is unclear how changes are related to changing sensitivity. For example, salt taste preference can be modified in infants as young as six months of age through experience. Children maintain a strong preference for sweet taste, certainly up to five years of age, but sweet taste preferences tend to decline between adolescence and adulthood.

The human olfactory system is anatomically complete before birth, and perception of odours via the amniotic fluid and early in infancy may play a role in later responses and preferences to odour stimuli. Studies have shown that new-borns can smell and can discriminate between different odours, although they do not demonstrate typical preferences initially. In fact, infants may like smells that are off-putting for adults, such as those of faecal odours. However, these infants soon learn to develop preferences that are more in keeping with their peers, and by the age of three tend to match those of adults. This and other evidence suggest that lifelong preferences and aversions for odours, as well as other strong emotional associations with odours, are formed during infancy and childhood.

Preferences for the flavours of fat also appear to be learned, as children may associate specific flavours with high-energy density. Children's preferences for fat are correlated with the BMI of their parents. This suggests a possible genetic link that may also be manifested as 'fat taste' sensitivity. However, the link may equally be due to the dietary habits these parents transmit to their children. In fact, given the evidence, variety of the diet in later life may be compromised by flavour preferences, or food experiences, learned during infancy and in early childhood. This relationship highlights a need to introduce variety in flavours linked to positive nutritional value early in life and there is evidence that this indeed can be beneficial. Relatively little is known about changes in sensitivity or hedonic response to chemesthesis that accompanies the change from infancy to adolescence and adulthood. Given the impact of chemical irritation on hedonic response, research work is now needed.

In a way that is similar to acquisition of odour preferences through familiarity and associative learning, early experience with texture variety will lead to a greater likelihood of intake of a new food texture. In addition, young children are less dominated by their visual sense, as the associations between appearance and colour and in-mouth perceptions of sensory attributes are not as well formed as those among adults. In many cases children can correctly identify flavour qualities when colours are varied from what is familiar (to an adult). In addition, due to having less formed associations between appearance and in-mouth perceptions, novel colours may be used to encourage tasting of a (expected) disliked food. This has been demonstrated with new coloured cultivars of vegetables, such as purple cauliflower and yellow beans.

Nutritional disorders may begin during childhood but are more commonly initiated during the transition from adolescence to adulthood and are established as adulthood progresses. Studies have been conducted to determine whether sensitivity and preferences for sweet taste, the flavour of fat and prevalence of obesity are related. No causal relationship was found, although obese people tend to prefer foods high in sugar and fat and select fat-rich taste stimuli in sensory tests. Recent evidence suggests a negative correlation between ability to perceive free fatty acids in the mouth and higher weight status because of greater consumption of energy in fatty foods. Additional factors such as income, socioeconomic status and the availability of sugars and fats were related to intake. On the other hand, women with the eating disorder anorexia nervosa seem to have dissociated taste responsiveness and eating patterns. Their taste preferences for sweet and fat do not differ from other people, but they use them to aid food avoidance rather than selection.

Although not conclusive, these studies may demonstrate that it is not sensitivity per se that is important, but how we have learned to use our sensory system as a guide to consumption.

In older people, and in particularly in the oldest old, a new sensory requirement emerges. Older people lose chemical sense function, in parallel with the loss of other biological functions. This loss is greatest for olfactory function, which is manifest as higher absolute odour thresholds, less ability to perceive differences between suprathreshold odour intensity levels and decreased ability to identify odours. Anatomically, decrement can be seen by morphological change to the olfactory bulb. Taste function, on the other hand, remains relatively intact. Thresholds for salt and bitter taste may increase, whereas sweet and sour thresholds show little change. There are also some intensity decrements to chemesthesis function, although little is known for certain. Chewing efficiency has been demonstrated to decrease among older people, especially when dental status is also compromised. This loss, and changes in dentition, give rise to troublesome-to-eat sensory attributes, such as hard and fibrous textures and the presence of peel or seeds. It has also been demonstrated that older people become less sensitive to particle size in mouth.

In general, adaptation proceeds more slowly, recovery from adaptation is lengthier and cross-adaptation is more severe in older people compared with the young. In addition, interactions within and between sensory modalities may be affected by a changed contribution of specific compounds or modalities to perception due to differential loss of function. For example, as smell declines at a faster rate, foods that are bitter but have pleasant odour may be experienced as just plain bitter by an older person with poor odour ability. There are also effects of ageing on saliva flow and composition. This change influences the ability to break down food, inhibits mixing, retards flavour release and makes swallowing difficult.

Loss of sensory function with ageing may cause older persons to lose interest in food and food-related activities such as cooking or dining out, leading to reduced energy intakes and a reduction in essential nutrient consumption. In addition, the motivation that sensory-specific satiation gives to seek variety may be reduced, leading to consumption of a monotonous diet, which can also lead to reduced intake of specific essential nutrients. Older people have also been found to be more food neophobic, which poses a barrier to adopting new dietary habits prompted by changes in their nutrient needs and/or dental issues. Losses in ability to sense saltiness can create problems in older hypertensive populations, as they are likely to put more salt in their food. The texture attributes that can become troublesome are often associated with fruits and vegetables and may lead to avoidance. However, the impact an age-related change in sensory function has on food preferences and food intake is unclear as few studies have demonstrated a causal relationship between sensory impairment, diminished hedonic response and altered food intake in the same group of older people.

In all ages, the perception of food in the mouth signals an innate biological response termed the cephalic phase response. This response stimulates saliva flow and gastric and pancreatic secretions, preparing the gastrointestinal tract for foods that are about to be ingested so that these may be processed efficiently by the digestive system. The extent to which cephalic phase responses change over the lifespan, and the importance of these responses to long-term nutritional status, has yet to be determined. On the other hand, changes in nutritional status, resulting from physiological disturbances that affect nutrient or energy balance, from normal fluctuations in energy status associated with hunger and satiety (sensory-specific satiety was discussed previously) or from modifications in diet (e.g. resulting in specific hunger for salt) can influence sensitivity and hedonic response, although these changes are generally short term.

Extreme cases of nutrient deficiency or toxicity can influence the function of the chemical senses. This effect has been observed for vitamins A, B_6 and B_{12}, and for the trace metals zinc and copper. For example, zinc deficiency has been associated with histological changes in taste buds as well as degeneration and loss of taste papillae. Vitamin A deficiency results in gradual, but reversible, loss of taste, although zinc may again be involved as it plays an important role in transporting vitamin A from the liver. Diabetes results in a general reduction in taste sensitivity and losses in ability to sense sweetness may cause diabetics to unwittingly add more sugar to their food. This can create additional nutritional problems, particularly in older people who develop late-onset diabetes and find it difficult to manage their diet. Although not directly related to nutrition per se, poor oral and dental health affect sensory function and can also cause dietary restriction if mechanical difficulties or pain are associated with eating some food types. Medications taken for chronic illness also contribute significantly to sensory loss, particularly in older people. The average older person in the USA is taking 3.7 different medications at any one time.

The visual system consists of specialised tissues that are vulnerable to nutritional insults. For example, vitamin A deficiency results in night blindness and xerophthalmia. Thiamin deficiency can result in ophthalmoplegia (paralysis of eye movement). The eye is also susceptible to damage by toxic material in food and drink. For example, the consumption of methanol results in the formation of formic acid in the body, which damages the retina and can cause blindness; cyanide toxicity can damage the optic nerves and cause amblyopia (dimness of vision without obvious defect or change in the eye). The outer tissues of the eye are susceptible to damage and infection if the production of tears and mucus is affected. The lens of the eye is at risk of developing opacities if the proteins from which it is made become oxidatively damaged. The extremely high oxygen tension within the retina and the high exposure of the eye to ionising radiation also make it susceptible to oxidative damage. To protect against free radical-mediated damage, the eye has several protective systems to mop up free radicals. With increasing age, the effectiveness of this system decreases and two major causes of blindness, cataract and age-related macular degeneration, are consequences.

9.10 Future perspectives

Earlier in this chapter, it was stated that the primary function of the human senses (associated with food) is to detect foods that would be bad for the body and should be rejected, and to identify foods that the body needs and that should be consumed. Nevertheless, 'what is bad for the body' can be interpreted as flavours and textures that we dislike for whatever personal factors. Conversely, 'what is good for the body' can be interpreted as those that we like for whatever personal factors. The factors determining personal likes and dislikes for sensory properties (especially flavours) are influenced by genetics, age, nutrition and health status, but most importantly by experience. Palatability is a strong determinant of food intake and food selection, and thus, individual nutritional status is determined by individual food likes and dislikes.

The aim of nutritional science is to improve dietary quality. Its focus is directed at improving the nutritional quality of foods, by provision of nutritionally enhanced foods and by recommendation of ideal eating habits. Relatively little regard is given to 'tastes' or 'taste preferences'. To address this, efforts should now be made to develop dietary strategies that take account of the sensory properties of food. Unacceptable flavour, or flavour that does not match individual likes or expectations, is an obstacle to compliance with a recommended change in diet, particularly now that consumers are becoming more affluent and more discerning. The potential nutritional value of functional ingredients may be compromised because they have tastes that do not meet acceptable criteria. For example, phytochemicals, linked with cancer prevention, have a bitter taste for which there is an innate dislike. Sugar, salt and fat replacers do not taste the same as the substances that they seek to replace.

Oral nutritional supplements (ONS) are often prescribed to combat malnutrition but liking for these high-protein ONS tends to be low and is the main reason for low intake. Reformulating foods poses a technological challenge but the relevance of the solution to that challenge is limited if the end consumer does not like its taste. Conversely, reformulating food to accommodate taste preferences may compromise its nutrient value. Apart from engineering nutrient-dense and preferred foods, some thought and effort need to be invested in engineering the consumers' attitudes towards certain foods. As many bitter-tasting compounds are associated with foods that have high nutritive benefit (e.g. vegetables such as bitter gourd), hedonic responses to these bitter tastes would need to be adjusted. On the other hand, macronutrients, such as sugar, salt and fat, are generally preferred. As these are linked with chronic disease when consumed in excessive quantities, hedonic response to these flavours needs to be adjusted to reduce consumption.

Nutritional scientists should also recognise the technological challenges facing the food manufacturing industry, as they seek to produce acceptable foods that contain new ingredients and formulations that differ markedly from tradition. In addition, dietary intervention strategies should consider the relationships between sensory likes and dislikes and demographic, economic and sociocultural factors. It is important to consider sensory and dietary needs in parallel, but at different stages in life. What is the contribution of sensory variety in infancy to lifelong variety-seeking behaviour, and do breast-fed and formula-fed infants have different hedonic potential? Although preference studies may show that older-age consumers are satisfied with foods currently available, these foods should be improved to compensate for specific sensory losses, which will, with time, lead to reacquaintance with an increased sensory variety. Given the strong link between sensory properties and digestive response, this compensation may have unforeseen benefits. However, it is important to recognise that the older-age population is very heterogeneous and individual preferences are likely to differ significantly. Responsible nutritional intervention and functional food use may also reduce the need for widespread pharmaceutical use as older age progresses. This will have the added benefit of delaying the adverse effects of medicines on ability to taste and smell, which may contribute to continued good eating habits.

Rather than attempting to play down the importance of the senses to food intake, by demonstrating that, say, price or availability are more significant determinants, it is now time to understand better the development of food palatability and preferences with positive nutrition in mind and to exploit sensory properties to increase intake of foods with high nutritive value. Future research should direct product development for individual nutrition and sensory/hedonic needs. The time seems ripe for personalised gastronomy.

Further reading

Etiévant, P., Guichard, E., Salles, C., and Voilley, A. (2016). Flavor: From Food to Behaviors, Wellbeing and Health. Cambridge: Woodhead Publishing.

Finger, T.E., Silver, W.L., and Restrepo, D. (2000). The Neurobiology of Taste and Smell, 2e. New York: Wiley–Liss.

Joseph, P.V. and Duffy, V.B. (2021). Sensory Science and Chronic Diseases: Clinical Implications and Disease Management. Cham: Springer.

Lawless, H.T. and Heymann, H. (1998). Sensory Evaluation of Foods, Principles and Practices. New York: Chapman & Hall.

10
The Gastrointestinal Tract

Mariano Mañas, Emilio Martínez de Victoria, Ángel Gil, María D. Yago, and John C. Mathers

Key messages

- The four basic processes that constitute the functions of the gastrointestinal tract are motility, secretion, digestion and absorption.
- Several nutritional factors can affect gastrointestinal function. Conversely, pathological modifications of this function lead to an altered nutritional state.
- The pancreas is a mixed gland containing both endocrine and exocrine portions. These two parts of the pancreas have a high level of interaction, both in health and in disease conditions.
- The most important digestive function of the liver is the secretion of bile, but it also plays a role in lipid, protein and carbohydrate metabolism as well as the excretion of bile pigments, the storage of several vitamins and minerals, and the processing of hormones, drugs and toxins.
- The intestinal epithelium is a complex system of multiple cell types which maintain precise inter-relationships. The regulatory mechanisms involved with developmental processes are at organism, cellular and molecular levels.
- The primary function of the large bowel is salvage of energy via bacterial fermentation of food residues. It also plays a role in the absorption of water and electrolytes, lipid metabolism, the synthesis of certain vitamins and essential amino acids, and the metabolism and absorption of phytochemicals. Bacterial fermentation in this organ may contribute to risk of obesity and influence other aspects of host metabolism and health.

10.1 Introduction

The overall function of the gastrointestinal system is to process ingested foods into molecular forms that can be transferred, along with salts and water, from the external environment to the internal environment of the body. The digestive processes are largely determined by the composition of food ingested. This fact determines the importance of the food and thus the diet, in most aspects of the physiology of the gastrointestinal system, including its regulation.

10.2 Structure and function of the gastrointestinal system

The **structure** of the gastrointestinal system (Figure 10.1) includes the gastrointestinal tract (GIT) and the accessory glands (salivary, exocrine pancreas and liver). The structure of the GIT varies greatly from region to region, but there are common features in the overall organisation of the tissue.

The layered structure of the wall of the GIT (Figure 10.2) includes, from the inside to outside:

Nutrition and Metabolism, Third Edition. Edited on behalf of The Nutrition Society by Helen M. Roche, Ian A. Macdonald, Annemie M.W.J. Schols and Susan A. Lanham-New.

Companion Website: www.wiley.com/go/nutrition/metabolism3e

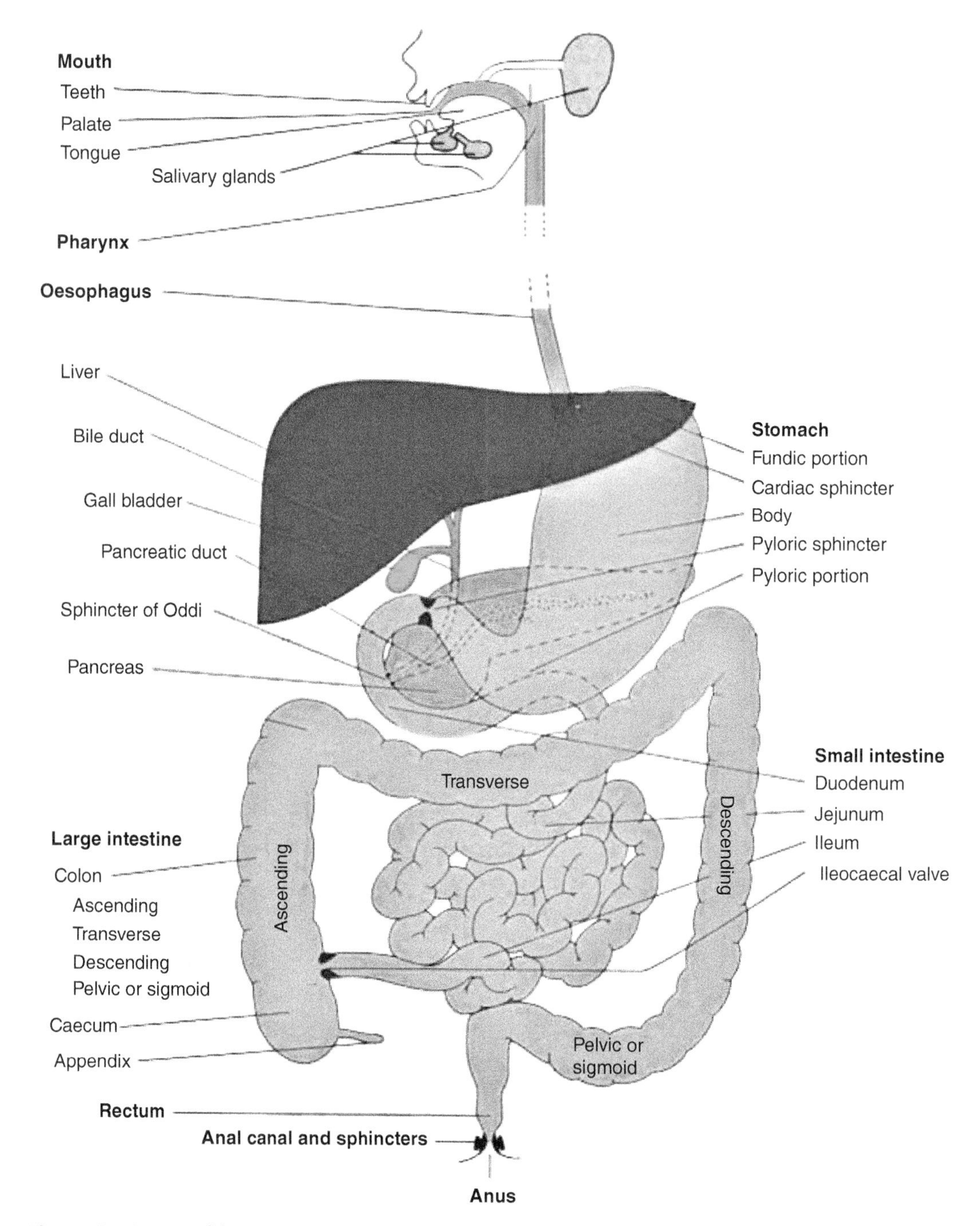

Figure 10.1 Structure of the gastrointestinal system, including the gastrointestinal tract and accessory glands (salivary, exocrine pancreas and liver).

- *Mucosa*: consists of an epithelium, the lamina propria and the muscularis mucosae
- *Submucosa*: consists largely of loose connective tissue with collagen and elastin fibres. Some submucosal glands are present in some regions. In this layer, there is a dense network of highly interconnected nerve cells called the submucosal plexus (Meissner's plexus)
- *Muscularis externa*: consists of two substantial layers of smooth muscle cells, an inner circular layer and an outer longitudinal layer. Between both layers there is another prominent network

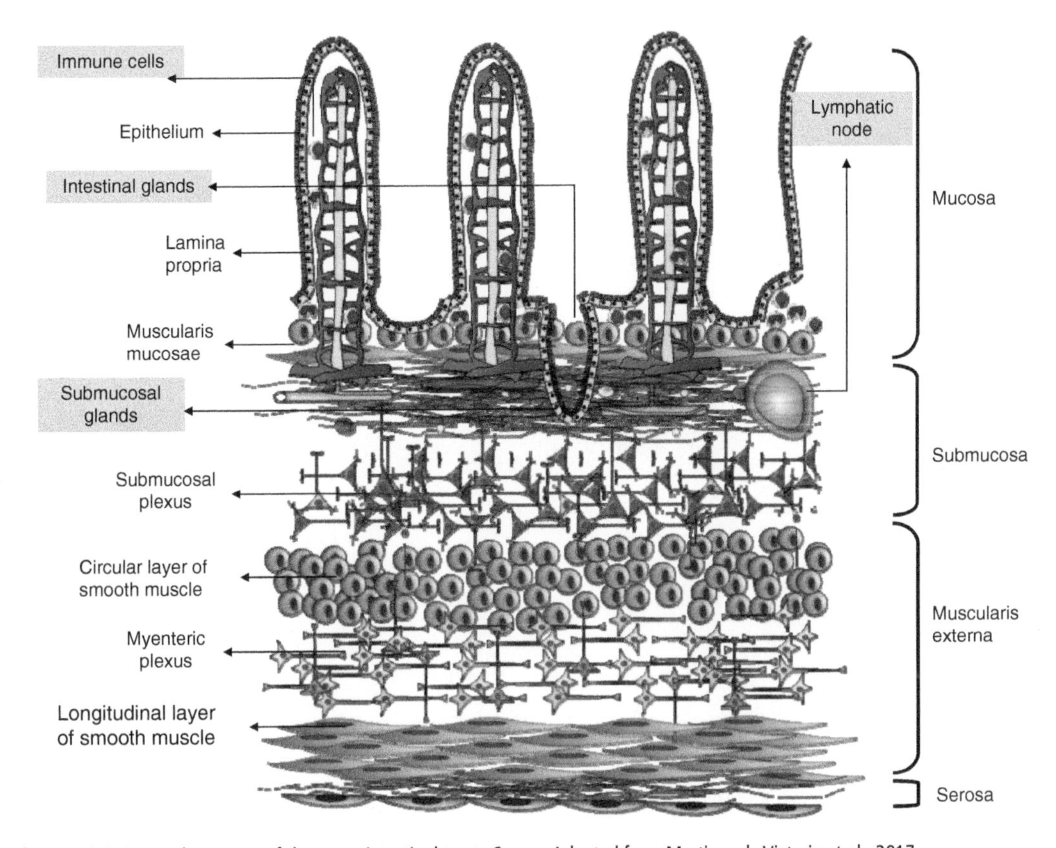

Figure 10.2 Layered structure of the gastrointestinal tract. *Source:* Adapted from Martinez de Victoria et al., 2017.

of highly interconnected nerve cells called the myenteric plexus (Auerbach's plexus). Both submucosal and myenteric plexuses (intramural plexuses) constitute, with the other neurons, the enteric nervous system. This innervation constitutes the intrinsic innervation (Figure 10.3)

- *Serosa*: the outermost layer consists mainly of connective tissue and mesothelial cells.

The gastrointestinal system receives innervations from neurons of the autonomic nervous system, both sympathetic and parasympathetic, which constitutes the extrinsic innervation. Both types of innervation are important for the regulation of the different functions of the GIT, mainly related to motility and secretion (exocrine and endocrine).

The oral cavity (or mouth) is the first portion of the GIT and is bounded by the lips (anteriorly), the fauces (posteriorly), the cheek (laterally), the palate (superiorly) and a muscular floor (inferiorly). The next structure is the pharynx, which is the common opening of both the digestive and respiratory systems. The pharynx can be divided into three regions: the nasopharynx, the oropharynx and the laryngopharynx. The first includes the uvula, while the second extends from the uvula to the epiglottis and communicates with the oral cavity through the fauces. The laryngopharynx extends from the tip of the epiglottis to the glottis and oesophagus.

The oesophagus is a segment of the GIT, 25 cm in length, that extends between the pharynx and stomach. It lies in the mediastinum, anterior to the vertebrae and posterior to the trachea. It passes through the oesophageal hiatus (opening) of the diaphragm and ends at the cardiac opening of the stomach. The oesophagus has two sphincters: the upper oesophageal sphincter between the pharynx and oesophagus, and the lower oesophageal sphincter between the oesophagus and stomach near the cardiac opening. Both sphincters regulate the movement of the ingested meal into and out of the oesophagus.

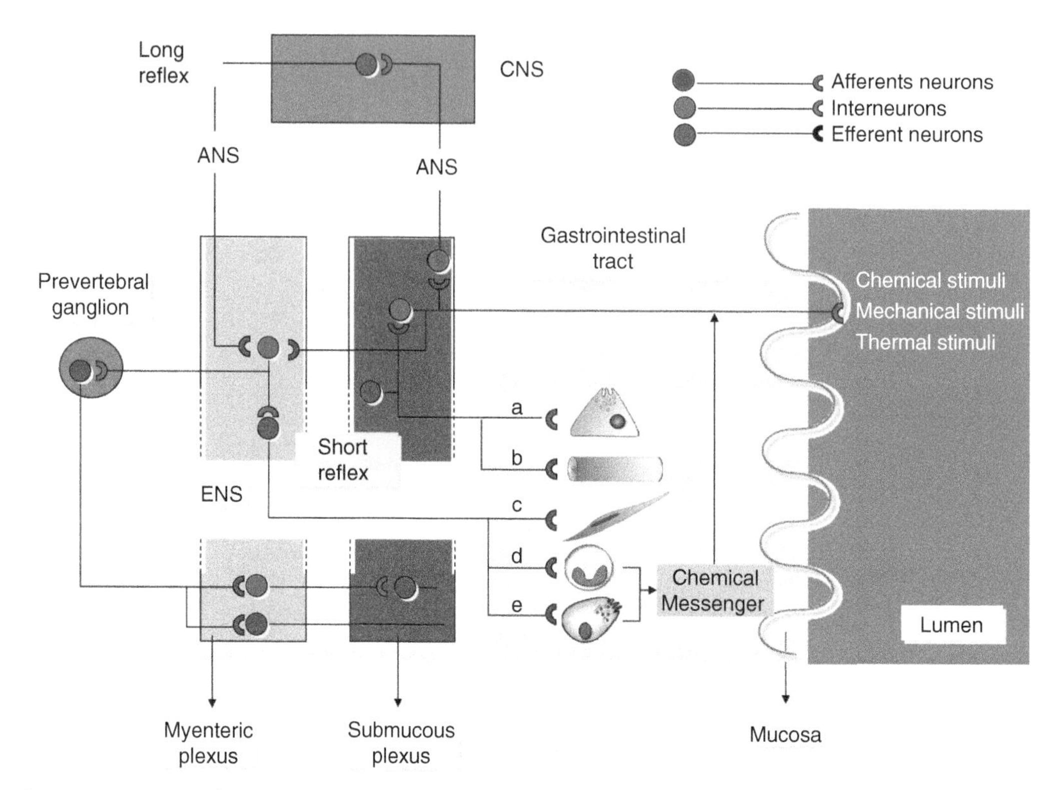

Figure 10.3 Intrinsic and extrinsic innervation and nervous regulation of the gastrointestinal tract. **a**: secretory exocrine cell; b: blood vesel; c: smooth muscle fibre; d: inmune cell; e: endocrine cell; ANS: Autonomous Nervous system; CNS: Central Nervous system; ENS Enteric Nervous system. *Source:* Adapted from Martinez de Victoria et al., 2017.

The stomach is an enlarged segment of the GIT in the left superior portion of the abdomen. The upper opening from the oesophagus is the gastro-oesophageal or cardiac opening. The region near this opening is called the cardiac region. In the left and upper parts of the cardiac region is the fundus. The body, the largest portion of the stomach, turns to the right, creating a greater curvature and a lesser curvature. The lower opening of the stomach, which communicates with the proximal segment of the small intestine (duodenal bulb), is the pylorus (or pyloric opening). The pylorus is surrounded by a thick ring of smooth muscle, the pyloric sphincter.

The small intestine is a long tube that consists of three portions: the duodenum, the jejunum and the ileum. The large intestine includes the caecum (most proximal), the colon (ascending, transverse, descending and sigmoid), rectum and anal canal. The longitudinal layer of the large intestine is incomplete and forms three bands that are called teniae coli. The contraction of the teniae coli causes pouches called haustra along the length of the colon.

The **function** of the GIT includes four general processes: motility, secretion, digestion and absorption (Figure 10.4).

- *Motility*: includes contractions of the smooth muscle of the GIT wall to mix the luminal contents with the various secretions and move them through the tract from mouth to anus. The components of motility are chewing, swallowing, gastric motility, gastric emptying, small and large intestinal motility, gall bladder contraction and defaecation.
- *Secretion*: the wall of the GIT and the accessory glands produce several secretions that contribute to the breakdown of ingested food into small molecules. These secretions consist of saliva, pancreatic juice and bile secreted by accessory glands (salivary, pancreas and liver, respectively) and gastric and intestinal juices secreted by the glands lining

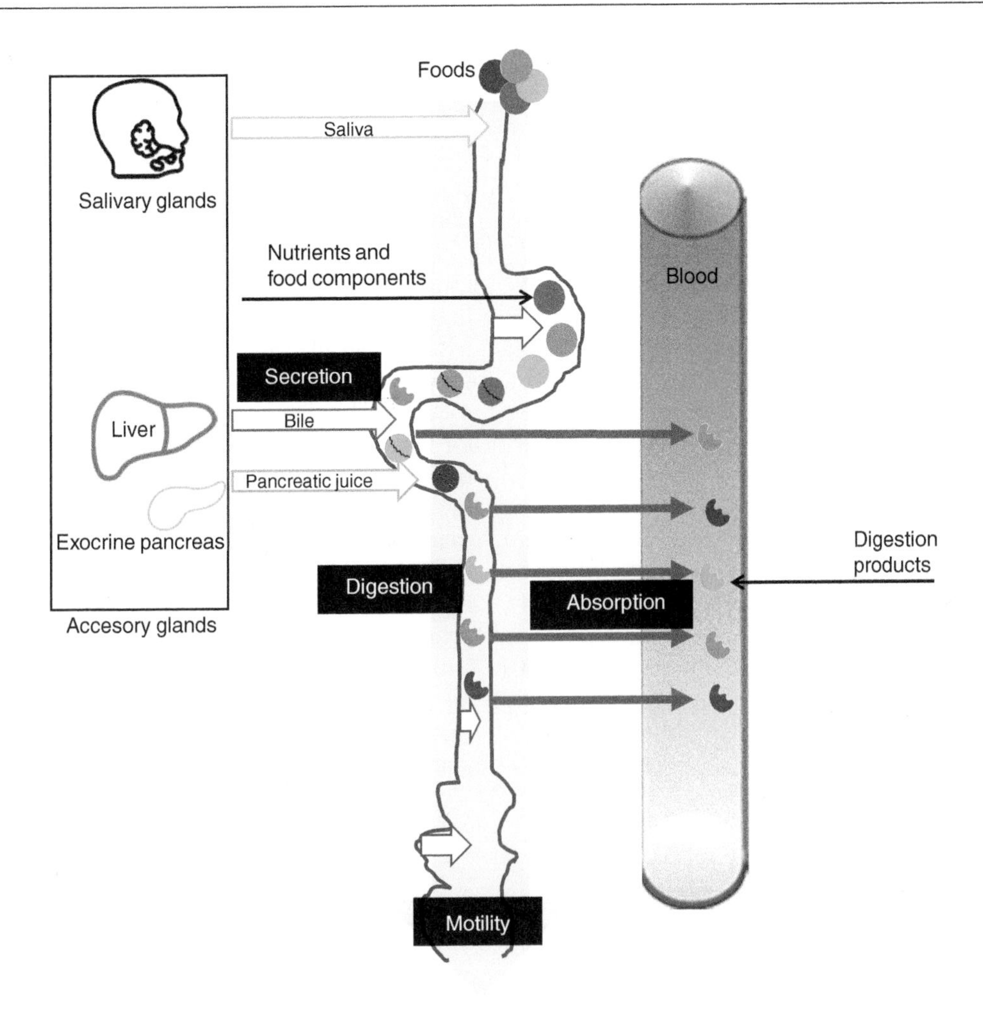

Figure 10.4 The four general functions of the gastrointestinal tract: motility, secretion, digestion and absorption.

the wall of the stomach and the small and large intestine.

- *Digestion*: most food is ingested as large particles containing macromolecules, such as proteins and polysaccharides, which are unable to cross the wall of the GIT. Before the ingested food can be absorbed, it must be broken down and dissolved. These processes of breakdown and dissolution of the food involve both gastrointestinal motility and secretion and are termed mechanical and chemical digestion.
- *Absorption*: the molecules produced by digestion then move from the lumen of the GIT across a layer of epithelial cells and enter the blood or lymph (internal environment).

These are the four basic processes that, with the mechanisms controlling them, constitute the functions of the gastrointestinal system.

10.3 Motility

Chewing can be carried out voluntarily, but normally is almost entirely under reflex control. Chewing serves to lubricate the food by mixing it with saliva, to start the starch digestion with ptyalin (α-amylase) and to subdivide the food.

Swallowing is a rigidly ordered reflex that results in the propulsion of food from the mouth to the stomach. The swallowing centre in the medulla and lower pons controls this reflex. There are three phases: oral, pharyngeal and oesophageal. The last phase is carried out by the motor activity of the oesophagus.

Gastric motility serves the following major functions:

- To allow the stomach to serve as a reservoir for the large volume of food that may be ingested at a single meal

- To fragment food into smaller particles and mix the luminal contents with gastric juice to begin digestion
- To empty gastric contents into the duodenum at a controlled rate.

This last function is closely regulated by several mechanisms so that the chyme is not delivered to the duodenum too rapidly.

The mixing and mechanical fragmentation of the luminal contents and gastric emptying are carried out by peristaltic waves that begin in the pacemaker zone in the middle of the body of the stomach and travel towards the pylorus and through the gastroduodenal junction. The gastroduodenal junction (pyloric sphincter) allows the carefully regulated emptying of gastric contents to the duodenum for optimal digestion and absorption of nutrients.

Gastric emptying is regulated by both neural and hormonal mechanisms. The signals come from duodenal and jejunal receptors that sense acidity, osmotic pressure and fat content (Figure 10.5).

If the pH of chyme is less than 3.5, the rate of gastric emptying is reduced by neural mechanisms (vagal reflex) and through the release of secretin, an intestinal hormone that inhibits the antral contractions and stimulates the contraction of the pyloric sphincter.

The chyme emptying into the duodenal bulb is usually hypertonic and becomes more hypertonic in the duodenum because of the action of digestive enzymes. The hypertonic duodenal contents slow gastric emptying. The luminal hypertonic solution releases an unidentified humoral factor (hormone) that diminishes the rate of gastric emptying. A neural component may also be involved. The presence of fat digestion products (mainly fatty acids and 2-monoglycerides) dramatically decreases the rate of gastric emptying due to an increase in the contractility of the pyloric sphincter. This inhibition of gastric emptying is mediated by both hormonal and neural mechanisms. The effect of unsaturated and long-chain (>14 carbons) fatty acids is greater than that of saturated and medium short-chain fatty

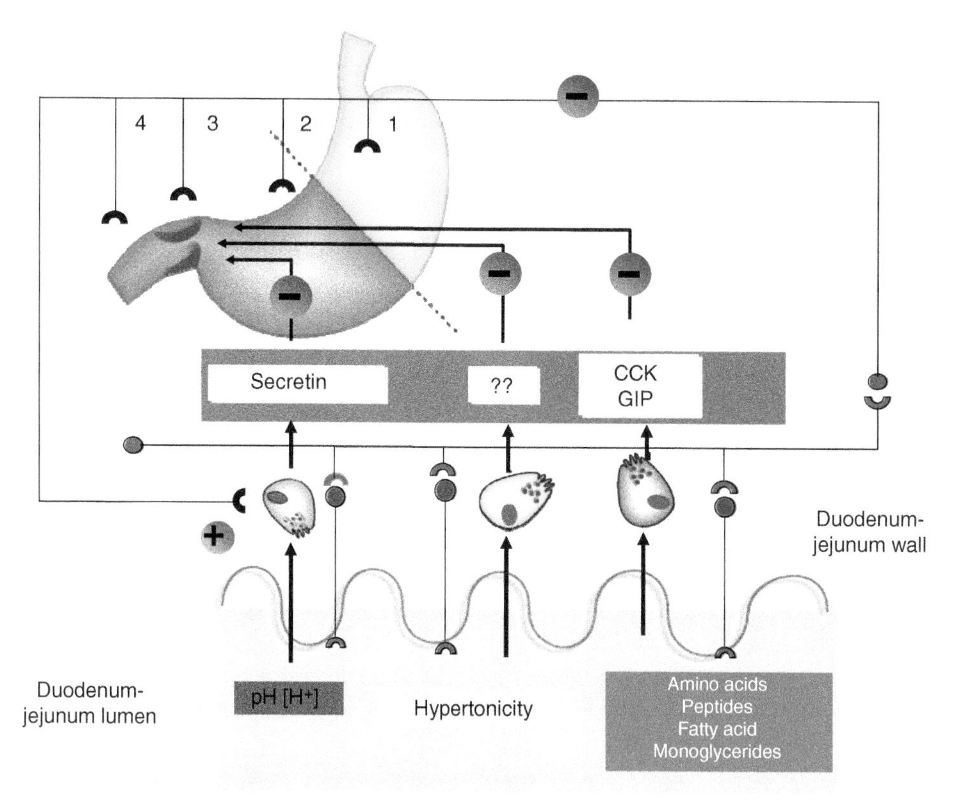

Figure 10.5 Neural and hormonal mechanisms involved in the regulation of gastric emptying. CCK, cholecystokinin; GIP, gastric inhibitory polypeptide. *Source:* Adapted from Martinez de Victoria et al., 2017.

acids. Cholecystokinin (CCK) is an intestinal hormone released by the presence of fatty acids and other fat digestion products. The net effect of CCK is to slow the rate of gastric emptying. Other gastrointestinal hormones, such as gastric inhibitory peptide (GIP) and peptide tyrosine-tyrosine (PYY), are implicated in the slowing of gastric emptying, seen later after food ingestion due to the presence of fatty acids in the ileum ('ileal brake'). The release of glucagon-like peptide-1 (GLP-1) also has an inhibitory role in gastric emptying.

Finally, the presence of proteins and peptides in the stomach also slows gastric emptying via the release of gastrin. This gastric hormone has a net effect of diminishing gastric emptying. Further inhibition is achieved by the presence of tryptophan, other amino acids and peptides in the duodenum, probably via CCK release.

Motility of the small intestine

The movements of the small intestine can be classified into three major patterns: segmentation and peristalsis, which occur in the postprandial period, and the migrating motor complexes (MMCs), which are seen during fasting. The postprandial motility involves alternating contractions of the intestine, which mix the chyme with digestive secretions, bringing fresh chyme into contact with the mucosal surface for absorption. The peristaltic pattern allows the progression of food throughout the intestine. The interdigestive period consists of bursts of intense electrical and contractile activity separated by longer quiescent periods. This pattern appears to be propagated from the stomach to the terminal ileum. The MMCs sweep the small intestine clean, emptying its contents into the caecum, avoiding bacterial overgrowth. The mechanisms that regulate the MMCs are both neural (vagal) and hormonal (motilin).

Motility of the colon

There are two major types of movement: segmentation (haustration), with a mixing function, and segmental propulsion, which allows the luminal contents to move in the distal direction.

Control of the contractile activities involves the central nervous system (long reflexes), the intrinsic plexuses of the gut (short reflexes) (see Figure 10.3), humoral factors (gastrin, CCK, secretin, etc.) and electrical coupling among the smooth muscle cells.

10.4 Secretion

Salivary secretion

Salivary secretion is produced by the three major salivary glands (parotids, submandibular and sublingual) and other minor glands in the oral mucosa. Saliva lubricates food for swallowing, facilitates speaking and begins the digestion of starch. The functional unit of the salivary gland is the acinus (secretory endpiece). The composition of saliva includes inorganic (salts) and organic (amylase, mucoproteins) components. The primary control of salivary secretion is by the autonomic nervous system, which regulates several gland effectors, including acinar cells, blood vessels, ductular cells and myoepithelial cells.

Gastric secretions

The major secretions of the stomach are hydrochloric acid (HCl), pepsinogens, intrinsic factor and mucus. These components arise from the many secretory glands in the wall of the gastric mucosa. These glands have different cellular types: mucous neck cells (which secrete mucus), parietal or oxyntic cells (which secrete HCl and intrinsic factor) and chief or peptic cells (which secrete pepsinogens). The regulation of gastric secretion is carried out by neural and humoral mechanisms. There are three phases in the regulation process: cephalic, gastric and intestinal. The first is elicited by the sight, smell and taste of food. The second is brought about by the presence of food in the stomach. The last phase is elicited by the presence of chyme in the duodenum.

In the past few years, it has been recognised that *Helicobacter pylori*, a Gram-negative microaerophilic flagellated urease-producing rod found in the gastric mucosa, is related to the development of peptic ulcer disease. Studies suggest that *H. pylori* can induce mucosal damage via both direct and indirect mechanisms. The direct mechanisms imply the production by the micro-organisms of urease, lipopolysaccharides and cytotoxins that induce mucosal inflammation. The indirect mechanisms include the release of gastrin and somatostatin. The importance of

H. pylori infection of gastric mucosa has led to the development of treatments of both gastritis and duodenal ulcers with anti-*H. pylori* regimens, such as antacids and antibiotics.

Intestinal secretions

The mucosa of all the segments of the GIT elaborates secretions that contain mucus, electrolytes and water.

Exocrine pancreatic secretion and bile secretion are covered in sections 10.8 and 10.10, respectively.

10.5 Digestion and absorption

Digestion and absorption of carbohydrates

The major source of carbohydrates in the diet is plant starch. The starch is hydrolysed by salivary and pancreatic α-amylase. The further digestion of the oligosaccharides is accomplished by enzymes in the brush-border membrane of the epithelium of the duodenum and jejunum. The major brush-border oligosaccharidases are lactase, sucrase, α-dextrinase and glucoamylase. The endproducts of the disaccharidases are glucose, galactose and fructose. Several mechanisms transport these products across the intestinal mucosa to the blood, including sodium (NA^+)-dependent and NA^+-independent active transport.

Digestion and absorption of proteins

The process of digestion of dietary protein begins in the stomach. In the gastric lumen, the pepsinogens secreted by the chief cells are activated by hydrogen ions to pepsins, which hydrolyse the proteins to amino acids and small peptides. In the duodenum and small intestine, the proteases secreted by the pancreas play a major role in protein digestion (trypsin, chymotrypsin and carboxypeptidase). The brush border of the small intestine contains a number of peptidases which reduce the peptides produced by pancreatic enzymes to oligopeptides and amino acids. The principal products of protein digestion are small peptides and amino acids. These products are transported across the intestinal mucosa via specific amino acid and oligopeptide transport systems.

Digestion and absorption of lipids

The digestion of dietary fat is carried out by lipolytic enzymes of the exocrine pancreas (lipase and colipase) and requires the presence of biliary phospholipid and bile salts. The absorption of lipids needs the formation of micelles with bile lecithin, bile salt and products of the fat digestion. Inside the intestinal epithelial cell, the lipids are reprocessed and the 2-monoglycerides are re-esterified, lysophospholipids are reconverted to phospholipids and most of the cholesterol is re-esterified. The reprocessed lipids, along with those that are synthesised de novo, combine with proteins to form chylomicrons, which enter the bloodstream at the thoracic vena cava, via the lymphatics draining the gut.

Absorption of water and electrolytes

Water movement into or out of the lumen of the GIT is passive. The water moves across the cell plasma membrane or in between cells via paracellular pathways composed of tight junctions and lateral intercellular spaces between epithelial cells. Water molecules follow the osmotic gradients created by electrolyte movement. Electrolytes are transported by both passive and active processes, and these ionic movements control water absorption and secretion.

Net intestinal fluid transport depends on the balance between intestinal water absorption and secretion. The processes are localised to specific regions of the intestinal epithelium. The cells of the villus tip are differentiated, so as to promote fluid and solute absorption, whereas those in the crypt region promote fluid and solute secretion. Figure 10.6 shows the transport mechanisms in the enterocytes of both regions (villus tip and crypts).

Na^+ entry into the epithelial cell from the intestinal lumen is passive. This occurs via two mechanisms. One is a Na^+/hydrogen (H^+) antiporter protein that exchanges one sodium ion (in) and one proton (out). The second is the Na^+-glucose co-transporter that mediates the coupled entry of Na^+ and glucose into the epithelial cell (1 Na:1 glucose or 2 Na:1 glucose). Na^+ moves down its electrochemical potential gradient and provides energy for moving the sugar into the epithelial cell against a concentration gradient. On the other hand, the presence of glucose in the intestinal lumen enhances Na^+ absorption, which is the basis for the use of

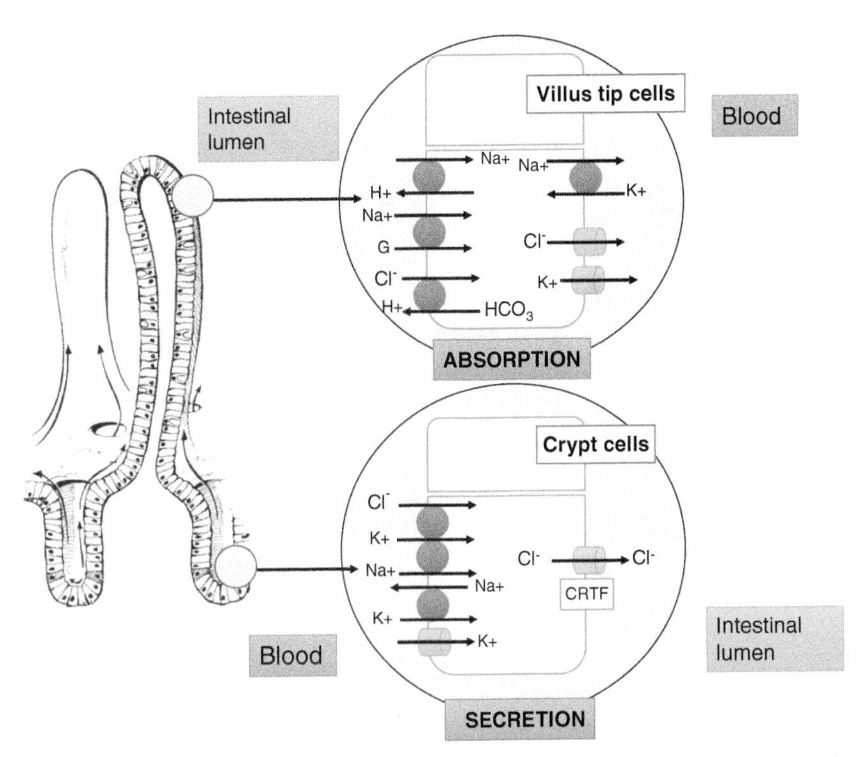

Figure 10.6 Transport mechanisms in the enterocytes of villus tips and crypts of the intestinal epithelium. CFTR, cystic fibrosis transmembrane conductance regulator.

glucose in rehydratation solutions during diarrhoea. The absorption of other sugars (galactose) and some neutral and acidic amino acids utilises a similar mechanism.

Chloride (Cl^-) is absorbed following different pathways. Some Cl^- is absorbed across a paracellular path, some through a cellular pathway composed of an apical Cl^-/bicarbonate ($^-HCO_3^-$) antiporter and possibly using an unidentified basolateral Cl^- selective channel or a potassium (K^+)/Cl^- co-transporter. There is a linkage between H^+ and $^-HCO_3^-$ via carbonic acid and carbonic anhydrase that provides a degree of coupling between the entry of Na^+ and Cl^-.

Potassium is passively absorbed in the small intestine (jejunum and ileum) when its luminal concentration rises because of absorption of water.

10.6 Water balance in the gastrointestinal tract

The water balance in the GIT depends on the input and output of water to and from the gastrointestinal lumen. The water inputs can be exogenous or endogenous. Average oral (i.e. exogenous) intake of water is about 2 l per day. The major components of the endogenous inputs are saliva (1.5 l), gastric juice (2.5 l), bile (0.5 l), pancreatic juice (1.5 l) and intestinal secretions (in normal state 1 l). Thus, the final daily water volume load in the duodenal lumen is 8–10 l. Most of the fluid is absorbed by the small intestine, and about 1–1.5 l reaches the colon, which continues to absorb water, reducing faecal water volume to about 100 ml per day (Figure 10.7).

The maximum absorptive capacity of the colon is only about 4 l per day, so if volumes of fluid greater than this enter the small intestine, diarrhoea will ensue despite normal colonic function. The normal intestinal secretion is about 1 l per day but can be as much as 20 l per day. This striking difference between the absorptive and secretory capacity of the small intestine and colon dictates that large-volume diarrhoea is most often caused by small intestinal dysfunction.

Diarrhoea

Diarrhoea is a problem of intestinal water and electrolyte balance. It results when excess water and electrolytes are actively transported into the lumen (secretory diarrhoea) or when water is

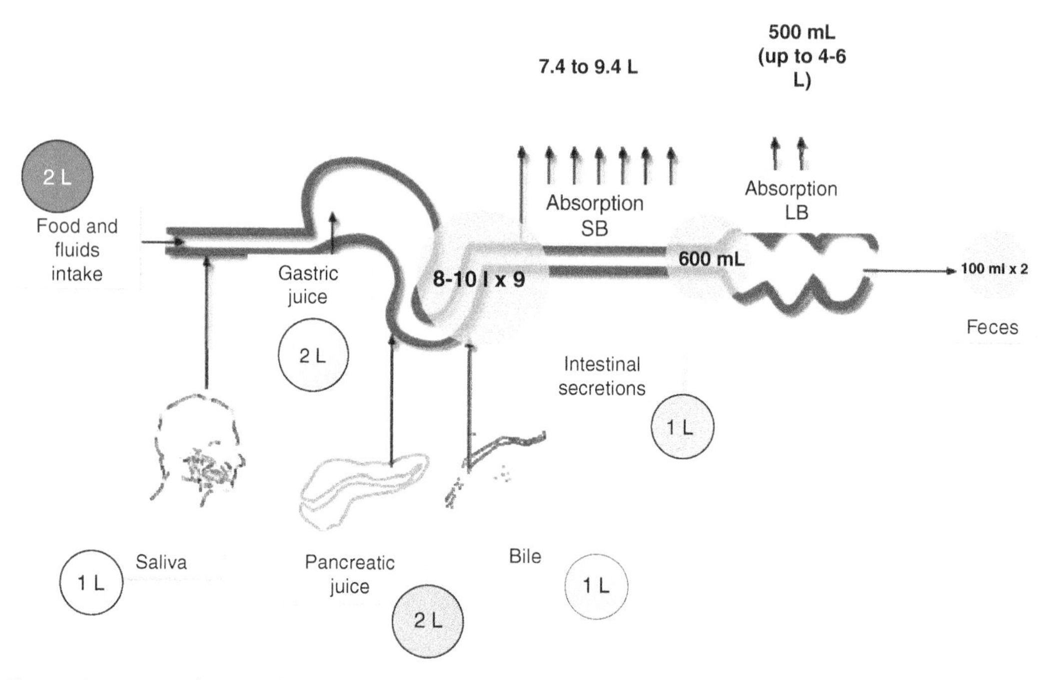

Figure 10.7 Inputs and outputs of water to and from the gastrointestinal lumen in a healthy adult. SB: Small Bowel; LB: Large Bowel. *Source:* Adapted from Martinez de Victoria et al., 2017.

retained in the lumen by osmotically active agents (osmotic diarrhoea). Another major contributor to diarrhoea is gastrointestinal motility. The rate of transit through the gut determines the time available for intestinal absorption of water and can result in diarrhoea.

Vomiting

Vomiting is a reflex behaviour controlled and co-ordinated by the vomiting centre in the medulla, and involving the somatic and autonomic nervous systems, the oropharynx, the GIT and the skeletal muscles of the thorax and abdomen. It is the expulsion of gastric (or sometimes gastric and duodenal) contents from the GIT via the mouth. This can result from irritation (overdistension or overexcitation) of both chemoreceptors and mechanoreceptors anywhere along the GIT. It is usually preceded by nausea and retching.

The events of the vomiting reflex are independent of the initiating stimulus and include the following.

- A wave of reverse peristalsis that sweeps from the middle of the small intestine to the duodenum.
- Relaxation of the pylorus and stomach.
- Forced inspiration against a closed glottis.
- Increase in the abdominal pressure owing to diaphragm and abdominal muscle contraction.
- Relaxation of the lower oesophageal sphincter and entry of gastric contents into the oesophagus.
- Forward movement of the hyoid bone and larynx, closure by approximation of the vocal cords and closure of the glottis.
- Projection of gastrointestinal contents into the pharynx and mouth.

Several physiological changes accompany vomiting, including hypersalivation, tachycardia, inhibition of gastric acid secretion and sometimes defaecation.

Certain chemicals, called emetics, can elicit vomiting. Their action is mediated by stimulating receptors in the stomach, or more commonly in the duodenum, or by acting in the central nervous system on the chemoreceptor trigger zone near the area postrema in the floor of the fourth ventricle (brainstem).

10.7 The exocrine pancreas

Anatomy and histology of the pancreas

The human pancreas is a pink, soft, elongated organ weighing less than 100 g, lying posterior to the greater curvature of the stomach. It consists of a head, located within the duodenal loop, a body and a tail. The latter extends towards the spleen, with which it has morphological relations. The pancreas is a mixed gland, containing both exocrine and endocrine portions (Figure 10.8). The major structural components responsible for the exocrine function of the pancreas are the acinar units (acini) and the duct system, accounting for about 86% of the gland mass.

The acini, the enzyme-secreting units of the exocrine pancreas, are round or oval structures composed of epithelial cells (acinar cells) bordering a common luminal space where enzymes are delivered. The acini form lobules that are separated by thin septa. The tiny ducts that drain the acini are called intercalated ducts. Within a lobule, a number of intercalated ducts empty into somewhat larger intralobular ducts. All intralobular ducts of a particular lobule then drain into a single extralobular duct; this duct in turn empties into still larger ducts. These larger ducts ultimately converge to form two ducts that drain the secretions into the small intestine. The main duct is called the pancreatic duct (duct of Wirsung), which joins the common bile duct, forming the hepatopancreatic ampulla (ampulla of Vater). The ampulla opens on an elevation of the duodenal mucosa known as the major duodenal papilla. The smaller of the two ducts, the accessory duct (duct of Santorini), leads from the pancreas and empties into the duodenum at the apex of the lesser duodenal papilla.

The endocrine portion of the pancreas is composed of cells organised into clusters called islets of Langerhans. Cells are of five different types (α, β, δ, PP and ε), with the hormones secreted being, respectively, glucagon, insulin, somatostatin, pancreatic polypeptide (PP) and ghrelin. Insulin and glucagon play a major role in the regulation of macronutrient metabolism. The main functions of these hormones are dealt with in other chapters. The functional interactions between the endocrine and exocrine pancreas are discussed in section 10.9.

The pancreas is innervated by vagal and sympathetic fibres. The former innervate both exocrine and endocrine cells, whilst the latter innervate the endocrine cells.

Composition of the pancreatic juice

The exocrine secretion of the pancreas is important in the digestion of foodstuffs. It is made up of an aqueous component and an enzyme component.

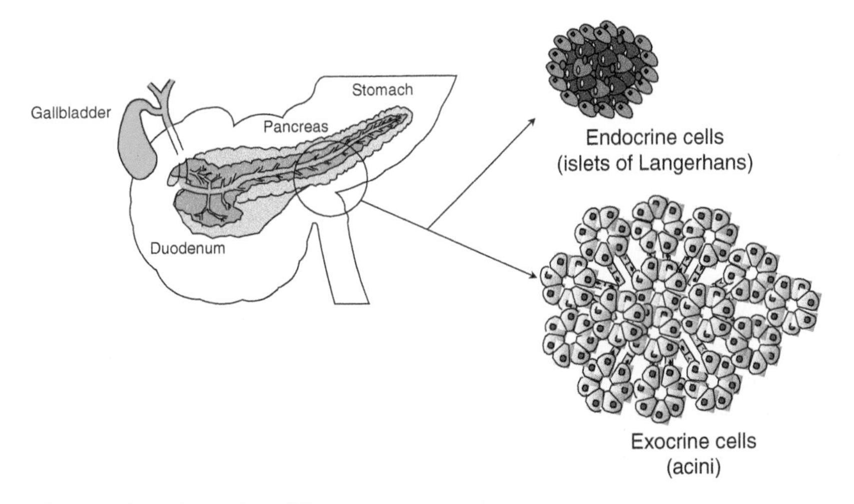

Figure 10.8 Endocrine and exocrine portions of the pancreas.

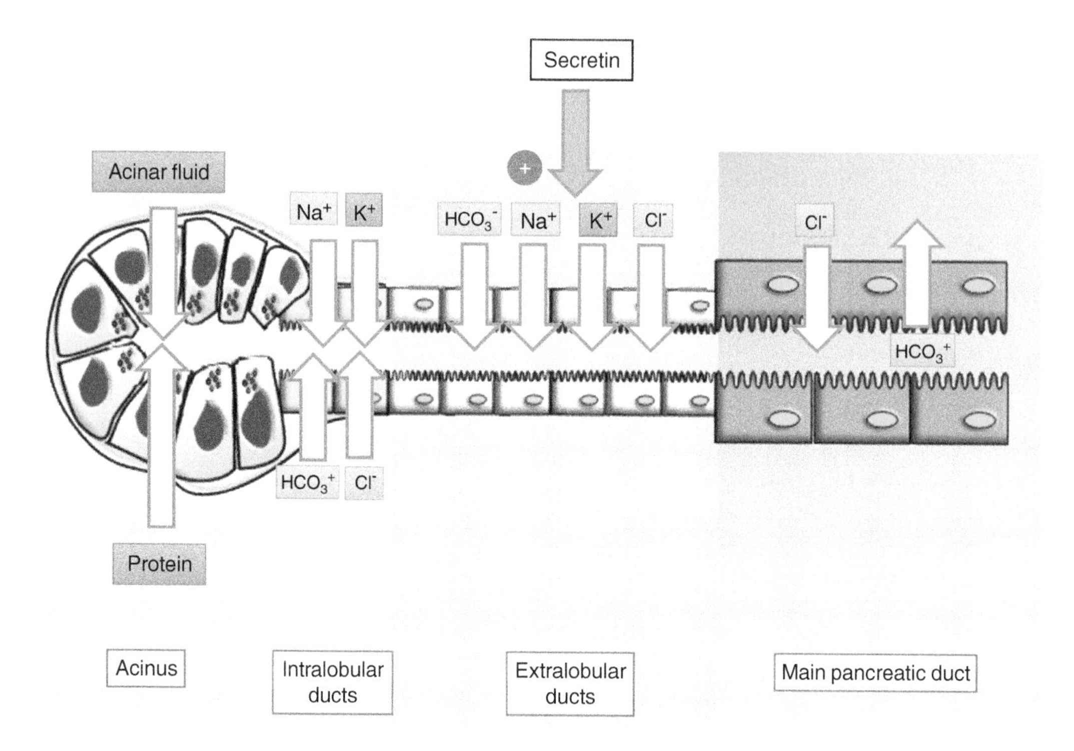

Figure 10.9 Composition and secretion of the aqueous component of the pancreatic juice.

Aqueous component of pancreatic juice

This is secreted by the epithelial cells lining the pancreatic ducts and is composed of water, sodium, potassium, bicarbonate and chloride (Figure 10.9). Bicarbonate and chloride are the major anions contained in pancreatic juice. The bicarbonate concentration increases and chloride concentration decreases reciprocally with the rate of secretion.

The primary pancreatic fluid is hypertonic to plasma. Pancreatic duct cells are water permeable and as this primary secretion flows through the ducts, water moves into the duct and makes the pancreatic juice isotonic to plasma.

The function of the pancreatic bicarbonate is to neutralise gastric acid entering the duodenum, creating the optimum pH for the action of digestive enzymes in the small intestine.

The hormone secretin is the primary stimulant for bicarbonate secretion. The fluid secreted under secretin stimulation has a higher bicarbonate concentration than that secreted under resting conditions (spontaneous secretion).

Since the function of pancreatic bicarbonate is to neutralise duodenal acidity, it is logical that the acidity of the chyme entering the duodenum stimulates the release of secretin.

Enzyme component of pancreatic juice

The acinar cells secrete the enzyme component of the pancreatic juice. The enzyme component in pancreatic juice includes enzymes for digesting carbohydrates, proteins, fats and nucleic acids (Table 10.1).

The proteases and phospholipases (enzymes that degrade phospholipids) are secreted in the form of zymogens, which are inactive precursors of the enzymes. In the duodenum, inactive pancreatic zymogens are converted to their active form. Enterokinase, an enzyme embedded in the membrane of duodenal epithelial cells, converts trypsinogen (inactive proteolytic zymogen) to

Table 10.1 Enzymes produced in pancreatic juice and the substrate they digest.

Enzyme	Substrate
Amylase	Starches (polysaccharides)
Lipase	Triacylglycerols
Phospholipase A2	Phospholipids
Carboxypeptidase	Proteins (exopeptidase)
Trypsin	Proteins (endopeptidase)
Chymotrypsin	Proteins (endopeptidase)
Elastase	Elastic fibres
Ribonuclease	Ribonucleic acids
Deoxyribonuclease	Deoxyribonucleic acids

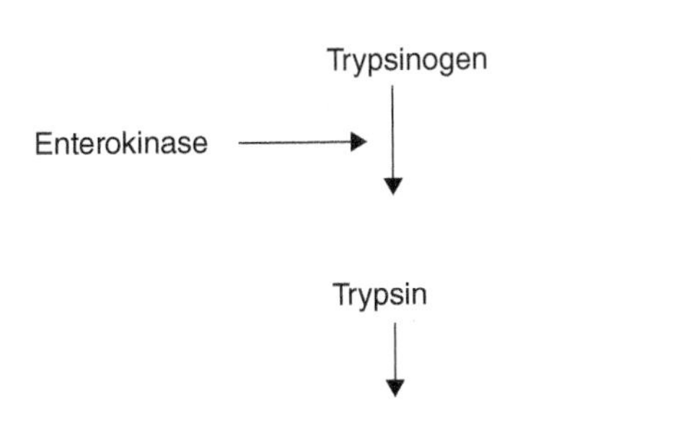

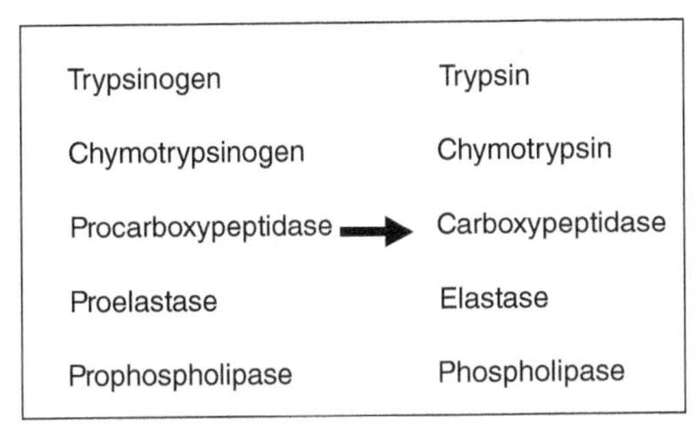

Figure 10.10 Process for converting inactive proteolytic enzymes to their active form.

trypsin (Figure 10.10). Trypsin then activates the other pancreatic zymogens by removing specific peptides from them.

A protein present in the pancreatic juice, the pancreatic secretory trypsin inhibitor (also known as serine protease inhibitor Kazal-type 1 or SPINK1), prevents the activation of inactive enzymes inside the pancreas. If the pancreatic duct is blocked, the concentration of endogenous trypsin can rise and when the concentration of trypsin inhibitor becomes insufficient, the pancreas begins to autodigest (acute pancreatitis).

The major pancreatic proteolytic enzymes are trypsin, chymotrypsin, elastase and carboxypeptidase.

Among the enzymes that are released by the pancreas in their active form, the most important are:

- *Amylase*: carbohydrate-digesting enzyme that degrades starch molecules into oligosaccharides
- *Ribonuclease and deoxyribonuclease*: nucleic acid-digesting enzymes
- *Lipase*: degrades triacylglycerols into fatty acids and monoacylglycerol.

The presence of fat and amino acids in the duodenum is the stimulus for the release of the intestinal hormone cholecystokinin. This hormone is the most powerful stimulus for the secretion of the enzymatic component of pancreatic juice. Thus, the nutrients in the diet initiate, via CCK release, the enzymatic secretion involved in their own digestion.

Regulation of the pancreatic exocrine secretion

Neural and hormonal reflexes regulate pancreatic exocrine secretion (Figure 10.11). The secretion of pancreatic juice occurs in three phases.

- *Cephalic phase*: the taste and smell of food lead to increased pancreatic secretion via the parasympathetic fibres to the pancreas. The vagal impulses stimulate the secretion of the enzyme component of the pancreatic juice.
- *Gastric phase*: the distention of the stomach by food transmits impulses via vagal afferents to the brain; then, efferent activity via the vagus nerves to the pancreas increases the enzyme secretion of the pancreas.

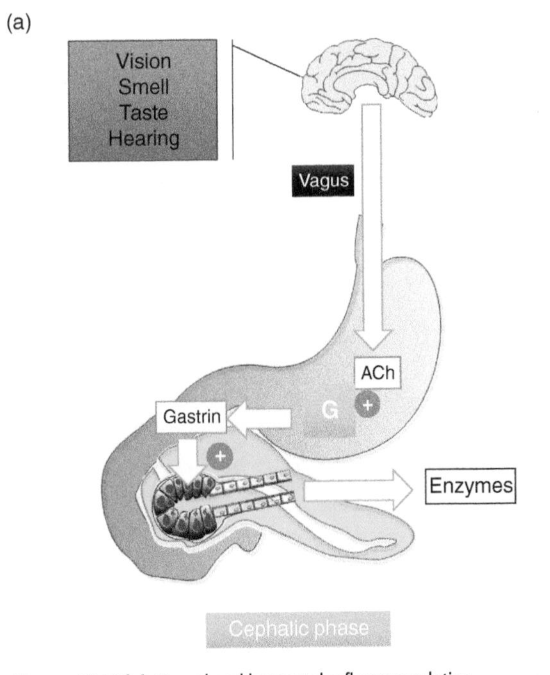

Figure 10.11(a) Neural and hormonal reflexes regulating pancreatic exocrine secretion during the cephalic (a), gastric (b) and intestinal (c) phases of digestion. ACh, acetylcholine; CCK, cholecystokinin. *Source:* Adapted from Martinez de Victoria et al., 2017.

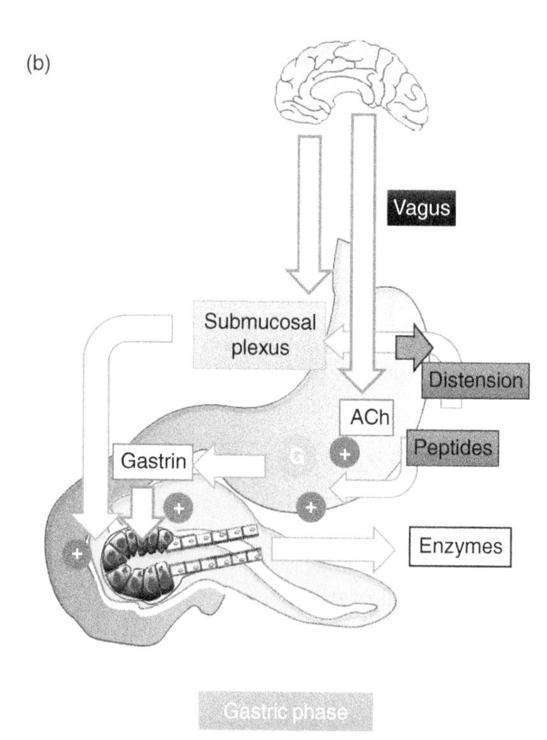

Figure 10.11(b) Neural and hormonal reflexes regulating pancreatic exocrine secretion during the cephalic (a), gastric (b) and intestinal (c) phases of digestion. ACh, acetylcholine; CCK, cholecystokinin. *Source:* Adapted from Martinez de Victoria et al., 2017.

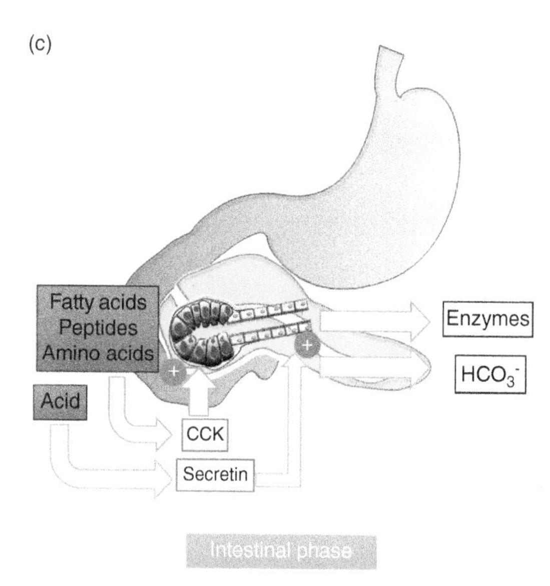

Figure 10.11(c) Neural and hormonal reflexes regulating pancreatic exocrine secretion during the cephalic (a), gastric (b) and intestinal (c) phases of digestion. ACh, acetylcholine; CCK, cholecystokinin. *Source:* Adapted from Martinez de Victoria et al., 2017.

During these two phases, vagal impulses, distention of the stomach and the presence of amino acids and peptides in the stomach evoke the release of the hormone gastrin, which also stimulates pancreatic secretion. Vagotomy reduces pancreatic exocrine secretion.

- *Intestinal phase*: the acidity of the chyme emptied into the duodenum stimulates the release of secretin. This hormone causes the ductular epithelial cells to secrete a solution high in bicarbonate and, in doing so, increases the pancreatic flow. Secretin produces a large volume of pancreatic juice with low protein concentration.

Cholecystokinin is released into the blood by fatty acids and polypeptides in the chyme and stimulates the secretion of digestive enzymes by the acinar cells. CCK potentiates the pancreatic stimulatory effects of secretin and vice versa. Secretin and CCK are released in the mucosa of the duodenum and upper jejunum by enteroendocrine cells and pass into the capillaries and thus into the systemic circulation.

10.8 Diet and exocrine pancreatic function

Adaptation of the exocrine pancreas to diet

The pancreas is very sensitive to a variety of physiological and pathological stimuli, the most important of which are nutritional in origin. Severe alterations in the diet leading to a state of malnutrition are associated with the occurrence of pancreatic injury. This section will focus on the functional changes that may occur when nutritional components in the diet are altered within physiological limits.

Pancreatic adaptation, a phenomenon first noted by Pavlov in the early twentieth century, refers to the ability of the pancreas to modify the volume and composition of its exocrine secretion in response to long-term changes in the levels of nutritional substrates available in the diet. Provided there is an adequate dietary protein supply, the synthesis and content of the major digestive enzymes (proteases, amylase and lipase) change proportionally to the amount of their respective substrates (protein, carbohydrate and fat) in the diet. This adaptation optimises the digestion and utilisation of such substrates.

The pancreatic synthesis and content of trypsinogen and chymotrypsinogen increase in response to high-protein diets. The quality of dietary protein also affects this adaptation. Increasing the intake of high-quality proteins such as casein or fish protein increases chymotrypsinogen, whereas increasing the intake of low-quality proteins such as gelatine or zein does not.

High-carbohydrate diets increase the pancreatic amylase content and synthesis. This effect occurs as the level of dietary carbohydrate increases at the expense of dietary fat or protein, which implies a primary response to the level of carbohydrate.

The intake of high-fat diets enhances the synthesis and content of pancreatic lipase, independently of whether the amount of dietary fat increases at the expense of protein or carbohydrate. Although lipase adapts to increasing dietary fat levels, there may be a threshold of fat content below which there is little adaptation and above which there is significant adaptation. Colipase, the protein co-factor required for lipase to efficiently hydrolyse triacylglycerols, may adapt to high dietary fat levels, but the response seems to be weaker than that of lipase. Considerable controversy exists over the effects of the type of fat (degree of saturation or major chain length) on the adaptation of pancreatic enzymes.

Most of the information on pancreatic adaptation to diet has been gathered from rats, and few studies have evaluated this topic in humans. Investigations in premature infants suggest that adaptive responses occur in humans in a manner similar to that observed in experimental animals. In healthy omnivore adults, a reduction and modification of protein intake due to a five-week vegan diet resulted in an adaptation of pancreatic protease secretion. In contrast, altering the quantity of carbohydrate, protein or fat in the diet of adult volunteers for 10 or 15 days failed to induce any changes in the ratios among the enzymes secreted. In humans adapted for 30 days to diets containing either olive oil (rich in oleic acid, a monounsaturated fatty acid) or sunflower oil (rich in linoleic acid, a polyunsaturated fatty acid) as the main source of dietary fat, no differences in the activity of proteases, amylase, lipase and colipase measured in duodenal contents were apparent after the administration of a liquid meal. Differences in diet composition, duration of the adaptation period and approach to pancreatic function testing may account for the discrepancies.

Is pancreatic adaptation mediated by specific hormones?

Although pancreatic adaptation is well recognised, the underlying mechanisms are still largely unknown. Adaptive changes in pancreatic enzymes have been postulated to be differentially mediated by specific hormones. In most cases, their release is increased by the nutrients whose digestion they regulate (Table 10.2).

The gastrointestinal hormone CCK, which is potently released by ingested protein, dramatically increases the synthesis and tissue level of proteases, so it has been proposed as the mediator of pancreatic adaptation to dietary protein. In many species, CCK release from endocrine I cells in the small intestine is regulated by a luminal negative feedback mechanism. This mechanism, which is important in the normal response of the pancreas to a meal, may also be involved in the pancreatic adaptation to high-protein diets. It is based on the existence of CCK-releasing peptides with the capacity to bind to intestinal endocrine I cells and enhance CCK release. However, because these peptides are sensitive to degradation by pancreatic proteases, only when proteolytic activity is removed (by administration of protease inhibitors or diversion of pancreatic juice) will they be able to exert this action. In the case of high-protein feeding, abundant protein in the lumen occupies protease active sites, thus preventing the inactivation of CCK-releasing peptides. Several CCK-releasing peptides have been identified in different species: luminal CCK-releasing factor (LCRF) and diazepam-binding inhibitor (DBI), both produced by the intestinal mucosa, and monitor peptide (MP), contained in the pancreatic juice. Figure 10.12 is a schematic diagram of the feedback mechanism.

Table 10.2 Hormonal mediators of pancreatic adaptation to increased nutritional substrates.

Substrate	Hormone	(Pro)enzyme synthesis and tissue content
Protein	CCK	↑ Trypsinogen and chymotrypsinogen
Triacylglycerol	Secretin, GIP	↑ Lipase
Starch	Insulin	↑ Amylase

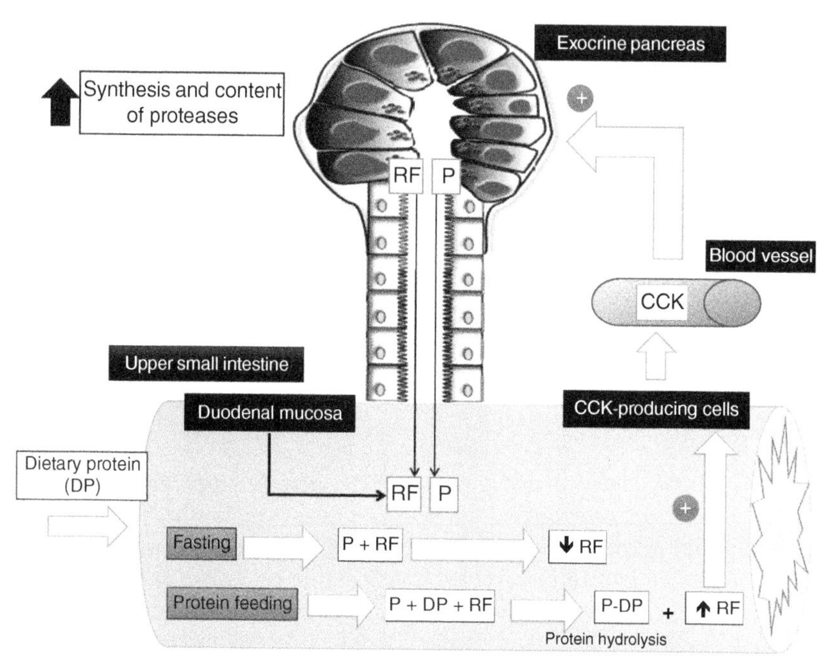

Figure 10.12 Role of luminal feedback regulation of CCK release in the pancreatic adaptation to high-protein diets. CCK-releasing peptides (either secreted by the exocrine pancreas or by the duodenal mucosa) are protease-sensitive factors that stimulate CCK release from endocrine I-cells in gut epithelium. They are represented by RF (releasing factors) in the diagram. Under conditions of protein fasting, RF are digested and inactivated by pancreatic proteases in the intestinal lumen. In this form, RF are not capable of stimulating CCK release. In contrast, proteases are bound to dietary proteins when these are ingested. RF survive and enhance the release of CCK into the blood circulation. DP, dietary protein; P, proteases.

A role for insulin in the adaptation of the pancreas to high-carbohydrate diets has been proposed. When animals are rendered diabetic, amylase content, synthesis and mRNA levels fall dramatically, and these parameters are restored following the administration of exogenous insulin. Section 10.9 includes a more detailed discussion about the effects of insulin on the exocrine pancreas.

Secretin and/or gastric inhibitory peptide (GIP), also known as glucose-dependent insulinotropic peptide, have been suggested to mediate pancreatic adaptation to the amount of lipids in the diet. Both are produced by cells of the upper intestine and their release is stimulated, among other factors, by the hydrolytic products of fat. Moreover, the administration of either secretin or GIP has proved to augment the synthesis and content of lipase and colipase in pancreatic acinar cells.

The type of dietary fat affects the circulating levels of hormones involved in the control of pancreatic secretion, as shown in a study conducted in dogs. After a six-month period on diets containing either virgin olive oil or sunflower oil as the source of dietary fat, plasma PYY and pancreatic polypeptide (PP) concentrations were shown to be higher in the olive oil group than in the sunflower oil group. The higher concentrations of inhibiting hormones such as PP and PYY in response to an olive oil diet can explain, at least in part, why long-term adaptation to olive oil leads in this species to an attenuation of the pancreatic response to food compared with the typical one in the animals fed sunflower oil.

Regulation of pancreatic gene expression by long-term dietary changes

Theoretically, diet can alter the expression of genes encoding specific pancreatic enzymes through mechanisms involving the gene (transcription), its mRNA (processing, extranuclear transport or cytoplasmic stability) and its translation into protein. However, adaptive changes in pancreatic enzymes in animals are mainly associated with changes in specific mRNA levels, suggesting an effect at the transcriptional level. In contrast, shorter-term meal-stimulated protein synthesis in the pancreas, during the postprandial phase, is regulated primarily at the translational level. Studies using administration of exogenous hormonal mediators support this view, since single periods of hormone stimulation (mimicking the postprandial situation) modify the efficiency of mRNA translation into protein, whereas repeated periods of stimulation (up to seven days, mimicking the situation after long-term intake of nutrients) lead to changes in mRNA levels. This illustrates the importance of the duration of the adaptation period to a dietary change.

The mechanism by which hormonal mediators regulate the expression of genes for specific enzymes after adaptation to high levels of nutrients has not been completely elucidated. In some cases, the genetic elements regulated in the promoter region have been identified, although the full intracellular pathway leading to their regulation is unknown.

Modification of membrane fatty acid composition as a mechanism for pancreatic adaptation to the type of dietary fat

Investigations of the in vivo pancreatic response after medium- or long-term intake of diets differing in the fat source have indicated that pancreatic adaptation to dietary fat type is mediated, at least in part, by changes in the circulating levels of some gastrointestinal hormones. However, the finding that pancreatic cell membranes are enriched in those fatty acids most abundant in the fat ingested suggests that, in addition to the above mechanism, in vivo data may reflect a direct modulatory effect of the type of dietary fat on the secretory activity at the cellular level.

Indeed, studies with viable pancreatic acinar cells of rats fed over eight weeks with diets containing either virgin olive oil or sunflower oil as the fat source have shown that dietary-induced changes in the fatty acid profile of pancreatic membranes is associated with modulation of pancreatic cell function as assessed by amylase release and intracellular calcium (Ca^{2+}) mobilisation in response to CCK-8. Such a direct mechanism for adaptation to dietary fat type seems reasonable when considering the role of membrane fatty acids in cellular signalling. Many steps of the stimulus–secretion coupling process in acinar cells are membrane dependent, as shown in Figure 10.13, which represents a very simplified scheme of signal transduction pathway triggered by CCK in the pancreatic acinar cell. Differential enrichment in certain fatty acids may influence the accessibility to secretagogue receptors, the interaction with G-proteins, the functionality of such enzymes as phospholipase C and protein kinase C that interact with cell membranes during their activation, or the fatty acyl composition of intracellular messengers like diacylglycerol, all of which may imply a different sensitivity of the gland for a given concentration of the hormonal secretagogue in blood.

The above results in fresh acinar cells from rats fed different dietary fats have been confirmed by more recent research using the AR42J pancreatoma cell line. These cells retain many characteristics of normal pancreatic acinar cells, and their membrane composition can be artificially modified. Supplementation of culture media with oleic acid, linoleic acid, or a mixture of n-3 long-chain polyunsaturated fatty acids has been reported to change membrane fatty acids in 72 h (Figure 10.14). For oleic and linoleic acids, the pattern and direction of changes were parallel to those found in rats fed diets enriched in olive or sunflower oil, respectively, and they were also accompanied by modulation of the secretory activity that involved Ca^{2+} signalling.

Moreover, exposure of native and modified cells to caerulein, a harmful compound that evokes acute pancreatitis when administered to rats in vivo, suggests that specific membrane changes relate not only to modified cell function, but also to differential protection against caerulein-induced injury, as measured by inflammatory, oxidative and cell death markers.

Effect of exocrine pancreatic insufficiency on nutrition

It is clear from the previous sections that a number of nutritional factors can affect exocrine pancreatic function. Conversely, pathological modifications of this function lead to an altered nutritional state.

Exocrine pancreatic insufficiency (EPI) can be defined as the inability of exocrine pancreatic secretion to carry out normal food digestion, usually due to reduced secretion of enzymes and bicarbonate. The most common causes are chronic pancreatitis and pancreatic cancer in adults, and cystic fibrosis in children. Among the three major macronutrients, fat digestion is

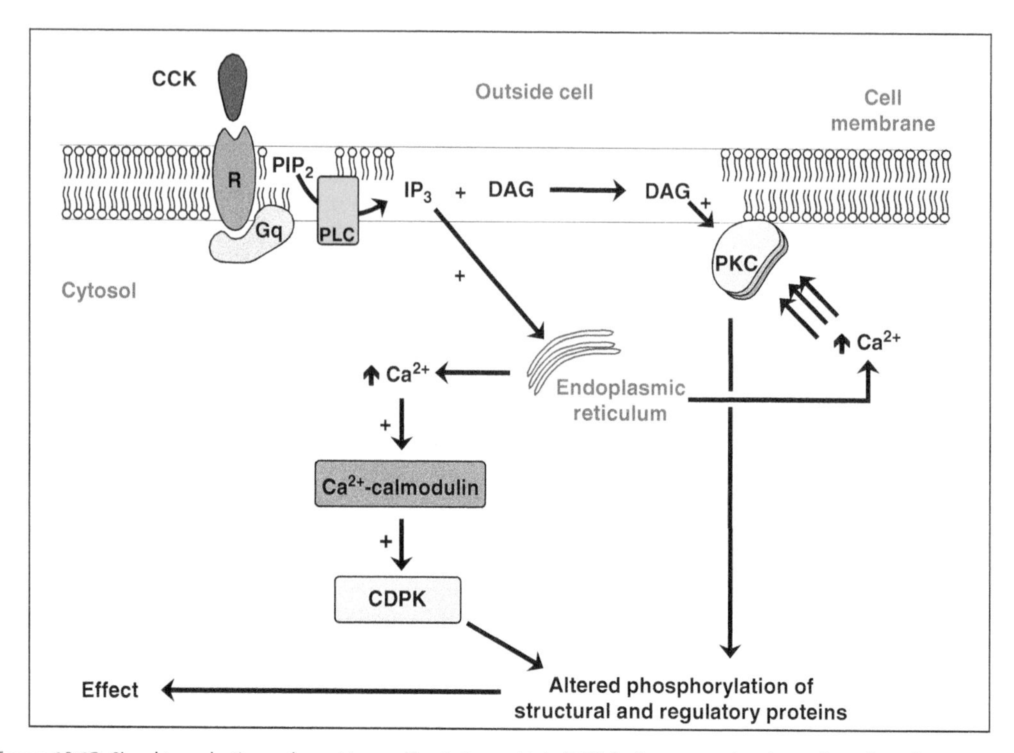

Figure 10.13 Signal transduction pathway triggered by cholecystokinin (CCK) in the pancreatic acinar cell. Binding of CCK to its specific receptor (R) activates the Gq class of heterotrimeric G proteins (Gq). The activated Gq then stimulates phospholipase C (PLC) to hydrolyse the membrane phospholipid phosphatidylinositol 4,5-bisphosphate (PIP_2). This cleavage releases inositol 1,4,5-trisphosphate (IP_3) and diacylglycerol (DAG). IP_3 binds to specific Ca^{2+} channels in the endoplasmic reticulum membrane causing Ca^{2+} to be released. The rise in cytosolic Ca^{2+} favours the formation of the complex Ca^{2+}-calmodulin, which in turn activates a number of calmodulin-dependent protein kinases (CDPK). In the presence of enhanced Ca^{2+}, protein kinase C (PKC) (which in unstimulated conditions is in the cytosol) also binds Ca^{2+}, and this causes PKC to migrate to the inner surface of the plasma membrane, where it can be activated by DAG. All PKCs and CDPKs phosphorylate key intracellular enzymes, leading to either activation or inactivation of downstream regulatory proteins and producing the cellular response.

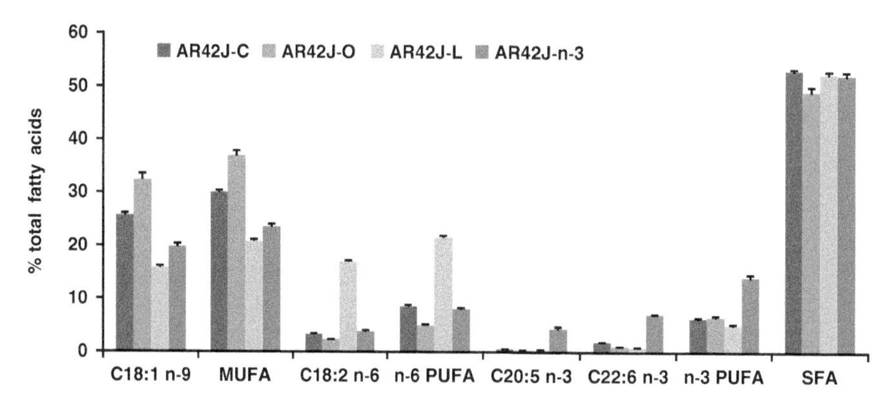

Figure 10.14 Fatty acid composition of crude membranes from differentiated AR42J cells cultured for 72 h in medium containing unmodified foetal calf serum (AR42J-C) or serum enriched in either C18:1 n-9 (AR42J-O), C18:2 n-6 (AR42J-L), or a 60:40 mixture of C20:5 n-3 plus C22:6 n-3 (AR42J-n-3). MUFA, monounsaturated fatty acids; PUFA, polyunsaturated fatty acids; SFA, saturated fatty acids.

much more severely impaired in EPI because it depends almost entirely on pancreatic lipase and colipase, while in carbohydrate digestion there is the contribution from salivary amylase and brush-border oligosaccharidases, and proteins can be partially digested by the proteolytic activity of the gastric juice. Even so, since the pancreas has a reserve capacity many times the physiological requirement, steatorrhoea (excessive amounts of fat in the faeces) occurs only when lipase secretion is reduced to less than 10% of normal.

Patients with severe and chronic exocrine pancreatic insufficiency are at high risk for developing different forms of malabsorption, the common endstage of which is protein-energy malnutrition. Apart from defective absorption of the three major nutrients, there may be other nutritional disorders.

- Deficiencies of fat-soluble vitamins (in patients with pancreatic steatorrhoea).
- In healthy subjects, a zinc-binding compound of pancreatic origin facilitates intestinal zinc absorption. In chronic pancreatitis, clinically significant malabsorption of zinc may ensue.
- Vitamin B_{12} deficiency: in the duodenum, vitamin B_{12} is bound to salivary and gastric R proteins. Cleaving of the vitamin from R proteins by pancreatic proteases allows B_{12} to bind to intrinsic factor for absorption in the ileum. In exocrine pancreatic insufficiency, the amount of vitamin B_{12} absorbed can decrease markedly.

Malnutrition in these patients is associated with the development of sarcopenia, osteopenia or osteoporosis, diminished immunocompetence and increased risk of cardiovascular events.

10.9 Interactions between the endocrine and exocrine pancreas

As described in previous sections (see 10.7), the pancreas is a complex organ with an endocrine part (islets of Langerhans) and an exocrine part which is composed of acini (enzyme secretion) and ducts (fluid and electrolyte secretion). The view of a pancreatic gland with functionally distinct compartments has been proven incorrect. It has been clearly established that there exist intensive interaction and integration between the exocrine and endocrine pancreas at morphological and physiological levels. These close links play a significant role in pancreatic function not only in normal but also in pathological conditions. Any disease affecting the endocrine pancreas will inevitably affect the exocrine function and vice versa, as described in the following paragraphs.

Influence of the endocrine pancreas on the exocrine pancreas

Studies aimed at examining the influence of the endocrine pancreas on the exocrine pancreas have been based on two pillars: on the one hand, on the effects of islet hormones on exocrine function and, on the other hand, on the morphological relationships between both components.

Regarding the first point, pancreatic growth and exocrine secretion seem to be, at least partially, under the control of islet hormones. Thus, although the gastrointestinal hormone CCK is considered to be the main trophic factor for the exocrine pancreas, insulin has been demonstrated to enhance protein synthesis, sensitivity to CCK and binding of insulin-like growth factor II (IGF-II) to acinar cells, thus promoting growth. Its contribution to the effects of CCK could be important in the case of adaptive growth (in response to high dietary protein intake), and essential in the case of regenerative growth following pancreatic damage.

In addition, islet hormones appear to play a role in the regulation of exocrine pancreatic secretion. Insulin enhances pancreatic enzyme synthesis and content, with emphasis on amylase, and potentiates the stimulatory action of nerves and hormones (CCK and secretin) on fluid and enzyme secretion. Saturable insulin-binding sites have been demonstrated on both acinar and duct cells in different species, so the effect of the hormone is probably direct. Evidence about the effects of glucagon is inconsistent, with reports of stimulatory and inhibitory effects on exocrine secretion. Somatostatin action on both acinar and duct cells is inhibitory, yet the underlying mechanism is still under debate. It may regulate secretagogue-evoked enzyme secretion via binding to its receptors on the acinar cells. Other proposals include inhibition of insulin release and the consequent decrease in its stimulating effect, as well as neurally mediated mechanisms, both central and peripheral. PP has an inhibitory effect on enzyme secretion. However, it has not been possible to confirm the presence of PP receptors in acinar cells, and neither is PP capable of inhibiting enzyme secretion in isolated pancreatic acini, so it has been postulated that its action is carried out through an indirect cholinergic mechanism.

From a morphological point of view, endocrine–exocrine interaction is supported by the fact that the peri-islet basement membrane, double-layered in human islets versus single-layered in rodent islets, is incomplete and presents occasional gaps that would allow paracrine signalling and even direct endocrine–exocrine cell contact. In addition, small ducts can penetrate the islets, making extensive contacts with core cells.

However, it is the proposed existence of an islet-acinar portal blood system that has provided the greatest support for the influence of islets on the exocrine pancreas. Such a system comprises feeding arterioles that reach the islets, form an intra-islet glomerulus-like network and leave the islets as efferent capillaries which subsequently perfuse the exocrine tissue. This means that a large part of the blood arriving to the acini has passed through the islets first. An important consequence is that, under physiological circumstances, the acinar cells are exposed to concentrations of islet hormones that may be much higher (20-fold) than those of peripheral blood. According to this model, only unidirectional blood flow from islets to exocrine tissues is possible.

This notion has recently been challenged by studies using intravital microscopy of the exteriorised mouse pancreas to individually track fluorescent-labelled red blood cells moving through islet capillaries in real time. Results obtained with this methodology suggest a new model of islet microcirculation where the islet capillaries are integrated with those of the exocrine pancreas, contradicting the assumption that microcirculation is confined to a closed structure within islets. In this new model, blood may flow not only from islets to acinar tissue but also from acinar tissue to islets in near-equal proportion, a major outcome being that islet-secreted products can reach acinar tissue and that molecules released by the acini can reach the islets.

Although in vivo recordings of capillary blood flow are not yet possible in the human pancreas, 3D structural analysis of vasculature by means of confocal microscopy and tissue clearing of human pancreatic slices indicates that, similar to rodents, the capillary structure in human islets is integrated with pancreatic microvasculature in its entirety. This new model still requires further experimental evidence and validation. Nonetheless, the implications in physically linking the endocrine and exocrine pancreas are particularly important in relation to clinical pathologies where both regions are affected.

Exocrine dysfunction secondary to diabetes mellitus

There is evidence that a significant proportion of diabetic patients have exocrine pancreatic insufficiency (EPI), the main features of which were described previously. Pathophysiology of these aspects has been more thoroughly investigated in insulin deprivation conditions (type 1 diabetes mellitus, T1D) than in states of insulin resistance and hyperinsulinaemia (type 2 diabetes, T2D).

General findings in the pancreas of T1D patients include acinar atrophy, fibrosis, arteriosclerosis and immune infiltration. Acinar atrophy is accompanied by reduced enzyme content and a marked decrease in pancreatic size. The absence of insulin, a growth and stimulatory factor otherwise reaching high local concentrations around the acinar cells, may lead to a smaller gland and impaired exocrine function. In support of this, acinar atrophy is most pronounced around insulin-deficient islets compared with insulin-expressing islets in autopsies of patients with T1D. Other islet hormones may be involved as well. Considering their regulatory role on exocrine function, a change in hormone balance with prevalence of inhibitory over activating signals probably contributes to the exocrine insufficiency.

While reduced pancreatic size is observed early in T1D, possibly reflecting an acute loss of islet hormone sensing, other contributing factors to EPI require a longer duration of diabetes.

- *Vascular complications*: long-standing hyperglycaemia diminishes pancreatic blood flow through arteriosclerosis, leading to acinar ischaemia. Hyperglycaemia and hypoxia lead to activation of pancreatic stellate cells, resulting in increased collagen production and fibrosis.
- *Diabetic neuropathy*: disruption of cholinergic input and enteropancreatic reflexes may potentiate exocrine dysfunction in T1D.

A last hypothesis proposed to explain EPI in T1D patients is autoimmunity, whereby autoantibodies raised against islet cells may cross-react with exocrine antigens.

Exocrine insufficiency is known to be present in 32% of T2D patients compared with 50% in those with T1D. Although evidence is not abundant, it has been reported to be also associated with fibrosis and inflammatory cell infiltration. Pancreatic size has been a controversial issue in T2DM patients, with several studies reporting a reduced size while other studies reported no difference compared with normoglycaemic controls. However, a recent study found that the pancreas is smaller in patients with mid- and long-term T2D, but not in recently diagnosed individuals, suggesting that pancreatic atrophy relates to the loss of insulin secretory capacity in these patients.

Influence of the exocrine pancreas on the endocrine pancreas

Endocrine dysfunction secondary to exocrine pancreatic disease

Pancreatogenic diabetes mellitus (also known as diabetes of the exocrine pancreas) is classified as a form (type 3c) of secondary or type 3 diabetes mellitus (T3cD) by the World Health Organization and the American Diabetes Association.

The most common underlying disease seems to be chronic pancreatitis, followed by pancreatic cancer. Less common pancreatic aetiologies include acute pancreatitis, cystic fibrosis and surgical pancreatic resection, among others. Recent data suggest that many T3cD patients are misdiagnosed and classified as T2D. Thus, T3cD is far more frequent than has been previously estimated.

Several mechanisms have been proposed to explain the pathogenesis of T3cD and, among the different aetiologies, they appear to contribute differentially to the diabetic phenotype.

- *Endocrine dysfunction*: sustained exocrine disease featuring inflammation and fibrotic replacement of exocrine tissue, as in chronic pancreatitis and cystic fibrosis, leads to islet damage and dysfunction. All cell subtypes within the islets can be affected, but β-cell failure plays a central role in the development of hyperglycaemia and diabetes. Different inflammatory mediators have been suggested to be involved in impaired insulin release and/or β-cell death. Over time, fibrosis contributes by progressively destroying the islets. In pancreatic cancer, there is no fibroinflammatory destruction of endocrine tissue. Rather, insulin deficiency is explained by release of β-cell toxic products by cancer cells. The new model of integrated, bidirectional pancreatic islet blood flow supports these hypotheses by providing a mechanism for delivery of such mediators to the islet cells.

Alterations in the glucagon and PP responses also appear to have important repercussions on glycaemic control in T3cD patients. Defective release of the counter-regulatory hormone glucagon, together with a reduced nutrient absorption due to exocrine insufficiency and poor compliance with medication and nutritional advice in some patients (e.g. alcohol-induced pancreatitis), makes serum glucose levels difficult to regulate, with alternately occurrence of hypoglycaemia and hyperglycaemia.

Regarding PP deficiency, it has been associated with the appearance of hepatic insulin resistance in patients with T3cD secondary to chronic pancreatitis, pancreatic ductal adenocarcinoma, pancreatic resection and cystic fibrosis. PP seems to improve liver cell sensitivity to insulin in normal conditions. Reduced PP secretion would cause hyperglycaemia through unrestrained hepatic glucose production. Indeed, diminished hepatic insulin receptor availability in chronic pancreatitis has been shown to be reversed by PP administration.

- *Impairment of the incretin system*: abnormal exocrine pancreatic function and defective fat hydrolysis may affect islet β-cell mass and function through disruption of the enteroinsular axis.
- *Peripheral insulin resistance*: appears to be present in some forms of T3cD, with different cytokines and paraneoplastic factors (in the case of pancreatic cancer) having been proposed as mediators. However, discrepancies between data presented by different studies exist. More research in this area is needed to ascertain the importance of changes in peripheral insulin sensitivity as contributors to diabetes secondary to exocrine pancreatic diseases.

In short, the available information leads us to increasingly view the pancreas as an integrated organ in which all the components are functionally related for well-orchestrated functional responses, a reasonable view considering that the exocrine portion secretes enzymes and bicarbonate, which affect the digestion and absorption of nutrients, whereas the endocrine part releases hormones that regulate the metabolism and disposal of breakdown products of food within the body.

10.10 Physiology of bile secretion and enterohepatic circulation

The liver of all vertebrates produces and secretes bile. This digestive secretion is elaborated from transport processes in the hepatocyte and is modified along the bile canaliculus and ducts. Finally, in most species, bile is stored in the gall bladder and then released into the common hepatic duct and duodenum after a meal.

The liver cells (hepatocytes) are arranged in many functional units called lobules. The lobules are sheets of hepatocytes organized around a central vein (Figure 10.15). Instead of capillaries, the liver has large spaces called sinusoids which receive blood containing absorbed nutrients from the hepatic portal vein. The hepatic arterial blood supply to the liver brings oxygen to the liver cells.

Branches of both the hepatic portal vein and the hepatic artery carry blood into the liver sinusoids, where nutrients and oxygen are taken up by the hepatocytes. The liver secretes back into the blood products synthesised by the hepatocytes and nutrients needed by other cells. Blood from both origins leaves the liver in the hepatic vein.

Functions of the liver

The most important digestive function of the liver is the secretion of bile, which is essential for lipid digestion in the intestine. Bile is secreted continuously into specialised ducts, called bile canaliculi, located within each lobule of the liver. These canaliculi empty into bile ducts at the periphery of the lobules. The bile ducts merge and form the right and left hepatic ducts that unite to form a single common hepatic duct. The common hepatic duct is joined by the cystic duct from the gall bladder to form the common bile duct, which empties into the duodenum.

Surrounding the common bile duct at the point where it enters the duodenum is the sphincter of Oddi. When the sphincter is closed, the dilute bile secreted by the liver passes into the gall bladder and becomes concentrated until it is needed in the small intestine.

After the beginning of a meal, the sphincter of Oddi relaxes and the gall bladder contracts, discharging concentrated bile into duodenum. The signal for gall bladder contraction is the intestinal hormone CCK. The presence of fat and amino acids in the duodenum is the stimulus for the release of CCK, a hormone that also causes relaxation of the sphincter of Oddi.

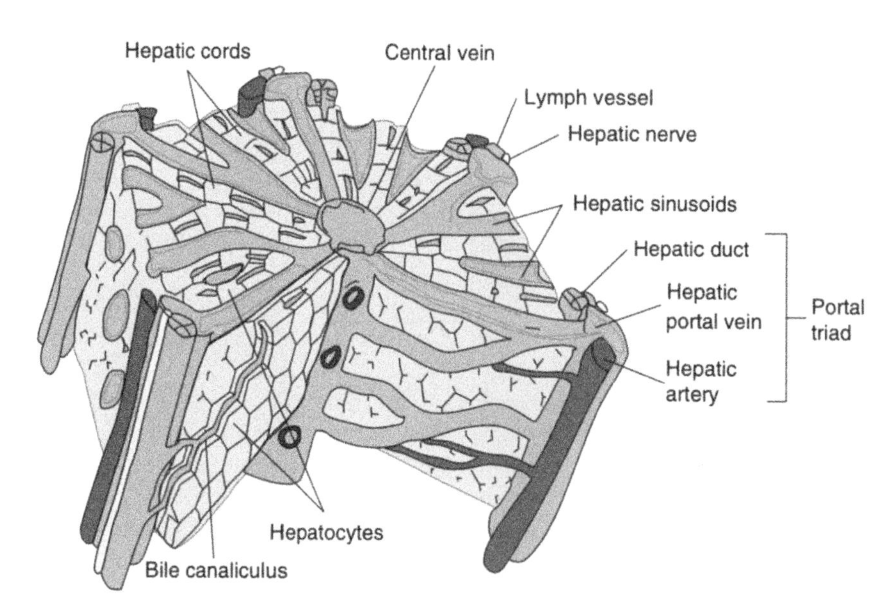

Figure 10.15 Hepatic lobule structure.

Apart from bile secretion the liver has other important functions.

- *Lipid metabolism*: the hepatocytes synthesise lipoprotein and cholesterol, store some triacylglycerols and use cholesterol to synthesise bile acids (BA).
- *Carbohydrate metabolism*: the liver is one of the major sites of glycogen storage in the body. This occurs after carbohydrate-containing meals and is stimulated by insulin. When blood glucose is low, glycogen is broken down to glucose (glycogenolysis) and the glucose is released into the bloodstream. Under these conditions, the liver also synthesises glucose from amino acids and lactic acid (gluconeogenesis).
- *Protein metabolism*: hepatocytes deaminate amino acids to form ammonia (NH_3), which is converted into a much less toxic product, urea. Urea is excreted in urine.
- *Protein synthesis*: the liver synthesises all the major plasma proteins (albumin, globulin, apoproteins, fibrinogen and prothrombin).
- *Excretion of bile pigments*: bile pigments are derived from the haem portion of haemoglobin in aged erythrocytes. The most important is bilirubin, which is absorbed by the liver from the blood and actively secreted into the bile. After entering the intestinal tract, bilirubin is metabolised by bacteria and eliminated in faeces.
- *Storage*: the liver stores some vitamins (A, D, E, K and B_{12}) and minerals (iron and copper) and participates in the activation of vitamin D.
- *Detoxification of hormones, drugs and toxins*: the liver converts these substances to inactive forms in the hepatocytes for subsequent excretion in bile or urine.

Role and composition of the bile

Bile is a mixture of substances synthesised by the liver. It has several components: bile acids, cholesterol, phospholipids (lecithin), bile pigments and small amounts of other metabolic endproducts, bicarbonate ions and other salts and trace elements. Bile acids are synthesised from cholesterol and before they are secreted are conjugated with the amino acids glycine or taurine. Conjugated bile acids are called bile salts. Bile salts emulsify lipids in the lumen of the small intestine to form micelles, increasing the surface area available to lipolytic enzymes. Micelles transport the products of lipid digestion to the brush-border surface of the epithelial cells, helping the absorption of lipids. Cholesterol is made soluble in bile by bile acids and lecithin.

The epithelial cells lining the ducts secrete a bicarbonate solution, similar to that produced by the pancreas, which contributes to the volume of bile living the liver. This salt solution neutralises acid in the duodenum.

Bile secretion

Bile secretion is one of the major functions of the liver, which serves two major purposes: (i) the excretion of hepatic metabolites – including bilirubin, cholesterol, drugs and toxins – and (ii) the facilitation of intestinal absorption of lipids and fat-soluble vitamins.

The hepatocytes produce a primary secretion isotonic to plasma that contains the substances that carry out the main digestive functions (bile acids, cholesterol and lecithin). The bicarbonate-rich fluid secreted by the epithelial cells of the ducts modifies the primary secretion of the liver (Figure 10.16).

Fraction of bile secreted by the hepatocytes

Bile acids, phospholipids, cholesterol, bile pigments and proteins are the most important substances secreted by the hepatocytes into the bile. Bile acids synthesised from cholesterol by the liver are called primary bile acids (cholic acid and chenodeoxycholic acid). The bacteria of the small intestine dehydroxylate primary bile acids to form secondary bile acids (deoxycholic acid and lithocholic acid).

Before they are secreted, bile acids are conjugated with the amino acids glycine or taurine to make them more water soluble. Bacteria in the intestine deconjugate bile salts into bile acids, which are absorbed in the ileum and go back to the liver, where they are converted into bile salts again and secreted along with newly synthesised bile acids. The uptake of bile acids by the liver stimulates bile salts release.

Hepatocytes secrete phospholipids, especially lecithin, which help to solubilise fat, cholesterol and bile pigments in the small intestine. The synthesis of bile acids from

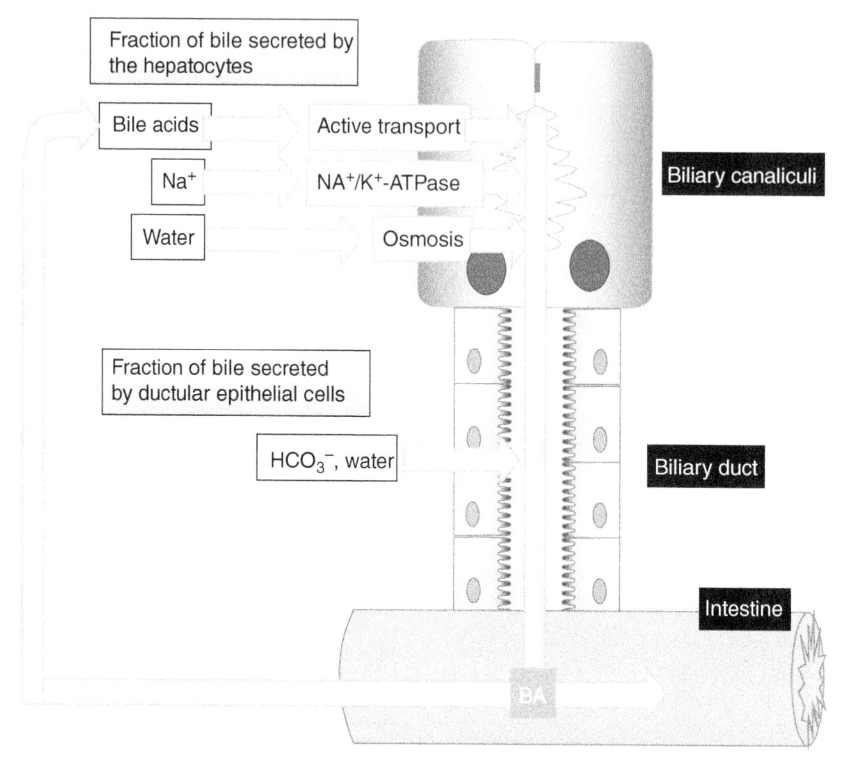

Figure 10.16 Process of bile secretion in the hepatocytes of the liver. *Source:* Adapted from Martinez de Victoria et al., 2017.

cholesterol and the excretion of cholesterol and bile acids in faeces are two of the mechanisms which maintain cholesterol homeostasis. Hepatocytes remove bilirubin from the bloodstream attached to glucuronic acid molecules. The resultant conjugated bilirubin is secreted into the bile.

Fraction of bile secreted by ductular epithelial cells

These cells secrete a bicarbonate-rich salt solution that accounts for about half of the total bile volume. The concentration of HCO_3 is greater than in the plasma and helps to neutralise the intestinal acid. This secretion is stimulated by secretin in response to the presence of acid in the duodenum.

Bile concentration and storage in the gall bladder

Between meals, the sphincter of Oddi is closed and the dilute bile secreted by the liver passes through the cystic duct into the gall bladder and becomes concentrated. The gall bladder mucosa absorbs water and ions (Na^+, Cl^-, HCO_3^-) and can concentrate bile salts 10-fold. The active transport of Na^+ into the interstitial space produces an elevated osmolarity in this region and the osmotic flow of water across the epithelium; HCO_3^- and Cl^- are transported to the interstitial space to preserve electroneutrality.

In the postprandial periods, the gall bladder is emptied, and bile acids reach the intestinal lumen in high concentration where they exert their functions in the digestion and absorption of dietary fat.

Then they are actively absorbed from the distal ileum (actively) and large intestine (passively), transported to the liver in portal venous blood, and then efficiently taken up by the hepatocyte. Bile acids traverse the hepatocyte and are actively secreted into canalicular bile, completing the enterohepatic cycle. This cycle is called the *enterohepatic circulation of bile acids* (EHC). The term enterohepatic circulation denotes the movement of bile acid molecules from the liver to the small intestine and back to the liver (Figure 10.17).

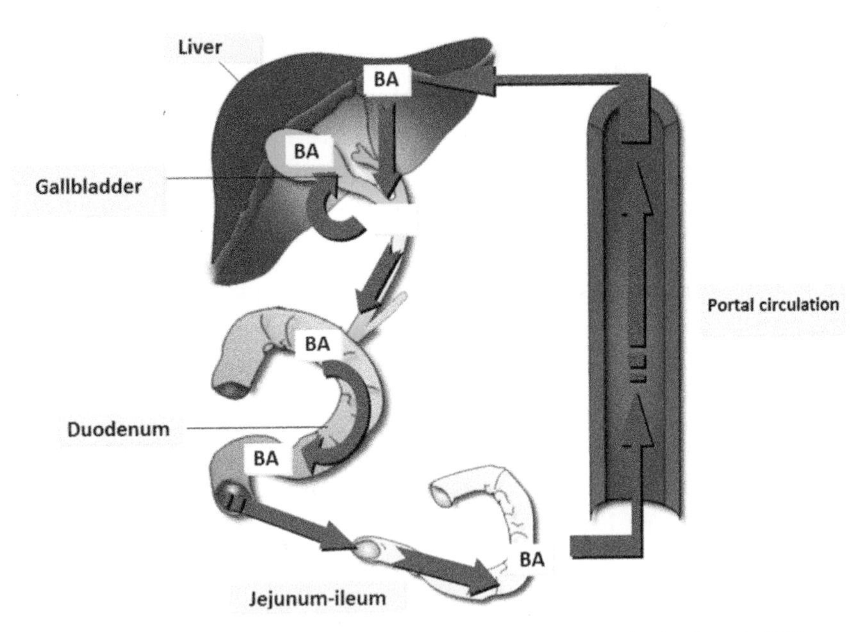

Figure 10.17 Enterohepatic circulation of bile acids (BA). *Source:* Adapted from Martinez de Victoria et al., 2017.

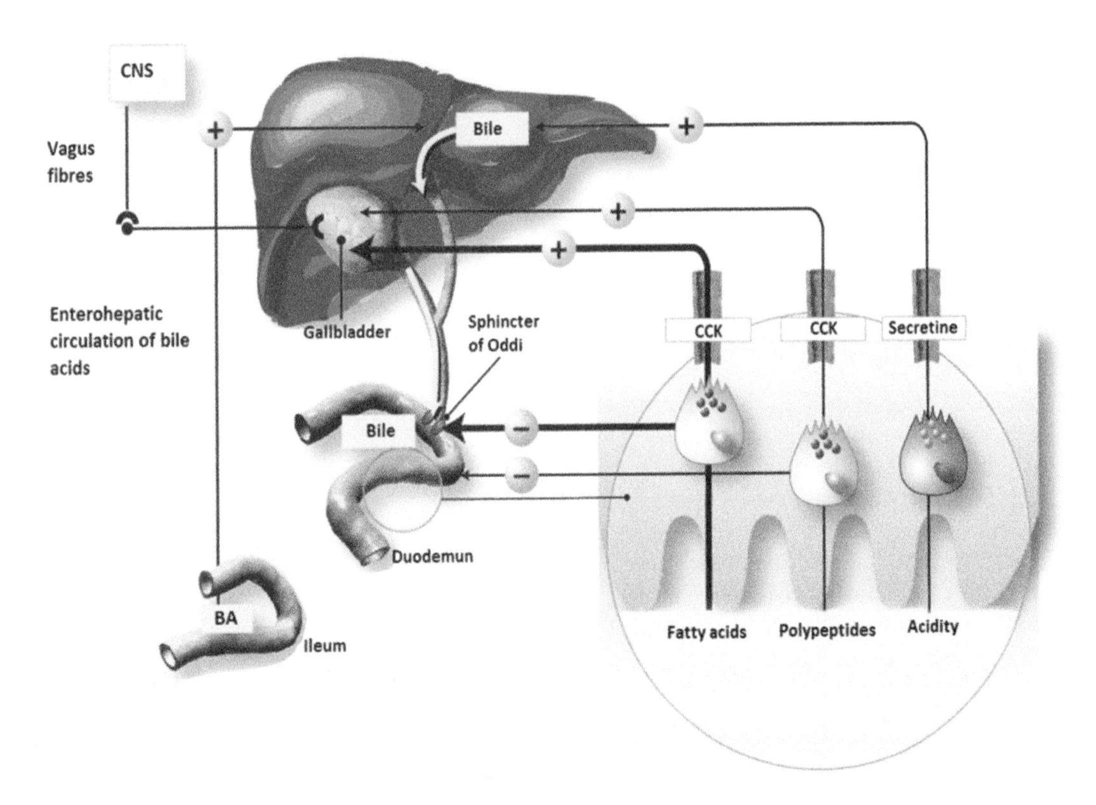

Figure 10.18 Factors controlling bile secretion. *Source:* Adapted from Martinez de Victoria et al., 2017.

Regulation of bile secretion

Several factors increase the production or release of bile (Figure 10.18).

- The rate of return of the bile acids to the liver via portal blood (enterohepatic recirculation of bile salts) affects the rate of secretion of bile acids. Bile acids stimulate bile production and secretion; in fact, there is a linear relationship between bile acid secretion rate and bile flow. Bile acids increase bile flow because they provide an osmotic driving force for filtration of water and electrolytes (choleretic effect).

There are other nervous and hormonal mechanisms, in addition to enterohepatic recirculation of bile salts, that affect the volume of bile that reaches the duodenum after ingestion of food.

- During interdigestive periods, most of the bile is found in the gall bladder where it is concentrated. The epithelium of the gall bladder wall absorbs electrolytes, creating an osmotic gradient that moves water into the blood. The concentration of bile acids can be of the order of 2–20 times.
- After ingestion of food, parasympathetic impulses along the vagus nerve (cephalic and gastric influences) cause contraction of the gall bladder and relaxation of the sphincter of Oddi.
- During the intestinal phase, the presence of acid and protein and, especially, fat digestion products in the duodenum induces an increase in hepatic bile secretion (choleretic effect) and a vesicular contraction (cholagogue effect).
- During a meal, CCK is released into the blood by fatty acids and polypeptides in the chyme. This causes contraction of the gall bladder and release of concentrated bile. CCK also relaxes the sphincter of Oddi, allowing the flow of bile, rich in bile salts, into the duodenum. These bile salts return via the enterohepatic circulation to the liver, further stimulating bile acid secretion and bile flow. CCK also directly stimulates the primary secretion by the liver.
- The acidity of the chyme emptying to the duodenum stimulates the release of secretin. This hormone causes the ductular epithelial cells to secrete a solution high in bicarbonate and in doing so increases the bile flow.

10.11 Adaptation of the biliary response to the diet

There is abundant information on adaptation of the pancreatic exocrine secretion in response to dietary composition. The physiological significance of this adaptation is to optimise the digestion and utilisation of macronutrients. Reports about the role of major dietary components in bile secretion, on the other hand, are relatively scarce, in spite of the fact that this secretion is essential for lipid digestion.

Bile salts are synthesised and secreted by the liver, stored and concentrated in the gall bladder, and delivered into the duodenum in response to a meal. Postprandial gall bladder contraction and sphincter of Oddi relaxation are mainly regulated by CCK. This hormone is released into the blood by the products of digestion of macronutrients. Changing the composition of macronutrients in the diet may evoke different plasma levels of CCK and thus modify the bile flow, the supply of bile salts into the duodenum, the enterohepatic circulation, the hepatic synthesis of bile acids and the amount excreted into the bile.

In healthy subjects, higher CCK release in response to food has been associated with stronger gall bladder emptying. There is a positive linear relationship between CCK release and gall bladder contraction. Meals differing in either the type or the amount of fat, protein or carbohydrates may be effective in stimulating CCK release and biliary secretion to varying degrees.

In addition, decreased postprandial gall bladder emptying has been suggested to play a major role in the development of gallstones in humans, so dietary factors may be important in the pathogenesis of gall bladder stasis.

Major dietary components, CCK release, and gall bladder contraction

Carbohydrates

Carbohydrates seem to have no influence on bile secretion when they constitute the standard proportion of the diet. Institution of a 95% carbohydrate diet reduces gall bladder emptying and diminishes bile salt pool size and bile acid secretion. Fat is the stronger stimulant for the release of CCK. Diets very rich in carbohydrates and with an adequate protein supply are very low in fat and these diets have been shown to be especially ineffective in the stimulation of CCK secretion and gall bladder contraction.

Protein

Several studies have shown that the amount and type of dietary protein influence biliary secretion. High-protein diets have been shown to increase bile acid secretion and the presence of bile salts in the lumen of the small intestine, and to stimulate plasma CCK secretion. The effect of low-protein diets on the postprandial emptying of the gall bladder has also been examined. A three-day low-protein/low-fat diet increased fasting gall bladder volume and significantly decreased fasting plasma CCK levels in humans. This suggests that CCK secretion regulates fasting gall bladder volume and that basal CCK release depends on diet composition. This diminished gall bladder stimulation affects bile flow and bile acid secretion in resting conditions and in response to food.

The quality of the dietary protein also affects CCK release. Animal studies have demonstrated this, although more human studies in this area are necessary before the relationship between biliary response to a meal and protein quality is fully understood. Intravenous infusion of amino acid solutions appears to have different effects on gall bladder contraction and CCK secretion, depending on the amino acid composition. Tryptophan and phenylalanine appear to be the most potent stimulants to CCK secretion and gall bladder contraction in humans.

Fat

The amount and type of dietary fat ingested also affect biliary secretion. In general, when the amount of fat ingested increases, the biliary response shows a marked rise in bile acid concentration and output. It has been suggested that this response is an adaptation process ensuring adequate fat digestion. In humans, CCK release and gall bladder contraction after a meal depend mainly on fat intake. Low-fat and fat-free diets decrease gall bladder emptying and CCK levels when compared to a normal mixed diet. High-fat diets produce opposite results. Some studies suggest that meals high in long-chain triacylglycerols (LCT) produce a postprandial increase in plasma CCK concentration while meals containing medium-chain triacylglycerols (MCT) inhibit CCK release.

The type of dietary fat also affects gall bladder emptying. Oleic acid, the major fatty acid in olive oil, is one of the most potent stimulators of CCK release known. In human subjects fed for 30 days diets containing either olive oil or sunflower oil as the main source of dietary fat, it was confirmed that the type of dietary fat affects the plasma levels of CCK. Postprandial CCK plasma levels were higher in the olive oil group, so a first option to explain the mechanisms of the biliary adaptation to dietary fat involves the existence of hormonal mediators. However, whether or not the secretory activity of liver cells is influenced directly by dietary fat alteration is unknown.

The biological effects of dietary fatty acids are partly due to their incorporation into the cellular structures of organs and tissues. In different tissues there is evidence that the lipid profile of the diet can influence the fatty acid composition of cell membranes, this being associated with a modification of cell function. This is not an unexpected finding since there is growing evidence that fatty acids, in addition to their role in determining membrane structure and fluidity, can participate in intracellular processes as diverse as signal transduction or the regulation of gene expression.

The plasma membranes of tissues and organs differ, however, in their adaptation to dietary fat type. The liver, for instance, is very sensitive to dietary changes. In contrast, other tissues, such as the brain, skeletal muscle and heart, seem to show relatively minor adaptive alterations in the lipid composition of their membranes. A second possibility to explain the mechanisms of biliary adaptation is that dietary fat composition may change the responsiveness of the liver to circulating secretagogues and/or to alter calcium signalling as a consequence of the modification of the fatty acid composition of the liver membranes, as explained in the pancreas section.

10.12 Growth, development, and differentiation of the gastrointestinal tract

The intestinal epithelium is a complex equilibrium system of multiple cell types undergoing continual renewal while maintaining precise inter-relationships. The crypt–villus axis is composed of a dynamic cell population in perpetual change from a crypt proliferative and undifferentiated stage to a mature villus stage. The migration of crypt cells is accompanied by cellular differentiation, which leads to morphological and functional changes.

The regulatory mechanisms involved with developmental processes are at organism, cellular and molecular levels. At the organism level, there is an expanded array of hormones and dietary effects that modulate the ontogeny of the intestine. The adult phenotype of the small intestine is established via a series of developmental transitions resulting from the interaction of visceral endoderm and mesoderm.

During development and in adult epithelium, cellular phenotypes are defined by the expression of specific sets of genes in individual cells. The regulation of those genes occurs mainly at the transcriptional level, although some genes may be regulated after initiation of the transcription during translation or even by late modifications of synthesised proteins in the endoplasmic reticulum and Golgi system. The particular set of genes expressed in a single cell type can be referred to as the 'transcriptome'. Intestinal epithelial cell transcriptomes shift in well-orchestrated patterns during the development, differentiation and adaptative processes of the intestinal mucosa. In addition to this intrinsic gene programme, intestinal epithelial cells respond to extrinsic signals, including nutrients and other dietary components, by producing various molecules. Using different experimental approaches, recent studies have further characterised intestinal epithelial cell biology and provided evidence of its polyvalent nature and important role in gut homeostasis.

Ontogeny of digestive and absorptive functions in humans

By the time of birth, all mammals must be able to digest all the nutritional components of their mother's milk. Lactose is the primary carbohydrate in the diet of the infant until the time of weaning. Lactase, sucrase and maltase activities appear between eight and nine weeks of gestational age in the jejunum. The greatest increase in lactase activity occurs during the third trimester, while at 20 weeks sucrase and maltase activities are already 50–75% of those found in term infants and adults. The appearance of lactase coincides with the appearance of microvilli. Unlike other species in which lactase activity declines at weaning, in the human it is maintained well beyond the weaning period. However, a decline occurs between three and five years of life. Sucrase and maltase activities are maintained throughout life.

Glucoamylase is the most important enzyme for digestion of complex carbohydrates. This enzyme is detectable by 20 weeks of gestational age and the greatest increase occurs between the period of foetal development and early postnatal development. Its activity in infants less than one year is comparable to that of adolescents.

Alkaline phosphatase is detectable by seven weeks of gestation but levels in newborns are substantially less than in adults.

As for sucrase and maltase, most brush-border peptidases appear to mature more rapidly in humans than in rats. γ-Glutamyl transpeptidase (γ-GT) activity increases more than twofold between 13 and 20 weeks of gestational age and its activity is higher in foetuses than in infants and adults. Aminopeptidase activity appears at eight weeks of gestation, and adult values are attained by 14 weeks. Oligoaminopeptidase, dipeptidylamino peptidase IV and carboxypeptidase increase in activity from eight to 22 weeks of gestational age and are present in infants. Lysosomal enzymes undergo little ontogenic change in the human.

Longitudinal distribution of the various enzymes along the small intestine often varies with the stage of development. Lactase and sucrase show a proximodistal gradient through adulthood, and glucoamylase and γ-GT activities are generally uniform throughout the small intestine.

Active transport of glucose is demonstrable by 10 weeks of life in the jejunum and by 12 weeks in the ileum. In vivo data suggest that little change occurs between 30 weeks and term. A high-affinity, low-capacity Na–glucose co-transport system along the length of the small intestine and a low-affinity, high-capacity system in the proximal intestine have been found in human foetuses. Glucose absorptive capacity seems to increase as a function of age from foetal life to adulthood.

Amino acid transport can also be demonstrated by 12 weeks of gestational age. Amino acid uptake is accomplished by sodium-dependent and -independent transport mechanisms. The Na-dependent pathways include the neutral brush border (NBB) and Phe systems for neutral amino acids, the Xag system for acidic amino acids, and the amino system for proline and hydroxyproline. The Na-independent pathways include the L system for neutral amino acids and y + system for basic amino acids. The NBB system, which transports leucine, is demonstrable in the entire small intestine of both infants and adults.

Macromolecules are also capable of crossing the small intestinal mucosa. The pathway of macromolecular uptake in the human is unclear, although endocytic and pinocytic processes can occur. The ability of the human intestinal mucosa to take up macromolecules is increased in infancy and during episodes of intestinal injury, such as diarrhoea and malnutrition. The intestinal closure is enhanced in breast-fed infants compared to those fed formula milk.

Calcium absorptive capacity in preterm infants is greater than 50% of adult levels. In relation to the needs of the growing infant, iron absorption is more efficient in infants than in adults, although iron bioavailability is dependent on dietary components, that is, the presence of lactoferrin in human milk.

By 8–11 weeks of gestational age, gastrin, secretin, motilin, gastric inhibitory peptide (GIP), enteroglucagon, somatostatin, vasoactive intestinal peptide (VIP), gastrin and cholecystokinin are demonstrable. The concentrations of these peptides increase throughout gestation and are at adult levels between 31and 40 weeks gestational age. Gastrin, secretin, motilin and GIP are localised to the duodenum and jejunum, whereas enteroglucagon, neurotensin, somatostatin and VIP are distributed throughout the small intestine.

Hormonal and dietary regulation of ontogenic changes in the small intestine

Administration of exogenous glucocorticoids during the first or second postnatal week causes precocious maturation of intestinal structure and function. In the absence of glucocorticoids, there is a dramatic reduction in the rate of maturation. Glucocorticoids cause intestinal maturation by transcriptional changes, but hormones may also have post-translational effects on proteins of the microvillus membrane, reflecting their capacity to alter membrane fluidity and glycosylation patterns. Glucocorticoids are capable of eliciting terminal maturation of the human intestine in a manner analogous to their effects in the rat. These hormones increase the lactase activity in vitro. In addition, retrospective and prospective studies have shown that prenatal corticosteroid treatment is associated with a significantly reduced incidence of necrotising enterocolitis in preterm infants.

Thyroxine (T_4) is a reasonable candidate for mediating the ontogeny of enzyme changes in the human small intestine, and total and free serum T_4 increase during the latter half of gestation. Insulin levels increase just prior to ontogenic changes in disaccharidase activities. For example, in human amniotic fluid, insulin directly reflects foetal insulin output, and the concentration increases during the last trimester. The rise in insulin precedes the prenatal increase in lactase activity that occurs in the third trimester.

Other hormones and substances have been proposed to play a role in the regulation of intestinal mucosal ontogeny. Many of them are known growth factors, but their direct effects on intestinal maturation are largely unknown.

In preterm infants, the provision of small amounts of milk enterally within the first week of life shortens the time it takes for infants to be able to tolerate all nutrients. Early introduction of feeding also appears to enhance the development of a more mature small intestinal motility pattern and improve enteral feeding tolerance. In addition, water feeding is not as effective as formula feeding in stimulating the release of gastrointestinal peptides in the premature infant.

Epithelial growth and differentiation

The small intestinal epithelium originates from stem cells that give rise to four different cell lineages.

- Absorptive enterocytes
- Mucus-producing goblet cells
- Enteroendocrine cells
- Paneth cells

The first three cell types migrate towards the villi, whereas Paneth cells migrate downwards to the crypt base. The intestinal epithelial cells form intimate contacts with T-lymphocytes, which in turn may regulate epithelial cell growth and barrier function. The intestinal epithelium provides unique opportunities for the study of cell differentiation and apoptosis. The gut epithelium along the crypt–villus axis is dynamic, with equilibrium between crypt cell production and the senescence and exfoliation of differentiated cells. Regulation of normal intestinal epithelial growth is thought to be dependent on mesenchymal–epithelial interactions as well as interactions with extracellular matrix proteins. Mesenchymal fibroblasts secrete growth factors that influence intestinal epithelial cell migration, proliferation and differentiation. As mature epithelial cells are produced, they express a variety of intestine-specific gene products involved in digestive, absorptive, cell migratory and cell protection functions.

Proteinase-activated receptors (PARs) are a family of G-coupled receptors for serine protease. PAR-2 has recently been found to be expressed in the villi and crypt region of the rat small intestine and in the basolateral and apical membrane of rat enterocytes. Trypsin at physiological concentrations activates PAR-2, which triggers the release of inositol triphosphate, arachidonic acid and prostaglandin E_2 and F_{1a}; therefore, luminal trypsin may serve as a signalling protein for enterocyte activation via PAR-2.

Protein kinase C (PKC) is involved in cell growth and differentiation in various cell types. PKCα expression is increased in Caco-2 spontaneous cell differentiation, although levels of other kinases remain unchanged in postconfluent cells. Concurrently, expression of PKCα-regulated cyclin-dependent kinase inhibitor $p21^{waf1}$ increases in differentiated cells, therefore the PKCα and $p21^{waf1}$ cascade could direct a signal that influences the differentiation status of intestinal epithelial cells. Paneth cells are likely to play a role in secretion and protection against microflora. Paneth cell allocation occurs early during epithelial cytodifferentiation and crypt formation and leads to the production of marker proteins such as phospholipase A2 (Pla2s), cryptidins and lysozyme.

Transforming growth factor (TGF)-α and TGF-β, as well as epidermal growth factor (EGF) and hepatocyte growth factor (HGF), are potent multifunctional growth factors that appear to play critical roles in the intestine under normal development and during injury. Using immunohistochemistry, TGF-β_1 has been localised to the smooth muscle cells of the muscularis propria in the human foetal intestine. In untreated adult rats, TGF-α expression has also been demonstrated in the epithelial cells at the villus tip, and TGF-β expression in epithelial cells in the upper crypt. TGF-α and TGF-β may accelerate wound healing in the intestine. Polyamines and ornithine decarboxylase-dependent cations have been demonstrated to be crucial for the migration of epithelial cells into regions of damaged tissue to promote early mucosal repair. Insulin-like growth factor (IGF) peptides are known to stimulate gastrointestinal growth in normal adult rats and IGF-I has been shown to selectively stimulate intestinal growth in developing rats.

Brush-border hydrolases

Lactase and sucrase isomaltase

The developmental expression of lactase and sucrase-isomaltase (SI) enzyme activity correlates with the change in diet at the time of weaning from predominantly lactose-containing milk to non-lactose-containing foods. The expression of the two enzymes appears to be largely genetically determined and transcriptionally regulated. However, the profiles are markedly different, with lactase mRNA appearing before SI mRNA, peaking earlier and then rapidly declining. SI mRNA peaks later and is then maintained at 70% of its peak value. A second major determinant of enzyme activity is the stability of the proteins, with lactase having a longer half-life than SI.

There is strong support for additional independent transcription factors to allow for the differential regulation of these two genes. Further support for differential regulation is provided by the observation that increased cyclic AMP levels increase lactase biosynthesis and mRNA levels but inhibit SI synthesis. Transgenic mouse experiments show that the patterns of SI gene expression are regulated by multiple functional *cis*-acting DNA elements. There are at least three groups of transcriptional factors involved in SI promoter activation, including hepatocyte nuclear factor 1 (HNF-1), Cdx proteins and nuclear proteins that interact with a GATA binding site. There is little correlation of SI transcription with the expression of these transcription factors, which is apparently due to the differential expression of co-activator proteins that thereby modulate the activation of the SI promoter.

Each of the intestinal epithelial cell lineages contains the necessary components for SI expression, and normal enterocyte-specific expression probably involves a repression mechanism in the goblet, enteroendocrine and Paneth cell lineages.

Peptidases

γ-GT catalyses the transfer of glutamic acid from a donor molecule such as glutathione to an acceptor molecule such as a peptide. The mouse γ-GT gene is unique since it has seven different promoters and contains six 5′-exons. The combination of different promoter usage along with alternative splicing generates tissue-specific mRNAs from this single gene.

Enterokinase is a hydrolase that activates latent hydrolytic enzymes in the intestinal lumen by proteolytic cleavage of the proenzymes. The cloning of the bovine enterokinase has revealed a complex protease with multiple domains. Its amino acid sequence suggests that the enterokinase is activated by an unidentified protease and therefore is not the first enzyme of the intestinal digestive hydrolase cascade. The regulation of the expression of the enterokinase gene is largely unknown.

Dipeptidyl peptidase IV (DPP-IV), a brush-border hydrolase, is identical to the T-cell activation molecule CD26. It is a member of the prolyl oligopeptidase family of serine proteases that plays an important role in the hydrolysis of proline-rich peptides. Differential expression of DPP-IV has been reported along the vertical villus axis, with high levels of expression in the villus cells and lower levels in the crypt cells and horizontal axis.

Alkaline phosphatase

Intestinal alkaline phosphatase (IAP) is a brush-border enzyme secreted into the serum and lumen, and it undergoes extensive postnatal developmental regulation of expression. The adult rat intestine expresses two IAP mRNAs, IAP-I and IAP-II, which have about 70% nucleotide sequence identity with complete divergence at the carboxy terminus.

Transporters

The Na-dependent glucose co-transporter (SGLT1) and seven facilitative glucose transporters (GLUT1 to GLUT7) have been cloned. SGLT1, GLUT2 and GLUT5 are expressed in the small intestine. SGLT1 is expressed in the apical membrane of the enterocytes, GLUT2 is anchored to the basolateral membrane and GLUT5 appears to function as an intestinal fructose transporter.

The Na-dependent uptake of bile is also developmentally regulated, with a marked increase at weaning. This co-transporter is regulated at the transcriptional level at the time of weaning with subsequent changes in the apparent molecular weight of the protein after weaning.

Regulatory peptides

Pituitary adenylate cyclase-activating peptide (PACAP) is a neuropeptide that exists in two functional forms and is a member of a group of regulatory peptides that includes secretin, glucagons, gastric inhibitory peptide, growth hormone-releasing factor and vasoactive intestinal peptide (VIP). PACAP receptors have tissue-specific affinity for both PACAP and VIP. PACAP nerves exist within the myenteric plexus, deep muscular plexus and submucosal plexus in the canine terminal ileum. In addition, PACAP receptors are present in deep muscular plexus and circular muscular fibres.

Uroguanylin and guanylin are endogenous proteins that regulate intestinal chloride secretion by binding to the guanylate cyclase C, and guanylin has also been shown to be involved in duodenal bicarbonate secretion. Uroguanylin is expressed throughout the intestinal tract and in the kidney. This gene is predominantly expressed in the villus. Guanilyn, however, is localised to the distal small intestine and proximal colon and is expressed in both the crypt and villus.

Intestinal cytoprotection

Cyclo-oxygenases

Two cyclo-oxygenase (Cox) isoforms, Cox-1 and Cox-2, catalyse the synthesis of prostaglandins. Cox-1 is expressed constitutively in most mammalian tissues and prostaglandins produced by Cox-1 are thought to play a major role in the maintenance of gastrointestinal homeostasis, including gastric cytoprotection and blood flow. Cox-2 is induced in monocytes and macrophages by proinflammatory cytokines, mitogens, serum and endotoxin. Prostaglandins synthesised by Cox-2 mediate the inflammatory response. Cox-2 is expressed at very low levels in the intestinal epithelium but is expressed at high levels in human colon cancers and adenomas and in spontaneous adenomas in mice that carry a mutant APC gene. In the normal colonic epithelium, there is no significant Cox-2 expression but in ulcerative colitis and Crohn's disease, Cox-2 is expressed in epithelial cells in upper crypts and on the surface but not in the lower crypts.

The area of Cox-1 expression in the small and large intestine corresponds to the area of epithelial proliferation in the crypt. As the epithelial cell migrates up out of the crypt and onto the villus, it differentiates and stops expressing Cox-1.

Trefoil peptides

Trefoil peptides contain a unique motif with six cysteine residues and three intrachain disulfide bonds, resulting in a three-loop structure. They are highly expressed in gastrointestinal mucosa and are resistant to protease digestion in the lumen, probably due to their unusual trefoil domain. Known members of the trefoil peptide family include pS2, spasmolytic peptide (SP) and intestinal trefoil factor (ITF), all cloned from several species. The stomach secretes pS2, the stomach and duodenum secrete SP and the small and large intestine secrete ITF. Although the exact function of these proteins has not yet been determined, they are found in high levels at the edges of healing ulcers and are believed to be involved in the protective function of the mucosal barrier.

Mucins

Mucins are a key component of the protective gel layer that coats the mucosal surface. Whether secretory or membrane bound, mucins contain diverse, highly *O*-glycosylated repeat amino acid residues. Mucin expression is known to be both cell-type and tissue specific as different mucins have been localised to specific regions of the gastrointestinal tract. Mucin precursors for mucins, MUC2–MUC6, have been identified from the stomach to the small and large intestine, MUC3 from the small intestine, MUC4 from the large intestine, and both MUC5AC and MUC6 from the stomach. Little is known about the specific function of individual MUCs, but the site-specific expression of these molecules suggests that the composition of MUC5AC and MUC6 is beneficial in protecting the stomach from acid-induced damage.

Human MUC2 is expressed almost exclusively in goblet cells and may be abnormally expressed in colon cancer. The gene for human MUC2 maps to chromosome 11p15 and analysis of the locus reveals two regulatory elements, an enhancer and an inhibitor, upstream from the MUC2 translation start site. MUC-1 protein expression has been located to microvilli on the luminal surface of the epithelial cell in the intestine.

Regulation of intestinal tract gene expression mediated by nutrients

Traditionally, it has been assumed that gene expression in higher eukaryotes was not directly influenced by food components but by the action of hormones, growth factors and cytokines. However, it is now known that diet is a powerful means of modifying the cellular environment of the gastrointestinal tract.

Dietary regulation of the genes expressed by the epithelium confers three fundamental advantages for mammals. It enables the epithelium to adapt to the luminal environment to better digest and absorb nutrients; it provides the means whereby breast milk can influence the development of the gastrointestinal tract; and when the proteins expressed by the epithelium act on the immune system, it constitutes a signalling mechanism from the intestinal lumen to the body's defences. Each of these mechanisms is amenable to manipulation for therapeutic purposes. Major nutrients, that is amino acids, glucose and fatty acids, as well as minor nutrients, that is liposoluble vitamins A and D, hydrosoluble vitamins B_1, B_6, C and biotin, and minerals (iron, zinc, copper and selenium), influence the expression of a number of genes directly or in a concerted action with hormones.

The regulation of genes by nutrients requires that some enzymes, transporters and membrane receptors interact with the particular nutrient, resulting in a series of cellular events that lead to modulation of the transcriptional or translational gene processes. Enzymes and transporters required for carbohydrate digestion and absorption are regulated by carbohydrate availability. Gamma delta ($\gamma\delta$) T-cells are immune cells best known for host barrier defences in epithelial tissues. They also have a key role in sensing nutrient uptake in the small intestine. Intestinal $\gamma\delta$ T cells regulate the expression of a carbohydrate transcriptional program by limiting interleukin-22 production from type 3 innate lymphoid cells. Hence, nutrition and barrier functions at the molecular levels allow the small intestine to adjust the balance between nutrient uptake and host defence in response to environmental change.

The enzymatic activity of disaccharidases in the small intestine is increased in response to dietary carbohydrates due to increased synthesis of their respective mRNAs. Lactase and sucrase promoters contain binding elements for transcription factors, CE-LPH1 and SIF1, which have a common consensus sequence, TTTTAT/C. To these elements binds Cdx-2, a transcription factor implicated in the transcriptional regulation of intestinal epithelial genes, which is of great importance in a wide variety of phenomena from early cell differentiation to the maintenance of the intestinal epithelial lining. Glucose, fructose, sucrose, galactose and glycerol increase the expression of lactase in rat jejunum homogenates, whereas high-starch diets increase its expression. In mice, the expression of the glucose–sodium co-transporter SGLT1 is increased by dietary carbohydrates and the effect is apparent only in the crypts. In addition, high-fructose diets increase the expression of GLUT5.

Production and secretion of insulin from the β-cells of the pancreas are crucial in maintaining normoglycaemia. This is achieved by tight regulation of insulin synthesis and exocytosis from the β-cells in response to changes in blood glucose levels. The synthesis of insulin is regulated by blood glucose levels at the transcriptional and post-transcriptional levels. Although many transcription factors have been implicated in the regulation of insulin gene transcription, three β-cell-specific transcriptional regulators, Pdx-1 (pancreatic and duodenal homeobox-1), NeuroD1 (neurogenic differentiation 1) and MafA (V-maf musculoaponeurotic fibrosarcoma oncogene homologue A), have been shown to play a crucial role in glucose induction of insulin gene transcription and pancreatic β-cell function. These three transcription factors activate insulin gene expression in a co-ordinated and synergistic manner in response to increasing glucose levels. It has been shown that changes in glucose concentrations modulate the function of these β-cells.

The transcriptional role of glucose has largely been investigated in liver, where involvement of the lipogenic pathway is prominent. In the presence of insulin, which is required to activate glucose metabolism in hepatocytes, high glucose concentrations induce expression of genes that encode glucose transporters and glycolytic and lipogenic enzymes (e.g. L-type pyruvate kinase [L-PK], acetyl-coenzyme A carboxylase [ACC] fatty acid synthase [FAS], malic enzyme and stearoyl-coenzyme A desaturase) and downregulate genes of gluconeogenesis, such as the phosphoenolpyruvate carboxykinase gene (PEPCK).

Recently, using microarrays, 283 glucose-responsive genes in rat hepatocytes that are essentially involved in carbohydrate metabolism and de novo lipogenesis have been identified. Promoter analysis of glucose-regulated genes (e.g. those encoding acetyl-coenzyme A carboxylase, FAS, L-PK and Spot 14) allowed investigators to identify a carbohydrate response element (ChoRE) that is constituted by two E-boxes separated by five nucleotides. A member of the basic helix–loop–helix leucine zipper family of transcription factors that can bind this DNA motif was purified and named ChoRE binding protein (ChREBP). ChREBP is abundant in tissues where lipogenesis is highly active, such as liver, small intestine and white and brown adipose tissue, whereas its expression is low in skeletal muscle. ChREBP functions as a glucose and glutamine sensor and its expression is induced by a carbohydrate-rich diet. Another protein named Max-like factor (Mlx) is needed for the regulation of gene expression dependent of ChREBP.

A growing number of reports clearly demonstrate that amino acids are able to control physiological functions at different levels, including the initiation of protein translation, mRNA stabilisation and gene transcription. Extensive studies on the asparagine synthetase (*ASNS*) and C/EBP homology binding protein (*CHOP*) genes allowed characterisation of specific responsive sequences in their promoter, which were named either nutrient-sensing response elements (NSRE) or amino acid responsive elements (AARE). Specific transcription factors involved in the amino acid response (AAR) pathway have also been identified, that are members of the basic region/leucine zipper superfamily of transcription factors. In parallel, some amino acids involved in many cellular functions, particularly glutamine and leucine, have been shown to exert a wide range of effects via the activation of different signalling pathways and transcription factors. Although the molecular details of these effects are not completely known, the heterogeneity of the involved factors might suggest multiple AAR pathways depending on the amino acid studied, the cell type used and the gene promoter configuration.

The glutamine-responsive genes and transcription factors involved correspond tightly to the specific effects of this amino acid in the inflammatory response, cell proliferation, differentiation and survival, and metabolic functions. Indeed, in addition to the major role played by nuclear factor-κB (NF-κB) in the anti-inflammatory action of glutamine, the stimulatory role of activating protein-1 and the inhibitory role of CHOP in growth promotion, and the role of c-myc in cell survival, many other transcription factors are also involved in the action of glutamine to regulate apoptosis and intermediary metabolism in different cell types and tissues. The signalling pathways leading to the activation of transcription factors suggest that several kinases are involved, particularly mitogen-activated protein kinases (MAPK).

The stimulatory effect of leucine on protein synthesis is mediated through upregulation of the initiation of mRNA translation. Several mechanisms, including phosphorylation of ribosomal protein S6 kinase, eukaryotic initiation factor (eIF)4E binding protein-1 (4E-BP) and eIF4G, contribute to the effect of leucine on translation initiation. These mechanisms not only promote global translation of mRNA but also contribute to processes that mediate discrimination in the selection of mRNA for translation.

A key component in a signalling pathway controlling these phosphorylation-induced mechanisms is the protein kinase, termed the mammalian target of rapamycin (mTOR). The activity of mTOR toward downstream targets is controlled in part through its interaction with the regulatory-associated protein of mTOR (known as raptor) and the G-protein β-subunit-like protein. Signalling through mTOR is also controlled by upstream members of the pathway such as the Ras homologue enriched in brain (Rheb), a GTPase that activates mTOR, and tuberin (also known as TSC2), a GTPase-activating protein which, with its binding partner hamartin (also known as TSC1), acts to repress mTOR. The Rag proteins, a family of four related small GTPases, interact with mTOR in an amino acid-sensitive manner and seem to be both necessary and sufficient for mediating amino acid signalling to mTOR. mTOR is known to play vital roles in protein synthesis. In particular, it controls the anabolic and catabolic signalling of skeletal muscle mass, resulting in the modulation of muscle hypertrophy and muscle wastage.

In addition to EGF, insulin and other growth factors and nutrients, human milk contains some nutrients that influence gene expression. Lactoferrin, a protein involved in Fe bioavailability, is a proliferative factor for lymphocytes, embryonic fibroblasts and human intestinal HT-29 cells. It also increases sucrase and alkaline phosphatase activities in intestinal cells. These data suggest that lactoferrin may affect intestinal growth and differentiation, probably by internalisation of the lactoferrin–receptor complex.

Soluble nucleotides are present in milk from various mammals, contributing up to 20% of the non-protein nitrogen. Although nucleotide deficiency has not been related to any disease as such, dietary nucleotides are reportedly beneficial for infants since they positively influence lipid metabolism, immunity and tissue growth, development and repair. Nucleotides are naturally present in all foods of animal and vegetable origin as free nucleotides and nucleic acids. Concentrations of RNA and DNA in foods depend mainly on their cell density, whereas the content of free nucleotides is species specific. Thus, meat, fish and seeds have a high content of nucleic acids, and milk, eggs and fruits have relatively lower levels. Milk has a specific free-nucleotide profile for each species. The nucleotide content of colostrum is qualitatively similar but quantitatively distinct in human and ruminants. In general, total colostrum nucleotide content increases immediately after parturition, reaches maximum levels from 24 to 48 h after birth and decreases thereafter with advancing lactation. Dietary nucleotides appear to modulate lipoprotein and fatty acid metabolism in human early life, and affect the growth, development and repair of the small intestine and liver in experimental animals. Moreover, dietary nucleotides modify the intestinal ecology of newborn infants, enhancing the growth of beneficial bifidobacteria and limiting that of potentially harmful enterobacteria, and have a role in the maintenance of the immune response in both animals and human neonates.

Erythrocytes and lymphocytes, enterocytes and glial cells have a common characteristic: their ability to synthesise nucleotides by the de novo pathway is very low. Thus, an external supply of nucleosides seems to be needed for optimal functioning. Several investigations support the hypothesis that the enterocyte is not fully capable of developing the de novo purine synthesis and that this metabolic pathway may be inactive unless it is induced by a purine-deficient diet. The purine salvage pathway, as measured by the activity of its rate-limiting enzyme hypoxanthine phosphoribosyl transferase (HGPRT), is highest in the small intestine relative to liver and colon; moreover, a purine-free diet lowers HGPRT activity.

A number of studies have demonstrated that dietary nucleotides in part regulate gene expression in the intestine (Figure 10.19). Intestinal mRNA levels of the enzymes HGPRT and APRT declined in response to nucleotide restriction in the diet. Nuclear 'run-on' assays both in nuclei isolated from the small intestine and from an intestinal epithelial cell line (IEC-18) demonstrated that dietary nucleotide restriction significantly altered the transcription rate. A 35 bp region (HCRE) in the promoter of the HPGRT gene has been identified as the element necessary to this response, and the protein that interacts with this region has been identified and purified. It has been shown that the addition of nucleosides to Caco-2 cells increased the expression and activity of the general transcription factors activating enhancer binding protein 2 α (TFAP2A) and CCAAT displacement protein (CUX1) and v-ets avian erythroblastosis virus E26 oncogene homologue 1 (ETS1), and SMAD family member 2. In contrast, these nucleosides decreased the expression and activity of the general upstream stimulatory factor 1 (USF1) and the cAMP response element-binding protein 1 (CREB), glucocorticoid receptor (NR3C1), nuclear factor erythroid (NFE2), NFκB and tumour protein p53. Recent studies have reported that dietary nucleotides markedly increased gene expressions of TLR-9, TLR-4 and TOLLIP, involved in mechanisms of intestinal innate immunity, and claudin-1, zonula occludens-1 (ZO-1) and occludin, typical structural proteins of epithelial tight junction.

Taken together, these data suggest that exogenous nucleosides affect the expression and activity of several transcription factors involved in cell growth, differentiation and apoptosis, and in immune response and inflammation.

Dietary fat plays a key role in the regulation of gene expression in many tissues controlling the activity or abundance of key transcription factors (Figure 10.20). Several transcription factors, including peroxisome proliferator receptors (PPAR), hepatocyte nuclear factor 4 (HNF4), liver X receptors (LXR), steroid response binding proteins (SREBP) 1a, 1c and 2, and NFκB,

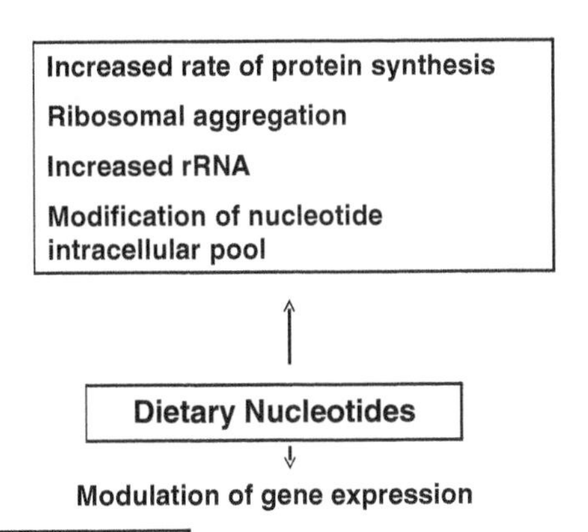

Figure 10.19 Modulation of gene expression by dietary nucleotides. TIMPs: HPRT, hypoxanthine phosphoribosyl transferase; APRT, aderine phosphoribosyl transferase; CNT, concentrative nucleoside transporters.

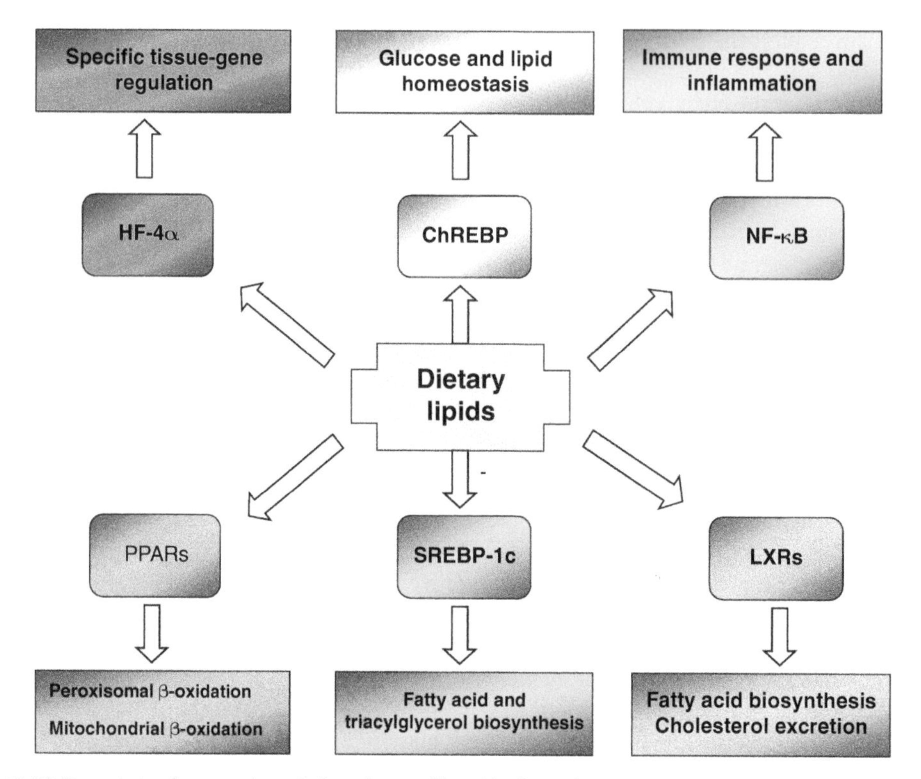

Figure 10.20 Transcription factors and metabolic pathways affected by dietary lipids. ChREBP, carbohydrate response element binding protein; HNF-4, hepatic nuclear factor-4; LXR, liver X receptors; NF-κB, nuclear factor κB; PPAR, peroxisome proliferator receptor; SREBP, sterol response element binding protein.

have been shown to be modulated by dietary fat components, particularly fatty acids and cholesterol, and some of their derivatives, namely eicosanoids and oxidised sterols.

In vivo studies established that PPAR-α- and SREBP-1c-regulated genes are key targets for polyunsaturated fatty acids (PUFA) control of hepatic gene expression. PUFA activate PPAR-α by direct binding, leading to induction of hepatic fatty acid oxidation. PUFA inhibit hepatic fatty acid synthesis by suppressing SREBP-1c nuclear abundance through several mechanisms, including suppression of SREBP-1c gene transcription and enhancement of proteasomal degradation and mRNA SREBP1c decay. Changes in intracellular non-esterified fatty acids (NEFA) correlate well with changes in PPAR-α activity and SREBP-1c mRNA abundance. Several mechanisms regulate intracellular NEFA composition, including fatty acid transport, acyl CoA synthetases and thioesterases, fatty acid elongases and desaturases, neutral and polar lipid lipases, and fatty acid oxidation. Many of these mechanisms are regulated by PPAR-α or SREBP-1c. Together, these mechanisms control hepatic lipid composition and affect whole-body lipid composition. It is assumed that δ-6 and δ-5 fatty acid desaturase genes which are key in the synthesis of long-chain polyunsaturated fatty acids from the essential parent fatty acids are regulated at the intestinal level in a similar fashion to that occurring in the liver.

Changes in diet greatly affect the mucosal immune system and alterations in the luminal environment of the intestine regulate the expression of genes in the enterocyte responsible for signalling to immune cells. Genes expressed by the epithelium orchestrate leukocytes in the lamina propria. For example, chemokine expression in the mouse intestinal epithelium, through transgenic means, induced the recruitment of neutrophils and lymphocytes into intestinal tissues. Diet alters the expression of the genes responsible for signalling by a variety of pathways. The introduction of a normal diet to a weanling mouse upregulates major histocompatibility complex (MHC) class II expression through a particular isoform of the class II transactivator, a protein that acts in the nucleus. Short-chain fatty acid (SCFA) concentrations in the intestinal lumen vary markedly with diet. SCFAs increase IL-8 and insulin-like growth factor binding protein-2 (ILGFBP-2) expression by inhibiting histone deacetylase activity in the enterocyte. Downregulation of gene expression by butyrate can act through acetylation of the inhibitory transcription factor Sp3. Myofibroblasts enhance enterocyte chemotactic activity by cleaving inactive precursors, and myofibroblast genes also are regulated by SCFA.

Vitamins A and D act as true hormones in the gut. Their hormonal action on gene expression of many targets occurs via binding to nuclear receptors in many organs, including the intestine (Figure 10.21). The biological actions of vitamin A, except those related to vision by 11-cis-retinal, depend primarily on interaction with specific nuclear receptors. There are mainly two types of nuclear receptors involved in the regulation of gene expression by retinoic acid (RA), called RAR (retinoic acid receptor) and RXR (retinoid X receptor). Although RAR can operate independently, high-affinity transcriptional activity requires the formation of heterodimers of a RAR-type unit with an RXR-type unit. RAR receptors are activated by trans-retinoic acid. Retinoid receptors recognise target sequences in DNA called RARE (retinoic acid response elements). Recognition of course requires additional co-factors and chromatin opening. The binding of RAR to the RA promotes association with co-activating factors, which are exchanged with the co-repressors to which it binds under basal conditions and constitutes the first step of transcriptional activation. Regulation of gene transcription depends on co-operation with a number of factors, such as histone acetyltransferases and deacetyltransferases, methyltransferases, etc. Indeed, vitamin D regulates the expression of numerous genes mainly through epigenetic mechanisms.

The mechanism of action of the active vitamin D 1,25(OH)2D3 (calcitriol) is mediated by the vitamin D receptor (VDR), which belongs to a subfamily of nuclear receptors that act as transcription factors into the target cells after forming a heterodimer with retinoid X receptor (RXR). Once dimerised, the complex binds to the VDR element, in the promoter regions of target genes or at distant sites, to positively or negatively regulate their expression. Besides 1,25(OH)2D3, the VDR-RXR dimer can associate with other molecules as the p160 co-activators family of steroid receptor co-activators 1, 2 and 3, that have histone acetylase (HAT) activity, and are primary co-activators that bind to the AF2 domain of liganded VDR. Members of p160 family recruit proteins as secondary

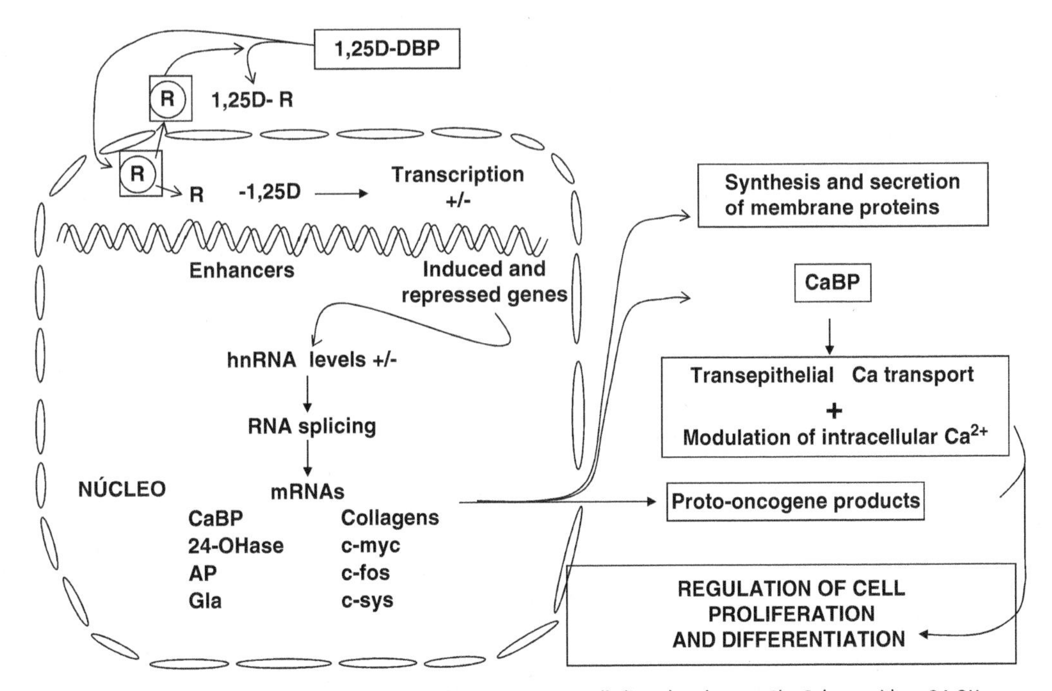

Figure 10.21 Regulation of gene expression mediated by vitamin D. AP, alkaline phosphatase; Gla, Galactooxidase; 24-OHase, 24-calciferol hydroxylase.

co-activators, such as CBP/p300, which also have HAT activity, resulting in a multi-subunit complex that modifies chromatin and destabilises histone/DNA interaction. There is increasing evidence that specific CAAT enhance binding protein (C/EBP) family members may be key mediators of 1,25(OH)2D3 action. Hence, as vitamin A, vitamin D regulates gene expression by epigenetic mechanisms

Other vitamins, such as biotin and vitamin B_6, are also involved in the regulation of gene expression, and it is well known that metals can influence the regulation of some genes. For example, the post-transcriptional regulation of ferritin and transferrin receptor by dietary iron levels is well recognised.

The intestinal microbiota contributes to the host homeostasis and rapidly changes in response to dietary changes. The gut microbiota modulates the immune system via the production of molecules with immunomodulatory and anti-inflammatory functions, mainly due to the interaction of commensal bacteria with epithelial cells, dendritic cells monocytes/macrophages and lymphocytes. Indeed, some intestinal bacteria and probiotics are capable of suppressing intestinal inflammation via the downregulation of TLR expression and other genes like cyclooxygenase-2. In addition, SCFAs produced in the gut by fermentation of some dietary fibres are key signalling molecules for the maintenance of gut health; they also enter the systemic circulation and interact with cell receptors in peripheral tissues. In fact, SCFAs have an important role in the regulation of energy homeostasis and metabolism.

10.13 The large bowel

Until the 1980s, the large bowel was a relatively neglected organ, once described as a 'sophisticated way of producing manure', which, unfortunately, was inclined to go wrong, resulting in common diseases and disorders such as constipation, ulcerative colitis and colorectal cancer. Fortunately, recent investigations on the aetiology, prevention and treatment of these diseases and disorders, and on the 'normal' physiology of this organ, have illuminated the structure and function of the large bowel. It is now clear that the colorectum is much more than just the last part of the gastrointestinal tract where stool is formed. From a nutritional perspective, recognition that the primary function of this organ is salvage of energy via bacterial fermentation of food residues has revolutionised understanding of this area of the body. The human body contains about 3×10^{13} cells and a roughly similar number of bacterial cells. Most of these bacteria are found in the colon where they dominate metabolic processes.

In the last decade, there has been intensive research on the bacteria, and other micro-organisms, in this part of the gastrointestinal tract stimulated by the development and widespread use of new laboratory approaches. As discussed later, there is emerging evidence that these microbes, their metabolic processes and their interaction with the immune system make important contributions to the physiology and health of the human host. Importantly, they may have implications not only for the health of the gut but also in influencing the risk of several common human diseases.

Structure

The large bowel forms the last 1.5 m of the intestine and is divided into the appendix, caecum, transverse colon, descending colon, sigmoid colon and rectum. It is about 6 cm in diameter, becoming narrower towards the rectum. With the exception of the rectum (the final 12 cm of the intestine), the large bowel has a complex mesentery supplying blood from the superior and inferior mesenteric arteries. Blood drained from the colon reaches the liver via the portal vein. As with the more proximal regions of the intestine, the large bowel has two layers of muscle, an outer longitudinal layer and an inner circular layer, which work together to sequester and move digesta along the tract. Although food residues travel through the small bowel in just a few hours, digesta may be retained in the large bowel for 2–4 days or longer.

The colon has a sacculated appearance composed of pouches known as haustra. These are formed by the organisation of the outer longitudinal smooth muscle layer into three bands called the taeniae coli (which are shorter than the other longitudinal muscles) and by segmental thickening of the inner circular smooth muscles.

The parasympathetic vagus nerve innervates the ascending colon and most of the transverse colon, with the pelvic nerves innervating the more distal regions. There are also connections to the colon from the lower thoracic and upper lumbar segments of the spinal cord via noradrenergic sympathetic nerves.

The mucosa of the colon consists of a single layer of columnar epithelial cells which are derived from stem cells located at or near the base of the crypts. Unlike the small bowel, there are no villi, so that the inner surface of the colon is relatively flat, which may have contributed to the misapprehension that this organ is unimportant in digestion and absorption. There is a higher density of goblet cells than in the small intestine, producing mucus that both protects the colonic surface and coats the increasingly more solid faeces as it is propelled through the large bowel. The water content of the gut decreases from about 85% in the caecum to 77% in the sigmoid colon. Immune protection extends into the large intestine with the presence of gut-associated lymphoid tissue (GALT) in nodules within the mucosa and extending into the submucosa.

The large bowel microbiome across the life-course

The large bowel microbiome describes the complex communities of micro-organisms including bacteria, archaea (single-celled prokaryotic organisms that are distinct from bacteria) and fungi (the mycobiome) that live in the distal part of the human gut. In addition, recent studies have shown that the human gut is also home to a vast array of viruses (the virome), most of which are bacteriophages, i.e. they infect, and replicate in, bacteria and archaea. Many of these micro-organisms colonise the undigested food particles that enter the large bowel from the terminal ileum whereas others are closely associated with the gut mucosa and its protective layer of mucus.

Immediately after a baby is born, the large bowel microbiome develops rapidly. However, the types of micro-organisms in the baby's gut depend on the mode of delivery. Babies born by vaginal delivery have a high abundance of lactobacilli, as a consequence of exposure to the lactobacilli in the mother's vagina, and develop a bacterial flora that resembles that of their mother. In contrast, the bacterial flora of babies born by caesarean section develops more slowly and, in the early days, may contain substantial numbers of facultative anaerobes such as *Clostridium* species.

Initially, the baby's large bowel microbiome has low microbial diversity and is dominated by two main phyla, Actinobacteria and Proteobacteria. In addition, it differs markedly between those babies who are breastfed exclusively and those who receive formula. At least in part, this occurs because breast milk contains a complex array of oligosaccharides that support the development of a healthy gut microbiome through stimulating the growth of beneficial bacteria such as *Bifidobacterium*. Similarly, the newborn's large bowel is virtually devoid of viruses, but viral abundance increases rapidly to to 10^9 per gram/stool by one month of age. Breastfeeding reduces the abundance of viruses that can infect human cells (as distinct from bacteriophages that infect bacteria) which may explain, in part, why breast milk can be protective against viral infections. Following weaning, the consumption of a wider range of foods leads to a more diverse range of undigested food materials entering the large bowel. In turn, this supports the development of an increasingly diverse microbial community. By 2–3 years of age, the composition, diversity and functional capability of the infant's microbiome are similar to those of the human adult (see Chapter 6).

During adulthood, the composition of each individual's large bowel microbiome stays relatively constant and has features that distinguish it from the microbiome of other adults. The same appears to be true of the large bowel virome but, to date, this is much less well investigated. However, the microbiome can be altered by changes in diet, by exposure to antibiotics and by disease. Some of these changes occur rapidly and are reversed when the individual returns to their usual diet or finishes a course of antimicrobial therapy.

Further changes to the large bowel microbiome occur during ageing. Often, these changes include an apparent simplification of the microbial communities with a reduction in diversity, an increase in the abundance of facultative anaerobes such as *Escherichia coli* and lower numbers of butyrate-producing bacteria (see below). The reasons for these microbial changes are not well understood but may include the effects of (i) reduced dietary diversity (including lower dietary fibre intake), (ii) chronic diseases and the treatments (including medications) used in their management, and (iii) age-associated changes in the immune system. There is some evidence that this age-related microbial dysbiosis may result in intestinal permeability, systemic inflammation and premature mortality. However, whether these microbial changes and their sequelae are a consequence or a driver of unhealthy ageing remains to be discovered.

Over the last decade, advances in DNA sequencing techniques, the availability of enhanced bioinformatics tools and rapidly expanding databases have enabled researchers to explore the human gut microbiome in increasing detail, and at scale. However, knowing what organisms are present is only the starting point. Understanding the roles that the microbiome plays in human physiology and health requires investigation of (i) interactions between members of the complex communities, (ii) the amounts of specific metabolites produced and absorbed into the host bloodstream, and (iii) interactions with the host immune system.

Functions of the large bowel

Energy salvage

Many animal species have evolved a symbiotic relationship with micro-organisms which allows them to extract energy from food materials that otherwise would be indigestible by the animal's enzymes. This is best exemplified by mammalian herbivores, which may obtain up to three-quarters of their energy as the by-product of anaerobic fermentation in specialised sacs before the gastric stomach (e.g. in ruminants) or in a massively enlarged large bowel (e.g. in equids, members of the horse family). In each case, commensal bacteria (and sometimes other micro-organisms, especially protozoa and fungi) proliferate by fermenting plant matter in specialised regions of the intestine that do not secrete digestive enzymes. The energy-containing end-products of this fermentation, principally SCFAs, are absorbed readily and metabolised. For those animals in which fermentation precedes the gastric stomach or where there is ingestion of faeces (best described for lagomorphs such as the rabbit), the bacterial biomass is also an important source of nutrients.

In humans consuming westernised diets, the equivalent of 50–60 g dry weight of food residues and endogenous material dissolved and suspended in approximately 1–1.5 l of fluid flows into the caecum from the ileum every day. For those eating low-fat diets with lots of plant foods rich in carbohydrates, which are poorly, or not at all, digested in the small bowel (e.g. non-starch polysaccharides and resistant starch), considerably more carbohydrate and other energy-containing food components escape into the colon. In a world with an increasing number of people with overweight or obesity, loss of this potential energy would not be a problem but, throughout human history, shortage of food has been much more common. Therefore, the ability to salvage as much energy as possible from food will have been an evolutionary advantage and may have made the difference between survival and death. This energy salvage is achieved by bacterial fermentation of carbohydrates and, to a lesser extent, proteins. Bacterial cells make up at least 50% by weight of colonic contents, where they are present at concentrations of about 10^{10}–10^{11} cells/g wet matter.

Dietary fibre is the major substrate for the large bowel microbiota. Dietary fibre includes plant-derived complex carbohydrates (mainly plant cells walls, resistant starch and non-digestible oligosaccharides) that have escaped digestion in the small bowel. These carbohydrates are subject to enzymatic hydrolysis by a complex array of glycoside hydrolases, polysaccharide lyases and carbohydrate esterases that, together, can degrade the polymers to simple sugars. To enable them to utilise the wide diversity of carbohydrates found in plant cell walls and other complex carbohydrates, large bowel bacteria have evolved a wide range of hydrolases. For example, the genome of *Bacteroides thetaiotaomicron* encodes 260 different glycoside hydrolases which maximise the potential to extract energy from food residues. The monosaccharides that are released are metabolised by multiple bacteria via the Embden–Meyerhof–Parnas, Entner–Doudoroff or pentose phosphate pathways to yield pyruvate, ATP and reduced nicotinamide adenine dinucleotide (NADH).

The virtual absence of oxygen in the anaerobic conditions in the lumen of the large bowel means that the micro-organisms cannot oxidise pyruvate using the Krebs cycle and oxidative phosphorylation. Instead, NADH, generated via the glycolytic pathway, is converted to NAD^+ and pyruvate is converted into succinate, lactate or acetyl-CoA. These intermediate compounds do not accumulate in large concentration in the colonic lumen because they are metabolised by other 'cross-feeding' micro-organisms to produce the SCFAs – acetate, propionate and butyrate. SCFAs are absorbed readily from the colonic lumen largely by transcellular mechanisms, including carrier-mediated transport and non-ionic diffusion. Only a very small proportion, usually much less than 5%, of SCFA production is lost in the faeces. Consequently, most of the energy present in the plant cell wall polysaccharides and resistant starch that escape small bowel digestion is absorbed across the colonic epithelial cells as SCFAs. The energy value to the host human of carbohydrates that have been fermented to SCFAs and absorbed as such is about 8 kJ/g compared with 16 kJ/g for carbohydrates that are digested in the small intestine.

The principal gaseous endproducts of large bowel fermentation are carbon dioxide, hydrogen and methane, with everyone generating carbon dioxide and hydrogen, but only 30–40% of people producing significant quantities of methane. These gases are released in the flatus (about 500 ml/day) or absorbed and excreted via the lungs.

Metabolism of other macromolecules

The colonic microflora preferentially ferment carbohydrates but they are capable of degrading proteins and other nitrogenous substrates in the more distal large bowel when accessible carbohydrates become exhausted. Degradation of the 10–12 g of nitrogenous compounds entering the colon from the small bowel daily yields a range of quantitatively minor endproducts, including branched-chain and longer-chain SCFA (isobutyrate, isovalerate, valerate and caproate), ammonia, hydrogen sulfide, indole, skatole and volatile amines. Several of these compounds are implicated in colonic diseases. For example, hydrogen sulfide is believed to contribute to the aetiology, or recurrence, of ulcerative colitis, while ammonia and sulfide may play a part in damaging colonocytes, leading to increased risk of colorectal cancer. Although not yet well understood, metabolism of the sulfur-containing, basic and aromatic amino acids generate proinflammatory, cytotoxic and neuroactive compounds with adverse effects on human health.

The most extensively studied aspect of lipid metabolism in the colon is the bacterial transformation of the up to 5% of bile salts that are not absorbed in the ileum. Bile salts are deconjugated (removal of glycine and taurine residues) and, depending on the pH of the digesta (which ranges from about 5.5 in the caecum to 6.5 in the distal colon), the primary bile acids may be converted to secondary bile acids by 7α-dehydroxylation – a process carried out by Gram-positive bacteria including Lachnospiraceae, Ruminococcaceae and *Blautia*. Secondary bile acids are more lipid soluble and are absorbed by passive diffusion (probably about 50 mg/day) then returned to the liver, where they have an important role in regulating bile acid synthesis. Most of the cholesterol entering the colon is excreted as such in faeces, but some is metabolised to the relatively insoluble derivatives coprostanol and coprostanone, which are not absorbed.

Normally, very little triacylglyceride reaches the large bowel, and any that does can be partially metabolised. For example, lactobacilli, enterococci, Clostridia and Proteobacteria can convert the glycerol backbone to 1,3-propanediol which may be metabolised to acetate by sulfate-reducing bacteria. The anaerobic nature of the colonic lumen does not allow the oxidation of long-chain fatty acids, which explains why diseases causing fat malabsorption in the small intestine result in greatly increased excretion of fat in the faeces (steatorrhoea). A similar effect is seen with antiobesity drugs such as orlistat (available on prescription as Xenical® or, in some countries, as the over-the-counter preparation sold as Alli®), which act by inhibiting gastric and pancreatic lipases. As a result, dietary triacylglycerols flow into the colon and give rise to the well-described side-effects of steatorrhoea and, in some cases, faecal incontinence. In the anaerobic conditions of the large bowel, unsaturated fatty acids may be biohydrogenated using the reducing power (H_2) generated as a by-product of carbohydrate fermentation.

Absorption of water and electrolytes

Daily faecal output is highly variable but is in the order of 100–150 g stool containing about 25 g dry matter. This means that the large bowel absorbs approximately 1 l of water every day. Water is absorbed down an osmotic gradient, mainly from the proximal colon, following the absorption of sodium and chloride ions and of SCFAs. Sodium uptake is driven by an active process powered by a Na^+/K^+-ATPase pump situated on the basolateral membrane of the colonocyte. Chloride ions are absorbed passively in exchange for bicarbonate ions. Water absorption is under both neural (via the enteric nerve plexus) and hormonal control. Aldosterone, angiotensin and the glucocorticoids stimulate water absorption, while the antidiuretic hormone vasopressin decreases water absorption.

An unpleasant side-effect of treatment with some antibiotics can be the development of diarrhoea as a result of suppression of colonic bacteria and subsequent reduction in SCFA production. However, such observations have led to improved therapies for diarrhoeal diseases. For example, addition of resistant starch to oral rehydration preparations has been shown to reduce the duration of diarrhoeal episodes. Although the mechanism responsible for this improved therapy remains to be confirmed, it seems likely that the absorption of the enhanced SCFA production from starch fermentation results in greater water absorption and so a shorter duration of loose stools.

Synthesis of vitamins and essential amino acids

For more than 50 years, it has been known that the bacteria in the human large bowel synthesise a wide range of vitamins, including the B vitamins folate, riboflavin, biotin, nicotinic acid, panthotenic acid, pyridoxine, vitamin B_{12} and thiamine and vitamin K, but understanding of the nutritional significance of this synthesis remains uncertain. In part, this is due to the considerable technical difficulty in making quantitative measurements of vitamin absorption from the human colon. The presence of menaquinones (bacterially derived forms of vitamin K) in human blood and tissues attests to uptake of these substances from the human gut, but there is only modest experimental evidence for the frequently cited assertion that about half of vitamin K needs can be obtained by colonic synthesis. Recent evidence suggests that certain bacteria, notably, *Bacteroides* and *Prevotella*, are associated with higher concentration of menaquinones in the human gut but the health implications of colonic-derived menquinones remain unclear.

The observation that low intakes of folate are associated with an increased risk of colorectal cancer and the accumulating evidence that low folate status can compromise the integrity of the genome are stimulating interest in factors (dietary or otherwise) that may modulate folate synthesis in the large intestine. Bacterial folate synthesis genes occur widely in the genomes of gut bacteria and, in one study, 13% of investigated genomes contained all genes required for complete de novo folate synthesis. There is growing evidence that folate biosynthesis by large bowel micro-organisms can contribute not only to folate supply to colonocytes but also, via the bloodstream, to the rest of the human body. Consequently, it is possible that bacterially synthesised folate could be absorbed and utilised by colonocytes (potentially affording protection against colorectal cancer) and, through effects on folate status, influence health elsewhere in the body.

In the same way that colonic bacteria synthesise vitamins for their own use, they synthesise amino acids, including the essential amino acids. Some of the nitrogen required for this synthesis is derived from urea circulating in the bloodstream, which diffuses readily into the bowel where it is hydrolysed by the gut microflora to ammonia. It has been proposed that intestinally synthesised amino acids may make a significant contribution to the amino acid needs of children and adults on very low protein intakes. However, quantifying this contribution remains a tough challenge. This is an important research area with potentially profound implications for the understanding of colonic physiology and for the derivation of human protein and amino acid needs.

Metabolism of phytochemicals and effects on the microbiome

The strong epidemiological evidence that those consuming diets rich in vegetables and fruits have a lower risk of developing several common non-communicable diseases, including cardiovascular disease, type 2 diabetes, respiratory disease and cancer, has stimulated research on the components of these plant foods that may be protective. There is particular interest in the non-nutrient secondary metabolites of plants, described collectively as phytochemicals, that have bioactivity when consumed by humans. Phytochemicals include polyphenols, carotenoids, phytosterols and phytostanols, lignans, glucosinolates and alkaloids. To be effective in protecting human cells against oxidative damage, for example, these compounds must be released from the plant tissue, absorbed across the intestinal mucosa and transported in sufficient concentrations to their site of action. Disruption of plant cells by cooking may help to release phytochemicals, but for uncooked foods degradation of plant cell walls and release of their phytochemical contents are aided greatly by bacterial fermentation in the colon.

Phytochemicals are chemically diverse, but a large proportion are found naturally as glycosides. In many cases, these must be converted to the aglycone derivatives (a process that is accomplished readily by colonic bacteria) before they can be absorbed.

Bacterial metabolism appears to be responsible for the production of the active derivatives of some phytochemicals. Among the best known are enterolactone and enterodiol, which are oestrogen-like compounds derived by bacterial degradation of the plant lignans matairesinol and secoisolariciresinol, respectively. There is some epidemiological evidence that these lignans may contribute to protection against hormone-related cancers and cardiovascular disease. In addition, phytochemicals may alter the abundance of certain bacteria, leading to higher bacterial diversity which is associated with better health outcomes (Figure 10.22).

Associations between the large bowel microbiome and human health

For decades, it has been suspected that the gut microbiota influences human health but more robust evidence for these links has emerged only recently. However, it is becoming clear that the large bowel micro-organisms (often via their metabolic endproducts) can affect many core body functions ranging from innate immunity to appetite and energy metabolism (Figure 10.23). As noted above, some of the beneficial effects of diets rich in vegetables and fruits may be mediated by metabolism of phytochemicals by the microbiota in the large bowel. On the other hand, adverse disturbances in the gut microbiota (described as dysbiosis) have been associated with a wide range of diseases including asthma, cancer, dementia, diabetes, obesity and poor mental health.

Colorectal cancer

Large-scale prospective cohort studies have provided strong evidence that higher intakes of dietary fibre reduce the risk of colorectal

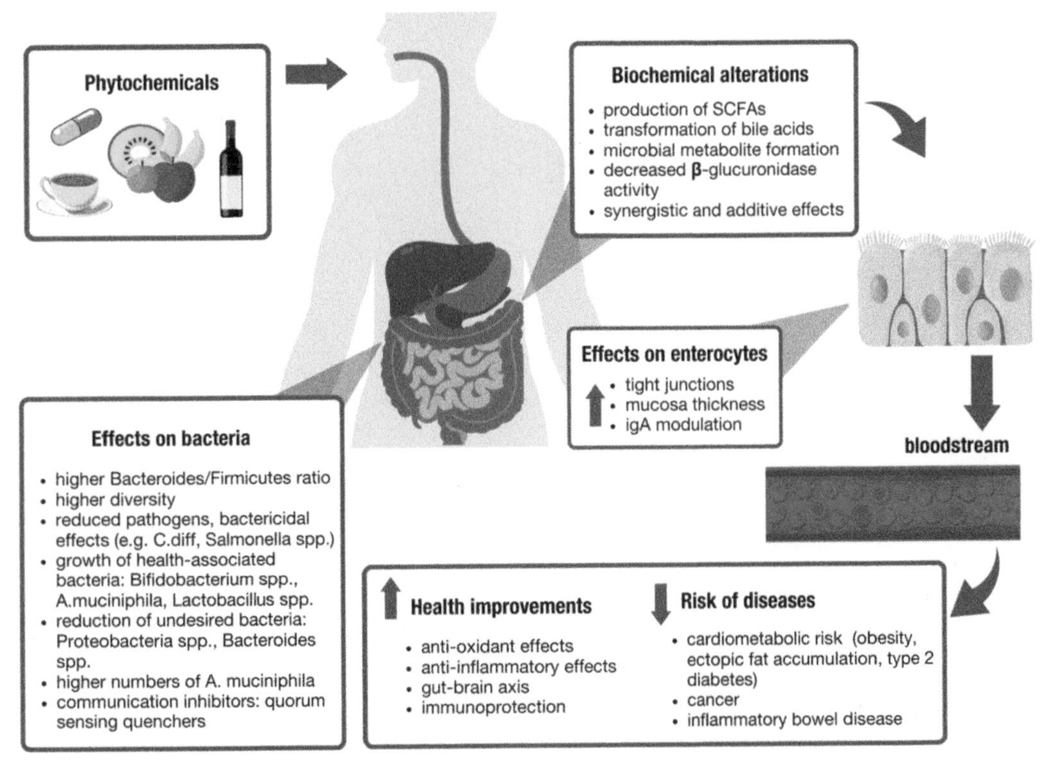

Figure 10.22 Metabolism of dietary phytochemicals by the gut microbiota and effects of phytochemicals on the microbiota and consequential implications for host health. *Source:* Dingeo G. et al. (2020) / The Royal Society of Chemistry.

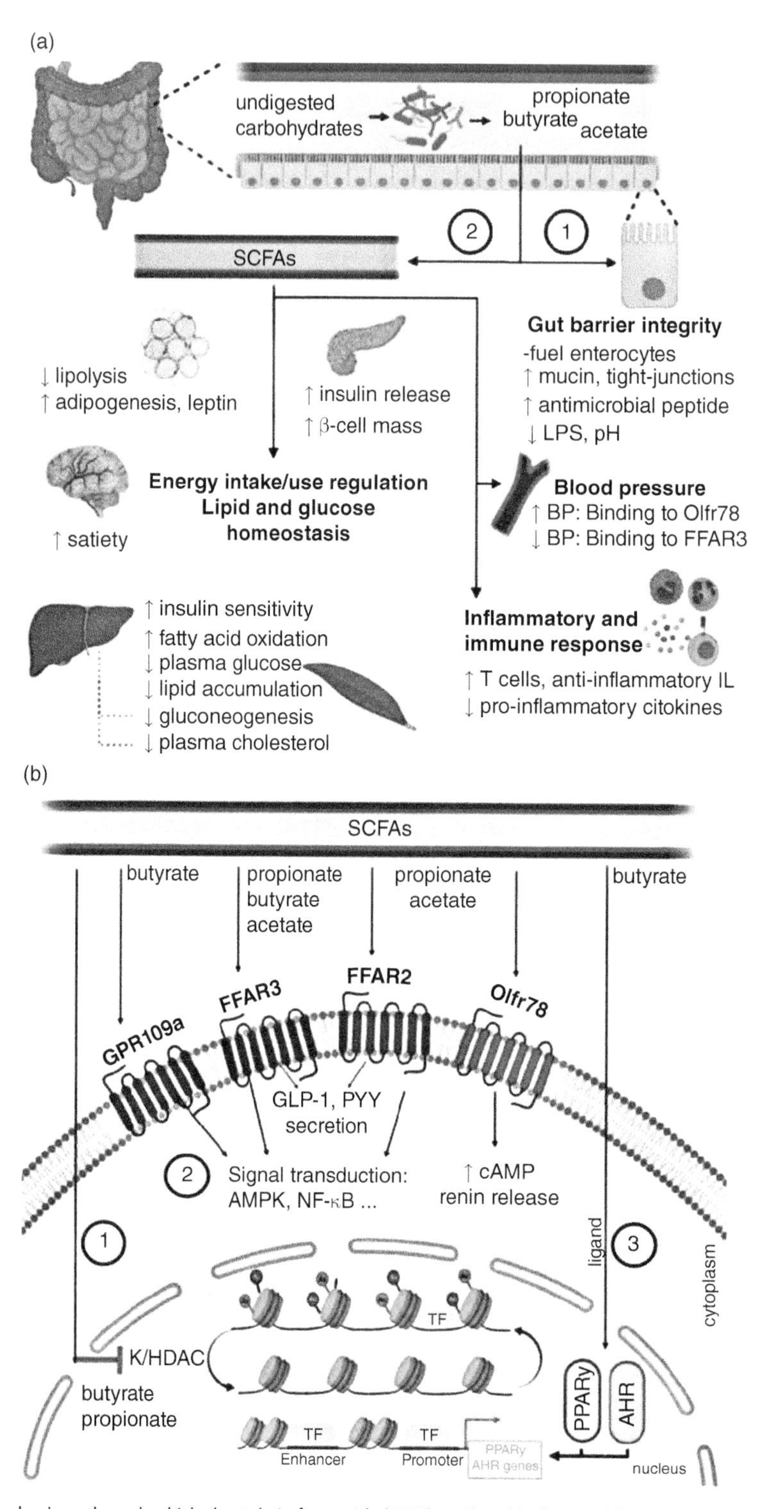

Figure 10.23 Mechanisms through which short-chain fatty acids (SCFA) produced by bacterial fermentation in the large bowel modulate a wide range of physiological processes. *Source:* Nogal et al., (2021) / Taylor and Francis / CC BY.

cancer. Such findings have stimulated extensive research on the role of the gut microbiome in mediating the effects of dietary fibre via the SCFAs (notably butyrate) produced from dietary fibre fermentation. In addition to being the preferred energy substrate for colonocytes (the cells lining the large bowel), butyrate is an inhibitor of enzymes known as histone deacetylases (EC

3.5.1.98; HDAC) that remove acetyl groups from histones, the globular proteins around which DNA is wrapped in the cell nucleus. The acetylation status of histones contributes to the constellation of epigenetic marks and molecules that regulate access to the genome and help to control gene expression. Cancer is characterised by dysregulated gene expression and there is evidence that butyrate may 'normalise' gene expression and help to maintain healthy colonocyte function.

Obesity

There is growing evidence that the mixture of bacterial species (the microbiome) in the colon of obese people differs from the microbiome of those who are leaner. Obesity is associated with a reduction in bacterial diversity with a lower proportion of Bacteroidetes and higher proportion of Actinobacteria. At first sight, one might imagine that these differences in the microbiome of obese individuals were a consequence of different dietary choices or of being obese per se but, most intriguingly, it now appears that the large bowel micobiota may play a causal role in the development of obesity. In studies of the large intestinal microflora of genetically obese *ob/ob* mice, lean *ob/+* and wild-type siblings all fed the same diet, the *ob/ob* mice had only half the proportion of Bacteroidetes but a greater abundance of Firmicutes than their lean siblings. In a large study of UK twins, bacteria from the genus *Christensenella* were rare in overweight people. In addition, low abundance of both *Christensenella* and *Akkermansia* correlates with lower visceral fat deposits. The mechanism(s) through which the microbiota may modulate adiposity are poorly understood but the dominant current hypothesis is that the flora associated with obesity are more efficient at harvesting energy from the food residues that flow into the large bowel.

Given the continuing rise in prevalence of overweight or obesity, the discovery that the intestinal microflora may contribute to obesity risk opens the way for research on innovative approaches for the prevention or treatment of obesity based on manipulation of the resident large bowel bacteria. This apparent link with risk of obesity (and other common diseases) is a considerable incentive for the development of better tools for studying the gut microbiota and the consequences of altered flora for human health and well-being. It is well established that differences in food intake are associated with alterations in the balance of species within the microbiota and in the patterns of endproducts of large bowel fermentation but there is little understanding of the underlying mechanisms, which limits the ability to use food (or food components) to change the flora to improve health. In contrast with the old idea that the colon is simply a 'sophisticated way of producing manure', the current revolution in knowledge about this organ places the colon and its associated micro-organisms centre stage in understanding diet and health and further discoveries of its importance are anticipated.

Manipulation of the large bowel to enhance or protect health

An unpleasant side-effect of treatment with some antibiotics can be the development of diarrhoea as a result of suppression of colonic bacteria and subsequent reduction in SCFA production. This is one of the most frequent, albeit unwanted, illustrations of the ease with which the colonic flora can be manipulated. The recognition that the balance of microflora within the large bowel may influence health has stimulated attempts to manipulate the flora in health-promoting ways using two main approaches.

Probiotics

A probiotic is a live microbial food ingredient that, when consumed in adequate amounts, is beneficial to health. Probiotics are often described as 'gut-friendly' bacteria and usually include *Bifidobacterium* and *Lactobacillus* species. Several probiotic products (mostly fermented milk-based products) are on the market in Japan, China, Europe and North America. Some potential probiotic species may help to overcome pathogenic bacteria such as *Clostridium difficile*, which can become dominant if the endogenous flora is disturbed, e.g. by the use of broad-spectrum antibiotics. Other potential probiotics are intended to improve immune defences, for example by direct interaction with the GALT or the production of 'protective' endproducts such as butyrate.

In addition to finding microbes that produce demonstrable benefits to human health in vivo, there have been challenges in the development of delivery systems (foods) that have a good shelf-life and can ensure that an appropriate and effective dose of bacteria reaches the large bowel (or other site of activity). Despite these challenges, there is now good evidence that probiotics are effective in the prevention of traveller's diarrhoea and they show promise in the prevention of antibiotic-associated diarrhoea in children.

Probiotics may also be beneficial in treating necrotising enterocolitis, acute upper respiratory tract infections, pulmonary exacerbations in children with cystic fibrosis and eczema in children. In addition, there is accumulating evidence that probiotics may have a role in alleviating some of the symptoms of irritable bowel syndrome (IBS) but more research is needed to determine the optimal dose of probiotics and the subgroups of IBS patients likely to benefit most.

Under EU food legislation, the term 'probiotic' represents a health claim and, to date, no health claims for probiotics have been approved by the European Food Safety Authority (EFSA).

Prebiotics

A prebiotic is defined as a substrate that is used selectively by the gut micro-organisms and may confer a health benefit. Given the difficulty of delivering enough 'gut-friendly' bacteria in foods in the form of probiotics, an alternative approach is to augment the numbers of the desired bacteria in the large bowel by supplying them with substrates that give them a competitive advantage.

All prebiotics developed to date are relatively small carbohydrates (degree of polymerisation 2–60), which are essentially indigestible by mammalian enzymes and so are delivered in digesta to the colon. Consequently, prebiotics are types of fermentable dietary fibre, but not all types of dietary fibre have prebiotic properties. There is good evidence that some of these oligosaccharides can produce significant shifts in the balance of bacterial species in the human colon, but how they do so is not yet certain. Prebiotics can have a substantial effect on the composition of the gastrointestinal microbiome, with specific species becoming enriched to constitute more than 30% of the faecal microbiota. Because many prebiotics occur naturally in commonly used foods such as onions and some cereal products, it is probable that such carbohydrates are safe in the amounts likely to be eaten. However, research to obtain sound evidence for the effectiveness of prebiotics in protecting or enhancing health is in its early stages.

As for probiotics, the term 'prebiotic' represents a health claim under EU legislation and, to date, the EFSA has not approved any health claims for prebiotics.

References

Dingeo G. et al. (2020) Food & Function 11: 8444. Phytochemicals as modifiers of gut microbial communities.

Martínez de Victoria E, Mañas M, Yago MD. (2017) Capítulo 2, Vol. I. Fisiología de la digestión. In: Gil A (ed.) Tratado de Nutrición, 3rd edn. Editorial Medica Panamericana, Madrid.

Nogal, A., Valdes, A.M., and Menni, C. (2021) The role of short-chain fatty acids in the interplay between gut microbiota and diet in cardio-metabolic health. Gut Microbes 13: 1897212.

Further reading

Beger, H.G., Warshaw, A.L., Hruban, R.H. et al. (2018). The Pancreas: An Integrated Textbook of Basic Science, Medicine, and Surgery, 3rd edn. Hoboken: Wiley-Blackwell.

Che, L., Hu, L., Liu, Y. et al. (2016). Dietary nucleotides supplementation improves the intestinal development and immune function of neonates with intra-uterine growth restriction in a pig model. PLoS One 11 (6): e0157314.

Dingeo, G., Brito, A., Samouda, H. et al. (2020). Phytochemicals as modifiers of gut microbial communities. Food & Function 11: 84444.

Dybala, M.P., Kuznetsov, A., Motobu, M. et al. (2020). Integrated pancreatic blood flow: bidirectional microcirculation between endocrine and exocrine pancreas. Diabetes 69: 1439–1450.

Gil, Á., Plaza-Diaz, J., Mesa, M.D., and Vitamin, D. (2018). Classic and novel actions. Annals of Nutrition and Metabolism 72 (2): 87–95.

Johnson, L.R. (2018). Gastrointestinal Physiology: Mosby Physiology Monograph Series, 9th edn.Philadelphia: Elsevier Mosby.

Macfarlane, G.T. and Gibson, G.R. (1995). Microbiological aspects of the production of short-chain fatty acids in the large bowel. In: Physiological and Clinical Aspects of Short-chain Fatty Acids (eds J.H. Cummings, J.L. Rombeau, and T. Sakata), 87–105. Cambridge: Cambridge University Press.

McBurney, M.I., Davis, C., Fraser, C. et al. (2019). Establishing what constitutes a healthy human gut microbiome: state of the science, regulatory considerations, and future directions. Journal of Nutrition 149: 1882–1895.

Plaza-Diaz, J., Ruiz-Ojeda, F.J., Gil-Campos, M., and Gil, A. (2019). Mechanisms of action of probiotics. Advances in Nutrition 10 (suppl_1): S49–S66.

Said, H.M. and Ghishan, F.K. (eds). (2018). Physiology of the Gastrointestinal Tract, 1 and 2, 6th edn. London: Academic Press.

Steidler, L., Hans, W., Schotte, L. et al. (2000). Treatment of murine colitis by Lactococcus lactis secreting interleukin-10. Science 289: 1352–1355.

Sullivan, Z.A., Khoury-Hanold, W., Lim, J. et al. (2021). gammadelta T cells regulate the intestinal response to nutrient sensing. Science 371 (6535): eaba8310.

Tilg, H., Moschen, A.R., and Kaser, A. (2009). Obesity and the microbiota. Gastroenterology 136: 1476–1483.

Valdes, A.M.,Walter, J., Segal, E., et al. (2018). Role of the gut microbiota in nutrition and health. BMJ 361: k2179.

11
The Cardiovascular System

Wendy L. Hall, Jayne V. Woodside, and J.A. Lovegrove

Key messages

- Dietary factors can modify the underlying causes of cardiovascular disease: high blood pressure, impaired cardiac electrophysiology and factors driving atherogenesis, such as inflammation, oxidative stress, platelet activation, endothelial function and lipoprotein infiltration of the artery wall and resultant lipid accumulation.
- Atherosclerosis can affect arteries from an early age and is largely preventable by lifelong maintenance of a healthy body weight, a balanced diet and regular physical activity.
- Replacing saturated fat intake with unsaturated fat or wholegrains can decrease low-density lipoprotein (LDL) cholesterol concentrations; lowering LDL cholesterol reduces cholesterol accumulation in atherosclerotic plaques and decreases risk of cardiovascular disease.
- The type of LDL particle is important: small, dense LDLs (and triacylglycerol-rich lipoprotein remnants) are particularly atherogenic. The atherogenic lipoprotein phenotype (more small dense LDL, reduced high-density lipoprotein cholesterol (HDL-C) and higher triacylglycerol) can be improved by weight loss and reducing intake of refined starches and sugars.
- Although dairy foods are rich in saturated fatty acids, evidence suggests that greater consumption of dairy foods is not associated with increased risk of cardiovascular disease, highlighting the importance of considering whole foods and dietary patterns in cardiovascular disease risk, rather than individual dietary components.
- Long-chain n-3 polyunsaturated fatty acids, derived from marine sources, can target many facets of atherosclerosis, including inflammation, thrombosis, blood pressure, vascular function, hypertriacylglycerolaemia and cardiac function, but supplementation in populations at risk or with existing cardiovascular disease has not convincingly demonstrated reductions in mortality or events.
- Lowering sodium intake from dietary salt lowers blood pressure, even in the context of a healthy, balanced diet and in normotensive individuals, and may improve endothelial function.
- Reductions in animal protein intake and a shift to a predominantly plant-based diet are recommended based on evidence that red and processed meat may be associated with increased risk of cardiovascular disease (possibly due to saturated fatty acid and sodium content) and the beneficial effects of higher intakes of fibre, potassium and other micronutrients, organic nitrate and phytochemicals in plant foods. Diets rich in fruits, vegetables, wholegrains, nuts and seeds may lower blood pressure and blood lipids, and improve endothelial function and glycaemic control.

11.1 Introduction

Diet is one of the key determinants of a healthy cardiovascular system. This chapter will provide an overview of the role of nutrition in the maintenance of a healthy cardiovascular system, with a particular focus on the impact of food groups and key macro- and micronutrients on well-established and emerging intermediary cardiometabolic risk factors.

Cardiovascular diseases (CVDs) are a collection of heart and circulatory conditions that cause half the deaths in Europe and are associated with the highest mortality globally (World Heart Federation 2021). There are many types of CVD, but the four main categories are: (i) coronary heart disease (CHD), (ii) stroke and transient ischaemic attacks (TIAs), (iii) peripheral arterial disease, and (iv) aortic disease, with CHD and stroke accounting for the majority of deaths in the

Nutrition and Metabolism, Third Edition. Edited on behalf of The Nutrition Society by Helen M. Roche, Ian A. Macdonald, Annemie M.W.J. Schols and Susan A. Lanham-New.

Companion Website: www.wiley.com/go/nutrition/metabolism3e

UK, and across the world. In brief, CHD occurs when there is a reduction or blockage in blood flow in the coronary arteries supplying the heart, resulting in angina (chest pain), heart attack or myocardial infarction (due to sudden blockage) or heart failure (insufficient blood delivery to the body). An ischaemic stroke follows insufficient blood supply to the brain via the carotid arteries, leading to death or brain damage, with a mini-stroke or TIA being a milder and temporary insufficiency in brain blood flow. A haemorrhagic stroke follows an arterial bleed in the brain. Peripheral arterial disease occurs due to blockage in arteries to limbs, predominantly the legs, causing pain, numbness or weakness in the legs or persistent ulcers. Finally, aortic disease includes a group of conditions that affect the aorta, the most common of which is aortic aneurysm, which is often symptomless, but could rupture, potentially causing life-threatening bleeding.

11.2 Factors involved in the development of atherosclerotic cardiovascular diseases

The heart is a powerful muscular organ that has evolved to be robust enough to beat around two to three billion times over an average global life span. Electrical signals originate in the sinoatrial node (the pacemaker) to set off a chain of electrical events resulting in the rhythmical contraction of the atria then ventricles, sending around 50–100 ml of blood from the left ventricle at each beat into the aorta. Blood is then distributed around the body, carrying oxygen, nutrients and hormones, and regulating temperature, via arteries, arterioles and capillaries, returning via venules and veins of increasing size until reaching the vena cava and returning to the heart. The thick elastic walls of the aorta need to withstand pressures of ~90 mmHg with every beat, and tiny capillaries sprout, withdraw and proliferate in response to fluctuating oxygen and nutrient demands of the body's organs and limbs. The endothelium, a single layer of endothelial cells lining blood vessels, maintains a finely tuned balance of secretion of homeostatic substances and regulation of exchanges between circulating blood and underlying tissue, maintaining vascular tone and avoiding thrombosis, and withstanding exposure to a variety of exogenous and endogenous circulating factors that may trigger inflammation. Considering the degree of mechanical and chemical stress endured by the cardiovascular system, it is not surprising that it can be prone to problems with ageing and that CVD is so prevalent worldwide.

Atherosclerosis

Atherosclerosis is a lifelong process that often starts in childhood, with early-stage lesions being common in young adults. The name atherosclerosis refers to the appearance of the lipid-filled lesions of atherosclerosis (*athere* derived from a word for 'porridge' in Greek), which eventually lead to hardening ('sclerosis') of the arteries and, ultimately, cardiovascular events, disability or sudden cardiac death. The original 'response to injury' theory for atherosclerosis development (Ross et al. 1999) proposed that injury to the artery due to trauma, infection, hypertension, reactive oxygen species (due to smoking or air pollution), hyperlipidaemia and chronic hyperglycaemia results in an inflammatory response that triggers initiation of an arterial lesion (Figure 11.1). This inflammatory response allows infiltration and trapping of apolipoprotein B-containing lipoproteins [low-density lipoproteins (LDLs) and triacylglycerol-rich remnant lipoproteins], triggering an inflammatory cascade where monocytes gather in the area and transform into macrophages, which engulf the trapped lipoprotein particles, accumulating cholesterol in the form of cholesteryl esters, thereby becoming foam cells. Eventually, multiplying and expanding foam cells coalesce and form a visible fatty streak. Ultimately, the lesion undergoes various stages of transformation involving migration of smooth muscle cells to the cap of the lesion, accumulation of cellular waste products, calcium and fibrin in the inner lining of the artery to become a 'plaque'. Atherosclerotic plaques may continue to expand and become more fibrous, causing remodelling and thickening of the artery wall and eventually narrowing of the lumen, restricting blood flow and causing further damage by turbulent blood flow. The plaque may eventually become unstable, with the fibrous cap thinning, macrophages dying and a necrotic core of free cholesterol and cellular waste forming within the plaque, culminating in rupture, thrombosis and possibly arterial occlusion and ischaemia of the surrounding tissue. The different stages of atherogenesis are shown in Figure 11.2.

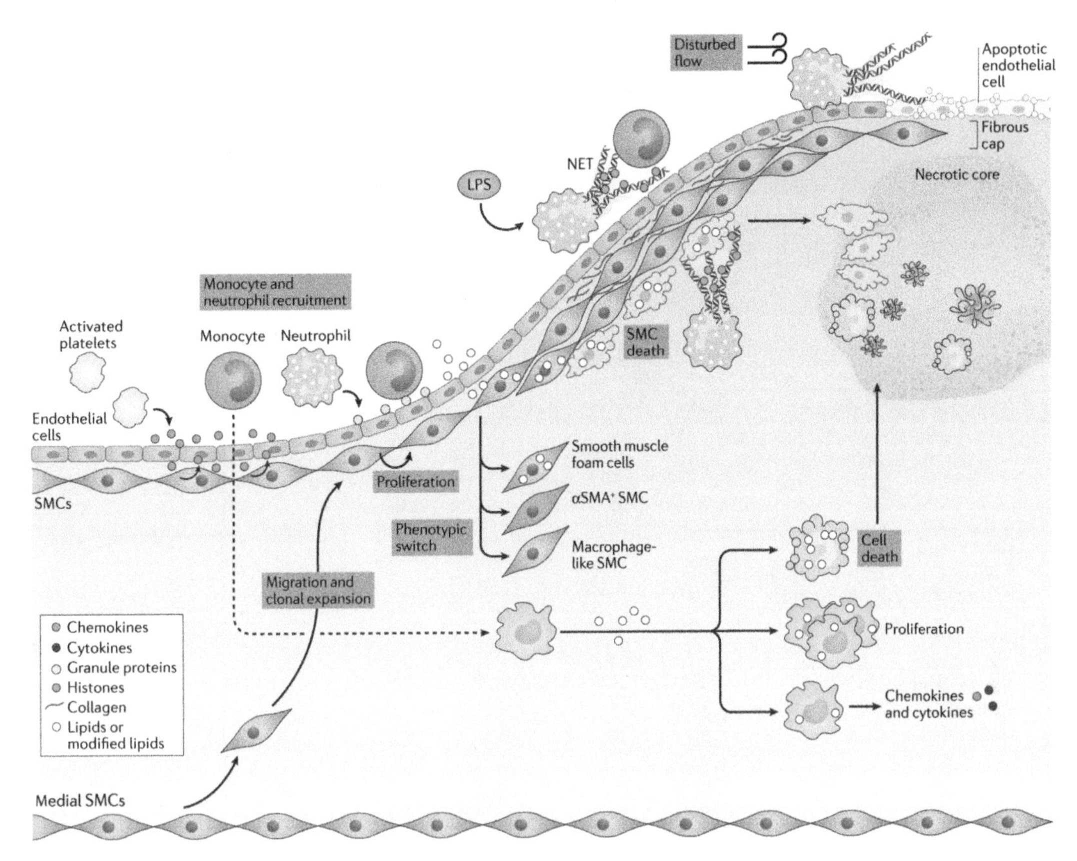

Figure 11.1 Inflammatory processes in atherosclerosis. In the early stages of atherosclerosis, activated platelets secrete chemokines that promote adhesion of monocytes and neutrophils. Neutrophils themselves secrete chemotactic granule proteins, thus paving the way for arterial monocyte infiltration. The chemokine milieu is supplemented by chemokines secreted by activated smooth muscle cells (SMCs). In progressing atherosclerotic lesions, medial SMCs migrate towards the developing fibrous cap where they undergo clonal expansion. SMC lipid loading triggers phenotype switching towards SMCs that express α-smooth muscle actin (αSMA+ SMCs), macrophage-like SMCs and smooth muscle foam cells. Heightened lipid loading of SMCs induces SMC apoptosis and – if not cleared quickly – necrosis. SMCs also undergo cell death after interaction with histone H4 presented in neutrophil extracellular traps (NETs). NET-associated cytotoxicity is observed during plaque erosion when NETs released at sites of disturbed flow induce endothelial cell desquamation. In systemic infections with Gram-negative organisms, which produce lipopolysaccharide (LPS), NET-associated histones promote the adhesion of monocytes, hence contributing to accelerated plaque growth under these conditions. Monocyte-derived macrophages ingest modified lipids and, in response, secrete inflammatory chemokines and cytokines. Excessive lipid uptake triggers macrophage proliferation or even cell death. *Source:* Reproduced with permission from Soehnlein & Libby, 2021/ Springer Nature.

Obesity

Obesity is frequently the unifying common factor underlying many of the risk factors for CVD: hypertension, impaired fasting glucose and/or glucose intolerance (pre-diabetes), or type 2 diabetes, and raised blood non-high-density lipoprotein (HDL) cholesterol and/or triacylglycerols. Substantial excess body fat, often associated with obstructive sleep apneoa, places additional demands on the heart, requiring increased heart rate and cardiac output, which may result in cardiac remodelling and eventually myocardial dysfunction. Abdominal obesity, where there is accumulation of fat around (visceral) and within (ectopic) the abdominal organs resulting in a higher waist circumference, is a major cause of atherosclerotic CVD, even in the absence of a raised body mass index (BMI). Visceral adiposity is associated with insulin resistance, dyslipidaemia, inflammation and hypertension, mediated through release of pro-inflammatory substances and free fatty acids from visceral adipose tissue. Higher amounts of visceral fat increase the likelihood of developing CVD risk factors over time, even after adjustment for BMI (Abraham et al. 2015), and weight loss can reverse CVD risk factors arising from abdominal obesity. Evidence

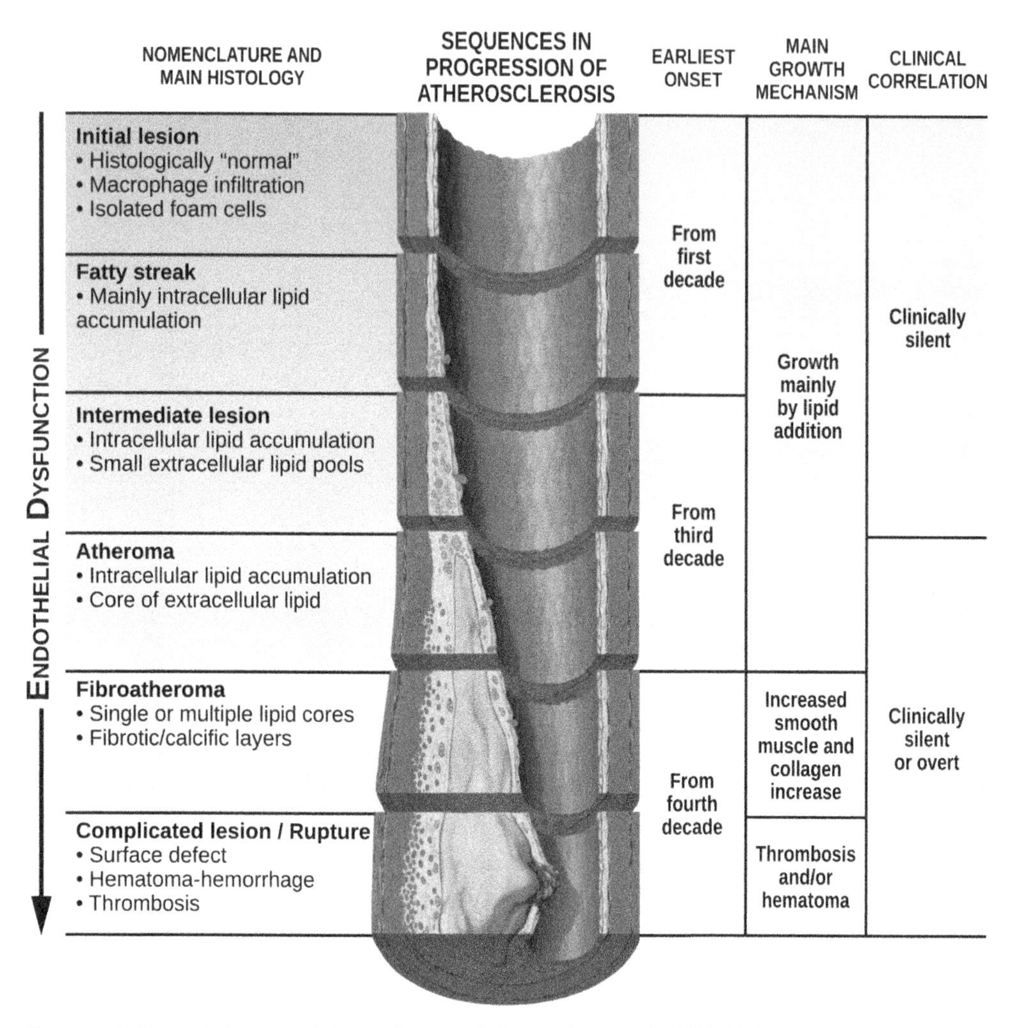

Figure 11.2 Stages of atherogenesis. *Source:* Unknown / Wikipedia Commons / CC BY-SA 3.0.

from primate models suggests that energy intake restriction can regress atherosclerosis, but only if hyperlipidaemia was present. In the DIRECT-Carotid human randomised controlled trial (Shai et al. 2010), energy-restricted diets led to a reduction in carotid artery thickening, an indicator of atherosclerosis, after two years in those participants who achieved a greater weight loss and larger reductions in systolic blood pressure.

Lipoprotein metabolism

Low density lipoproteins are produced predominantly from metabolism of hepatically derived very low-density lipoproteins (VLDLs). Their main purpose is to transport cholesterol to peripheral tissues for the formation of cell membranes and synthesis of steroid hormones. Reduction of elevated circulating LDL cholesterol (LDL-C) significantly reduces the incidence of myocardial infarction and death from cardiovascular causes. As shown in Figure 11.3, LDL-C increases the progression of atherosclerosis in a dose-dependent manner (Ference et al. 2017) with an estimated 1 mmol/l reduction in LDL-C associated with 23–54% proportional reduction in the risk of CHD with evidence from randomised controlled trial and Mendelian randomisation studies, respectively.

While all LDL is potentially atherogenic, small, dense LDL particles show greater atherogenicity than larger particles, and confer a greater risk for CVD, due to their increased ability to move into the vascular intima, where they undergo oxidation, and contribute to the atherosclerotic plaque. By contrast, a low concentration of serum HDL-C is related to an increased risk of CVD. HDL particles are involved in 'reverse cholesterol transport', a process in which

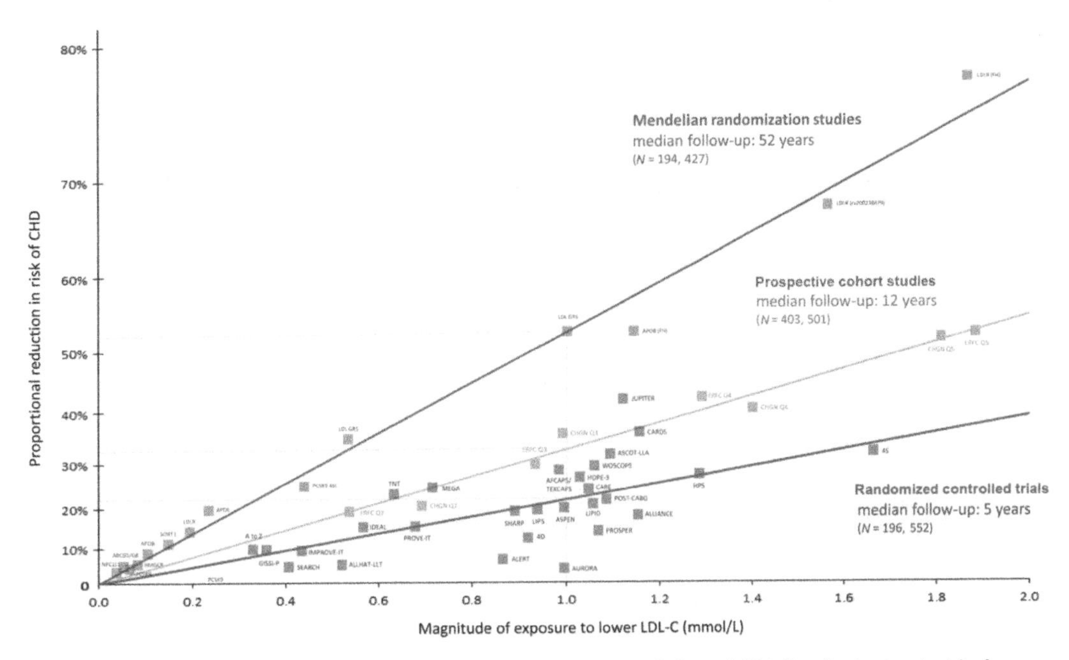

Figure 11.3 Associations between exposure to lower low-density lipoprotein cholesterol (LDL-C) and reductions in risk of coronary heart disease from meta-analyses of Mendelian randomisation studies, prospective cohort studies and randomised controlled trials. Log-linear association per unit change in LDL-C and the risk of cardiovascular disease as reported in meta-analyses of Mendelian randomisation studies, prospective epidemiological cohort studies and randomised trials. The increasingly steeper slope of the log-linear association with increasing length of follow-up time implies that LDL-C has both a causal and a cumulative effect on the risk of cardiovascular disease. *Source:* Ference et al. 2017 / Oxford University Press / CC BY 4.0.

cholesterol is removed from tissues and organs and returned to the liver for metabolism. However, recent evidence has shown that increasing serum HDL-C, by use of drugs, may not result in the anticipated reduction in CVD risk, which seems to be more closely related to the functionality, rather than the cholesterol content of HDL particles. A more sensitive and specific CHD risk prediction is the total cholesterol (TC):HDL-C or LDL-C:HDL-C ratios, which take account of the contrasting contributions of these two lipoproteins to CVD risk.

Elevated fasting, and more recently postprandial, circulating triacylglycerol concentrations have been recognised as independent risk factors for CVD. Triacylglycerol-rich hepatically (VLDL) and dietary (chylomicrons, CMs)-derived lipoprotein particles have been implicated in the development of atherosclerosis. These lipoproteins are involved with 'neutral lipid exchange' where triacylglycerol is exchanged for cholesterol esters from LDL and HDL, resulting in small, dense LDL and HDL particles and relatively cholesterol ester-enriched-VLDL and CM remnants. These HDL particles are more efficiently cleared, resulting in lower circulating HDL-C, whereas the small, dense LDL particles have lower affinity with hepatic receptors and persist in the circulation. Furthermore, apoB48, the apolipoprotein found exclusively on dietary-derived CMs and their remnants, has been found in atherosclerotic plaques, demonstrating the atherogenic potential of CM remnants. Details of endogenous and exogenous lipoprotein metabolism are shown in Figure 11.4.

Hypertension and vascular function

There is a strong, continuous relationship between blood pressure and risk of CVD even at the lower, 'normal' end of the scale for both systolic and diastolic blood pressures. The biggest risk of having high blood pressure is the greater likelihood of stroke, but risk of CHD and heart failure also increases steadily with increasing blood pressure. Many randomised controlled trials have demonstrated that treatment of hypertension can effectively reduce incidence of CVD and mortality. Lifestyle changes are key for the optimal treatment of high blood pressure through maintenance of a healthy body weight, limiting salt intake and increasing consumption of fruits and vegetables. Untreated hypertension

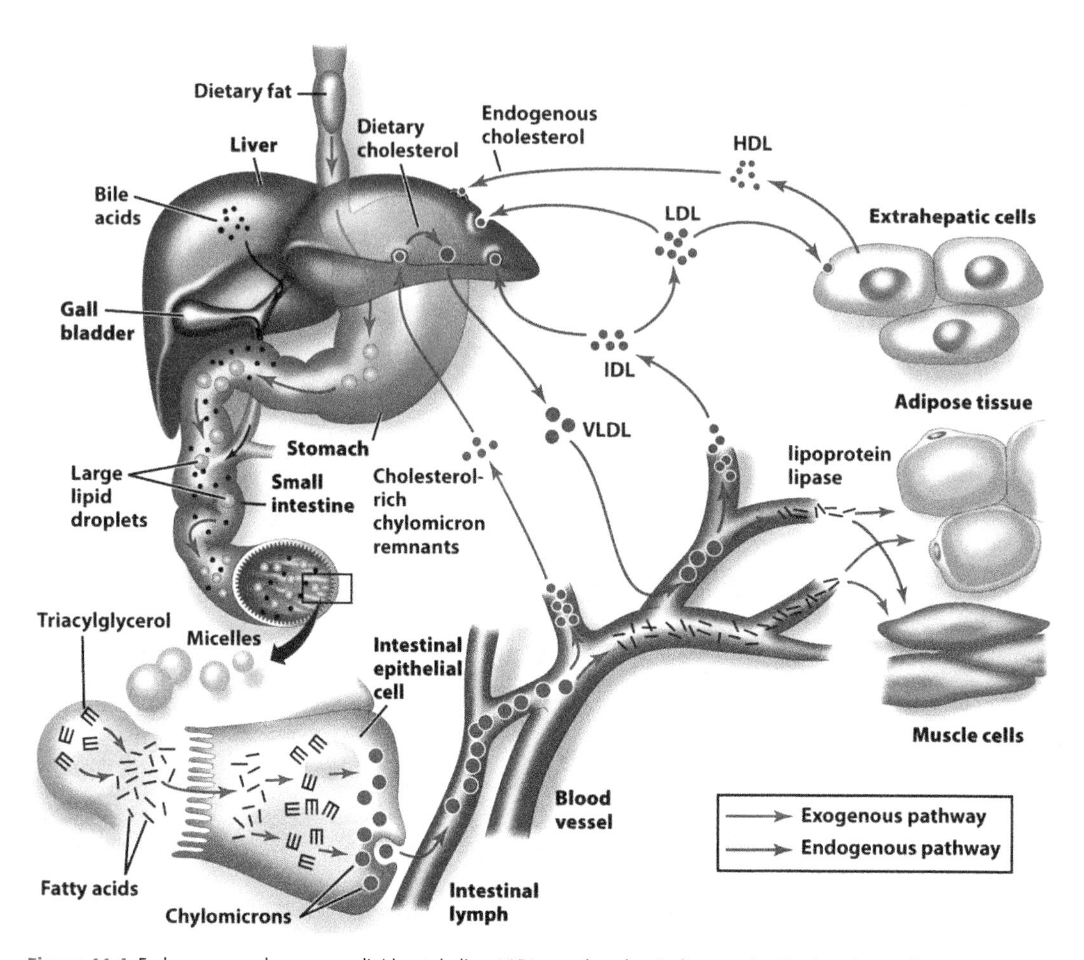

Figure 11.4 Endogenous and exogenous lipid metabolism. VLDL, very low density lipoprotein; LDL, low-density lipoprotein; IDL, intermediate density lipoprotein; HDL, high-density lipoprotein (https://www.slideshare.net/422459/voet-chap-20-synthesis-and-degradation-of-lipids).

can lead to damage to blood vessels, the heart and the kidneys, thereby exacerbating the progression of arterial stiffening, heart failure, atherosclerosis and chronic kidney disease. Arterial stiffening with ageing is one of the main drivers of raised systolic blood pressure, as arterial stiffness increases blood flow velocity and causes the reflected pressure wave due to peripheral vascular resistance to arrive in the aorta earlier during systole, thereby augmenting systolic blood pressure. However, hypertension itself can then cause further structural damage to arteries that leads to further arterial stiffening as well as endothelial dysfunction through turbulent blood flow at branches in the arterial tree. Endothelial dysfunction results in impaired vasodilation through impaired nitric oxide availability, increasing permeability to lipoproteins, monocyte adhesion and a procoagulatory state, which combine to promote atherogenesis. Arterial stiffness, aortic stiffness and endothelial function can be measured using a range of non-invasive techniques and, through these approaches, they have been demonstrated to be independent predictors of cardiovascular events.

Inflammation

Inflammation is a complex adaptive defence response associated with each of the modifiable risk factors for CVD: dyslipidaemia, hypertension, diabetes, obesity and smoking. It is a feature of all stages of atherosclerosis from initiation to progression to plaque rupture, and comprises a complicated cascade of pathways that mediate the effects of CVD risk factors on arteries. Where damage to the arteries has been initiated, blood flow turbulence, pro-inflammatory cytokines and endotoxins, oxidised LDL and advanced glycation endproducts (AGE) activate endothelial cells, triggering NF-κβ-induced endothelial expression of cell adhesion molecules, e.g. vascular cell adhesion molecule-1 (VCAM-1), chemokines, e.g. monocyte chemoattractant protein-1 (MCP-1) and interleukin-1 (IL-1), and prothrombotic agents such as tissue factor and plasminogen activator inhibitor. This facilitates the movement of monocytes, neutrophils and lymphocytes from the circulation into the subendothelial space, where transformation of monocytes into macrophages and interaction with trapped lipoproteins and deranged smooth muscle cells results in a self-propagating process of accumulating cytokine-secreting immune cells, promoting further inflammation and production of reactive oxygen species, and resulting in degraded and necrotic tissue within the plaque. Due to the highly localised and complex nature of atherosclerotic inflammation, it has been challenging to demonstrate direct effects of diet on precise markers of atherothrombotic inflammatory pathways in human randomised controlled trials. However, circulating high-sensitivity C-reactive protein (hs-CRP), an acute phase inflammatory protein secreted mainly by the liver (but also in the vasculature) in response to IL-6, is an independent predictor of CVD risk that has been shown to decrease following a cardioprotective diet (Reidlinger et al. 2015).

Dysglycaemia

It is estimated that more than 450 million people globally are currently living with diabetes, which is around 9% of the world's adult population, with almost half of these undiagnosed. The figure is expected to rise to 642 million people living with diabetes worldwide by 2040. Diabetes is associated with microvascular complications such as retinopathy, nephropathy and neuropathy, leading to increased risk of sight loss, chronic kidney disease and amputation, as well as a much greater risk of CHD, peripheral vascular disease and stroke; consequently, CVD is the leading cause of death in those with diabetes. Hyperglycaemia is a key driver of the increased risk of CVD in type 2 diabetes. Chronic uncontrolled elevated blood glucose concentrations, particularly pronounced postprandial 'spikes', cause damage to the endothelium by increasing oxidative stress and inflammation, resulting in atherogenic stimuli in arteries and ischaemic injury to organs and other tissues. Sustained hyperglycaemia can also lead to the formation of advanced glycation end-products (AGEs), which are proteins or lipids that can become modified (glycated) by chronic exposure to aldose sugars, often in the presence of oxidative stress, thereby altering cellular structures and function with tissues, including blood vessel walls. In pre-diabetics, where impaired glucose tolerance may increase oxidative stress and inflammation in the vasculature, frequent pronounced elevations in blood glucose may also promote atherosclerosis.

11.3 The global burden of diet-related cardiovascular disease and dietary recommendations

Usual diet varies markedly between countries and therefore the impact of dietary patterns on cardiovascular outcomes may differ depending on geographical location and local food culture. The Global Burden of Diseases, (GBD) attempted to quantify the burden of mortality and disability attributable to specific dietary risks, specifically evaluating the consumption of major foods and nutrients across 195 countries and focusing on adults aged ≥25 years (GBD 2017 Diet Collaborators 2019).

Globally, in 2017, consumption of nearly all healthier foods and nutrients, but particularly nuts and seeds, milk and wholegrains, was suboptimal, whereas that of all unhealthy items, including sugar-sweetened beverages, sodium and red and processed meats, was higher than recommended. The analysis found that 11 million (95% uncertainty interval 10–12) deaths, the majority of which were due to CVD, and 255 million (234–274) disability-adjusted life-years (DALYs) were attributable to dietary risk factors. High intake of sodium, low intake of wholegrains and low intake of fruits were the leading risk factors for mortality and DALYs among dietary factors, accounting for more than half of all diet-related deaths and two-thirds of diet-related DALYs, and there was a disproportionate burden in low-income settings, as shown in Figure 11.5.

This study clearly illustrates the potential impact of suboptimal diet on CVD morbidity and mortality, and highlights the need to improve diet across nations.

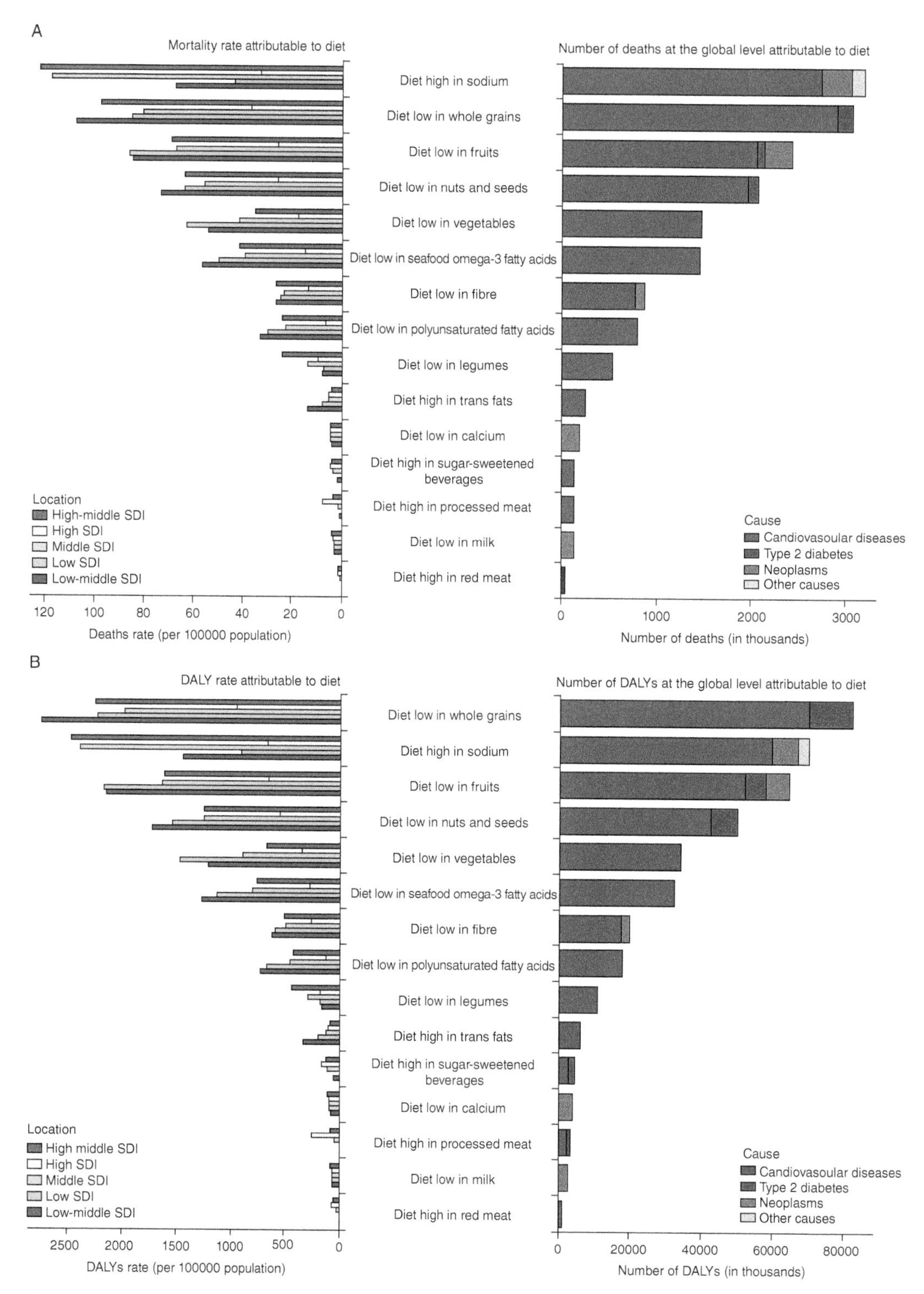

Figure 11.5 Number of deaths and DALYs and age-standardised mortality rate and DALY rate (per 100 000 population) attributable to individual dietary risks at the global and Socio-demographic Index (SDI) level in 2017. *Source:* Reproduced with permission from GBD 2017 Diet Collaborators 2019 / Elsevier / CC BY 4.0.

Population dietary guidance for CVD prevention was proposed in the USA as far back as 1961, which included guidance to reduce saturated fatty acids (SFAs). Similar UK nutrient recommendations on dietary SFA reductions to less than 10% total energy were proposed by the National Advisory Committee on Nutritional Education (NACNE) in 1983. Since then, consistent recommendations have been published on SFAs (SACN 2019) and on numerous other nutrients. However, nutrient-based dietary recommendations can be challenging to translate into practical advice for the general public. Furthermore, as total diet quality is recognised as a key contributor to health and disease, food-based recommendations are being increasingly valued, as exemplified by the UK Eatwell Guide and the US Dietary Guidelines for Americans. Another example of this shift is the dietary guidance proposed by the Netherlands in 2015, which was exclusively food-based. This included general guidelines to eat a more plant-based dietary pattern, to limit consumption of animal-based food, and 15 specific guidelines on salt, vegetables, fruits, fish, legumes, nuts, meat, dairy produce, cereal products, fats and oils, tea, coffee, and sugar-containing and alcoholic beverages (Kromhout et al. 2016).

11.4 Diet, cardiovascular risk factors and cardiovascular outcomes

11.4.1 Fats and oils

There is convincing evidence that dietary SFAs and trans-fatty acids increase the risk of CVD events, and that replacement with unsaturated fats, polyunsaturated (PUFAs) and monounsaturated fatty acids (MUFAs), but not simple carbohydrates, reduce risk. This evidence has led to the formulation of key dietary recommendations on fat intake in the UK (Table 11.1) and around the world.

Saturated fatty acids. Randomised controlled trials performed in humans since 1950s have consistently shown that SFAs (which are derived predominantly from animal products) are the key dietary factor responsible for increased plasma cholesterol, and that this effect is largely due to increased LDL-C, particularly in those with higher baseline concentrations of LDL-C. Individual SFAs have differential effects on plasma cholesterol fractions. The most hypercholesterolaemic SFAs are C12:0 (lauric acid), C14:0 (myristic acid) and C16:0 (palmitic acid), while stearic acid (C18:0) appears to be relatively neutral. Odd-chain SFAs (particularly C:17 and, to a lesser extent, C:15), found predominantly in dairy fats, have been associated with a lower risk of type 2 diabetes, which may reflect the impact of other components in dairy foods (see Section 11.4.2) in studies that explore circulating fatty acids and disease risk. The LDL-C-raising capacity of SFAs has been proposed to be due to the regulatory effects of LDL receptor expression on hepatic cell membranes. Exposure to high concentrations of SFAs have been shown to downregulate the expression of the LDL receptor gene by a mechanism that lowers the amount of free cholesterol within cells. This reduces the amount of LDL receptors on the cell surface and the rate of removal of LDL from the circulation, effectively raising plasma cholesterol. The more limited effects of stearic acid may be due to its high rate of conversion to 20:1 monounsaturated fatty acid or lower bioavailability in foods. Increasing evidence from controlled intervention studies supports detri-

Table 11.1 UK Dietary Reference Values (DRVs) for fats for adults as a percentage of total energy intake (unless otherwise stated).

	Individual minimum	Population mean	Individual maximum
SFAs		10%	
cis-PUFAs	n-6 PUFAs 1.0%	6%	10%
	n-3 PUFAs 0.2%	LC n-3 PUFAs 0.45g/day	
cis-MUFAs		12%	
trans fatty acids		2%	
Total fatty acids		30%	
Total fat		33%	

SFAs, saturated fatty acids; PUFAs, polyunsaturated fatty acids; LC n-3 PUFAs, long-chain n-3 polyunsaturated fatty acids; MUFAs, monounsaturated fatty acids.
Source: Adapted from Mensink, 2016.

mental effects of SFAs on vascular risk (including arterial reactivity, measured by the gold standard flow-mediated dilation, circulating cell adhesion molecules, endothelial progenitor cells and microparticles), blood pressure and glycaemic control, mainly in 'at risk' populations, with benefit from replacement with unsaturated fatty acids (found in plant seeds and foods). However, more studies are required to investigate and confirm these effects further.

Unsaturated fatty acids. A large body of evidence from strictly controlled metabolic ward and population dietary intervention studies has consistently demonstrated the similar hypocholesterolaemic effects of *cis*-MUFAs (found in plants and oils, e.g. rapeseed or olive, and animal foods) and *cis*-n-6 PUFAs (found in plants and oils, e.g. corn and sunflower), particularly as a replacement for SFAs as shown in Table 11.2 (Mensink 2016). Mechanistically, unsaturated fatty acids increase the expression of the LDL receptor protein, causing a greater rate of removal of LDL from the circulation. However, there is also a modest lowering of HDL-C when SFAs are replaced by unsaturated fats, although it is not clear whether this is associated with detrimental effects on CVD risk. *Cis*-n-6 PUFAs have also been associated with beneficial effects on other risk factors such as blood pressure and vascular function, but differential effects on inflammation. Numerous *in vitro* cell studies have indicated that *cis*-n-6 PUFAs are relatively pro-inflammatory in comparison to long-chain n-3 PUFAs, especially anti-inflammatory eicosapentaenoic acid (EPA). However, pro-inflammatory effects of n-6 PUFAs in human intervention studies are not consistent with these in vitro studies.

Trans-unsaturated fatty acids. There are two distinct types of *trans*-fatty acids (*t*FAs): those produced commercially by partial hydrogenation of vegetable oil for preparation of spreads and margarines or from repeated heating of unsaturated oils (e.g. elaidic acid); and ruminant-*t*FAs produced in the rumen from microbiome fermentation (found in meat and dairy products) (e.g. vaccenic acid). Metabolic studies on *t*FAs have consistently shown a significant increase in LDL-C and decrease in HDL-C concentrations with an estimated 1% energy replacement of unsaturated fatty acids, with *t*FA inducing an approximately 0.05 mmol/l (2 mg/dl) increase in LDL-C. However, contrary to SFAs, this replacement is associated with a significant decrease in HDL-C concentrations, leading to an increase in the LDL:HDL-C ratio, linked with an increasing CHD risk. Due to this marked effect of total *t*FAs on CVD risk, commercial *t*FAs have been almost totally removed from the food chain by industry, leading to a significant reduction *in* *t*FA intake

Table 11.2 Estimated multiple regression equations for the mean changes in serum lipids when 1% of dietary energy from SFAs is isoenergetically replaced by carbohydrates, *cis*-MUFAs or *cis*-PUFAs.

Lipid	SFAs for CHOs	SFAs for *cis*-MUFAs	SFAs for *cis*-PUFAs	No.[1]
Change TC[2] (mmol/l)	**−0.041**	**−0.046**	**−0.064**	177/74
95% CI	−0.047 to −0.035 *P* <0.001	−0.051 to −0.040 *P* <0.001	−0.070 to −0.058 *P* <0.001	
Change LDL-C (mmol/L)	**−0.033**	**−0.042**	**−0.055**	165/69
95% CI	−0.039 to −0.027 *P* <0.010	−0.047 to −0.037 *P* <0.001	−0.061 to −0.050 *P* <0.001	
Change HDL-C (mmol/L)	**−0.010**	**−0.002**	**−0.005**	163/68
95% CI	−0.012 to −0.008 *P* <0.011	−0.004 to −0.000 *P* = 0.014	−0.006 to −0.003 *P* <0.001	
Change in TAG (mmol/L)	**0.011**	**−0.004**	**−0.010**	172/72
95% CI	0.007 to 0.014 *P* <0.001	−0.007 to −0.001 *P* = 0.022	−0.014 to −0.007 *P* <0.001	
Change in TC:HDL-C ratio	0.001	**−0.027**	**−0.034**	159/66
95% CI	−0.006 to 0.007 *P* = 0.842	−0.033 to −0.022 *P* <0.001	−0.040 to −0.028 *P* <0.001	

SFAs, saturated fatty acids; CHOs, carbohydrates; *cis*-MUFAs, *cis*-monounsaturated fatty acids; *cis*-PUFAs, *cis*-polyunsaturated fatty acids; CI, confidence interval; HDL-C, high-density lipoprotein cholesterol; LDL-C, low-density lipoprotein cholesterol.

[1] Number of diets/number of studies.

[2] The 95% confidence intervals (CI) refer to the regression coefficients on the line directly above.

Source: Adapted from Mensink (2016).

in Western countries, with the mean UK intake below the recommendation of 0.2% total energy/day. With this marked reduction in commercial *t*FAs, a greater proportion of total *t*FAs are from ruminant sources. However, recent evidence has supported a relative lack of effect on CVD risk of ruminant *t*FAs compared with industrial *t*FAs, with some evidence for a negative association with risk of type 2 diabetes in prospective cohort studies, although confirmation of these differential effects of *t*FAs is required.

Dietary cholesterol, sterols and stanols. Dietary cholesterol (found in animal foods, particularly egg yolks and shellfish) is generally associated with SFAs in foods and has an additive or even synergistic effect on LDL-C. These effects are proposed to be mediated through LDL receptor-related mechanisms, similar to those from SFAs, which occur at an intake threshold of about 400 mg/day of dietary cholesterol. High intakes of dietary cholesterol have been associated with increased risk of CVD in observational studies. Yet these results have not been substantiated in high-quality intervention studies in healthy groups or individuals with type 2 diabetes or hyperlipidaemia, where dietary cholesterol (primarily from eggs) had no clinically relevant effects on serum cholesterol. Evidence suggests that up to seven eggs per week can be consumed, although, in patients with CVD or type 2 diabetes, only with special emphasis on a healthy lifestyle (Geiker et al 2017).

Plant sterols (found naturally in small quantities in some plant foods and as foods additives) and their saturated derivatives, stanols, have a similar structure to cholesterol and consistently reduce LDL-C concentrations (with an intake of 2 g/day resulting in an estimated 5–10% reduction). There are a number of proposed mechanisms of action, but primarily they inhibit dietary cholesterol absorption by competitively binding to receptors.

Dairy foods

Dairy products are the greatest contributor to dietary SFAs in the habitual diets in the UK, and in many European countries, contributing an average 23% of SFA intake. Yet data from meta-analyses of prospective cohort studies suggest no associations of dairy with risk of all-cause mortality, CHD and CVD, with a possible role for fermented dairy (including cheese and yoghurts) in lower CVD risk (Guo et al. 2017) and some evidence for greater benefits from low-fat dairy. Dairy products are complex foods and are rich in proteins (caseins ~ 82%, whey ~18%), bioactive peptides and micronutrients, with some fermented foods also containing probiotics (see Section 11.4.12). Despite the recognised detrimental effects of SFAs on CVD events and risk factors (see Sections 11.4.1), other components within dairy foods may attenuate this effect, thus reducing the overall detrimental effect of dairy foods on CVD risk.

Significant hypotensive effects of dairy-derived proteins (predominantly whey) and peptides (e.g. lactotripetides) are evident in healthy and mildly hypertensive patients. Bioactive peptides, derived from dairy protein either chemically, enzymatically or as a consequence of endogenous protein digestion, have inhibitory effects on angiotensin-converting enzyme (ACE), required for the conversion of angiotensin I to II, the latter of which is a key mediator of vascular constriction and elevated blood pressure. In vitro evidence for the effect of dairy proteins on ACE inhibition is strong, with further confirmatory studies required for in vivo effects. Furthermore, dairy is a good source of calcium and magnesium, which have also been implicated in reducing blood pressure.

There are a wide variety of dairy foods which have differential effects on risk. Butter (containing only fat) has a significant LDL-C-raising capacity, yet the equivalent fat intake from a hard cheese has substantially lower LDL-C-raising effects. This has been termed the 'matrix effect'. Dietary calcium present in high amounts in cheese can bind SFAs and cholesterol-containing bile acids and salts, reducing their absorption via the enterohepatic circulation, leading to faecal loss, which could contribute to a reduction in circulating LDL-C concentrations, relative to butter. The protein matrix of cheese may also reduce SFA absorption by reducing digestive enzyme access, although these effects and mechanisms need confirmation.

Higher dairy intake is consistently associated with lower risk of type 2 diabetes, a finding supported by the previously mentioned negative relation between circulating dairy-derived odd-chain SFAs and type 2 diabetes in epidemiological studies. While the mechanistic link for this association is unclear, there is emerging evidence to suggest that dairy proteins, mainly whey-derived, may beneficially affect glycaemic control, which may therefore contribute to lower type 2 diabetes risk.

Fish

There have been many decades of research on the health effects of consuming seafood and the n-3 fatty acids that they contain, which began with associations between dietary seafood and low incidences of CHD among Inuit populations. Fish are rich in long-chain (LC) n-3 PUFAs, iodine, selenium, vitamin D and protein, which may contribute to their association with reduced risk of CVD, as demonstrated in epidemiological studies. Many national dietary guidelines include a recommendation to consume any kind of fish, with the UK government and some other authorities specifying inclusion of oily fish for prevention of CVD. However, estimates of fish consumption suggest intakes are lower than recommendations, and fish may be completely absent from some diets. It is relevant to note that global fish stocks are unlikely ever to be large enough to meet the recommended intake for every human being on the planet. The most intensively researched nutrients in fish, particularly oily fish, are LC n-3 PUFAs: namely EPA and docosahexaenoic acid (DHA). Oily fish (e.g. salmon, trout, herring, mackerel, sardines) are a concentrated source of LC n-3 PUFAs due to the progressive concentration up the food chain from the original de novo synthesisers of EPA and DHA: marine single-celled microalgae. Vegan diets are devoid of dietary LC n-3 PUFAs, and so long-term consumers of vegan diets rely on endogenous conversion of dietary alpha-linolenic acid (ALA) to EPA and then, to a very limited extent, further conversion to DHA. Vegetarians, however, consume small amounts of EPA and DHA through consumption of dairy foods and eggs.

Although epidemiological evidence strongly suggests that fish consumption is associated with reduced risk of CVD, randomised controlled trials of fish oil supplements have yielded mixed results, probably due to methodological limitations. Meta-analyses of randomised controlled trials in adults have shown that higher intakes of LC n-3 PUFAs (mainly as oil supplements) had no overall effect on CVD, although there is low-certainty evidence for reductions in relative risk of CHD (Abdelhamid et al. 2020). Evidence from clinical and basic science studies indicates that LC n-3 PUFAs modify cardiovascular risk factors through modulation of gene expression, cell signalling, membrane fluidity, and by conversion to specialised pro-resolving mediators. Supplementation with 0.8–3 g LC n-3 PUFAs/day can reduce fasting triacylglycerol levels in a dose-dependent manner by 8–30% in both normolipidaemic and hyperlipidaemic individuals by reducing VLDL secretion from the liver, as well as lowering postprandial lipaemia by reducing chylomicron secretion and increasing chylomicron clearance rates. Mechanisms include decreased production and increased catabolism of apoB48 and apoB100, inhibition of hepatic fatty acid synthesis, higher rates of fatty acid peroxisomal beta-oxidation, and reduced adipose tissue lipolysis and release of circulating free fatty acids, further reducing supply for hepatic triacylglycerol synthesis. There is little evidence of an effect of LC n-3 PUFAs on total cholesterol or HDL cholesterol. However, high-dose DHA-containing supplements can slightly increase HDL cholesterol and LDL cholesterol concentrations in some individuals, particularly those not on LDL-lowering medication with high initial plasma triacylglycerol concentrations, due to enhanced conversion of VLDL to larger, buoyant LDL particles (which are less atherogenic than small dense LDL).

A high proportion of deaths from CHD are caused by sudden cardiac death, which is due to abnormal heart rhythms (arrhythmia) and can be triggered by a myocardial infarction. LC n-3 PUFAs have been shown in animal studies to have arrhythmic effects, and reduced risk of CHD mortality associated with fish oil supplementation, is attributed in part to prevention of lethal arrhythmias based on early secondary prevention trials such as the GISSI-Prevenzione study, demonstrating a reduction in risk of sudden cardiac death. Fish consumption and, to some extent, fish oil supplementation are associated with reductions in heart rate and increased heart rate variability, predictors of risk of arrhythmia and sudden cardiac death. Mechanisms include influencing neuronal function in the hypothalamus, sympathetic and parasympathetic neurons, and the intrinsic cardiac nervous system by neurotrophic and neuroprotective anti-inflammatory and haemodynamic mechanisms, and by incorporation into myocardial cell membranes influencing membrane ion channels. Beneficial effects on autonomic function, together with increased production of vasodilatory prostaglandin derivatives, may result in blood pressure lowering, particularly with high-dose LC n-3 PUFA supplementation, although evidence from randomised controlled trials that fish consumption or LC n-3 PUFA supplementation lowers blood pressure suggests that high doses in groups with hypertension are required for a measurable effect, with some studies suggesting that DHA is more effective than EPA. Improvements in vascular function may be responsible to some extent for the anti-hypertensive properties, with decreased arterial stiffness and improved endothelial function reported in some studies, but not all. Importantly, LC n-3 PUFAs are likely to have beneficial effects on inflammation and haemostasis through their role as precursors to eicosanoids that have anti-inflammatory, anti-platelet properties or result in fewer pro-inflammatory or pro-platelet aggregatory effects, thereby reducing the likelihood of thrombosis and improving plaque stability. LC n-3 PUFAs may also lower inflammatory burden by inhibiting expression of VCAMs and cytokines, and through the inflammation-resolving properties of their oxygenated metabolites, specialised pro-resolving mediators including E- and D-resolvins, protectins and maresins (Innes & Calder 2020).

Meat

Meat falls into three distinct groups: red meat (e.g. beef, pork and lamb), processed meat (e.g. ham, sausages, bacon, salami; defined by the World Health Organization as meat that has been transformed through salting, curing, fermentation, smoking or other processes to enhance flavour or improve preservation), and poultry (e.g. chicken and turkey). Until recently, any association between meat and CHD was considered as reflecting the fat content of the meat, but this now seems to be more complex.

A summary of cohort studies concluded that each additional 100 g/day of red meat is associated with an increased risk of CHD of 15%, while each extra 50 g/day of processed meat is associated with an increased risk of 27% (Bechthold et al. 2019). Therefore, intake of processed meat is more strongly associated with cardiovascular health than that of red meat. However, the evidence for causality for red and processed meat has been judged as of low or very low certainty, due to the fact that the studies conducted are largely observational and therefore prone to residual confounding.

The composition of processed meat is likely to contribute to this stronger observed positive association, in that processed meat products have high amounts of saturated fat, but also preservatives (used during meat processing) and sodium and these may impact blood pressure and lipid profile, two key risk factors for CVD.

Studies on poultry and its association with CVD are inconsistent, but it is likely that this type of meat has little or no association with CVD risk, at least compared with other meat types.

Plant-based protein foods are increasingly consumed, as plant-based diets are growing in popularity based on concerns regarding climate change and sustainable diets. Human consumption of plant-based protein products such as tofu, tempeh and seitan dates back to ancient times in different parts of the world but, increasingly, these sources are being supplemented by commercial production of plant-based meat analogues which, to varying degrees, aim to approximate the organoleptic properties of traditional meat; the relative health benefits of these different plant proteins is as yet less well understood. However, a shift from animal to plant proteins is likely to reduce levels of total and LDL cholesterol and, consequently, CVD risk.

In the context of a healthy diet pattern, moderate intake of lean cuts of red meat can be a good source of protein, iron and other micronutrients; randomised controlled trial evidence has demonstrated that lean and unprocessed red meat intake in the context of a Mediterranean diet did not increase cardiovascular risk and improved 10-year cardiovascular risk score. Most European dietary guidelines recommend consuming lean meat in moderation.

In the UK, the mean intake of both red and processed meat is around 56 g/day, which is within the recommended levels; however, these numbers do include non-consumers of meat, and therefore intake of consumers is likely to be higher. Current recommendations also suggest consideration of the whole diet, limiting the intake of red meat while having other protein sources in the diet such as fish, poultry and plant-based proteins, consumption of which would help to reduce environmental impacts. When moderately consuming red meat, it is recommended to choose lean cuts and to try to use healthier cooking options. Considering the current evidence, it is recommended to limit intake of processed meat.

Salt

There is a well-established link between salt and high blood pressure, or hypertension. As hypertension is a major risk factor for CVD, a reduction in salt intake is strongly recommended across all major dietary guidelines. Independent of its blood pressure-raising effect, studies in normotensive experimental animals and human subjects have revealed that dietary salt also reduces vascular nitric oxide (NO) bioavailability, which limits endothelium-dependent dilation. A meta-analysis of cohort studies that included 617 000 people showed that, compared with individuals with low sodium intake, those with a relatively high intake had a 19% greater incidence of CVD (Wang et al. 2020). Moreover, there is a dose-dependent association: the risk of CVD increased up to 6% for every 1 g increase in dietary sodium intake (Wang et al. 2020). Consistent with these observations, randomised controlled trials have shown a reduced risk of CVD when subjects with elevated blood pressure were placed on a diet with a reduced salt content. High-quality self-reported sodium intake can be challenging to collect, but studies examining sodium excretion in urine support the findings of the dietary intake studies.

When source of salt in dietary intake is considered, it is not only salt added during cooking and at the table that should be taken into account. Most salt intake in North America and Europe comes from processed, restaurant and fast foods, which tends to go unnoticed by the population and is more difficult to consciously reduce for individuals. Ultra-processed foods are recently receiving attention; these are formulations of low-cost ingredients, many for non-culinary use, that result from a sequence of industrial processes, although definitions and examples of such foods do vary. These foods tend to have higher amounts of salt, and also SFAs, and sugar, than minimally processed foods. These nutritional features suggest they are likely to increase the risk of CVD. Nutritional intervention for CVD prevention should thus include education on food labels, encourage the reduction of intake of processed foods high in salt and promote replacement of these with unprocessed or minimally processed foods, in addition to control of the use of table salt and addition of salt during cooking.

Sugars

Carbohydrate intake is a key component of a balanced diet, but there is no dietary requirement for sugars. UK recommendations are to lower intakes of free sugars to below 5% total energy and minimise consumption of sugar-sweetened beverages (SSBs). This level of intake is exceeded by the vast majority of the UK, European and USA populations, probably due to the desirability of sugary foods and drinks, which can result in their overconsumption, particularly in foods that are also rich in fats. Sugar intake has not been independently associated with risk of CVD, CHD, stroke or type 2 diabetes, but it has been associated with higher BMI, particularly when consumed as a SSB (SACN 2015). Furthermore, a positive relationship between SSBs and risk from type 2 diabetes has been identified. High intake of free sugars (>20% dietary energy), particularly fructose, has been linked to metabolic abnormalities characteristic of the metabolic syndrome and type 2 diabetes, key risk factors for CVD. These include non-alcoholic fatty liver disease (>5% of hepatic lipids in the absence of a high intake of alcohol) and elevated serum triacylglycerol, which is often associated with lower HDL-C concentrations. These effects are proposed to be either through the direct lipogenic effects of sugars on liver fat and VLDL metabolism, and/or via the indirect effects of the energy from sugars on body weight over a longer time period.

Cereals, wholegrains and prebiotics

According to the UK Government's Eatwell Guide (https://www.gov.uk/government/publications/the-eatwell-guide), cereals and starches should contribute 37% of the total food intake and make the greatest contribution to dietary carbohydrates. In relation to disease risk, total carbohydrate consumption has little relation to cardiometabolic risk, although the quality of carbohydrate-rich foods is related, with wholegrains and low glycaemic index foods associated with lower risk. Wholegrains are complex foods and consists of endosperm (starch), bran (fibre, protein, B vitamins, minerals, flavonoids, tocopherols) and germ (protein, fatty acids, phytochemicals) and they can be intact (eg, quinoa) or partially intact (eg, rolled oats, stone-ground bread). The bran protects the starchy endosperm from digestion, resulting in reduced glycaemic responses, although this is less so in finely milled wholegrain products (e.g. most wholegrain breads, breakfast cereals). Refined grains (including white bread, crackers) consist of mainly starch with high glycaemic response and contains little fibre, minerals or other nutrients.

Data from the 2019 Global Burden of Disease study indicates that dietary risks, primarily those low in wholegrains, accounted for the largest proportion of ischaemic heart disease burden in low- and middle-income countries (Figure 11.5; GBD 2017 Diet Collaborators 2019). Further evidence from meta-analyses of prospective cohort studies has shown that higher intakes of wholegrains are associated with a reduction in risk of CHD mortality [relative risk (RR) = 0.66, 95% confidence interval (CI) 0.56–0.77] and incidence (RR = 0.80, 95% CI 0.7–0.91), stroke mortality (RR = 0.74, 95% CI 0.58–0.94) and type 2 diabetes (RR = 0.67, 95% CI 0.5–0.78) compared with lower intakes, although evidence is graded as low, due to the probability of residual confounding (Reynolds et al. 2019). This is supported by results from randomised controlled trials indicating that adults with or without CVD have significantly improved CVD risk markers, including total cholesterol, LDL cholesterol, haemoglobin A1c (longer-term marker of glycaemic control), CRP and BMI when consuming wholegrains rather than refined grains.

There are many components of wholegrains that could contribute to the observed reduction in CVD risk. Total dietary fibre, and more strongly cereal fibres, are associated with lower risk of CVD, coronary events and stroke with the evidence graded at moderate to adequate. In 2015 UK dietary recommendations for adults increased to 30 g/day of dietary fibre (measured by the Association of Analytical Chemists). Prebiotics are specific fibres defined as a selectively fermented ingredient that results in specific changes, in the composition and/or activity of the gastrointestinal microbiome, thus conferring benefit(s) upon host health. There is no evidence for associations between prebiotics and CVD disease, although some intervention studies illustrate benefits of certain prebiotic supplements on lipid profiles, including total, LDL and HDL cholesterol and triacylglycerol.

Fruits and vegetables

Fruits and vegetables are a major food group; diets rich in fruit and vegetables have been linked with a reduced risk of chronic disease, whereas low fruit and vegetable consumption has been linked to poor health. Fruit and vegetables are complex foods: they have low energy density and are micronutrient- (e.g. folate, potassium and magnesium) and fibre-rich, as well as containing a range of beneficial non-nutrient components, including plant sterols, flavonoids and other bioactives. Consuming a variety of fruit and vegetables will help to ensure an adequate intake of many of these essential nutrients. Fruits and vegetables are, therefore, recommended across all dietary guidelines. The World Health Organization suggests consuming >400 g/day of fruits and vegetables to improve overall health and reduce the risk of certain non-communicable diseases, and this has been widely translated as five portions of fruit and vegetables per day, although the exact number of portions recommended and portion size descriptions vary between countries. Currently, the UK population is not meeting the five-a-day target, with an average intake of around four portions of fruit and vegetables per day.

Cohort studies have consistently shown that higher consumption of fruit and vegetables has a strong protective association with risk of CHD. A recent meta-analysis found a lower risk of CHD incidence associated with the highest versus the lowest intakes of fruits and vegetables (RR = 0.88, 95% CI 0.83–0.92), fruits alone (RR = 0.88, 95% CI 0.84–0.92), and vegetables alone (RR = 0.92, 95% CI 0.87–0.96), especially green leafy vegetables (RR = 0.82, 95% CI 0.76–0.89). There was a consistent linear dose–response association between fruits and vegetables and CHD, with a maximum daily intake of seven fruit and seven vegetable servings showing a risk reduction of ≈20% and 30% in CHD incidence and mortality, respectively (Zurbau et al. 2020).

Fruits and vegetables are suggested to improve cardiovascular health via a range of possible mechanisms, including their antioxidant, anti-inflammatory, anti-platelet, blood pressure-lowering, lipid and glucose-altering properties, as well as having positive effects on endothelial function. Although usually grouped together within the dietary guidelines, fruits and vegetables have their own nutrient profile, which varies depending on the part of the plant they represent. For example, the stronger association observed for dark green leafy vegetables is likely to be due to their rich dietary nitrate content.

The edible dried seeds of legumes, such as chickpeas, lentils, beans and peas (also called pulses), are included in UK dietary guidelines under the fruit and vegetable group, although they are also protein-rich, and therefore may be a good replacement for meat intake. A recent umbrella review compared low versus high intake of dietary pulses and found a significant inverse association with CVD and hypertension. The benefits they offer may be explained by their high amounts of fibre, plant protein, micronutrients and phytochemicals, while also being low in fat and glycaemic index.

Although the evidence linking increased fruit and vegetable intake with reduced risk of chronic disease is relatively strong, it is difficult to measure fruit and vegetable intake accurately, and there are some uncertainties regarding the importance of variety, the optimum number of portions and whether different types of fruit and vegetables have different health effects. The effect of storage, processing and cooking methods on nutrient content and therefore health benefits is also uncertain. Regardless, populations worldwide are not meeting current guidelines.

Nuts and seeds

Dietary patterns characterised by higher intakes of nuts and seeds are associated with reduced risk of CVD. Tree nuts can differ from each other markedly in their nutrient profiles. For example, walnuts are notable for their high alpha-linolenic acid content, almonds contain a lot of vitamin E, and brazil nuts are an extremely rich source of selenium. Greater satiety, features of a healthy microbiome and reduced blood LDL cholesterol and triacylglycerol concentrations have been attributed to the favourable macronutrient profile of tree nuts, which are rich in fibre, protein and unsaturated fatty acids. Their relatively high micronutrient and phytochemical density may also act in combination with physico-chemical properties of nuts to further lower lipids (e.g. phytosterols) and blood pressure (e.g. folate, riboflavin and phenolic compounds), and modulate inflammatory and oxidative pathways (e.g. selenium, vitamin E, proanthocyanidins, flavonoids and phenolic acids).

The Prevención con Dieta Mediterránea (PREDIMED) dietary intervention trial investigated effects of a Mediterranean diet (also see Section 11.4.13) with either added almonds, walnuts and hazelnuts or added extra virgin olive oil, compared with a low-fat control diet, which resulted in a 30% reduction in incidence of CVD over 5.5 years following both Mediterranean diets (Estruch et al. 2018), but interestingly a small sub-study showed that only the arm featuring Mediterranean diet plus nuts slowed progression of atherosclerosis (Sala-Vila et al. 2014). The Portfolio diet, which includes 50 g/day nuts plus other plant foods, has been shown to markedly lower LDL cholesterol, triacylglycerol, apoB, blood pressure, a marker of inflammation (CRP), and estimated 10-year CHD risk calculated from an algorithm, even though body weight was not significantly affected (Chiavaroli et al. 2018). Although nuts are relatively energy-dense, due to their high-fat content, they do not contribute to weight gain. In vitro and in vivo experiments have demonstrated that this was due to low macronutrient bioaccessibility from nuts as well as being a consequence of increased satiety; fat is located within plant cell walls that are inefficiently broken down during digestion, resulting in reductions of as much as 20–25% of available energy in comparison to what would be expected using standard equations for calculation of metabolisable energy. Prospective cohort studies show significant inverse associations between nut consumption and risk of both CHD and CVD mortality. These reductions in risk are unlikely to be solely attributable to lipid-lowering, but randomised controlled trials have also demonstrated other beneficial cardiovascular responses to nut consumption, including improved endothelial function.

Alcohol

Heavy alcohol consumption is an important cause of death and disability, but the association between moderate drinking and CVD is complex. CVD itself is complex and the patterns of association with alcohol intake may differ; observational studies generally show that alcohol consumption is positively associated with risk of atrial fibrillation, heart failure and haemorrhagic stroke, whereas moderate drinking is associated with lower risk of CHD and ischaemic stroke. Data relating alcohol consumption to other types of CVD, including venous thromboembolism, peripheral artery disease, aortic valve stenosis and abdominal aortic aneurysm are limited or inconsistent. As stated for other dietary factors, observational studies are unable to fully account for confounding and reverse causation bias, and therefore causality in the associations of alcohol consumption with different CVDs remains uncertain. Furthermore, and again this is common to other dietary risk factors, but is perhaps particularly true for alcohol intake, self-reported alcohol consumption may be underestimated, leading to measurement error in the assessment of alcohol consumption, which may result in altered and likely attenuated risk estimates.

Based on the totality of the evidence, however, which mirrors dietary guidelines, the lowest risk of CHD, and probably CVD in general, is seen at an alcohol intake of about one to two drinks per day (equivalent to approximately 8–16 g of pure alcohol). As alcohol intake increases to above four drinks/day, so does the risk of CVD, especially stroke and heart failure. The pattern of consumption is also important, with the benefits of moderate consumption being seen with consumption across multiple days per week, not with high levels on a few days. For that reason, guidelines also suggest the avoidance of binge drinking (defined as more than five standard drinks in a single sitting) and to have several drink-free days each week.

There has been much speculation that certain types of alcohol may offer more protection against CVD when consumed moderately than others, for example that red wine is more potent than other forms of alcoholic beverages. However, the epidemiological evidence currently indicates that all forms of alcoholic beverages – beer, spirits, and wine, both red and white – have similar patterns of association.

Likely mechanisms by which alcohol may mediate its effects on CVD outcomes, is also complex. Moderate alcohol consumption increases HDL cholesterol, insulin sensitivity and adiponectin levels while decreasing inflammation, all of which have positive effects on the risk for CVD, and may also affect haemostatic factors. Heavier consumption has serious physiological effects, including mitochondrial dysfunction and changes in circulation, inflammatory responses, oxidative stress and programmed cell death, as well as anatomical damage to the cardiovascular system, including the heart itself.

Tea and coffee

The role of non-alcoholic beverages in influencing CVD risk has been largely overlooked until relatively recently. However, tea and coffee are widely consumed around the world and may make a significant contribution to the total impact of diets on risk of CVD. For instance, tea and coffee were the main sources of dietary polyphenols in the UK, contributing 60–70% of total intakes in adults aged 19–64 years, with tea being the main source of flavonoids. Flavonoid-rich foods (tea, cocoa, chocolate, apples, berries) have been shown to reduce blood pressure and improve endothelial function, lower lipids and have anti-platelet and anti-inflammatory effects in preclinical and human dietary intervention studies. Epidemiological studies have reported associations between moderate flavonoid intakes and risk of CVD mortality, and a meta-analysis of prospective cohort studies revealed a 4% lower risk of CVD mortality associated with each additional cup of tea (Chung et al. 2020). Although the evidence on tea and CVD has not been widely translated to public health advice, the 2015 Dutch dietary guidelines included a recommendation to drink three cups of tea (black or green) daily, based on evidence that tea consumption reduced risk of stroke and reduced blood pressure, as well as being associated with reduced risk of type 2 diabetes (Kromhout et al. 2016).

Coffee is also a significant source of dietary polyphenols, although unfiltered coffee has been shown to increase LDL cholesterol and triacylglycerols due to the diterpenes, kahweol and cafestol, that are extracted from the coffee bean. Paradoxically, these diterpenes also inhibit inflammation in activated macrophages and animal models, and coffee consumption is associated with reduced risk of CVD in observational studies. As well as diterpenes, coffee is rich in polyphenols called chlorogenic acids (which inhibit oxidative stress, platelet aggregation and vascular inflammation), and contains a range of other bioactive phytochemicals, including caffeine, unless decaffeinated. Although tea is a significant source of caffeine, coffee contains the greatest amount, and high intakes could increase CVD risk by impairing sleep quality, increasing anxiety and acute blood pressure-raising effects. In the absence of causal evidence, it remains unclear whether coffee consumption is cardioprotective, harmful or has no net effect on risk of CVD (Nordestgaard 2021).

Probiotics

Probiotics are non-pathogenic live microbes, which, when ingested, beneficially affect the host by improving the intestinal microbial balance and its properties. There are many diverse probiotics, with lactobacilli and bifidobacteria being common examples. Probiotics are found in fermented, live or 'bio' foods, such as some yoghurts or cheeses, and they can be added to foods or taken as supplements. To exert their effects, probiotics need to be able to reach and persist in a viable state in the lower gut and colon. Intervention studies in those with high, borderline or normal cholesterol concentrations suggest hypocholesterolaemic effects of a daily consumption of some probiotic-containing products. Potential mechanisms include the hepatic effects of some short-chain fatty acids (produced from fibre fermentation by probiotic bacteria) on endogenous cholesterol metabolism. Some probiotics (via bile salt hydrolysation) can convert primary to secondary bile acids, which are poorly absorbed, leading to increased net faecal cholesterol excretion and subsequent hepatic LDL-C removal from the circulation. However, reported hypocholesterolaemia effects of probiotics are inconsistent, particularly in normocholesterolaemic groups, and mechanisms need confirmation.

Whole dietary patterns

Historically, in the field of nutritional epidemiology, research was structured to analyse the effects of individual nutrients, foods or food groups on health. However, food consumption leads to the intake of a complex combination of macronutrients, micronutrients and other bioactive components, which also have synergistic effects. Moreover, changing the dietary intake of one specific food or food group may affect the ingestion of many dietary components, directly or indirectly (through substitution of one food for another) (Cespedes and Hu 2015).

Considering that people do not consume isolated dietary components and that focusing on single nutrients or specific foods may not reflect the full complexity of the association between diet and diseases, a change of focus away from nutrients towards the examination of whole foods, food groups and dietary patterns has been suggested when conducting research and advising individuals on dietary changes to improve cardiovascular outcomes.

A number of whole dietary patterns have been extensively studied in terms of their links with CVD risk. The Mediterranean diet is the diet pattern with the strongest evidence supporting a positive effect on cardiovascular health. The diet varies from one country to another around the Mediterranean sea, but the traditional pattern is typical of many regions in Greece and southern Italy in the early 1960s. Typical features include a high content of plant foods (i.e. fruit, vegetables, cereals, legumes, nuts and seeds); moderate amounts of fish, poultry, milk and other dairy products; and small amounts of red and processed meat. The diet includes olive oil as the main fat in food preparation; and low-to-moderate alcohol consumption (especially red wine, consumed mainly at meals). Meta-analyses of cohort studies have consistently reported significant inverse associations between adherence to the Mediterranean diet and CHD risk (Rosato et al. 2019), with a 19% reduction in the relative risk of CVD for the highest versus the lowest category of the Mediterranean Diet Score, an indicator of adherence to the Mediterranean diet. Furthermore, the PREDIMED study, a multicentre randomised controlled trial, found that, compared with the control diet, the Mediterranean diet with extra-virgin olive oil or with nuts reduced the incidence rates of major cardiovascular events by 31% and 28%, respectively (Estruch et al. 2018). These findings suggest a protective effect of the Mediterranean diet on the risk of CVD, and the Lyon Diet Heart Study showed similar effects in an earlier secondary prevention trial.

How these dietary patterns protect the cardiovascular system is not fully understood, but those who follow the diet more closely seem to have a healthier lipid profile, lower blood pressure, lower insulin resistance and less inflammation. Emerging evidence also suggests that eating a Mediterranean diet may have additional benefits in relation to overall longevity and other chronic diseases, such as diabetes, cancer at certain sites and Alzheimer's disease. The health benefits offered by a Mediterranean diet appear to be attributable to interactions between different food components rather than the effects of single nutrients.

Thus the Mediterranean diet has been proposed as an alternative and palatable lifestyle change that is beneficial to health. What is not yet clear is whether non-Mediterranean populations can adopt and maintain dietary behaviours consistent with a traditional Mediterranean diet, because of the required changes in foods consumed, eating patterns and food culture. There are also concerns that Mediterranean populations are changing their dietary habits towards a Western-style diet.

Another dietary pattern that has been explored is the Dietary Approaches to Stop Hypertension (DASH) diet. This was originally developed with the goal of lowering elevated blood pressure, but studies have now shown other benefits for cardiovascular health. The DASH diet has a similar pattern to the Mediterranean diet in terms of food groups, but with extra attention to the intake of salt, added sugar, wholegrains, fat-free/low-fat dairy and dietary cholesterol. It promotes the use of vegetable oils rather than SFAs, but not necessarily olive oil. The DASH dietary pattern was associated with decreased incident CVD (RR = 0.80, 95% CI 0.76–0.85), CHD (RR = 0.79, 95% CI 0.71–0.88), stroke (RR = 0.81, 95% CI 0.72–0.92) and type 2 diabetes (RR = 0.82, 95% CI 0.74–0.92) in the analysis of prospective cohort studies (Chiavaroli et al. 2019).

Many plant-based diet patterns look promising in terms of offering a protective role against CVD. Meta-analyses of trials with lipid outcomes have demonstrated a decrease in total and LDL cholesterol, and also in HDL cholesterol, but no effect on triacylglycerol. Historically those following a vegetarian diet had lower CVD morbidity and mortality than those who ate followed traditional dietary patterns, but these cohorts are now old and do not reflect modern eating habits and food availability. Given the recent increasing adoption of plant-based foods and availability of plant-based meat alternatives, more studies are necessary to confirm the benefits of modern plant-based dietary patterns (Jafari 2021).

11.5 Future perspectives

Although much is known about the dietary factors that can affect CVD risk factors and disease outcomes (via a range of mechanisms), there are still a number of uncertainties, including areas where novel approaches and future research can contribute. Progressing our understanding is critical, given the demographic shifts now seen in our population towards an ageing population with increased non-communicable disease burden, with the accompanying healthcare and societal burden that this will bring, including through CVD morbidity.

Although we know that improving diet is likely to reduce CVD risk in both primary and secondary settings, achieving effective behaviour change in the long term is challenging. Personalised nutrition, a research field that leverages human individuality to drive nutrition strategies that prevent, manage and treat disease as well as optimising health, offers potential avenues to promote more effective behaviour change or reduction in CVD risk. Advice can be personalised based either on biological evidence of differential responses to foods/nutrients dependent on genotypic or phenotypic characteristics (including nutritional status) or on current behaviour, preferences, barriers and objectives that motivate and enable individuals to make appropriate changes to their eating patterns. Behavioural change is important in terms of non-communicable disease outcomes, but recently dietary changes are becoming increasingly recognised as vital to reduce the impact on planetary health and it is accepted that a diet that promotes heart health is also likely to be more sustainable.

While achieving behavioural change towards what we already know is a 'heart-healthy' diet is important, there is also still much to learn regarding nutrients, foods, dietary patterns and their effects on metabolism, CVD risk factors and hard clinical cardiovascular outcomes. Systems biology is an emerging approach that attempts to address the problems posed by complex biological processes. It has already helped in making progress towards understanding CVD, but promises more. Much of the research described in this chapter is based on the consideration of the link between an individual's diet and their future CVD risk, yet the Barker hypothesis considers that maternal nutrition is also vitally important and that under-nutrition in middle-to-late gestation leads to disproportionate foetal growth and programmes increased risk of CHD later in life. Such a concept requires further study. Finally, as the pharmacological management of CVD continues to develop rapidly, the possibility of diet–pharmacological interactions must be studied to ensure that harmful effects of such interactions on disease outcomes are minimised.

References

Abdelhamid, A.S., Brown, T.J., Brainard, J.S., Biswas, P., Thorpe, G.C., Moore, H.J., Deane, K.H., Summerbell, C.D., Worthington, H.V., Song, F., and Hooper, L. (2020). Omega-3 fatty acids for the primary and secondary prevention of cardiovascular disease. *The Cochrane Database Syst Rev* 3 (2): CD003177. doi: 10.1002/14651858.CD003177.pub5.

Abraham, T.M., Pedley, A., Massaro, J.M., Hoffmann, U., and Fox, C.S. (2015 27). Association between visceral and subcutaneous adipose depots and incident cardiovascular disease risk factors. *Circulation* 132 (17): 1639–1647.

Bechthold, A., Boeing, H., Schwedhelm, C. et al. (2019). Food groups and risk of coronary heart disease, stroke and heart failure: a systematic review and dose-response meta-analysis of prospective studies. *Crit Rev Food Sci Nutr* 59: 1071–1090.

Cespedes, E.M. and Hu, F.B. (2015). Dietary patterns: from nutritional epidemiologic analysis to national guidelines. *Am J Clin Nutr* 101 (5): 899–900.

Chiavaroli, L., Nishi, S.K., Khan, T.A., Braunstein, C.R., Glenn, A.J., Mejia, S.B., Rahelić, D., Kahleová, H., Salas-Salvadó, J., Jenkins, D.J.A., Kendall, C.W.C., and Sievenpiper, J.L. (2018). Portfolio dietary pattern and cardiovascular disease: a systematic review and meta-analysis of controlled trials. *Prog Cardiovasc Dis* 61 (1): 43–53.

Chiavaroli, L., Viguiliouk, E., Nishi, S.K., Blanco Mejia, S., Rahelić, D., Kahleová, H., Salas-Salvadó, J., Kendall, C.W., and Sievenpiper, J.L. (2019). DASH dietary pattern and cardiometabolic outcomes: an umbrella review of systematic reviews and meta-analyses. *Nutrients* 11 (2): 338.

Chung, M., Zhao, N., Wang, D., Shams-White, M., Karlsen, M., Cassidy, A., Ferruzzi, M., Jacques, P.F., Johnson, E.J., and Wallace, T.C. (2020). Dose-response relation between tea consumption and risk of cardiovascular disease and all-cause mortality: a systematic review and meta-analysis of population-based studies. *Adv Nutr* 11 (4): 790–814.

Estruch, R., Ros, E., Salas-Salvadó, J., Covas, M.I., Corella, D., Arós, F., Gómez-Gracia, E., Ruiz-Gutiérrez, V., Fiol, M., Lapetra, J., Lamuela-Raventos, R.M., Serra-Majem, L., Pintó, X., Basora, J., Muñoz, M.A., Sorlí, J.V., Martínez, J.A., Fitó, M., Gea, A., Hernán, M.A., and Martínez-González, M.A. (2018). PREDIMED study investigators. Primary prevention of cardiovascular disease with a Mediterranean diet supplemented with extra-virgin olive oil or nuts. *N Engl J Med* 378 (25): e34.

Ference, B.A., Ginsberg, H.N., Graham, I., Ray, K.K., Packard, C.J., Bruckert, E., Hegele, R.A., Krauss, R.M., Raal, F.J., Schunkert, H., Watts, G.F., Borén, J., Fazio, S., Horton, J.D., Masana, L., Nicholls, S.J., Nordestgaard, B.G., van de Sluis, B., Taskinen, M.R., Tokgözoglu, L., Landmesser, U., Laufs, U., Wiklund, O., Stock, J.K., Chapman, M.J., and Catapano, A.L. (2017). Low-density lipoproteins cause atherosclerotic cardiovascular disease. 1. Evidence from genetic, epidemiologic, and clinical studies. A consensus statement from the European atherosclerosis society consensus panel. *Eur Heart J* 38: 2459–2472.

GBD 2017 Diet collaborators (2019). Health effects of dietary risks in 195 countries, 1990-2017: a systematic analysis for the global burden of disease study 2017. *Lancet* 393 (10184): 1958–1972.

Geiker, N.R.W., Larsen, M.L., Dyberg, J., Stender, S., and Atrup, A. (2017). Egg consumption, cardiovascular diseases and type 2 diabetes. *Eur J Clin Nutr* 72 (1): 44–56. doi: 10.1038/ejcn.2017.153. Epub 2017 Sep 27. PMID: 28952608.

Guo, J., Astrup, A., Lovegrove, J.A., Gijsbers, L., Givens, D.I., and Soedamah-Muthu, S.S. (2017). Milk and dairy consumption and risk of cardiovascular diseases and all-cause mortality: dose-response meta-analysis of prospective cohort studies. *Eur J Epidemiol* 32 (4): 269–287.

Innes, J.K. and Calder, P.C. (2020). Marine Omega-3 (N-3) fatty acids for cardiovascular health: an update for 2020. *Int J Mol Sci* 21 (4): 1362.

Jafari, S., Hezaveh, E., Jalilpiran, Y., Jayedi, A., Wong, A., Safaiyan, A., and Barzegar, A. (2021 May). Plant-based diets and risk of disease mortality: a systematic review and meta-analysis of cohort studies. *Crit Rev Food Sci Nutr* 6: 1–13.

Kromhout, D., Spaaij, C.J.K., de Goede, J., and Weggemans, R.M. for the committee dutch dietary guidelines 2015 (2016). The 2015 Dutch food-based dietary guidelines. *Eur J Clin Nutr* 70 (8): 869–878.

Mensink, R.P. (2016). Effects of Saturated Fatty Acids on Serum Lipids and Lipoproteins: A Systematic Review and Regression Analysis. Geneva: WHO.

Nordestgaard, A.T. (2021 July 28). Causal relationship from coffee consumption to diseases and mortality: a review of observational and Mendelian randomization studies including cardiometabolic diseases, cancer, gallstones and other diseases. *Eur J Nutr* doi: 10.1007/s00394-021-02650-9. Epub ahead of print.

Reidlinger, D.P., Darzi, J., Hall, W.L., Seed, P.T., Chowienczyk, P.J., and Sanders, T.A.; (2015). Cardiovascular disease risk REduction study (CRESSIDA) investigators. How effective are current dietary guidelines for cardiovascular disease prevention in healthy middle-aged and older men and women? A randomized controlled trial. *Am J Clin Nutr* 101 (5): 922–930.

Reynolds, A., Mann, J., Cummings, J., Winter, N., Mete, E., and Te Morenga, T. (2019). Carbohydrate quality and human health: a series of systematic reviews and meta-analyses. *Lancet* 393: 434–445.

Rosato, V., Temple, N.J., La Vecchia, C., Castellan, G., Tavani, A., and Guercio, V. (2019). Mediterranean diet and cardiovascular disease: a systematic review and meta-analysis of observational studies. *Eur J Nutr* 58: 173–191.

Ross, R. (1999). Atherosclerosis–an inflammatory disease. *N Engl J Med* 340: 115–126.

Sala-Vila, A., Romero-Mamani, E.S., Gilabert, R., Núñez, I., de la Torre, R., Corella, D., Ruiz-Gutiérrez, V., López-Sabater, M.C., Pintó, X., Rekondo, J., Má, M.-G., Estruch, R., and Ros, E. (2014). Changes in ultrasound-assessed carotid intima-media thickness and plaque with a Mediterranean diet: a substudy of the PREDIMED trial. *Arterioscler Thromb Vasc Biol* 34 (2): 439–445.

Scientific Advisory Committee on Nutrition (2015 July). SACN report on carbohydrates and health [Internet]. Available from: https://www.gov.uk/government/publications/sacn-carbohydrates-and-health-report (accessed 25 August 2023).

Scientific Advisory Committee on Nutrition (2019 July). SACN report on saturated fats and health. [Internet] Available from: https://www.gov.uk/government/publications/saturated-fats-and-health-sacn-report (accessed 25 August 2023).

Shai, I., Spence, J.D., Schwarzfuchs, D., Henkin, Y., Parraga, G., Rudich, A., Fenster, A., Mallett, C., Liel-Cohen, N., Tirosh, A., Bolotin, A., Thiery, J., Fiedler, G.M., Blüher, M., Stumvoll, M., and Stampfer, M.J. DIRECT Group (2010). Dietary intervention to reverse carotid atherosclerosis. *Circulation* 121 (10): 1200–1208.

Soehnlein, O. and Libby, P. (2021). Targeting inflammation in atherosclerosis - from experimental insights to the clinic. *Nat Rev Drug Dis* 20 (8): 589–610.

Wang, Y.J., Yeh, T.L., Shih, M.C., Tu, Y.K., and Chien, K. (2020). Dietary sodium intake and risk of cardiovascular disease: a systematic review and dose-response meta-analysis. *Nutrients* 12: 2934.

World Heart Federation (2021). *Cardiovascular Diseases – Global Facts and Figures*. Available from: https://world-heart-federation.org/resource/cardiovascular-diseases-cvds-global-facts-figures.

Zurbau, A., Au-Yeung, F., Blanco Mejia, S. et al. (2020). Relation of different fruit and vegetable sources with incident cardiovascular outcomes: a systematic review and meta-analysis of prospective cohort studies. *J Am Heart Ass* 9: e017728.

Further reading

Mozaffarian, D. (2016). Dietary and policy priorities for cardiovascular disease, diabetes, and obesity: A comprehensive review. *Circulation* 133: 187–225. https://www.ahajournals.org/doi/10.1161/CIRCULATIONAHA.115.018585.

Riccardi, G., Giosuè, A., Calabrese, I., Vaccaro, O. (2022). Dietary recommendations for prevention of atherosclerosis. *Cardiovasc Res* 118: 1188–1204. doi: 10.1093/cvr/cvab173.

Stanner, S., Coe, S., and Frayn, K. (eds.) (2018). *Cardiovascular Disease: Diet, Nutrition and Emerging Risk Factors*, 2e. Wiley Blackwell.

12
The Skeletal System

John M. Pettifor, Ann Prentice, Kate A. Ward, and Ansuyah Magan

Key messages

- Bone provides important supportive and protective functions for the body, and its constituent cells are in close relationship with those of the bone marrow, from which they are derived.
- Bone is a dynamic tissue, which is continually being resorbed and replaced (the process of remodelling). The rate of remodelling is influenced by a number of different factors, including circulating hormones such as sex steroids and parathyroid hormone. Bone formation and resorption can be measured by various different biochemical assays, which may be of use in assessing bone and mineral metabolism.
- Bone plays an essential role as a calcium store in maintaining serum calcium homeostasis.
- Osteocytes, the mechanical strain sensing cells of bone, also play a vital role in controlling phosphate and vitamin D metabolism by secreting fibroblast growth factor 23 (FGF23).
- Bone mass can be measured in vivo by several techniques, including dual-energy X-ray absorptiometry (DXA) and computed tomography. Each method has its advantages and disadvantages, which may complicate the interpretation of the results.
- Peak bone mass, which is reached in early adulthood, is influenced by different factors, including heredity, sex, anthropometry, nutrition, hormonal status and lifestyle patterns. Peak bone mass may influence the prevalence of fragility fractures occurring in later life.
- Important metabolic bone diseases include osteomalacia/rickets and osteoporosis. The former is common in infants, young children and the elderly in a number of countries owing predominantly to vitamin D deficiency, while the latter is a severe problem in the ageing population of developed countries.
- With an increase in ageing populations across the globe, osteoporotic fractures and their treatment are significant healthcare burdens. Importantly, there remains a concerning treatment gap between those at high risk of fracture and those who actually receive treatment.
- Sarcopenia is a progressive age-related loss of skeletal muscle mass and strength impacting bone health through increased fall, and hence fragility fracture, risk. It shares many common risk factors with osteoporosis. Sarcopenia also contributes to bone loss through reduced loading.
- Several nutritional factors play permissive roles in ensuring optimal bone health. Among the most important of these are vitamin D, calcium and phosphorus, but many other nutrients and dietary factors, singly or in combination, may influence bone and mineral homeostasis.
- Dental diseases share modifiable risk factors, such as consumption of free sugars, excessive alcohol intake and smoking with other non-communicable diseases, highlighting the need for an interdisciplinary approach to health management.
- Dental caries and periodontitis are chronic and progressive disorders and may track through the life course. For example, caries experience in childhood can influence oral and general health through adolescence and into adulthood.

12.1 Introduction

Rickets/osteomalacia and osteoporosis are two diseases of bone which have major impacts on the state of health and quality of life of young and old in many countries of the world. Nutritional factors play important roles in determining the prevalence of these two diseases. This chapter sets the scene for the reader by providing an overview of the structure and physiology of bone, and of the factors that determine bone and tooth growth and development. The

Nutrition and Metabolism, Third Edition. Edited on behalf of The Nutrition Society by Helen M. Roche, Ian A. Macdonald, Annemie M.W.J. Schols and Susan A. Lanham-New.

Companion Website: www.wiley.com/go/nutrition/metabolism3e

physiological changes that occur during pregnancy and lactation and with ageing are discussed and the factors that may influence these changes are described. This chapter should be read in conjunction with other chapters in other books in this series.

12.2 Bone architecture and physiology

The skeletal system plays a number of important physiological roles, thus its integrity must be maintained for the normal function of the human body. These physiological functions include:

- *Support* for the body: in this role the skeleton is responsible for posture, for allowing normal joint movement and muscle activity through providing the levers on which muscles act, and for withstanding functional load bearing
- *Protection* of organs, such as the brain and lungs
- *Providing a reservoir* of calcium to maintain serum calcium homeostasis
- *Acting as a buffer* to maintain normal acid–base balance
- Through its close relationship with bone marrow, *maintaining a normal haemopoietic and immune system*
- *Controlling vitamin D and phosphorus homeostasis*, through the role of osteocytes.

Bone may be divided into compact (cortical) bone, which provides mainly the supporting and protective functions of bone and makes up approximately 85% of bone tissue, and trabecular, cancellous or spongy bone, which is composed of thin calcified trabeculae enclosing the haemopoietic bone marrow and adipose tissue and comprises only 15% of the skeleton.

Trabecular bone is considered to be physiologically more active than cortical bone because of its larger surface area, despite it forming only a small fraction of total mineralised bone. The internal trabecular structure of bones such as that found in the vertebral bodies and femoral neck plays an important supportive role, preventing collapse or fracture. Figure 12.1 schematically depicts a long bone such as the tibia and highlights the various anatomical regions of the bone.

Bone composition

Unlike most other tissues, which are composed mainly of cells, bone is mainly composed of an extracellular matrix. The cells that maintain this matrix are relatively sparse, being present only on the various surfaces of the calcified matrix and scattered within the matrix that makes up cortical and trabecular bone. Bone thus consists of a non-cellular calcified matrix and the cells that maintain this matrix. As shown in Table 12.1, mineral is the major constituent of bone by weight.

Bone matrix

The matrix is made up of both *collagenous* and *non-collagenous* proteins, with type 1 collagen making up 90% of total bone protein (Table 12.1). The fibrous nature of collagen provides elasticity and flexibility to bone as well as the scaffolding on which mineralisation can occur. The collagen fibres are oriented in directions influenced by the stresses and strains experienced by the developing bone through weight-bearing, and by the attachments of muscles, tendons and ligaments. Type 1 collagen is a triple-helical molecule containing two identical α_1 (I) chains and an α_2 (I) chain. These chains are rich in lysine and proline, which undergo post-translational modifications, including hydroxylation of lysyl and prolyl residues (which requires vitamin C), glycosylation of lysyl and hydroxylysyl residues, and the formation of intramolecular and intermolecular covalent cross-links (Figure 12.2). These post-translational modifications are important in ensuring the linkage of the collagen molecules into fibrils, thus increasing the strength of the collagen network. The collagen molecules are linked end to end and side to side in a staggered pattern, resulting in gaps between the molecules, where mineral deposition occurs. Measurement of these various collagen cross-links in serum and urine has been used successfully to assess the rate of bone formation and resorption in the clinical situation.

The non-collagenous proteins make up only approximately 10% of total bone protein and can be divided into three major groups:

- proteoglycans
- glycosylated proteins
- γ-carboxylated proteins.

The *proteoglycan* molecules are some of the most abundant in the body and are found in many

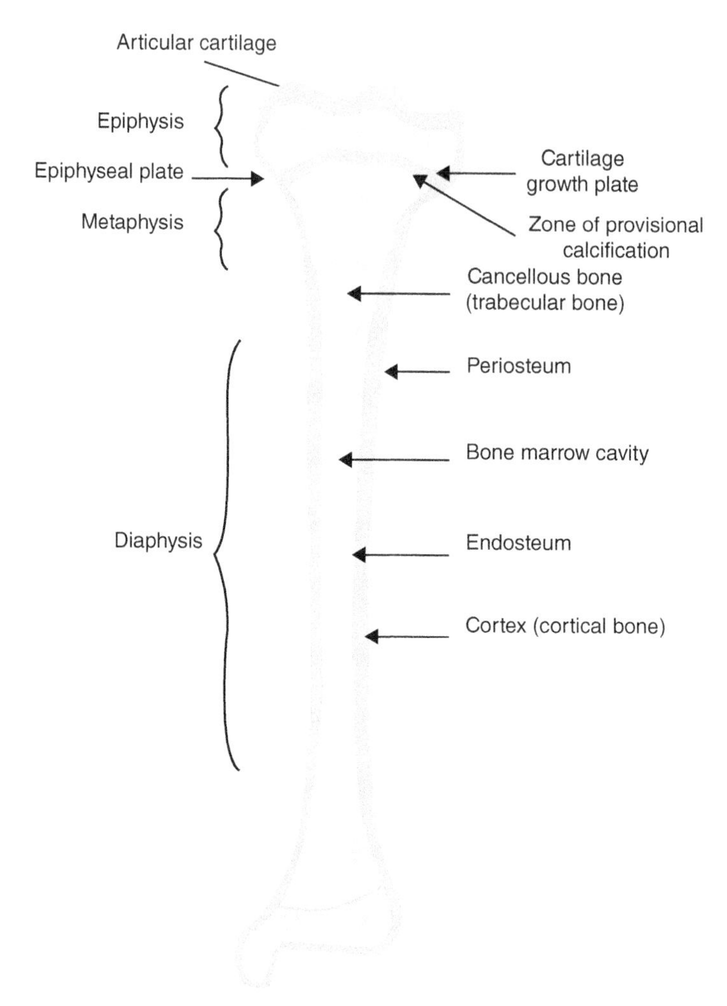

Figure 12.1 Schematic representation of a long bone with the important anatomical areas labelled.

Table 12.1 Composition of bone of an adult.

Skeletal weight	6.5–13.4 kg (height, sex and ethnicity dependent)
Mineral	60–70% dry weight
of which calcium	~ 1000 g (32% of mineral)
Organic matrix	20–30% (90% type 1 collagen, ~5% non-collagenous proteins, ~2% lipids by weight)
Water	5–10%
Lipids	3%

tissues including bone and cartilage. They may be important regulators of bone formation, but their physiological functions in bone have not been clearly elucidated. The *glycosylated proteins*, such as alkaline phosphatase, osteonectin, osteopontin, fibronectin and bone sialoprotein, probably play a number of different roles, which for many of the proteins have not been well established. They may be important in regulating matrix mineralisation, bone cell growth and proliferation, and in osteoblast differentiation and maturation. Alkaline phosphatase, which is a zinc-containing metalloenzyme, is an essential enzyme for the normal mineralisation of bone. Zinc deficiency is associated with low circulating levels of alkaline phosphatase. Inherited defects in the molecule result in the condition of hereditary hypophosphatasia, which in its most severe form may be lethal in infancy without treatment.

The *γ-carboxylated proteins* osteocalcin, matrix-Gla-protein and protein S are post-translationally modified by the action of vitamin K-dependent γ-carboxylases to form dicarboxylic glutamyl (Gla) residues, which enhance calcium binding. The actual role of these proteins in bone is to regulate properties of bone mineral. Osteocalcin is produced by osteoblasts and acts not only on bone but also as a hormone involved in glucose metabolism and potentially on

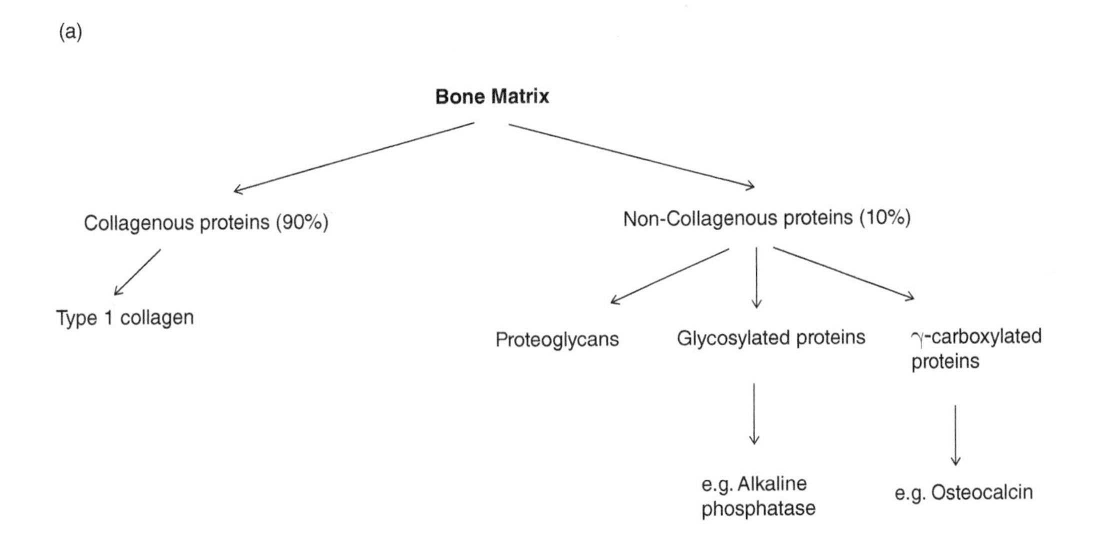

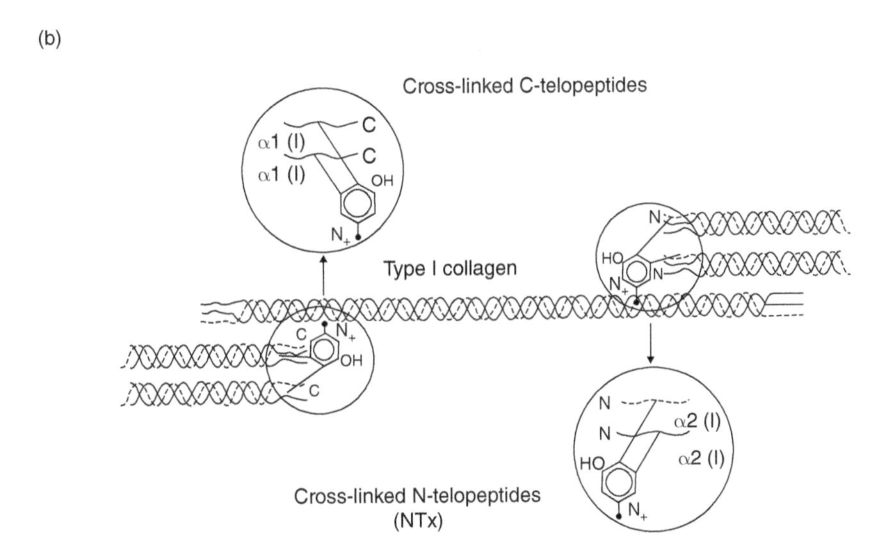

Figure 12.2 (a) The major components of bone matrix, produced by osteoblasts. Collagenous proteins make up about 90%. (b) Schematic diagram of type 1 collagen, highlighting its linkages with other fibres. *Source:* Reproduced with permission from Calvo MS et al. © 1996 Oxford University Press.

cognition and reproductive function. Clinical vitamin K deficiency reduces the number of carboxylated glutamic acid residues per molecule of osteocalcin.

The *mineral component* of bone is mainly in the form of hydroxyapatite [$Ca_{10}(PO_4)_6(OH)_2$], which provides stiffness and load-bearing strength to bone. The crystals of hydroxyapatite are approximately 200 Å in length, and form in and around the collagen fibrils, where they grow both by increasing in size and by aggregation. The mechanism by which mineralisation occurs is not well understood, although it is believed that extracellular matrix vesicles (produced by osteoblasts) initiate mineralisation by removing inhibitors of mineralisation, such as pyrophosphate and adenosine triphosphate (ATP) present in the matrix, and by increasing calcium and phosphate concentrations locally to allow crystallisation to occur. Key enzymes involved in phosphate production include alkaline phosphatase and bone-specific cytosolic phosphatase, Orphan 1 (PHOSPHO1). Both seem essential for normal mineralisation; an example is the severe mineralisation defect seen in children with hypophosphatasia (see earlier).

The hydroxyapatite crystal may take up dietary cations and anions into its lattice. Magnesium or strontium may replace calcium in the crystal lattice, resulting in smaller, less perfect crystals, while fluoride incorporation increases crystal size and decreases solubility. Bisphosphonates, a family of antiresorptive agents used in the treatment of osteoporosis and prevention of fragility fractures, bind to the surface of apatite crystals, preventing/reducing resorption. Tetracycline, an antibiotic, also binds avidly to newly formed apatite crystals, resulting in fluorescence under UV light of newly deposited bone and tooth mineral. This characteristic of tetracycline is used clinically to measure bone mineralisation rates and the extent of bone surface undergoing mineralisation in histological sections. If taken during the formation of teeth, tetracyclines result in staining of teeth through their incorporation in the mineralising enamel.

Bone cells

The important bone cells are *stromal osteoprogenitor cells*, *osteoblasts*, *osteocytes*, *lining cells* and *osteoclasts* and their precursors. However, it is becoming increasingly apparent that there is a close inter-relationship between the various haemopoietic cells and adipocytes in the bone marrow and bone cells, not only because precursors of bone cells may reside within the marrow but also because of the cross-talk between the various cell types.

Osteoprogenitor cells are found in the periosteum and bone marrow. These mesenchymal stem cells may develop through appropriate stimuli into two very different cell types: adipocytes (fat cells) or osteoblast precursors. Various growth factors, cytokines and hormones (including transforming growth factor-β_1 [TGF-β_1], fibroblast growth factors [FGFs], a number of bone morphogenetic proteins [BMPs] and parathyroid hormone [PTH]) are responsible for controlling the proliferation and differentiation of these mesenchymal cells into preosteoblasts, osteoblasts and osteocytes. Both PTH and the active form of vitamin D, 1,25-dihydroxyvitamin D [$1,25(OH)_2D$], are important in controlling the proliferation and differentiation of these bone-forming cells. The control of the development of these cells is complex and beyond the scope of this chapter; the reader is referred to a number of reviews for further information.

Osteoblasts are mesenchymal stem cells present on bone surfaces and are responsible for secretion of the organic constituents of the bone extracellular matrix and for the production of matrix vesicles, which probably initiate mineralisation of the preformed osteoid. During maturation of the osteoblast, the cell initially secretes collagen and is rich in alkaline phosphatase, but later produces osteocalcin and other matrix proteins, such as osteopontin. As the osteoblast becomes encircled by matrix, it transforms into an osteocyte. Osteoblasts also produce an osteoclastogenic protein, receptor activator of nuclear factor κ-β ligand (RANKL), which regulates bone resorption. Evidence also suggests that osteoblasts co-ordinate physiological processes in skeletal and extraskeletal tissues (Figure 12.3).

Osteocytes are known as the mechanosensing cells and constitute 90–95% of all bone cells within the adult. They have cytoplasmic extensions which lie in canaliculi throughout the bone matrix. These extensions connect with adjacent cells (osteocytes, osteoblasts and lining cells) and extend into the bone marrow. Osteocytes sense changes in strain within the matrix and signal to both osteoclasts and osteoblasts to stimulate bone resorption or formation. These mechanosensing cells are thus vital to the process of bone remodelling. The exact mechanism by which the osteocyte senses strain changes in vivo is not fully understood, but is thought to be related to changes in fluid flow shear stress at the cell membrane and/or through microvillous transduction of changes in pressure. Factors secreted by the osteocytes, such as RANKL, BMPs, FGF23 and sclerostin, have endocrine, paracrine and autocrine actions, such as controlling the renal handling of phosphate and osteoblast formation through the WNT signalling pathway. Antibodies of RANKL (denosumab) and sclerostin (romosuzamab) have been used as new pharmaceutical products to increase bone density and reduce fragility fractures (Figure 12.3).

Osteoclasts (bone-resorbing cells) are derived from haemopoietic stem cells, which have the potential to become macrophages or multinucleated osteoclasts depending on the stimuli received during development. Osteoclasts lie in contact with the mineralised trabecular surface or in Howship's lacunae. These cells are rich in endoplasmic reticulum and Golgi complexes. The cell membrane adjacent to the bone matrix is

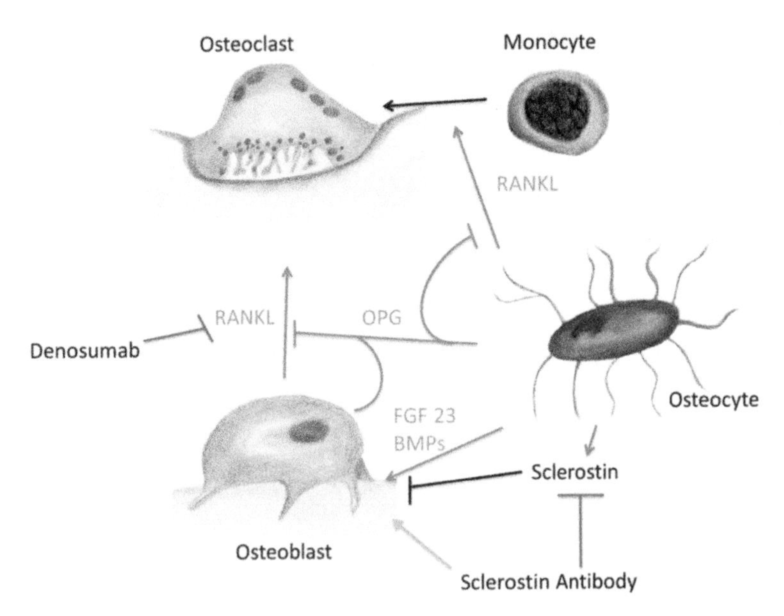

Figure 12.3 Osteocyte control of the local bone environment. Osteoclasts orchestrate bone resorption and bone formation by controlling osteoclast and osteoblast activity. RANKL antibodies (denosumab) and sclerostin antibodies (romosuzamab) are two pharmacological agents used to reduce osteoclastic bone resorption and stimulate osteoblastic activity respectively. BMPs, bone morphogenetic proteins; FGF 23, fibroblast growth factor 23; OPG, osteoprotegerin; RANKL, receptor activiator of nuclear factor κ-β ligand.

characterised by a peripheral sealing zone that is rich in integrins, and a central ruffled border that forms a space between the osteoclast and the bone matrix into which lysosomal enzymes (such as tartrate-resistant acid phosphatase and cathepsin K), proteinases (such as collagenase) and hydrogen ions are secreted. The hydrogen ions and enzymes dissolve the mineral and digest the demineralised matrix, the products of which are internalised into the osteoclast, transported across the cell into the extracellular fluid or released through the sealing zone. Key osteoclastogenic cytokines are RANKL and monocyte colony-stimulating factor. The physiological inhibitor of RANKL is osteoprotegerin (OPG) which binds to the RANKL receptor RANK with higher affinity than RANKL. Hormones that stimulate osteoclast number and activity include PTH, $1,25(OH)_2D$ and a number of proinflammatory cytokines, while OPG and calcitonin reduce osteoclastic activity. There are many remaining unknowns concerning osteoclast function and biology.

Bone remodelling

Bone is not a dead organ; rather, it is continually being resorbed and replaced. This process is known as bone remodelling (Figure 12.4) and it not only helps to maintain bone in optimal condition through repairing microfractures but is also important in serum calcium homeostasis. Bone turnover occurs in discrete packages throughout the skeleton, and the process of resorption followed by replacement is tightly coupled so that in the healthy young adult there is no net loss or gain in bone. The balance between bone formation and resorption is essential in maintaining bone mass. As one grows older, and particularly in the post-menopausal period in women, progressive bone loss occurs as a result of incomplete replacement of the bone that has been resorbed. The control of the rate of bone remodelling, the mechanisms by which the osteoclasts are activated and how the whole process is coupled are areas of intensive investigation.

Recent studies have provided a better understanding of how the various bone cells interact, with osteocytes playing a crucial role in determining the rate of bone remodelling and interaction between cells (see Figure 12.3). For example, osteoblasts control osteoclast precursor development through the production of RANKL, which binds to RANK on osteoclast precursors and stimulates osteoclast proliferation. Osteoprotegerin, a circulating binder of RANKL, reduces the amount of RANKL available to bind to osteoclast precursors, thus reducing osteoclast production.

The whole process of remodelling of a particular package of bone takes about 3–4 months, with

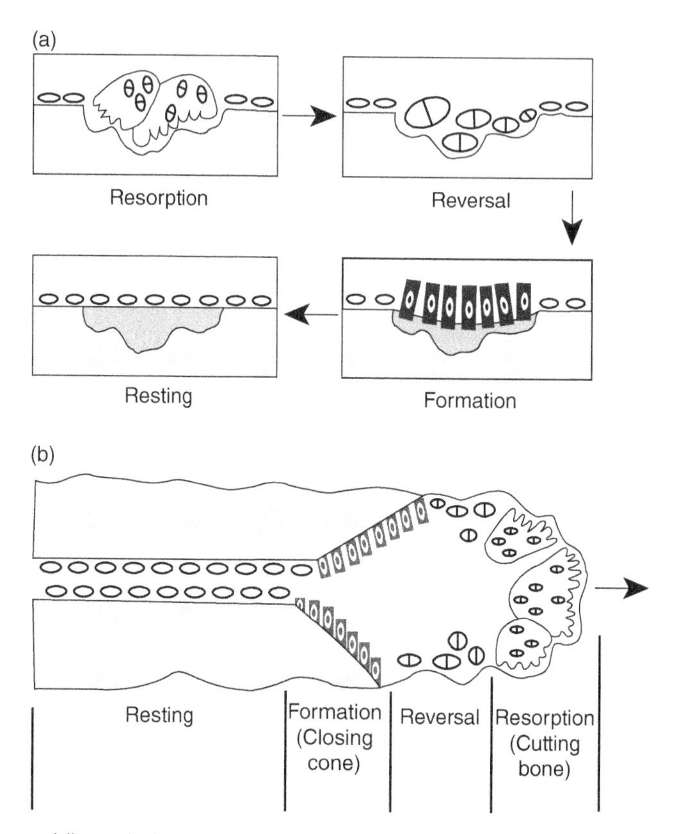

Figure 12.4 The bone remodelling cycle. (a) Remodelling at the trabecular bone surface. (b) Remodelling occurring in cortical bone. *Source:* John P. Bilezikian, 2018/Reproduced with permission of John Wiley & Sons.

the resorption process lasting about 10 days and the filling in of the cavity taking about three months (Figure 12.4). Systemic hormones, such as PTH, $1{,}25(OH)_2D$ and calcitonin, alter bone remodelling mainly through inducing more or fewer remodelling sites in the skeleton. Oestrogen impacts bone remodelling, acting directly on osteocytes, osteoblasts and osteoclasts in men and women. Oestrogen deficiency, particularly at the time of the menopause, causes bone loss through excess bone resorption and incomplete filling in of the resorption cavity that develops during the remodelling cycle. Excessive bone resorption, either through the formation of excessively deep resorption cavities or through the progressive thinning of trabecular bone with each remodelling cycle, results in weakening of the trabecular bone structure owing to the loss of connectivity between the various rods and plates that make up the trabecular scaffolding, resulting in progressive risk of minimal trauma fractures. States of very high bone turnover, which may occur in conditions such as Paget disease or severe primary or secondary hyperparathyroidism, result in the formation of woven rather than lamellar bone. In woven bone, the collagen fibres are laid down in a disorganised fashion, resulting in loss of the normal lamellar structure with a resultant loss in bone strength.

Biochemical assessment of bone remodelling

Since the early 1990s, great strides have been made in the development of assays for the measurement of both serum and urine markers of bone formation and resorption, and these allow non-invasive assessment of bone turnover. Each indicates a different stage in bone turnover physiology (Lorentzon et al. 2019).

Markers of bone formation

Table 12.2 lists the biochemical markers currently available to assess bone turnover.

Total alkaline phosphatase has until recently been the only marker of osteoblastic activity available, but as serum total alkaline phosphatase also reflects that produced at other sites, in particular the liver and placenta, elevated total levels do not necessarily reflect an increase in

Table 12.2 Biochemical markers of bone turnover.

Formation
• Serum/plasma
– Total alkaline phosphatase (TALP)
– Bone-specific alkaline phosphatase (BALP)
– Osteocalcin (OCN)
– Carboxy-terminal propeptide of type 1 collagen (PICP)
– Amino-terminal propeptide of type 1 collagen (PINP)
Resorption
• Serum/plasma
– C-telopeptide of collagen cross-links (CTx-Ia)
– Cross-linked C-telopeptide of type 1 collagen (ICTP)
– Tartrate-resistant acid phosphatase 5b (TRAP-5b)
– N-telopeptide of collagen cross-links (NTx)
• Urine
– N-telopeptide of collagen cross-links (NTx)
– C-telopeptide of collagen cross-links (CTx-Ia)

osteoblastic activity. Alkaline phosphatase concentrations are characteristically elevated in all forms of osteomalacia and rickets, in hyperparathyroidism and in Paget disease.

The measurement of *bone-specific alkaline phosphatase* (BALP) utilises specific monoclonal antibodies to detect the bone component of the tissue non-specific alkaline phosphatase isoenzyme and more accurately reflects an increase in osteoblastic activity. Depressed levels of alkaline phosphatase may occur in zinc deficiency and protein–energy malnutrition, and in inherited disorders of the alkaline phosphatase gene (hereditary hypophosphatasia).

Osteocalcin is a non-collagenous protein secreted by the mature osteoblast. Thus, serum levels should reflect osteoblastic activity and bone formation. In general, BALP and osteocalcin values correlate but some studies suggest that osteocalcin levels may be normal in patients with rickets despite markedly elevated alkaline phosphatase values. The reasons for this dissociation are unclear.

The *amino- and carboxy-terminal propeptides of type 1 collagen* (PINP and PICP as shown in Table 12.2 respectively) are cleaved and excreted during collagen biosynthesis. Assays for the determination of the propeptides at both the N- and C-terminals have been developed, but they do not appear to be as useful as either BALP or osteocalcin in assessing bone formation and osteoblastic activity.

Markers of bone resorption

The assessment of bone resorption is possible using serum, plasma and urine. The most useful markers are those derived from the breakdown of type 1 collagen, which occurs during the resorption of bone matrix. Assays have also been developed to measure the *N- and C-terminal telopeptides of collagen* in urine and C-terminal telopeptides in serum/plasma.

A reasonably specific marker of osteoclast activity is *tartrate-resistant acid phosphatase* (TRAP), which is secreted into serum. Its measurement is limited by its instability in serum and by the fact that it is not entirely specific for osteoclasts.

Clinical usefulness of biochemical markers of bone turnover

Bone markers are useful in assisting in the diagnosis of metabolic bone disease, such as osteomalacia, rickets and hyperparathyroidism. While they do help in understanding the pathogenesis of osteoporosis and monitoring the response to therapy, they do not accurately diagnose osteoporosis, nor are they useful tools to predict future bone loss (Compston et al. 2017). Guidelines from the UK, USA and International Osteoporosis Foundation recommend the use of β-CTx-I and P1NP for the purposes of osteoporosis treatment monitoring and exploring underlying causes for osteoporosis. These two bone turnover markers have also been shown to slightly improve fracture risk prediction independently of bone mineral density (BMD) and other risk factors, though stronger evidence is still required.

There are a number of factors that influence measurements of bone turnover markers and that has perhaps limited their adoption into routine clinical practice. A number of the markers have a circadian rhythm, thus specimens collected at different times of the day or night may vary markedly. Recommendations are to standardise collection to the morning and if they are being collected with specimens to assess calcium homeostasis, they should be collected in a fasting state. Other factors which may affect bone turnover include diet and exercise, menopause status, ethnicity, age, season, fracture and glucocorticoid use.

Bone turnover markers are probably most useful as a research tool to assess rates of bone turnover between groups of subjects, and in clinical medicine to determine the bone response to an intervention such as drug therapy. They are less helpful in determine the bone turnover status of an individual.

Assessment of bone mass

One of the techniques used to assess bone health is the measurement of the amount of mineral present in the bone. Before the advent of more advanced techniques, the usual method of assessing the amount of mineral in bone was to use radiographs of the lumbar spine or hips and judge whether the radiographic density appeared to be normal, increased or decreased. The problem with this technique is that one needs to have lost some 30% of bone mineral before it becomes obvious on the routine radiograph and the interpretation is open to considerable subjectivity. The loss of radio-density on radiographs is termed *osteopenia*, which could be due to a loss of bone (matrix and mineral; osteoporosis) or a failure of mineralisation of normal amounts of matrix (osteomalacia).

Advances in technology since the early 1980s have resulted in the availability of rapid and accurate methods of assessing bone mass. A number of different techniques have been used (Table 12.3), but dual-energy X-ray absorptiometry (DXA) is the most widely accepted (Figure 12.5). Measurement of bone mass is relevant to assess the degree of osteoporosis/osteopenia, to determine possible fracture risk and to assess response to therapy or the effect of disease on bone mass.

Dual-energy X-ray absorptiometry is the only method that can be used for diagnosis of osteoporosis in adults or children. Importantly, in children the definition of osteoporosis also requires the presence of prior fracture(s) and/or other risk factors. DXA measures bone mineral content (BMC) in grams and bone area in cm^2. From these two measurements, the so-called bone mineral density (BMD) in g/cm^2 is calculated. It should be noted that the BMD is an areal bone density and does not measure true volumetric density of bone. Thus, unlike true density, DXA-derived BMD is influenced not only by the true density of bone but also by the volume of bone (Figure 12.6). With improvements in detector quality, the greatest advance in bone assessment by DXA has been the ability to visualise lateral projections of the spine, which allows assessment of vertebral fractures. DXA lateral vertebral assessment (LVA) performs at least as well as radiographic diagnosis of fractures in both adults

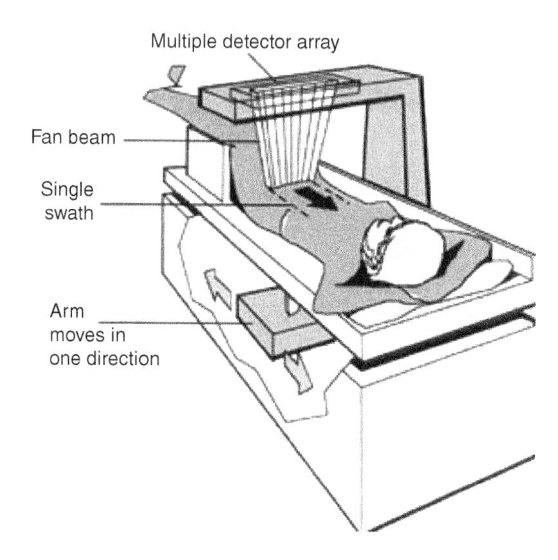

Figure 12.5 Schematic diagram of a dual-energy X-ray densitometer. Note the collimated X-ray beam moves through the body from inferior to superior. Using this method, whole body, lumbar spine, femoral and radial bone mineral density and area may be measured and bone mineral content calculated. Newer scanners are able to image the lateral spine, which can be used for assessment of vertebral shape and fractures.

Table 12.3 Current techniques used to measure bone mass.

Technique	Sites measured	Precision (% CV)	Radiation dose (μSv)
Dual-energy X-ray densitometry (DXA)[a]	Spine, hip, forearm, whole body, lateral distal femur, lateral vertebral assessment	1–2.5 depending on site	~1–11
Quantitative computed tomography (QCT)	Spine	2–4	~50
Peripheral QCT (pQCT) single slice	Radius, tibia, femur	~1–2	~1
High-resolution pQCT	Radius and tibia	<3%	<3
Quantitative ultrasound (QUS)	Calcaneus, tibia, phalanx	0.3–5	0

[a] Earlier versions were single- and dual-photon absorptiometry, which were superseded due to poorer precision and technical limitations such as decay of isotope sources.

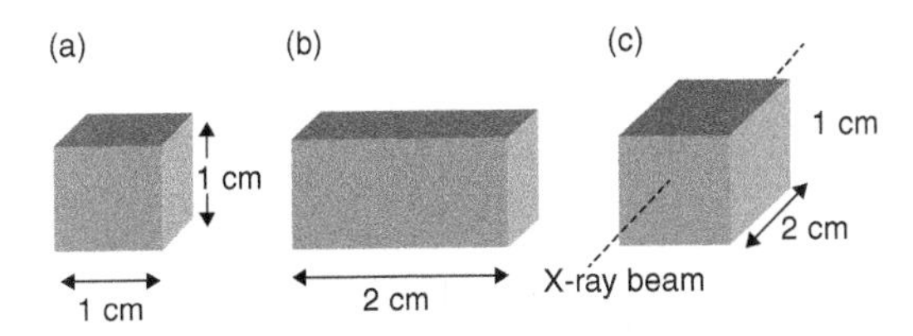

Figure 12.6 The concept of areal bone mineral density (BMD). The three blocks (a), (b) and (c), represent three blocks of bone of the same volumetric density of x g/cm^3. The projected area of block (a) is 1 cm^2, block (b) 2 cm^2 and block (c) 1 cm^2. Thus the BMD measured by DXA would be x g/cm^2 for blocks (a) and (b), but $2x$ g/cm^2 for block (c) as it is twice as thick as blocks (a) and (b). Note that BMD measured by DXA is dependent not only on the true bone density (volumetric) but also on the thickness of the bone, which is not measured by DXA. The dashed line represents the path of the X-ray beam through the three blocks.

and children. Additional measures that can be obtained on DXA include the trabecular bone score and hip structural analysis.

An understanding of the concept of areal BMD compared to volumetric BMD is important, as bones with the same volumetric bone density but different volumes will have different areal BMD as assessed by DXA (Figure 12.6). Areal BMD will increase with increasing bone size without a change in the volumetric bone density. Furthermore, DXA-assessed BMD reflects the average areal bone density of all the constituents of the particular bone being measured, thus the value will depend on amount of cortical as well as trabecular bone. Quantitative computed tomography (QCT) has an advantage in that it measures the volumetric density of bone. Furthermore, regions of bone, such as the cortical or trabecular regions, can be selected, thus allowing an assessment of the differential effects of a disease or treatment on the different types of bone. Central QCT, using general CT scanners, is not more widely promoted and used because of the practicalities of accessing hospital scanners, the costs involved and the relatively high radiation dose. Peripheral devices (pQCT) are available, which measure bone geometry and strength, and the volumetric BMD of cortical and trabecular bone in the appendicular skeleton (radius, tibia and femur). More recently, high-resolution pQCT (HRpQCT) has allowed in vivo assessment of cortical and trabecular bone microarchitecture and strength at the distal radius and tibia; the newer scanners also allow high-resolution scanning of the radius and tibia diaphysis. Both pQCT and HRpQCT assess the skeleton at very low radiation doses (Fuggle et al. 2019).

Bone growth

Bone growth during fetal, childhood and adolescent periods involves not only elongation of long bones by proliferation of the growth plate cartilage, but also growth at the periosteal surfaces of both the long bones and membranous bones. Thus, as bones enlarge they undergo modelling, which involves the resorption of bone in one area and the deposition of bone in another, allowing individual bones to retain their shape despite growth enlargement (Figure 12.7). From birth to closure of the epiphyses and cessation of growth during adolescence, the skeleton increases in length approximately three-fold, during which time quite marked changes in skeletal proportions occur, with limb length increasing more than trunk length, resulting in changes in the upper body (trunk and head) to lower body (lower limb) ratio.

Skeletal growth rates are not constant throughout childhood. After birth, there is a marked deceleration in growth until three years of age, when the growth rate plateaus until the onset of puberty. Girls have their adolescent growth spurt approximately two years before boys and fuse their epiphyses earlier than boys (see Figure 7A.4).

Hormonal control of skeletal growth

Postnatally, the most important hormones regulating skeletal growth are growth hormone, insulin-like growth factor-1 (IGF-1), thyroid hormone and the sex steroids (Table 12.4).

The control of fetal growth is less clearly understood, but considerable interest in this area has been generated by the important role that fetal growth has not only in adult size but also in the future risk for cardiovascular disease, hypertension and non-insulin-dependent diabetes mellitus. Although genetic factors play a role in determining fetal growth, the predominant factor is the nutritional, oxygen and hormonal milieu in which the fetus develops. Both IGF-1 and IGF-2 are necessary to achieve normal fetal growth, but the effect of IGF-1 deficiency during fetal life is more severe than that of IGF-2 deficiency.

Growth hormone (GH) probably induces most of its effect on growth through the stimulation of

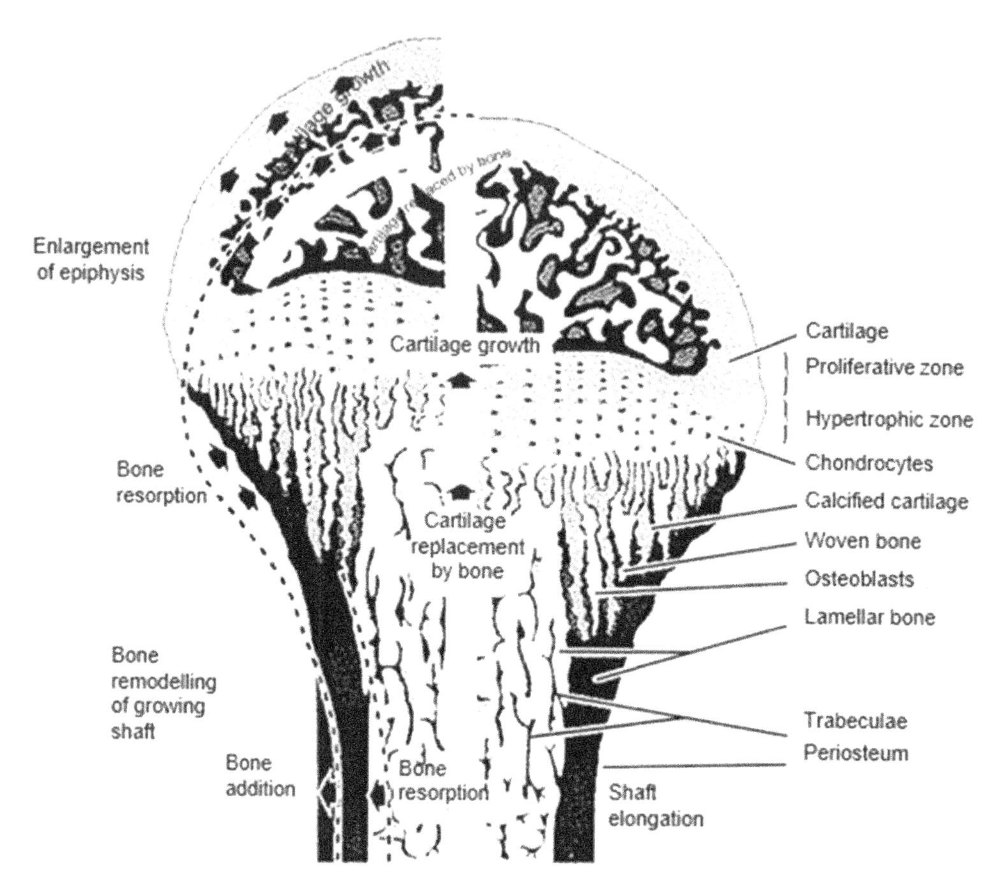

Figure 12.7 Changes associated with longitudinal growth of a long bone. Note the modelling that needs to take place in order for the bone to be the same shape but larger. *Source:* Feldman D. et al. 2007/With permission of Elsevier.

Table 12.4 Important hormonal regulators of bone growth.

- Growth hormone
- Insulin-like growth factor-1
- Thyroid hormone
- Sex hormones, especially oestrogen

IGF-1 secretion both from the liver and in the growth plate, where it stimulates the proliferation of cartilage cells, resulting in an elongation of long bones. It is possible that the paracrine actions of IGF-1 in the growth plate are more important than the hormonal effects of IGF-1 produced in the liver, as studies using a targeted knockout of the IGF-1 gene in the liver did not show a reduction in the growth-promoting actions of growth hormone. Unlike congenital deficiency of IGF-1, which manifests during fetal development, congenital GH deficiency manifests postnatally with a falling off of growth velocity around 12 months of age. Nutritional deprivation in infants and children probably causes a reduction in growth velocity through reducing IGF-1 production. Recent evidence suggests that undernutrition might act through altering the gut microbiota, which plays an important role modulating GH and IGF-1 circulating levels, possibly through altering ghrelin and leptin secretion (De Vadder 2021).

It also appears that IGF-1 may be an important mediator of the effect of sex steroids on bone growth during puberty, as both growth hormone and IGF-1 levels rise during puberty. The rise in growth hormone during puberty is a result of an increase in circulating oestrogen concentrations in both boys and girls. In boys, it is likely that the increase in oestrogen is through the aromatisation of testosterone.

Sex steroids play an essential role not only in the growth spurt that occurs during puberty but also in the cessation of growth through the closure of the epiphyses. It appears that at physiological levels, circulating oestrogen stimulates growth, while at pharmacological levels it suppresses longitudinal growth, possibly through reducing IGF-1 levels. Oestrogen is responsible for epiphyseal closure in both boys and girls. Sex steroids have effects not only on the growth plate but also on the endosteal and periosteal bone surfaces, resulting in an increase in cross-sectional diameter of the long bones. It appears that androgens and oestrogen may have different effects on these bone surfaces. Both androgens and oestrogen cause an increase in periosteal new bone formation, with resultant widening of the bone, but only oestrogen has an effect on the endosteal bone surface, with a consequent increase in cortical bone thickness. Thus after puberty, boys have wider bones than girls, and girls have a narrower endosteal diameter due to bone deposition on the endosteal surface (Figure 12.8).

Thyroid hormone is essential for the normal proliferation and maturation of the growth plate. Cretinism or congenital thyroid deficiency is associated with abnormalities of the epiphyses and marked short stature. From a nutritional standpoint, severe iodine deficiency in the mother and infant will manifest with growth failure and delayed neurological development.

Bone mass accumulation during childhood and adolescence

The skeleton increases in calcium content from birth to the end of adolescence by about 40 times (from 25 to 1000 g). A large proportion of this

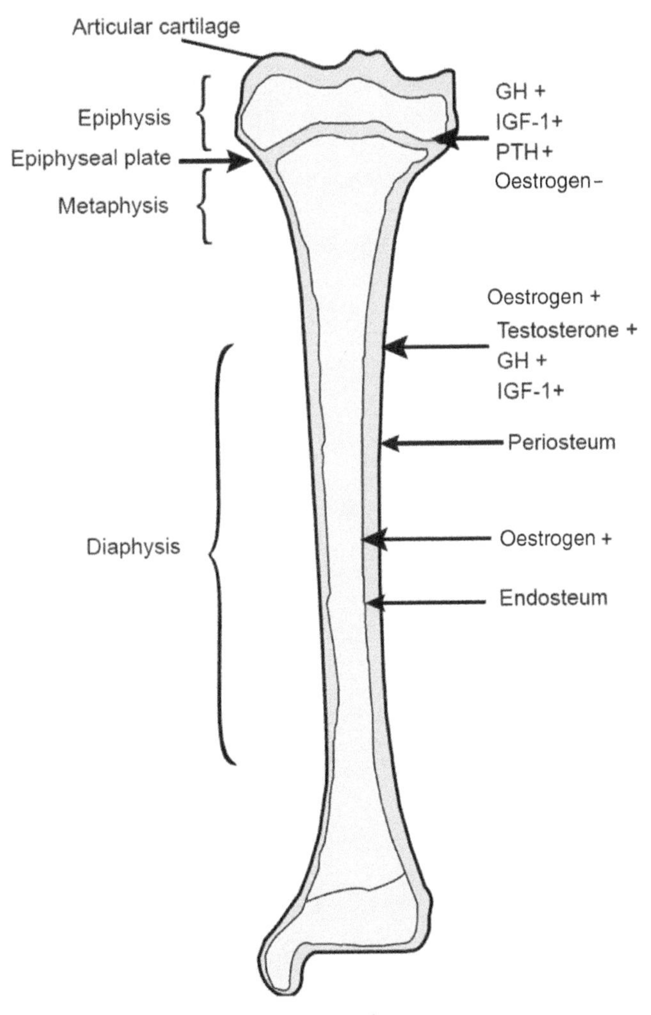

Figure 12.8 Sites of hormone actions on the growing long bone. Stimulation of bone growth is indicated by +, while inhibition of growth is indicated by – . GH, growth hormone; IGF-1, insulin-like growth factor-1; PTH, parathyroid hormone.

deposition occurs during puberty. It is estimated that in both boys and girls, over 35% of final bone mineral in the whole body is laid down during the four years around the age of peak height velocity during adolescence, despite the eventually greater final bone mass in males than females due to the later onset of puberty in boys. In healthy children, maximal gain in bone mass occurs between 11 and 14 years of age in girls and between 13 and 17 years of age in boys. In children with chronic or acute disease, the timing of puberty is often later, with respect to population norms. For example, timing of puberty in children living with HIV or cystic fibrosis may be later than population-matched norms, and is slightly delayed when compared to the ages of peak height velocity.

The pattern of bone mass accumulation during childhood is similar to that of growth in height, as bone growth accounts for the majority of bone mass accumulation. There is a sequential pattern whereby height increases first, followed by lean mass (a surrogate for muscle mass), bone area and finally consolidation with accrual of bone mineral content and consequently BMD. The dissociation between the rates of statural growth and mineral mass accumulation during puberty could be viewed as a period of relatively low bone mass (Figure 12.9) and may account for the increase in fracture incidence that occurs round this time. As linear growth slows, bone mass accumulation continues to occur and the deficit is made up.

Peak bone mass

Although longitudinal growth of bone ceases with fusion of the epiphyses, bone mass continues to accumulate as bone consolidates and periosteal new bone formation continues. The concept of peak bone mass, defined as the amount of bone tissue present at the end of skeletal maturation, has assumed considerable significance as there is good evidence that the risk of osteoporotic fractures in later life is inversely related to the amount of bone accumulated during maturation (Table 12.5). Skeletal maturation is considered to occur early in the third decade of life, although the timing of the achievement of peak bone mass may vary slightly between different skeletal sites.

Factors known to influence peak bone mass include heredity, sex, race, nutrition, hormonal

Table 12.5 Factors influencing peak bone mass.

- Genetic
- Race and ethnicity
- Sex
- Nutrition
- Physical exercise
- Hormonal status
- Body weight

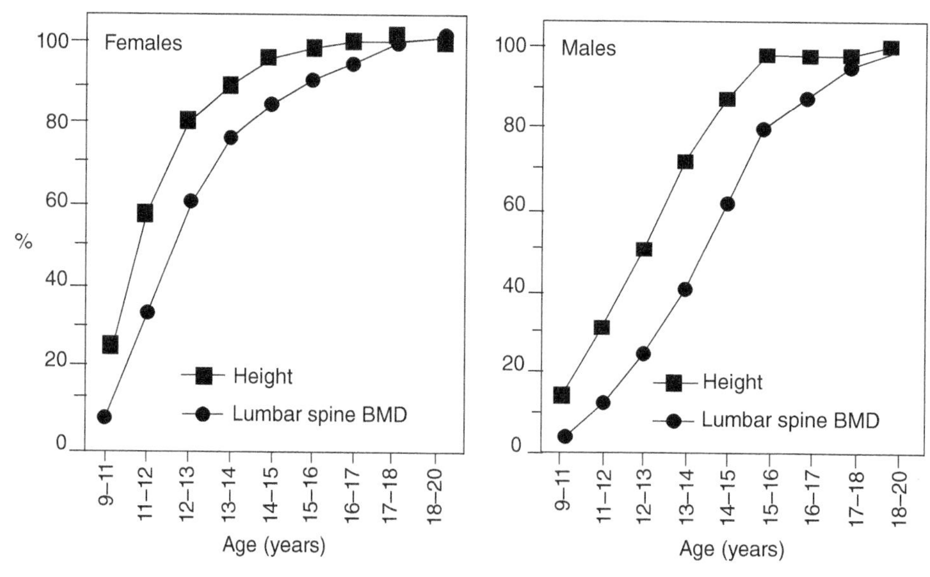

Figure 12.9 Comparison in cumulative gain between lumbar spine bone mineral density (BMD) and standing height during adolescence. The values are expressed as the percentage difference between 18–20-year-old and 9–11-year-old groups. *Source:* Reproduced with kind permission from Fournier P.E. et al. © 1997 Springer Nature.

status (in particular the sex steroids and IGF-1), exercise and body weight (lean mass and fat mass), height and weight growth rates and pubertal timing throughout childhood and adolescence (Ward et al. 2018). By far the most important is genetic influence, with 50–85% of the variance in peak bone mass at different sites being accounted for by heritability. The role of nutrition and in particular dietary calcium intake during childhood in influencing peak bone mass has been an area of intense investigation and will be discussed later in this chapter. However, a definitive answer on the role of dietary calcium is not yet available. Exercise has been shown to have a modulating influence on peak bone mass, with weight-bearing exercises being the most effective, but the effect of exercise is relatively small (3–5%). Although small, on a population basis an increase in peak bone mass of this magnitude might have a considerable influence on fracture rates in later life.

There is an increasing gradient of fracture risk with decrease in bone mass, such that, for example, a BMC or BMD at the femoral neck of one standard deviation below the young adult mean increases the relative risk of hip fracture by 2.6. Consequently, both peak bone mass and the rate of subsequent bone loss are major determinants of osteoporotic fracture risk in later life. The maximisation of peak bone mass by optimising environmental factors that influence skeletal development during childhood and adolescence is regarded as an important preventive strategy against future fractures.

Conceptually, peak bone mass, as defined by BMC or BMD, contains elements related to the size of the skeleton, to the amount of bony tissue contained within it, to the mineral content of that tissue, and to the degree to which the bony tissue is actively undergoing remodelling. It is, as yet, unclear which of these aspects is most influential in determining future fracture risk. Furthermore, the measurement of BMC or BMD by DXA does not assess all the factors that might be responsible for fracture risk, such as the microarchitecture of bone. However, a computer program which generates the trabecular bone score from DXA scans of the lumbar vertebrae captures information related to trabecular microarchitecture and may provide extra information (Shevroja et al. 2017), as may pQCT and HRpQCT-derived parameters of bone microarchitecture and strength.

Sex and ethnic differences in bone mass

Peak bone mass is significantly greater in adult males than females (Figure 12.10). This difference is due to a greater bone size in males than females, rather than due to a difference in true bone density. At birth, boys and girls have similar bone masses that begin to deviate prior to the pubertal growth spurt, gradually increasing as puberty proceeds. Sex differences start appearing early in childhood and accelerate during puberty due to the delayed and more prolonged growth spurt in boys.

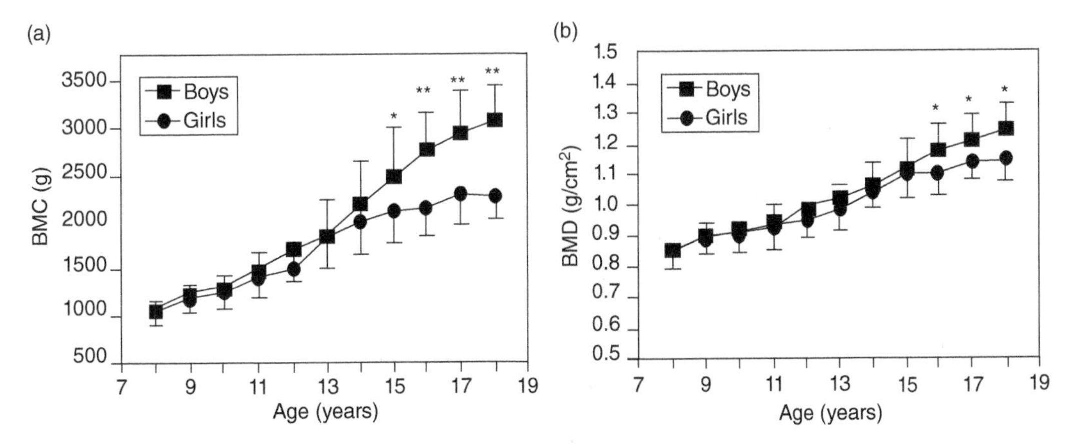

Figure 12.10 (a) Total body bone mineral content (BMC) and (b) areal bone mineral density (BMD) in Caucasian boys and girls from eight to 18 years of age. Note the divergence of values between boys and girls developing after the onset of puberty. The effect is more marked for BMC than BMD as true volumetric bone density is similar in boys and girls, the difference in BMD being mainly due to a difference in bone size. *Source:* Reproduced with permission from Maynard L.M. et al. © 1998 Oxford University Press.

Ethnic differences in bone mass have been consistently found in studies conducted in the USA, with African-Americans having approximately 5–15% greater bone mass at all measured sites than white Americans, after adjusting for body size (Ettinger et al. 1997). In keeping with these bone mass differences, fracture rates are lower in African-American children and adults. In Africa itself, there are few data available assessing either bone mass or fracture rates in sub-Saharan populations. However, recent studies indicate that rates of both fractures in childhood and adult hip fractures are very much lower in the indigenous populations compared to North American and European whites (Figure 12.11). In studies from East, West and southern Africa, lumbar spine BMD has been reported to be similar to or lower than that of African or European whites. These differences are present before the onset of puberty. Consequently, the recommendation for clinical assessment of black populations from outside the USA is that a white reference norm should be used for calculation of T- and Z-scores to ensure that overestimation of deficits in BMD does not occur through the comparison to the African-American population. Studies of other ethnic groups have been less detailed, but it appears that Asian women from the Indian subcontinent have similar or lower bone mass than American or European Caucasian women.

Skeletal changes during pregnancy and lactation

Pregnancy and lactation are associated with alterations in calcium and bone metabolism that temporarily affect the mineral content of the skeleton (Table 12.6). Both the fetus and the nursing infant increase calcium demands on the pregnant and lactating woman. These changes are evident from early to mid-gestation and continue through and beyond lactation. In pregnancy, bone resorption and formation rates are increased by 50–200%, and maternal intestinal calcium absorption efficiency and urinary calcium excretion are elevated. In lactation, calcium absorption efficiency returns to normal, but there is evidence of renal conservation of calcium in some women. Bone turnover continues to be elevated and differences in the timing of the skeletal response in terms of resorption and formation favour the release of calcium from the skeleton during early lactation, with restitution during and after weaning.

Direct studies of changes in BMC using densitometry have been largely restricted to lactating women because of the small radiation dose involved. Such studies have demonstrated striking reductions in BMC after 3–6 months of lactation, particularly in axial regions of the skeleton, such as the lumbar spine and femoral neck, where decreases average 3–5% (Figure 12.12).

Lactation-associated reductions in BMC are remarkable given that the rate of bone loss is 2–5 times faster than that found in postmenopausal bone loss, which is typically 1–3% per year. BMC

Table 12.6 Skeletal and mineral changes during pregnancy.

- Increased bone turnover
- Increased intestinal calcium absorption
- Increased urinary calcium loss
- Increased serum $1,25(OH)_2D$ and PTHrP, decreased PTH
- Variable changes in bone mineral content

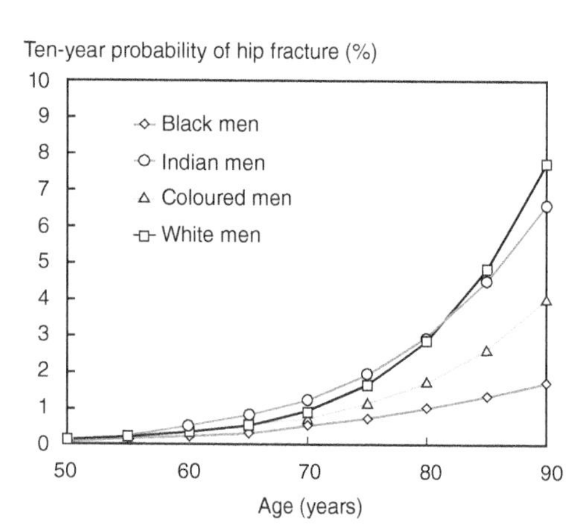

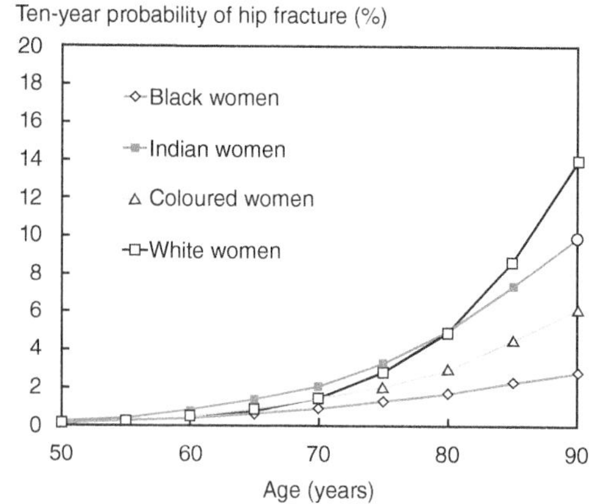

Figure 12.11 Ten-year probability of hip fracture in South African men and women by ethnic group. *Source:* Johansson H. et al. 2021/Springer Nature/Licensed under CC BY 4.0.

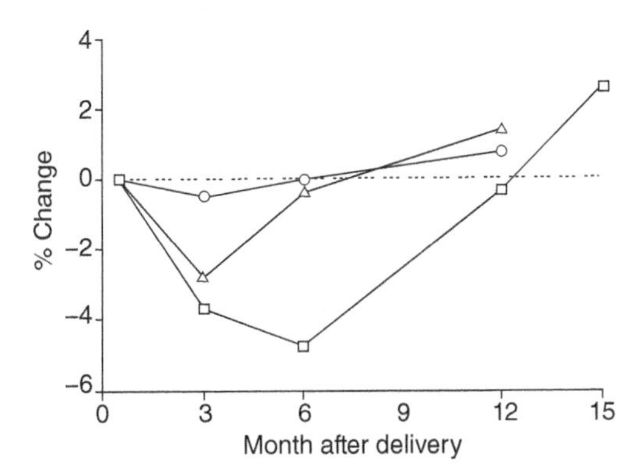

Figure 12.12 Changes in bone mineral content of the lumbar spine during lactation and after weaning. Data are adjusted for scanned bone area. □, Mothers who breast-fed for >9 months (n = 20); Δ, Mothers who breast-fed for 3–6 months (n = 13); ○: Mothers who formula-fed (n = 11). *Source:* Data from Laskey M.A., Prentice A. 1999.

recovers later in lactation or after weaning, and at some sites exceeds that measured after parturition. Lactational amenorrhoea, the length of lactation and other aspects of infant-feeding behaviour influence the magnitude and temporal pattern of the skeletal response experienced by breastfeeding women. However, the final outcome following lactation appears to be similar irrespective of duration of lactation, or indeed whether the woman breast-fed or not.

Studies in women in which measurements were made before conception and after delivery also demonstrate reductions in whole body and regional BMC sufficient to make a sizeable contribution to maternal and fetal calcium economy. In contrast, substantial increases have been reported in women entering pregnancy while still breastfeeding, suggesting that skeletal changes during pregnancy may depend on the status of the maternal skeleton before conception.

The mechanisms underlying the effects of pregnancy and lactation on calcium and bone metabolism are not fully understood. Originally, it was considered that the observed metabolic changes were due to physiological hyperparathyroidism, driven by the inability of the dietary calcium supply to meet the high calcium requirement for fetal growth and breast milk production. For this reason, women in the past were advised to increase their calcium intake during pregnancy and lactation. It is, however, now generally accepted that this is not the case – first, because PTH concentrations are not elevated during pregnancy and lactation, with parathyroid hormone-related peptide (PTHrP) dominating instead; second, because, in pregnancy, the metabolic changes precede the increased requirement for calcium; and third, because the effects during lactation appear to be independent of calcium intake as shown through supplementation studies in populations accustomed to low and medium-to-high calcium intakes. As a consequence, it is no longer considered necessary for a woman to increase her calcium intake during pregnancy and lactation, and this is reflected in recent dietary recommendations, for example by the Department of Health (UK), the Institute of Medicine Food and Nutrition Board (NAS) (US/Canada) and the European Food Standards Agency (European Union).

The World Health Organization (WHO 2018) recommends women in populations with a low calcium intake to take high-dose calcium supplements (1500–2000 mgCa/d) during pregnancy to reduce the risk of pre-eclampsia and its complications, based on the outcome of several randomised controlled trials (RCTs). No assessments of maternal skeletal health were made in these trials. However, an RCT in Gambian mothers with low calcium intakes has demonstrated that 1500 mgCa/d taken from 20 weeks of pregnancy to term accentuated the metabolic and skeletal changes of lactation and reduced postlactational bone mineral recovery, with deficits in bone mineral still measurable after five years (Figure 12.13). The pregnancy supplement was also associated with sex-specific differences in childhood growth and bone development of the offspring, discussed more fully in a later section. The long-term skeletal

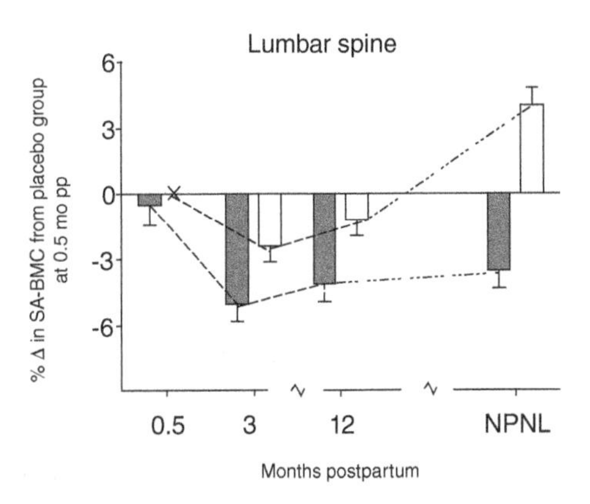

Figure 12.13 Effects on size-adjusted bone mineral content of the lumbar spine (L1-4) of maternal calcium supplementation during pregnancy in Gambian mothers. Dark bars, calcium supplemented group; light bars, placebo group, NPNL-not pregnant, not lactating. *Source:* Prentice A, 2021/Cambridge University Press/Licensed under CC BY 4.0.

health consequences for the mother and her children of the use of high-dose calcium supplements during pregnancy are as yet unknown.

No convincing protective or deleterious link between fracture risk in later life and either pregnancy or lactation has been shown in retrospective studies where women self-report the number of pregnancies and lactation periods during their reproductive lives.

Osteoporosis with fractures is a rare complication of pregnancy and lactation. When it does occur, the condition frequently involves the hip or spine, is more common in the first pregnancy and usually resolves spontaneously. The cause is unknown, and most cases are either idiopathic or secondary to warfarin or corticosteroid therapy. There is no evidence that osteoporosis of pregnancy and lactation is a consequence of nutrient deficiencies or that it can be prevented by changes in diet and lifestyle.

12.3 Bone disease and associated challenges

Nutritional rickets

Nutritional rickets occurs in the paediatric age range, especially in the infant and young child. Although rickets has a number of different aetiologies, nutritional causes are by far the most important globally and constitute a major public health problem in a number of countries. Despite the ease of its prevention and the rapidity of its response to treatment, the disease does not receive the attention it should from international agencies and national governments, so children are being left physically handicapped and stunted unnecessarily. In this section, the causes, presentation and biochemical and radiological changes associated with nutritional rickets are presented. A brief overview of its treatment and prevention is also provided.

Rickets is a clinical syndrome characterised by a delay in or failure of mineralisation of the cartilaginous growth plates in the growing child (Figure 12.14). These abnormalities result in deformities at the growth plates of rapidly growing bones. Accompanying these changes, there is also a delay in mineralisation of newly formed osteoid at the endosteal and periosteal bone surfaces, resulting in an increase in osteoid seam width. These latter abnormalities are features of osteomalacia, which is also characterised by an increase in osteoid surface and volume, an increase in the mineralisation lag time and a decrease in tetracycline uptake at the mineralisation front. Thus, the disease of rickets is associated with clinical features related to both the

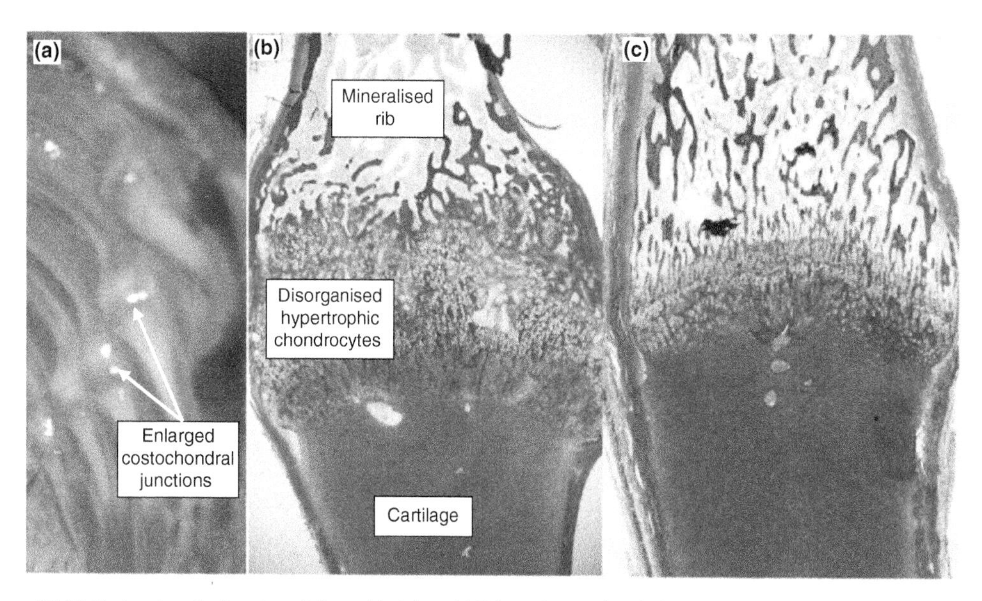

Figure 12.14 Postmortem findings in an infant with rickets. (a) Enlarged costochondral junctions of the ribs viewed from the inner surface; (b) histology of the costochondral junction in the child displaying marked increase and disorganisation of the hypertrophied cartilage; (c) histology of a normal costochondral junction. The hypertrophied cartilage is narrower than in (b) and more organised. The zone of provisional calcification is arranged in rows and more delicate than in (b). *Source:* Reproduced from Uday et al. 2018/ Springer Nature/CC BY 4.0.

growth plate abnormalities of rickets and those of osteomalacia at the various bone surfaces. Once growth plates fuse in late puberty, rickets can no longer occur, but osteomalacia may still develop at the trabecular, endosteal and periosteal bone surfaces (see Section 12.6).

The differences between osteoporosis and rickets are listed in Table 12.7.

Causes of rickets

As mentioned above, rickets is primarily a disease characterised by a failure of mineralisation of preformed growth cartilage matrix. For bone or cartilage matrix to mineralise, the concentrations of both calcium and phosphorus at the mineralisation site must be sufficient for the growth and aggregation of the hydroxyapatite crystals, the formation of which is facilitated by extracellular matrix vesicles (from osteoblasts), which actively accumulate calcium and phosphorus ions and remove inhibitors of crystal formation.

Thus, the causes of rickets may be broadly classified into three large groups: those related to an inability to maintain adequate calcium concentrations at the mineralising bone surface or growth plate (calciopenic rickets), those related to an inability to maintain appropriate phosphorus concentrations (phosphopenic rickets) and those that directly inhibit the process of mineralisation (Table 12.8). This classification is useful clinically to help decide the most appropriate initial investigations required to establish a diagnosis (Carpenter et al. 2017); however, laboratory studies indicate that despite the cause, it is the consequent hypophosphataemia that is primarily responsible for the failure of mineralisation and impaired chondrocyte apoptosis which is essential for mineralisation to occur (Sabbagh, Carpenter and Demay 2005).

From Table 12.8, it is clear that there are numerous causes of rickets, the majority of which are not nutritional in origin and thus will not be considered in this chapter. Nevertheless, nutritional rickets remains the most common form of rickets globally. Until recently, nutritional rickets was considered to be synonymous with vitamin D deficiency rickets, but studies now suggest that although vitamin D deficiency is an important cause, so too is a low dietary calcium intake, and the two may act synergistically to exacerbate the risk of developing rickets. Another nutritional cause of rickets is dietary phosphorus deficiency, but this condition occurs almost exclusively in breast-fed very low birth-weight infants during the first several months of life (Chinoy, Mughal and Padidela 2019).

Nutritional rickets: a global perspective

Despite advances in our understanding of calcium homeostasis (see *Introduction to Human Nutrition*, Chapter 9) and vitamin D metabolism (see *Introduction to Human Nutrition*, Chapter 8), nutritional rickets remains a major cause of morbidity and mortality in children in many parts of the world.

Although rickets has been thought of as a disease that originated with the Industrial Revolution in Europe, descriptions of rickets have been attributed to both Homer (900 BC) and Soranus Ephesius (AD 130). Before the Industrial Revolution, rickets in England was a disease of the children of the aristocracy, as they were frequently kept indoors and excessively clothed if venturing outside. With the rapid growth of cities associated with urban

Table 12.7 Differences between osteoporosis and rickets.

	Osteoporosis	Nutritional rickets
Age of presentation	Older adults and postmenopausal women	Infants and young children
Pathology	Loss of bone (mineral: matrix normal)	Failure to mineralise bone matrix (mineral: matrix decreased)
Speed of onset/development	Slow	Rapid
Presentation	Pathological fractures	Limb deformities
Diagnosis	Densitometry assessment or biopsy	Radiographs
Biochemistry	Typically normal	Hypocalcaemia, hypophosphataemia and elevated alkaline phosphatase and parathyroid hormone[a]

[a]Hyperparathyroidism is variable in dietary calcium deficiency. In stages 2 and 3 of vitamin D deficiency PTH concentrations are consistently elevated.

Table 12.8 Classification of the causes of rickets.

Calciopenic rickets	Phosphopenic rickets	Inhibition of mineralisation
Vitamin D deficiency	Dietary deficiency, e.g. breast-fed very low-birthweight infant	Aluminium toxicity
Increased vitamin D metabolite catabolism 25-hydroxylase deficiency	Inhibition of intestinal phosphate absorption	Fluoride excess
1α-hydroxylase deficiency Renal failure	Increased renal loss of phosphate: FGF-23 related: – X-linked hypophosphataemia – Tumour associated	Hereditary hypophosphatasia First-generation bisphosphonates
Abnormalities of the vitamin D receptor	Kidney disease related: – Fanconi syndrome – Distal renal tubular acidosis	
Dietary calcium deficiency		

The table is not a comprehensive list, but it does highlight the numerous causes of rickets.

migration, which accompanied the Industrial Revolution, urban slums rapidly developed. The excessive pollution, narrow streets and overcrowding resulted in most young children living in these appalling conditions receiving little or no sunshine. Their mothers likewise were often vitamin D deficient and suffering from osteomalacia, resulting in obstructed labour and neonates being born with no vitamin D stores. In the late nineteenth and early twentieth century several studies described the almost universal prevalence of rickets in young children in Europe and north-eastern USA. The importance of ultraviolet light in the prevention and treatment of rickets was appreciated in the first quarter of the twentieth century, and with the discovery of vitamin D and its role in the aetiology of rickets, the stage was set for the almost complete elimination of vitamin D deficiency rickets from a number of developed countries through programmes of food fortification, vitamin D supplementation and education.

During World War II, the UK introduced vitamin D fortification of a number of foods including milk, cereals and bread. The incidence of rickets fell dramatically, but the programme fell into disrepute because of uncontrolled fortification, which was thought to have resulted in an increase in idiopathic hypercalcaemia in young infants, although this association has not been conclusively proven. Since then, the incidence of nutritional rickets has risen in the UK and is a problem particularly among Asian and other immigrants with dark skin living in the northern cities.

In North America, the prevention of rickets through fortification has been more successful. The universal fortification of milk with vitamin D at 400 IU/quart (32 fl oz, approximately 900 ml) since the 1930s has almost eradicated the disease in young children, except in those who are exclusively breast-fed or who are on milk-free diets. However, since the 1980s there have been reports of a resurgence of rickets in at-risk families, in particular in infants who are breast-fed for prolonged periods, in vegan and vegetarian families, and in infants of African-American mothers.

In central Europe and Algeria, rickets has been successfully prevented in infants and young children by the administration of high doses of vitamin D every 3–5 months for the first two years of life (*stosstherapie*), while in Turkey the introduction of free vitamin D drops from health clinics for infants 0–12 months of age has dramatically reduced the prevalence of rickets in infants and young children from 6% to 0.1%.

Despite advances in the prevention of rickets in young children in many developed countries, the disease remains a health problem in a number of these countries. Not surprisingly, nutritional rickets is more common in those countries at the extremes of latitude, as ultraviolet light exposure is limited for large parts of the year and the cold weather necessitates wearing clothes that cover the majority of the skin surface. Thus, the prevalence of clinical rickets in Tibet has been estimated to be over 60% in children, and similar figures have been obtained from surveys in Mongolia and northern China.

Nutritional rickets remains a problem in many countries, even those closer to the equator with larger amounts of sunshine than the countries mentioned above. These countries include Sudan, Ethiopia, Nigeria, Algeria, Saudi Arabia, Kuwait and other countries in the Middle East, India, Pakistan and Bangladesh. Furthermore, a high prevalence of rickets has been described in young children in areas of Greece. Thus, there must be other factors beside high latitude that contribute to predisposing a child to rickets. A number of studies have highlighted poverty, malnutrition, maternal vitamin D deficiency, prolonged breastfeeding and overcrowding as important risk factors (Table 12.9).

Cultural and social customs may also aggravate the problem. Several countries have large Muslim communities, which practise 'purdah', thus increasing the risk of vitamin D deficiency in mothers and their young children.

Although the typical age of presentation of vitamin D deficiency rickets is during the first 18 months, reports from a number of developing countries describe clinical signs of rickets in children over the age of two years. It is possible that many of these children are suffering from the residual effects of earlier active, but now healed, rickets. However, studies from Nigeria, South Africa, The Gambia, Myanmar and Bangladesh have suggested that children outside the infant age group with clinical signs of rickets may be suffering from the combined effects of a relatively poor vitamin D status in association with extremely low dietary calcium intakes (Sempos et al. 2021) (Figure 12.15). The mean calcium intakes of affected children are remarkably similar in the different countries, estimated to be approximately 200 mg/day. The characteristic features of the diets are high cereal content

Table 12.9 Factors predisposing a child to vitamin D deficiency rickets.

- Living in a country at high latitude
- Lack of sunlight exposure through overcrowding, social customs, clothing or pollution
- Infants whose mothers lacked vitamin D in pregnancy
- Prolonged breastfeeding without vitamin D supplementation or appropriate UVB exposure
- Low dietary calcium intake
- Increased melanin pigmentation

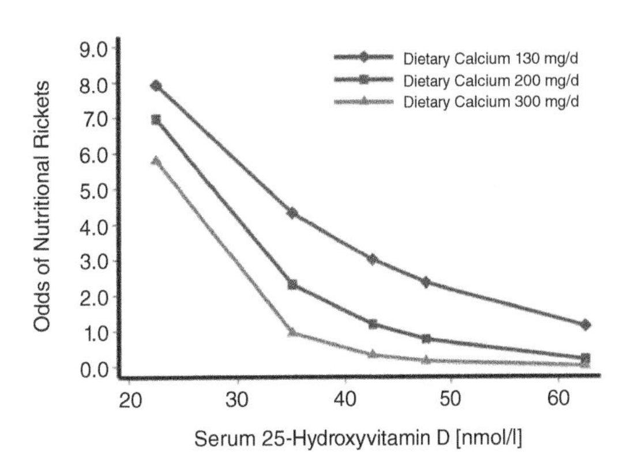

Figure 12.15 The odds of having rickets at three different calcium intakes is increased by a declining vitamin D status (circulating 25-hydroxyvitamin D concentration). *Source:* Sempos C.T. et al., 2021, with permission of Oxford University Press.

and lack of variety and of dairy product intake. It is unclear how widespread this form of nutritional rickets is, but low dietary calcium intakes may also increase the requirements for vitamin D, thus exacerbating the severity of rickets by increasing the catabolism of vitamin D.

The burden that nutritional rickets places on children globally is unclear, but in those countries in which there is a high prevalence of rickets, a significant proportion of the infant mortality may be directly or indirectly due to the disease. This has clearly been shown in Ethiopia, where it has been estimated that children with rickets have a 13 times higher incidence of pneumonia, which is associated with a 40% mortality. The long-term sequelae of severe rickets are difficult to quantitate, but short stature, residual limb deformities resulting in early osteoarthritis and pelvic deformities in women leading to a high incidence of obstructed labour are some of the problems.

In addition to rickets, recent cross-sectional evidence suggests that vitamin D deficiency may also have associations with wider complications, such as impairing immunity and increasing tuberculosis infection, increasing the risk of type 1 diabetes, increasing cancer risk and possibly increasing allergic diseases in children. Randomised controlled trials have in general not confirmed these associations (Giustina et al. 2020).

Vitamin D deficiency rickets

As mentioned in the section above, the most common cause of nutritional rickets is vitamin D deficiency. To maintain vitamin D sufficiency, people are largely dependent on the conversion in the skin of 7-dehydrocholesterol to vitamin D_3 through the photochemical action of ultraviolet light, as the normal unfortified diet is generally deficient in vitamin D (see *Introduction to Human Nutrition*, Chapter 10). Thus, vitamin D deficiency is most common at the two extremes of life: in young infants who are unable to walk and in older people who are infirm and unable to go out of doors.

Vitamin D deficiency rickets is most prevalent in infants and toddlers between the ages of three and 18 months, although it may occur at any age if social customs, skin pigmentation and living in countries at extremes of latitude combine to prevent adequate ultraviolet B (UV-B) exposure of the skin. It is uncommon under three months of age because the newborn infant is provided with some vitamin D stores as 25-hydroxyvitamin D (25OHD) crosses the placenta, umbilical cord values being approximately two-thirds of maternal concentrations. However, the half-life of 25OHD is only 3–4 weeks, so values fall rapid after birth unless the infant is exposed to ultraviolet light or receives a dietary source of vitamin D. Furthermore, these stores may be inadequate if the mother is vitamin D deficient, resulting in an earlier presentation of rickets or, in rare cases, of the neonate being born with congenital rickets.

Human breast milk from unsupplemented mothers typically contains very little of the parent vitamin D or its metabolites, 25OHD being the major component (90%) of breast milk antirachitic activity. It has been estimated that normal human milk contains approximately 1.3–2.1% of maternal circulating 25OHD concentrations, which equates to an infant median daily intake of ~77 IU/d of antirachitic activity or <20% of an infant's recommended daily intake of 400 IU/d (vio Streym et al. 2016). Vitamin D activity in breast milk has been shown to be dependent on maternal circulating levels of 25OHD, which in turn are dependent on maternal sunlight exposure, dietary vitamin D intake and supplementation. Serum levels of 25OHD in exclusively breast-fed infants correlate with their sunshine exposure, highlighting the importance of UV-B exposure, rather than diet, in preventing rickets in breast-fed infants.

The amount of sunlight needed to maintain normal serum concentrations of 25OHD in the infant appears to be very little, although it will vary according to latitude and season. In Cincinnati, USA, during summer, it has been estimated to vary between 20 minutes a week for an infant in a nappy only to two hours a week for an infant fully clothed but without a cap. The reliance on sunlight exposure and the intradermal formation of vitamin D for the maintenance of vitamin D sufficiency in exclusively breast-fed infants has resulted in the presentation of clinical rickets having a strong seasonality in most studies. Rickets is most prevalent in late winter and the early spring months (Figure 12.16).

Because of the difficulty of ensuring vitamin D sufficiency in infants in general and breast-fed infants in particular, many national authorities, including the UK, recommend routine vitamin D supplementation (usually 10 μg/d [400 IU/d]) to breast-fed infants less than a year of age. Some authorities recommend that all infants should be supplemented irrespective of their mode of feeding (Munns et al. 2016).

Prior to the development of formula milks for the feeding of non-breast-fed infants, breastfeeding was noted to reduce the risk of rickets in young infants compared to those fed diluted cow's milk (the then current alternative to breast milk). More recent studies, however, consider breastfeeding to be a risk factor for vitamin D deficiency and rickets in infants. An explanation for this apparent paradox is that breast milk substitutes are now all required to be vitamin D fortified, thus protecting against vitamin D deficiency. Before the introduction of modified cow's milk formulae, non-breast-fed infants were fed unmodified or diluted cow's milk which, like breast milk, contains very little vitamin D, unless fortified. Furthermore, the calcium:phosphorus ratio of 1:1 and the high phosphate content of cow's milk both adversely affect calcium homeostasis in the relatively vitamin D-deficient infant, resulting in a higher prevalence of clinical rickets.

Although vitamin D deficiency is the ultimate cause of rickets in the Asian community in the UK, it is likely that several of the factors mentioned earlier combine to exacerbate the low vitamin D concentrations and the development of rickets (Figure 12.17). These factors include decreased vitamin D formation in the skin owing to increased skin coverage by clothing and a greater degree of melanin pigmentation than the Caucasian population, a low vitamin D intake, a low dietary calcium content of the mainly vegetarian diet, and poor intestinal absorption of calcium because of the high phytate content of the diet. Low calcium intakes and impaired calcium absorption have been shown to increase vitamin D

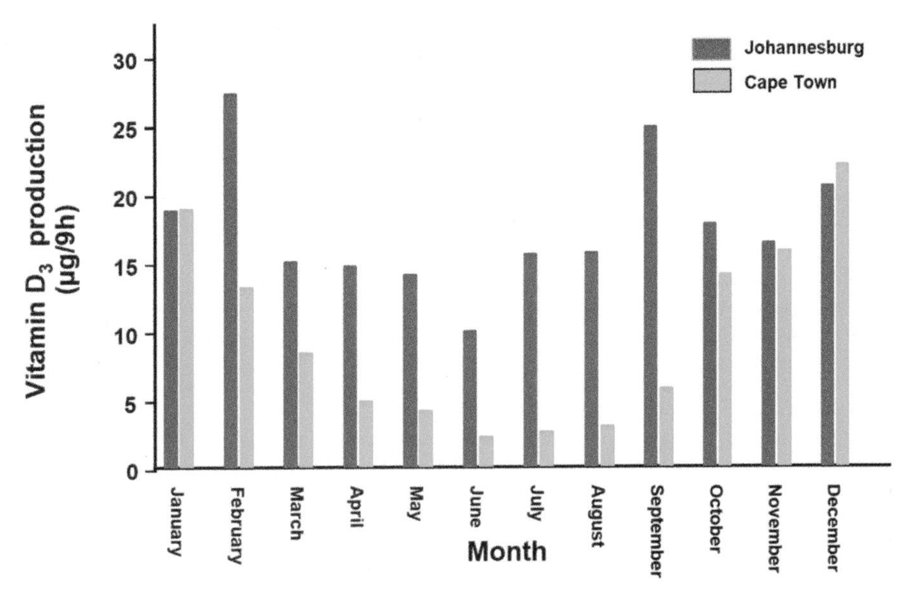

Figure 12.16 Seasonal variation in the production by sunlight of vitamin D_3 from 7-dehydrocholesterol in vitro in two cities at different latitudes in the southern hemisphere (Johannesburg 26°S and Cape Town 32°S). Note the production of vitamin D in the two cities during the spring and summer months (October through January) is similar, but during the autumn and winter months (March–September) vitamin D production is minimal in the more southerly city of Cape Town. *Source:* Reproduced from Pettifor J.M. et al., 1996/Health & Medical Publishing Group.

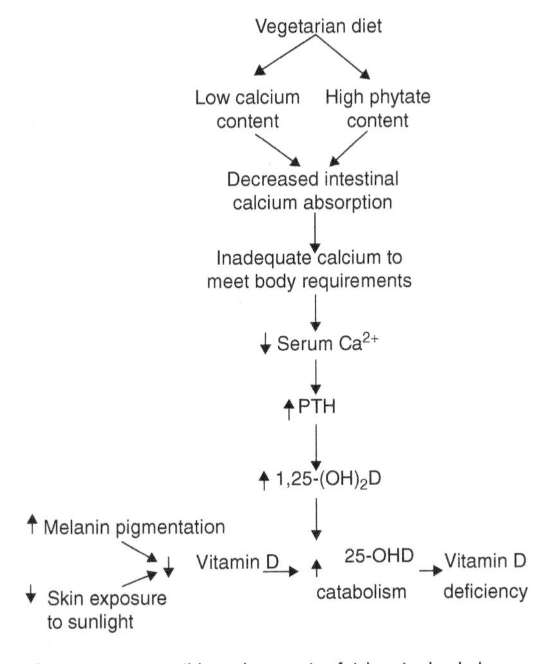

Figure 12.17 Possible pathogenesis of rickets in the darker skinned immigrant community in northern Europe. The disease is thought to be the result of a combination of relative vitamin D insufficiency and inadequate intestinal calcium absorption. A similar mechanism may also be responsible for nutritional rickets in children after weaning in Africa and south Asia. PTH, parathyroid hormone.

requirements through an increase in its catabolism as a result of the elevated $1,25(OH)_2D$ concentrations. Thus, these mechanisms push mild vitamin D insufficiency into frank vitamin D deficiency and rickets unless vitamin D intake or skin exposure to sunlight is increased (Table 12.10). Recommended vitamin D intakes are summarised in Table 12.11. Possible strategies to prevent vitamin D deficiency are summarised in Table 12.10.

Biochemical changes associated with vitamin D deficiency rickets

Vitamin D plays an important role in maintaining normal calcium homeostasis, mainly through optimising intestinal calcium absorption (as discussed in Chapter 8 of *Introduction to Human Nutrition*). The biochemical hallmark of vitamin D deficiency is a low serum/plasma concentration of 25OHD, which is the major circulating form of the vitamin. In the majority of studies of the vitamin D status in untreated vitamin D deficiency rickets, 25OHD concentrations are reported to be less than 4 ng/ml (<10 nmol/l). Vitamin D deficiency has been based by several authorities on a plasma concentration of 25(OH)D below which the risk of poor musculoskeletal health increases; in the UK this level is <10 ng/ml (<25 nmol/l) and in North America it is <12 ng/ml (<30 nmol/l). However, there is considerable debate around whether or not there should be a zone of vitamin D insufficiency between clear vitamin D deficiency and sufficiency.

Table 12.10 Possible strategies to prevent vitamin D deficiency rickets.

- Ensure regular skin exposure to sunlight
- Regular vitamin D supplementation
- Intermittent high-dose supplementation (*stosstherapie*)
- Food fortification

Table 12.11 Dietary reference values for vitamin D (μg/d) recommended by British, American and European expert committees.

United Kingdom[a]		US/Canada[b]		European Union[c]	
Lifestage	Males and females	Lifestage	Males and females	Lifestage	Males and females
Children					
0–6 mo*	8.5–10	0–6 mo$	10	0–6 mo^	–
7–11 mo*	8.5–10	6–12 mo$	10	7–11 mo	10
1–3 y*	10	1–3 y	15	1–3 y	15
4–6 y	10	4–8 y	15	4–10 y	15
7–10 y	10				
Adolescents					
11–14 y	10	9–13 y	15	11–17 y	15
15–18 y	10	14–18 y	15	18–24 y	15
Adults					
19–50 y	10	19–50 y	15	≥25 y	15
50 + y	10	51–70 y	15		
		>70 y	15		
Pregnancy and lactation					
Pregnancy	+0	Pregnancy	+0	Pregnancy	+0
Lactation	+0	Lactation	+0	Lactation	+0

Unit conversion for vitamin D: 1 μg = 40 IU; mo – months, y – years.
[a]Reference Nutrient Intake (except *Safe Intake), developed and published by the Scientific Committee on Nutrition (2016) based on maintaining a population protective level of serum 25-hydroxyvitamin D of ≥25 nmol/l when UVB exposure is minimal in 97.5% of the UK population.
[b]Recommended Dietary Allowance (except $Adequate Intake), developed and published by the Institute of Medicine (2011), based on coverage of the needs for nearly all the population at a serum 25-hydroxyvitamin D of 50 nmol/l.
[c]Adequate Intake, developed and published by the European Food Safety Authority (2016), based on the consideration that the majority of the population should achieve a serum 25-hydroxyvitamin D concentration near or above the target of 50 nmol/l (^no DRV was established for infants 0–6 mo).

The UK has calculated its RNI for vitamin D based on ensuring 25OHD levels ≥25 nmol/l (vitamin D sufficiency) in the general population. The IOM, however, has defined vitamin D sufficiency for North America as ≥50 nmol/l, being the level that would protect nearly all the population from the musculoskeletal effects of vitamin D deficiency. Higher thresholds, for example 75 nmol/l, have been proposed by other bodies, such as the Endocrine Society, but at present there is no consensus.

As 1,25(OH)$_2$D is considered to be the physiologically active metabolite of vitamin D, low levels would be expected to be reported in active vitamin D deficiency rickets. However, serum 1,25(OH)$_2$D concentrations are not consistent and low, normal or raised levels have been found in patients. It is suggested that the elevated levels are due to a slight increase in substrate, perhaps through sunlight exposure or dietary intake, causing a transient elevation in 1,25(OH)$_2$D concentrations. Rapid rises in 1,25(OH)$_2$D to supraphysiological levels have been documented in vitamin D-deficient subjects after the administration of very small doses of vitamin D or exposure to sunlight.

The biochemical progression of vitamin D deficiency in infants and young children has been divided into three biochemical stages, although there are no sharp boundaries between

Table 12.12 Progression of the biochemical and radiological changes in untreated vitamin D deficiency.

	Stage I	Stage II	Stage III
X-ray changes	Nil to slight osteopenia	Mild to moderate changes of rickets	Severe changes of rickets
Serum calcium	Low	Low to normal	Very low
Serum phosphorus	May be normal	Low	Low
Alkaline phosphatase	Normal to mildly elevated	Raised	Very raised
Parathyroid hormone	Normal to mildly elevated	Raised	Very raised

the stages (Table 12.12). Stage I is the earliest stage and is characterised by hypocalcaemia with normal serum phosphorus, alkaline phosphatase and PTH values. During this phase, the infant may present with features of hypocalcaemia without clinical or radiological signs of rickets being found. Frequently, this stage is not detected clinically as the infant may pass through this stage to stage II without clinical symptoms. In stage II, secondary hyperparathyroidism develops and partially corrects the hypocalcaemia, so serum calcium concentrations may be within the low normal range, but serum phosphorus values are low and alkaline phosphatase concentrations are elevated. In this stage, the typical radiographic features of rickets are found. Stage III is associated with severe clinical and radiological rickets, with hypocalcaemia once again occurring and alkaline phosphatase values reaching even higher levels.

Other biochemical features found in vitamin D deficiency rickets include decreased urinary calcium excretion, decreased renal tubular reabsorption of phosphorus, increased urinary adenosine monophosphate (cAMP) excretion, generalised aminoaciduria and impaired acid excretion, which are all features of hyperparathyroidism. Bone turnover markers are typically increased, with the effect being more marked on bone resorption markers. Characteristically, alkaline phosphatase values are elevated, but serum osteocalcin has been reported to be within the normal range in untreated rickets.

Histological changes of bone

Histological changes in vitamin D deficiency rickets occur both at the growth plate and at the endosteal and periosteal bone surfaces. At the growth plate, there is a failure of calcification of the longitudinal septa surrounding cartilage cells in the lower hypertrophic zone with the chondrocytes failing to undergo apoptosis, while cells in the proliferative zone continue to divide (Figure 12.14). Blood vessels invading the zone of provisional calcification also cease to proliferate. Thus, the growth plate widens and with the effect of weight bearing and continued proliferation, the longitudinal rows of cartilage cells and thus the growth plate become distorted and splayed. The underlying mechanism of the growth plate abnormalities is thought to be as a result of hypophosphataemia secondary to the induced hyperparathyroidism (Tiosano and Hochberg 2009).

At the endosteal and periosteal bone surfaces, newly formed osteoid produced during modelling and remodelling fails to mineralise so the trabecular surface covered by unmineralised osteoid increases and osteoid seams widen. Although these features are typical of osteomalacia, they are not pathognomonic unless they are accompanied by the finding of an increase in the mineralisation lag time (the time taken for newly laid osteoid to mineralise).

Radiological changes of bone

The radiological changes of rickets are typically seen best at the growth plates of rapidly growing bones, so the distal radius and ulna, and the femoral and tibial growth plates at the knee are the usual sites examined (Figure 12.18). The characteristic changes include loss of the provisional zone of calcification and thus blurring of the demarcation between the metaphysis and the cartilaginous growth plate, widening of the growth plate (in the vertical dimension), and cupping and splaying of the distal metaphysis. The epiphyses typically are poorly mineralised and underdeveloped, resulting in a delay in the bone age compared with chronological age. In older children and adolescents, features of osteomalacia, such as Looser zones or pseudofractures, may predominate.

The type and position of deformities of the long bones vary depending on the age of the

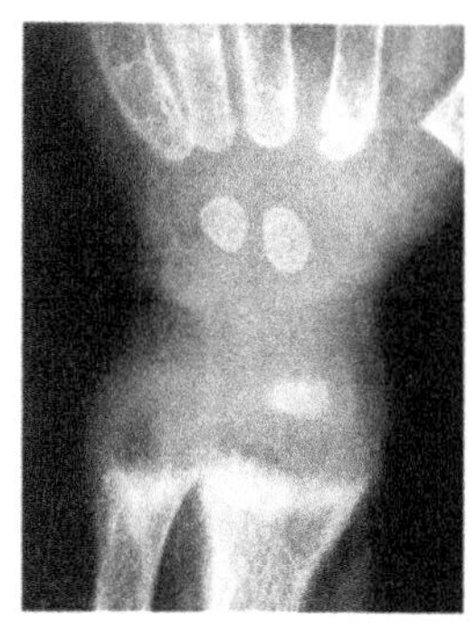
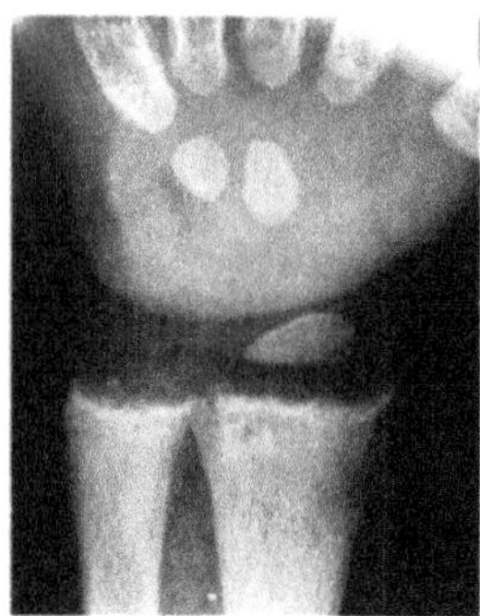
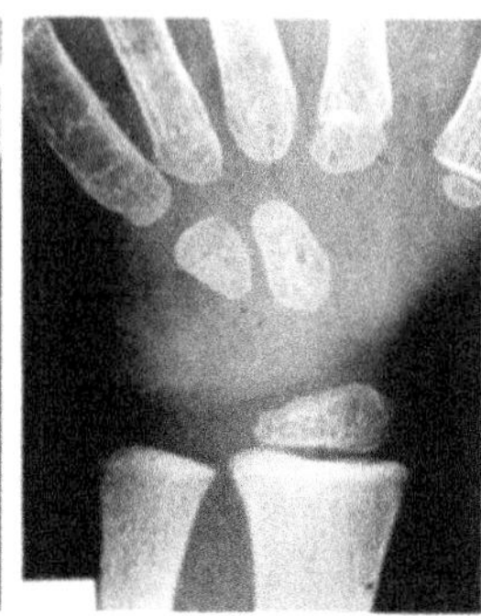

Figure 12.18 Radiographic changes of rickets. Note the progressive improvement in the radiographic picture from left to right in response to vitamin D treatment over a period of three months. *Source:* Reproduced with permission from Pettifor 1994, © NestecInc.com.

child, the degree and direction of weight bearing and the severity of the rickets. If hypotonia is severe, deformities might not develop until weight bearing occurs with the commencement of treatment. Genu varum (bow legs) tends to occur in young children who develop rickets, as this age group tends to have a normal physiological bowing. In the older child, genu valgum (knock knees) or wind-swept deformities of the legs are more common.

Once treatment has commenced, the initiation of radiological healing may be seen within three weeks, with the appearance of a broad band of increased density occurring at the position of the provisional zone of calcification at the ends of the metaphyses (see Figure 12.18). Over the following weeks, the growth plates narrow, the epiphyses mineralise and cortices thicken. Modelling of deformities occurs so that in many situations, apparently severe deformities disappear over a period of months to a few years. The spontaneous correction of deformities tends to occur more frequently in children under five years of age than in older children.

Clinical presentation

The clinical picture of rickets depends to a certain extent on the age of the child at presentation. Clinical features of hypocalcaemia may occur at any age, but are more common in the young infant, when they may present with apnoeic attacks, tetany, convulsions, stridor or features of dilated cardiomyopathy. Delay in motor milestones, enlargement of the ends of long bones and costochondral junctions, lower limb deformities and hypotonia are common presentations. Typically, the infant is floppy and sweating and has a protuberant abdomen (Figure 12.19). In severe rickets, hepatosplenomegaly may be present, partially pushed down by the flattened diaphragm but also enlarged by extramedullary erythropoiesis.

The clinical picture of hypocalcaemia and rickets is covered in greater detail in the *Clinical Nutrition* textbook of this series.

Dietary calcium deficiency

Studies conducted since the 1970s in a number of countries, including South Africa, Nigeria, The Gambia, Bangladesh and India, have highlighted the importance of low dietary calcium intakes in the pathogenesis of nutritional rickets in vitamin D-replete children outside the infant age group. Prior to these studies, it was believed that low dietary calcium intakes were not responsible for rickets in children, except in a few very unusual situations, when young children with gastrointestinal problems were placed on very restricted diets that were very low in calcium (Figure 12.20).

The biochemical hallmark of rickets due to dietary calcium deficiency rather than vitamin D deficiency is the finding of 25OHD values within the normal range and elevated $1{,}25(OH)_2D$ concentrations in children who are older than those normally at risk of developing vitamin D deficiency (Table 12.13). Further dietary calcium intakes are characteristically low, at about 200 mg/day, and diets are devoid of dairy products and high in phytates and oxalates. In South Africa, the children with dietary calcium deficiency are usually aged between six and 16 years and live in rural parts of the country. Unlike those with vitamin D deficiency, the

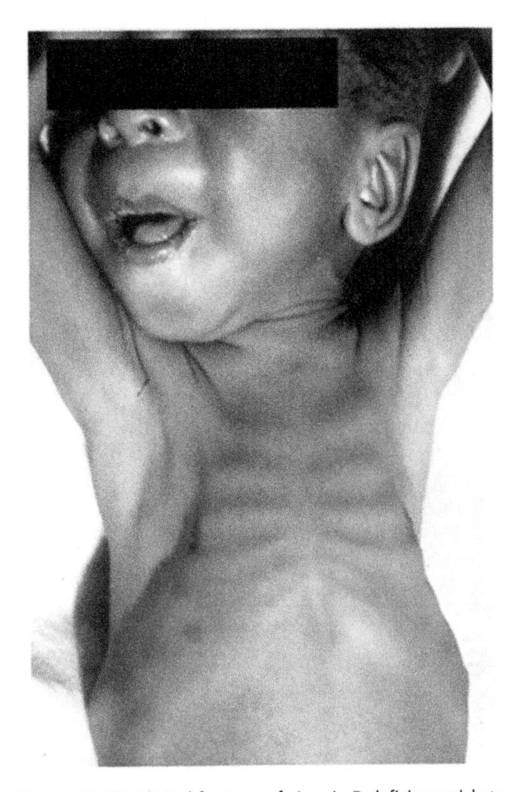

Figure 12.19 Clinical features of vitamin D deficiency rickets in an eight-month-old child. Note the marked chest deformities and the rather protuberant abdomen. *Source:* Reproduced with permission from Pettifor et al. 2018, © Elsevier.

children with dietary calcium deficiency do not have muscle weakness. In Nigeria, the children are younger than those described from South Africa. They tend to present around four years of age, having had symptoms for approximately two years. In both countries, medical assistance is generally sought because of progressive lower limb deformities.

Support for the hypothesis that dietary calcium deficiency plays an important role in the pathogenesis of rickets in these children comes from the therapeutic response to calcium supplements alone. Reanalysis of the data from the original studies in Nigeria has highlighted the synergistic effect of the combination of low dietary calcium intakes and relatively low vitamin D status on increasing the odds of having rickets in young children (see Figure 12.15). It is likely that many children with rickets suffer from varying combinations of poor vitamin D status and low calcium intakes.

It is not known how widespread dietary calcium deficiency is. It is likely that children with dietary calcium deficiency have been and continue to be diagnosed as vitamin D deficient and treated with vitamin D and calcium supplements. On this regimen, the bone disease will respond, and thus the incidence of dietary calcium deficiency will be severely underestimated. Nevertheless, the apparent scarcity of the clinical disease needs to be confirmed by more detailed studies in developing countries where maize (corn) or rice forms a major part of the cereal staple. Although in both Nigeria and South Africa maize is the staple, in Bangladesh rice is the cereal staple, and it has been suggested that the low dietary calcium intakes associated with its almost exclusive consumption may be responsible for an apparent increase in the incidence of rickets in children since the 1980s (Figure 12.21).

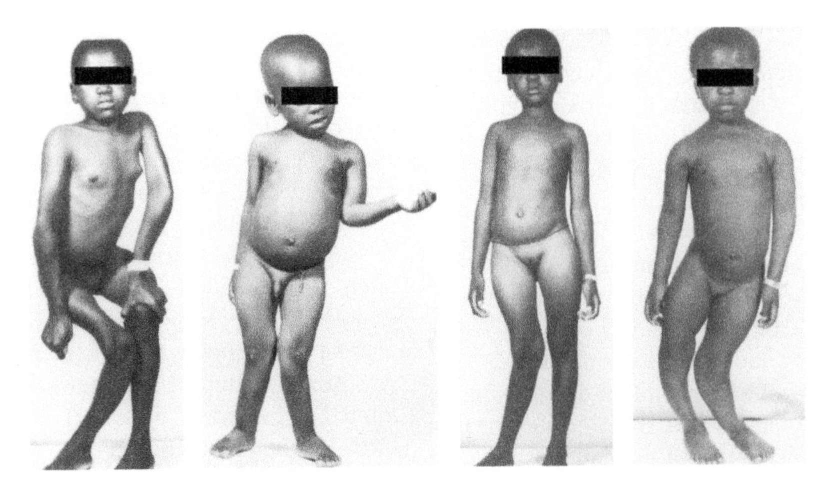

Figure 12.20 Lower limb deformities in children in South Africa with dietary calcium deficiency rickets. *Source:* Reproduced with permission from Pettifor 1994, © NestecInc.com.

Table 12.13 Differentiating features between vitamin D deficiency and dietary calcium deficiency rickets.

	Vitamin D deficiency	Dietary calcium deficiency
Onset	Usually between six and 18 months	Usually after weaning (>2 years)
Hypotonia	Present	Usually not present
Biochemistry:		
25OHD	Usually <10 nmol/l	Usually >25 nmol/l
$1,25(OH)_2D$	Variable (may be low, normal or elevated)	Elevated
Dietary calcium intake	Variable (usual close to RDA)	Low (usually ~200 mg/day)

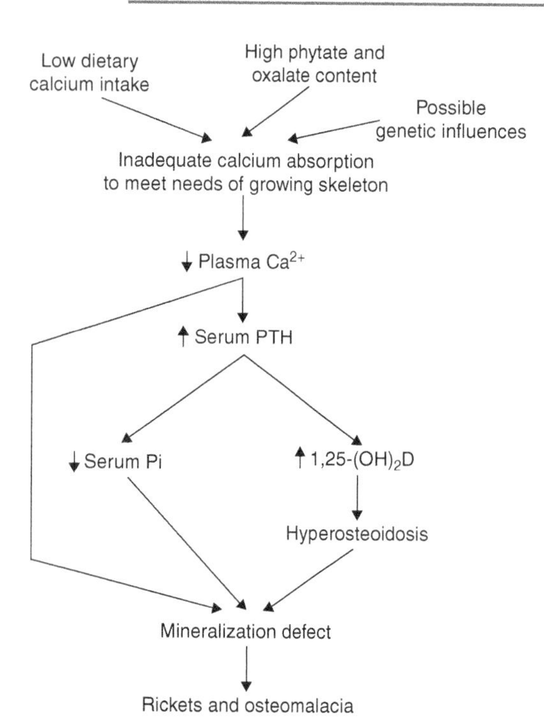

Figure 12.21 Proposed pathogenesis of dietary calcium deficiency rickets. Note that a combination of factors may play a role in the pathogenesis. PTH, parathyroid hormone; Pi, inorganic phosphate; $1,25(OH)_2D$, 1,25-dihydroxyvitamin D.

Table 12.14 Typical biochemical features of metabolic bone disease in preterm infants.

- Hypophosphataemia
- Normal or increased serum calcium levels
- Elevated $1,25(OH)_2D$ concentrations
- Raised alkaline phosphatase concentrations
- Normal parathyroid hormone concentrations
- Hypercalciuria
- Reduced urinary phosphate excretion

Rickets of prematurity

Over the past few decades, rapid advances in the management of very low birth-weight infants (<1500 g) have resulted in a marked improvement in survival rates, particularly for extremely low birth-weight infants (weighing <1000 g). Associated with the increased survival has been an increase in clinical and biochemical evidence of metabolic bone disease in these infants (Table 12.14). It is estimated to affect some 50% of infants <1000 g and 23–32% of infants <1500 g, who manifest some features of the disease, which become apparent 4–12 weeks after birth. Metabolic bone disease of prematurity encompasses a range of radiographic abnormalities from osteopenia to frank rickets and pathological fractures and laboratory perturbations, the most common of which are hypophosphataemia and elevated alkaline phosphatase levels. The radiographic features of the disease at the wrist include loss of the provisional zone of calcification at the metaphysis, increased submetaphyseal lucency, thinning of the cortices, and fraying, splaying and cupping of the metaphyses. The major risk factor associated with the development of the disorder is severe immaturity (<1000 g) accompanied by breast milk feeding, diuretic and steroid administration and prolonged illness (Rustico et al. 2014). Although the pathogenesis is multifactorial, phosphorus depletion due to an inadequate intake to meet the needs of the rapidly growing infant is considered to be mainly responsible.

Supplementation of the breast-fed premature infant with phosphorus rapidly improves the retention of both calcium and phosphorus, and corrects the biochemical picture of the phosphorus depletion syndrome. Once an adequate phosphorus intake is assured, the calcium content of breast milk may become a limiting factor, so simultaneous calcium supplementation is often recommended. However, care should be exercised as supplemental calcium may precipitate out before being administered to the baby. As vitamin D stores in the newborn infant are limited and many of these very premature low birth-weight infants spend prolonged periods in intensive care units and neonatal nurseries, an adequate vitamin D intake needs to be ensured.

It is recommended that vitamin D 800–1000 IU/daily should be provided as a supplement.

Osteomalacia

Osteomalacia and rickets have the same aetiologies, with vitamin D deficiency being the most common in both conditions. The characteristic feature of the two conditions is a failure of mineralisation of either cartilage or bone matrix; the difference between osteomalacia and rickets is in the organs involved, in that rickets involves mineralisation of the growth plates (endochondral bone formation) with the participation of chondrocytes and cartilage, while osteomalacia involves bone formation at the endosteal and periosteal bone surfaces (intramembranous bone formation) through the participation of osteoblasts. In children with open growth plates, rickets and osteomalacia occur simultaneously, while in adults only osteomalacia can occur because the growth plates are fused. In children, the majority of the clinical signs are related to rickets, although osteomalacia may also contribute, for example in producing bone pain and fractures, while in adults the symptoms are related to osteomalacia alone.

Nutritional osteomalacia occurs typically in the elderly and infirm, who are no longer self-sufficient and find it difficult to get out of doors, thus limiting their exposure to sunlight. Symptoms are frequently vague and non-specific; general aches and pains, difficulty in getting out of chairs and in climbing stairs, and a proximal myopathy. Depression is also considered to be a non-specific feature of vitamin D deficiency. These symptoms may be associated with an increased risk of falls and fractures. The major condition from which osteomalacia must be differentiated is osteoporosis which, as discussed later (see section Osteoporosis: a global perspective), is also typically a disease of the elderly, and may present with back pain associated with vertebral fractures.

Radiographically, osteomalacia may be difficult to diagnose, as once again features are often absent or may be similar to those of osteoporosis. The bones look osteopenic with thin cortices and reduced trabecular bone pattern. As opposed to osteoporosis, the trabecular pattern may be coarse and pseudofractures (Looser zones) may be visible, particularly on the pelvic rami and the medial cortical surfaces of the femurs (Adams 2018).

The biochemical changes associated with nutritional osteomalacia are similar to those found in nutritional rickets. Unlike postmenopausal osteoporosis, in which serum calcium, phosphorus, parathyroid hormone and alkaline phosphatase values are typically normal, osteomalacia is characterised by hypocalcaemia, hypophosphataemia and elevated alkaline phosphatase and parathyroid hormone concentrations. Serum 25OHD levels are typically in the vitamin D deficiency range.

Osteoporosis: a global perspective

The WHO has defined osteoporosis as a condition characterised by low bone mass and microarchitectural deterioration of bone tissue that leads to enhanced bone fragility and a consequent increase in fracture risk. These fractures occur most commonly in the wrist, spinal vertebrae and hip, but can occur elsewhere in the skeleton. Osteoporosis is further defined in postmenopausal women, and men aged 50 years and over, by the WHO as BMD, measured by DXA, of more than 2.5 standard deviations below the young adult mean (DXA T-score). Fractures are the clinically important manifestation of osteoporosis, whereas low BMD classifies those at risk.

Osteoporosis is a major health problem among older adults. In 2017, in the UK, over 3 million people were affected, with 536 000 fragility fractures a year; approximately 79 000 hip fractures, 66 000 clinically diagnosed vertebral fractures and 69 000 wrist fractures occur each year, estimated to cost over £4.4 billion/year in 2012. Clinically diagnosed vertebral fractures are the most prevalent type of osteoporotic fracture and hip fractures most associated with morbidity and mortality. More than one in two British women and one in six men over the age of 50 years can expect to experience an osteoporotic fracture during their remaining years.

Similarly high fracture rates occur in white populations of northern Europe, the USA and Australasia. However, in other populations, such as those of Africa and China, the incidence is much lower, at least in terms of hip fracture, which is the most reliable statistic available (see Figure 12.11). In countries where hip fracture rates are high, women are at greater risk than men. Within countries, there are differences between ethnic groups in hip fracture risk. For example, black,

Indian and mixed-race South Africans are at lower risk than their white counterparts, while the incidence of hip fractures in Singapore is highest in the Indian community. However, within all regions and ethnic groups, hip fracture incidence is increasing because of the rise in the number of people surviving to an age when hip fractures are more common. Populations projected to see the steepest rise in older people are in low- and middle-income countries. It has been estimated that the global incidence will increase from 1.66 million in 1990 to 6.26 million in 2050 because of the change in demographic profile (Table 12.15). This inevitable rise in fragility fractures will create high healthcare demand and burden, particularly in countries where resources are poor.

Fracture risk assessment tools have been developed to be used in conjunction with DXA or standalone; the Fracture Risk Assessment Tool (FRAX), Garvan and Q-Fracture are the most widely applied. FRAX is now available in over 66 countries globally, one of the most recent being South Africa, covering around 80% of the world's population. By taking a combination of the most common risk factors for osteoporosis, FRAX provides an individual with a predicted 10-year risk of any osteoporotic fracture or a hip fracture (Fuggle et al. 2019). Risk factors included in FRAX are age, sex, height, weight, fracture history, parental hip fracture history, smoking, glucocorticoids, rheumatoid arthritis, secondary osteoporosis, alcohol >3 units per day, femoral neck BMD (optional).

The pathogenesis of osteoporosis is still a matter of intensive research. Factors that have been implicated include low sex hormone concentrations, especially oestrogen withdrawal at the menopause, compromised supply or metabolic handling of calcium and other nutrients, poor vitamin D status and physical inactivity. Low bone mass and propensity to fracture are heritable traits in the pathogenesis of the disease. Genome-wide association studies in large consortia or biobanks indicate that osteoporosis is a complex polygenic disease.

There are no accepted explanations for the geographical and ethnic differences in fracture incidence (Cauley et al. 2014). Effects of differences in diet, sunlight exposure and activity levels have been suggested, as well as genetic influences. With increased longevity in people living wih HIV, now a chronic disease of ageing, this is an important risk factor to consider, particularly in populations with a high prevalence of HIV. The variation in fracture incidence is unlikely to be related to low bone mass per se, since Asian and African people generally have lower BMC than their western counterparts. This is partly, but not entirely, related to their smaller body, and therefore bone size. Global variations in osteoporosis cannot be explained by differences in calcium intake because hip fracture incidence is greatest in countries with the highest calcium intakes, such as those of northern Europe. Differences in skeletal anatomy, metabolism and microarchitecture may be important, such as the shorter hip axis length of Asian and African people relative to their height, and the higher bone turnover and greater trabecular width in the spines of black South Africans.

Loss of muscle mass and strength (sarcopenia) with ageing also contributes to bone loss and risk of osteoporotic fracture. Sarcopenia was given an ICD-10 code in 2016, and has several working definitions which use combinations of thresholds for grip strength and walking speed, with or without lean mass (Cruz-Jentoft and Sayer 2019). Given the lack of access to DXA in many ageing populations, and the limitations of DXA-associated lean mass and vastly different body composition across the globe, it may perhaps be most appropriate to find definitions which do not rely on DXA. To date, a limitation for sarcopenia is the lack of a single definition. However, what is clear is that the working definitions do not necessarily apply across populations. Sarcopenia and osteoporosis share many common risk factors, including age, sex, poor nutrition including low 25(OH)D, poor diet quality, low protein intakes and reduced physical activity (see later sections on diet and physical activity). Sarcopenia may also be a causal factor in age-related bone loss and increased fracture risk. This may be through the associated loss of type II muscle fibres resulting in reduced ability

Table 12.15 Important facts about osteoporosis.

- Major public health problem in many countries
- Prevalence varies with ethnicity
- Incidence rising in most countries creating inevitable demands on healthcare systems
- Lifestyle factors probably play an important role in the pathogenesis
- BMD combined with clinical risk scores, including FRAX, can be used to predict which individuals are at high risk of fracture
- There is a concerning treatment gap between those who are at increased risk of fracture versus those receiving treatment

to arrest a fall, and with falls comes a consequent increased risk of fracture. Secondly, lower muscle strength reduces loading to bones, and with that reduction in loading comes bone loss.

Postmenopausal osteoporosis

After the attainment of peak bone mass during the second to third decades of life, bone tissue is gradually lost from the skeleton in both men and women. Periosteal bone growth continues, and the tissue that is lost is from endosteal surfaces. In women, this bone loss is accentuated in the 10–15 years that follow oestrogen withdrawal at the menopause, when approximately one-third of cortical bone tissue and one-half of cancellous bone tissue are lost.

Bone loss is the net consequence of imbalances between bone resorption and formation resulting from changes in the recruitment, activity and lifespan of osteoclasts and osteoblasts. Two types of bone loss are often distinguished, postmenopausal (type I) and age-related (type II) osteoporosis, the first being characterised by a relative increase in osteoclast numbers, the latter by a relative decrease in osteoblast numbers. However, the classification of osteoporosis is not clear-cut; both types may co-exist or may represent parts of the same continuum. The features of postmenopausal osteoporosis are summarised in Table 12.16.

Postmenopausal, or type I, osteoporosis affects individuals aged 50–75 years and is more common in women than in men. Accelerated bone loss is associated with reductions in calcium absorption and in circulating concentrations of PTH and $1,25(OH)_2D$. Fractures occur most commonly at the distal radius (Colles fracture of the wrist) and in the spinal vertebrae, regions rich in cancellous bone. Oestrogen deficiency is regarded as the main pathogenetic factor; of type I osteoporosis at the cellular level, oestrogen withdrawal produces a greater number of active osteoclasts, more bone remodelling sites and deeper excavation of resorption cavities. Trabeculae become perforated and may disappear altogether, which reduces the strength of cancellous bone to withstand compressive forces. A number of antiresorptive agents are available for the treatment of postmenopausal osteoporosis, primarily bisphosphonates.

Table 12.16 Features of postmenopausal osteoporosis.

- 45% of 50-year-old women will have an osteoporotic fracture in their lifetime
- It is estimated that 96% of fractures occur in women in Europe without low BMD which is why population-level screening with DXA is not recommended
- Occurs between 50 and 75 years of age
- Associated with fractures of the distal forearm and vertebrae
- Associated with oestrogen withdrawal
- Associated with excessive osteoclast activity

Age-related osteoporosis

Age-related, or type II, osteoporosis affects women and men over the age of about 70 years (sex ratio 2:1). In general, bone loss is not accelerated and bone turnover may be reduced. Calcium absorption is reduced, $1,25(OH)_2D$ concentrations may be low but PTH levels may be increased. Fractures occur commonly in the proximal femur (hip) and spinal vertebrae. They are associated with trabecular thinning in cancellous bone and the presence of giant resorption cavities in cortical bone.

Ageing is associated with several factors that may result in bone loss and increased bone fragility. These include decreases in physical activity and muscle strength that have direct effects on the skeleton, and also indirect factors via impaired neuromuscular protective mechanisms, which increase the likelihood of falling and of a fall resulting in fracture. Age-related low oestrogen and growth hormone concentrations may impair the function and senescence of osteoblasts, resulting in declining bone turnover and osteocyte numbers, which may reduce the ability to repair fatigue microdamage. Secondary hyperparathyroidism caused by declining renal function and compromised vitamin D status may also be important, via direct effects on the skeleton and indirect effects on calcium handling. The combination of calcium and vitamin D supplementation has been shown to reduce the incidence of non-vertebral fractures in older men and women (Harvey et al. 2017). The features of age-related osteoporosis are summarised in Table 12.17.

Low circulating levels of endogenous oestrogens are produced in women even though they are many years postmenopause and these may continue to play an important role in the bone health of older women. Current oestrogen use in women aged 65 and older, for example, has been associated with a decreased risk of non-vertebral fractures, especially at the wrist. Similarly in men, some of the skeletal

Table 12.17 Features of age-related osteoporosis.

- Occurs generally in people over 70 years of age
- Female:male ratio 2:1
- Incidences of age-specific vertebral, forearm and hip fractures are increasing due to the growing ageing population across the globe
- There is a wide global variation by geography, ethnicity and
- socioeconomic status
- Associated with impaired osteoblast function
- Associated with decreased intestinal calcium absorption, hyperparathyroidism and poor vitamin D status

effects of testosterone appear to be mediated through its conversion by aromatase to oestradiol, and this may be one factor involved in male osteoporosis associated with low testosterone concentrations. It has been proposed that oestrogen deficiency is the underlying cause of the late, slow phase of bone loss in postmenopausal women and the continuous phase of bone loss in ageing men in addition to the early, accelerated bone loss post menopause.

Role of vitamin D deficiency

Poor vitamin D status may have a role in the pathogenesis of age-related bone loss and fracture, although frank osteomalacia is rarely implicated (Box 12.1). It is thought that the mechanism for the increase in bone loss and fractures is via the resulting secondary hyperparathyroidism, although the muscle weakness and depression associated with vitamin D deficiency may also be important.

Vitamin D is obtained either from the diet or by endogenous production in the skin by the action of sunlight (see Chapter 8 in *Introduction to Human Nutrition*). The intermediate metabolite 25OHD, produced from vitamin D in the liver, is a useful status marker, being responsive to changes in dietary vitamin D and to sunlight exposure. Plasma 25OHD concentrations decline with age, as a result of decreased endogenous production of the vitamin caused by reduction in the exposure to sunlight, in the tissue content of the precursor, 7-dehydrocholesterol, and in the efficiency of the synthetic process.

Associations have been reported between 25OHD concentration and BMD in middle-aged and older women. However, vitamin D intervention trials of older people with either bone loss or fracture as outcome have met with mixed success, possibly reflecting differing degrees of vitamin D insufficiency in the various study populations. Trials of calcium and vitamin D together have resulted in a decreased incidence of non-vertebral fracture. Calcium plus vitamin D supplementation is recommended for patients with osteoporosis at high risk of developing vitamin D or calcium insufficiency, and in those taking anti-osteoporosis medications.

Box 12.1 Vitamin D and older people

- High prevalence of poor vitamin D status in older people
- Poor vitamin D status associated with decreased sunlight exposure, low 7-dehydrocholesterol levels in the skin, poor vitamin D intakes, decreased kidney and liver function for conversion of vitamin D to active metabolites and obesity
- Vitamin D plus calcium supplements modestly reduce fracture risk
- Reduced kidney/liver function affects vitamin D status in older people fractures, but not consistently the attenuation of bone loss
- Vitamin D may play a role in reducing falls in older people, particularly those with very low 25OHD

Traditionally, a serum 25OHD concentration of 25 nmol/l has been used to define the lower end of the normal range, a value higher than that at which clinical osteomalacia is usually seen. Using this cut-off, a significant prevalence of vitamin D insufficiency has been recognised in the older populations of UK and elsewhere in Europe, particularly those in residential accommodation. A recent meta-analysis of data from Africa notes the paucity of information in adolescents and older adults. Several studies have reported inverse relationships between plasma concentrations of 25OHD and PTH, suggesting that the rise in PTH that accompanies declining vitamin D status may occur at plasma 25OHD concentrations higher than the traditional cut-off, especially in older age groups. There are calls to increase the lower cut-off of normality for 25OHD to match a putative threshold below which the concentration of PTH would be expected to rise. Such a change would substantially increase the numbers of people classified as vitamin D deficient and would, perhaps, prompt greater action to improve vitamin D status. However, on a population basis, there is a wide variation in PTH concentration at any given 25OHD level and such a threshold has yet to be defined with any certainty. In addition, in the context of age-related bone loss and fragility fractures, it is currently unclear what concentrations of PTH, and hence of 25OHD, are optimal for long-term bone health.

12.4 Effects on bone health of specific nutrients and dietary factors

In this section, the effects of specific nutrients on bone health will be discussed. It should be borne in mind that isolated nutrient deficiencies are unusual; rather, they reflect an unbalanced diet and therefore may be associated with other less obvious nutrient or energy deficiencies, which could mask or aggravate the apparent effects on bone. However, all dietary reference values (DRV) are developed on the basis that they are for healthy individuals and that intake of all other nutrients is sufficient.

Calcium

Calcium is one of the main bone-forming minerals (99% of the body's approximately 1000 g of calcium is in bone), so an appropriate supply to bone is essential at all stages of life. Only approximately one-third of dietary calcium from a western-style diet is absorbed, and calcium is lost from the body by urinary excretion and in dermal and gastrointestinal secretions. Most estimates of calcium requirements are based upon either a factorial approach, where calculations of skeletal accretion and turnover rates are combined with typical values for calcium absorption and excretion, or a variety of methods based on experimentally derived balance data. Reference calcium intakes current in the UK, US/Canada and European Union are shown in Table 12.18.

Calcium intake and skeletal growth

Direct evidence from supplementation studies has shown that increases in maternal calcium intake during pregnancy and lactation have no effect on the growth and bone development of the offspring during fetal or infant life. Increases in calcium intake of children and adolescents by supplementation with calcium salts, such as calcium carbonate, have been associated with higher BMD but these are considered to reflect a temporary decrease in bone remodelling rate that is not sustained after the intervention ceases. Supplementation with milk appears to increase bone mineral by promoting skeletal growth, but this may involve components of milk other than calcium.

Whether these interventions ultimately alter peak bone mass and, if so, whether later fracture

Table 12.18 Dietary reference values for calcium (mg/d) recommended by UK, US/Canadian and European Union expert committees.

United Kingdom[a]			US/Canada[b]			European Union[c]		
Lifestage	Males	Females	Lifestage	Males	Females	Lifestage	Males	Females
Children								
0–6 mo	525	525	0–6 mo[$]	200	200	0–6 mo^	–	–
7–12 mo	525	525	6–12 mo[$]	260	260	7–11 mo[¶]	280	280
1–3 y	350	350	1–3 y	700	700	1–3 y	450	450
4–6 y	450	450	4–8 y	1000	1000	4–10 y	800	800
7–10 y	550	550						
Adolescents								
11–14 y	1000	800	9–13 y	1300	1300	11–17 y	1150	1150
15–18 y	1000	800	14–18 y	1300	1300	18–24 y	1000	1000
Adults								
19–50 y	700	700	19–50 y	1000	1000	≥25 y	950	950
50+ y	700	700	51–70 y	1000	1200			
			>70 y	1200	1200			
Pregnancy and lactation								
Pregnancy		+0	Pregnancy		+0	Pregnancy		+0
Lactation*		+550*	Lactation		+0			+0

[a] Reference Nutrient Intake, developed and published by the Committee on the Medical Aspects of Food Policy 1991 and re-evaluated and endorsed in 1998; [b] Recommended Dietary Allowance (except [$]Adequate Intake), developed and published by the Institute of Medicine 2011; [c] Population Reference Intake (except ^ no DRV was established for 0–6 mo and Adequate Intake), developed and published by the European Food Safety Authority 2015.
* Lactation for 0–4 months and >4 months; indicated in 1998 that the increment 'may not be necessary'. mo – months, y – years
Source: Adapted from Prentice A., 2021.

risk is reduced has yet to be determined. However, all three of the DRV committees in Table 12.18 considered that there was no evidence of benefit of calcium intakes above those determined from estimates based on calcium retention rates and balance studies. Very low calcium intakes appear to predispose children to rickets, especially if combined with poor vitamin D status and deficiencies of other nutrients such as iron. This is discussed more fully in an earlier section.

Boys and girls differ in their growth, skeletal development and maturity rates, especially during puberty, but currently no distinction is made in the reference intakes for calcium in childhood and pre-adolescence. However, sex differences in the effects of calcium supplementation have been reported in some studies. For example, in rural Gambia, a population with a very low calcium intake, girls born to mothers who had been supplemented with calcium carbonate during pregnancy were shorter, lighter with smaller and less dense bones at age 8–12 years than girls born to mothers consuming a placebo, whilst the converse trend was seen in boys (Ward et al. 2017). This was associated with similar disparities in plasma IGF1 earlier in childhood, suggesting that the pregnancy supplement may have altered the timing of the pubertal growth spurt through sex-specific in utero programming of the growth hormone–IGF1 axis. In a separate Gambian trial, supplementation of children with calcium carbonate for 12 months at age 8–12 years, prior to the onset of physical signs of puberty, resulted, in boys, in an earlier pubertal growth spurt and effects on growth (Prentice et al. 2012). The lack of an effect in girls may have been due to their age at intervention because a trial of a calcium phosphate salt extracted from milk consumed for a year by 6–9-year-old Swiss girls was shown to have advanced the age of menarche (Chevalley et al. 2005).

More investigations are required to confirm and add to these findings, but these studies suggest that there may be unexpected consequences of calcium supplementation in populations with a low calcium intake that could be a cause for concern.

Calcium intake and peak bone mass

Although linear growth ceases with the fusion of the long bone epiphyses, bone growth continues at a very low rate into adulthood, peaking at different ages depending on the region of the skeleton, the rate of maturation and the sex of the individual. Correlations between calcium intake and young adult BMC/BMD have been reported in a large number of cross-sectional and retrospective studies, although there are many other studies where no such association has been observed. Meta-analyses have concluded that calcium intake is a significant predictor of adult BMD, but the magnitude of the effect is small, about 1% of the population variance. Some studies suggest that an adequate calcium intake is required to optimise the effects of physical activity in childhood and adolescence on peak bone mass. Interpretation of these associations is difficult, however, because few studies have adjusted adequately for the confounding effects of body size and, in calcium supplementation trials, for bone turnover rate. Calcium intake and the bone health of women during pregnancy and lactation are discussed in an earlier section.

Calcium and the peri- and postmenopausal periods

Studies evaluating current calcium intake as a risk factor for low BMD in 40–60-year-old women have shown inconsistent results. Calcium supplementation given to women around the time of the menopause has little or no effect on the BMD of cancellous regions of the skeleton, where the greatest loss of bone is occurring at that time, but may cause a modest increase in regions rich in cortical bone because of its effects in reducing bone resorption.

Calcium intake and bone loss during ageing

Studies of an association between customary calcium intake and BMD in older individuals have produced inconsistent results. Longitudinal studies, which follow people prospectively over time, have shown a relationship between low customary calcium intake and bone loss in some studies but not others. Case–control studies in Britain, Australia and Canada, populations with medium to high average calcium intakes, have reported no relationship between customary calcium intake and the risk of hip fracture, whereas studies in Hong Kong and southern Europe, where average calcium intakes are lower, have observed an increase in hip fracture risk with declining calcium intake. Cohort studies gave similar results. While meta-analyses have suggested an increase in hip fracture incidence with declining calcium intake, this effect appears to be strongest in populations with a comparatively low average calcium intake and suggests that there is a threshold of increasing risk below around 400–500 mg/day. Many populations, especially in Africa, have average customary calcium intakes at or below this level but with lower age-adjusted hip fracture rates than in countries with a higher calcium intake (Figure 12.22). However, more recent studies suggest that the degree of bone loss and osteoporosis prevalence in these populations may be higher than previously recognised.

Calcium supplementation in older women is associated with a higher BMD, by around 1–3%, with reductions in bone loss in the first 1–2 years after supplementation is started, although it does not prevent some loss from occurring. Calcium has antiresorptive properties, and the increase in BMD that accompanies calcium supplementation is thought to reflect a reduction in the activation of new bone remodelling sites, the infilling of current resorption cavities and an increase in the reversible calcium space. Once this process is complete and a new steady state has been achieved, no further increase in BMD occurs and bone loss continues at a similar rate to before. There have been only a few calcium supplementation trials with fracture as an endpoint. Results have been modest and inconsistent, but sample sizes have been small. Larger trials of calcium and vitamin D supplementation in older people have demonstrated reductions in the incidence of non-vertebral fractures.

In general, the effect of customary calcium intake on the outcome of calcium supplementation has not been investigated. In those studies where it has, no relationship has been noted, except in a study of American women where the effect on BMD and bone loss was limited to those with a daily calcium intake below 400 mg/day. Taken together with the observational data, this suggests, for older adults living in western-style environments, that customary calcium intakes below the UK lower reference nutrient intake (LRNI) of 400 mg/day may not be compatible with long-term bone health. There is, however, no evidence of beneficial skeletal effects of a customary calcium intake above the current UK reference nutrient intake (RNI) for any age group, although calcium supplementation is a recognised adjunct in the treatment of bone loss and the prevention of fracture in vulnerable individuals.

Figure 12.22 Global map of average calcium intake among adults. Each country with available data is coloured based on an estimate of mean or median dietary calcium intake. Bright red <400 mg/day, dark red 400–499 mg/day, orange 500–599 mg/day, brown 600–699 mg/day, yellow 700–799 mg/day, moss green 800–899 mg/day, light green 900–999 mg/day, and dark green ≥1000 mg/day. *Source:* Reproduced with permission from Balk E.M. et al., 2017/Springer Nature.

Recent committees reviewing the evidence have felt unable to use bone health and fracture outcomes as criteria for calculating RNI values for older people, although the US/Canada RDA for calcium includes an increment for women aged 51–70 years and for all adults aged over 70 years 'to ensure public health protection and err on the side of caution' (see Table 12.18).

Concerns have been expressed about potential adverse effects of long-term calcium supplementation on cardiovascular outcomes. However, recent systematic reviews have shown that findings of adverse effects have been inconsistent, while conversely there is evidence for a small, protective lowering of systolic blood pressure in hypertensive individuals and in pregnant women.

Phosphorus

Phosphorus is an essential bone-forming element and, as with calcium, an adequate supply of phosphorus to bone is necessary throughout life. Both calcium and phosphorus are required for the appropriate mineralisation of the skeleton, and a depletion of serum phosphate leads to impaired bone mineralisation and compromised osteoblast function. However, there is little evidence that, in healthy individuals, the dietary intake of phosphorus limits good bone health, except in the special case of very low birthweight infants. Although there is a set proportion of calcium and phosphorus in bone (Ca:P = 10:6), the ratio of calcium to phosphorus in the diet can vary over a wide range with no detectable effects on the absorption and retention of either mineral. Reports of correlations between phosphorus intake and BMD are inconsistent and likely to be affected by size-confounding. Concerns have been expressed about the possible adverse effects of the increasingly high intake of phosphorus in western-style diets, especially in relation to the consumption of phosphate-based carbonated drinks. At present, it seems unlikely that high phosphorus intakes have consequences for long-term bone health.

Magnesium

Magnesium is involved in bone and mineral homeostasis and is important in bone crystal growth and stabilisation. In a number of studies, magnesium intake and serum magnesium concentration have been reported to be positively associated with both BMD and excretion of bone resorption markers in teenagers, middle-aged and older women and men. In a UK cohort study (EPIC-Norfolk), magnesium intake was linked with lower incidence of hip, spine and wrist fractures when combined in an index with potassium intake (Hayhoe et al. 2015). The converse was observed in a large US cohort of postmenopausal women (Women's Health Initiative) in which no such relation was seen for hip fracture but a high magnesium intake was associated with wrist fractures, possibly confounded by differences in physical activity (Orchard et al. 2014). Short-term increases in BMD have been observed with magnesium supplementation. However, the influence of magnesium nutrition, within the range of normal customary intakes, on long-term bone health is unknown.

Magnesium is one of a number of nutrients found in fruit and vegetables which contribute to an alkaline environment and which may promote bone health by a variety of mechanisms (see Section Vegetarianism, veganism and other dietary regimes), making it difficult to examine the effects of magnesium alone.

Protein

High protein intake

Dietary protein is necessary for healthy bone maintenance in adults. Many populations, especially those with a western-style diet, have protein intakes that match or exceed dietary reference values. On a worldwide basis, high protein intakes have been linked with hip fracture because the consumption of protein, particularly as meat and dairy products, is greatest in countries where hip fractures are common.

Protein intake is a determinant of urinary calcium excretion, and animal protein, which is rich in sulfur-containing amino acids, contributes to an acidic environment. This has led to concerns that high protein intakes are inadvisable for long-term bone health. This is not supported by the evidence from systematic reviews of studies in such populations. These report either no effect or a positive association between protein intake and BMD in adults, with no apparent detriment or benefit of intakes higher than those of current dietary reference values.

The failure of the high protein hypothesis to explain global differences in hip fracture incidence may be partly explained by the alkalising effect of fruits and vegetables in a balanced diet (see later in this section) which offsets the acid load of a high protein intake. Also, when meat is the protein source, the hypercalciuric effect of protein is offset by the hypocalciuric effect of meat phosphorus. Although the data are limited, the importance of protein to bone health does not appear to be dependent on whether the source of protein is derived from animals or plants.

Dietary protein is necessary for bone growth in childhood and adolescence, mediated through the stimulation of IGF-1 and other growth-promoting hormones. Clinical and cohort studies have reported positive associations between protein intake and bone variables related to skeletal size and mineral content. Studies of an association between protein intake and skeletal growth generally relate to milk and meat consumption and are therefore confounded by intakes of accompanying nutrients, especially calcium and phosphorus. Protein also promotes muscle growth and there are studies reporting a positive interaction between physical activity and protein intake on bone growth. In infants and children aged under two years, there are studies suggesting that a high protein intake early in life promotes weight gain and childhood obesity risk, although no differences in growth to two years were detected in a European trial where healthy term infants were randomized to high- or lower-protein formula milks from approximately one to six months of age (Kouwenhoven et al. 2021).

Protein deficiency

Protein deficiency very seldom occurs as an isolated nutrient deficiency, thus it is difficult to separate out the effects of protein deficiency from those of other nutrient deficiencies that occur in protein–energy malnutrition. In children, protein–energy malnutrition is associated with low bone mass and decreased bone growth, which manifests clinically as stunting, a decrease in cortical thickness and a loss of trabecular bone. There is also a delay in bone age due to a delay in the mineralisation of the ossification centres, but the epiphyseal plates are narrowed owing to a reduction in cartilage cell proliferation.

In protein–energy malnutrition, the biochemical markers of bone turnover are reduced, as evidenced by a reduction in serum alkaline phosphatase levels and urinary hydroxyproline excretion. A similar reduction in bone turnover is seen at the histological level, with a reduction in osteoblastic and osteoclastic surfaces. Total serum calcium concentrations are often low, in keeping with the reduction in serum albumin values, and serum phosphorus levels may also be markedly reduced. The pathogenesis of the latter changes is ill understood, although there is evidence of poor renal conservation of phosphate in the child with kwashiorkor. The long-term consequences of protein deprivation in infancy and childhood on bone health have not been studied, but it appears as though there are few detrimental consequences, as in most countries that have high malnutrition rates, osteoporosis and fragility fractures in adulthood are uncommon.

Malnutrition can also occur in the frail elderly. Protein supplements have been shown to benefit the frail elderly and patients recovering from hip fracture, especially when calcium and vitamin D requirements are met.

Vitamin C

Vitamin C (ascorbic acid) acts as a co-factor in the hydroxylation of lysine and proline, which are major constituents of collagen. Hydroxylation is important in the formation of cross-links between the collagen fibres and the formation of mature collagen. A severe nutritional deficiency of vitamin C leads to scurvy (Figure 12.23), but the florid syndrome is rarely seen today. A systematic review of studies in adolescents and adults, mostly of case–control or longitudinal cohort design, has reported positive correlations between vitamin C intakes or serum concentrations with BMD and fracture risk. However, interpretation of these findings is complex. Dietary intakes of vitamin C are strongly associated with fruit and vegetable intake, and serum concentrations are affected by smoking and other factors.

Vitamin K

Vitamin K plays a major role in the γ-carboxylation of glutamic acid residues on a number of proteins (the vitamin K-dependent proteins). The best known of these proteins are the vitamin K-dependent coagulation proteins, but there are also three such proteins in bone: osteocalcin (OCN), matrix Gla protein and protein S. The best studied of these is OCN which, unlike the other two bone vitamin K-dependent proteins, is

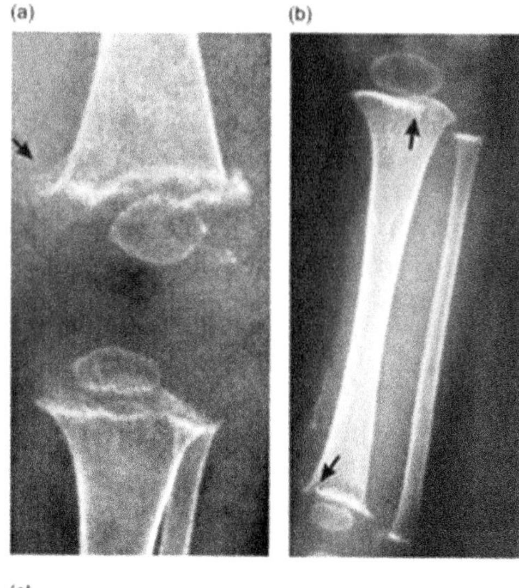

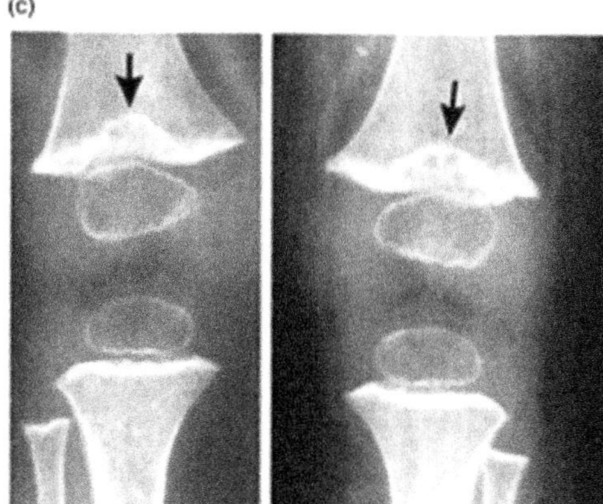

Figure 12.23 Radiographic changes associated with scurvy in children. Advanced scurvy with fractures of thickened, brittle provisional zones of calcification. (a) Multiple infractions in the provisional zone, with peripheral spurring and early subperiosteal calcification. (b) Longitudinal fractures of the provisional zones of calcification. (c) Crumpling fractures of the provisional zones of calcification. *Source:* Reproduced with permission from Silverman et al. 1985, © Elsevier.

produced only in mineralised tissue, synthesised by osteoblasts and odontoblasts. The characteristic of the γ-carboxylated proteins is their calcium-binding activity, thus osteocalcin is found bound to mineralised matrix.

The physiological role of these bone Gla proteins is unclear, but it has been suggested that osteocalcin may inhibit hydroxyapatite formation or control bone resorption. Vitamin K is required for the synthesis of the fully carboxylated form of osteocalcin; the proportion of under-carboxylated osteocalcin (ucOCN) is considered as a marker of vitamin K deficiency. Osteocalcin has wider metabolic functions beyond bone, especially in its under-carboxylated form, acting on the pancreas to increase insulin secretion, on muscle and white adipose tissue to promote glucose and lipid metabolism, and in the regulation of the hypothalamic–pituitary–gonadal axis. The extent to which these non-skeletal functions relate to vitamin K intakes is not clear.

Prospective cohort studies and RCTs have linked vitamin K intake with lower serum ucOCN concentration, higher BMD and reduced fracture risk. The interpretation of these findings in nutritional terms is difficult. Vitamin K_1 (phylloquinone) is the major dietary form, found predominantly in fruits and leafy green vegetables. The reported correlations between dietary intakes and bone health are therefore confounded by the other nutrients and lifestyles associated with fruit and vegetable intake. Many of the RCTs involved vitamin K_2 (menaquinones) at doses far in excess of normal dietary intakes and were in patients with osteoporosis. As such, these are more correctly considered as pharmacological interventions.

Other nutrients and dietary factors

Many other nutrients and dietary factors may be important for long-term bone health. Among the essential nutrients, plausible hypotheses for involvement with skeletal health, based on biochemical and metabolic evidence, can be made for zinc, copper, manganese, boron, vitamin A, B vitamins, vitamin E, potassium and sodium. Evidence from physiological and clinical studies is largely lacking.

Zinc is important in infant growth, and associations with BMD have been noted in middle-aged premenopausal women. Copper supplementation, usually together with other micronutrients such as zinc, has been variously reported as having no effect on BMD or reducing bone loss in perimenopausal women. High vitamin A intake as retinol is associated with increased hip fracture risk in older people, whereas intakes of provitamin A carotenoids are associated with lower fracture risk. High intakes of preformed vitamin A in pregnancy, especially in the first trimester, are not advised because of potential risks of birth defects but, within the usual range of vitamin A intakes, serum retinol in pregnancy has been positively associated with offspring BMD in early adulthood. A higher potassium intake has been associated with a higher BMD, along with other nutrients associated with fruit and vegetable intake, and supplementation with potassium salts reduces bone resorption markers and urinary calcium excretion, but with no apparent effect on BMD. Sodium is intimately involved in the excretion of calcium through the renal tubules, and a direct relationship is found between urinary sodium and calcium excretion in free-living populations. A high sodium intake may reduce BMD and exacerbate age-related bone loss, but the evidence is not conclusive.

Other components of the diet could also influence bone health, and there is growing interest in a range of compounds found in plants. These include phytoestrogens, naturally occurring plant chemicals with weak oestrogenic properties. Phytoestrogens include isoflavones derived from soy or red clover, such as genistein and daidzein, lignans derived from cereals, fruit and vegetables, and coumestans found in a variety of beans and sprouts. In recent years there have been a number of trials of phytoestrogen supplementation in peri- and postmenopausal women, with mixed results. For example, supplementation with isoflavone-containing soy protein or isolated isoflavones increased BMD in relatively short-term studies but longer term studies report no effect on BMD or bone turnover markers.

Among the array of other plant-derived compounds being investigated for potential bone-protective effects are flavinoids, including quercitin and kaempferol, polyphenols, such as resveratrol found in red grapes and berries, non-digestible dietary fibre compounds with prebiotic properties, and botanicals found in herbs. The extent to which sufficient quantities of these phytochemicals can be consumed through dietary means to elicit their putative bone effects, rather than by pharmaceutical intervention, is still to be determined.

Vegetarianism, veganism, and other dietary regimes

It is suggested that the skeleton may act as a reservoir of alkaline salts for the maintenance of adequate acid–base homeostasis, and foods that promote an alkaline environment, such as fruits and vegetables, may diminish the demand for skeletal salts to balance acid generated from foods such as meat. Positive associations have been shown between BMD and nutrients found in fruits and vegetables, such as magnesium, vitamin C and potassium (see earlier). A regular intake of fruits and vegetables has been linked to a reduction in hip fracture risk in observational studies. Dietary regimes that are rich in fruits and vegetables are likely to have greater alkalising properties than the standard western diet, and may affect bone health even when designed for another purpose. Examples include regimes designed to protect cardiovascular health such as the DASH Diet (Dietary Approaches to Stop Hypertension), a diet rich in potassium, magnesium and calcium and low in sodium, and Mediterranean diets that emphasise fruits, vegetables, nuts and olive oil. Both of these regimes include some animal protein and dairy products. No significant associations between BMD or fracture risk with either of these dietary programmes were reported in the US Women's Health Initiative, except for a lower hip fracture risk in women with a greater adherence to a Mediterranean-style diet (Haring et al. 2016).

More generally, there is increasing interest in and adherence to plant-based regimes such as vegetarian diets, which generally exclude fish and meat but include some dairy products, and vegan diets that exclude all animal-derived foods. There have been relatively few investigations of individuals consuming vegetarian, vegan or macrobiotic diets, but there is evidence that these are associated with low BMD in adults and slower skeletal growth in children. Infants weaned onto macrobiotic diets are at risk of growth retardation and rickets. In the UK EPIC-Oxford cohort, a prospective longitudinal study with a large proportion of non-meat eaters among this adult cohort, a greater hip fracture risk was reported in vegans, and to a lesser extent in vegetarians and fish-eaters, than meat eaters, and vegans also had higher risk of total fractures and fractures at other main sites, before and after adjustment for BMI, calcium and protein intakes (Tong et al. 2020).

The extent to which differences in acid–base balance underlie any effects of diet composition on bone health is unclear and controversial. A high potential renal acid load (PRAL), predominantly reflecting high protein and low potassium intakes, has been associated with higher fracture risk in some studies. However, interpretation of studies in this area is difficult because of other differences that may be associated with adherence to a specific dietary regime or adopting a vegetarian or vegan lifestyle which may affect bone health, such as body weight and composition, smoking habits, physical activity patterns and socioeconomic status.

12.5 Lifestyle factors and bone health

Alcohol

Ethanol use increases the risk of fracture through several direct and indirect mechanisms, for example through an increased risk of falls, reduced nutrient intakes with associated malnutrition, increased prevalence of heavy smoking, hypogonadism and a direct inhibition of bone formation. Thus, studies of patients suffering from alcoholism have shown reduced bone density and histological evidence of osteoporosis. Biochemically, the characteristic feature is a reduction in serum osteocalcin, but reduced serum calcium and vitamin D levels have also been reported.

Whether social drinking (1–2 units/day) is associated with an effect on bone is unclear and inconsistent, with some studies suggesting a positive effect and others suggesting negative effects. There is some evidence that the association is U-shaped, with the lowest risk of future fracture correlated with low-to-moderate alcohol consumption compared to non-use and high alcohol consumption.

Smoking

Smoking is associated with a small but significant reduction in BMD, particularly in older men and women when compared with age-matched controls in the majority of studies. The mechanisms by which smoking reduces bone density are unclear, although a number of possibilities exist, including depression of the vitamin D–PTH system, increased alcohol consumption, decreased exercise, reduced free oestradiol and decreased body mass. Smoking in middle age has also been associated with an increased future fracture risk, especially hip fracture in men.

Physical activity and exercise

Physical activity is an important modulator of bone mass, not only during childhood but also during adulthood and in older people. Complete immobilisation may result in a loss of some 40% of bone mass. However, exercise added to the normal routine of daily activity may only add a few per cent onto the average bone mass of children and adults. In a meta-analysis of randomised and non-randomised controlled trials conducted over a 30-year period, exercise training programmes were found to prevent or reverse bone loss of almost 1% per year in both the lumbar spine and femoral neck in pre- and postmenopausal women. Leisure-time physical activity during adulthood has been associated with greater BMD and reduced fracture risk in early old age. The osteogenic effects of exercise are specific to the anatomical sites at which the mechanical strain occurs. Thus, tennis players have greater bone density in the dominant than in the non-dominant arm. The type of exercise may also make a difference, as swimmers have lower bone density at axial sites than other athletes.

Eating disorders and malabsorption syndromes

Eating disorders (anorexia nervosa and bulimia) are major problems among the adolescent and young adult population of the most developed nations of the world. Besides the obvious adverse effects on nutritional status, these disorders have varying adverse effects on bone. Studies suggest that anorexia nervosa carries with it more severe consequences for bone health than bulimia. Bone loss occurs in the majority of anorexic subjects, with over 50% having bone densities more than two standard deviations below age-matched controls. Non-spinal fractures are reported to have a sevenfold increase. It appears that trabecular bone is more severely affected than cortical bone. The pathogenesis of bone loss in anorexia nervosa is multifactorial (Table 12.19), but oestrogen deficiency associated with amenorrhoea probably plays a central role. Other contributing

Table 12.19 Possible factors responsible for bone loss in anorexia nervosa.

- Oestrogen deficiency
- Dehydroepiandrosterone deficiency
- Undernutrition with associated protein, mineral and vitamin deficiencies
- Insulin-like growth factor-1 deficiency
- Increased cortisol levels
- Excessive exercise

factors include malnutrition, IGF-1 deficiency, dehydroepiandrosterone deficiency, increased cortisol levels and possibly associated excessive exercise.

As mentioned earlier, bulimia appears to affect bone mass less severely than anorexia nervosa. In general, women with bulimia are not as wasted or undernourished as anorectic subjects, nor do they have the same prevalence of amenorrhoea. They also appear to be less obsessive about the need for exercise, as in one study only 30% of bulimics exercised regularly, compared with 100% of anorexics. Bulimic subjects who exercise maintain their BMD at weight-bearing sites better than sedentary peers, and have higher BMD than anorexic subjects.

Malabsorption syndromes such as cystic fibrosis (see *Clinical Nutrition* textbook) and coeliac disease are associated with an increased risk of osteoporosis through associated malnutrition, and calcium and vitamin D malabsorption. As life expectancy in subjects with cystic fibrosis increases with the advances that have been made in the control of respiratory complications, minimal trauma fractures are becoming more common.

Inflammatory bowel disease is also associated with an increase in osteoporosis, with approximately 50% of patients estimated to suffer from osteopenia. Factors aggravating the condition include undernutrition, high levels of circulating inflammatory cytokines, lack of exercise and the use of corticosteroids.

Lactose intolerance

Lactose has long been thought to enhance the absorption of calcium, based on animal studies. However, data from human investigations have been inconsistent and recent work suggests that lactose does not have an effect on either calcium absorption or excretion in healthy humans. Lactose intolerance is associated with a low calcium intake because of avoidance of milk and milk products and is regarded as a likely risk factor for osteoporosis. Studies with fracture or bone loss as an outcome have produced an inconsistent picture, with some suggesting a modest risk for those with lactose intolerance, but not others.

Body weight, body composition, and obesity

Body weight and height are major determinants of BMC and BMD. Small build is a risk factor for vertebral fracture, while tall individuals are more prone to hip fracture (see Section 12.3). Anatomical variations between adults may reflect the impact of environmental effects at different stages of skeletal development and these may influence later predisposition to fractures. Size in infancy predicts adult BMC, suggesting that the environment in utero and early life may be an important modulating factor.

Low body weight, especially in connection with anorexia nervosa and the frailty of old age, is associated with an increased risk of fractures, and being overweight with reduced risk. In contrast, obesity is a recognised risk factor for fracture in children, with obese individuals having inappropriately low BMC/BMD for their weight. Obesity is associated with a higher prevalence of poor vitamin D status, and this may contribute to any negative effect of fat mass on bone mass. Although obesity is generally considered as protective against fracture in older people, it is a recognised risk factor for certain types of fracture, especially of the ankle and upper leg. Weight loss by overweight and obese persons through dietary means or bariatric surgery is associated with decreases in BMD and increases in bone resorption markers and fracture risk.

In young people and men, studies of body composition suggest that, for the same body weight, leanness (a higher lean-to-fat ratio) is associated with higher bone mineral mass, whereas in postmenopausal women, it is fatness (a lower lean-to-fat ratio) that is positively related to bone mineral mass. Various interpretations have been put forward to explain this dichotomy, including the osteogenic effects of muscle in younger people, the shock-absorbing effects of adipose tissue in older people and the possible endogenous production of oestrogens by adipose tissue, which may be particularly important in women after the menopause. Animal studies have suggested a role for leptin and other adipokines in the control of bone mass. Situations associated with poor leptin signalling are associated with high bone mass.

12.6 Teeth and supporting tissues

Anatomy and development

Humans have two dentitions. The primary or deciduous dentition, known colloquially as 'milk teeth', comprises 20 teeth in four quadrants each with two incisors, one canine and two molars. The permanent dentition contains 32 teeth; each quadrant has two incisors, one canine, two premolars and three molars. The incisors, canines and premolars succeed the overlying primary teeth. The permanent molar teeth have no predecessors and develop distal to the premolars as the jaws grow.

The teeth develop from two of the three primary germ layers, ectoderm and mesoderm. Ectoderm gives rise to the enamel of teeth; mesoderm provides the dentin, pulp, cementum and periodontal ligament. By the 37th day of development there is a horseshoe-shaped epithelial thickening in each jaw (the dental lamina). At intervals along this, thickenings develop (the tooth germs). The bottom part of these invaginates to form a bell-shaped structure around the inner tissue, called the dental papilla. A double layer of cells lining the inner aspect of the bell develops into the outer ameloblasts that lay down enamel from the inside out, and the inner odontoblasts that form dentin as they retreat towards the dental papilla, which is the future pulp area. As enamel and dentin are deposited, the cell layers move away from each other, the gap being filled with the developing crown. Once the crown has formed, root development begins (Figure 12.24).

The sequence of tooth eruption is different in the primary and permanent dentitions. In general, the permanent mandibular teeth erupt before the corresponding maxillary teeth. It is important to take care of and retain the primary dentition until it naturally exfoliates, as this guides the development of the permanent dentition, thereby minimising the risk of malocclusion. Notice in Table 12.20 that calcification of the crowns of primary teeth begins in utero. There is great variation in the timing of tooth eruption between individuals. Like bone, tooth eruption is affected by nutrition, hormones, ethnicity, height, weight and sex of the individual. Improved diet has led to secular changes in dental maturation in many populations, with tooth eruption occurring at an earlier age. Conditions such as diabetes and obesity have also been shown to be associated with earlier tooth eruption, highlighting the role of nutrition on dental development.

The anatomy of a molar tooth in sagittal section is shown in Figure 12.25. There are three dental hard tissues: enamel, dentin and cementum. Enamel is the outermost layer of the anatomical crown while cementum covers the anatomical root of the tooth. The clinical crown is that part of the anatomical crown which is visible in the mouth. The layer internal to the enamel and cementum is dentin. Dentin forms the bulk of the tooth and surrounds the central soft tissue

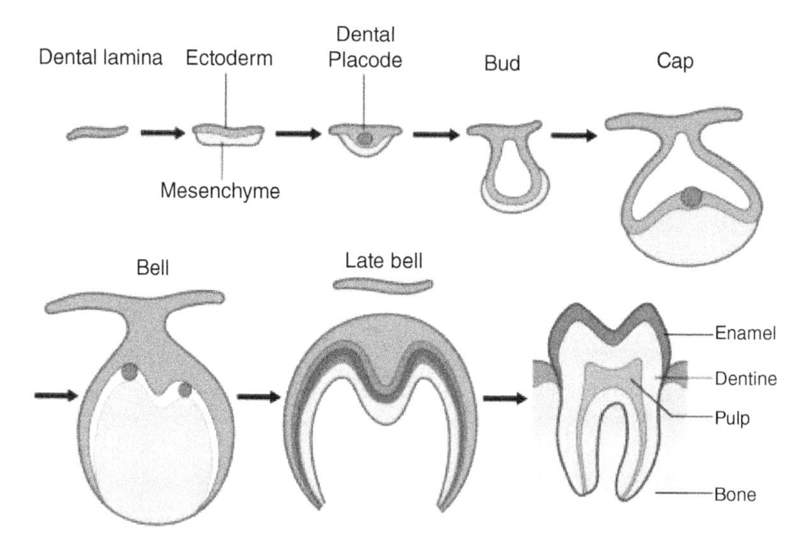

Figure 12.24 Diagram of the developmental stages of a tooth. *Source:* Adapted from Thesleff I., 2003.

Table 12.20 Timings of tooth formation and eruption.

Dentition	Jaw	Order of eruption	Calcification first seen	Crown completed	Eruption	Root completed
Primary	Mandible and maxilla	Central incisor	14 weeks[a]	1½–2½ months	8–10 months	1½ months
		Lateral incisor	16 weeks[a]	2½–3 months	11–13 months	1½–2 months
		First molar	15½ weeks[a]	5½–6 months	16 months	2¼–2½ months
		Canine	17 weeks[a]	9 months	19–20 months	3¼ months
		Second molar	19 weeks[a]	10–11 months	27–29 months	3 months
Permanent	Mandible		Postnatal			
		First molar	At birth	2½–3 years	6–7 years	9–10 years
		Central incisor	3–4 months	4–5 years	6–7 years	9 years
		Lateral incisor	3–4 months	4–5 years	7–8 years	10 years
		Canine	4–5 months	6–7 years	9–10 years	12–14 years
		First premolar	1¼–2 years	5–6 years	10–12 years	12–13 years
		Second premolar	2¼–2½ years	6–7 years	11–12 years	13–14 years
		Second molar	2½–3 years	7–8 years	11–13 years	14–15 years
		Third molar	8–10 years	12–16 years	17–21 years	18–25 years
	Maxilla		Postnatal			
		First molar	At birth	2½–3 years	6–7 years	9–10 years
		Central incisor	3–4 months	4–5 years	7–8 years	10 years
		Lateral incisor	10–12 months	4–5 years	8–9 years	11 years
		First premolar	1½–1¾ years	5–6 years	10–11 years	12–13 years
		Second premolar	2–2¼ years	6–7 years	10–12 years	12–14 years
		Canine	4–5 months	6–7 years	11–12 years	13–15 years
		Second molar	2½–3 years	7–8 years	12–13 years	14–16 years
		Third molar	7–9 years	12–16 years	17–21 years	18–25 years

[a] In utero.
Source: Reproduced with permission from Nelson S.J. and Ash Jr. MM. (2010) @ Elsevier.

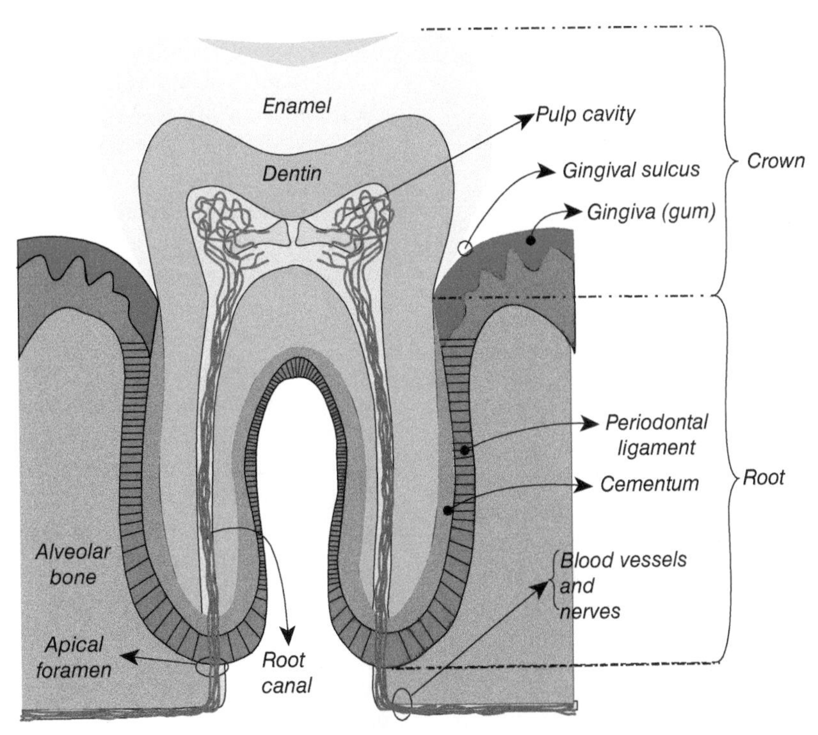

Figure 12.25 Diagram of a sagittal section of a molar tooth and its supporting tissues. *Source:* Reproduced from ScientificStock/ Adobe Stock.

or pulp which is made up of loose connective tissue, nerves and blood vessels. The tooth is suspended in its socket by connective tissue fibres that extend from the cementum to the surrounding alveolar bone. Additional support for the tooth is provided by the gingiva (gum). The supporting tissues (cementum, alveolar bone, periodontal ligament and gingiva) of the tooth are collectively known as the periodontium.

Composition of enamel, dentin, and cementum

Enamel is synthesised by ectodermal cells (ameloblasts) that disappear once enamel formation is complete. Therefore, enamel is not able to repair or regenerate. Enamel is highly calcified, with approximately 96% inorganic material, 0.8% organic and 3.2% water, arranged as crystals of calcium hydroxyapatite (see Section on bone matrix). If ameloblasts are disturbed, they produce poorer quality enamel matrix (enamel hypoplasia) which presents as pits or fissures on the enamel surface. If calcification has been altered, hypocalcification will be seen as opaque, white areas ranging in size from a spot to an entire surface. Highly calcified enamel is transparent, allowing the yellow colour of the underlying dentin to show through; less calcified areas lack this transparency.

The position of enamel hypoplasia or hypocalcification of individual or multiple teeth, in both primary and permanent dentitions, can give a reasonable indication of when development was abnormal. The cause of the defect cannot be easily determined. Excess fluoride is the only nutritional cause that can be diagnosed with reasonable certainty; mostly, enamel defects are due to transient infections, chronic illness or metabolic conditions such as rickets, but a definitive link to the cause is rarely possible.

Dentin and cementum are similar in composition to bone, and are less calcified than enamel, being approximately 68% inorganic material, 22% organic and 10% water. Dentin is formed by mesodermal cells (odontoblasts) that retreat towards the pulp as dentin is laid down. The odontoblasts remain viable after tooth formation is complete and so can form a reparative dentin in response to an irritant such as thermal, chemical or mechanical trauma or dental caries. Dentin itself contains no cells but has fine tubules that contain a portion of the odontoblasts (odontoblast process) which form as the cells retreat towards the pulp while secreting dentin. Reparative dentin may contain trapped cells and does not have tubules. Cementum has a similar appearance to bone as it contains entrapped cells (cementocytes). Cementum does not undergo remodelling; however, the alveolar bone which supports the teeth is able to remodel, thereby facilitating tooth movement during orthodontic treatment.

Tooth function and importance

Teeth function primarily in the chewing and breaking down of food (mastication) to aid digestion. They also play an important role in speech and provide support for the lips and cheeks. Diseases of the teeth and/or supporting tissues lead to loss of function and pain, and may thereby contribute to malnutrition. Also, there may be associated social and psychological impairment. The two most important diseases of the oral cavity are dental caries (tooth decay) and periodontal disease (periodontitis). Caries is the destruction of the hard tissues of the tooth. Periodontitis is an inflammatory condition in which there is destruction of the hard and soft supporting tissues of the tooth (cementum, alveolar bone, gingiva and periodontal ligament).

Caries and periodontitis are among the most common of all diseases globally. Both diseases are non-communicable and preventable, and share modifiable risk factors with more than 100 other non-communicable diseases. These risk factors include a high-sugars diet, lifestyle factors such as tobacco use, excessive alcohol consumption and unfavourable socioeconomic factors. Due to a worldwide rise in these risk factors, non-communicable diseases are rated in the top 10 of the most critical threats to global health by the WHO. Notably, caries experience is cumulative and caries in childhood may have direct or indirect skeletal, cardiac and psychological effects throughout an individual's life (Figure 12.26). Both overnutrition and undernutrition contribute to the development of dental diseases.

Dental caries aetiology and pathogenesis

Dental caries occurs once teeth have erupted into the mouth. Its aetiology is multifactorial (Figure 12.27), consisting of the interaction of

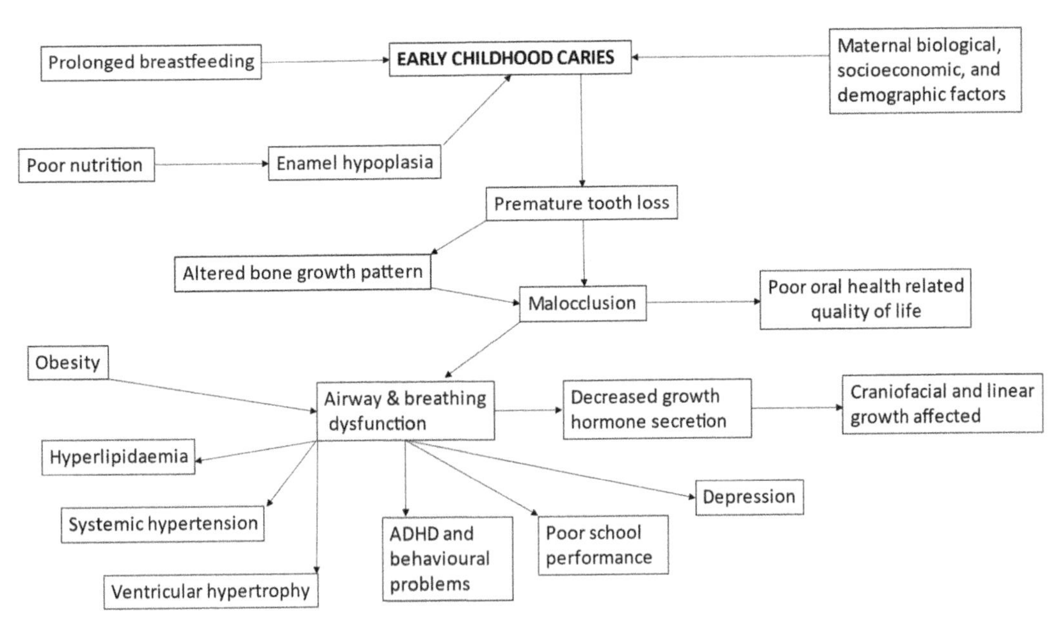

Figure 12.26 Associations between poor nutrition, early childhood caries and oral and general health through the life course.

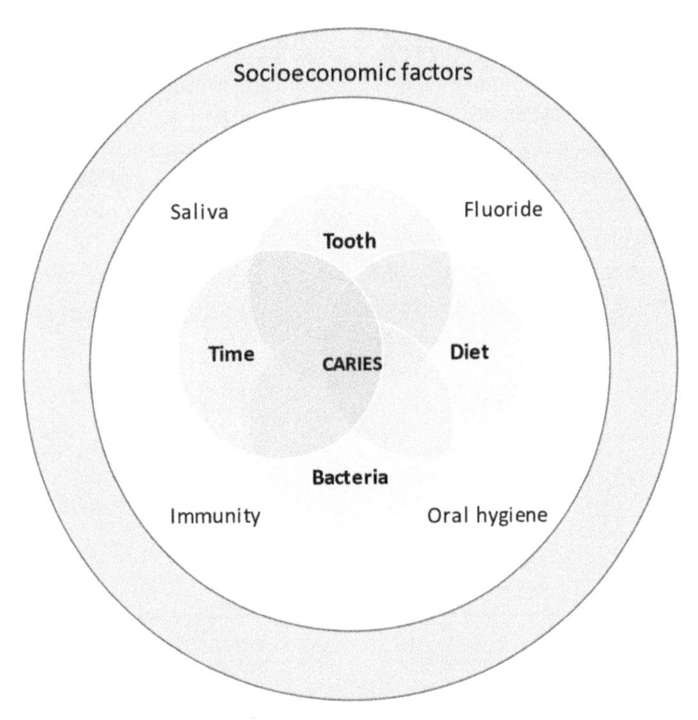

Figure 12.27 Four key interactions necessary for the development of dental caries – fermentable carbohydrates, cariogenic oral bacteria, the tooth surface and time.

four factors, namely, fermentable carbohydrates, cariogenic oral bacteria, a tooth surface and time, which results in demineralisation of the hard tissues of the tooth. The primary pathogen associated with caries is *Streptococcus mutans* but several other bacterial species including *Lactobacillus*, *Actinomyces* and non-*mutans* streptococci are associated with the various stages of the carious process.

Initial caries appears as white spots on the teeth and is reversible. If the unfavourable conditions in the oral cavity persist, then there is destruction of tooth structure accompanied by pain, compromised function and ultimate loss of teeth. Some individuals are at greater risk for caries development due to factors such as poor oral hygiene, a low salivary flow rate, inadequate access to dental care and low fluoride exposure (in drinking water or toothpaste). Dental caries is therefore also greatly influenced by economic, social and environmental factors.

Bacteria in dental biofilm (plaque) metabolise fermentable carbohydrates into organic acids which demineralise the hard tissues of the teeth on contact. The critical pH, that is, the highest pH at which mineral loss occurs, is 5.5 for enamel and 6.0 for dentin. Subsequently, proteolytic enzymes break down the organic component. Remineralisation, aided by saliva and fluoride, may occur in the early stages before proteolysis. The more rapidly a food is fermented and the longer a tooth is exposed to a pH below critical levels, the worse is the potential damage. The length of exposure is influenced by a food's inherent retention in the mouth. For example, liquids are cleared from the mouth more rapidly than solids, and foods that stimulate salivary flow through their consistency or chemical properties are cleared more rapidly than bland foods.

In 1943, Robert Stephan first described the typical pattern of change in the pH of 24-hour-old dental biofilm following a 10% sucrose rinse, depicted in Figure 12.28. After a one-minute rinse with 10% sucrose, there is a rapid drop in pH within three minutes to below the critical level at which demineralisation of enamel occurs. Thereafter, there is a gradual rise back to resting levels over the next 30 minutes. The return to resting levels is produced by the flow of saliva, the buffering action of the salts in saliva and removal of the food from the mouth. Non-fermentable foods that stimulate saliva flow do not drop the pH below 5.7 and may increase it, as in the case of peanuts (Figure 12.28). Since all meals are mixtures of many foods, the pH response is complex. Traditional research has reported on eating of single foods and on sequences of food, but not on food mixtures.

Frequency of food intake and contact time with teeth affect demineralisation, therefore it is advisable to restrict the intake of sticky, fermentable foods to three meals and two snacks per day. Carbonated beverages and wine are known to cause demineralisation and should be avoided; however, contact of liquids with tooth surfaces can be minimised by drinking through a straw. Dental biofilm removal should be performed twice a day to reduce contact of cariogenic micro-organisms with teeth. It is recommended that biofilm removal should not be done immediately after consuming acidic foods since the demineralised tooth surfaces are less hard, rougher and more easily eroded by the abrasive action of the toothbrush, leading to accelerated tooth wear.

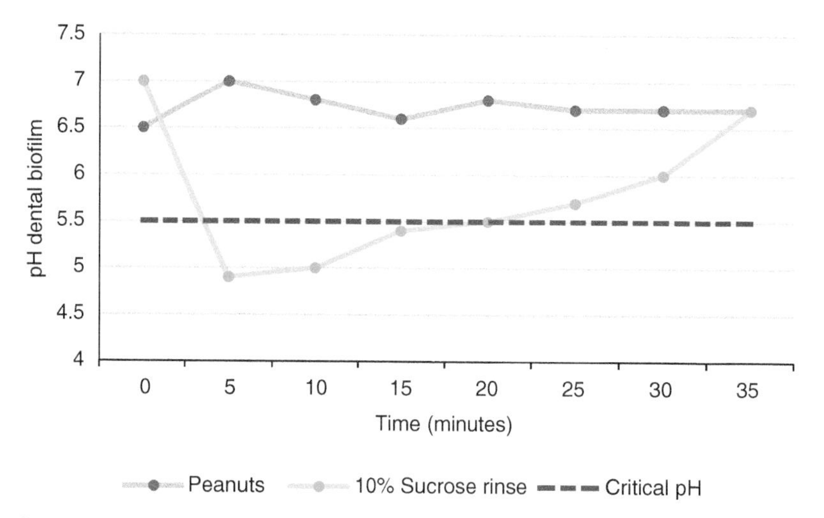

Figure 12.28 Changes in the pH of 24-hour-old dental biofilm following a 10% sucrose rinse, and consuming peanuts. *Source:* Adapted from Geddes, DAM et al., 1977.

Periodontitis aetiology and pathogenesis

Periodontitis is a chronic inflammatory non-communicable disease of the supporting tissues of the teeth (alveolar bone, cementum, periodontal ligament and gingiva). The disease occurs in response to micro-organisms in the dental biofilm that accumulates on teeth which cause a gingivitis (inflammation of the gingiva) that is reversible in its initial stages provided that adequate biofilm control measures are implemented. Healthy gingivae appear pink, firm and taut (Figure 12.29A), but when inflamed they are red, swollen and bleed easily (Figure 12.29B). Persistence of the gingivitis leads to destruction of the periodontal ligament coupled with loss of the associated alveolar bone (Figure 12.29C). As the disease progresses, there is an increase in tooth mobility, pain and compromised function (Figure 12.29D).

Nutrition and developing teeth

Poor nutrition contributes to developmental defects in tooth enamel. This hypoplastic enamel is thinner, more easily colonised by bacteria, retains more biofilm, and dissolves more easily in an acidic environment than normal enamel, making the teeth more susceptible to decay. Since the calcification of teeth begins in utero and continues into adolescence (see Table 12.20), maternal nutritional status affects development of the primary dentition and those permanent teeth whose development begins during gestation.

Calcium and phosphorus are the most important minerals for developing teeth and their levels are modulated by vitamin D. Both calcium and vitamin D are able to cross the placenta easily to reach the fetus. Low maternal calcium intake is associated with enamel hypoplasia in the developing primary teeth, but there is a lack of evidence on the association between adequate calcium in a balanced diet and tooth development after birth.

Maternal vitamin D deficiency poses a risk for fetal enamel hypoplasia which leads to caries susceptibility after eruption postnatally. The exact underlying mechanisms by which vitamin D influences tooth mineralisation are not clearly understood. While mineralisation may be affected through calcium and phosphorus regulation, vitamin D may also act directly on the ameloblasts and odontoblasts, as shown in an animal study (Zhang et al. 2009).

Fluoride is not essential for healthy tooth development but its incorporation into tooth mineral renders the tooth more resistant to demineralisation (see Section Fluoride). Fluoride does not cross the placenta easily. Maternal fluoride ingestion must be very high, >3 ppm/day, to produce even mild fluorosis (mottled staining) of the primary dentition, and if the mother ingests the usual upper therapeutic supplemental dose of 0.5–1.0 mg per day (see Table 12.10 and Section Fluoride), this does not raise fluoride levels in the primary teeth developing in utero.

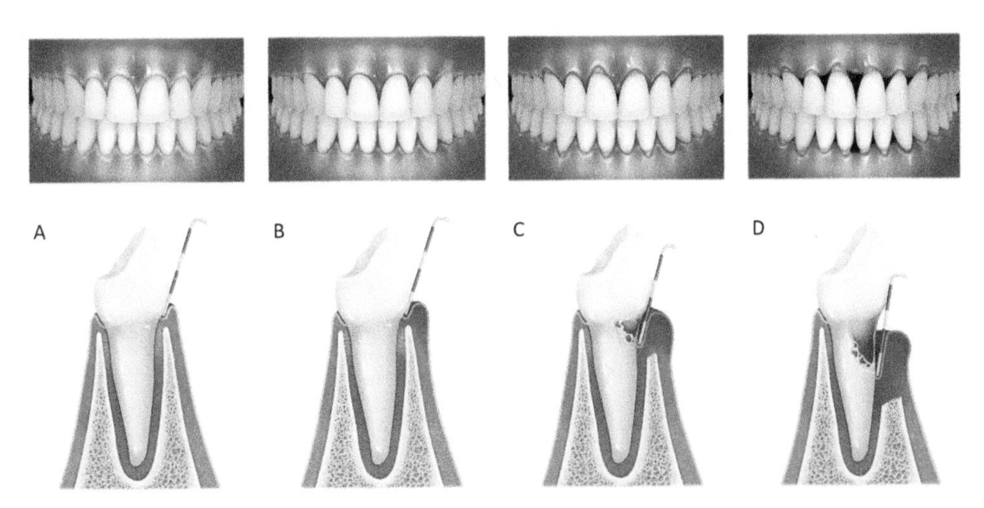

Figure 12.29 Schematic representation of a healthy periodontium and the progression of gingivitis to periodontitis. A – healthy periodontium; B – gingivitis; C – mild periodontitis; D – advanced periodontitis. *Source:* Reproduced from Instituto Maxilofacial/www.institutomaxilofacial.com/wp-content/uploads/2016/04/periodontitis.

Nutrition and erupted teeth

Infant feeding practices

Maternal breastfeeding practices influence caries development in the infant. Early childhood caries (Figure 12.30) is defined as the presence of at least one primary tooth that is decayed, missing (due to caries) or has a filled surface, in a child younger than six years. The WHO recommends that in the first six months of life, infants should be exclusively breast-fed; thereafter, breastfeeding should continue until the age of two years and should be complemented with appropriate nutrient-rich foods. In infants aged 6–12 months, breastfeeding has a protective effect against caries. The risk of infant feeding practices on caries is dependent on the duration and frequency of breastfeeding. Prolonged breastfeeding beyond 12 months or feeding inappropriate foods, coupled with poor oral hygiene practices, leads to rampant early childhood caries.

An additional risk for early childhood caries is bottle feeding with milk or juice during the night; the combination of slow and prolonged sucking on the teat, with the presence of residual liquid in the mouth after the child falls asleep, and a decreased salivary flow rate which occurs during sleep creates a highly unfavourable oral environment.

The risk of caries development in infants in low fluoride areas may be reduced by oral fluoride supplementation (see Table 12.10). Those infants living in areas of high water fluoride concentration should have their formula reconstituted with low fluoride (less than 0.5 ppm) bottled water to prevent fluorosis. In addition, tooth brushing or at least the wiping down of teeth with gauze should be implemented as soon as they appear in the mouth.

Early childhood caries can have long-term consequences for both oral and general health.

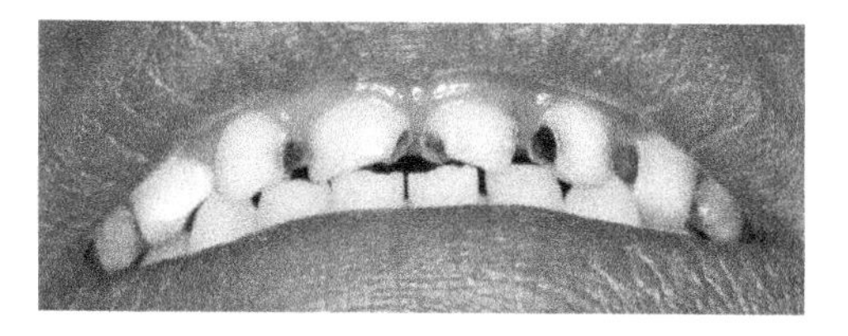

Figure 12.30 Clinical presentation of early childhood caries in a four-year old child. *Source:* Dr Younus Seedat.

Some of these potential consequences are shown in Figure 12.26. Notably, several recent studies have found an association between longitudinal growth and oral health. In a randomised controlled trial, children who had severe early childhood caries treated had greater height after a six-month follow-up compared to their peers who had no dental treatment (Yengopal 2017). Both direct and indirect (immune, endocrine and metabolic) mechanisms have been proposed to explain the relationship between untreated caries and impaired growth. Early childhood caries should therefore be explored as a risk factor in the high prevalence of stunting seen in many low-income countries.

Carbohydrates and dental diseases

Sugars are one of the main risk factors for dental caries. In the last 50 years, there has been a threefold increase in the consumption of free sugars, particularly in low- to middle-income countries. Free sugars are sugars which have been added to foods during preparation, and those sugars which occur naturally in honey, syrups and fruit juices. They do not include sugars in dairy products and in intact fruit and vegetables. According to the WHO guideline, the daily intake of free sugars should not exceed 10% (or 50 g = around 12 teaspoons) of total energy intake in both adults and children. In the UK, a population average free sugars intake that does not exceed 5% of total energy is recommended for age groups older than two years. Furthermore, reducing consumption to less than 5% (or 25 g = around 6 teaspoons) of total energy intake would not only minimise the risk of dental caries, but probably also decrease the risk for noncommunicable diseases such as diabetes and obesity throughout the life course.

It is widely accepted that sugars, as fermentable carbohydrates, play an aetiological role in dental caries but given the complexity of the disease, the precise relationship remains unclear. Epidemiological studies have shown a decline in the prevalence of dental caries over the past 40 years in industrialised countries even though fermentable carbohydrate intake has either remained the same or increased. This reduction may be attributed to the protective effect of fluoride in toothpastes and improved dental care.

Systematic reviews of published evidence allow for a more objective evaluation of the association between dental caries and the intake of sugars. Recently, the Grading of Recommendations Assessment, Development and Evaluation method (GRADE) was used to assess the quality of evidence obtained from systematic reviews that were used by trustworthy organisations to produce guidelines on the intake of sugars (Erickson et al. 2017). The outcome of this GRADE assessment was that guidelines for sugars intake have mostly been based on low-quality evidence. Also, the rigour of studies used to devise guidelines should be carefully considered. For example, the WHO guideline for sugars intake and its risk for dental caries included studies of total sugars, yet its recommendations are based on free sugars.

A fair summary of the relationship between fermentable carbohydrates and dental caries is that intake of these in a person at high risk of dental caries plays a more powerful role than in someone with low caries risk. The problem is that risk of dental caries (or resistance to the disease) is a much talked about but ill-understood concept. What is clear from many studies around the world is that there has been a shift in caries prevalence, with a decline amongst the affluent and an increase in socially deprived communities, in both developed and developing countries. Targeting of such high-risk persons should be the most cost-effective way of preventing the disease, but identification of high-risk people before dental caries develops remains elusive.

High carbohydrate consumption is causally linked to the development of type 2 diabetes mellitus, which in turn has a bidirectional relationship with periodontitis. The relationship between periodontitis and diabetes was first reported more than 50 years ago, yet its importance has only gained recognition recently. Numerous studies since have shown that diabetic patients are at greater risk of developing periodontitis, and that periodontitis may contribute to the development of diabetes. There is a growing body of evidence that shows improved blood glycaemia in diabetic patients in response to treatment for periodontitis (Baeza et al. 2020). Inappropriate carbohydrate consumption clearly has implications not only for the health of teeth but also for their supporting tissues, but the specific mechanistic pathways of the latter are complex and poorly understood.

Calcium and dental diseases

The importance of calcium intake for fully developed teeth and the supporting alveolar bone is not fully understood. Risk for bone diseases such as osteoporosis is known to be determined in adolescence when peak bone mass is attained. Periodontitis and osteoporosis are both bone-resorptive conditions that commonly occur in older individuals. The bone loss in periodontitis is localised, while in osteoporosis it is systemic. Although current research has shown a strong correlation between the two conditions, there is no clear evidence for a causal relationship. However, they share multiple risk factors such as age, genetics, hormones, smoking, calcium intake and vitamin D status. Older women, particularly those who are postmenopausal or who have osteoporosis, are at greater risk for developing periodontitis.

It has been suggested that inflammatory cytokines such as interleukin-1, interleukin-6, and tumour necrosis factor-α, which are associated with bone-resorptive activity, may be the common mediators of the two diseases. Additional controlled studies are needed to give more insight into the relationship between alveolar bone loss and skeletal bone mineral density.

Fluoride

Fluoride is an important nutrient that reduces susceptibility to dental caries; however, its role in bone health remains controversial. In many areas in the world, such as India, South Africa and Tanzania, endemic fluorosis has been reported due to the chronic ingestion of borehole water with high fluoride concentrations (8–20 ppm). Fluoride is deposited in bones and teeth, where it displaces hydroxyl ions in the hydroxyapatite crystals to form fluorapatite. Sources of fluoride intake are drinking water and food, fluoridated salt, milk and fluoride-containing dental products such toothpastes or gels that are applied topically to teeth. Of the ingested fluoride, children and adults retain approximately 50% and 36%, respectively, with the remainder excreted in the urine.

Fluoride and teeth

The protective mechanism of fluoride on erupted teeth is primarily topical through the formation of calcium fluorapatite in the outer few microns of enamel. Fluoride also interferes with the metabolism of cariogenic organisms present in the dental biofilm, making the teeth more resistant to demineralisation. Fluoride may be incorporated into enamel and dentin during tooth development postnatally, but the level of protection that this confers to teeth once they erupt is difficult to measure precisely, since multiple factors contribute to caries development.

Excess fluoride exposure, especially during the first three years of life, produces dental fluorosis, which presents as mottling and staining and surface pitting of the permanent teeth (Figure 12.31b); the severity of the mottling increases with rising fluoride intake. At a water fluoride ion concentration of 1 ppm, about 10% of individuals will show mild mottling (occasional white patches on enamel); the rate is about 20% at a concentration of 2 ppm, thereafter it rises exponentially to 100% at 3 ppm. When very severe mottling is present, enamel is particularly brittle and flakes off the underlying dentin. Caries in such teeth is more frequent than at lower levels of dental fluorosis.

The most cost-effective source of fluoride, in public health terms, is from drinking water yet there is no general agreement on the ideal concentration. The WHO recommends a maximum of 1.5 ppm of fluoride in drinking water, which is double that proposed by the United States Health Department. Much depends on the climate, with greater water consumption in hot compared to temperate conditions and intake from other sources such as fluoridated toothpastes. Guidelines for dietary supplementation of fluoride are based on water fluoride levels, and are shown in Table 12.21. Recommendations for the optimal intake of fluoride should be considered in the context of an amount that would offer protection against caries but not have adverse effects such as dental fluorosis. There also needs to be judicious use of recommended values due to individual variation in fluoride exposure from sources other than water, and genetic variation in the ability to metabolise fluoride that would predispose to fluorosis.

Fluoride and bone

Excess systemic fluoride over time can result in endemic fluorosis which may present with generalised bone abnormalities (osteosclerosis, osteomalacia or osteoporosis), and deformities suggestive of rickets (see Section 12.5) (Figure 12.31a). Its incorporation in the apatite crystal reduces the ability of the crystal to dissolve but there is little evidence that at physiological concentrations, fluoride has any effect on bone mass. At pharmacological doses, fluoride (approximately 50 mg of sodium fluoride) increases BMD by stimulating bone formation. At higher doses, bone formation is abnormal, with features of woven bone appearing. Despite the increase in BMD, particularly at the lumbar vertebrae, there is no conclusive evidence that fracture rates are reduced, although studies from India and China have reported an increased risk of bone fracture in individuals living in areas with a water fluoride concentration of 4.5 ppm.

Fluoride has a differential action on bone cells which is dose dependent (Jiang et al. 2020). It may induce the destruction of osteoblasts but promote the formation of osteoclasts; however, its action on osteocytes through the expression of the glycoprotein sclerostin is the most important in the pathogenesis of skeletal fluorosis. In vitro studies show that osteocytes are better able to withstand fluoride toxicity than the other bone cells.

Table 12.21 Recommended daily dietary fluoride supplement dosage according to drinking water fluoride concentration.

	Water fluoride concentration		
Age	<0.3 ppm	0.3–0.6 ppm	>0.6 ppm
Birth–6 months	0.25	0	0
6 months–3 years	0.25	0	0
3–6 years	0.50	0.25	0
6–16 years	1.00	0.50	0

Recommended dosages are in milligrams of fluoride per day (2.2 mg sodium fluoride = 1 mg fluoride ion).ppm, parts per million (1 ppm = 1mg/L)
Source: Adapted from Rozier, R.G. et al., 2010.

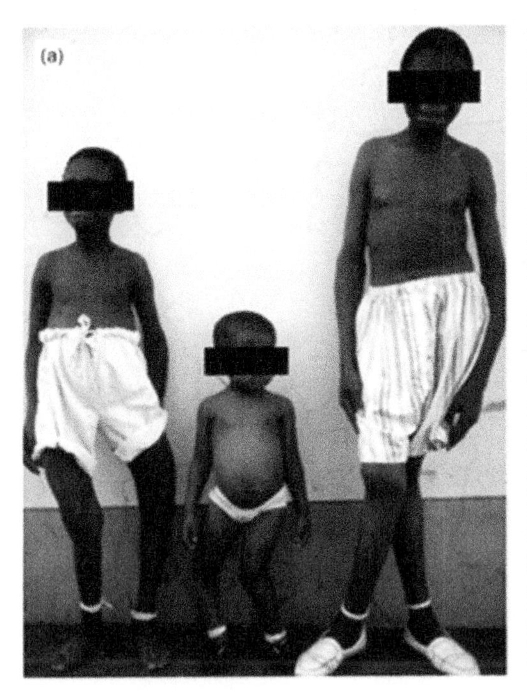

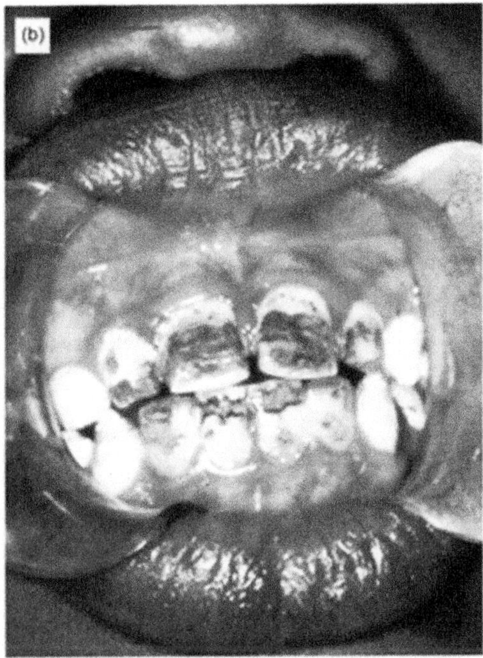

Figure 12.31 (a) Limb deformities in children living in an area associated with endemic fluorosis. (b) Teeth staining due to endemic fluorosis.

Typically, the features of endemic fluorosis include joint stiffness, limb deformities and staining of the teeth (see Figure 12.31). In children, features of rickets may be seen. Radiologically, there may be osteopenia of the distal ends of the long bones, but the axial skeleton shows osteosclerosis with ligamentous calcification around the joints. The biochemical changes are usually minimal, although hypocalcaemia and elevated alkaline phosphatase and parathyroid concentrations have been reported.

The relationship between fluoride and bone is paradoxical and unclear, in contrast to teeth where it plays a pivotal role in reducing caries. Fluoride's beneficial effect on dental health is predominantly topical, and exposure is necessary throughout life to sustain this. Dental fluorosis should alert clinicians to assess for excess fluoride exposure and skeletal fluorosis.

12.7 Conclusions and future perspectives

Bone research has progressed in leaps and bounds since the 1980s and in the next decade it is likely that a much greater understanding of the genetic and biochemical factors that are important in controlling bone mass and thus the incidence of fragility fractures will develop. These developments will probably offer health professionals an enormous spectrum of therapeutic possibilities for the optimisation of bone mass and the treatment of osteoporosis, which will be required in a global population in which the elderly are becoming a larger and larger proportion.

From a more nutritional point of view, a lot more needs to be learnt about the factors responsible for bone mass accretion during childhood and adolescence, the development of peak bone mass and bone health in older people. Furthermore, the interaction of nutrients, nutritional status, lifestyle, sarcopenia and paracrine and endocrine factors on bone homeostasis needs to be explored in greater depth at all stages of the human life cycle. Of increasing interest is the role of micronutrients and non-nutritional compounds in fruit and vegetables and plant-based diets on bone metabolism and resilience. Although the evidence of a link between diet and bone health is not sufficiently secure for firm dietary recommendations, the accumulating picture suggests that current healthy eating advice to consume more fresh fruit and vegetables, eat less meat and reduce sodium (salt) intake is unlikely to be detrimental to bone health and may be beneficial.

Diet and nutritional status play a central role in oral and skeletal health and since both are integral to general health, cross-collaboration between nutrition science and different health disciplines should be adopted, particularly during pregnancy, early life and adolescence, to mitigate the development of adverse health outcomes through the life course. Given the nutritional and epidemiological transitions occurring in many countries, conceptual frameworks in future research should consider not only biological but also environmental and social contexts.

References

Adams, J.E. (2018). Radiology of rickets and osteomalacia. In: Vitamin D, 4e, (eds) FeldmanD. et al., 975–1006. San Diego: Academic Press.

Baeza, M. et al. (2020). Effect of periodontal treatment in patients with periodontitis and diabetes: systematic review and meta-analysis. Journal of Applied Oral Science 28: e20190248.

Carpenter, T.O. et al. (2017). Rickets. Nature Reviews Disease Primers 3: 17101.

Cauley, J.A. et al. (2014). Geographic and ethnic disparities in osteoporotic fractures. Nature Reviews Endocrinology 10 (6): 338–351.

Chevalley, T. et al. (2005). Interaction between calcium intake and menarcheal age on bone mass gain: an eight-year follow-up study from prepuberty to postmenarche. Journal of Clinical Endocrinology and Metabolism 90: 44–51.

Chinoy, A., Mughal, M.Z., and Padidela, R. (2019). Metabolic bone disease of prematurity: causes, recognition, prevention, treatment and long-term consequences. Archives of Disease in Childhood Fetal Neonatal Edition 104 (5): F560–F566.

Compston, J. et al. (2017). UK clinical guideline for the prevention and treatment of osteoporosis. Archives of Osteoporosis 12 (1): 43.

Cruz-Jentoft, A.J. and Sayer, A.A. (2019). Sarcopenia. Lancet 393 (10191): 2636–2646.

De Vadder, F., Joly, A., and Leulier, F. (2021). Microbial and nutritional influence on endocrine control of growth. Journal of Molecular Endocrinology 66: R67–R73.

Erickson, J. et al. (2017). The scientific basis of guideline recommendations on sugar intake: a systematic review. Annals of Internal Medicine 166: 257–267.

Ettinger, B. et al. (1997). Racial differences in bone density between young adult black and white subjects persist after adjustment for anthropometric, lifestyle, and biochemical differences. Journal of Clinical Endocrinology and Metabolism 82 (2): 429–434.

Feldman, D., Glorieux, F.H., and Pike, J.W. (2005).Vitamin D, 1e. Elsevier Science.

Fuggle, N.R. et al. (2019). Fracture prediction, imaging and screening in osteoporosis. Nature Reviews Endocrinology 15 (9): 535–547.

Giustina, A. et al. (2020). Controversies in vitamin D: a statement from the third international conference. JBMR Plus 4 (12): e10417.

Haring, B. et al. (2016). Dietary patterns and fractures in postmenopausal women: results from the Women's Health Initiative. JAMA Internal Medicine 176: 645–652.

Harvey, N.C. et al. (2017). The role of calcium supplementation in healthy musculoskeletal ageing: an expert consensus meeting of the European Society for Clinical and Economic Aspects of Osteoporosis, Osteoarthritis and Musculoskeletal Diseases (ESCEO) and the International Foundation for Osteoporosis (IOF). Osteoporosis International 28 (2): 447–462.

Hayhoe, R.P. et al. (2015). Dietary magnesium and potassium intakes and circulating magnesium are associated with heel bone ultrasound attenuation and osteoporotic fracture risk in the EPIC-Norfolk cohort study. American Journal of Clinical Nutrition 102: 376–384.

Jiang, N. et al. (2020). Different effects of fluoride on the three major bone cell types. Biological Trace Element Research 193: 226–233.

Kouwenhoven, S.M.P. et al. (2021). Long-term effects of a modified, low-protein infant formula on growth and body composition: follow-up of a randomized, double-blind, equivalence trial. Clinical Nutrition 40: 3914–3921.

Lorentzon, M. et al. (2019). Algorithm for the use of biochemical markers of bone turnover in the diagnosis, assessment and follow-up of treatment for osteoporosis. Advances in Therapy 36 (10): 2811–2824.

Munns, C.F. et al. (2016). Global consensus recommendations on prevention and management of nutritional rickets. Journal of Clinical Endocrinology and Metabolism 101: 394–415.

Orchard, T.S. et al. (2014). Magnesium intake, bone mineral density, and fractures: results from the Women's Health Initiative observational study. American Journal of Clinical Nutrition 99: 926–933.

Prentice, A. et al. (2012). The effect of pre-pubertal calcium carbonate supplementation on the age of peak height velocity in Gambian adolescents. American Journal of Clinical Nutrition 96: 1042–1050.

Rustico, S.E. et al. (2014). Metabolic bone disease of prematurity. Journal of Clinical and Translational Endocrinology 1: 85–91.

Sabbagh, Y., Carpenter, T.O., and Demay, M.B. (2005). Hypophosphatemia leads to rickets by impairing caspase-mediated apoptosis of hypertrophic chondrocytes. Proceedings of the National Academy of Sciences USA 102: 9637–9642.

Sempos, C.T. et al. (2021). Serum 25-hydroxyvitamin D requirements to prevent nutritional rickets in Nigerian children on a low-calcium diet-a multivariable reanalysis. American Journal of Clinical Nutrition 114: 231–237.

Shevroja, E. et al. (2017). Use of trabecular bone score (TBS) as a complementary approach to dual-energy X-ray absorptiometry (DXA) for fracture risk assessment in clinical practice. Journal of Clinical Densitometry 20: 334–345.

Tiosano, D. and Hochberg, Z. (2009). Hypophosphatemia: the common denominator of all rickets. Journal of Bone and Mineral Metabolism 27: 392–401.

Tong, T.N.Y. et al. (2020). Vegetarian and vegan diets and risks of total and site-specific fractures: results from the prospective EPIC-Oxford study. BMC Medicine 18: 353.

vio Streym, S. et al. (2016). Vitamin D content in human breast milk: a 9-mo follow-up study. American Journal of Clinical Nutrition 103: 107–114.

Ward, K.A. et al. (2017). Long-term effects of maternal calcium supplementation on childhood growth differ between males and females in a population accustomed to a low calcium intake. Bone 103: 31–38.

Ward, K.A. et al. (2018). Nutrition and bone health during childhood and adolescence a global perspective. In: Osteoporosis: A Lifecourse Epidemiology Approach to

Skeletal Health, (eds) Harvey, N.C. and Cooper, C., 65–79. Boca Raton: CRC Press.
World Health Organization (2018). WHO Recommendation: Calcium Supplementation during Pregnancy for the Prevention of Pre-eclampsia and its Complications. Geneva: World Health Organization.
Yengopal, V. (2017). The effect of dental treatment on weight gain in children in South Africa. PhD thesis. University of the Western Cape.
Zhang, X. et al. (2009). Different enamel and dentin mineralization observed in VDR deficient mouse model. Archives of Oral Biology 54: 299–305.

Further reading

Anil, S. and Anand, P.S. (2017). Early childhood caries: prevalence, risk factors, and prevention. Frontiers in Pediatrics 5: 157.
Bhadada, S.K. and Rao, S.D. (2021). Role of phosphate in biomineralization. Calcified Tissue International 108: 32–40.
Bishop, N. et al. (2014). Fracture prediction and the definition of osteoporosis in children and adolescents: the ISCD 2013 pediatric official positions. Journal of Clinical Densitometry 17: 275–280.
Brondani, J.E. et al. (2019). Fruit and vegetable intake and bones: a systematic review and meta-analysis. PLOS One 14: e0217223.
Buzalaf, M.A.R. (2018). Review of fluoride intake and appropriateness of current guidelines. Advances in Dental Research 29: 157–166.
Compton, J.T. and Lee, F.Y. (2014). A review of osteocyte function and the emerging importance of sclerostin. Journal of Bone and Joint Surgery American Volume 96: 1659–1668.
Crabtree, N.J. et al. (2014). Dual-energy X-ray absorptiometry interpretation and reporting in children and adolescents: the revised 2013 ISCD pediatric official positions. Journal of Clinical Densitometry 17: 225–242.
Deane, C.S. et al. (2020). Animal, plant, collagen and blended dietary proteins: effects on musculoskeletal outcomes. Nutrients 12: 2670.
Department of Health (1991) Dietary Reference Values for Food Energy and Nutrients for the United Kingdom. Report of the Panel on Dietary Reference Values of the Committee on Medical Aspects of Food Policy. London: HMSO.
Department of Health (1998). Nutrition and bone health: with particular reference to calcium and vitamin D. Report of the Subgroup on Bone Health, Working Group on the Nutritional Status of the Population of the Committee on Medical Aspects of the Food and Nutrition Policy.
European Food Standards Authority Panel on Dietetic Products, Nutrition and Allergies. (2015). Scientific opinion on dietary reference values for calcium. EFSA Journal 13: 4101.
European Food Standards Authority Panel on Dietetic Products, Nutrition and Allergies. (2016). Scientific opinion on dietary reference values for vitamin D. EFSA Journal 14: 4547.
Gregson, C.L. et al. (2019). Fragility fractures in sub-Saharan Africa: time to break the myth. Lancet Global Health 7: e26–e27.
Harvey, N.C. et al. (2017). The role of calcium supplementation in healthy musculoskeletal ageing: an expert consensus meeting of the European Society for Clinical and Economic Aspects of Osteoporosis, Osteoarthritis and Musculoskeletal Diseases (ESCEO) and the International Foundation for Osteoporosis (IOF). Osteoporosis International 28 (2): 447–462.
Hörnell, A. et al. (2013). Protein intake from 0 to 18 years of age and its relation to health: a systematic literature review for the 5th Nordic nutrition recommendations. Food and Nutrition Research 57.
Institute of Medicine Food and Nutrition Board. (2011). Dietary Reference Intakes for Calcium and Vitamin D. Washington, DC: National Academies Press.
Koo, W.W. et al. (1982). Skeletal changes in preterm infants. Archives of Disease in Childhood 57: 447–452.
Kovacs, C.S. (2016). Maternal mineral and bone metabolism during pregnancy, lactation, and post-weaning recovery. Physiological Reviews 96: 449–547.
Lambert, M.N.T., Hu, L.M., and Jeppesen, P.B. (2017). A systematic review and meta-analysis of the effects of isoflavone formulations against estrogen-deficient bone resorption in peri- and postmenopausal women. American Journal of Clinical Nutrition 106: 801–811.
Li, T., Li, Y., and Wu, S. (2021). Comparison of human bone mineral densities in subjects on plant-based and omnivorous diets: a systematic review and meta-analysis. Archives of Osteoporosis 16: 95.
Mogire, R.M. et al. (2020). Prevalence of vitamin D deficiency in Africa: a systematic review and meta-analysis. Lancet Global Health 8: e134–e142.
Olausson, H. et al. (2012). Calcium economy in human pregnancy and lactation. Nutrition Research Reviews 25: 40–67.
Popa, D.S., Bigman, G., and Rusu, M.E. (2021). The role of vitamin K in humans: implications in aging and age-associated diseases. Antioxidants 10: 566.
Prentice, A. (2004). Diet, nutrition and the prevention of osteoporosis. Public Health Nutrition 7: 227–243.
Rogmark, C., Fedorowski, A., and Hamrefors, V. (2021). Physical activity and psychosocial factors associated with risk of future fractures in middle-aged men and women. Journal of Bone and Mineral Research 36: 852–860.
Schousboe, J.T. et al. (2013). Executive summary of the 2013 International Society for Clinical Densitometry position development conference on bone densitometry. Journal of Clinical Densitometry 16: 455–466.
Schroth, R.J. et al. (2021). Prenatal and early childhood determinants of enamel hypoplasia in infants. Journal of Pediatrics Perinatology and Child Health 1: 5–17.
Scientific Advisory Committee on Nutrition (2005). Review of Dietary Advice on Vitamin A. London: Scientific Advisory Committee on Nutrition.
Scientific Advisory Committee on Nutrition (2016). Vitamin D and Health. https://assets.publishing.service.gov.uk/government/uploads/system/uploads/attachment_data/file/537616/SACN_Vitamin_D_and_Health_report.pdf
Shams-White, M.M. et al. (2017). Dietary protein and bone health: a systematic review and meta-analysis from the National Osteoporosis Foundation. American Journal of Clinical Nutrition 105: 1528–1543.
Takayanagi, H. (2019). Osteoclast biology and bone resportion. In: Primer on the Metabolic Bone Diseases and Disorders of Mineral Metabolism, 9e, (eds) Bilezikian, J.P., et al. Hoboken: John Wiley and Sons Inc.
Tonetti, M.S. et al. (2017). Impact of the global burden of periodontal diseases on health, nutrition and wellbeing of mankind: a call for global action. Journal of Clinical Periodontology 44: 456–462.
Uusi-Rasi, K., Karkkainen, M.U., and Lamberg-Allardt, C.J. (2013). Calcium intake in health maintenance. Food and Nutrition Research 57: 21082.

Wall, C.R. et al. (2016). Vitamin D activity of breast milk in women randomly assigned to Vitamin D3 supplementation during pregnancy. American Journal of Clinical Nutrition 103: 382–388.
Wang, J. et al. (2019). The role of the fibroblast growth factor family in bone-related diseases. Chemical Biology and Drug Design 94: 1740–1749.
Wheeler, B.J. et al. (2019). A brief history of nutritional rickets. Frontiers in Endocrinology 10: 795.
Winter, E.M. et al. (2020). Pregnancy and lactation, a challenge for the skeleton. Endocrine Connections 9: R143–R157.
Wu, A.M. et al. (2014). The relationship between vitamin A and risk of fracture: meta-analysis of prospective studies. Journal of Bone and Mineral Research 29: 2032–2039.
Xu, S. et al. (2021). Associations between osteoporosis and risk of periodontitis: a pooled analysis of observational studies. Oral Diseases 27: 357–369.
Zibellini, J. et al. (2015). Does diet-induced weight loss lead to bone loss in overweight or obese adults? A systematic review and meta-analysis of clinical trials. Journal of Bone and Mineral Research 30: 2168–2178.

13
The Immune and Inflammatory Systems

Parveen Yaqoob and Philip C. Calder

Key messages

- There is a bidirectional interaction between nutrition, infection and immunity, whereby undernutrition decreases immune defences against invading pathogens, making an individual more susceptible to infection, but the immune response to an infection can itself impair nutritional status and alter body composition.
- Practically all forms of immunity are affected by protein–energy malnutrition, but non-specific defences and cell-mediated immunity are more severely affected than humoral (antibody) responses.
- Obesity impairs many aspects of immunity and is associated with increased low-grade inflammation.
- Micronutrients are required for an efficient immune response and deficiencies in one or more of these nutrients diminish immune function, providing a window of opportunity for infectious agents. However, excessive intakes of some micronutrients may also impair immune function.
- Essential fatty acids have an important role to play in the regulation of immune and inflammatory responses, since they provide precursors for the synthesis of bioactive lipid mediators, including eicosanoids.
- Deficiencies in essential amino acids are likely to impair immune function, and some non-essential amino acids (e.g. arginine and glutamine) may become conditionally essential in stressful situations.
- The gut microbiota seems to be important in regulating the immune and imflammatory responses.
- Probiotic bacteria have been shown to enhance immune function in laboratory animals and may do so in humans. Prebiotics may also have these effects but research in this area is not so conclusive.
- Breast milk has a composition that promotes the development of the neonatal immune response and helps protect against infectious diseases.
- There appears to be a range of nutrient intakes over which the immune system functions optimally, but the exact nature of this range is not clear. It is often assumed when defining the relationship between nutrient intake and immune function that all components of the immune system will respond in the same dose-dependent fashion to a given nutrient. This is not correct, at least as far as some nutrients are concerned, and it appears likely that different components of the immune system show an individual dose–response relationship to the availability of a given nutrient.

13.1 Introduction

Associations between famine and epidemics of infectious disease have been noted throughout history; Hippocrates recognised that poorly nourished people were more susceptible to infectious disease as early as 370 BC. In general, undernutrition impairs the immune system, suppressing immune functions that are fundamental to host protection against pathogenic organisms. Undernutrition leading to impairment of immune function can be due to insufficient intake of energy and macronutrients and/or deficiencies in specific micronutrients (vitamins and minerals). Often these occur in combination; this is particularly notable for protein–energy malnutrition and deficiencies in micronutrients such as vitamin A, iron, zinc and iodine. Clearly, the impact of

Nutrition and Metabolism, Third Edition. Edited on behalf of The Nutrition Society by Helen M. Roche, Ian A. Macdonald, Annemie M.W.J. Schols and Susan A. Lanham-New.

Companion Website: www.wiley.com/go/nutrition/metabolism3e

undernutrition is greatest in low- and middle-income countries, but it is also important in high-income countries, especially among the elderly, individuals with eating disorders, patients with certain diseases, premature babies and those born small for gestational age.

The precise effects of individual nutrients on different aspects of immune function have been notoriously difficult to study in humans. However, it is becoming clear that many nutrients have defined roles in the immune response and that each nutrient has a distinct range of intakes and concentrations over which it supports immune function. Lowering the level of the nutrient below this range or increasing it in excess of the range can impair immune function. Thus, the functioning of the immune system is influenced by nutrients consumed as normal components of the diet and appropriate nutrition is required for the host to maintain adequate immune defences towards bacteria, viruses, fungi and parasites.

Unfortunately, the immune system can become dysfunctional or dysregulated, resulting in inappropriate activation of some components. In some individuals, the immune system recognises a host (self) antigen and then proceeds to direct its destructive activities against host tissues. These diseases are termed autoimmune diseases and always involve chronic inflammation. Examples are rheumatoid arthritis, psoriasis, systemic lupus erythmatosus, multiple sclerosis, ulcerative colitis and Crohn's disease. In other individuals, the immune system becomes inappropriately sensitised to a normally benign antigen, termed an allergen, and so reacts vigorously when that antigen is encountered. These are the atopic diseases, which include allergies, asthma and atopic eczema. It is now recognised that atherosclerosis has an immunological component and some cancers arise and develop as a result of diminished immunosurveillance. Thus, modulation of immune function by dietary components might be an effective means for altering the course of these diseases. Furthermore, diet may underlie the development of some of these diseases.

This chapter begins with an overview of the key components of the immune system, concentrating on the cells that participate in immune responses, the mechanisms by which they communicate and how the system operates in health and disease. The role of nutrients in the immune system is examined using specific examples, and the cyclic relationship between infection and nutritional status is discussed. The main body of the chapter is devoted to an evaluation of the influence of individual macronutrients and micronutrients on immune function.

13.2 The immune system

The immune system acts to protect the host from infectious agents that exist in the environment (bacteria, viruses, fungi, parasites) and from other noxious insults. The immune system has two functional divisions: the innate (or natural) immune system and the acquired (also termed specific or adaptive) immune system (Figure 13.1).

Innate immunity

Innate immunity consists of physical barriers, soluble factors and phagocytic cells, which include granulocytes (neutrophils, basophils, eosinophils), monocytes and macrophages (Table 13.1). Innate immunity has no memory and is therefore not influenced by prior exposure to an organism. Phagocytic cells, the main effectors of innate immunity, express surface receptors that recognise certain structures on microbes. These receptors are termed 'pattern recognition receptors' and include the toll-like receptors; the microbial structures they recognise are termed 'microbe-associated molecular patterns'. Binding of bacteria to pattern recognition receptors triggers phagocytosis and subsequent intracellular destruction of the pathogenic micro-organism by toxic chemicals, such as superoxide radicals and hydrogen peroxide. Natural killer cells also possess surface receptors and destroy their target cells by the release of cytotoxic proteins.

In this way, innate immunity provides a first line of defence against invading pathogens. However, an immune response often requires the co-ordinated actions of both innate immunity and the more powerful and flexible acquired immunity.

Acquired immunity

Acquired immunity involves the specific recognition of molecules (antigens) on an invading pathogen which distinguish it as being foreign to the host. Lymphocytes, which are classified into T- and B-lymphocytes, effect this form of

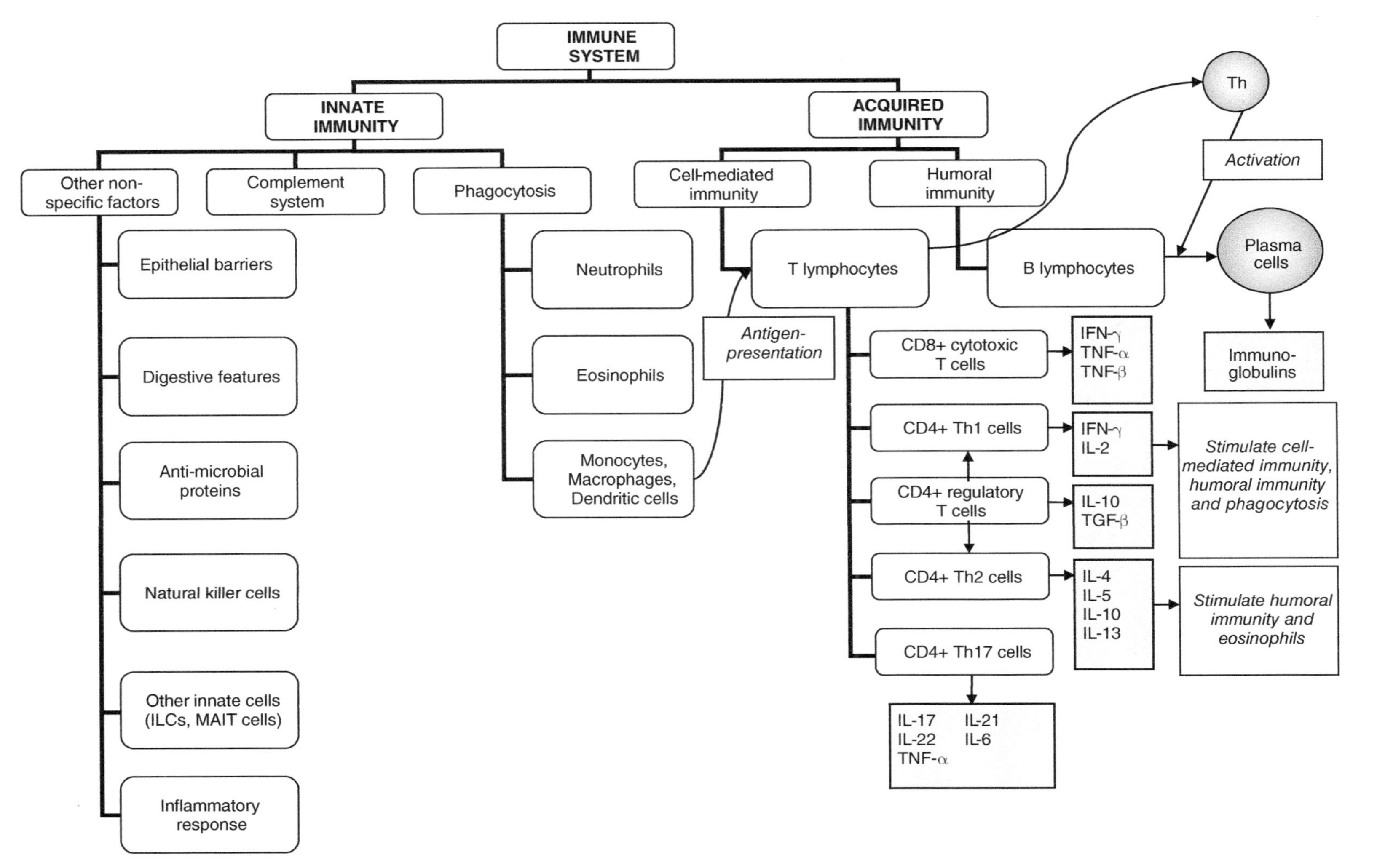

Figure 13.1 The components of the immune system and their division into innate and acquired immunity. IFN, interferon; IL, interleukin; ILCs, innate lymphoid cells; MAIT, mucosal associated invariant T; TGF, transforming growth factor; TNF, tumour necrosis factor. *Source:* P.C. Calder (2021)/Springer Nature/CC BY 4.0.

Table 13.1 Components of innate and acquired immunity.

	Innate	Acquired
Physicochemical barriers	Skin	Cutaneous and mucosal immune systems
	Mucous membranes	Antibodies in mucosal secretions
	Lysozyme	
	Stomach acid	
	Commensal bacteria	
Circulating molecules	Complement	Antibodies
Cells	Granulocytes	Lymphocytes (T and B)
	Monocytes and macrophages	
	Natural killer cells	
Soluble mediators	Macrophage-derived cytokines	Lymphocyte-derived cytokines

immunity (see Figure 13.1). All lymphocytes (indeed, all cells of the immune system) originate in the bone marrow. B-lymphocytes undergo further development and maturation in the bone marrow before being released into the circulation, while T-lymphocytes mature in the thymus. From the bloodstream, lymphocytes can enter peripheral lymphoid organs, which include lymph nodes, the spleen and mucosal lymphoid tissue (e.g. gut-associated lymphoid tissue). Immune responses occur largely in these lymphoid organs, which are highly organised to promote the interaction of immune cells with one another and with invading pathogens.

The acquired immune system is highly specific, since each lymphocyte carries surface receptors for a single antigen. However, acquired immunity is extremely diverse; the lymphocyte repertoire in humans has been estimated as recognition of approximately 10^{11} antigens. The high degree of specificity, combined with the huge lymphocyte repertoire, means that only a relatively small number of lymphocytes will be able to recognise any given antigen. The acquired immune system has developed the ability for clonal expansion to deal with this. Clonal expansion involves the proliferation of a lymphocyte once an interaction with its specific antigen has occurred, so that a single lymphocyte gives rise to a clone of lymphocytes, all of which have the ability to recognise the antigen causing the initial response. This feature of acquired immunity has often been likened to building up an army to fight a foreign invasion. The acquired immune response becomes effective over several days after the initial activation, but it also persists for some time after removal of the initiating antigen. This persistence gives rise to immunological memory, which is also a characteristic feature of acquired immunity. It is the basis for the stronger, more effective immune response to re-exposure to an antigen (i.e. reinfection with the same pathogen) and is the basis for vaccination. Eventually, the immune system will re-establish homeostasis using self-regulatory mechanisms which involve communication between cells.

The major features of the acquired immune response are summarised in Box 13.1.

B- and T-lymphocytes

B-lymphocytes are characterised by their ability to produce antibodies (these are soluble antigen-specific immunoglobulins) which confer antigen specificity to the acquired immune system (i.e. the antibodies produced by B-lymphocytes are specific for individual antigens). This form of protection is termed humoral immunity. B-lymphocytes carry immunoglobulins, which are capable of binding an antigen, on their cell surfaces. Binding of immunoglobulin with antigen causes proliferation of the B-lymphocyte and subsequent transformation into plasma cells, which secrete large amounts of antibody with the same specificity as the parent cell.

Immunoglobulins (antibodies) are proteins consisting of two identical heavy chains and two identical light chains. Five different types of heavy chain give rise to five major classes of immunoglobulin, IgA, IgD, IgG, IgM and IgE, each of which elicits different components of the humoral immune response.

Box 13.1 Features of the acquired immune response

- Specificity
- Diversity
- Memory
- Self-regulation

Antibodies work in several ways to combat invading pathogens. They can 'neutralise' toxins or micro-organisms by binding to them and preventing their attachment to host cells, and they can activate complement proteins in plasma, which in turn promote the destruction of bacteria by phagocytes. Since they have binding sites for both an antigen and for receptors on phagocytic cells, antibodies can also promote the interaction of the two components by forming physical 'bridges', a process known as opsonisation. The type of phagocytic cell bound by the antibody will be determined by the antibody class; macrophages and neutrophils are specific for IgM and IgG, while eosinophils are specific for IgE. In this way, antibodies are a form of communication between the acquired and innate immune responses; they are elicited through highly specific mechanisms, but are ultimately translated to a form that can be interpreted by the innate immune system, enabling it to destroy the pathogen.

Humoral immunity deals with extracellular pathogens. However, some pathogens, particularly viruses but also some bacteria, infect individuals by entering cells. These pathogens will escape humoral immunity and are instead dealt with by cell-mediated immunity, which is conferred by T-lymphocytes. T-lymphocytes express antigen-specific T-cell receptors (TCRs) on their surface, which have an enormous antigen repertoire. However, unlike B-lymphocytes, they are only able to recognise antigens that are presented to them on a cell surface (the cell presenting the antigen to the T-lymphocyte is termed an antigen-presenting cell); this is the distinguishing feature between humoral and cell-mediated immunity. Activation of the TCR results in entry of T-lymphocytes into the cell cycle and, ultimately, proliferation. Activated T-lymphocytes also begin to synthesise and secrete the cytokine interleukin (IL)-2, which further promotes proliferation and differentiation by autocrine mechanisms. Thus, the expansion of T-lymphocytes builds up an army of T-lymphocytes in much the same way as that of B-lymphocytes.

Effector T-lymphocytes have the ability to migrate to sites of infection, injury or tissue damage. There are three principal types of T-lymphocytes: cytotoxic T-cells, T helper cells and regulatory T-cells. Cytotoxic T-lymphocytes carry the surface protein marker CD8 and kill infected cells and tumour cells by secretion of cytotoxic enzymes, which cause lysis of the target cell, or secretion of the antiviral interferons (IFNs), or by inducing apoptosis (suicide) of target cells. Helper T-lymphocytes carry the surface protein marker CD4 and eliminate pathogens by stimulating the phagocytic activity of macrophages and the proliferation of, and antibody secretion by, B-lymphocytes. Regulatory T-cells ($CD4^{+}CD25^{+}FoxP3^{+}$) suppress the activities of B-cells and other T-cells and prevent inappropriate activation.

While effective immune responses are highly desirable, some aspects of immunity have undesirable consequences. For example, bactericidal and inflammatory mediators secreted by macrophages are toxic not only to pathogens but also to host tissues, resulting in unavoidable tissue damage. For this reason, immune responses, and macrophage responses in particular, need to be tightly controlled and the self-regulatory properties of the immune system need to be highly effective; regulatory T-cells are an important part of this self-regulatory activity. Loss of regulatory T-cell function is associated with many immune-mediated diseases.

The gut-associated immune system

The immune system of the gut (sometimes termed gut-associated lymphoid tissue, GALT) is extensive and includes the physical barrier of the intestine, as well as components of innate and adaptive immune responses. The physical barrier includes acid in the stomach, peristalsis, mucus secretion and tightly connected epithelial cells, which collectively prevent the entry of pathogens. The cells of the immune system are organised into specialised lymphoid tissue, termed Peyer's patches, which are located directly beneath the gut epithelium in the lamina propria. This also contains M cells, which sample small particles from the gut lumen. Other lymphocytes are also present in the lamina propria, as well as associated with the epithelium itself.

Because the gut-associated immune system is fairly inaccessible and requires invasive techniques for study, it is relatively poorly understood and much of our understanding of the influence of nutrition on this aspect of immunity comes from animal studies. Endoscopy and biopsy of gut-associated immune tissues is possible, but data from this type of study are very limited. The most useful practical marker of the gut-associated immune system is secretory IgA, although this relies on the assumption that secretion of Ig in saliva reflects that of the rest of the gut-associated system. There are also some markers available for measuring inflammation in the intestine; these include components of neutrophils in stools and markers of protein loss. New technologies employ capsules which travel through the gut transmitting images and sampling tissue from selected sites. Such techniques are likely to revolutionise our understanding of gut immunity in humans.

Communication within the immune system: cytokines

Communication within the acquired immune system and between the innate and acquired systems is brought about by direct cell-to-cell contact often involving pairs of adhesion molecules on each cell type and by the production of chemical messengers, which send signals from one cell to another. Chief among these chemical messengers are proteins called cytokines, which can act to regulate the activity of the cell that produced the cytokine or of other cells (i.e. in either an autocrine or paracrine manner). Each cytokine can have multiple activities on different cell types. Cytokines act by binding to specific receptors on the cell surface and thereby induce changes in the growth, development or activity of the target cell.

Tumour necrosis factor (TNF), IL-1 and IL-6 are among the most important cytokines produced by monocytes and macrophages. These cytokines activate neutrophils, monocytes and macrophages to initiate bacterial and tumour cell killing, increase adhesion molecule expression on the surface of neutrophils and endothelial cells, stimulate T- and B-lymphocyte proliferation, and initiate the production of other proinflammatory cytokines (e.g. TNF induces production of IL-1 and IL-6, and IL-1 induces production of IL-6). Thus, TNF, IL-1 and IL-6 are mediators of both natural and acquired immunity and are an important link between them. In addition, these cytokines mediate the systemic effects of infection and inflammation such as fever, weight loss and acute-phase protein synthesis in the liver. Production of appropriate amounts of TNF, IL-1 and IL-6 is clearly important in response to infection, but inappropriate production or overproduction can be dangerous and these cytokines, particularly TNF, are implicated in causing some of the pathological responses that occur in chronic inflammatory conditions (e.g. rheumatoid arthritis and psoriasis) and in sepsis. IL-10 opposes the action of TNF and, as such, is an important anti-inflammatory cytokine.

Helper T-lymphocytes have traditionally been subdivided into two broad categories according to the pattern of cytokines they produce, although a third category has more recently been identified. Helper T-cells that have not previously encountered antigen produce mainly IL-2 upon the initial encounter with an antigen. These cells may differentiate into a population, sometimes referred to as Th0 cells, which differentiate further into either Th1 or Th2 cells (Figure 13.2). This differentiation is regulated by cytokines: IL-12 and IFN-γ promote the development of Th1 cells, while IL-4 promotes the development of Th2 cells (see Figure 13.2). Th1 and Th2 cells have relatively restricted profiles of cytokine production: Th1 cells produce IL-2 and IFN-γ, which activate macrophages, natural killer cells and cytotoxic T-lymphocytes and are the principal effectors of cell-mediated immunity. Interactions with bacteria, viruses and fungi tend to induce Th1 activity. Since Th1 cytokines activate monocytes and macrophages, these cytokines may also be regarded as proinflammatory. Th2 cells produce IL-4, which stimulates IgE production, IL-5, an eosinophil-activating factor, and IL-10, which together with IL-4 suppresses cell-mediated immunity (see Figure 13.2). Th2 cells are responsible for defence against helminthic parasites, which is due to IgE-mediated activation of mast cells and basophils. Since Th2 cytokines suppress Th1 responses, these cytokines may be regarded as anti-inflammatory, although they are involved in allergic responses.

The patterns of cytokine secretion by Th1 and Th2 lymphocytes were first demonstrated in mice. It has subsequently been demonstrated that while human helper T-lymphocytes do show differences in cytokine profile, the divisions are not clear and while some cells have a typical Th1 or Th2 profile, the majority secrete a mixture of Th1 and Th2 cytokines in differing proportions. Thus, the terms 'Th1 dominant' and 'Th2 dominant' are commonly used to describe the cytokine profiles of these cells. An

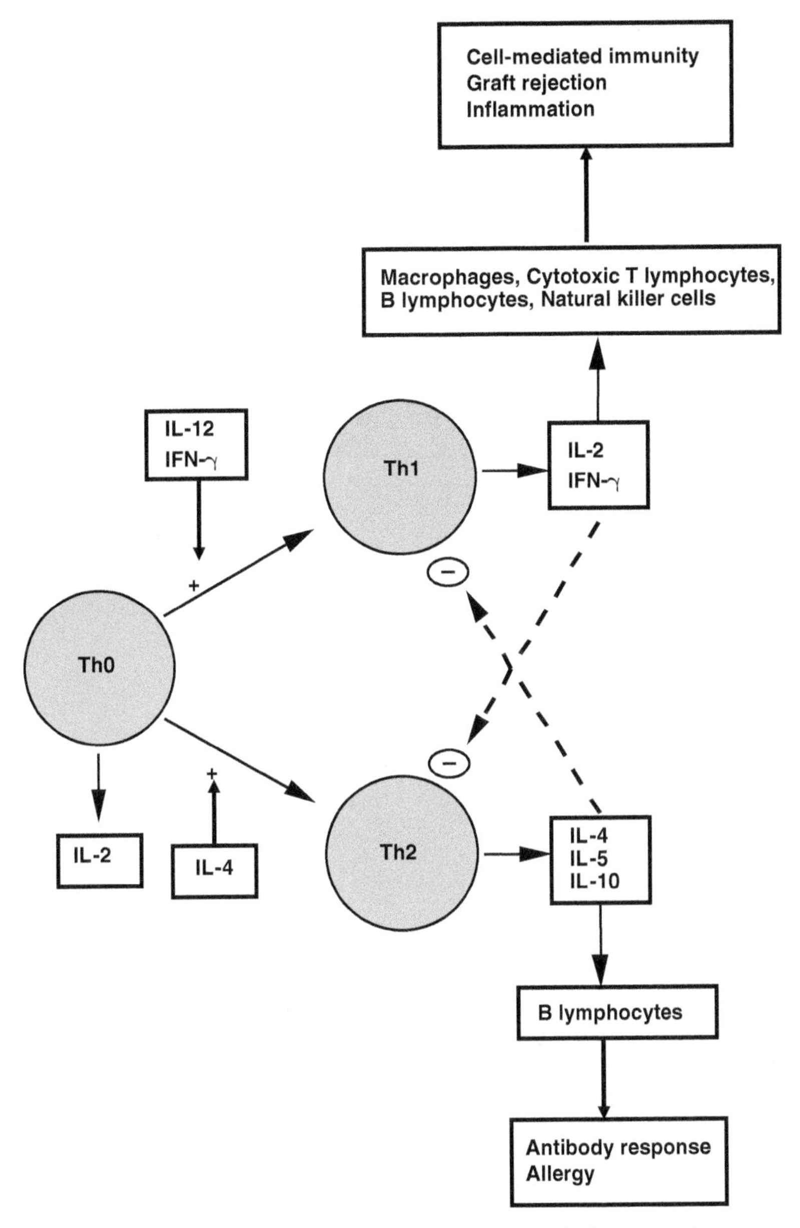

Figure 13.2 Development and cytokine profiles of Th1 and Th2 lymphocytes. IL, interleukin; IFN, interferon.

interesting feature of Th1/Th2 dominance is that once a pattern of cytokine secretion has been established, the dominant arm is able to self-amplify and to antagonise the non-dominant arm. In this way, once a helper T-lymphocyte response has been established, it becomes increasingly polarised towards the dominant phenotype (inflammatory conditions for Th1 and allergy for Th2).

The third, most recently characterised category of T helper cells is Th17, which appears to play an important role in autoimmunity (where the immune system attacks host tissues inappropriately).

Inflammation

Inflammation is the body's immediate response to infection or injury. It is typified by redness, swelling, heat and pain. These occur as a result of increased blood flow, increased permeability across blood capillaries, which permits large molecules (e.g. complement, antibodies, cytokines) to leave the bloodstream and cross the endothelial wall, and increased movement of leucocytes from the bloodstream into the surrounding tissue. Thus, inflammation is an integral part of the innate immune response. However, if inflammation is uncontrolled or does not resolve, it can be damaging to the host and a cause of disease. It is now recognised that resolution of inflammation is an active process promoted by some cytokines and bioactive lipid mediators. In particular, mediators produced from long-chain n-3 fatty acids seem to be important to the resolution of inflammation and these have been collectively termed 'specialised pro-resolving mediators'.

The immune system in health and disease

Clearly, a well-functioning immune system is essential to health and serves to protect the host from the effects of ever-present pathogenic organisms. Cells of the immune system also have a role in identifying and eliminating cancer cells. There are, however, some undesirable features of immune responses.

First, in developing the ability to recognise and eliminate foreign antigens effectively, the immune system is responsible for the rejection of transplanted tissues.

Second, the ability to discriminate between 'self' and 'non-self' is an essential requirement of the immune system and is normally achieved by the destruction of self-recognising T- and B-lymphocytes before their maturation. However, since lymphocytes are unlikely to be exposed to all possible self-antigens in this way, a second mechanism termed clonal anergy exists, which ensures that an encounter with a self-antigen induces tolerance. In some individuals, there is a breakdown of the mechanisms that normally preserve tolerance; a number of factors contribute to this, including a range of immunological abnormalities and a genetic predisposition in some individuals. As a result, an inappropriate immune response to host tissues is generated and this can lead to autoimmune and inflammatory diseases, which are typified by ongoing chronic inflammation and a dysregulated Th1 response. Examples of this type of disease include psoriasis, multiple sclerosis and rheumatoid arthritis.

Third, the immune system of some individuals can become sensitised to usually benign antigens from the environment and can respond inappropriately to them. Such antigens can include components of foods or of so-called allergens (e.g. cat or dog fur, house dust mite, some pollens), such that this response can lead to allergies, asthma and related atopic diseases. Although these diseases are often termed chronic inflammatory diseases and are typified by inappropriate recognition of and/or responses to antigens, they have a different immune basis from the diseases described above. Atopic diseases are typically initiated by the production of IgE by B-lymphocytes in response to exposure to an antigen for the first time (this can be present in a variety of forms, e.g. pollen, dust or in foods). Binding of IgE to specific receptors on the surfaces of mast cells and basophils then occurs (termed sensitisation). If the antigen is reintroduced, it will interact with the bound IgE, leading to activation of the cells and the release of both preformed and newly synthesised inflammatory mediators (particularly histamine and the Th2 cytokines, IL-4, IL-5 and IL-10) that cause the symptoms of an allergic reaction.

13.3 Factors that influence immune function

Many factors influence immune function and resistance to infection, leading to great variability within the normal adult population. These factors include genetics, sex, early life events, age and hormonal status. Immunological 'history' also plays a role in the form of previous exposure to pathogens, vaccination history and chronic disease burden (accumulating conditions over time). Environmental factors influencing immune function include stress (environmental, physiological and psychological), exercise (acute and chronic), body composition (e.g. obesity), smoking, alcohol consumption, gut microbiota and nutritional status.

In a newborn baby, immunological competence is gained as the immune system encounters new antigens and so matures and expands. Some of these early encounters with antigen play an important role in assuring tolerance and a breakdown in this system can lead to increased likelihood of childhood atopic diseases and perhaps also to certain inflammatory conditions later in life. At the other end of the life course, some older people experience a progressive dysregulation of the immune system, leading to decreased cell-mediated immunity and a greater susceptibility to infection. This age-related immune decline has been termed 'immunosenescence'. Innate immunity appears to be less affected by ageing; indeed, there is a progressive increase in chronic inflammation during ageing, which is termed 'inflammageing'.

13.4 The impact of infection on nutrient status

Undernutrition decreases immune defences against invading pathogens and makes an individual more susceptible to infections. However, the immune response to an infection can itself impair nutritional status and alter body composition. Thus, there is a complex interaction between nutrition, infection and immunity (Figure 13.3).

Infection impairs nutritional status and alters body composition in the following ways.

Infection is characterised by anorexia

Reduction in food intake (anorexia) following infection can range from as little as 5% to almost complete loss of appetite. This can lead to nutrient deficiencies even if the host is not deficient before the infection and may make apparent existing borderline deficiencies.

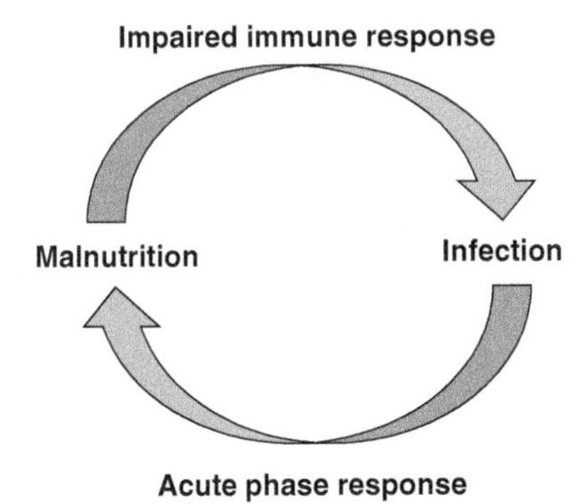

Figure 13.3 Cyclic relationship between malnutrition and infection. Undernutrition impairs immune defences, lowering resistance to invading pathogens. In turn, infection alters nutrient status and contributes to the undernourished state.

Infection is characterised by nutrient malabsorption and loss

The range of infections associated with nutrient malabsorption is wide and includes bacteria, viruses, protozoa and intestinal helminths. Infections that cause diarrhoea or vomiting will result in nutrient loss. Apart from malabsorption, nutrients may also be lost through the faeces as a result of damage to the intestinal wall caused by some infectious organisms and their toxins.

Infection is characterised by increased resting energy expenditure

Infection increases the basal metabolic rate during fever: each 1 °C increase in body temperature is associated with an approximate 13% increase in metabolic rate, which significantly increases energy requirements. This places a significant demand on nutrient supply, particularly when coupled with anorexia, diarrhoea and other nutrient losses (e.g. in urine and sweat).

Infection is characterised by altered metabolism and redistribution of nutrients

The acute-phase response is the name given to the metabolic response to infections and it includes the onset of fever and anorexia, the production of specific acute-phase reactants, and the activation and proliferation of immune cells. This catabolic response occurs with all infections, even when they are subclinical, and serves to redistribute nutrients away from skeletal muscle and adipose tissue and towards the host immune system. This redistribution is mediated by the production of proinflammatory cytokines (e.g. IL-6) by leucocytes and associated endocrine changes. Amino acids, mobilised from skeletal muscle, are used by the liver for the synthesis of acute-phase proteins (e.g. C-reactive protein) and by leucocytes for the synthesis of immunoglobulins and cytokines. The average loss of protein over a range of infections has been estimated to be 0.6–1.2 g/kg body weight per day.

It is clear that inflammatory cytokines mediate many of the effects that lead to compromised nutritional status following an infection, including anorexia, increased energy expenditure and redistribution of nutrients, while malabsorption

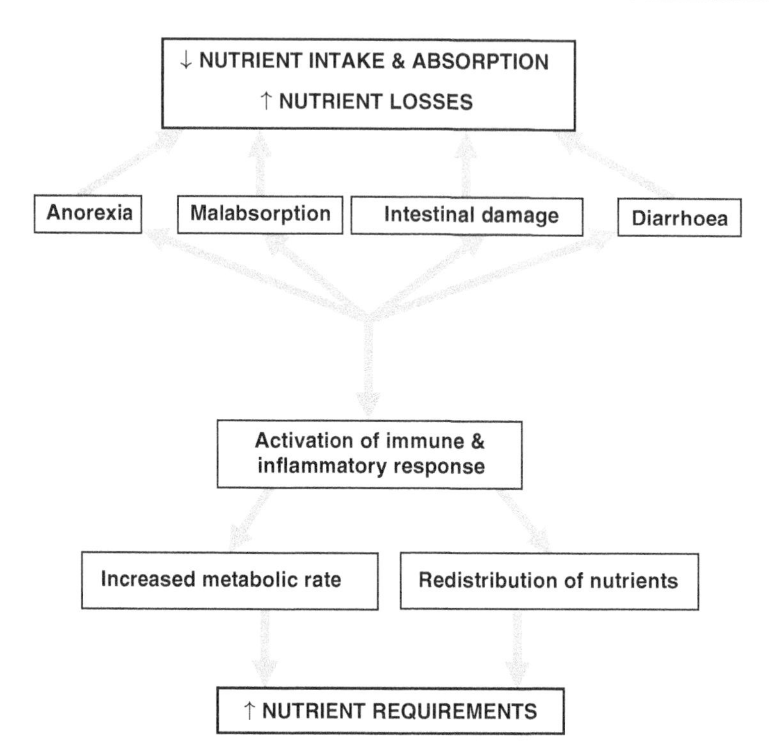

Figure 13.4 Mechanisms by which infection impairs nutritional status. Infection increases nutrient requirements to support the immune response but at the same time can decrease nutrient intake and absorption and increase nutrient losses.

and maldigestion are brought about by the pathogen itself. The result is that an increased nutrient requirement coincides with reduced nutrient intake, reduced nutrient absorption and nutrient losses (Figure 13.4).

13.5 Why should nutrients affect immune function?

Although the immune system is functioning at all times, specific immunity becomes activated when the host is challenged by pathogens. This activation is associated with a marked increase in the demand of the immune system for substrates and nutrients to provide a ready source of energy, which can be supplied from exogenous sources (i.e. from the diet) and/or from endogenous pools. The cells of the immune system are metabolically active and are able to utilise glucose, amino acids and fatty acids as fuels, although different immune cells exhibit preferences for non-oxidative and oxidative metabolism. It is now recognised that there is a link between immune cell metabolism and function.

Energy generation involves electron carriers, which are nucleotide derivatives, for example nicotinamide adenine dinucleotide (NAD), flavin adenine dinucleotide (FAD), and a range of coenzymes. The electron carriers and coenzymes are usually derivatives of vitamins: thiamine pyrophosphate is derived from thiamin (vitamin B_1), FAD and flavin mononucleotide from riboflavin (vitamin B_2), NAD from nicotinate (niacin), pyridoxal phosphate from pyridoxine (vitamin B_6), coenzyme A from pantothenate, tetrahydrofolate from folate (vitamin B_9) and cobamide from cobalamin (vitamin B_{12}). In addition, biotin is required by some enzymes for activity. The final component of the pathway for energy generation (the mitochondrial electron transfer chain) includes electron carriers that have iron or copper at their active site.

Activation of the immune response gives rise to the increased production of proteins (immunoglobulins, cytokines, cytokine receptors, adhesion molecules, acute-phase proteins) and lipid-derived mediators (prostaglandins, leukotrienes). To respond optimally, there must be the appropriate enzymic machinery in place (for RNA synthesis and protein synthesis and their regulation) and ample substrate available (nucleotides for RNA synthesis, the correct mix of amino acids for protein synthesis, polyunsaturated fatty acids [PUFAs] for eicosanoid synthesis).

An important component of the immune response is the oxidative (sometimes called respiratory) burst, during which superoxide anion radicals are produced from oxygen in a reaction linked to the oxidation of NADPH. The reactive oxygen species produced can be damaging to host tissues and thus antioxidant protective mechanisms are necessary. Among these are the classic antioxidant vitamins, α-tocopherol (vitamin E) and ascorbic acid (vitamin C), glutathione, a tripeptide composed of glutamate, cysteine and glycine, the antioxidant enzymes superoxide dismutase and catalase, and the glutathione recycling enzyme glutathione peroxidase. Superoxide dismutase has two forms, a mitochondrial form and a cytosolic form; the mitochondrial form includes manganese at its active site, whereas the cytosolic form includes copper and zinc. Catalase contains iron at its active site, whereas glutathione peroxidase contains selenium.

Cellular proliferation is a key component of the immune response, providing amplification and memory; before division, there must be replication of DNA and then of all cellular components (proteins, membranes, intracellular organelles, etc.). In addition to energy, this clearly needs a supply of nucleotides (for DNA and RNA synthesis), amino acids (for protein synthesis), fatty acids, bases and phosphate (for phospholipid synthesis), and other lipids (e.g. cholesterol) and cellular components. Although nucleotides are synthesised mainly from amino acids, some of the cellular building blocks cannot be synthesised in mammalian cells and must come from the diet (e.g. essential fatty acids, essential amino acids, minerals). Amino acids (e.g. arginine) are precursors for the synthesis of polyamines, which have roles in the regulation of DNA replication and cell division. Various micronutrients (e.g. iron, folic acid, zinc, magnesium) are also involved in nucleotide and nucleic acid synthesis.

Thus, the roles for nutrients in immune function are many and varied and it is easy to appreciate that an adequate and balanced supply of these is essential if an appropriate immune response is to be mounted.

13.6 Assessment of the effect of nutrition on immune function

There is a wide range of methodologies by which to assess the impact of nutrients on immune function. Assessments can be made of cell functions ex vivo (i.e. of the cells isolated from animals or humans subjected to dietary manipulation and studied in short- or long-term culture) or of indicators of immune function *in vivo* (e.g. by measuring the concentrations of proteins relevant to immune function in the bloodstream or the response to an immunological challenge). Table 13.2 lists some examples of methods used for the assessment of immune function.

However, the biological relevance of these markers of immune function remains unclear. An Expert Task Force on Nutrition and Immunity in Man published a report in 2005 (see Further reading), which classified markers into three categories with high, medium or low suitability for assessment of immune function. Those in the 'high suitability' category include the delayed-type hypersensitivity response, vaccination response and production of secretory IgA. These were deemed to be suitable because they are *in vivo* measures, which involve no manipulation of cells outside the body. Those in the 'medium suitability' category include ex vivo markers, such as natural killer cell activity and cytokine production. The Task Force concluded

Table 13.2 Measurements to assess the effect of nutrition on immune function.

In vivo measures
Size of lymphoid organs
Cellularity of lymphoid organs
Numbers and types of immune cells circulating in bloodstream
Cell surface expression of molecules involved in immune response (e.g. antigen presentation)
Circulating concentrations of Ig specific for antigens after an antigen challenge
Concentration of secretory IgA in saliva, tears and intestinal washings
Delayed-type hypersensitivity (DTH) response to intradermal application of antigen
Response to challenge with live pathogens (mainly animal studies; outcome usually survival)
Incidence and severity of infectious diseases (can be used in human studies)
Ex vivo measures
Phagocytosis by neutrophils and macrophages
Oxidative burst by neutrophils and macrophages
Natural killer cell activity against specific target cells (usually tumour cells)
Cytotoxic T-lymphocyte activity
Lymphocyte proliferation (following stimulation with an antigen or mitogen)
Production of cytokines by lymphocytes and macrophages
Production of immunoglobulins by lymphocytes
Cell surface expression of molecules involved in cellular activation

that no single marker should be used to draw conclusions about modulation of the whole immune system, except for those studies which have infection as a clinical outcome, and that combining markers with high and medium suitability is currently the best approach to measure immunomodulation in human nutritional studies.

13.7 Protein–energy malnutrition and immune function

Protein–energy malnutrition, although often considered a problem solely of low- and middle-income countries, has been described in even the most affluent of countries. Moderate malnutrition in high-income countries is encountered among the elderly, those with anorexia or bulimia, premature babies, hospitalised patients, and patients with various disease conditions (e.g. cystic fibrosis, acquired immunodeficiency syndrome and some cancers). It is important to recognise that protein–energy malnutrition often co-exists with micronutrient deficiencies, and poor outcome of intervention can result from a lack of awareness of multiple deficiencies.

Practically all forms of immunity may be affected by protein–energy malnutrition but non-specific defences and cell-mediated immunity are more severely affected than humoral (antibody) responses. Protein–energy malnutrition causes atrophy of the lymphoid organs (thymus, spleen, lymph nodes, tonsils) in laboratory animals and humans. There is a decline in the number of circulating lymphocytes, which is proportional to the extent of malnutrition, and the proliferative responses of T-lymphocytes to mitogens and antigens is decreased by malnutrition, as is the synthesis of IL-2 and IFN-γ and the activity of natural killer cells. Production of cytokines by monocytes (TNF, IL-6 and IL-1β) is also decreased by malnutrition, although their phagocytic capacity appears to be unaffected. The *in vivo* skin delayed-type hypersensitivity response to challenge with specific antigens is reduced by malnutrition. However, numbers of circulating B-lymphocytes and immunoglobulin levels do not seem to be affected or may even be increased by malnutrition; it has been suggested that underlying infections may influence these observations.

13.8 Obesity and immune function

Obesity can be associated with a loss of immune competence, with impairments of the activity of helper T-lymphocytes, cytotoxic T-lymphocytes, B-lymphocytes and natural killer cells, and reduced antibody and IFN-γ production. Thus, compared with healthy-weight individuals, those living with obesity can have increased susceptibility to a range of bacterial, viral and fungal infections, and poorer responses to vaccination. During the 2009 H1N1 influenza A virus pandemic, individuals living with obesity showed delayed and weakened antiviral responses to infection and showed poorer recovery from disease compared with healthy-weight individuals. Animal studies and case studies in humans show that obesity is associated with prolonged shedding of influenza virus, indicating an impairment in viral control and killing, and the emergence of virulent minor variants. Compared with healthy-weight individuals, vaccinated individuals with obesity have twice the risk of influenza or influenza-like illness, indicating poorer protection from vaccination. Paradoxically, obesity is also linked to an increase in blood concentrations of many inflammatory mediators, a state of chronic low-grade inflammation, that is considered to contribute to an increased risk of chronic conditions of ageing and may predispose to mounting an excessive inflammatory response when infected.

13.9 The influence of individual micronutrients on immune function

Much of what is known about the impact of single nutrients on immune function comes from studies of deficiency states in animals and humans, and from controlled animal studies in which the nutrients are included in the diet at known levels. There is now overwhelming evidence from these studies that particular nutrients are required for an efficient immune response and that deficiencies in one or more of these nutrients diminish immune function and provide a window of opportunity for infectious agents. It is logical that multiple nutrient deficiencies might have a more significant impact on immune function, and therefore resistance to infection, than a single nutrient deficiency. What is also apparent is that excess amounts of some nutrients also impair immune function and decrease resistance to pathogens. Thus, for some nutrients there may be a relatively narrow range of intake that is associated with optimal immune function.

Vitamin A

The vitamin A (or retinoid) family includes retinol, retinal, retinoic acid and esters of retinoic acid. Not long after its discovery, vitamin A was described as the 'anti-infective vitamin'. This is probably not entirely accurate, however, because vitamin A tends to enhance recovery from infection rather than prevent it. The discovery and characterisation of nuclear receptors for vitamin A in the 1980s did much to help elucidate its functions. There is a transitory decrease in serum retinol during the acute-phase response that follows trauma or infection, and which is likely to be due to decreased synthesis of, and therefore decreased release of, retinol-binding protein (RBP) by the liver, and also to increased vascular permeability at sites of inflammation, allowing leakage into the extravascular space. For these reasons, serum retinol cannot be used as an indicator of vitamin A status in individuals with an active acute-phase response.

Vitamin A is essential for maintaining epidermal and mucosal integrity; vitamin A-deficient mice have histopathological changes in the gut mucosa consistent with breakdown in gut barrier integrity and impaired mucus secretion, both of which would facilitate entry of pathogens through this route. One of the key changes caused by vitamin A deficiency is the loss of mucus-producing goblet cells; the resulting lack of mucus diminishes resistance to infection by pathogens that would otherwise be trapped and washed away. Vitamin A regulates keratinocyte differentiation, and vitamin A deficiency induces changes in skin keratinisation, which may explain the observed increased incidence of skin infection.

Many aspects of innate immunity, in addition to barrier function, are affected by vitamin A. It modulates gene expression to control the maturation of neutrophils; in vitamin A deficiency there are increased neutrophil numbers but decreased phagocytic function. Macrophage-mediated inflammation is increased by vitamin A deficiency, but the ability to ingest and kill bacteria is impaired. Vitamin A deficiency may therefore lead to more severe infections, coupled with excessive inflammation. Natural killer cell activity is diminished by vitamin A deficiency. The impact of vitamin A on acquired/specific immunity is less clear, but there is some evidence that deficiency alters the Th1/Th2 balance, decreasing the Th2 response, often without affecting the Th1 response. This area requires further research. There is very little evidence for the effects of vitamin A supplements on the proliferation or activation of B-lymphocytes.

The impact of vitamin A deficiency on infectious disease has been studied widely in the developing world. Vitamin A deficiency is associated with increased morbidity and mortality in children, and appears to predispose to respiratory infections, diarrhoea and severe measles. Although vitamin A deficiency increases the risk of infectious disease, the interaction is bidirectional such that infections can lead to vitamin A deficiency; diarrhoea, respiratory infections, measles, chickenpox and human immunodeficiency virus (HIV) infection are all associated with the development of vitamin A deficiency.

Replenishment of vitamin A in deficient individuals by provision of supplements decreases mortality by approximately 30% in children aged six months to five years in areas of the world where deficiency is a problem. In general, frequent, small doses tend to decrease mortality more dramatically than infrequent high doses. Vitamin A supplements improve recovery from measles and decrease the duration, risk of complications and mortality from the disease. Because measles is an acute, immunosuppressive viral infection, it is often associated with secondary, opportunistic bacterial infections, and it is not clear whether vitamin A improves recovery from the measles itself, the secondary infection, or both. The ability of vitamin A to promote the regeneration of damaged mucosal epithelium and phagocytic activity of neutrophils and macrophages results in a reduction in the incidence and duration of diarrhoea, which may be of particular benefit to infants who are not breastfed.

The effect of vitamin A on respiratory infections is particularly interesting. In community studies, vitamin A supplements increased the risk of respiratory infection, while in most clinical studies vitamin A has simply failed to improve recovery. There is some evidence that low-dose supplements given frequently can reduce the severity of respiratory infections in children with underlying protein–energy malnutrition but not in those who are not malnourished; indeed, several studies indicate that vitamin A increases the severity of respiratory infections in the non-malnourished. The apparently undesirable effects of vitamin A supplements in community and clinical studies are puzzling and cannot be easily explained, but may be related to induction of excessive inflammation in the airways.

Carotenoids

The carotenoids are a group of over 600 naturally occurring pigmented compounds that are widespread in plants, although fewer than 30 carotenoids occur commonly in human foodstuffs. Some of the carotenoids possess vitamin A activity and most effectively quench free radicals, which could be useful in counteracting the damaging effects of reactive oxygen species (ROS) generated by respiratory burst. Early studies reported that providing carotenoids to children improved severe ear infections, and that dietary carotenoids appeared to protect against lung and skin cancers. However, detailed studies investigating the effects of carotenoids on parameters of immune function (e.g. lymphocyte proliferation, natural killer cell activity, cytotoxic T-cell activity, expression of cell surface molecules on monocytes, DTH) have generally not been consistent, with some studies showing benefit and others showing no effect. Further confusion in the area arose when several major intervention trials showed no benefit or an increase in lung cancer in smokers receiving β-carotene supplementation. This area requires further research.

Folic acid and other B-vitamins

Folic acid and the other B vitamins participate as coenzymes in the synthesis of nucleic acids and proteins, which is crucial for many aspects of immune function. There is some suggestion that folate supplementation of elderly individuals improves immune function, and in particular, natural killer cell activity, although this is not entirely conclusive. However, elderly subjects supplemented with a combination of folic acid, vitamin E and vitamin B_{12} were reported to have fewer infections. Elderly individiuals tend to be at risk of vitamin B_{12} deficiency, and studies have shown that subjects aged >65 years with low serum vitamin B_{12} concentrations have impaired antibody responses to a vaccine. Patients with vitamin B_{12} deficiency also have decreased numbers of lymphocytes and suppressed natural killer cell activity, which may be reversed with supplementation.

Vitamin B_6 deficiency in laboratory animals causes thymus and spleen atrophy and decreases lymphocyte proliferation and the DTH response. In a study in healthy elderly humans, a vitamin B_6-deficient diet (3 μg/kg body weight per day or about 0.17 and 0.1 mg/day for men and women, respectively) for 21 days resulted in a decreased percentage and total number of circulating lymphocytes, decreased T- and B-cell proliferation in response to mitogens and decreased IL-2 production. Repletion at 15 or 22.5 μg/kg body weight per day for 21 days did not return the immune functions to starting values; however, repletion at 33.75 μg/kg body weight per day (about 1.9 and 1.1 mg/day for men and women, respectively) returned immune parameters to starting values. Providing a larger dose of 41 mg vitamin B_6/day for four days caused a further increase in lymphocyte proliferation and IL-2 production. This comprehensive study indicates that vitamin B_6 deficiency impairs human immune function, that the impairment is reversible by repletion, and that lymphocyte functions are enhanced at levels of vitamin B_6 above those typical of habitual consumption.

Vitamin C

Vitamin C is a water-soluble antioxidant found in high concentrations in circulating leucocytes and appears to be utilised during infections. High circulating levels of vitamin C are associated with enhanced antibody responses, neutrophil function and antiviral activity in animal studies. In humans, supplementation studies have mainly been conducted in athletes and populations with chronic illnesses. The interest in athletes is due to the fact that exercise induces an increase in the numbers of neutrophils and their capacity to produce ROS, which, if prolonged, can be immunosuppressive and reduce neutrophil activity in the recovery period following exercise. Since neutrophils form an important part of the defence against viruses, the suppression of neutrophil activity after strenuous exercise may explain why upper respiratory tract infections are often noted to coincide with this. Because of its antioxidant capacity, vitamin C could potentially counteract the exercise-induced generation of ROS and limit the postexercise immunosuppression. However, randomised controlled trials (RCTs) have often been limited by lack of statistical power and do not conclusively show an effect of vitamin C on cell numbers, neutrophil function or ROS production. Similarly, RCT evidence for effects of vitamin C on immune function in chronic illnesses is lacking, although it does support a role for reducing trauma-induced increases in serum concentrations of proinflammatory cytokines.

Several studies suggest a modest benefit of vitamin C supplementation, at doses ranging from 1000 to 8000 mg/day, in reducing the duration, but not the incidence, of respiratory infections. However, the incidence of common colds and pneumonia has been shown to be reduced by vitamin C in those individuals who regularly engage in strenuous physical activity, and also in those who live in crowded conditions. The potential benefits and risks of vitamin C supplementation at doses above 8000 mg/day, and the role of vitamin C in non-respiratory infections, have not been well investigated.

Vitamin D

The active form of vitamin D is 1,25-dihydroxy vitamin D_3 and is referred to here as vitamin D.

Vitamin D receptors have been identified in most immune cells, suggesting that it has immunoregulatory properties, while some immune cells (including dendritic cells and macrophages) can synthesise the active form of vitamin D; this latter observation indicates important roles of vitamin D in immunity. There are reports suggesting immune defects in vitamin D-deficient patients and experimental animals, and anecdotal evidence that individuals with rickets are particularly prone to infection. However, there is, paradoxically, a large body of literature supporting an immunosuppressive role of vitamin D and related analogues. The current view is that under physiological conditions, vitamin D probably facilitates immune responses, but that it may also play an active role in the prevention of autoimmunity and that there may be a therapeutic role for vitamin D in some immune-mediated diseases.

Vitamin D acts by binding to its receptor and regulating gene expression in target cells. Its effects include promotion of phagocytosis, superoxide synthesis and bacterial killing, but it is also reported to inhibit T-cell proliferation, production of Th1 cytokines and B-cell antibody production, highlighting the paradoxical nature of its effects. Vitamin D promotes the synthesis of some antimicrobial proteins including cathelcidin and β-defensins.

The role of vitamin D in autoimmunity is particularly interesting. There is increasing evidence, mainly from animal studies, that vitamin D deficiency is linked with autoimmune diseases such as multiple sclerosis, rheumatoid arthritis and inflammatory bowel disease. The inhibition of Th1 activity by vitamin D is thought to be key to this link. In addition, a polymorphism in the vitamin D receptor gene has been associated with increased risk of Crohn's disease.

Taken together, the evidence suggests that vitamin D is a selective regulator of immune function and the effects of vitamin D deficiency, vitamin D receptor deficiency or vitamin D supplementation depend on the immunological situation of the individual (e.g. health, infectious disease, autoimmune disease).

Vitamin E

Vitamin E is the major lipid-soluble antioxidant in the body and is required for protection of membrane lipids from peroxidation. Free radicals and lipid peroxidation are immunosuppressive; thus, it is considered that vitamin E should act to optimise and even enhance the immune response. Except in premature infants and the elderly, clinical vitamin E deficiency is rare in humans, although many individuals have vitamin E intakes below the recommended daily intake in many countries. Cigarette smoking imposes free radical damage and smokers have increased levels of indicators of free radical damage to lipids, low levels of lung and serum vitamin E, increased numbers of neutrophils and macrophages in the lung, increased reactive oxygen species production by phagocytes and depressed immune responses. Thus, cigarette smokers have a higher vitamin E requirement than non-smokers.

A positive association exists between plasma vitamin E levels and DTH responses, and a negative association has been demonstrated between plasma vitamin E levels and the incidence of infections in healthy adults over 60 years of age.

There appears to be particular benefit of vitamin E supplementation for the elderly, with studies demonstrating enhanced Th1 cell-mediated immunity at high doses. However, RCTs do not consistently support a role for vitamin E supplementation in reducing the incidence, duration or severity of respiratory infections in elderly populations or smokers, although one study did show benefit specifically for upper respiratory tract infections in older people in nursing homes. A comprehensive study demonstrated increased DTH responses in elderly subjects supplemented with 60, 200 and 800 mg vitamin E/day, with maximal effect at a dose of 200 mg/day. This dose also increased the antibody responses to hepatitis B, tetanus toxoid and pneumococcus vaccinations. This 'optimal' dose of 200 mg vitamin E/day is well in excess of a typical recommended dose; thus, it appears that adding vitamin E to the diet at levels beyond those normally recommended enhances some immune functions above normal, and it has even been argued that the recommended intake for vitamin E is not adequate for optimal immune function. However, as with many other micronutrients, doses that are hugely in excess of normal requirements may suppress the immune response; indeed, the 800 mg vitamin E/day supplement decreased some of the antibody responses to below those of the placebo group.

Zinc

Zinc is important for DNA synthesis and in cellular growth, differentiation and antioxidant defence, and is therefore an important candidate for potential modulation of immune function. Zinc also has specific antiviral actions. Zinc deficiency impairs many aspects of innate immunity, including phagocytosis by macrophages and neutrophils, natural killer cell activity, respiratory burst and complement activity, all of which could be important contributors to increased susceptibility to infection. Zinc deficiency has a marked impact on bone marrow, decreasing the number of nucleated cells and the number and proportion of cells that are lymphoid precursors. In patients with zinc deficiency related to sickle cell disease, natural killer cell activity is decreased, but can be returned to normal by zinc supplementation. In acrodermatitis enteropathica, which is characterised by reduced intestinal zinc absorption, thymic atrophy, impaired lymphocyte development and reduced lymphocyte responsiveness and DTH are observed. Moderate or mild zinc deficiency or experimental zinc deficiency (induced by consumption of <3.5 mg zinc/day) in humans results in decreased thymulin activity, natural killer cell activity, lymphocyte proliferation, IL-2 production and DTH response; all can be corrected by zinc repletion.

Low plasma zinc levels can be used to predict the subsequent development of lower respiratory tract infections and diarrhoea in malnourished populations. Indeed, diarrhoea is considered a symptom of zinc deficiency and several studies show that zinc supplementation decreases the incidence, duration and severity of childhood diarrhoea. However, most (but not all) studies fail to show a benefit of zinc supplementation in respiratory disease in malnourished populations and available trials on the effect of zinc on the common cold in non-malnourished populations report conflicting results. Very high zinc intakes can result in copper depletion, and copper deficiency impairs immune function (see below).

Copper

Although overt copper deficiency is believed to be rare in humans, modest deficiency is likely to be present among some populations. Zinc and iron impair copper uptake, so that taking high doses of these might induce mild copper deficiency. Copper deficiency has been described in premature infants and in patients receiving total parenteral nutrition. The classic example of copper deficiency is Menkes syndrome, a rare congenital disease which results in the complete absence of caeruloplasmin, the copper-carrying protein in the blood. Children with Menkes syndrome have increased susceptibility to bacterial infections, diarrhoea and pneumonia.

Copper participates as a co-factor in the formation of ROS and appears to have an important role in innate immunity, particularly in respiratory burst. However, as with many other micronutrients, high intakes over long periods can have a negative impact on immune function.

Iron

Iron deficiency has multiple effects on immune function in laboratory animals and humans. However, the relationship between iron deficiency and susceptibility to infection remains controversial. Furthermore, evidence suggests that infections caused by organisms that spend part of their life cycle intracellularly, such as plasmodia, mycobacteria and invasive salmonellae, may actually be enhanced by iron therapy.

In the tropics, in children of all ages, at doses of >2 mg/kg/day, iron has been associated with increased risk of malaria and other infections, including pneumonia. For these reasons, iron intervention in malaria-endemic areas is not advised, particularly high doses in the young, those with compromised immunity (e.g. HIV) and during the peak malaria transmission season. Iron treatment for anaemia in a malarious area must be preceded by effective antimalarial therapy and should be oral, rather than parenteral.

The detrimental effects of iron administration may occur because micro-organisms require iron and providing it may favour the growth and replication of the pathogen. Indeed, it has been argued that the decline in circulating iron concentrations that accompanies infection is an attempt by the host to 'starve' the infectious organism of iron. There are several mechanisms for witholding iron from a pathogen in this way. Lactoferrin has a higher binding affinity for iron than do bacterial siderospores, making bound iron unavailable to the pathogen. Furthermore, once lactoferrin reaches 40% saturation with iron, it is sequestered by macrophages. It is notable that breast milk contains lactoferrin, which may protect against the use of free iron by pathogens transferred to an infant. It is important to note that oral iron supplementation has not been shown to increase risk of infection in non-malarious countries.

Selenium

Selemium is essential for an effectively functioning immune system. Deficiency in laboratory animals affects both innate and adaptive immunity, particularly neutrophil function. It also increases susceptibility to bacterial, viral, fungal and parasitic challenges. Lower selenium concentrations in humans have also been linked with increased virulence, diminished natural killer cell activity, increased mycobacterial disease and HIV progression. A study conducted in the UK demonstrated that selenium supplementation to adults with low selenium status improved some aspects of their immune response to a poliovirus vaccination.

Low dietary intakes of selenium are often associated with concurrent vitamin E deficiency, which has led to the conclusion that selenium deficiency has a significant impact on oxidant/antioxidant processes. This occurs mainly through glutathione peroxidases, and explains why neutrophils and macrophages from selenium-deficient animals are able to ingest pathogens, but are unable to generate the respiratory burst to destroy them.

Supplementation with micronutrients in combination

The roles that micronutrients play in immunity are summarised in Box 13.2.

Supplementation with combinations of vitamins may be particularly beneficial to athletes, since the combined antioxidant activity could enhance protection against free radical damage and reduce exercise-induced immunosuppression. Studies have indeed shown that supplementation with combined antioxidants increased respiratory burst, upregulated the antioxidant system of neutrophils and reduced the production of proinflammatory cytokines from exercising muscles. In contrast, in elderly individuals, combinations of vitamins do not appear to affect innate or adaptive immunity, DTH or the incidence of respiratory or urogenital infections. However, it should be noted that in the

Box 13.2 Key points: micronutrients

- Micronutrient deficiencies are common in low- and middle-income countries, but also occur in high-income countries, especially among the elderly, premature infants and patients with certain diseases
- Micronutrient deficiencies impair immune function: all aspects of immunity can be affected
- Micronutrient deficiencies make the host more susceptible to infections
- Providing micronutrients to deficient individuals can restore immune function and resistance to infections in some situations
- Increasing the intake of some micronutrients (vitamin A, vitamin E, zinc) above the levels normally recommended may enhance immune function
- Excess amounts of some micronutrients (vitamin A, zinc, iron) can impair immune function
- Some diseases are characterised by oxidant stress and this will be compounded by micronutrient deficiencies
- Insufficient intake of micronutrients may contribute to the progression of diseases that have a strong oxidative stress component

combined supplementation studies, the doses of vitamins used were lower than in the single vitamin studies showing benefits in respiratory infections. This is also the case for studies using combined multivitamin and multimineral supplements, which have tended to be small and to vary greatly in the doses of micronutrients provided.

A systematic review concluded that there was no effect of such an approach on episodes of infection, number of days of infection, or antibiotic use in those supplemented compared with those not supplemented, but given the lack of consistent methodology and study design, this remains an important topic for study. It is also pertinent to note that most studies investigating the effects of multivitamin supplements on immune function and infection have been conducted in elderly or immunocompromised individuals and there is little or no information from studies conducted in children or populations in developing countries.

13.10 Dietary fatty acids and immune function

Essential fatty acid deficiency and immune function

Polyunsaturated fatty acids are important components of the membranes of cells involved in the immune and inflammatory responses. They contribute to the physical properties of the membrane through regulation of membrane fluidity and influence the environment in which membrane proteins function. The formation of signalling platforms, termed lipid rafts, within the membrane of stimuated cells is influenced by the fatty acid make-up of those membranes and this seems to be an important way in which fatty acids affect immune cell function. As described in further detail below, omega-6 (n-6) and omega-3 (n-3) PUFAs released from cell membrane phospholipids upon cell stimulation act as substratres to produce bioactive lipid mediators (oxylipins) involved in regulation of immunity and inflammation; importantly, many of these oxylipins have direct proinflammatory actions while others are involved in the resolution of inflammation. Hence, there are multiple important roles for PUFAs within immune cell membranes. Typically in humans, the content of n-6 PUFAs in cell membranes is much higher than n-3 PUFAs; arachidonic acid (20:4n-6) is often present in relatively high amounts. The content of n-3 PUFAs present can be increased by eating more fish or by taking omega-3 supplements.

Animal studies have shown that deficiencies in the essential fatty acids (linoleic acid [18:2n-6] and α-linolenic acid [18:3n-3]) result in decreased thymus and spleen weight, lymphocyte proliferation, neutrophil chemotaxis, macrophage-mediated cytotoxicity and DTH response compared with animals fed diets containing adequate amounts of these fatty acids. Thus, the immunological effects of essential fatty acid deficiency appear to be similar to the effects of single micronutrient deficiencies, although there are no human studies to confirm this (essential fatty acid deficiency is very rare in humans). However, essential fatty acid deficiency would be expected to have a similar effect, because of the vital roles that PUFAs play in immune cell membranes and as oxylipin precursors.

Oxylipins: an important link between fatty acids, immunity and inflammation

Oxylipins are oxidised PUFAs. The best known oxylipins are the eicosanoids produced from the n-6 PUFA arachidonic acid; these include prostaglandins, thromboxanes and leukotrienes. Such oxylipins provide a key link between fatty acids and immune function, including the inflammatory component.

The membranes of most cells contain fairly large amounts of arachidonic acid, which is the principal precursor for eicosanoid synthesis. Arachidonic acid in cell membranes can be released by various phospholipase enzymes, most notably phospholipase A_2, and the free arachidonic acid can subsequently act as a substrate for cyclo-oxygenase (COX) enzymes, forming prostaglandins (PGs) and related compounds, for one of the lipoxygenase (LOX) enzymes, forming leukotrienes (LTs) and related compounds, or for cytochrome P450 (CYP) enzymes (Figure 13.5).

There are many different compounds belonging to each class of eicosanoid and they are each formed in a cell-specific manner. For example, monocytes and macrophages produce large amounts of PGE_2 and PGF_2, neutrophils produce moderate amounts of PGE_2 and mast cells produce PGD_2. The LOX enzymes have different

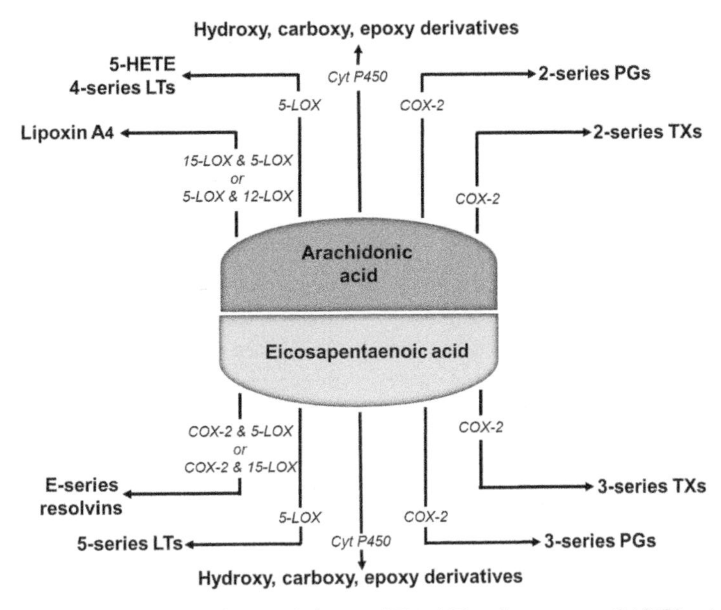

Figure 13.5 Summary of eicosanoid synthesis from arachidonic and EPA. COX, cyclo-oxygenase; Cyt P450, cytochrome P450; HETE, hydroxyeicosatetraenoic acid; LOX, lipoxygenase; LT, leukotriene; TX, thromboxane. *Source:* P.C. Calder (2020) /Cambridge University Press.

tissue distributions, with 5-LOX being found mainly in mast cells, monocytes, macrophages and granulocytes, and 12- and 15-LOX mainly in epithelial cells.

Prostaglandins are involved in modulating the intensity and duration of inflammatory and immune responses. For example, PGE_2 has a number of proinflammatory effects, including induction of fever and erythema, increasing vascular permeability and vasodilation and enhancing pain and oedema caused by other agents such as histamine. However, PGE_2 also suppresses lymphocyte proliferation and natural killer cell activity, and inhibits the production of TNF, IL-1, IL-6, IL-2 and IFN-γ; thus, in these respects, PGE_2 is immunosuppressive and anti-inflammatory. LTB_4 increases vascular permeability, enhances local blood flow, is a potent chemotactic agent for leucocytes, induces the release of lysosomal enzymes, enhances the generation of reactive oxygen species, inhibits lymphocyte proliferation, promotes natural killer cell activity and can enhance the production of inflammatory cytokines. PGD_2, LTC_4, D_4 and E_4 are produced by the cells that mediate lung inflammation in asthma, such as mast cells, and are believed to be the major mediators of asthmatic bronchoconstriction.

Thus, arachidonic acid gives rise to a range of mediators that may have opposing effects to one another, so the overall physiological effect will be governed by the nature of the cells producing the eicosanoids, the concentrations of the mediators, the timing of their production and the sensitivity of target cells to their effects. Many drugs have been developed that target arachidonic acid metabolism as a way to control inflammation.

Consumption of increased amounts of the n-3 PUFAs eicosapentaenoic acid (EPA) and docosahexaenoic acid (DHA) results in a decrease in the amount of arachidonic acid in the membranes of most cells in the body, including those involved in inflammation and immunity. This means that there is less substrate available for synthesis of eicosanoids from arachidonic acid. Furthermore, EPA competitively inhibits the oxygenation of arachidonic acid by COX. Thus, n-3 PUFAs result in decreased capacity of immune cells to synthesise eicosanoids from arachidonic acid. In addition, EPA is itself able to act as a substrate for both COX and 5-LOX (see Figure 13.5), giving rise to eicosanoid derivatives that have a different structure from those produced from arachidonic acid (i.e. 3-series PGs and thromboxanes and 5-series LTs). Thus, the

EPA-induced suppression of the production of arachidonic-acid derived eicosanoids is mirrored by elevation of the production of EPA-derived eicosanoids. The eicosanoids produced from EPA are often less biologically potent than the analogues synthesised from arachidonic acid, although the full range of biological activities of these compounds has not been investigated.

The reduction in the generation of arachidonic acid-derived mediators that accompanies n-3 PUFA consumption has led to the idea that n-3 PUFAs are anti-inflammatory and may affect immune function in general. In accordance with this, supplementation of the diet with fish oil as a source of EPA and DHA has often been shown to result in a decrease in markers and indicators of inflammation (e.g. decreased movement of leucocytes towards sites of inflammatory activity, decreased binding of leucocytes to endothelial cells and their movement from the bloodstream to the subendothelial space, and decreased cellular activation and the release of chemoattractants, cytokines, eicosanoids and reactive species). However, fairly high doses of EPA+DHA (probably at least 2 g/day) are required to elicit these effects and thus not all studies are consistent in their findings.

Omega-3 PUFAs and resolution of inflammation

One exciting development in this field has been the recognition that resolution of inflammation is an active process and that oxylipins produced from EPA and DHA are central to this. The oxylipins involved in resolution of inflammation have been termed 'specialised pro-resolution mediators' or SPMs. There are several families of SPMs, including E- and D-series resolvins produced from EPA and DHA, respectively, and protectins (sometimes called neuroprotectins) and maresins produced from DHA (Figure 13.6). SPMs are synthesised using COX and LOX enzymes operating within the same pathway, whereas for eicosanoid synthesis the COX and LOX pathways are distinct. Furthermore, synthesis of many SPMs is promoted by aspirin and different epimers are produced in the presence and absence of aspirin. Both aspirin-triggered and non-aspirin-triggered SPMs have biological activity. SPMs are produced by human cells (e.g. monocytes, neutrophils) and have been reported in human blood, adipose tissue, breast milk and synovial fluid. Increased intake of EPA and DHA has been shown to result in higher concentrations of some SPMs in human blood. The effects of SPMS have been widely studied in cell culture and in animal models and all SPMs tested to date have anti-inflammatory and inflammation-resolving actions. The potent activity of SPMs may explain many of the documented actions of EPA and DHA in inflammation.

The roles that dietary fatty acids play in immunity are summarised in Box 13.3.

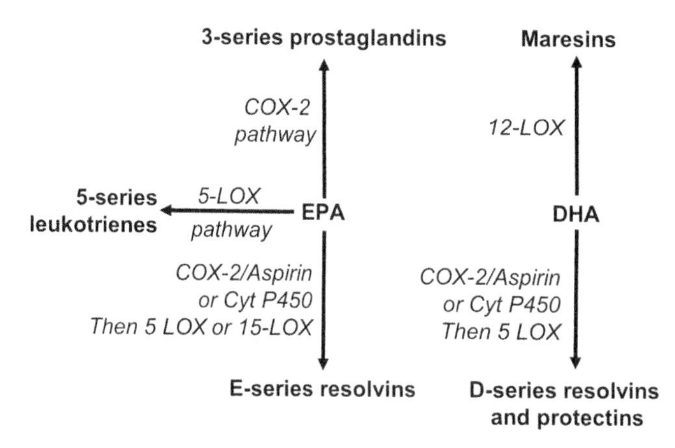

Figure 13.6 Summary of the roles of eicosapentaenoic acid (EPA) and docosahexaenoic acid (DHA) as precursors for biosynthesis of different lipid mediator (oxylipin) families. COX, cyclo-oxygenase; Cyt P450, cytochrome P450; LOX, lipoxygenase.

Box 13.3 Key points: fatty acids and immune function

- Key roles of polyunsaturated fatty acids are as components of membranes and as precursors for hormone-like compounds termed oxylipins (these include eicosanoids)
- Essential fatty acid deficiency impairs cell-mediated immunity
- Eicosanoids synthesised from arachidonic acid influence inflammation and immunity
- The n-3 fatty acids EPA and DHA replace arachidonic acid in cell membranes and therefore can oppose some of the effects of arachidonic acid
- The n-3 fatty acids EPA and DHA are anti-inflammatory
- SPMs produced from EPA and DHA have key roles in resolution of inflammation

13.11 Dietary amino acids and related compounds and immune function

Sulfur amino acids and related compounds

Sulfur amino acids are essential in humans. Deficiency in methionine and cysteine results in atrophy of the thymus, spleen and lymph nodes and prevents recovery from protein–energy malnutrition. When combined with a deficiency of isoleucine and valine, which are also essential amino acids, sulfur amino acid deficiency results in severe depletion of gut lymphoid tissue, very similar to the effect of protein deprivation.

Glutathione is an antioxidant tripeptide that consists of glycine, cysteine and glutamate (from glutamine). Glutathione concentrations in the liver, lung, small intestine and immune cells fall in response to infection and inflammatory stimuli (probably as a result of oxidative stress), and this fall can be prevented in some organs by the provision of cysteine in the diet. Although the limiting precursor for glutathione biosynthesis is usually cysteine, the ability of sulfur amino acids to replete glutathione stores is related to the protein level of the diet. Glutathione can enhance the activity of human cytotoxic T-cells, while depletion of intracellular glutathione diminishes lymphocyte proliferation and the generation of cytotoxic T-lymphocytes.

Taurine is a sulfonated β-amino acid derived from methionine and cysteine metabolism, but is not a component of proteins. Taurine is present in high concentrations in most tissues and particularly in cells of the immune system; in lymphocytes, it contributes 50% of the free amino acid pool. The role of taurine within lymphocytes is not clear. In neutrophils, taurine appears to play a role in maintaining phagocytic capacity and microbicidal action through interaction with myeloperoxidase, an enzyme involved in respiratory burst. Taurinechloramine is formed by complexing of taurine with hypochlorous acid (HOCl) produced by myeloperoxidase. Hypochlorous acid, although toxic to bacteria, causes damage to host tissues and it has been proposed that the formation of taurinechloramine is a mechanism to protect the host from this damage. Although taurine appears not to affect mediator production by macrophages, taurinechloramine decreases PGE_2, TNF and IL-6 production by macrophages.

Arginine

Arginine is a non-essential amino acid in humans and is involved in protein, urea and nucleotide synthesis, and adenosine triphosphate (ATP) generation. It also serves as the precursor of nitric oxide, a potent immunoregulatory mediator that is cytotoxic to tumour cells and to some micro-organisms. In laboratory animals, arginine decreases the thymus involution associated with trauma, promotes thymus repopulation and cellularity, increases lymphocyte proliferation, natural killer cell activity and macrophage cytotoxicity, improves DTH, increases resistance to bacterial infections, increases survival to sepsis and burns, and promotes wound healing. There are indications that arginine may have similar effects in humans, although these have not been tested thoroughly.

There is particular interest in the inclusion of arginine in enteral formulae given to patients hospitalised for surgery, trauma and burns, since it appears to reduce the severity of infectious complications and the length of hospital stay. However, in many of the clinical studies carried out in these patients, the enteral formulae used have contained a variety of nutrients with immunomodulatory actions, so it has been difficult to ascribe beneficial effects to any one nutrient alone.

Glutamine

Glutamine is the most abundant amino acid in the blood and in the free amino acid pool in the body; skeletal muscle is considered to be the most important glutamine producer in the body. Once released from skeletal muscle, glutamine acts as an interorgan nitrogen transporter. Important users of glutamine include the kidney, liver, small intestine and cells of the immune system. Plasma glutamine levels are lowered (by up to 50%) by sepsis, injury and burns, and following surgery. Furthermore, the skeletal muscle glutamine concentration is lowered by more than 50% in at least some of these situations. The lowered plasma glutamine concentrations that occur are likely to be the result of demand for glutamine (by the liver, kidney, gut and immune system) exceeding supply, and it has been suggested that the lowered plasma glutamine contributes, at least in part, to the impaired immune function that accompanies such situations. It has been argued that restoring plasma glutamine concentrations in these situations should restore immune function. As with arginine, there are animal studies to support this.

Clinical studies, mainly using intravenous infusions of solutions containing glutamine, have also reported beneficial effects for patients undergoing bone marrow transplantation and colorectal surgery, patients in intensive care and low-birthweight babies, all of whom are at risk from infection and sepsis. In some of these studies, improved outcome was associated with improved immune function. In addition to a direct immunological effect, glutamine, even provided parenterally, improves gut barrier function in patients at risk of infection. This would have the benefit of decreasing the translocation of bacteria from the gut and eliminating a key source of infection.

The roles that dietary amino acids play in immunity are summarised in Box 13.4.

Box 13.4 Key points: amino acids and immunity

- Deficiencies in essential amino acids are likely to impair immune function
- A key role of sulfur amino acids is in maintaining levels of the antioxidant glutathione and so in preventing oxidative stress
- Some classically non-essential amino acids (arginine, glutamine) may become essential in stress situations

13.12 Gut microbiota, probiotics, and immune function

Indigenous bacteria are believed to contribute to the immunological protection of the host by creating a barrier against colonisation by pathogenic bacteria. This barrier can be disrupted by disease and by the use of antibiotics, so allowing easier access to the host gut by pathogens. It is now believed that this barrier can be maintained by providing supplements containing live 'desirable' bacteria; such supplements are termed *probiotics*. Probiotic organisms are found in fermented foods, including traditionally cultured dairy products and some fermented milks. The organisms included in commercial probiotics include lactic acid bacteria (lactobacilli) and bifidobacteria. These organisms only colonise the gut temporarily, making regular consumption necessary.

In addition to creating a barrier effect, some of the metabolic products of probiotic bacteria (e.g. lactic acid and a class of antibiotic proteins termed bacteriocins produced by some bacteria) may inhibit the growth of pathogenic organisms. Probiotic bacteria may also compete with pathogenic bacteria for nutrients and may enhance the gut immune response to pathogenic bacteria.

Probiotics have various routes for contact with the gut epithelium and the underlying immune tissues; it is through these interactions that probiotics are thought to be able to influence immune function. The interactions involve both physical contact and metabolites, such as short-chain fatty acids produced by the bacteria. Our understandings of these interactions is incomplete and a focus of current research.

A number of studies have examined the influence of various probiotic organisms, either alone or in combination, on immune function, infection and inflammatory conditions in humans. Probiotics appear to enhance innate immunity (particularly phagocytosis and natural killer cell activity), but have lesser effects on adaptive immunity. In children, probiotics have been shown to reduce the incidence and duration of diarrhoea, although the effects depend on the nature of the condition. Some studies demonstrate a reduction in the risk of traveller's diarrhoea in adults and in antibiotic-associated diarrhoea in older people in hospital. The effects

Box 13.5 Key points: probiotics, prebiotics, and immunity

- Probiotics is the name given to desirable bacteria that colonise the human gut creating a health benefit
- Probiotic bacteria help to maintain the gut barrier and prevent colonisation by pathogenic bacteria
- Probiotic bacteria enhance some aspects of immune function in laboratory animals and may do so in humans
- Probiotic bacteria may decrease the incidence and severity of diarrhoea.
- Prebiotics increase the numbers of beneficial bacteria in the gut and may affect immune function, but it is not clear whether their effect is direct or indirect

on other infectious outcomes are much less clear. There is also some evidence that probiotics could be beneficial in ulcerative colitis, irritable bowel syndrome and allergy, but this is not entirely consistent. One of the difficulties in interpreting results is that there may be significant species and strain differences in the effects of probiotics, as well as doses used, duration of treatment, subject characteristics, sample size and technical considerations.

Prebiotics are selectively fermented by the gut microbiota, leading to increased numbers of beneficial bacteria within the gut, which interact with other members of the gut microbiota and the host immune system. Although there is growing evidence for potential immunomodulatory effects of prebiotics, it is not clear whether they are direct effects or simply manifested through alteration of the gut microbiota. Furthermore, it is notable that while there is considerable supporting literature for the immunomodulatory effects of probiotics, the data for prebiotics alone remain inconsistent.

The effects of probiotics and prebiotics on immune function are summarised in Box 13.5.

13.13 Breastfeeding and immune function

Composition of breast milk

The human neonate is born with an immature immune system. Before birth, the fetus is protected by passive immune factors (e.g. antibodies) passed across the placenta from the mother. After birth, the neonate must continue to receive immune support from the mother until its immune system develops sufficiently. Such support comes from breast milk, which is the best example of a foodstuff with immune-enhancing properties.

Milk contains a wide range of immunologically active components, including cells (macrophages, T- and B-lymphocytes, neutrophils), immunoglobulins (IgG, IgM, IgD, sIgA), lysozyme (which has direct antibacterial action), lactoferrin (which binds iron, so preventing its uptake by bacteria), cytokines (IL-1, IL-6, IL-10, IFN-γ, TNF, transforming growth factor-β), growth factors (epidermal growth factor, insulin-like growth factor), hormones (thyroxin), fat-soluble vitamins (vitamins A, D, E), amino acids (taurine, glutamine), fatty acids, amino sugars, oligosaccharides (human milk oligosaccharides or HMOs), gangliosides and nucleotides. Breast milk also contains factors that prevent the adhesion of some microorganisms to the gastrointestinal tract and so prevents bacterial colonisation. It contains factors that promote the growth of useful bacteria (e.g. bifidobacteria) in the gut. The content of many factors varies among milks of different species and is different between human breast milk and many infant formulae.

Breastfeeding, immune function, and infection

Breastfeeding is thought to play a key role in the prevention of infectious disease, particularly diarrhoea and gastrointestinal and lower respiratory infections, in infants in all countries. In addition to preventing infectious disease, breastfeeding enhances the antibody responses to vaccination. Several studies have examined the effect of breastfeeding versus formula feeding on risk of death due to infectious diseases in low- and middle-income countries. A meta-analysis of these studies suggested that infants who are not breastfed have a sixfold greater risk of dying from infectious diseases in the first two months of life than those who are breastfed. However, it appears that this protection decreases steadily with age, as infants begin complementary feeding, so that by 6–11 months the protection afforded by breastfeeding is no longer apparent. Breastfeeding may provide better protection against diarrhoea (up to six months of age) than

Box 13.6 Key points: breast milk and immunity

- Breast milk contains a variety of immunologically active factors
- The factors in breast milk appear designed to protect the neonate from infection and to promote development of the neonatal immune response
- Human breast milk has a different composition from other milks
- Breastfeeding protects against some infectious diseases

against deaths due to respiratory infections. There are also geographical influences on the protection afforded by breastfeeding; in some continents, protection can be observed throughout the first year of life, whereas in others it is much more short-lived.

The roles that breastfeeding plays in immunity are summarised in Box 13.6.

13.14 Future perspectives

Deficiencies of total energy or of one or more essential nutrients, including vitamins A, B_6, B_{12}, C and E, folic acid, zinc, iron, copper, selenium, essential amino acids and essential fatty acids, impair immune function and increase susceptibility to infectious pathogens. This occurs because each of these nutrients is involved in the molecular, cellular and metabolic responses to challenge of the immune system. Providing these nutrients to deficient individuals restores immune function and improves resistance to infection. For several nutrients, the dietary intakes that result in greatest enhancement of immune function are greater than recommended intakes. However, excessive intake of some nutrients (e.g. zinc, iron) also impairs immune responses. Thus, four potential general relationships between the intake of a nutrient and immune function appear to exist (Figure 13.7).

It is often assumed when defining the relationship between nutrient intake and immune function that all components of the immune system will respond in the same dose-dependent fashion to a given nutrient. This is not correct, at least as far as some nutrients are concerned, and it appears likely that different components of the immune system show an individual dose–response relationship to the availability of a given nutrient (Figure 13.8).

It is now appreciated that the supply of nutrients that are not considered to be essential according to traditional criteria may also influence immune function; this is particularly notable for the amino acids glutamine and arginine, and may indicate that a re-evaluation of the definitions of essentiality, nutrient requirements and nutrient status is required for some dietary components.

An early point of contact between nutrients and other food components (e.g. plant polyphenols, fibre) and the immune system occurs within the intestinal tract. Relatively little is known about the relationship between nutrient status and the function of the gut-associated immune system. This is of particular relevance when considering adverse reactions to foods; the role of immunoregulatory nutrients in responses to food components and in sensitisation to food-borne allergens is largely unknown. An understanding of the interaction between nutrients, the types of bacteria that inhabit the gut, and gut-associated and systemic immune responses is only now beginning to emerge.

The term 'optimal immune function' is often used in the literature without careful thought about its definition. An optimal immune response to any given nutrient measured by one marker will not necessarily be optimal according to a second marker of immune function (see Figure 13.8). Furthermore, the effect of a given nutrient on immune response may be altered by levels of other nutrients. For these reasons, the natural desire to 'optimise' the immune response may not be realistic. At best, it is reasonable to expect that correction of marginal deficiencies will improve immunity, but further enhancement using, for example, micronutrient supplements cannot be guaranteed and in excessive doses may be detrimental. At the other extreme, there is interest in the potential therapeutic effect of nutrients in diseases involving dysregulation of the immune system (e.g. n-3 fatty acids in rheumatoid arthritis). In some but by no means all cases, there is supportive evidence for this approach. Between these extremes, there are many unanswered questions, but it is clear that the study of the modulation of immune function by nutrients has important implications.

The appreciation that chronic low-grade inflammation contributes to risk of cardiovascular, metabolic and cognitive disease and that these diseases are linked to ageing and to diet

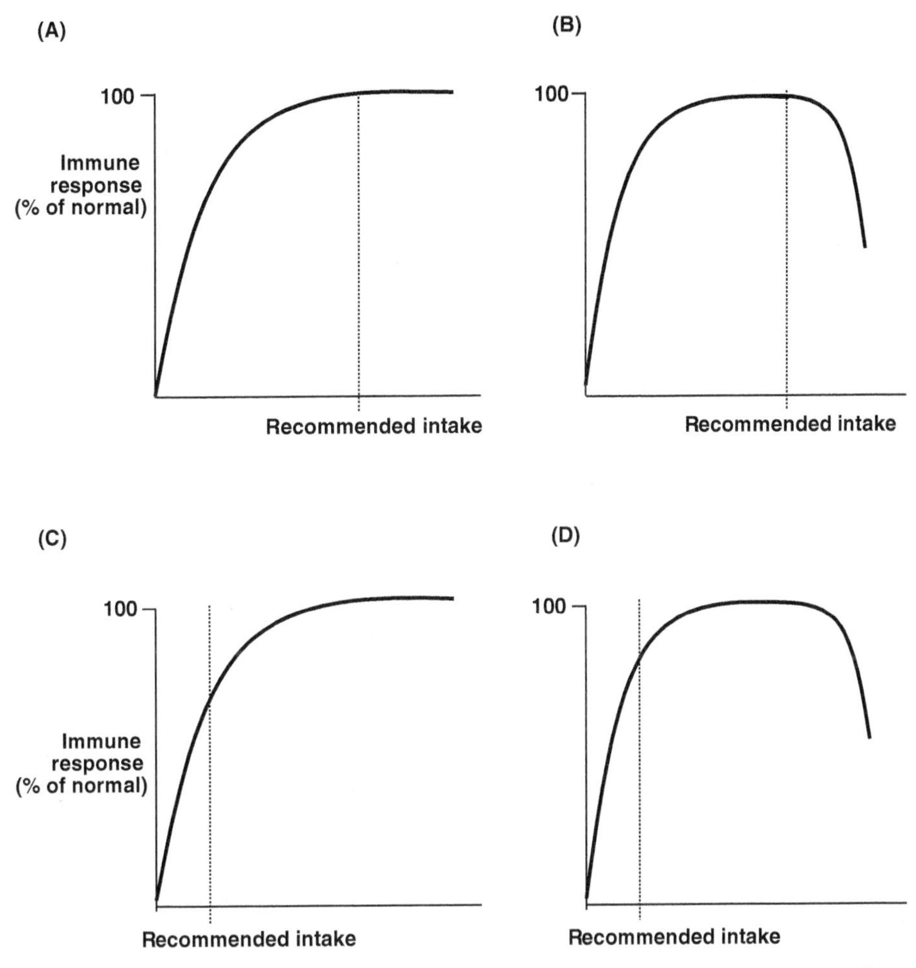

Figure 13.7 Four possible relationships between nutrient intake and immune function. All patterns assume that a deficiency of the nutrient impairs the immune response. In pattern (A) the immune response is maximal, in terms of relationship to the intake of the nutrient under study, at the recommended level of intake and intakes somewhat above the recommended intake do not impair immune function. In pattern (B) the immune response is maximal, in terms of relationship to the intake of the nutrient under study, at the recommended level of intake and intakes somewhat above the recommended intake impair immune function. In pattern (C) the immune response is submaximal, in terms of relationship to the intake of the nutrient under study, at the recommended level of intake and intakes somewhat above the recommended intake increase and do not impair immune function. In pattern (D) the immune response is submaximal, in terms of relationship to the intake of the nutrient under study, at the recommended level of intake and intakes somewhat above the recommended intake impair immune function.

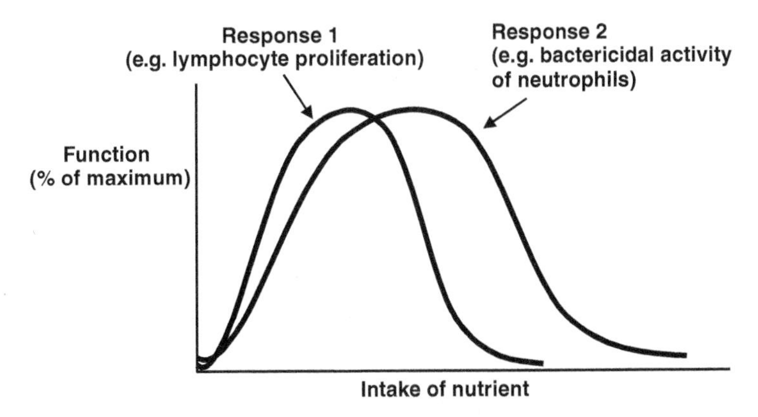

Figure 13.8 Dose–response relationships of different immune functions to the same nutrient may not be identical.

has led to attempts to classify diets according to the extent to which they are likely to promote inflammation. A number of such classification systems have been proposed. The 'dietary inflammatory index' (DII) is a literature-derived nutrient-based index based on published associations of 45 dietary components (energy, macronutrients, fibre, micronutrients, plant polyphenols, herbs, spices) which are each weighted (positively for increasing inflammation and negatively for decreasing inflammation) and six inflammatory biomarkers (CRP, TNF, IL-1β, IL-6, IL-4 and IL-10); a higher score indicates a more proinflammatory diet. The 'dietary inflammation score' (DIS) is a literature-derived food and supplement-based score based on biological plausibility and published associations of intakes of 18 foods or food groups groups (e.g. leafy greens and cruciferous vegetabes, tomatoes, apples and berries, deep yellow or orange vegetables and fruit, legumes, fish, high-fat dairy, low-fat dairy, nuts, red meat, added sugars, refined grains and starchy vegetables) and micronutrient supplement use which are each weighted (positively for increasing inflammation and negatively for decreasing inflammation) and four inflammatory biomarkers (CRP, IL-6, IL-8 and IL-10); a higher score indicates a more proinflammatory diet. The 'anti-inflammatory dietary index' (AIDI) is a literature-derived food-based score based on the associations of intakes of 18 foods or food groups (e.g. total fruits and vegetables, whole-grain breads, dry fruits, legumes, nuts, unprocessed red meat) which are each weighted (negatively for increasing inflammation, positively for decreasing inflammation) and plasma CRP concentrations; a higher score indicates a more anti-inflammatory diet. The 'empirical dietary inflammatory pattern score' (EDIP) is based on regression models predicting the association of intakes of foods and food groups which are each weighted (positively for increasing inflammation and negatively for decreasing inflammation) and three inflammatory biomarkers (CRP, IL-6, TNF receptor 2); a higher score indicates a more proinflammatory diet.

The EDIP, DIS and AIDI are all based on whole foods, although they are derived differently, while DII is mainly nutrient based. Although high correlations have been reported between DII and DIS in at least three cohorts, there are low-to-moderate correlations between EDIP and DIS and low correlations between EDIP and DII. EDIP and AIDI seem to be similar although they have not been correlated in a single study.

The emergence of systemic acute respiratory distress syndrome coronavirus 2 (SARS-CoV-2) and the disease it causes (COVID-19) focused attention on the role of nutrition in supporting the immune system and in helping protect individuals from infection and from infections becoming severe. In keeping with their adverse impact on immunity, both frailty and obesity were clearly demonstrated to increase the likelihood of SARS-CoV-2 infection and the risk of more severe COVID-19. Many case–control and other association studies linking low status of several micronutrients, especially vitamin D and zinc but also vitamin C and selenium, with increased risk of more severe COVID-19 were published. Although the findings of such studies are consistent with the importance of these micronutrients in supporting the immune system and suggest that low intakes and status are a risk, reverse causality cannot be excluded and more definitive intervention trials, especially RCTs, have been inconclusive.

Finally, although outside the scope of this chapter, it is important to consider the role of hormones in regulating immune function during malnutrition. An inadequate supply of nutrients to the body may cause physiological stress, leading to an elevation in the circulating concentrations of glucocorticoids and catecholamines. Both classes of hormones have an inhibitory effect on immune function and may therefore be important factors when considering the relationship between nutrient supply and immunological outcome.

Further reading

General reviews of nutrition, immunity, and infection

Calder, P.C. (2013). Feeding the immune system. Proceedings of the Nutrition Society 72: 299–309.

Calder, P.C. and Jackson, A.A.. (2000). Undernutrition, infection and immune function. Nutrition Research Reviews 13: 3–29.

Scrimshaw, N.S. and SanGiovanni, J.P.. (1997). Synergism of nutrition, infection, and immunity: an overview. American Journal of Clinical Nutrition 66: 464S–477S.

Wu, D., Lewis, E.D., Pae, M., and Meydani, S.N. (2019). Nutritional modulation of immune function: analysis of evidence, mechanisms, and clinical relevance. Frontiers in Immunology 9: 3160.

Assessment of the impact of nutrition on immunity and inflammation

Albers, R., Antoine, J.-M., Bourdet-Sicard, R. et al.(2005). Markers to measure immunomodulation in human nutrition intervention studies. British Journal of Nutrition 94: 452–481.

Albers, R., Bourdet-Sicard, R., Braun, D.et al. (2013). Monitoring immune modulation by nutrition in the general population: identifying and substantiating effects on human health. British Journal of Nutrition 110 (suppl. 2): S1–S30.

Calder, P.C., Ahluwalia, N., Albers, R. et al.(2013). A consideration of biomarkers to be used for evaluation of inflammation in human nutritional studies. British Journal of Nutrition 109 (Suppl. 1): S1–S34.

The role of inflammation in chronic disease

Calder, P.C., Albers, R., Antoine, J.M. et al.(2009). Inflammatory disease processes and interactions with nutrition. British Journal of Nutrition 101: S1–S45.

Calder, P.C., Ahluwalia, N., Brouns, F. et al.(2011). Dietary factors and low-grade inflammation in relation to overweight and obesity. British Journal of Nutrition 106 (Suppl. 3): S5–S78.

Calder, P.C., Bosco, N., Bourdet-Sicard, R. et al.(2017). Health relevance of the modification of low grade inflammation in ageing (inflammageing) and the role of nutrition. Ageing Research Reviews 40: 95–119.

Protein–energy malnutrition, immunity, and infection

Michael, H., Amimo, J.O., Rajashekara, G., Saif, L.J., and Vlasova, A.N.. (2022). Mechanisms of kwashiorkor-associated immune suppression: insights from human, mouse, and pig studies. Frontiers in Immunology 13: 826268.

Schaible, U.E. and Kaufmann, S.H.. (2007). Malnutrition and infection: complex mechanisms and global impacts. PLoS Medicine 4: e115.

Obesity, immunity, and infection

Huttunen, R. and Syrjänen, J. (2013). Obesity and the risk and outcome of infection. International Journal of Obesity 37: 333–340.

Milner, J.J. and Beck, M.A. (2012). The impact of obesity on the immune response to infection. Proceedings of the Nutrition Society 71: 298–306.

Micronutrients, immunity, and infection

Calder, P.C., Ortega, E.F., Meydani, S.N. et al.(2022). Nutrition, immunosenescence and infectious disease: an overview of the scientific evidence on micronutrients and on modulation of the gut microbiota. Advances in Nutrition 13: 1S-26S.

Gombart, A.F., Pierre, A., and Maggini, S. (2020). A review of micronutrients and the immune system – working in harmony to reduce the risk of infection. Nutrients 12: 236.

Maggini, S., Pierre, A., and Calder, P.C. (2018). Immune function and micronutrient requirements change over the life course. Nutrients 10: 1531.

Fatty acids, immunity, and inflammation

Calder, P.C. (2020). Eicosanoids. Essays in Biochemistry 64: 423–441.

Calder, P.C.. (2020). n-3 PUFA and inflammation: from membrane to nucleus and from bench to bedside. Proceedings of the Nutrition Society 79: 404–416.

Chiang, N. and Serhan, S.N. (2020). Specialised pro-resolving mediator network: an update on production and actions. Essays in Biochemistry 64: 443–462.

Gutierrez, S., Svahn, S.L., and Johansson, M.E. (2019). Effects of omega-3 fatty acids on immune cells. International Journal of Molecular Science 20: 5028.

Amino acids and immunity

Li, P., Yin, Y.L., Li, D., Kim, S.W. and Wu, G. (2007). Amino acids and immune function.British Journal of Nutrition 98: 237–252.

Gut microbiota, probiotics, and immunity

Ahern, P.P. and Maloy K.J. (2020). Understanding immune–microbiota interactions in the intestine. Immunology 159: 4–14.

Mowat, A.M. (2003). Anatomical basis of tolerance and immunity to intestinal antigens. Nature Reviews Immunology 3: 331–341.

Thomas, C.M. and Versalovic, J. (2010). Probiotics–host communication: modulation of signaling pathways in the intestine. Gut Microbes 1: 148–163.

Breast milk and immunity

Andreas, N.J., Kampmann, B. and Mehring Le-Doare, K. (2015). Human breast milk: a review on its composition and bioactivity. Early Human Development 91: 629–635.

Ogra, P.L. (2020). Immunology of human milk and lactation: historical overview. Nestle Nutrition Institute Workshop Series 94: 11–26.

Weström, B., Arévalo Sureda, E., Pierzynowska, K., Pierzynowski, S.G. and Pérez-Cano, F.J. (2020). The immature gut barrier and its importance in establishing immunity in newborn mammals. Frontiers in Immunology 11: 1153.

Dietary inflammatory scores

Byrd, D.A., Judd, S.E., Flanders, W.D., Hartman, T.J., Fedirko, V. and Bostick, R.M. (2019). Development and validation of novel dietary and lifestyle inflammation scores. Journal of Nutrition 149: 2206–2218.

Hébert, J.R., Shivappa, N., Wirth, M.D., Hussey, J.R. and Hurley, T.G. (2019). Perspective: The Dietary Inflammatory Index (DII) – lessons learned, improvements made, and future directions. Advances in Nutrition 10: 185–195.

Kaluza, J., Harris, H., Melhus, H., Michaëlsson, K. and Wolk, A. (2018). Questionnaire-based anti-inflammatory diet index as a predictor of low-grade systemic inflammation. Antioxidants and Redox Signaling 28: 78–84.

Shivappa, N., Steck, S.E., Hurley, T.G., Hussey, J.R. and Hébert, J.R. (2014). Designing and developing a literature-derived, population-based dietary inflammatory index. Public Health Nutrition 17: 1689–1696.

Tabung, F.K., Smith-Warner, S.A., Chavarro, J.E. et al. (2017). An empirical dietary inflammatory pattern score enhances prediction of circulating inflammatory biomarkers in adults. Journal of Nutrition 147: 1567–1577.

Tabung, F.K., Smith-Warner, S.A., Chavarro, J.E. et al. (2016). Development and validation of an empirical dietary inflammatory index. Journal of Nutrition 146: 1560–1570.

Nutrition, immunity, and COVID-19

Calder, P.C. (2020). Nutrition, immunity and COVID-19. BMJ Nutrition Prevention and Health 3: e000085.

Calder, P.C. (2021). Nutrition and immunity: lessons for COVID-19. European Journal of Clinical Nutrition 75: 1309–1318.

14
Phytochemicals

Gary Williamson and Jeremy P.E. Spencer

Key messages

- Phytochemicals are found in fruits, vegetables, tea, coffee and cocoa.
- Examples of phytochemicals are polyphenols, carotenoids, glucosinolates, monoterpenes, sterols, stanols and sulfur-containing compounds.
- Phytochemicals confer particular properties on the plant such as colour, taste and nutritional properties.
- The pathways of absorption and metabolism of polyphenols and carotenoids are well understood, and partially involve the gut microbiota.
- Phytochemicals, especially polyphenols, can have beneficial effects on metabolic health and insulin resistance through effects on digestion and cell signalling.
- There is a strong evidence base to support the actions of phytochemical intake, in particular flavonoids, on protection against the development and progression of cardiovascular disease.
- Phytochemicals may affect human cognitive function and the development of neurodegenerative disease, through their potential to improve cerebrovascular blood.
- There are early indications that phytochemicals may affect the growth and diversity of the gut microbiota, in a similar manner as that seen for other non-nutrients such as fibre.

14.1 Introduction

The word 'phytochemical' refers to plant-derived secondary metabolites and means 'chemicals from plants'. In nutrition, it is associated with several classes of compounds which can provide benefits to health and well-being over and above basic essential nutrition. The well-known health benefits of increased dietary intake of fruits and vegetables are derived from high fibre content, low calorie density and the presence of phytochemicals. Phytochemicals are present in most foods and beverages containing plant-based products. Each plant species has a characteristic phytochemical profile, but the amount is highly dependent on growing conditions, variety, degree of ripeness and after harvest, on storage conditions and degree of processing. Ultra-processed foods often retain very little of the phytochemicals originally present in the raw ingredients.

The most abundant classes in the diet are polyphenols and carotenoids, and these components impart colour, taste, sensory characteristics and nutritional properties. The term also includes glucosinolates, alkaloids, terpenes, isoflavones and phytosterols. These phytochemicals confer unique characteristics to the plant; for example, glucosinolate breakdown products, formed by chewing, cooking or processing, are pungent compounds which give *Brassica* vegetables their unique flavour.

This chapter will focus mainly on the bioavailability and biological effects in humans of the most widely distributed and abundant phytochemicals, polyphenols and carotenoids, and indicate mechanism of action where known.

Nutrition and Metabolism, Third Edition. Edited on behalf of The Nutrition Society by Helen M. Roche, Ian A. Macdonald, Annemie M.W.J. Schols and Susan A. Lanham-New.

Companion Website: www.wiley.com/go/nutrition/metabolism3e

14.2 Chemistry, intake, and nomenclature

The main classes of phytochemicals are shown in Table 14.1, together with some of the most commonly studied examples. The nomenclature is complicated, as many compounds have often been named after the plant from which they were originally isolated. For example, quercetin was named in the nineteenth century after the oak tree, genus *Quercus*, and bears no relation to the chemical structure. Here we focus on polyphenols and carotenoids, the most abundant and well-studied classes of phytochemicals. We do not cover terpenoids, phytosterols or glucosinolates in detail, but further information can be obtained from the reviews indicated in the further reading list.

Polyphenols

Polyphenols can be roughly divided into several classes (see Table 14.1). It is often claimed that thousands of flavonoids are present in plants, but the relevant number present in the diet and with potential effects on health is much lower. Flavonoids and isoflavones have a basic three-ring chemical structure, whereas phenolic acids contain just one phenolic ring, together with a side chain and a terminal carboxyl group (Figure 14.1). Other polyphenols such as ellagitannins and curcumin consist of multiple phenolic rings linked together. Many flavonoids have one or more sugars attached, and phenolics acids are ester-linked to an organic acid, which increases stability and water solubility and stops the polyphenol from passively crossing

Table 14.1 Classification and nomenclature of phytochemicals.

Phytochemical class	Division	Class	Selected common examples	Most common food sources
(Poly)phenols[a]	Flavonoids	Flavonols	Quercetin	Onion, apples, tea
	Flavonoids	Flavanols	(Epi)catechin	Green tea, fruits, cocoa
	Condensed tannins	Proanthocyanidins (flavanol oligomers)	Procyanidin B_2, prodelphinidins	Cocoa, cranberry, grapeseed
	Flavonoids	Anthocyanins	Cyanidin, pelargonidin	Berries
	Flavonoids	Flavanones	Hesperidin	Citrus
	Flavonoids	Flavones	Apigenin	Herbs
	Isoflavonoids	Isoflavones	Genistein, daidzein	Soy
	Phenolic acids	Hydroxy-cinnamates	Ferulic acid, caffeic acid (chlorogenic acids)	Coffee, fruits
	Hydrolysable tannins	Ellagitannins	Ellagic acid, punicalagin	Strawberry, pomegranate
Terpenoids	Monoterpenoids, iridoids, di- and triterpenoids		Carvone, menthol, thymol	Plant essential oils
	Tetraterpenoids (carotenoids)	Carotene	β-Carotene (provitamin A)	Carrots
	Tetraterpenoids (carotenoids)	Carotene	Lycopene	Tomatoes, watermelon
	Tetraterpenoids (carotenoids)	Xanthophyll	Lutein	Green leafy vegetables
	Tetraterpenoids (carotenoids)	Xanthophyll	Zeaxanthin	Corn, peppers
Sulfur-containing compounds		Glucosinolates	Glucoraphanin (→ sulforaphane)	*Brassica* vegetables
		Organosulfur	Allicin	Garlic
Sterols and stanols		Phytosterols	β-Sitosterol, sitostanol, campesterol, campestanol	Soy, legumes, nuts, plant oils

[a]The term (poly)phenols is used here to include all classes including phenolic acids.

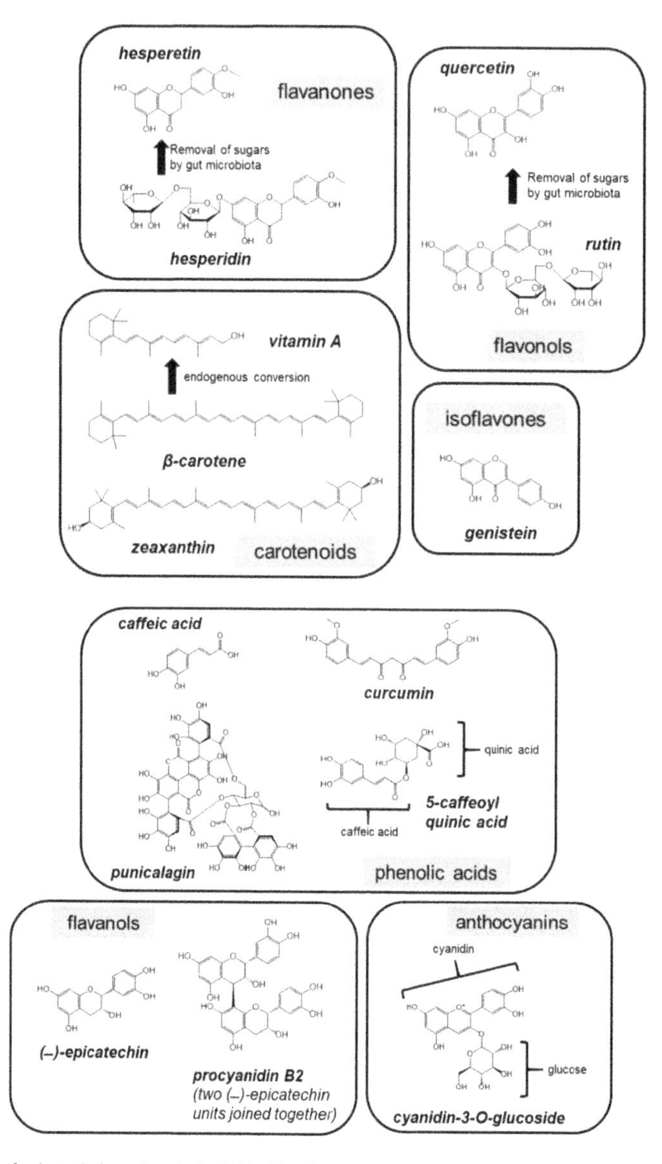

Figure 14.1 Structures of selected phytochemicals divided by class. Important structural features are indicated as well as some metabolic conversions by gut microbiota.

biological membranes. Without an attached sugar, the aglycone forms of flavonoids are less water soluble and more readily able to diffuse across biological membranes. Without an attached organic acid, phenolic acid solubility is highly dependent on pH, but they are more soluble under physiological conditions.

Polyphenols generated interest originally through their antioxidant activity, since they are redox active and able to scavenge free radicals in vitro. However, it has become apparent that their activity is more subtle and complex than originally thought. Recent research has identified several molecular targets where specific interactions are important, rather than a general redox activity. These aspects are discussed below.

Carotenoids

Carotenoids are hydrophobic molecules found widely in plants, but especially in vegetables. There are two main types: xanthophylls, which contain oxygen atoms, and carotenes, hydrocarbon structures with no oxygen atoms (see Table 14.1). Carotenoids such as α- and β-carotene may be converted to vitamin A, whereas zeaxanthin and lutein are not. Carotenoids are present in many vegetables and some fruits in low milligram amounts, are very lipid soluble and almost insoluble in aqueous solution and have an intense colour.

Recommended intakes

Although there are very few specific recommendations for phytochemical intake, many countries recommend increasing fruit and vegetable intake, leading to campaigns such as Five-a-day. Typical diets that include one cup of tea and one cup of coffee a day, together with five portions of fruit/vegetables, are shown in Figure 14.2. It should be noted that these values are based on raw, unprocessed fruit and vegetables. Minimally processed foods should be eaten to maximise intake of phytochemicals, which can be lost during processing.

14.3 Absorption and metabolism

Polyphenols

Unlike vitamins and minerals, polyphenols are not stored in the body, but are metabolised by pathways common to many drugs. Depending on the habitual diet, up to several grams of polyphenols can be consumed every day as part of a diet rich in fruits, vegetables and polyphenol-rich beverages.

The processes and pathways of absorption and metabolism in general are well known and understood for many types of polyphenols. The chemical attachment of sugars or organic acids to flavonoids and phenolic acids has a dramatic effect on their absorption. For example, quercetin is attached to a glucose moiety in onions, but to glucose and rhamnose in tea. For the quercetin aglycone to be absorbed, the sugars must first be hydrolysed, as occurs in all carbohydrate digestive processes. The milk sugar, lactose, is hydrolysed into constituent glucose and galactose by the brush border enzyme lactase phlorizin hydrolase (LPH). The same enzyme hydrolyses glucose from quercetin, and the resulting quercetin product is passively absorbed into the enterocytes, where it is conjugated and enters blood circulation. However, if a rhamnose is also attached to the quercetin glucoside (as in quercetin rhamnoglucoside, or rutin) then the molecule is no longer a substrate for LPH as it cannot fit into the active site (Figure 14.3). Rutin is too large and too hydrophilic to

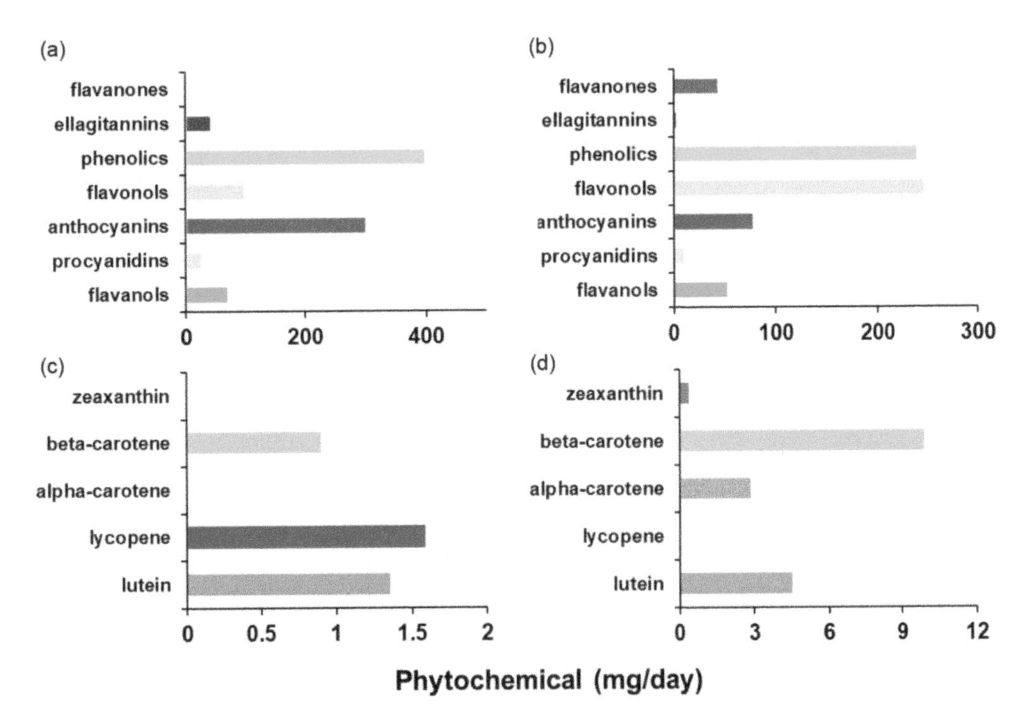

Figure 14.2 Phytochemical content of different diets. Each diet includes one cup of coffee and one cup of tea per day plus fruit and vegetables as indicated. Variability in fruit and vegetable cultivars, strength and number of cups of tea and coffee means that the values are only an indication of intake. Heavy tea or coffee drinkers, for example, will have several times higher intake of phenolics and flavanols. (a) Polyphenol content or (c) carotenoid content from one portion each of blackberries, blueberries, tomato, broccoli and apple, together with tea and coffee. (b) Polyphenol content or (d) carotenoid content from one portion each of strawberries, red onions, orange, carrot and spinach, together with tea and coffee.

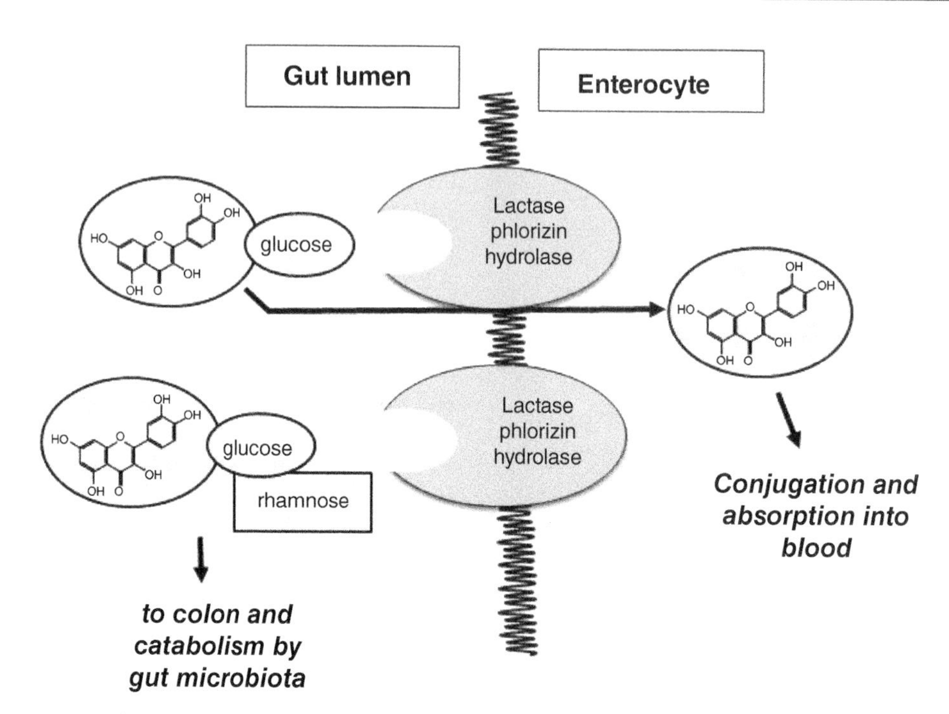

Figure 14.3 Key absorption step: the attached sugar defines the pathway of absorption. The attachment of a glucose or rhamnose determines whether the polyphenol is absorbed intact, or reaches the colon where it is catabolised by gut microbiota.

be passively absorbed into enterocytes, and so it is untouched in the small intestine and reaches the colon. In the colon, the gut bacteria hydrolyse the glucose and rhamnose. Some of the resulting quercetin is absorbed, whereas the majority is broken down into smaller phenolics including phenolic acids, which are then also absorbed and conjugated.

The total bioavailability of some phytochemicals is estimated from a sum of the parent and catabolic products. There are no known specific transporters to allow rapid absorption of polyphenols, unlike nutrients like glucose or vitamin C which have specific transporters. Despite this, at least two-thirds of quercetin from onions and of the phytochemicals in apple are absorbed in the small intestine. For polyphenols which are not attached to a sugar or organic acid, such as epicatechin, the molecule is absorbed across the enterocyte brush border at a rate which depends on molecular size, hydrophobicity and rate of conjugation. The latter generates a concentration gradient of the parent compound, enhancing absorption. Various factors can further influence the absorption of polyphenols, including the food matrix, fat content and the presence of other nutrients, but, unlike for carotenoids (see below), the overall effect on polyphenol absorption is usually quite modest.

The total bioavailability of polyphenols has been studied using radiolabelled compounds in animal models. When rats consumed either radiolabelled epicatechin or procyanidin B_2, more than 80% of the dose was absorbed. It is important to note that this value includes the parent compound, its metabolites and the gut microbiota catabolites as described above. The latter are often found in plasma at a higher concentration than the parent compounds and their conjugates, but with a later maximum time of peak concentration in the blood. The circulating metabolites and gut microbiota catabolites are in the form of sulfates or glucuronides, either methylated or unmethylated. The exact chemical nature of many of the conjugates is known. For example, the main constituents in the blood after epicatechin consumption are 3′-methyl-epicatechin-5-*O*-sulfate, epicatechin-3′-*O*-sulfate and epicatechin-3′-*O*-β-D-glucuronide.

Once absorbed into the blood, polyphenols and their catabolites are eventually excreted into the urine via the kidney. Urinary excretion can be used as a measure of bioavailability. For polyphenols, most of the intact parent compound is in urine as conjugates, together with microbial metabolites and conjugates. The percentage of parent compound in the urine depends on the polyphenol, and can be used to compare interindividual differences in bioavailability, but it is not a measure of absolute absorption as some polyphenols are excreted in the bile rather than urine. Specific urinary markers can be used to indicate habitual consumption in intervention and population studies. In this respect, apple consumption is correlated with urinary phloretin, grapefruit consumption with urinary naringenin and orange consumption with urinary hesperetin.

Carotenoids

Carotenoids are mostly insoluble in water, and so must be in the lipid phase for absorption in the small intestine. This implies that some fat is essential for efficient absorption amd so carotenoid absorption is highly dependent on the food matrix. In some plants, the carotenoids are containined within oil droplets but in others, the carotenoids are in a more crystalline form which hinders absorption. Xanthophylls are sometimes attached to fatty acids through hydroxyl groups and are readily incorporated into micelles. The fatty acid moieties are removed during digestion in the gastrointestinal tract. About a third of β-carotene from spinach and lutein is absorbed in the small intestine provided that some fat is present.

During absorption, carotenoids in micelles enter enterocytes by passive absorption, with some active transport. After packaging into chylomicrons, the carotenoids are released into the lymphatic system and plasma. After processing by the liver, carotenoids enter the blood as very low-density lipoprotein particles. Ultimately, carotenes are mostly associated with low-density lipoprotein, whereas xanthophylls are higher in high-density lipoprotein, but both classes are taken up into liver and adipose tissue. In general, the concentration of carotenoids in the plasma depends on the habitual diet, and the half-life in the body is generally much longer than for polyphenols. As for polyphenols, carotenoids are also catabolised by the gut microbiota; as an example, for lycopene, about one-fifth of the dose is found in urine as catabolites derived through β-oxidation and microbial metabolism.

14.4 Protection against cardiovascular disease

The first observations of the interactions of flavonoids with the human cardiovascular system were in the 1930s during Albert Szent-Györgyi's work on elucidating the mode of action of the micronutrient vitamin C. It was noted that vitamin C alone was not as effective at preventing scurvy as the crude yellow extract derived from citrus fruits. They attributed the increased activity of this extract to the other substances in this mixture, which they referred to as 'citrin' (referring to citrus) or 'vitamin P', a reference to its effect on reducing the permeability of capillaries. These components of the extract were defined as the flavanones, hesperidin, eriodictyol, hesperidin methyl chalcone and neohesperidin. Whilst these phytochemicals were found not to fulfil the criteria of a vitamin, that of being 'essential' nutrients, these observations were the first to highlight the physiological activity of phytochemicals, in this case in the human vascular system. Figure 14.4 illustrates some of the historical developments in phytochemical research.

Recently, the Cocoa Supplement and Multivitamin Outcomes Study (COSMOS), the largest nutritional study investigating the impact of flavanols on health and the risk of developing heart disease, stroke or cancer, provided evidence of a beneficial effect from daily flavanol supplementation on cardiovascular health. The five-year, randomized, double-blind, placebo-controlled trial, involving more than 21 000 healthy US participants aged 60 and older, investigated whether daily flavanol supplementation (500 mg of cocoa flavanols, including 80 mg of the flavanol (–)-epicatechin) indicated reductions in all cardiovascular events. Notably in the per-protocol analysis the results indicated: (i) a statistically significant 15% percent reduction in the number of all cardiovascular events (from heart attacks and strokes to vascular surgeries); (ii) a statistically significant 39% reduction in cardiovascular death; and (iii) a statistically significant 24% reduction in the total number of major cardiovascular events. This is the first dietary intervention trial to study the effects of a bioactive or non-essential nutrient on cardiovascular disease and death at a such a large scale that the findings are relevant for public health.

In support of these large-scale observations, there have been many short-term, small-scale, human intervention studies designed to test the

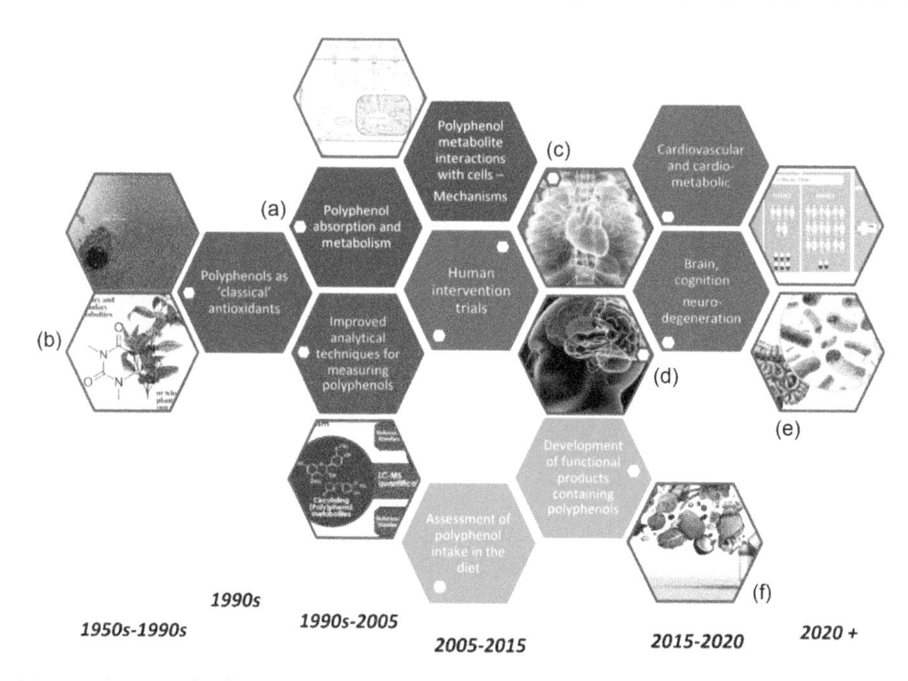

Figure 14.4 Approximate timeline for the evolution of research for phytochemicals. *Source:* (a) John Wiley & Sons, (b)Interfoto Scans/AGE Fotostock, (c) Yodiyim /Adobe Stock, (d) V. Yakobchuk / Adobe Stock, (e) and (f) Freshidea / Adobe Stock.

effect of flavonoid-rich foods on well-characterized, medically significant CVD risk factors, including hypertension, endothelial dysfunction, lipid metabolism and platelet activation. There is good evidence to suggest that many phytochemical-rich foods (cocoa, tea, coffee, berries, orange, red wine, etc.), extracts (green tea, cocoa, berries) and individual compounds (epicatechin, anthocyanins) have blood pressure-lowering effects in hypertensive individuals. As hypertension is one of the primary prognostic risk factors for the development of CVD (and indeed for dementia), a capacity to reduce elevated blood pressure can be regarded as cardioprotective. The COSMOS trial was the first truly large-scale study to examine the effect of flavonoids on cardiovascular risk factors (Figure 14.5).

With regard to endothelial function, a well-known surrogate marker for vascular health, several flavonoids have been shown to acutely improve endothelium-dependent vasodilation assessed by flow-mediated dilation (FMD; a gold-standard measure of human vascular health) in healthy individuals and in patients with coronary artery disease, hypertension or diabetes. These vascular improvements, and indeed blood pressure-lowering effects, correlate temporally with levels of flavonoid metabolites in circulation, suggesting a 'cause-and-effect' relationship between such compounds and vascular improvements. Such effects on blood flow assessed by FMD are often rapid (within hours) and return to fasting levels after around 4–6h: however, sustained increases in baseline FMD levels and additional acute-on-chronic increases in FMD response have also been reported after chronic consumption (weeks/months).

With respect to other risk factors of CVD, flavonoid interventions have been suggested to improve insulin resistance, and may also affect platelet function, inhibiting platelet activation. The latter is supported by chronic intervention studies in which flavonoids have been shown to improve platelet function after 28 days of supplementation, and although less effective than aspirin in this regard, seem capable of inhibiting platelet aggregation, a detrimental event (thrombosis/haematoma) which occurs towards the latter stages of atherosclerosis, the event which leads to heart attack and stroke.

14.5 Effects on metabolic syndrome, insulin resistance and risk of type 2 diabetes

Metabolic health is an important determinant of quality of life. Obesity, insulin resistance, metabolic syndrome and type 2 diabetes are mainly driven by habitual diet and sedentary lifestyles, as described in more detail in Chapters 16 and 17.

COSMOS study on cocoa flavanols and cardiovascular risk

	Reduction in events (all participants) over 5 years	**Reduction in events (compliant participants) over 5 years**
Total count of all cardiovascular events (primary endpoint)	10	15*
Deaths from cardiovascular disease (secondary endpoint)	27*	39*
Total number of all cardiovascular events	16*	24*

>21,000 healthy male and female volunteers (>60 years old) were given cocoa flavanols for 5 years in a randomised, placebo-controlled, double-blinded study. Significant change indicated by *.

Sesso et al, 2022. Effect of cocoa flavanol supplementation for the prevention of cardiovascular disease events: the COcoa Supplement and Multivitamin Outcomes Study (COSMOS) randomized clinical trial, AJCN 115, 1490–1500.

Figure 14.5 Main outcomes of the COSMOS trial on cocoa flavanols and cardiovascular health.

In healthy individuals, after a meal containing sugars and carbohydrate, the glucose released during digestion is absorbed in the gut and triggers multiple processes, including insulin release from the pancreas (Figure 14.6). Insulin stimulates the uptake of glucose into tissues such as muscle, by translocating sugar transporters to the surface of the muscle cells. This results in uptake of glucose into the tissue and the blood level of glucose returns to normal within a couple of hours. Insulin resistance occurs when the glucose transporter translocation process in tissues is impaired. Oxidative stress and impaired cell signalling prevent the full translocation of the glucose transporters, leading to a much slower rate of glucose uptake. Since the glucose remains in the blood for longer, the pancreatic β-cells continue to produce insulin. However, this released insulin has a weaker effect, and so the elevated blood glucose continues to signal to the β-cells to produce more insulin. Eventually, the β-cells become burnt out and less efficient at producing insulin. The entire process is accompanied by increased low levels of inflammation through reactive oxygen species and NF-κB activation.

One of the factors leading to insulin resistance is the repeated high consumption of sugars. Polyphenols can affect this process in several ways; some can slow down the digestion of sugars and carbohydrates by inhibiting the digestive enzymes and the transport of glucose into the blood, which can blunt the postprandial glucose spikes in blood. Certain polyphenols can also attenuate chronic inflammation through interaction with targets in cells, affecting cell signalling and so improving the response to insulin (see Figure 14.6).

One of the parameters used to define insulin resistance is HOMA-IR (Homeostatic Model Assessment for Insulin Resistance, and variations) which takes into account both fasting blood glucose and insulin. Intervention studies have shown that long-term intake of some polyphenol-rich foods can improve HOMA-IR. For

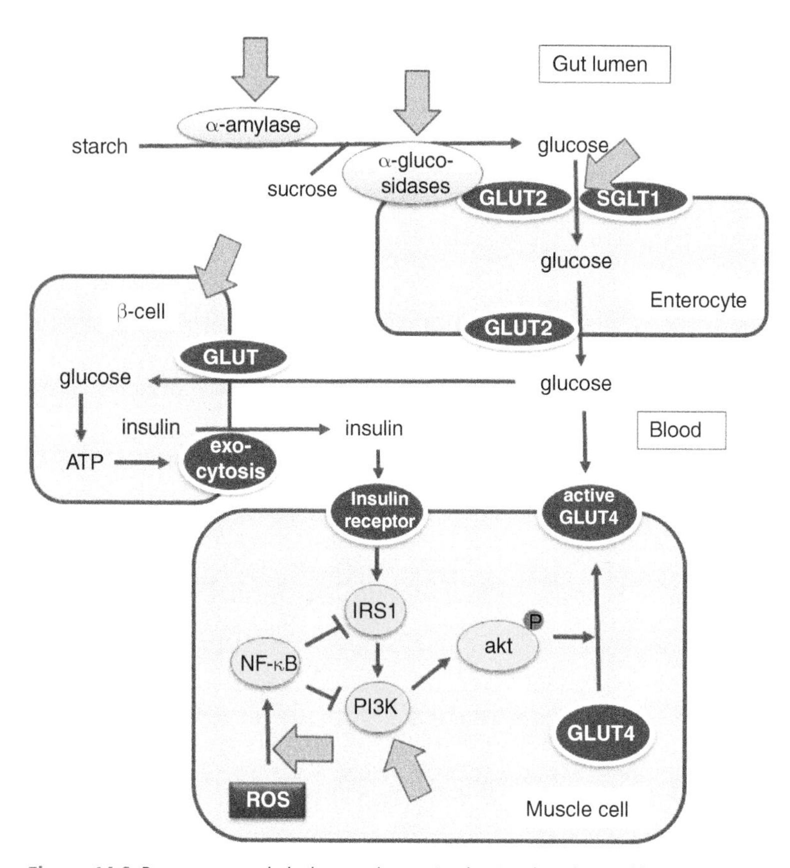

Figure 14.6 Response to carbohydrate and sugar intake. Starch is digested by α -amylase and α -glucosidases including sucrase/isomaltase and maltase/glucoamylase on the outside surface of the enterocyte. The product, glucose, is absorbed into the enterocyte and the blood through transporters. Glucose stimulates pancreatic β-cells to secrete insulin. Insulin acts on muscle and other cells to increase phosphorylation of Akt leading to translocation of the glucose transporter GLUT4 to the cell surface, and clearance of glucose from the blood. Polyphenols can interact at various points shown by yellow arrows. ROS, reactive oxygen species.

example, anthocyanins and anthocyanin-rich foods improve HOMA-IR, lower blood pressure, decrease inflammatory markers and lower the risk of type 2 diabetes. Quercetin and quercetin-containing foods do not directly affect HOMA-IR but lower systolic blood pressure, and only reduce fasting plasma glucose at higher intakes.

There is more evidence in relation to the effect of polyphenols, rather than carotenoids, on metabolic conditions. Nevertheless, some intervention studies involving consumption tomatoes showed reduction in the inflammatory marker interleukin-6 (IL-6), while lycopene supplementation reduced systolic blood pressure. Although higher intake and blood concentrations of the xanthophyll lutein, present in spinach and kale, is associated with better cardiometabolic health, there appears to be no effect on risk of diabetes.

Epidemiological studies on phytochemical-rich foods such as fruits and vegetables often support protective effects against metabolic conditions. However, determining the role of individual phytochemicals is a much more difficult undertaking, and so most data are derived from intervention and mechanistic studies.

14.6 Impact on brain function and cognitive performance

The projected increase in incidence and prevalence of neurodegenerative disorders, such as Alzheimer disease (AD), vascular dementia and Parkinson disease (PD), highlights a need for a more comprehensive understanding of how different aspects of lifestyle, particularly exercise and diet, may affect neural function and consequently cognitive performance as people age.

The precise mechanisms by which the brain degrades during typical or atypical ageing remain unclear, although several likely modes of action have been postulated (Figure 14.7). Notably, the degeneration of select domains of neurons within the brain (e.g. nigral neurons in PD and hippocampal neurons in AD) is probably impacted by many physiological disruptions, including a reduction in cerebrovascular blood flow, an increase in oxidative stress within select brain areas and the accumulation of toxic protein adducts. Furthermore, these detrimental processes can manifest globally or within specific brain regions. However, as well as neurodegeneration of the brain, such as that which leads to the classic neurodegenerative disorders, acute modulation of processes such as cerebrovascular blood flow may also provide immediate cognitive enhancements, such as that seen with methylxanthines, such as caffeine, which are known to enhance cognition without meaningful change to brain architecture or physiology. Thus, it is important to consider acute enhancement of cognitive processes in addition to an ability to attenuate neurodegeneration.

Flavonoids and other phytochemicals have been identified as promising bioactive compounds capable of influencing different aspects of brain function, including cerebrovascular blood flow and synaptic plasticity, resulting in improvements in memory and neurocognitive performance in humans and other mammalian species. For example, data from human intervention trials have demonstrated that the consumption of flavonoid-rich foods is associated with cognitive benefits, particularly memory

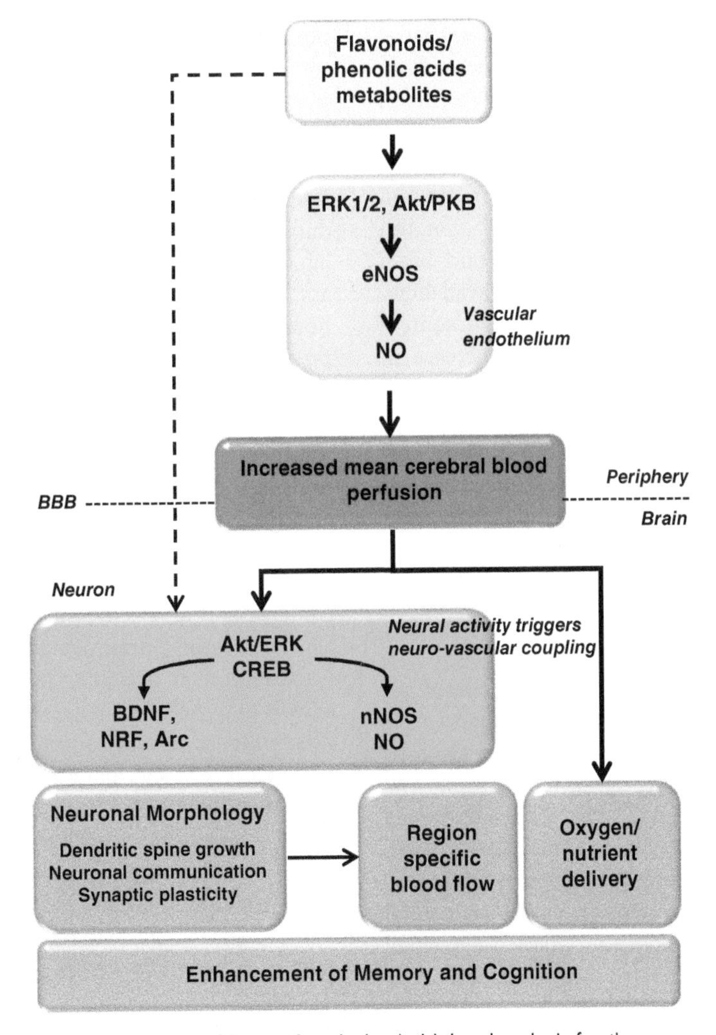

Figure 14.7 Proposed mechanisms of action of flavonoids and other (poly)phenols on brain function.

and learning, executive function and complex and sustained attention, while it has little impact on domains such as motor function.

The mechanisms by which flavonoids exert these actions on cognitive performance are being elaborated, with evidence suggesting that they may induce increased cerebrovascular blood flow, both generally and specifically, within the brain, and may interact with neuronal signalling pathways crucial in modulating neuronal function and survival through the enhanced expression of genes and proteins involved in synaptic plasticity, neuronal repair, long-term potentiation (LTP) and neuronal spine density. Whether these effects are mediated directly (i.e. within the brain) or from the periphery is currently unknown. However, there is considerable evidence suggesting that neuronal activity during cognitive performance is tightly coupled to increases in regional blood flow, a process known as cerebrovascular coupling, which is primarily mediated by nitric oxide (NO) generated by the activation of both endothelial nitric oxide synthase (eNOS in endothelium) and neuronal nitric oxide synthase (nNOS in neurons). In particular, NO derived from eNOS activation contributes to cerebral arterial dilation, by migrating to vascular smooth cells and increasing blood flow at the blood–brain interface. This overall increase in bioavailability of NO from neuronal and endothelial sources may result in vasodilation and an increase in functional, regional vascularisation and blood delivery during neuronal activity. As discussed above, flavonoids are known to affect blood flow through the modulation of nitric oxide levels and thus this may represent a likely candidate for their biological function in the brain.

14.7 Gut immune system

The microbiota can be categorised as being either beneficial or potentially pathogenic due to their metabolic activities and fermentation end-products. Health-promoting effects of the microbiota may include immunostimulation, improved digestion and absorption, vitamin synthesis, inhibition of the growth of potential pathogens and lowering of gas distension, while detrimental effects include carcinogen production, intestinal putrefaction, toxin production, diarrhoea/constipation and intestinal infections.

There has been much interest in the dietary modulation of the microbiota, particularly via the use of probiotics and prebiotics. To be effective, prebiotics must have low digestibility and/or bioavailability, must be selective for the growth and metabolism of commensal bacteria and must alter the microbiota to a healthy composition. Presently, the most widely used prebiotics are fructo-oligosaccharides and galacto-oligosaccharides, which enhance the growth of bifidobacteria, positively influence mineral absorption and lipid metabolism, stimulate innate immunity and demonstrate anticancer effects. Recently, it has been proposed that an imbalance between the two dominant phyla of bacteria, the Bacteroidetes and Firmicutes, plays a role in the progression of metabolic disorders such as obesity and diabetes, whilst bacterial lipopolysaccharides may control the tone of the innate immune system, thus regulating the general inflammatory status, insulin resistance and adipose tissue plasticity.

As outlined in the absorption and metabolism section, most ingested flavonoids escape absorption in the upper gastrointestinal tract (stomach, duodenum, jejunum and ileum), passing to the large intestine where interaction between them and the microbiota leads to their further degradation/metabolism. These interactions, as well as producing an array of smaller phenolic acids (akin to short-chain fatty acids for fibres), lead to changes in the profile and levels of bacteria residing there. The full significance of these interactions between flavonoids and other polyphenols and the large gut microbiota is yet to be fully determined, but it is likely that they have a variety of effects, including increasing the growth of some probiotic species, such as Bifidobacteria, and decreasing detrimental groups, such as Clostridia. For example, flavanols have been shown to significantly increase the growth of the *Clostridium coccoides-Eubacterium rectale* group, *Bifidobacterium* spp. and *Escherichia coli*, and significantly decrease the growth of the *Clostridium histolyticum* group. Furthermore, tea phenolics have been shown to affect the growth of pathogenic bacteria such as *Clostridium perfringens*, *Clostridium difficile* and *Bacteroides* spp. more than commensal anaerobes like *Bifidobacterium* spp. and *Lactobacillus* spp.

Current data suggest that the consumption of flavanol-rich foods may have the potential to support gut health through their ability to exert prebiotic-like activity, which most likely manifests either by directly inducing the growth of beneficial groups of bacteria (that are also involved in their degradation/metabolism) or by inhibiting the growth of deleterious groups, thus allowing probiotic bacteria to exploit the niche left behind.

Recent interest has centred on the gut–brain axis and how modulation of the gut microbiota may affect the brain. There are various ways in which the large gut and the microbiota that reside there might influence the brain, despite their physiological isolation from each other. First, the brain may influence the function of the gut directly through the hypothalamic–pituitary–adrenal axis and the autonomic nervous system, and the gut may modulate brain physiology via release of a variety of microbiota-derived metabolites and products, including neurotransmitters, such as GABA, which reach the brain through the circulation. Modulation of the gut microbiota is also likely to affect gut hormone release and immune modulation, ultimately influencing brain function via the circulatory and/or immune systems.

Currently, there are observations that flavonoid intake can affect cognitive processes in the brain and change gut microbiota profiles, but evidence is yet to emerge which links these two directly.

14.8 Future perspectives

Flavonoids are classed as non-nutrients, and thus not essential for life; however, like fibre, their regular intake can impart significant effects on health. Indeed, flavonoids, in a similar way to fibre, have been shown to impart significant protection with respect to the development of chronic diseases, in particular cardiovascular disease and cognitive function. Notably, the effects of regular consumption have been shown to reduce cardiovascular deaths (recent COSMOS study), lower blood pressure in hypertensive individuals and improve blood flow to the brain, thereby affecting cognitive performance. While the precise mechanisms for this activity are yet to be fully established, it is likely that once absorbed, their metabolite forms act to optimise the physiological function of several organ systems through interactions with intracellular signalling in various cell types. As such, lacking these components in the diet is likely to result in more rapid ageing of physiological systems and more rapid development of chronic diseases.

It is important to provide evidence-based information to individuals regarding intake amounts so that these benefits can be realised on a population level. Presently this is difficult, although data have emerged to suggest that intakes of specific flavonoids (flavanols) between 0.5 and 1.0 mg/kg body weight appear to be effective at inducing beneficial vascular effects. Furthermore, if individuals adhere to current UK nutritional intake guidelines, then overall intake of these bioactives in the diet could be as high as 0.5–1.0 g per day. However, it is too early to define a precise value for the Reference Nutrient Intake (RNI) of these non-nutrients, and whether such values would be different for individual phytochemical components. One may question whether such a RNI is appropriate for non-nutrients, but we should remember that much effort was expended in defining such a value for the non-nutrient fibre, which has been equally linked to beneficial effects on human physiological health, without a precise mode of action being established. Further large-scale clinical trials, conducted over longer durations, will be required to fully establish realistic RNIs for either groups of phytochemicals or individual compounds.

When such intake recommendations are established, it will be important to provide better information as to where such plant components can be found within the diet and whether there is a need to develop supplements. Currently, there are a huge array of both micronutrient and non-nutrient supplements available on the market, although whether one needs such supplements is debatable if a healthy diet according to nutritional intake guidelines is followed. With respect to all dietary components, but particularly non-nutrient phytochemicals, there is likely to be an optimum intake, which if exceeded may have no additional benefit, or in the worst-case scenario may be detrimental to health (e.g. vitamin A is essential but at high doses is highly detrimental to health). Thus, the best way to achieve reasonable, physiologically relevant amounts of phytochemicals is via the diet, specifically by increasing the intake of a variety of fruits, vegetables and foods/beverages derived from these.

Further reading

Polyphenols

Del Rio, D., Rodriguez-Mateos, A., Spencer, J.P. et al. (2013). Dietary (poly)phenolics in human health: structures, bioavailability, and evidence of protective effects against chronic diseases. Antioxidants and Redox Signaling 18: 1818–1892.

Kay, C.D., Clifford, M.N., Mena, P. et al. (2020). Recommendations for standardizing nomenclature for dietary (poly)phenol catabolites. American Journal of Clinical Nutrition 112: 1051–1068.

Frank, J., Fukagawa, N.K., Bilia, A.R. et al. (2020). Terms and nomenclature used for plant-derived components in nutrition and related research: efforts toward harmonization. Nutrition Reviews 78: 451–458.

Williamson, G., Kay, C.D., and Crozier, A. (2018). The bioavailability, transport, and bioactivity of dietary flavonoids: a review from a historical perspective. Comprehensive Reviews in Food Science and Food Safety 17: 1054–1112.

Carotenoids

Cheng, H.M., Koutsidis, G., Lodge, J.K., Ashor, A.W., Siervo, M., and Lara, J. (2019). Lycopene and tomato and risk of cardiovascular diseases: a systematic review and meta-analysis of epidemiological evidence. Critical Reviews in Food Science and Nutrition 59: 141–158.

Leermakers, E.T., Darweesh, S.K., Baena, C.P. et al.(2016). The effects of lutein on cardiometabolic health across the life course: a systematic review and meta-analysis. American Journal of Clinical Nutrition 103: 481–494.

Muller, L., Caris-Veyrat, C., Lowe, G., and Bohm, V. (2016). Lycopene and its antioxidant role in the prevention of cardiovascular diseases – a critical review. Critical Reviews of Food Science and Nutrition 56: 1868–1879.

Other phytochemicals

Barba, F.J., Nikmaram, N., Roohinejad, S.et al. (2016). Bioavailability of glucosinolates and their breakdown products: impact of processing. Frontiers in Nutrition 3: 24.

Ludwiczuk, A., Skalicka-Wozniak, K., and Georgiev, M.I. (2017). Terpenoids. In: Pharmacognosy, 233. St Louis: Elsevier Inc.

Salehi, B., Quispe, C., Sharifi-Rad, J., et al. (2020). Phytosterols: from preclinical evidence to potential clinical applications. Frontiers in Pharmacology 11: 599959.

15
The Control of Food Intake

Graham Finlayson and Mark Hopkins

Key messages

- Food intake is a behaviour determined by internal and external factors. The psychology of eating therefore involves the interaction of our physiology and environment.
- Important descriptors of eating behaviour include meal size, meal frequency, meal timing and food choice. These patterns of behaviour are episodic and cyclical.
- Important motivational constructs include satiation and satiety (inhibitory) and hunger, liking and wanting (stimulatory).
- Physiological signals affecting motivation to eat reflect nutrient and energy composition of ingested food and the requirements of body tissues and energy stores.
- In 'obesogenic' environments where palatable, energy-dense food is ubiquitous, intake can easily exceed energy requirements partly due to motivation underpinned by food reward.
- Food intake is subject to conscious control/loss of control and moderated by individual factors such as cognitive restraint and disinhibition (reflective and reactive traits).

15.1 Introduction

Food intake is a complex human behaviour, subject to many influences internal and external to the body. Internal physiological mechanisms, which influence when and how much we eat, have evolved to help regulate energy balance and ensure our survival. External environmental factors guide food selection, shape the development of eating habits and affect the types and amounts of foods consumed. Studying the interactions among the internal and external controls of food intake is needed to understand malnutrition and the management of body weight. Such studies can help us deal with diet-related diseases and the global obesity epidemic. However, deciding whether the motivation to eat owes more to physiology, psychology or to the food environment is no simple task.

Until recently, the prevailing view of food intake was one of stimulation and inhibition of motivation to eat by physiological signals acting on hypothalamic neurons in the brain to promote stability of the internal environment (energy stores, tissue needs). However, it has become clear that eating behaviour does not simply support regulation of internal energy requirements, but also reflects psychological and environmental factors that influence the amount and timing of foods eaten and the type of foods selected. Placing these complex interactions in a proper context requires a theoretical framework. Looking at how human food intake is organised according to cycles is one place to start. Food intake in humans and in most animals is intermittent: people eat in episodes of varying duration, and then stop eating for varying periods of time. How individual eating episodes are distributed over time is an important emerging area of study.

15.2 The cycles of eating behaviour

The nineteenth-century French physiologist Claude Bernard noted that any living organism must continuously be able to satisfy a number of needs

Nutrition and Metabolism, Third Edition. Edited on behalf of The Nutrition Society by Helen M. Roche, Ian A. Macdonald, Annemie M.W.J. Schols and Susan A. Lanham-New.

Companion Website: www.wiley.com/go/nutrition/metabolism3e

in order to maintain an independent life and survive. While the needs for energy, nutrients and water are continuous, the behaviours required to satisfy them are episodic and cyclical.

The circadian cycle

The alternation between day and night defines the eating behaviour of humans and many animals. The circadian cycle is species specific: although humans mostly eat during the day, other species eat mostly or exclusively at night. Little or no food intake occurs during the inactive resting phase. Disturbances in the circadian cycle can signal an imbalance between energy consumption and energy needs.

The modern food environment encourages intake of food at any time of day. A seminal study conducted by Gill and Panda (2015) found that onset, termination and frequency of human eating episodes within a 24-hour cycle range enormously between individuals and rarely conform to a 'three meals per day' structure (Figure 15.1). As food intake acts as a synchroniser of the body's biological clock, this may promote food consumption in excess of energy requirements and disrupt metabolic processes reliant on endogenous circadian rhythms.

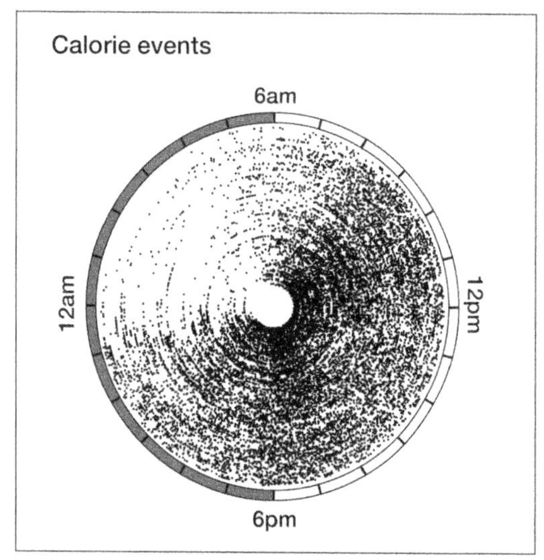

Figure 15.1 Polar plot of all energy-containing (≥5 kcal) eating episodes of 156 individuals plotted against the time of day (radial axis) in each concentric circle. The onset, termination and frequency of eating events range enormously between individuals. Median number of eating occasions in a day was 6 (decile range 3–10). *Source:* Gill & Panda (2015)/ With permission of Elsevier.

Most interventions for the prevention and treatment of obesity focus on 'what' and 'how much' to eat, but not 'when'. It is now recognised that patterns of daily energy intake, including meal timing and frequency, influence the management of body weight and cardiometabolic health. Evening food intake is associated with lower satiety, greater energy density of chosen foods and increased energy intake. Certain night-eating symptoms can predict weight gain in adults. In contrast, an earlier window of food intake during the day is associated with better body weight outcomes. The factors responsible for altering the circadian patterns of food consumption in humans are not clear. Neuroendocrine disturbances brought about by depression and stress as well as irregular work patterns and sleep habits may threaten the circadian cycle of food consumption (James et al. 2017).

There are many other ways in which underlying biology can be in conflict with the demands of contemporary society. Humans are social animals that live in complex cultures. The circadian distribution of meals is a behavioural adaptation to energy needs, but one that is shaped by culture and the food environment. In response to cultural demands, humans learn to be hungry at specific times of day, corresponding to mealtime social norms. Laboratory animals can also learn to adapt to a fixed meal pattern and eat adequate amounts at each meal in order to minimise hunger between meals and meet their energy needs. Even physiological signals such as glucose and insulin levels adapt to fixed meal patterns. However, such learning does not take place when food is constantly available. By providing easy access to palatable food at all times of day, the obesogenic environment is lacking the fixed pattern of three meals per day within an 8–10-hour time-window in favour of constant grazing and snacking. If the control of food intake is partly learned, then a weakening of the regulatory mechanisms is likely to follow.

The meal cycle

During waking hours, food is consumed in discrete episodes. The total time spent eating over any 24h period can be highly variable for humans, depending on age, custom and culture. Early studies in physiology suggested that energy deficits were a direct signal to start a meal.

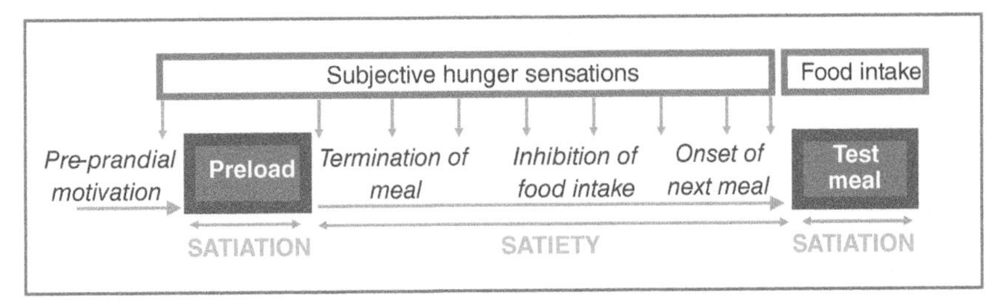

Figure 15.2 The typical preload-test meal study design used to test the impact of different foods on subsequent motivational ratings (subjective hunger) and food intakes. Satiation is measured from the within-meal inhibition of hunger and cessation of food intake. Satiety is measured from the between-meal inhibition of hunger until the onset of the next meal.

However, a look at human eating habits suggests that such a notion is too simple to account for the many complexities of the timing and duration of meals.

First, hunger is not the only trigger for eating. People (and animals) clearly eat even though they are not hungry and may not stop eating when full. Second, meals are generally finished well before the energy contained in the ingested food has been absorbed or metabolised to cover energy needs. Third, eating episodes can be timed to satisfy not only current but also future energy needs. At culturally determined times of day, people eat meals and snacks in anticipation of future energy expenditures. Fourth, the perceived deprivation of pleasure/satisfaction from eating can be just as potent a stimulus for further intake or food seeking as an actual deprivation of energy.

Immediately following the meal, a complex succession of events, described as a 'satiety cascade', inhibits further eating for a certain time, until the onset of the next meal. How long the period of satiety will last depends on the amount of energy consumed and on the volume and nutrient content of the foods just eaten. The state of satiety is influenced by physiological processes such as stomach distension and emptying, but can also be affected by the sensory characteristics of the food itself and by the attitudes and beliefs of the consumer.

Studying how eating behaviours adapt to the provision of too much or too little energy is one way to understand the control of food intake, although it is important to note that studies looking at the responses to short-term food deprivation or overconsumption tell us little about the long-term regulation of body weight. The usual expectation is that people will eat more in response to a reduction in energy but will adjust their food consumption downwards if energy is increased. Studies based on the 'preload-test meal paradigm' have examined the impact of different foods on subjective sensations of hunger and satiety inferred by the amount of food eaten during the next meal (Figure 15.2).

Such studies reveal that humans are much better at upward adjustment of energy intakes when faced with acute reductions in energy intake. Reducing preload energy was often followed by more food being consumed at the next eating occasion. Although results varied, growing children appeared to compensate for the missing energy better than adults and younger people better than older people. Arguably, responding to inadequate intakes by eating more is an important survival mechanism. By contrast, there are fewer control mechanisms to deal with episodes of overeating. Indeed, the control of food intake appears to be asymmetrical, with prolonged periods of food restriction and weight loss met with stronger compensatory responses than periods of excess intake and weight gain. This asymmetry may help account for the ease with which people gain weight but are typically unable to sustain reductions in body weight over the long term following intentional weight loss. While energy compensation after weight loss may occur partly through mechanisms of energy expenditure, changes in motivation to eat and control of food intake following prolonged energy deficit appear more important (Casanova et al. 2019).

The pleasure cycle

In societies with easy access to palatable energy-dense foods, the pleasure response may be responsible for most episodes of overeating. For the majority of people, food is a reliable source

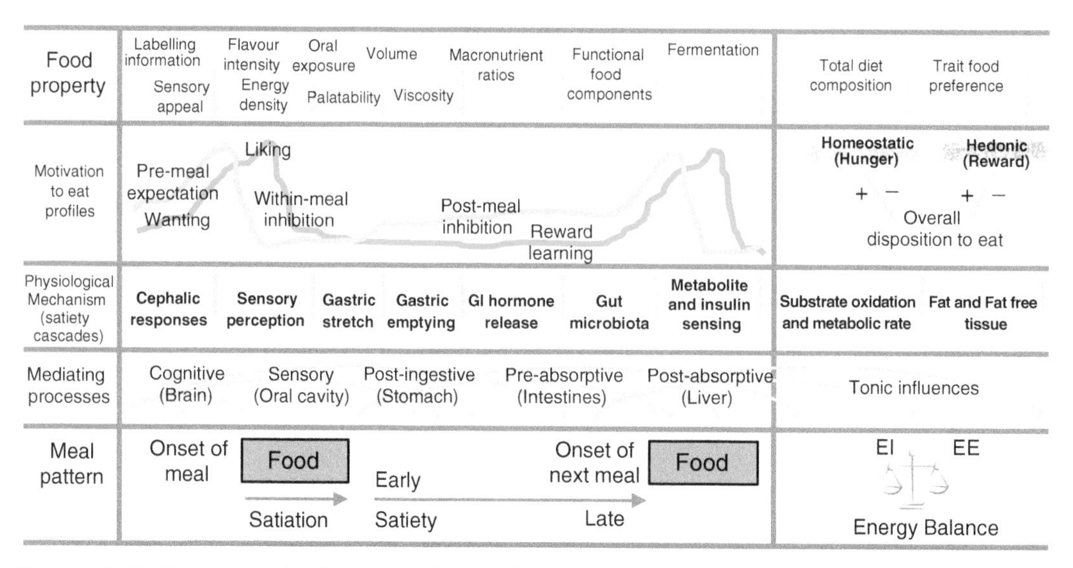

Figure 15.3 Multilevel model of food intake control based on the satiety cascade. The model incorporates tonic and episodic signalling and the influence of specific food-based properties and physiological mechanisms on the temporal overlap of homeostatic and hedonic drives. *Source:* From Dalton & Finlayson (2013)/With permission of Elsevier.

of pleasure, and food reward plays an important part in controlling food intake with a role in the initiation, maintenance and cessation of an eating episode. While the present theoretical framework talks about food reward, other researchers have talked about desire to eat, cravings or even food addictions. In scientific terms, the reward response to modern foods can interact with homeostatic mechanisms to shape eating patterns in a way that no longer balances food consumption with energy expenditure.

Recent research has distinguished between 'liking' and 'wanting' responses to food as subcomponents of food reward. Liking and wanting are now considered two important but distinct aspects of the motivation to eat. Both can depend on homeostatic or hedonic mechanisms. The constructs of food liking and wanting are separable in their everyday meaning for consumers and have separable neural correlates. Their relative functions in food intake control can be distinguished temporally, according to the pattern of eating episodes over the course of a day.

In the typical consumer, the overall hedonic experience of food begins with a thought or cue relating to a palatable food that guides attention towards obtaining that food and prepares for its consumption. Once obtained, the pleasure elicited by the oro-sensory properties of the food as it is consumed becomes primary. Next, if the oro-sensory stimulation is sustained, the pleasure will eventually subside and the motivation to consume the food will diminish until a new thought or sensation starts the process again. These distinct anticipatory, consummatory and postconsummatory phases have been termed the 'food pleasure cycle' (Kringelbach et al. 2012) and help to illustrate how the operation of wanting, liking and subsequent reward learning might oscillate in an overlapping but time-dependent manner in relation to each discrete eating episode. As mentioned above, the cyclical nature of food intake is also conceptualised by the satiety cascade which depicts satiety as a time-dependent process. It therefore becomes useful to map the temporal action of liking and wanting with those of hunger (Figure 15.3). In this way, it is possible to integrate emerging knowledge on food liking and wanting with the established theoretical principles and mechanisms identified in the satiety cascade (Dalton and Finlayson 2013).

15.3 Internal controls of food intake

The observation that adult body weights are relatively stable over long periods of time suggests that some critical signalling mechanism informs the brain of the current energy status of the body and accordingly sets the basal motivation to eat.

Early theories proposed that a deviation in body weight from a physiological 'set-point' would be met with changes in food intake to restore weight to its optimal level. Such theories depended on some regulated physiological parameter and the existence of a feedback loop sending information about the nutritional status of the body from the periphery to the brain.

A number of such physiological systems have been identified over the years, based around the oxidation or storage of specific macronutrients in the body (Watts et al. 2022). The glucostatic theory posited that eating was triggered by a decrease in the availability of glucose to tissues, with 'metabolic hypoglycaemia' acting as a cue for meal initiation. The lipostatic theory suggested that body fat was the key substance that determined eating behaviour. Linking to body fat stores, the lipostatic theory also tried to account for the fact that the amount of body fat was regulated and weight loss attempts were most often followed by weight regain until the pre-diet body fat was restored. The aminostatic theory suggested that serum amino acids not directed towards protein synthesis were the crucial agents determining satiety. However, it has become clear that food intake does not simply reflect internal energy or nutrient requirements, with the oxidation or storage of individual macronutrients failing to exert strong feedback on motivation to eat in humans.

Long-term signals

The long-term 'tonic' system influencing control over food intake is a function of the size of the body's fat stores (energy storage) and the body's demand for energy (energy turnover). This feedback has traditionally centred on the action of leptin, but it is now increasingly recognised that the energy expenditure of metabolically active tissues may also influence energy intake. Leptin is a hormone secreted by adipose tissue and provides feedback to the brain concerning stored energy in adipose tissue. Changes in circulating leptin concentrations are thought to promote changes in both energy intake and energy expenditure (Andermann and Lowell 2017).

More recently, a conceptual model of food intake that incorporates the energetic demands of metabolically active tissues has been proposed, in which fat-free mass and resting metabolic rate form the basis of a tonic drive to eat that helps energy intake track the basal energy requirements of tissues and metabolic processes (Blundell et al. 2020). This is based on a series of studies reporting positive associations between fat-free mass, but not fat mass, and subjective hunger, food intake and total daily energy intake. It has also been shown that the effect of fat-free mass on energy intake could be mediated by resting metabolic rate (Hopkins et al. 2019), suggesting that energy expenditure per se may influence daily energy intake. When considered alongside the negative feedback models from energy and macronutrient storage, additional mechanisms linking the energetic demands of metabolically active tissues to energy intake may provide a stronger account of the long-term internal signals that influence food intake (Figure 15.4).

Short-term signals

Short-term internal determinants of the control over food intake come in the form of gastric and hormonal 'episodic' signals released in response to the passage of nutrients through the gastrointestinal tract. The concentration of ghrelin, an orexigenic hormone secreted mainly in the stomach when nutrients are absent from the gastrointestinal tract, is high before meals and falls after intake and is thought to promote meal initiation and food intake. Other hormones, such as cholecystokinin, glucagon-like peptide 1 and peptide YY, are generally viewed as satiety signals that reduce intake following a meal. The secretion of these hormones and the distension of the stomach provide preabsorptive feedback to the brain, allowing it to predict the volume, energy content and macronutrient composition of nutrients consumed (Morton et al. 2006).

It is worth noting that different foods may produce the same effect on motivation to eat but display quite distinctive profiles of postprandial hormones. This suggests there is no single hormone or unique pattern of hormones that defines satiety. Indeed, marked individual variability exists in the postprandial concentration of these hormones following meals, but whether such variability underlies individual differences in food intake control is unclear (Gibbons et al. 2016).

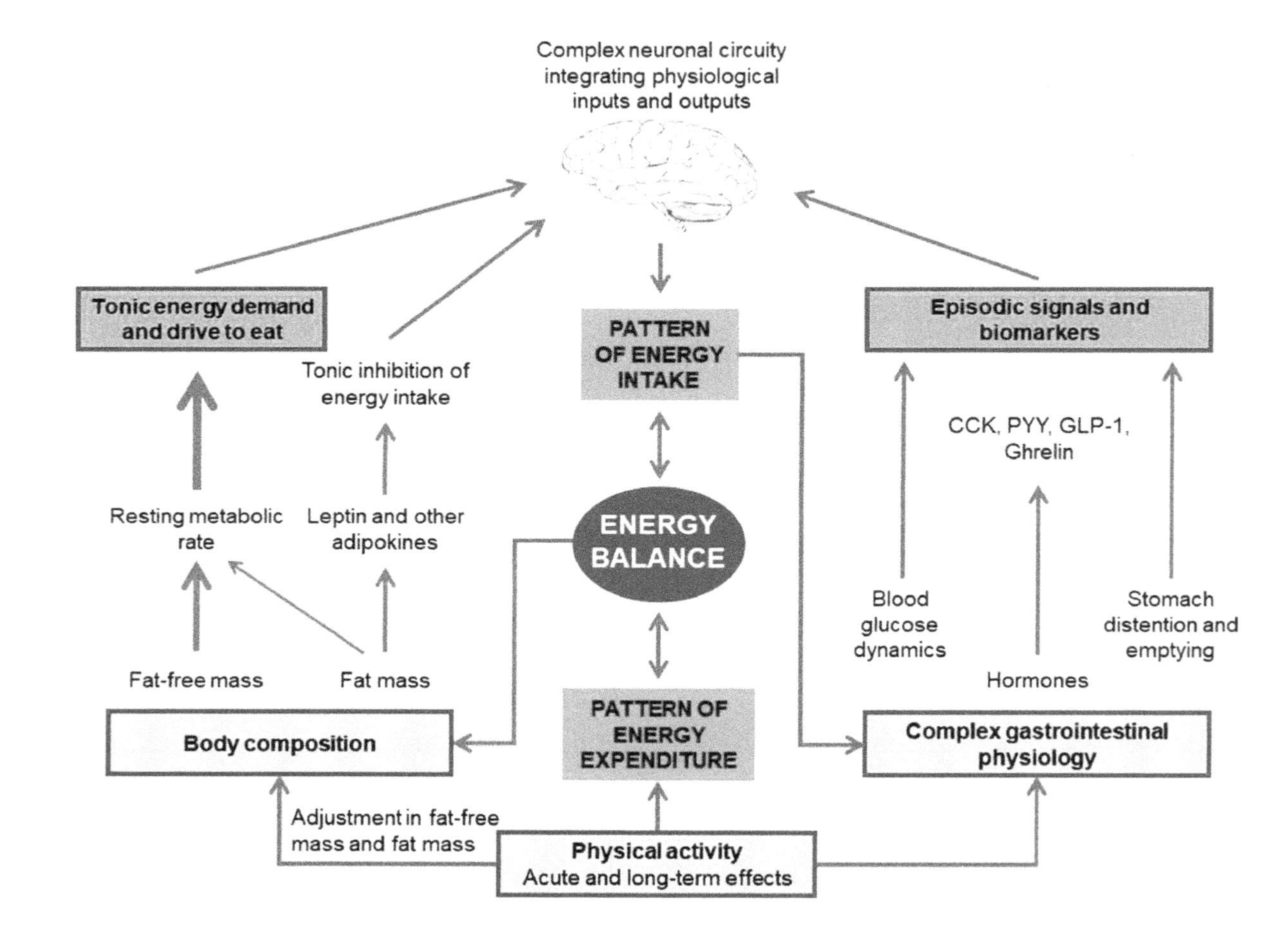

Figure 15.4 This model illustrates the interplay between tonic and episodic processes of appetite control, together with drive and inhibitory pathways. A major feature of the model is that body composition influences appetite control via both a drive and an inhibitory system. It is proposed that fat-free mass acts via resting metabolic rate reflecting energy needs of vital organs and constitutes a drive to eat. In contrast, fat mass acts as a tonic inhibitory influence on eating. Both of these processes achieve an expression in behaviour through transformation and integration in complex neuronal processes. In turn, these tonic influences on eating are periodically interrupted by episodic biological signals emanating from complex gastrointestinal physiology arising from food consumption. These signals are also represented in neuronal processes. The model illustrates how energy expenditure is related to energy intake. A further feature of this relationship is the contribution from physical activity energy expenditure, which influences both tonic and episodic processes. *Source:* From Hopkins & Blundell (2016)/With permission of Elsevier.

15.4 External controls of food intake

Availability

In modern societies, many food-related cues are present in the environment. The availability of foods at all times of day is an obvious stimulus of intake. The presence of late-night food outlets and vending machines in the environment may trigger spontaneous food intake at any time of day or night. Environmental cues that have been associated with past eating bouts can reinstate consumption even in satiated children and adults. The experience of restriction imposed by well-meaning parents on the intake of favourite foods seems to be one potent facilitator of eating in the absence of hunger observed in young girls when the forbidden food becomes available (Campbell et al. 2007).

Distraction

Not all environmental stimuli need to be food related. The presence of television or music in the environment can increase meal size or meal frequency. As a general rule, the presence in the environment of factors that distract the consumer from the act of ingestion are likely to increase intake.

One very potent factor in free-living conditions is the presence of others, especially friends and family. Social stimulation is a phenomenon found in free-living subjects and is a function of the number of persons present. Under laboratory or controlled environments, however, the stimulation effect is less clear. Eating with strangers can sometimes decrease intake. In free-living individuals who most often eat with familiar companions, a clear facilitation effect is observed, perhaps mediated by the duration of meals consumed in company (Robinson et al. 2013).

Portion size and energy density

The food environment can also determine portion size. Food consumption studies among children and adults show that intake is a direct function of the amount of food or drink that is served. Larger portions stimulate larger intake, regardless of the intensity of hunger, but do not lead to a corresponding increase in postingestive satiety. Large portion sizes can induce what has been called 'passive overconsumption', which exceeds bodily needs and potentially facilitates weight gain. Large portion sizes combined with energy-dense foods are especially problematic. Unaware of their high energy density, the consumer is likely to overeat, given the weight of the food is low. Portion size and energy density have additive effects, so that large portions of high energy density foods are said to have low satiating power (Ello-Martin et al. 2005).

Variety

The variety of foods presented within a meal can also have an impact on total consumption. Sensory variety, regardless of nutrient content, can stimulate food consumption. The motivation to eat is said to be sensory specific, such that the liking for the just-eaten food decreases relative to uneaten foods with differing sensory properties. The attractiveness of foods that share similar sensory characteristics (taste, aroma, colour, shape) also decreases. This is why selecting or presenting foods with varied sensory characteristics is associated with larger intake. Interestingly, the motivation for uneaten foods can depend on the sensory quality of the eaten food. For example, motivation for sweet food increases strongly after consuming a savoury meal (the so-called 'dessert stomach'), but motivation for savoury foods is relatively unchanged after a sweet meal (Griffioen-Roose et al. 2010).

Palatability

Pleasure is at the forefront of a normal eating pattern but palatability and processes of food reward can lead to loss of control over food intake. This raises the question of whether liking and wanting are viable targets for restoring control in people living with obesity. It is reasonable to suggest that in instances where processes of liking and wanting dissociate or become either enhanced or attenuated, the resulting impact on behaviour may resemble certain forms of disordered eating.

A particularly relevant behaviour marked by loss of control over food intake is binge eating. Binge eating behaviour is associated with increased cravings for sweet, energy-dense foods and excessive consumption of food accompanied by a perceived loss of control over eating, and guilt following a binge. Using a psychometric assessment of binge eating severity, studies have shown that high scorers had an enhanced reported liking for all types of food but an increased wanting specific for high-fat sweet foods only. This elevated wanting was measured by subjects responding faster to images of high-fat sweet foods, suggesting that these foods had an enhanced motivational value compared to other food categories. Furthermore, the elevated wanting for sweet foods in binge eaters coincided with them consuming 50% more high-fat sweet foods in an ad libitum test meal (Dalton and Finlayson 2014).

15.5 The psychology of eating

A vast literature has addressed the psychological determinants of food intake. However, early attempts to find out whether people with obesity shared a common obesogenic eating style were largely unsuccessful. Eating behaviour traits have been extensively studied in an attempt to identify potential markers that detect a tendency to overconsume. These eating behaviours influence energy intake through choices about what type of food to eat, when to start and stop eating and where to eat. However, there is wide individual variability in eating styles, which has led to the development of several measures and constructs which aim to capture individual differences in eating behaviour. These constructs are assessed using psychometric, self-report scales, for example the Three Factor Eating Questionnaire and its sub-scales of cognitive restraint, susceptibility to hunger and disinhibition (Stunkard and Messick 1985).

Restraint as a concept was developed to describe the state of chronic dietary concern, experienced by individuals who believe they need to regulate their weight. It therefore relates to a cognitive intention to restrict food intake, which is not necessarily reflected in actual adherence to a weight-reducing diet. Some experts now distinguish rigid restraint, which can indeed be associated with poor appetite control, from flexible restraint, which is a beneficial attitude facilitating dieting and weight loss.

Based on responses to multiple questionnaires, studies have categorised people by such supposedly permanent psychological traits as disinhibition. People likely to lose control over eating or eat 'opportunistically' in a variety of circumstances are thought to show high levels of disinhibition. High disinhibition is associated with overweight and obesity, as well as with metabolic disorders such as type 2 diabetes. A review of the literature found that disinhibition had the largest amount of evidence linking it to prospective weight gain, with 10 out of 11 studies demonstrating positive associations between disinhibition and BMI or weight gain (French et al. 2012). Moreover, disinhibition could influence weight regain, with studies showing that participants who were able to maintain clinical weight loss had lower disinhibition scores.

15.6 Conclusion

The control of food intake in humans involves physiology, the environment and psychology. Motivation to eat is an important factor influencing food intake if the environment does not constrain behaviour. Food intake is not necessarily solely related to situations of nutritional depletion and can be influenced by a number of physiological and non-physiological factors. Appetites are often learned and frequently sensory specific. The term 'control of food intake' is often used without specifying quite what is being controlled. Does eating behaviour change to maintain some constancy and motivation to eat? Does appetite change in order to control food intake around some central tendency? Does appetite change with the aim of changing food intake to regulate energy balance? Often, it is the latter view that is implied but not clearly articulated. The implication of asymmetrical energy balance regulation is that food intake is under stronger physiological control in relation to negative energy balances, whereas in a state of energy balance or positive energy balance, the linkages between physiological signalling and subjective motivation to eat are weaker. In obesogenic environments where palatable, energy-dense food is ubiquitous, intake can easily exceed energy requirements due to motivation underpinned by food reward.

References

Andermann, M.L. and Lowell, B.B. (2017). Toward a wiring diagram understanding of appetite control. Neuron 95: 757–778.

Blundell, J.E., Gibbons, C., Beaulieu, K. et al. (2020). The drive to eat in homo sapiens: energy expenditure drives energy intake. Physiol Behav 219: 112846.

Campbell, K.J., Crawford, D.A., Salmon, J., Carver, A., Garnett, S.P. and Baur, L.A. (2007). Associations between the home food environment and obesity-promoting eating behaviors in adolescence. Obesity 15: 719–730.

Casanova, N., Beaulieu, K., Finlayson, G. and Hopkins, M. (2019). Metabolic adaptations during negative energy balance and their potential impact on appetite and food intake. Proceedings of the Nutrition Society 78: 279–289.

Dalton, M. and Finlayson, G. (2013). Hedonics, Satiation and Satiety. Satiation, Satiety and the Control of Food Intake. Philadelphia: Elsevier.

Dalton, M. and Finlayson, G. (2014). Psychobiological examination of liking and wanting for fat and sweet taste in trait binge eating females. Physiology and Behavior 136: 128–134.

Ello-Martin, J.A., Ledikwe, J.H. and Rolls, B. (2005). The influence of food portion size and energy density on energy intake: implications for weight management. American Journal of Clinical Nutrition 82: 236S–241S.

French, S.A., Epstein, L.H., Jeffery, R.W., Blundell, J.E. and Wardle, J. (2012). Eating behavior dimensions. Associations with energy intake and body weight. A review. Appetite 59: 541–549.

Gibbons, C., Finlayson, G., Caudwell, P. et al. (2016). Postprandial profiles of CCK after high fat and high carbohydrate meals and the relationship to satiety in humans. Peptides 77: 3–8.

Gill, S. and Panda, S.J.C.M. (2015). A smartphone app reveals erratic diurnal eating patterns in humans that can be modulated for health benefits. Cell Metabolism 22: 789–798.

Griffioen-Roose, S., Finlayson, G., Mars, M., Blundell, J.E. and De Graaf, C. (2010). Measuring food reward and the transfer effect of sensory specific satiety. Appetite 55: 648–655.

Hopkins, M., Blundell, J. (2016) Energy balance, body composition, sedentariness and appetite regulation: pathways to obesity. Clinical Science 130(18):1615–1628.

Hopkins, M., Finlayson, G., Duarte, C. et al. (2019). Biological and psychological mediators of the relationships between fat mass, fat-free mass and energy intake. International Journal of Obesity 43: 233–242.

James, S.M., Honn, K.A., Gaddameedhi, S.and and Van Dongen, H. (2017). Shift work: disrupted circadian rhythms and sleep – implications for health and well-being. Current Sleep Medicine Reports 3: 104–112.

Kringelbach, M.L., Stein, A. and Van Hartevelt, T.J. (2012). The functional human neuroanatomy of food pleasure cycles. Journal of Physiology 106: 307–316.

Morton, G., Cummings, D., Baskin, D., Barsh, G. and Schwartz, M. (2006). Central nervous system control of food intake and body weight. Nature 443: 289–295.

Robinson, E., Aveyard, P., Daley, A. et al. (2013). Eating attentively: a systematic review and meta-analysis of the effect of food intake memory and awareness on eating. American Journal of Clinical Nutrition 97: 728–742.

Stunkard, A. and Messick, S. (1985). The three-factor eating questionnaire to measure dietary restraint, disinhibition and hunger. Journal of Psychosomatic Research 29: 71–83.

Watts, A.G., Kanoski, S.E., Sanchez-Watts, G. and Langhans, W. (2022). The physiological control of eating: signals, neurons, and networks. Physiology Reviews 102: 689–813.

16
Overnutrition

Albert Flynn

Key messages

- For a number of vitamins and minerals, adverse health effects can result when the capacity for homeostasis is exceeded by continuing high dietary intakes.
- The tolerable upper intake level (UL) is a dietary reference standard that may be used to evaluate the risk of excessive intakes of vitamins and minerals in groups and as a guide to individuals for the maximum level of the usual intake of micronutrients.
- The UL for nutrients may be derived based on the principles of risk assessment, and ULs have been established for several vitamins and minerals by authorities in a number of countries.

16.1 Introduction

The increased availability of fortified foods and the increased use of dietary supplements in many countries have led to increased interest in the adverse health effects that may arise from overconsumption of vitamins and minerals. It is well recognised that such adverse effects can occur with high intakes of some vitamins and minerals. While it is generally considered that the risk of such effects is low, the incidence of such occurrences in different populations is generally not known with any certainty. In some regions, for example in the European Union and North America, the need to protect consumer health through regulating the addition of vitamins and minerals to foods and nutritional supplements has led to the establishment of scientifically based upper limits of intake for these micronutrients.

16.2 Adverse effects of vitamins and minerals: concepts

Failure of homeostasis

Vitamins and essential minerals are subject to homeostatic control, through which body content is regulated. This reduces the risk of depletion of body pools, which could lead to deficiency when intakes are low. It also reduces the risk of excessive accumulation in tissues that could lead to adverse effects when intakes are high. A measure of protection against potential adverse effects of high intakes is provided by adaptation of homeostatic control mechanisms, for example limiting absorption efficiency, adaptation of metabolic processes or enhancing excretion in faeces, urine, skin or lungs (Box 16.1).

However, for a number of micronutrients, the capacity for homeostasis may be exceeded by

Nutrition and Metabolism, Third Edition. Edited on behalf of The Nutrition Society by Helen M. Roche, Ian A. Macdonald, Annemie M.W.J. Schols and Susan A. Lanham-New.

Companion Website: www.wiley.com/go/nutrition/metabolism3e

Box 16.1 Adaptive mechanisms that protect against adverse effects of continuing exposure to high dietary intakes of vitamins and minerals

- Reduced absorption efficiency with increasing dietary intake or body stores; for example, for iron, zinc and calcium, the mediated transport route of absorption in the small intestine mucosa is downregulated.
- Regulation of excretory pathways where excretion increases with increasing dietary intake or body stores; for example, for calcium, increasing calcium ion concentration in plasma depresses parathyroid hormone and reduces the efficiency of calcium ion reabsorption in the renal tubule, leading to increased urinary calcium excretion; for selenium, high intakes lead to increased excretion of selenium as the volatile dimethyl selenide through the lungs or as the water-soluble trimethylselenonium ion in urine.
- Increased endogenous secretion into the intestine occurs for some nutrients when intakes are high, for example increased secretion in pancreatic fluid (zinc) and bile (manganese), or increased losses by sloughing off of intestinal mucosal cells (iron, zinc).

continuing high dietary intakes. This can lead to abnormal accumulation in tissues or overloading of normal metabolic or transport pathways.

Threshold dose

For nutrients, no risk of adverse health effects is expected unless a threshold dose (or intake) is exceeded. Thresholds for any given adverse effect vary among members of the population. In general, for nutrients there are insufficient data to establish the distribution of thresholds in the population for individual adverse effects.

Variation in the sensitivity of individuals

Sensitivity to the adverse effects of micronutrients is influenced by physiological changes and common conditions associated with growth and maturation that occur during an individual's lifespan. Even within relatively homogenous lifestage groups, there is a range of sensitivities to adverse effects (e.g. sensitivity is influenced by body weight and lean body mass). Some subpopulations have extreme and distinct vulnerabilities owing to genetic predisposition or certain metabolic disorders or disease states (Box 16.2).

Box 16.2 Subpopulations with extreme sensitivities to the adverse health effects of vitamins and minerals

- Haemochromatosis: individuals who are homozygous for the high iron *Fe* (HFE) gene accumulate excessive levels of iron in body stores, leading to organ (e.g. liver, pancreas) damage.
- Wilson disease: an autosomal recessive disease of copper storage in which copper accumulates in the liver, brain and cornea of the eye.
- Renal disease: reduced renal function increases susceptibility to adverse effects of nutrients that are excreted by this route, for example phosphorus and calcium.
- Glucose-6-phosphate dehydrogenase deficiency: increased sensitivity to the adverse effects of vitamin C.

Effect of bioavailability

The bioavailability of a nutrient relates to its absorption and utilisation, and may be defined as its accessibility to normal metabolic and physiological processes. Bioavailability influences the usefulness of a nutrient for physiological functions at physiological levels of intake, and the nature and severity of adverse effects at excessive intakes. There is considerable variation in nutrient bioavailability in humans; for instance, the chemical form of a nutrient may have a large influence on bioavailability. Other modulating factors include the nutritional status of the individual, nutrient intake level and interaction with other dietary components and the food matrix (e.g. consumption with or without food). For example, high zinc intakes increase the synthesis of intestinal mucosal cells of metallothionein, a protein that avidly binds copper and reduces its absorption.

16.3 The tolerable upper intake level

While dietary reference standards have been established over many years for evaluation of the nutritional adequacy of dietary intakes, it is only in recent years that the need for dietary reference standards for evaluating and managing the risk of excessive intakes of vitamins and minerals has been recognised. Such reference standards have been established for a number of vitamins and minerals and are referred to as tolerable upper intake levels (sometimes also called upper safe levels).

The tolerable upper intake level (UL) is the maximum level of total chronic daily intake of a nutrient (from all sources, including foods, water, nutrient supplements and medicines) judged to be unlikely to pose a risk of adverse health effects to almost all individuals in the general population. 'Tolerable' means a level of intake that can be tolerated physiologically by humans. ULs may be derived for various lifestage groups in the population (e.g. adults, pregnant and lactating women, infants and children). The UL is not a recommended level of intake but is an estimate of the highest (usual) level of intake that carries no appreciable risk of adverse health effects.

The UL is meant to apply to all groups of the general population, including sensitive individuals, throughout the lifespan. However, it is not meant to apply to individuals receiving the nutrient under medical supervision or to those with predisposing conditions that render them especially sensitive to one or more adverse effects of the nutrient, such as those with genetic predisposition or certain disease states.

The term 'adverse health effect' may be defined as any significant alteration in structure or function or any impairment of a physiologically important function that could lead to an adverse health effect in humans.

16.4 Derivation of tolerable upper intake level

Risk assessment

The tolerable upper intake level can be derived for nutrients using the principles of risk assessment that have been developed for biological and chemical agents. Risk assessment is a systematic means of evaluating the probability of occurrence of adverse health effects in humans from an excess exposure to an environmental agent (in this case, nutrients in food and water, nutrient supplements and medicines). The hallmark of risk assessment is the requirement to be explicit in all the evaluations and judgements that must be made to document conclusions.

In general, the same principles of risk assessment apply to nutrients as to other food chemicals, but it is recognised that vitamins and minerals possess some characteristics that distinguish them from other food chemicals (Box 16.3).

Box 16.3 Characteristics that distinguish vitamins and essential minerals from other food chemicals

- They are essential for human well-being within a certain range of intakes and there is a long history of safe consumption of nutrients at the levels found in balanced human diets.
- For some nutrients, there may be experience of widespread chronic consumption (e.g. from dietary supplements) at levels significantly above those obtained from endogenous nutrients in foods without reported adverse health effects.
- Many nutrients are subject to homeostatic regulation of body content through adaptation of absorptive, excretory or metabolic processes, and this provides a measure of protection against exposures above usual intakes from balanced diets.
- Data on the adverse effects of nutrients are also often available from studies in humans, which helps to reduce the scientific uncertainties associated with extrapolation from the observed data to human populations.

The steps involved in the application of risk assessment principles to the derivation of ULs for vitamins and minerals are summarised in Figure 16.1 and explained in more detail below.

Hazard identification

This involves the collection, organisation and evaluation of all information pertaining to the adverse health effects of a given nutrient, and summarises the evidence concerning the capacity of the nutrient to cause one or more types of adverse health effect in humans. Human studies provide the most relevant data for hazard identification and, when they are of sufficient quality and extent, are given the greatest weight. Other experimental data (from experimental animals and in vitro studies) may also be used.

Key issues that are addressed in the data evaluation include the extent to which there is evidence of adverse health effects on humans and whether or not the relationship established by the published human data is causal, mechanisms of adverse effects, quality and completeness of the database, and identification of distinct and highly sensitive subpopulations.

Hazard characterisation

As intake of a nutrient increases, a threshold is reached above which increasing intake increases the risk of adverse health effects. This is illustrated diagrammatically in Figure 16.2.

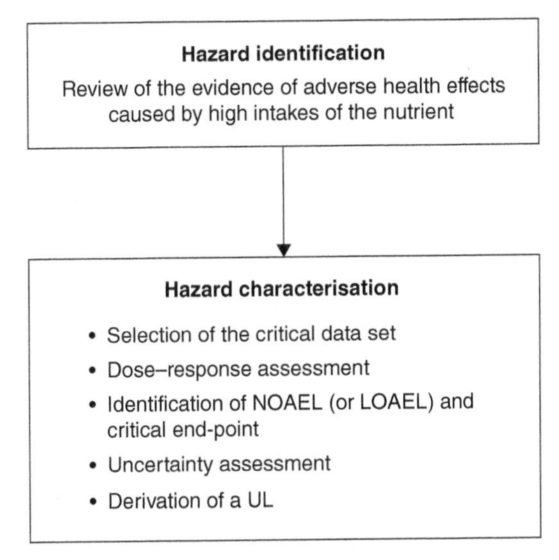

Figure 16.1 Steps in the development of the tolerable upper intake level (UL). NOAEL, no observed adverse effect level; LOAEL, lowest observed adverse effect level.

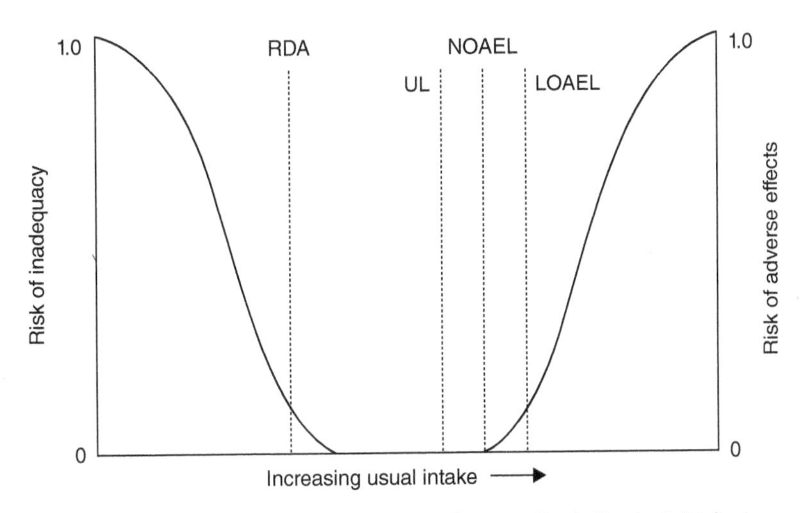

Figure 16.2 Theoretical description of health effects of a nutrient as a function of level of intake. As intakes increase above the tolerable upper intake level (UL), the risk of adverse effects increases. RDA, recommended daily allowance; NOAEL, no observed adverse effect level; LOAEL, lowest observed adverse effect level.

Hazard characterisation involves the qualitative and quantitative evaluation of the nature of the adverse health effects associated with a nutrient. This includes a dose–response assessment, which involves determining the relationship between nutrient intake (dose) and adverse health effect (in terms of frequency and severity). Based on these evaluations, a UL is derived, taking into account the scientific uncertainties in the data. ULs may be derived for various lifestage groups within the population.

Dose–response assessment

The dose–response assessment involves a number of key components.

Selection of the critical dataset

The data evaluation process results in the selection of the most appropriate or critical dataset(s) for deriving the UL. The critical dataset defines the dose–response relationship between intake and the extent of the adverse health effect known to be most relevant to humans. Data on bioavailability need to be considered and adjustments in expressions of dose–response are made to determine whether any apparent differences in dose–response between different forms of a nutrient can be explained. The critical dataset should document the form of the nutrient investigated, the route of exposure, the magnitude and duration of intake and the intake that does not produce adverse health effects, as well as the intake that produces adverse health effects.

Identification of the no observed adverse effect level or lowest observed adverse effect level and critical endpoint

The no observed adverse effect level (NOAEL) is the highest intake of a nutrient at which no adverse effects have been observed. The NOAEL can be identified from evaluation of the critical dataset. If there are no adequate data demonstrating a NOAEL, then a lowest observed adverse effect level (LOAEL, the lowest intake at which an adverse effect has been demonstrated) may be used. Where different adverse health effects (or endpoints) occur for a nutrient, the NOAELs (or LOAELs) for these endpoints will differ. The critical endpoint is the adverse health effect exhibiting the lowest NOAEL (i.e. the most sensitive indicator of a nutrient's adverse effects). The derivation of a UL based on the most sensitive endpoint will ensure protection against all other adverse health effects.

Uncertainty assessment

There are usually several scientific uncertainties associated with extrapolation from the observed data to the general population, and several judgements must be made in deriving uncertainty factors to account for the individual uncertainties. The individual uncertainty factors may be combined into a single composite uncertainty factor for each nutrient and applying this (composite) uncertainty factor (UF) to a NOAEL (or LOAEL) will result in a value for the derived UL that is less than the experimentally derived NOAEL, unless the uncertainty factor is 1.0. The larger the uncertainty, the larger the UFs and the lower the UL.

Uncertainty factors are used to account for imprecision of the data, lack of data and adequacy of data on variability between individuals. There are several potential sources of uncertainty, including:

- interindividual variation and sensitivity with respect to the adverse effect
- extrapolation from experimental animal data to humans
- if a NOAEL is not available, a UF may be applied to account for the uncertainty in deriving a UL from the LOAEL
- using a subchronic NOAEL to predict chronic NOAEL.

The UFs are lower with higher quality data and when the adverse effects are extremely mild and reversible. For example, for magnesium, a UF of 1.0 may be used since the adverse effect (osmotic diarrhoea) is relatively mild and reversible, there is a sufficiently large amount of data available relating magnesium intake level to this adverse effect in humans to cover adequately the range of interindividual variation in sensitivity, and a clear NOAEL can be established. In contrast, for vitamin B_6 a UF of 4 may be used since the adverse effect (neurotoxicity) is potentially severe, there are only limited data, mainly from inadequate studies of insufficient duration relating vitamin B_6 intake level to this adverse effect in humans, and no clear NOAEL can be established.

In the application of UFs, there should be cognisance of nutritional needs; for example, the derived UL should not be lower than the recommended intake.

16.5 Determining tolerable upper intake levels

The UL is derived by dividing the NOAEL (or LOAEL) by the (composite) UF. ULs are derived for different lifestage groups using relevant data. In the absence of data for a particular lifestage group, extrapolations are made from the UL for other groups on the basis of known differences in body size, physiology, metabolism, absorption and excretion of a nutrient. For example, when data are not available for children and adolescents, extrapolations are usually made on the basis of body weight using the reference weights for adults and children in the population.

Where possible, ULs are derived for separate lifestage groups (e.g. infants, children, adults, the elderly and women during pregnancy or lactation). Although within relatively homogeneous

Table 16.1 Examples of tolerable upper intake levels (ULs) for vitamins and minerals.

Nutrient	UL (adults)	Adverse effect
Retinol	3000 μg	Teratogenicity, hepatotoxicity
Vitamin D	100 μg	Hypocalcaemia
Vitamin E	1000 mg	Haemorrhagic effects
Vitamin B_6	25 mg	Sensory neuropathy
Folic acid	1000 μg	Progression of neuropathy or masking of anaemia in B_{12} deficiency
Vitamin C	2000 mg	Osmotic diarrhoea, gastrointestinal disturbances
Calcium	2500 mg	Hypercalciuria, nephrolithiasis (kidney stones)
Magnesium	250 mg (supplemental)	Osmotic diarrhoea
Phosphorus	4000 mg	Hyperphosphataemia
Iron	45 mg (supplemental)	Gastrointestinal effects
Zinc	40 mg	Reduced copper status
Copper	10 mg	Hepatotoxicity
Manganese	11 mg	Neurotoxicity
Selenium	300 μg	Clinical selenosis: brittle nails, hair loss
Iodine	1100 μg	Thyrotoxicosis

Source: Adapted from US Food and Nutrition Board, European Food Safety Authority and EU Scientific Committee on Food.

lifestage groups there is a range of sensitivities to adverse health effects due, for example, to differences in body weight and lean body mass, it is generally not possible to make a distinction between males and females in establishing ULs for adults or children.

The derivation of ULs for the normal healthy population, divided into various lifestage groups, accounts for normally expected variability in sensitivity, but it excludes subpopulations with extreme and distinct vulnerabilities due to genetic predisposition or other considerations. (Including these would result in ULs that are significantly lower than are needed to protect most people against the adverse effects of high intakes.) Subpopulations needing special protection are better served through the use of public health screening, healthcare providers, product labelling or other individualised strategies. The extent to which a subpopulation becomes significant enough to be assumed to be representative of a general population is an area of judgement and risk management.

It should be noted that derivation of a UL does not take into account possible adverse effects of acute bolus dosages. In general, adverse effects from acute or short-term intake require much greater intake levels than those arising from long-term or chronic exposure.

Tolerable upper intake levels have been derived for a number of vitamins and minerals (Table 16.1) and examples of the derivation of UL for selenium and vitamin D are given in Boxes 16.4 and 16.5. Adverse effects associated with high intakes of vitamin C, β-carotene and vitamin A are outlined in Boxes 16.6–16.8. While ULs are usually based on total intake of a nutrient, in some cases adverse health effects have been associated with intake from a particular source, such as supplements, rather than total intake, and in such cases the UL is based on intake from these sources only. For many nutrients, there are no systematic studies of the adverse health effects of high intakes and the UL is derived from limited data. Experience has shown that it is not always possible to establish a UL for a micronutrient using the science-based risk-assessment approach. Such a situation can arise for different reasons:

- evidence of the absence of any adverse health effects at high intakes
- lack of evidence of any adverse effect; this does not necessarily mean that there is no potential for adverse effects resulting from high intake
- evidence of adverse effects but insufficient data on which to base a dose–response assessment.

16.6 Use of tolerable upper intake levels as dietary reference standards

The UL is a dietary reference standard that can be used for both the evaluation and management of risk of excessive intake of vitamins or

Box 16.4 Derivation of UL for selenium

- Chronic selenosis has been described in regions of high soil selenium in China. Symptoms include hair and nail brittleness and loss, skin rash, mottled teeth, garlic breath odour and neurological disturbances, including peripheral anaesthesia, with numbness, convulsions, paralysis and motor disturbances in more severe cases.
- The high prevalence of selenosis in Enshi, South China, provided an opportunity for the study of the dose–response relationship of selenium intake to selenosis in 350 adults during the 1980s. Toxic effects occurred with increasing frequency as selenium intake increased above 850 μg/day. Lowering selenium intake in affected individuals led to the disappearance of symptoms. The LOAEL for clinical symptoms of selenosis is about 900–1000 μg Se/day and a NOAEL of 850 μg/day can be established.
- A UF of 3 may be used to cover uncertainties in the data (the NOAEL used was derived from a study on a large number of subjects and is expected to include sensitive individuals; although the toxic effect is not severe, it may not be readily reversible). This leads to a UL of 300 μg/day for adults.
- This UL may also be applied to pregnant and lactating women as there is no evidence to indicate that there is increased sensitivity of the fetus or breastfed infant to high levels of maternal selenium intake.
- Owing to insufficient data on the adverse effects of selenium in children, extrapolation from the UL for adults to children on a body weight basis may be performed to derive a UL for children.

Box 16.5 Derivation of UL for vitamin D

- Vitamin D toxicity has been described in individuals consuming excessive amounts of vitamin D-containing supplements. The effects observed include damage to kidney and other soft tissues, such as blood vessels, heart and lungs, owing to calcification. This results from hypercalcaemia caused by a vitamin D-induced increase in calcium absorption in the gastrointestinal tract and calcium resorption from bone. Elevated serum 25-hydroxyvitamin D (an indicator of vitamin D status) levels may lead to hypercalcaemia, which is considered the critical effect of excess intake of vitamin D.
- Data from controlled studies of the response of serum calcium to increasing intakes of vitamin D in human subjects may be used to establish the dose–response relationship. A NOAEL of 250 μg/day can be established as the highest intake that does not result in elevation of serum calcium above the upper end of the reference range.
- A UF of 2.5 may be used to cover uncertainties in the data (i.e. to cover the range of interindividual sensitivity in the population). This leads to a UL of 100 μg/day for adults. This UL may also be applied to pregnant and lactating women as there is evidence to suggest that there is no increased sensitivity of the fetus or breastfed infant to this level of maternal vitamin D intake.
- Although there are insufficient data for the adverse effects of vitamin D in children, a UL of 100 μg/day may be used for children of 10 years and older, and for children aged 1–9 years extrapolation from the UL for adults on a body weight basis may be performed to derive a UL.

Box 16.6 Vitamin C high intakes: myths and facts

- Vitamin C is often taken in large amounts (500 mg/day or more) for presumed health benefits. Evidence to date does not support a role for high doses of vitamin C in prevention of the common cold, although there is a moderate benefit in terms of duration and severity of episodes in some groups. Although there is epidemiological evidence to suggest a protective effect of vitamin C against cardiovascular disease, some cancers (e.g. stomach, lung) and cataracts, such benefits have not been established unequivocally.
- It has been suggested that antioxidants such as vitamin C may act as pro-oxidants at high intake levels. This is partly based on in vitro observations that ascorbic acid may interact with free iron or copper to promote lipid peroxidation or oxidative damage to DNA. However, this hypothesis has not been substantiated in vivo, where both iron and copper are normally tightly bound to transport and storage proteins and are unavailable to participate in such redox reactions.
- Reports of increased urinary oxalate excretion and kidney stone formation with high intakes of vitamin C have not been confirmed. Such effects are considered unlikely given the limited intestinal absorption of vitamin C at doses greater than 200 mg/day.
- A UL of 2000 mg/day may be derived based on osmotic diarrhoea and related gastrointestinal disturbances that arise from the osmotic effect of unabsorbed vitamin C passing through the intestine when the capacity of the saturable absorption system has been exceeded.

Box 16.7 β-carotene supplementation and smoking

- Prospective epidemiological studies have shown that higher consumption of carotenoid-containing fruits and vegetables, and higher plasma levels of carotenoids, including β-carotene, are associated with a lower risk of a number of cancers and cardiovascular disease.
- Several long-term intervention studies with supplemental β-carotene have not shown a reduction in chronic disease risk. Indeed, two of the studies carried out in the 1990s reported an increase in lung cancer associated with supplemental β-carotene (20–30 mg/day for 4–8 years) in current smokers. However, another study reported that supplementation with 25 mg/day of β-carotene for up to 12 years (including smokers and non-smokers) showed no excess of lung cancer.
- No other adverse effects have been described for β-carotene. For example, no toxic effects have been reported from the therapeutic use of β-carotene at high doses (about 180 mg/day) over many years for the treatment of erythropoietic protoporphyria, a photosensitivity disorder. Carotenodermia, a yellowish discoloration of the skin, is a harmless and well-documented effect of excess intake of β-carotene and other carotenoids.
- At present, the data on the potential for β-carotene to produce increased lung cancer are conflicting and not sufficient for a dose–response assessment and derivation of a UL. However, caution is advised in the use of β-carotene supplements.
- This does not conflict with dietary advice for fruit and vegetable consumption (e.g. five or more servings per day) since this provides much lower intakes of β-carotene (3–6 mg/day), which is also of lower bioavailability, than the supplemental intake levels associated with possible adverse effects (20 mg/day or more).

Box 16.8 Vitamin A toxicity

- High intakes (in excess of 3000 μg retinol equivalents [RE]/day) of preformed vitamin A (retinol and related compounds) appear to be teratogenic in humans, resulting in craniofacial malformations and abnormalities of the central nervous system (except for neural tube defects), thymus and heart. Evidence for this includes epidemiological studies on vitamin A intake level and pregnancy outcome, established teratogenicity in humans of therapeutic use of retinoic acid (a vitamin A metabolite), as well as established teratogenicity of vitamin A in a number of animal species.
- Hepatotoxicity is one of the most severe outcomes of chronic intake of high dosages of vitamin A (generally in excess of about 15000 μg RE/day in supplements or in rich food sources such as liver and liver oils), arising from overload of the storage capacity of the liver for the vitamin.
- Some epidemiological studies have indicated that intakes of vitamin A in excess of 1500 μg RE/day may reduce bone mineral density and increase the risk of osteoporosis and hip fractures in some population groups, particularly in pre- and postmenopausal women. However, other studies have not found this effect and this conflicting evidence is not adequate to establish a causal relationship or a UL.
- A UL of 3000 μg RE/day may be derived for women of reproductive age based on epidemiological data on the occurrence of birth defects in infants of exposed mothers. For other adults, a UL of 3000 μg RE/day may be derived based on case reports of hepatoxicity arising from prolonged overconsumption of vitamin A.

minerals in individuals or populations. The UL is not in itself an indication of risk. However, it can be applied to data derived from intake assessment (e.g. the distribution of usual total daily nutrient intakes among members of the general population) to identify those individuals or population groups at risk and the circumstances in which risk is likely to occur. In general, risk is considered to be the probability of an adverse effect (and its severity) and will depend on the fraction of the population exceeding the UL and the magnitude and duration of such excesses.

Individuals

It is not possible to identify a single risk-free intake level for a nutrient that can be applied with certainty to all members of a population. However, if an individual's usual nutrient intake remains below the UL, there is no appreciable risk of adverse effects from excessive intake. At usual intakes above the UL, the risk of adverse effects increases with increasing level of intake. However, the intake at which a given individual will develop adverse health effects as a result of taking large amounts of a nutrient is not known with certainty. The UL can be used as a guide for the maximum level of usual intake of individuals, although it is not a recommended intake. Occasional excursions above the UL are possible without an appreciable risk of adverse health effects.

Groups

The proportion of the population with usual intakes below the UL is likely to be at no appreciable risk of adverse health effects due to overconsumption, while the proportion above the UL may be at some risk, the magnitude of which depends on the magnitude and duration of the excess. The UL is derived to apply to the most sensitive members of the general population. Thus, many members of the population may regularly consume nutrients at or even somewhat above the UL without experiencing adverse effects. However, because it is not known which individuals are most sensitive, it is necessary to interpret the UL as applying to all individuals.

In the management of risk of excessive intake of vitamins or minerals in individuals or populations, the UL can be used in several ways. For example, it may be used for setting maximum levels of addition of vitamins and minerals to foods or nutritional supplements such that the risk of excessive intake in the population is minimised. It is also a useful basis for providing information and advice to individuals or groups on maximum intake levels of nutrients, such as in the labelling of nutritional supplements.

16.7 Perspectives on the future

The application of science-based risk assessment methods to the establishment of ULs for vitamins and minerals represents a significant advance in providing dietary reference standards for evaluating and managing the risk of overconsumption of these nutrients. These reference standards will be increasingly used in nutritional surveillance as studies on nutrient intakes in populations pay greater attention to the possibility of overconsumption of micronutrients resulting from the wider use of nutritional supplements and consumption of fortified foods in some countries.

This field is still limited by a lack of good data on dose–response relationships for the adverse health effects of micronutrients in humans. To improve the understanding of such effects, research is needed to provide better knowledge of the physiological effects of micronutrients at the molecular, cellular and whole-body levels, and of the kinetics of their absorption, metabolism and excretion at different dietary levels of intake.

Further reading

European Food Safety Authority (2006). Tolerable Upper Intake Levels for Vitamins and Minerals. Parma, Italy: European Food Safety Authority. www.efsa.europa.eu/sites/default/files/efsa_rep/blobserver_assets/ndatolerableuil.pdf.

Food and Agricultural Organization of the UN/World Health Organization, Expert Consultation (2006). A model for establishing upper levels of intake for nutrients and related substances: report of a joint FAO/WHO technical workshop on nutrient risk assessment. Geneva, Switzerland: 2–6 May 2005. https://apps.who.int/iris/handle/10665/43451

Food and Nutrition Board, Institute of Medicine, National Academy of Sciences (1997–2001). Dietary Reference Intakes. Washington, DC: National Academy Press

Food and Nutrition Board, Institute of Medicine, National Academy of Sciences (2001). Dietary reference intakes: applications in dietary assessment: a report of the subcommittees on interpretation and uses of dietary reference intakes and upper reference levels of nutrients, and the standing committee on the scientific evaluation of dietary reference intakes, food and nutrition board. Washington, DC: National Academy Press. www.nap.edu/catalog/9956.html.

17 Obesity

Gijs H. Goossens and Ellen E. Blaak

Key messages

- The prevalence of obesity is increasing throughout the world in both developed and developing countries, and obesity has become a worldwide health problem since it acts as a gateway to a range of other complications and non-communicable diseases.
- Body mass index (BMI) is the measure most widely used to assess obesity, yet phenotyping beyond BMI is needed to better assess disease risk.
- Body fat mass, body fat distribution and adipose tissue (dys) function are important determinants of cardiometabolic health.
- Obesity is the result of an energy imbalance where energy intake exceeds energy expenditure, and characterised by impairments in substrate utilisation during fasting, post-prandial and exercise conditions.
- The aetiology of obesity is complex and may be a result of genetic, metabolic, socioeconomic, behavioural and environmental factors. Therefore, this complex multifactorial chronic disease should be prevented and treated using a holistic approach, taking these different factors into account.

17.1 Introduction

The prevalence of obesity is increasing throughout the world. Globally, more than 1.9 billion adults are overweight, of whom at least 650 million are obese. According to current estimates, the global prevalence of obesity will further increase in the near future (WHO 2021). The aetiology of obesity is complex and may be attributed to genetic, metabolic, socio-economic, behavioural and environmental factors. In this chapter, we will discuss the identification and assessment of obesity, the concept of obesity as an adiposity-based chronic disease, how energy imbalance can lead to obesity, the impairments in substrate metabolism often present in people with obesity, potential factors contributing to the development of obesity and the consequences of obesity. Finally, we will discuss future perspectives with respect to research as well as the prevention and management of obesity.

17.2 Diagnosis

Overweight and obesity are defined as abnormal or excessive fat accumulation that presents a risk to health. Body fat can be measured precisely by laboratory methods such as dual-energy X-ray absorptiometry (DEXA), total body water and hydrodensitometry. These methods give an estimate of total body fat, or adipose tissue, in the body. Because people of different sizes will have different amounts of body fat, identification of obesity cannot be made on the basis of absolute amounts of fat.

For example, a person who has 20 kg of body fat and weighs 60 kg has a percentage body fat of 33%:

$$\frac{20}{60} \times 100 = 33\%$$

Nutrition and Metabolism, Third Edition. Edited on behalf of The Nutrition Society by Helen M. Roche, Ian A. Macdonald, Annemie M.W.J. Schols and Susan A. Lanham-New.

Companion Website: www.wiley.com/go/nutrition/metabolism3e

A second person who has 25 kg of fat but weighs 100 kg will have 25% body fat:

$$\frac{25}{100}\times 100 = 25\%$$

The first person has less total fat but a higher percentage body fat than the second person.

Although obesity is defined as an excess of body fat, laboratory techniques for directly assessing body fatness (or body fat percentage) are not available for clinical use. Furthermore, there are currently no agreed cut-off points or defined criteria for identifying obesity from measures of body fatness. Thus, identifying obesity in clinical practice is not presently based on an individual's percentage of body fat but rather on measures that correlate with body fat. The most common methods used in clinical assessment are body mass index (BMI), skinfold thickness and bioelectrical impedance. Several methods used to assess body composition and body fat mass are discussed below.

Body mass index

Adults

Body mass index is the recommended international screening tool used to identify overweight and obesity (Table 17.1) (WHO 1995, 2000). BMI is defined as weight in kilograms divided by height in metres squared.

A man who is 172 cm tall and weighs 89 kg will have a BMI of 30.1:

$$\frac{89}{(1.72)^2} = 30.1\,\text{Kg/m}^2$$

Another man who is 172 cm tall and weighs 73 kg will have a BMI of 24.7:

$$\frac{73}{(1.72)^2} = 24.7\,\text{Kg/m}^2$$

Studies have shown that BMI, although not a direct measure of fatness, is significantly correlated with body fat as measured by laboratory methods. One of the major advantages of using BMI to identify obesity is that it can be derived from measures of height and weight, which are much easier to obtain than measures of skinfold thickness. However, BMI cannot distinguish whether the excess weight is due to fat or muscle mass. Thus, the use of BMI to identify people living with obesity may result in misclassification for individuals with increased muscle mass, such as body builders or wrestlers.

Furthermore, recent studies suggest that percentage body fat at a specific BMI may be different in certain (sub)populations, and seems to be related to age, sex and ethnicity. For example, women in general have a higher fat mass for a given BMI. In addition, body composition changes occur with ageing. The age-related loss of muscle mass and strength or physical

Table 17.1 International classification of adult underweight, overweight and obesity according to BMI.

	BMI (kg/m^2)	
Classification	Principal cut-off points	Additional cut-off points
Underweight	<18.50	<18.50
Severe thinness	<16.00	<16.00
Moderate thinness	16.00–16.99	16.00–16.99
Mild thinness	17.00–18.49	17.00–18.49
Normal range	18.50–24.99	18.50–22.99
		23.00–24.99
Overweight	≥25.00	≥25.00
Pre-obese	25.00–29.99	25.00–27.49
		27.50–29.99
Obese	≥30.00	≥30.00
Obese class I	30.00–34.99	30.00–32.49
		32.50–34.99
Obese class II	35.00–39.99	35.00–37.49
		37.50–39.99
Obese class III	≥40.00	≥40.00

performance (that is, sarcopenia) is often accompanied by increased amounts of adipose tissue, a condition referred to as sarcopenic obesity. In other words, older individuals often have a lower muscle mass and higher fat mass for a given BMI compared to young adults. Finally, ethnicity seems to affect the relationship between BMI and body fat percentage. For example, African-Americans generally have a lower body fat percentage, while South Asians have a higher body fat percentage for a given BMI. Therefore, the WHO has defined Asian-specific obesity cut-off values (Table 17.2) (WHO Expert Consultation 2004), and research is under way to determine and implement cut-points for BMI specific to various populations.

Children and adolescents

Body mass index is currently used to identify children living with overweight and obesity. Because BMI changes with age in children, there is no specific numerical cut-off point for identifying overweight and obesity in children as there is for adults. Thus, BMI growth charts that are age and gender specific are used to identify obesity.

Cut-off points for overweight and obesity in children differ among countries. Several countries, such as the USA, France, Japan, the UK and Italy, have specific growth charts developed from populations within their countries. While many countries use their own reference population to define obesity, others may use reference populations from other countries. An international task force on obesity (IOTF) developed reference standards to identify obesity in children. These standards are based on pooled data from six countries, including Brazil, the UK, Hong Kong, the Netherlands, Singapore and the USA.

Measurements for the assessment of body fat mass and fat distribution

Quantifying the amount and distribution of adipose tissue and its related components is integral to the study and treatment of human obesity. The determination of clinical measures of overweight and obesity that accurately reflect body fat has been an area of much research and debate. Several field measurements as well as laboratory methods are available to determine body fat mass and body fat distribution. Currently, studies to identify obesity based on percentage body fat are in progress.

Field methods

Bioelectrical impedance Bioelectrical impedance (BIA) can be used in the clinical setting to estimate body composition. BIA is based on the body's resistance to a harmless electrical current, which is related to total body water. The measure of resistance, height and weight is used to determine the percentage body fat from predictive equations. One concern with BIA is that the predictive equations were created based on data from specific populations. It is important that the predictive equations used to determine percentage body fat are derived from a population similar to the population being measured.

Skinfold thickness Skinfold methods provide an estimation of subcutaneous adipose tissue mass by measuring skinfold thickness using a caliper that has been designed for this purpose. Measures of triceps skinfold (TSF) thickness provide an index of subcutaneous fatness (fat underneath the skin).

Studies have shown that measures of TSF thickness correlate well with body fatness in both children and adults. However, two

Table 17.2 Asian-specific classification of underweight, overweight and obesity in adults according to BMI.

	BMI (kg/m^2)	
Classification	Cut-off points	Risk of co-morbidities
Underweight	<18.50	
Normal range	18.50–22.99	Increasing but acceptable risk
Overweight	≥23.00	
Pre-obese	23.00–27.49	Increased risk
Obese	≥27.50	High risk

assumptions are made when TSF thickness measures are used to assess body fatness. The first is that the TSF is representative of subcutaneous fatness and the second is that subcutaneous fatness is correlated with total body fat. However, these assumptions are not always valid. In people living with certain diseases, or in individuals with significant amounts of visceral fat (fat around the internal organs), changes in body fat distribution may alter the relationship of TSF thickness with total body fat.

Investigators have measured skinfold thickness at several subcutaneous sites, including the subscapular, suprailiac, abdominal, bicep and tricep areas. Equations have been developed to calculate body fatness from the sum of several skinfolds, with an error of ~3.5–5%. Although the use of these equations is probably more representative of subcutaneous fat than TSF thickness alone, skinfold measurements cannot identify individuals with significant amounts of visceral fat.

A significant limitation of using skinfold thickness to measure body fatness is that extensive training is required to perform the measurement correctly. Furthermore, it is more difficult to measure skinfold thickness accurately in people with excess fat mass. In other words, the reliability of these measures decreases with increasing fatness. These limitations make this method less useful for identifying obesity in a clinical setting. Skinfold thicknesses have been used to describe the distribution of subcutaneous fat. This aspect of adipose tissue distribution has been called 'fat patterning'.

Circumferences Waist circumference and waist-to-hip ratio reflect body fat distribution, including visceral adipose tissue, and have been widely used to assess abdominal obesity. Waist circumference seems a better index of visceral adipose tissue than waist-to-hip ratio. As discussed in Chapter 16 in more detail, fat storage in the abdominal region conveys the largest risk for major chronic diseases and mortality. It should be noted, however, that muscle mass and bone influence waist and hip circumferences. Furthermore, minor changes in the volume of visceral adipose tissue cannot be detected by waist circumference. Since waist circumference can be measured at different sites (i.e. at the level of the umbilicus, smallest circumference on the torso below the sternum, lower margin of the ribs, and iliac crests), standardisation of measurements is very important.

Laboratory measurements

Body volume-based body composition measures Body volume measurements combined with assessment of body mass allow calculation of body density. Body volume measurement techniques are based on the assumption that our body consists of two main components: fat mass (FM) and fat-free mass (FFM). FFM can be further divided into total body water, protein and minerals. All these separate components have known densities. The two-component model assumes stable densities for FM and FFM, and body fat percentage can thus be calculated from body density. Importantly, volume-based measures should be adjusted for lung volume, which can be assessed with spirometry (i.e. helium dilution technique).

Hydrodensitometry, frequently called underwater weighing, is one of the oldest ways of assessing body composition in humans. This technique is based on the Archimedean principle that applies measured body density to determine fat mass and fat-free mass based on the assumed densities of these tissues. Although this method is very accurate, a limitation of underwater weighing is that it requires substantial co-operation of study participants or patients to perform the measurement, and it may therefore not be suitable for young children, disabled individuals or elderly. In addition, the procedure takes more time than several other methods.

Air displacement plethysmography determines body volume by assessing air displaced when an individual enters a controlled environment. The individual should wear a tightly fitting bathing suit and swim cap. After body weight has been determined, the individual takes a seat in a BOD POD® unit (BOD POD Body Composition System, Cosmed, Rome, Italy) to measure body volume. An advantage of this technique is that the measurement only takes a few minutes to perform and in contrast to hydrodensitimetry, this method can be relatively easy to use in children, disabled and elderly individuals.

Dual-energy X-ray absorptiometry Dual-energy X-ray absorptiometry (DEXA) technology uses a filtered X-ray source that produces two main energy peaks. Soft tissue and bone have different attenuation coefficients. The difference in attenuation between the two energy peaks on X-ray has a typical signature for fat tissue, lean tissue and bone mineral. Although DEXA was originally developed to determine bone mineral density, the technique has substantially improved, allowing quantifation of fat and lean tissue mass. DEXA is frequently used to accurately assess body composition with high reproducibility (i.e. coefficient of variation of ~0.5%–4% for body fat percentage). An advantage of this technique is that different body compartments are measured, and as such body fat distribution can be determined. This is of particular importance given the relationship between body fat distribution and cardiometabolic health, as outlined in more detail in Section 17.6.

Imaging methods More recently, several imaging methods have become available to assess body fat mass, fat distribution and lipid content in different organs. These include computed tomography (CT), magnetic resonance imaging (MRI) and magnetic resonance spectroscopy. Although these techniques provide accurate measurements of body fat mass and lipid deposition within organs, they are expensive and require highly trained technicians to execute the measurements.

17.3 Energy balance

The maintenance of energy balance is one of the most vital processes for individuals and species. It is evident that obesity results from a failure of homeostatic mechanisms that regulate body weight. Obesity is the result of a long-term energy imbalance where energy intake exceeds energy expenditure (Table 17.3). This leads to a positive energy balance where the excess energy is stored in the form of triglycerides or triacylglycerols, predominantly in adipose tissue.

The first law of thermodynamics states that energy can neither be created nor destroyed, but can only be transformed from one form into another. Although incontrovertible, this law describes the overall handling of energy. Small energy imbalances maintained over a long period can result in significant weight gain or loss. Calculation of how a small energy excess or deficit would lead to quantitative changes in the body fat stores is, however, an oversimplification of the concept of energy balance. It is of the utmost importance to understand the complexities of human energy balance and obesity, which requires a more integrated and holistic approach. Various factors need to be considered, for example that both energy intake and expenditure may be influenced by changes in body energy stores; that voluntary energy intake may rise with increasing levels of physical activity; and that food intake itself may affect energy expenditure. These aspects are discussed below.

Energy intake

Humans, like other mammals, are meal feeders. The amount of food ingested varies on average from 8 to 13 megajoules (MJ) a day, but may vary considerably depending on the degree of physical activity and hyperphagia. Research on food intake is very complicated due to difficulties in assessing it reliably. A considerable amount of research has examined whether people with obesity consume more energy than people without obesity. Most studies of energy intake in children with obesity or normal weight and adults with obesity do not show differences in their energy intakes. The findings of these studies have been questioned because of the uncertainty about whether the subjects' self-reported energy intakes are accurate.

It was not until the doubly labelled water (DLW) method for measuring energy expenditure was developed that it became possible to assess unobtrusively the validity of dietary reporting. Using the DLW technique, measures of daily energy expenditure can be made simultaneously with measures of energy intake. If an individual is in energy balance, energy expenditure should equal energy intake. However, studies using DLW to measure energy expenditure have revealed that people with and without obesity substantially

Table 17.3 Energy balance.

Energy balance	Energy intake = Energy expenditure	Maintenance of fat stores
Positive energy balance	Energy intake > Energy expenditure	Increase in fat stores
Negative energy balance	Energy intake < Energy expenditure	Decrease in fat stores

under-report their energy intakes, and energy and fat intake under-reporting may increase with increasing BMI. Thus, comparisons of self-reported energy intake between groups are often inaccurate and cannot be used to determine whether or not individuals with obesity eat more than individuals with a normal body weight (Ravelli and Schoeller 2020).

Globally, the availability in dietary energy per capita per day has risen from 2358 kcal in 1964–1966 to 2940 kcal in 2015 (www.fao.org/3/ac911e/ac911e05.htm). Although the waste of food has increased substantially, it probably did not do so at the same rate as the increase in production. Consequently, there appears to be a discrepancy between the increase in food production and the reported energy intake, for which under-reporting is one of the most likely explanations.

Thus, at a population level, overconsumption of food occurs and is likely to contribute to the spread of obesity. The neuroendocrine factors that control food intake are briefly addressed below. Human feeding behaviour is also heavily influenced by social, psychological, environmental and economic factors which may override the physiological mechanisms.

Control of energy intake

Impaired (regulation of) energy intake is one of the key factors contributing to the development of a positive energy balance and to overweight or obesity. The main brain areas responsible for regulating energy balance are the hypothalamus and brainstem, which receive numerous central and peripheral inputs, including nutritional signals from the body that convey information about current energy stores and fat mass.

As reviewed recently, multiple studies indicate that the neuroendocrine processes involved in the regulation of energy intake and appetite are disturbed in people suffering from overweight or obesity (Miller 2019). This includes, for example, the release of gastrointestinal satiety hormones in response to meal ingestion, which is suggested to be impaired in individuals with overweight and obesity (Mishra et al. 2016). In addition to disturbances in homeostatic pathways, the hedonic pathway regulating energy intake is also suggested to be altered in individuals with overweight and obesity, leading to an impaired food reward response and an addiction-like perception of food in some individuals (Berthoud et al. 2017). The current notion is that gut microbiota also influences the bidirectional communication between the gastrointestinal tract and the brain, also known as the gut–brain axis. Studies on the microbiota and gut–brain axis show that the microbiota might play a key role in the regulation of energy intake and satiety signalling (Torres-Fuentes et al. 2017).

Energy expenditure

The overall process by which fuel energy is converted to adenosine triphosphate (ATP) and subsequently used in diverse metabolic processes or lost as heat is referred to as energy expenditure. There are important differences in the amount of oxygen utilised and the amount of energy released when different macronutrients are used as metabolic fuels. The oxidation of fat yields over twice as much energy (9 kcal/g) as the oxidation of carbohydrates or proteins (4 kcal/g). Furthermore, for a given amount of energy expenditure, more oxygen is needed when using fat as a substrate, compared with carbohydrate or protein. Also, the ratio of carbon dioxide produced to oxygen consumed – the respiratory quotient (RQ) – differs between the macronutrient substrates, being highest for carbohydrate (1.0) and lowest for fat (0.7).
Energy expenditure comprises four components.

- Resting metabolic rate (RMR), which reflects core body functions and is measured at complete rest and without food.
- Diet-induced thermogenesis (DIT), which is the energy required to digest, absorb and assimilate ingested food.
- Activity thermogenesis, which is the energy expended by skeletal muscle contraction, which can be divided into exercise and non-exercise activity thermogenesis (NEAT).
- Energy required for growth.

A recent study analysed a large, diverse database of total expenditure measured by the DLW method for males and females aged eight days to 95 years. Energy expenditure (adjusted for FFM) accelerates rapidly in neonates to ~50% above adult values at ~1 year; declines slowly to adult levels by ~20 years; remains stable in adulthood (20–60 years), even during pregnancy; and then declines in older adults (Rimbach et al. 2022).

Resting metabolic rate accounts for a major proportion of daily energy expenditure. Its contribution to energy expenditure varies among individuals, but in sedentary adults it accounts for between 60% and 70% of daily energy expenditure. The remaining 30–40% of daily energy expenditure in sedentary adults is the energy spent on activity and DIT. As physical activity increases, RMR's proportion of total energy expenditure decreases.

Variability in energy expenditure

Since the 1980s, many studies have been conducted to compare RMR, DIT and daily energy expenditure among individuals with and without obesity. Studies have not shown significant differences in RMR between adults or children with and without obesity. Studies on the TEF are equivocal, with some suggesting that energy expenditure may be decreased in individuals with obesity and others showing no differences between people with and without obesity. Although the DIT contributes only a small percentage to daily energy expenditure (6–10%), small differences over a long period could potentially lead to a significant energy imbalance.

Studies of daily energy expenditure that include RMR, DIT and physical activity among people with and without obesity have not found differences in total energy expenditure between the two groups. Although these studies have not found lower energy expenditure in people with obesity, it has been hypothesised that differences may exist among individuals before obesity develops (pre-obese state). When an individual gains weight, RMR and daily energy expenditure increase, due to an increase in both fat and FFM. It has been hypothesised that lower energy expenditure in individuals predisposed to obesity may be normalised with weight gain. Earlier studies (Ravussin et al. 1988) suggested that low energy expenditure may lead to increased weight gain as an adaptive mechanism to increase energy expenditure. However, most recent longitudinal studies in children and adults do not support the hypothesis that lower energy expenditure is associated with an increase in body fat.

Interestingly, a recent study used repeated DLW measurement for total energy expenditure (TEE) in 348 adults and 47 children. The authors concluded that having a 'fast' or 'slow' metabolism is a repeatable, durable trait for adults that is consistent over years. However, no evidence was found that individuals with lower adjusted TEE are at increased risk of gaining body fat, nor that higher adjusted TEE protects against weight gain. The causes and consequences of variation in metabolic rate in humans remain a critical focus for future investigation (Rimbach et al. 2022).

Diet-induced thermogenesis

In the 1980s, considerable research focused on the role of diet-induced thermogenesis (DIT) in maintaining energy balance. DIT refers to an increase in energy expenditure in response to eating. If energy expenditure was increased in response to an excess intake of energy, the excess energy would be dissipated as heat and obesity could be prevented. It has been hypothesised that lean individuals compensate for excess energy intake by burning the excess energy as heat, whereas individuals with obesity are metabolically more efficient and store excess energy as fat. Results of overfeeding studies in which energy expenditure was measured have found little evidence for adaptations in DIT (adaptive thermogenesis) in individuals with or without obesity.

Stock (1999) suggested that the role of DIT in regulating energy balance is secondary to its role in eliminating non-essential energy for individuals consuming low nutrient-dense diets and that in our society, only recently food has become so accessible that the role of DIT in maintaining energy balance has become significant. Stock calculated the theoretical cost of weight gain in many of these overfeeding studies that reported little evidence of an increased DIT and argued that 84% of subjects in these studies exhibit some increase in DIT. Individual variability in adaptive thermogenesis has been hypothesised as a reason why some individuals are more susceptible to weight gain and obesity in response to overeating.

Several mechanisms have been proposed for adaptive thermogenesis. The most widely studied is mitochondrial uncoupling through the action of uncoupling proteins (UCPs). UCPs provide heat without generating ATP by uncoupling oxidation and phosphorylation. Thus, when food is oxidised, the energy is dissipated as heat rather than converted to ATP. In animals, DIT occurs in brown adipose tissue (BAT). Until recently, it was thought that adult humans do not have significant amounts of BAT, and it was assumed that UCPs were not important for humans. However, BAT has been found in adults (van Marken Lichtenbelt et al. 2009). Further research is required to identify the quantitative significance of BAT in human body weight regulation, to elucidate the role of BAT volume and activity in cardiometabolic health, and to develop strategies to activate BAT in humans (Hanssen et al. 2016; Saari et al. 2020; Shamsi et al. 2021).

Other possible mechanisms for adaptive thermogenesis that are presently under investigation include futile calcium cycling, protein turnover and substrate cycling.

Cold-induced thermogenesis

In many lower mammals, chronic exposure to cold is associated with an increase in energy expenditure. In the cold-exposed rat, energy expenditure is initially increased through shivering, but is soon replaced by non-shivering thermogenesis due to SNS-induced activation of BAT and other heat-producing tissues. Present-day humans rarely need to increase their heat production for the purposes of thermal regulation, because they can wear appropriate clothing or seek a warmer environment.

As with energy intake, large interindividual differences in energy expenditure in response to cold have been reported. It has been shown that short cold acclimatisation may induce BAT in humans and that cold-induced fatty acid uptake may be impaired in obesity (Saari et al. 2020). These adaptations as well as putative disturbances in individuals with obesity may have therapeutic implications.

Activity thermogenesis

Non-exercise activity thermogenesis (NEAT)

Non-exercise activity thermogenesis describes the low level of involuntary physical activity which is spontaneous and subconscious, and is distinct from voluntary exercise. Several factors affect NEAT, such as occupation, age, genetic background, body composition and seasonal variations.

The importance of NEAT in weight regulation is highlighted by findings that, even when subjects are confined to a metabolic chamber, the 24h energy expenditure attributed to NEAT (as assessed by radar systems) was found to vary between 94 and 700 kcal (0.4 and 3 MJ), and also predicted subsequent weight gain in the longer term. High-effect NEAT movements (fidgeting, walking and standing) have been suggested to result in up to an extra 2000 kcal of expenditure per day beyond the basal metabolic rate, depending on body weight and level of activity (Villablanca et al. 2015). When positive energy balance is imposed through overfeeding, NEAT increases, which may account for more than 60% of the rise in total daily energy expenditure; again, the change in NEAT was found to be the most significant predictor of increase in fat mass (Levine et al. 1999).

Energy expenditure through physical activity

Physical activity makes a highly variable contribution to daily energy expenditure, representing up to 70% of the total for competitive athletes or those involved in heavy manual labour. However, most individuals in developed populations are now so sedentary that physical activity accounts for only ~30% of total daily energy expenditure.

17.4 Substrate metabolism

Substrate oxidation

The three macronutrients, carbohydrate, protein and fat, are oxidised to carbon dioxide and water and can provide energy for metabolic needs.

According to a model proposed by Flatt (1995), the long-term stability of body weight requires not only that energy expenditure equals intake, but also that the relative proportions of fuels oxidised are the same as those consumed. Because the human body has limited storage capacity for excess carbohydrate, a lack of storage capacity for excess protein and a large capacity to store excess fat, excess energy intakes of carbohydrate and protein will be oxidised preferentially to fat. Fat oxidation is less well regulated and only slowly adapts to an increase in intake and is not tightly regulated by an increase in fat stores. The failure to adjust fat oxidation to a high-fat diet may lead to a depletion of glycogen stores due to increased carbohydrate oxidation and the depleted glycogen stores may be a trigger stimulating energy intake. In the theory of Flatt (1995), an increase in fat mass may be required to stimulate fat oxidation and achieve again a condition of fat and energy balance.

The interpretation of this concept of nutrient balance is that food intake may be controlled by both (short-term) 'glycogenic' and (long-term) 'lipostatic' mechanisms and that glycogen stores and fat mass both act as regulatory signals. This nutrient balance theory has been challenged by others showing that a very low-carbohydrate, high-fat diet to deplete glycogen stores leads to an increase in fat oxidation to meet energy needs and has no effect on appetite (Stubbs et al. 2004). Furthermore, it may be too simplistic to consider a separate regulation of macronutrient balances in view of their complex interactions.

Despite this, there are indications that a reduced ability to use fat as a fuel may predispose towards obesity, insulin resistance and type 2 diabetes mellitus. A subgroup of people with obesity may have a reduced capacity to increase fat oxidation after a high fat load (Blaak et al. 2006), which may persist after weight reduction. Further studies are needed to characterise these subgroups with a limited flexibility to regulate fat oxidation ('metabolic inflexibility') (Corpeleijn et al. 2009; Kelley 2005).

17.5 Determinants of obesity

Obesity is a complex multifactorial disease representing a major public health problem. It is driven by an excess of available energy. As already mentioned, the availability of dietary energy per capita per day has risen over time, indicating that at population level energy intake has increased. Besides that, in the developed world the level of physical activity is low and a sedentary lifestyle is very prevalent. Beside genetic, environmental and physiological factors, human energy balance is heavily influenced by social, psychological, environmental and economic factors which may over-ride the physiological mechanisms. These include, among other factors, the continuous availability and extensive offering of foods and the higher price of healthy compared to unhealthy foods.

Medical conditions

Although obesity is rarely the result of a metabolic disorder, there are pathological syndromes in which this is the case. These syndromes account for only a small percentage of the obese population and include Prader–Willi syndrome, Laurence–Moon–Bardet–Biedl syndrome, hypothyroidism, Alsrom–Hallgren syndrome, Carpenter syndrome, Cohen syndrome, Cushing syndrome, growth hormone deficiency and polycystic ovary syndrome.

Genetic factors

Heritability of body fatness

Evidence suggests that there is a genetic component to obesity (Loos and Yeo 2022). Studies examining fatness similarities among twins and other biological siblings have shown that similarities are strongest for monozygotic twins (MZ), followed by dizygotic (DZ) twins and then other siblings. The heritability of fatness has been examined by measuring BMI in MZ and DZ twins living in similar or different environments and in adopted children separated from their biological parents. Studies of twins and adoptees both found heritability of obesity between 40% and 70%. This observation supports a strong genetic component to fatness. In addition, overfeeding studies in twins have shown that genetics may influence the amount of weight that is gained as well as the amount of weight that is lost in response to exercise.

Genetic mutations and obesity

Recent improvements in genetic screening, as well as reduction in cost of the screens themselves, have advanced efforts to identify genetic mutations related to obesity. Research has uncovered at least 11 single-gene mutations that may result in obesity. Many of these have been found through genetic screening of individuals with severe obesity. Approximately 5% of people with obesity are estimated to have a single-gene mutation, for example in the melanocortin pathways (MC4R, POMC), congenital leptin deficiency and in the gene encoding the leptin receptor.

The discovery of genes that influence polygenic obesity, which is common in the general population, started off slowly with candidate gene studies and genome-wide linkage studies, but only had marginal impact on the progression of gene discovery for common obesity outcomes.

The gene discovery for common diseases accelerated with progress in genome-wide association studies (GWAS). To date, 60 GWAS have identified more than 1100 independent regions (loci) associated with a range of obesity traits. The first GWAS identified the fat mass and obesity (FTO) associated gene, an example of a region of genes shown to have significant effect on BMI. A recent GWAS for BMI, which included nearly 800 000 individuals, identified more than 750 loci, and explained 6% of the variation in BMI (Yengo et al. 2018).

Single-gene mutations are much rarer but typically have a larger effect on how much excess weight the person has. Polygenes are more common, but the effect size (i.e. the amount of excess weight) is smaller and can probably be mediated by many other factors.

Genetic factors influence energy intake and energy expenditure

A genetic basis of energy balance would be expected to show its effect on energy intake, energy expenditure or both. As previously mentioned, the lipostatic theory indicates that signals stimulating satiety and energy expenditure would be released from adipose tissue in proportion to the body's adipose tissue stores. Therefore, if body fat stores increased, a compensatory mechanism would be initiated that would result in a decrease in energy intake.

The discovery of leptin in 1994 provides a mechanism to support this theory of body weight regulation (Zhang et al. 1994). Leptin is a hormone secreted by adipose tissue that is involved in weight regulation through a series of complex interactions with other neurohormones that affect energy intake and energy expenditure. Leptin plays a key role in a negative feedback loop to maintain adipose tissue stores. When adipose tissue decreases, leptin levels are reduced, resulting in an increase in food intake and a decrease in energy expenditure. Subsequent increases in adipose tissue are associated with an increase in leptin levels and a decrease in food intake. This homeostatic mechanism results in the maintenance of adipose tissue.

Studies in mice have shown that a defect in the gene for leptin is associated with obesity, resulting in increased food intake and reductions in metabolic rate. Although rare, mutations in the gene for leptin and the leptin receptor in humans are also associated with obesity, as indicated above. Plasma levels of leptin are positively associated with fat mass. In general, the more adipose tissue an individual has, the higher their leptin level will be. This suggests that human obesity is not associated with a decrease in leptin. It is hypothesised that obesity may be associated with a resistance to leptin. Thus, individuals with leptin resistance may have normal or high leptin levels, but may not be responsive to these levels of leptin.

Ethnic differences and obesity prevalence

Ethnic differences in the prevalence of obesity have led to speculation that they may be related to differences in energy expenditure. Because of the high prevalence of obesity reported in black individuals, many studies have been conducted to compare energy expenditure between blacks and whites. These studies have examined RMR, total energy expenditure and substrate oxidation. Although most studies comparing RMR between the two groups report a reduced RMR (adjusted for lean and fat mass) in blacks, some do not. Even if these differences are established, these are cross-sectional studies so they cannot be interpreted to suggest that a low metabolic rate causes obesity. Longitudinal studies to date do not support the association between a low RMR and weight gain among blacks compared with whites. There are limited data on the relationship between RMR and weight gain in other ethnic groups.

Gene–environment interaction and epigenetic alterations

Although genetic factors may increase the susceptibility of an individual to obesity, the significant increase in the prevalence of obesity in the past 30 years suggests that genetic factors alone could not be responsible for the obesity epidemic. Thus, it is likely that an interaction between genetics and an environment that promotes increased energy intake and/or decreased energy expenditure contributes to the development of obesity.

Twin studies take into account both familial and environmental factors, but the relationship is complicated. The degree of heritability of obesity varies among studies, so it is unknown how much influence one's genes and the environment have. Because families share the same environment, it is difficult to determine how much of the similarity in fatness is due to genetic factors and how much is due to environmental factors.

In an environment where energy intake is increased and energy expenditure is reduced, genetic factors that predispose individuals to obesity may result in a positive energy balance. A good example of this can be seen in the Pima Indians living in Arizona, whose BMI is significantly higher than the BMI in a related group of Pima Indians living in Mexico. The diet of the Arizona Pima Indians is typical of a western diet, as it is higher in fat and lower in complex carbohydrates than the traditional diet of the Mexican Pimas. In addition, the Mexican Pimas engage in strenuous physical activity, whereas the Arizona Pimas are less physically active. This combination of a high-fat diet and decreased physical activity in this genetically susceptible population has contributed to a high prevalence of obesity.

Gene–environment interactions can also mediate the effect of a genetic predisposition to obesity. The aforementioned FTO gene region significantly increases the risk for overweight. However, emerging research suggests that this effect is nullified when persons with the FTO genotype are more physically active than average adults (Rampersaud et al. 2008), and the effect is intensified (i.e. there is a gain in adipose tissue) when they engage in little daily physical activity (Andreasen et al. 2008). Nevertheless, genome-wide searches to discover novel loci that interact with lifestyle have proven to be challenging. These studies often do not have sufficient statistical power, suggesting that interaction effects are probably small and/or the precision and accuracy with which non-genetic (lifestyle) factors are assessed are low.

Finally, recent advances in the study of DNA and its alterations have considerably increased our understanding of the function of epigenetics in regulating energy metabolism and expenditure in obesity and metabolic diseases. Epigenetics may provide an additional explanation for the increasing obesity prevalence over the past few decades without necessitating a radical change in the genome. The known epigenetic mechanisms include DNA methylation, histone modifications and miRNA-mediated regulation. In future, both genetic and epigenetic causes of obesity should be considered.

Factors affecting energy intake

The obesogenic environment, including increased accessibility to food outside the home, increases in portion size and increased availability of high-energy, low-nutrient foods, may lead to increased energy intake. The obesity epidemic has been fuelled in large part by increased energy from greater availability of highly rewarding and energy-dense food (high-fat, high-sugar foods, sugar-sweetened beverages).

In a 13-year follow-up study on 3000 young people, those who consumed much more fast food were found to weigh an average of ~6 kg more and have larger waist circumferences than those with the lowest fast food intake (Duffey et al. 2007). Obesogenic marketing to promote beverages or foods that are high in sugar and fat negatively modulates human behaviour. Studies have shown that advertising these foods during typical television viewing hours increased children's requests for the specific food items. The availability of meals and food outside the home may also contribute to excess energy intake. Portion sizes in restaurants are often large and fast-food restaurants generally provide high-fat meals, resulting in increased energy intakes. As discussed above, impaired regulation of energy intake and an impaired brain reward response to food intake may contribute to increased food intake in certain individuals (Lin and Li 2021).

Factors affecting energy expenditure

Physical inactivity

The decline in physical activity has undoubtedly contributed to the rising prevalence of obesity, and low levels of physical activity are also an independent risk factor for the development of diabetes and cardiovascular disease. Attempts to increase physical activity are an important strategy in public health efforts to reduce obesity and diabetes. Recent WHO data suggest that 60% of the world's population does not meet the recommended physical activity guidelines. Physical inactivity is a significant independent risk factor for obesity and chronic disease.

Environmental factors have resulted in decreased activity among individuals in both developed and developing countries. Technological advances in computers, labour-saving devices for food or goods production, and online shopping have decreased the time spent in activity at home and in the workplace. Automobile travel has reduced the amount of walking time, even for short distances. Living in an unsafe area can prohibit walking to and from work or school as well as outdoor play and exercise. All these changes have contributed to a decrease in daily energy expenditure.

Sedentary behaviour

Television viewing is a major sedentary behaviour in our current lifestyle. Much of the research on sedentary time has been on screen time (i.e. use of computers, video games and other electronic media) or television viewing. Total screen time has jumped in recent years, adding to time spent in sedentary activity. Studies suggest that increased time spent in sedentary behaviour increases the risk of obesity, but the mechanism is still unknown. Research on behaviour while watching television suggests that energy consumption (e.g. snacking) during viewing is increased, resulting in positive energy imbalance and potentially weight gain. One postulate is that individuals may fail to recognise or register their satiety signals while watching television, thus resulting in higher levels of energy intake. Furthermore, some studies have shown that food-related television commercials result in increased total energy intake during television viewing.

A recent study suggested that sedentary time, physical activity and cardiorespiratory fitness should all be targeted to optimally address prevention of metabolic syndrome and type 2 diabetes (van der Velde et al. 2018).

Impaired regulation of energy expenditure

As discussed above (Section 17.3), in (subgroups of) individuals predisposed to obesity, disturbances have been reported in adaptive thermogenesis (non-exercise-induced thermogenesis, cold-induced thermogenesis and diet-induced thermogenesis) as well as in the regulation of substrate metabolism (metabolic inflexibility), which may have a more or less genetic basis. On the other hand, in a recent study using merged DLW data to measure TEE, no evidence was found that individuals with lower adjusted TEE are at increased risk of gaining body fat. The causes and consequences of variation in metabolic rate in humans remain an important research focus.

Circadian rhythms, stress, and sleep patterns

Our modern lifestyle is increasingly characterised by a poor sleep quality, stress and a decoupling of the sleep/wake cycle from the light/dark cycle (as also seen in shiftwork), with prolonged waking times disrupting circadian rhythmicity and the central and peripheral clocks. Interrupting the circadian system has physiological (metabolic stress) and psychological consequences (impaired mental health), which increase obesity, T2D and cardiometabolic disease risk, and reduce quality of life (Stenvers et al. 2019).

Epidemiological evidence is accumulating to suggest an association between short sleep duration and increased body weight. Cross-sectional studies in both children and adults suggest an association between sleep duration and obesity. Experimental and observational studies have suggested several pathways to explain the link between short sleep duration and weight gain. Mechanisms currently under investigation are those that influence hunger and appetite; alterations in key energy intake hormones such as leptin and ghrelin are thought to play a role in the association of obesity with sleep quality and duration. The relationship of obesity to reduced sleep duration, the quality of sleep and the mechanisms responsible for this association are currently an area of active research.

Evidence is mounting that stress, and particularly an increase of the glucocorticoid stress hormone cortisol, plays a role in the development of obesity. Chronic stress, a disturbed circadian rhythmicity, a reduced amount of sleep and a high energy-dense diet may all enhance cortisol production. This is most pronounced in individuals who have an increased glucocorticoid exposure or sensitivity. More insight into exact mechanisms may lead to more effective and individualised obesity treatment strategies (van der Valk et al. 2018).

17.6 Beyond BMI: obesity as an adiposity-based chronic disease

Although BMI has retained its value in epidemiology, this anthropometric surrogate measure of adiposity does not seem to be a good predictor of morbidity and mortality risk at an individual level (Goossens 2017). Research over the past decade has demonstrated that body fat distribution and adipose tissue function are key determinants of obesity-related complications.

Body fat distribution

In humans, we can distinguish subcutaneous and visceral white adipose tissue, with each of these depots being subdivided into specific fat depots. Subcutaneous adipose tissue is stored directly underneath the skin, and is mainly located in the abdominal (upper body) and gluteal-femoral (lower body) region. In contrast, visceral adipose tissue (also called intra-abdominal adipose tissue) refers to fat that is dispersed between the abdominal organs. Individuals with upper body obesity often have increased amounts of both abdominal subcutaneous adipose tissue and visceral adipose tissue. Waist circumference is correlated with the amount of visceral fat in the body and appears to be a good marker for upper body obesity. As early as the 1950s, an association between upper body fat obesity and metabolic complications such as diabetes and cardiovascular diseases was recognised. In the 1980s, numerous studies showed a significant relationship between upper body obesity, assessed by the ratio of waist to hip circumference, and risk factors for cardiovascular diseases and type 2 diabetes such as glucose intolerance, insulin resistance, hypertension and abnormal triglyceride and cholesterol levels.

The limitation of BMI in predicting risk of future complications is illustrated by studies demonstrating that mortality risk and the risk for myocardial infarction were positively associated with waist-to-hip ratio within each BMI category, with individuals with normal BMI but a high waist-to-hip ratio being at greater risk for these complications compared to people with obesity with a low waist-to-hip ratio (Bowman et al. 2017; Yusuf et al. 2005). These observations are indicative of the importance of body fat distribution, rather than BMI or total fat mass, in the development of cardiometabolic complications (Goossens 2017; Goossens et al. 2021; Karpe and Pinnick 2015). In agreement with these observations, differences in body fat distribution between men and women (i.e. women store relatively more adipose tissue in the lower part of the body, predominantly on the hips and thighs) seem to explain sexual dimorphism in the risk of developing cardiometabolic diseases (Goossens et al. 2021). Therefore, it is recommended to measure waist circumference in addition to BMI to identify people at risk of obesity-related complications. Waist circumferences greater than 102 cm in men and 88 cm in women are associated with an increased risk of cardiometabolic complications.

Adipose tissue function

Adipose tissue is a highly dynamic, metabolically active organ involved in a multitude of physiological processes. As such, the adipose organ exerts a fundamental role in both health and disease (Goossens 2017). The classic function of adipose tissue is to provide a long-term energy reserve, which can be used during food deprivation by mobilising fatty acids to be utilised by other organs. Other benefits of healthy adipose tissue include protection of delicate organs and certain regions of the body that are exposed to high levels of mechanical stress, thermal insulation properties and preservation of normal reproductive function.

Lipid accumulation in adipocytes and other cells in metabolic organs influences the normal function of these tissues. The functioning of adipose tissue seems to be more important than adipose tissue mass. This lack of a direct relationship between adipose tissue mass and metabolic health becomes evident in patients with (partial) lipodystrophy (i.e. complete or partial lack of subcutaneous adipose tissue), who are severely insulin resistant; the lack of change in insulin sensitivity following liposuction (i.e. surgical removal of subcutaneous adipose tissue); and increased insulin sensitivity following treatment with insulin-sensitising thiazolidinediones, despite increased subcutaneous adipose tissue mass (Goossens 2017).

Research in the past two decades has provided substantial evidence that adipose tissue dysfunction – characterised by adipocyte hypertrophy, mitochondrial dysfunction, inflammation and insulin resistance of adipocytes – contributes to dyslipidaemia, ectopic fat storage in skeletal muscle, liver, visceral adipose tissue, pancreas and heart, systemic low-grade inflammation and atherosclerotic plaque formation. Thus, epidemiological, clinical and translational studies over recent decades have provided strong evidence that body fat distribution and adipose tissue dysfunction play a pivotal role in the aetiology of many chronic cardiometabolic complications (Fuster et al. 2016; Goossens 2008, 2017; Goossens et al. 2021; Rosen and Spiegelman 2014). It is increasingly recognised that different adipose depots have distinct and sex-specific functions (Goossens et al. 2021).

17.7 Consequences of obesity

Physiological effects of excess fat

The obesity pandemic poses a major public health issue. Although obesity should be considered as a disease in its own right, it is also a key risk factor for many non-communicable diseases, including type 2 diabetes (T2DM), cardiovascular diseases (i.e. coronary heart disease, hypertension and stroke), non-alcoholic fatty liver disease (NAFLD), gall bladder disease, respiratory diseases (i.e. obstructive sleep apnea, COPD), several types of cancer and Alzheimer disease (Fruhbeck et al. 2013; Kopelman 2000). Obesity is also associated with osteoarthritis, gout and low back pain (thus limiting mobility and reducing independence in activities of daily living), reproductive problems and impaired mental health, together affecting quality of life. Furthermore, obesity has a significant impact on life expectancy (Prospective Studies Collaboration 2009).

Children and adolescents with obesity show elevations in cardiovascular disease risk factors such as cholesterol levels, blood glucose and blood pressure. These risk factors tend to cluster in individuals and track from childhood to adulthood. The incidence of type 2 diabetes, once thought to be an adult disease, is still increasing in children and adolescents. This is concerning given the increasing prevalence of childhood obesity in many countries and the complications associated with both obesity and type 2 diabetes.

Not all people with obesity will develop these diseases. The risk that a person with obesity will develop any of the health consequences associated with obesity may be influenced by many factors. The detrimental health consequences of excess body fat mass seem to be determined by (epi)genetics, body fat distribution, adipose tissue function, obesity history and lifestyle factors such as physical (in)activity, diet and behaviours such as drinking alcohol or smoking.

In addition to obesity acting as a gateway to many non-communicable diseases (NCDs) (Fruhbeck et al. 2013; Kopelman 2000), and people with obesity in general having an elevated risk for hospitalisation, serious illness and mortality, the rising prevalence of obesity has increased awareness of its relationship with communicable disease. Recent studies have demonstrated that obesity has a significant impact on communicable diseases. For example, evidence suggests that obesity is an independent risk factor for worse clinical outcomes in patients with COVID-19. Moreover, obesity-related complications are associated with clinical outcomes in patients with COVID-19. People with obesity were more likely to require acute care and admission to the intensive care unit, intubation and mechanical ventilation, especially among patients younger than 60 years (Docherty et al. 2020; Goossens et al. 2020; Popkin et al. 2020).

Psychological effects

The psychosocial effects of obesity on both children and adults can be substantial. People with obesity are often viewed as lazy or lacking the willpower to diet and exercise. This misunderstanding of the causes of obesity leads to negative attitudes towards individuals with obesity. Children and adolescents with obesity often experience discrimination and teasing from other children. Research has documented that people living with obesity are discriminated against in employment, housing, educational, earning and marital opportunities. There is considerable literature about the negative effects of stigma and the coping strategies that stigmatised individuals employ, which may include isolation, withdrawal, decreased achievement motivation, low self-esteem and depression.

17.8 Future perspectives

Body mass index is a simple, easy and inexpensive way of identifying obesity. However, it is well established that obesity defined by BMI alone is a heterogeneous condition with varying cardiometabolic complications across individuals, which is related to age, ethnicity and sex, amongst other factors. Therefore, BMI does not reflect the complexity of the disease. Both the degree of obesity and the associated health risk of individuals should be taken into consideration, implying a better evaluation of adipose tissue is warranted. The reshaping of obesity as an adiposity-based chronic disease (ABCD), which includes the total amount, distribution and function of adipose tissue, may help to improve the diagnostic criteria for obesity based on aetiology, degree of adiposity and health risks (Fruhbeck et al. 2019; Mechanick et al. 2017).

The study of obesity is a rapidly changing field, and research in the past decade has significantly contributed to a better understanding of disease pathophysiology and progression. The aetiology of obesity is complex and may be a result of genetic, metabolic, socioeconomic, behavioural and environmental factors. The obesogenic environment of our modern 24-hour society and associated poor sleep quality, stress and disruptions of circadian rhythmicity pose a major public health challenge. Future studies are needed to identify what factors increase an individual's susceptibility to obesity and obesity-related complications. Identifying the key biological and external factors that influence food intake and energy expenditure will help to inform interventions to prevent and treat obesity. Accumulating evidence indicates that early factors in growth and development affect the risk for obesity and related complications. The role of epigenetic factors in susceptibility to obesity later in life, and what factors during pregnancy, infancy and early childhood may impact the risk for obesity and obesity-related diseases through the life course needs further clarification. In addition, the mechanisms underlying interindividual differences in body fat distribution and related metabolic abnormalities are not fully understood. Further research in this area is warranted to obtain more knowledge on the mechanisms linking abdominal fatness and the associated cardiometabolic abnormalities. Detailed characterisation of patients is crucial to optimise strategies to prevent and treat obesity-related complications for different subgroups of patients, taking age, sex, body fat mass and distribution and the metabolic phenotype into account, amongst other factors. Many studies are currently ongoing to investigate the potential of personalised nutrition to optimise health outcomes.

Obesity is a complex, relapsing chronic disease that generates an increasingly heavy burden on individuals, healthcare systems and society at large. The global obesity epidemic in children and adults will worsen without evidence-based targeted interventions that address both the individual and society as a whole, including the environmental (socio-political-economic) context. Public health efforts are needed to promote a healthy lifestyle and to provide opportunities for all individuals across our society to adopt a healthy lifestyle. For example, efforts to increase the accessibility and affordability of fresh fruits and vegetables have been shown to improve health indicators. Using these foods to displace fast-food meals may have an important impact on energy intake and obesity. Taken together, social and physical environments that promote health and well-being, and reduce health inequalities, should be created. Clearly, a multidisciplinary, holistic approach is needed to prevent and treat obesity.

Further reading

Andreasen, C.H. et al. (2008). Low physical activity accentuates the effect of the FTO rs9939609 polymorphism on body fat accumulation. Diabetes 57: 95–101.

Berthoud, H.R. et al. (2017). Blaming the brain for obesity: integration of hedonic and homeostatic mechanisms. Gastroenterology 152: 1728–1738.

Blaak, E.E. et al. (2006). Fat oxidation before and after a high fat load in the obese insulin-resistant state. Journal of Clinical Endocrinology and Metabolism 91: 1462–1469.

Bowman, K. et al. (2017). Central adiposity and the overweight risk paradox in aging: follow-up of 130,473 UK Biobank participants. American Journal of Clinical Nutrition 106: 130–135.

Corpeleijn, E. et al. (2009). Metabolic flexibility in the development of insulin resistance and type 2 diabetes: effects of lifestyle. Obesity Reviews 10: 178–193.

Docherty, A.B. et al. (2020). Features of 20 133 UK patients in hospital with covid-19 using the ISARIC WHO Clinical Characterisation Protocol: prospective observational cohort study. BMJ 369: m1985.

Duffey, K.J. et al. (2007). Differential associations of fast food and restaurant food consumption with 3-y change in body mass index: the coronary artery risk development in young adults study. American Journal of Clinical Nutrition 85: 201–208.

Flatt, J.P. (1995). Body composition, respiratory quotient, and weight maintenance. American Journal of Clinical Nutrition 62: 1107S–1117S.

Fruhbeck, G. et al. (2013). Obesity: the gateway to ill health – an EASO position statement on a rising public health, clinical and scientific challenge in Europe. Obesity Facts 6: 117–120.

Fruhbeck, G. et al. (2019). The ABCD of obesity: an EASO position statement on a diagnostic term with clinical and scientific implications. Obesity Facts 12: 131–136.

Fuster, J.J. et al. (2016). Obesity-induced changes in adipose tissue microenvironment and their impact on cardiovascular disease. Circulation Research 118: 1786–1807.

Goossens, G.H. (2008). The role of adipose tissue dysfunction in the pathogenesis of obesity-related insulin resistance. Physiology and Behavior 94: 206–218.

Goossens, G.H. (2017). The metabolic phenotype in obesity: fat mass, body fat distribution, and adipose tissue function. Obesity Facts 10: 207–215.

Goossens, G.H. et al. (2020). Obesity and COVID-19: a perspective from the European Association for the Study of Obesity on immunological perturbations, therapeutic challenges, and opportunities in obesity. Obesity Facts 13: 439–452.

Goossens, G.H. et al. (2021). Sexual dimorphism in cardiometabolic health: the role of adipose tissue, muscle and liver. Nature Reviews Endocrinology 17: 47–66.

Hanssen, M.J. et al. (2016). Short-term cold acclimation recruits brown adipose tissue in obese humans. Diabetes 65: 1179–1189.

Karpe, F. and Pinnick, K.E. (2015). Biology of upper-body and lower-body adipose tissue – link to whole-body phenotypes. Nature Reviews Endocrinology 11: 90–100.

Kelley, D.E. (2005). Skeletal muscle fat oxidation: timing and flexibility are everything. Journal of Clinical Investigation 115: 1699–1702.

Kopelman, P.G. (2000). Obesity as a medical problem. Nature 404: 635–643.

Levine, J.A. et al. (1999). Role of nonexercise activity thermogenesis in resistance to fat gain in humans. Science 283: 212–214.

Lin, X. and Li, H. (2021). Obesity: epidemiology, pathophysiology, and therapeutics. Frontiers in Endocrinology 12: 706978.

Loos, R.J.F. and Yeo, G.S.H. (2022). The genetics of obesity: from discovery to biology. Nature Reviews Genetics 23: 120–133.

Mechanick, J.I. et al. (2017). Adiposity-based chronic disease as a new diagnostic term: the American Association of Clinical Endocrinologists and American College of Endocrinology position statement. Endocrine Practice 23: 372–378.

Miller, G.D. (2019). Appetite regulation: hormones, peptides, and neurotransmitters and their role in obesity. American Journal of Lifestyle Medicine 13: 586–601.

Mishra, A.K. et al. (2016). Obesity: an overview of possible role(s) of gut hormones, lipid sensing and gut microbiota. Metabolism 65: 48–65.

Popkin, B.M. et al. (2020). Individuals with obesity and COVID-19: a global perspective on the epidemiology and biological relationships. Obesity Reviews 21: e13128.

Prospective Studies Collaboration. et al. (2009). Body-mass index and cause-specific mortality in 900 000 adults: collaborative analyses of 57 prospective studies. Lancet 373: 1083–1096.

Rampersaud, E. et al. (2008). Physical activity and the association of common FTO gene variants with body mass index and obesity. Archives of Internal Medicine 168: 1791–1797.

Ravelli, M.N. and Schoeller, D.A. (2020). Traditional self-reported dietary instruments are prone to inaccuracies and new approaches are needed. Frontiers in Nutrition 7: 90.

Ravussin, E. et al. (1988). Reduced rate of energy expenditure as a risk factor for body-weight gain. New England Journal of Medicine 318: 467–472.

Rimbach, R. et al. (2022). Total energy expenditure is repeatable in adults but not associated with short-term changes in body composition. Nature Communications 13: 99.

Rosen, E.D. and Spiegelman, B.M. (2014). What we talk about when we talk about fat. Cell 156: 20–44.

Saari, T.J. et al. (2020). Basal and cold-induced fatty acid uptake of human brown adipose tissue is impaired in obesity. Scientific Reports 10: 14373.

Shamsi, F. et al. (2021). The evolving view of thermogenic adipocytes – ontogeny, niche and function. Nature Reviews Endocrinology 17: 726–744.

Stenvers, D.J. et al. (2019). Circadian clocks and insulin resistance. Nature Reviews Endocrinology 15: 75–89.

Stock, M.J. (1999). Gluttony and thermogenesis revisited. International Journal of Obesity and Related Metabolic Disorders 23: 1105–1117.

Stubbs, R.J. et al. (2004). Rate and extent of compensatory changes in energy intake and expenditure in response to altered exercise and diet composition in humans. American Journal of Physiology Regulatory Integrative and Comparative Physiology 286: R350–358.

Torres-Fuentes, C. et al. (2017). The microbiota-gut-brain axis in obesity. Lancet Gastroenterology and Hepatology 2: 747–756.

van der Valk, E.S. et al. (2018). Stress and obesity: are there more susceptible individuals? Current Obesity Reports 7: 193–203.

van der Velde, J. et al. (2018). Which is more important for cardiometabolic health: sedentary time, higher intensity physical activity or cardiorespiratory fitness? The Maastricht study. Diabetologia 61: 2561–2569.

van Marken Lichtenbelt, W.D. et al. (2009). Cold-activated brown adipose tissue in healthy men. New England Journal of Medicine 360: 1500–1508.

Villablanca, P.A. et al. (2015). Nonexercise activity thermogenesis in obesity management. Mayo Clinic Proceedings 90: 509–519.

WHO Expert Consultation. (2004). Appropriate body-mass index for Asian populations and its implications for policy and intervention strategies. Lancet 363: 157–163.

World Health Organization (WHO). (1995). Physical status: the use and interpretation of anthropometry. Report of a WHO Expert Consultation. WHO Technical Report Series Number 854. Geneva.

World Health Organization (WHO). (2000). Obesity: preventing and managing the global epidemic. Report on a WHO Consultation on Obesity, Geneva, 3–5 June, 1997. Technical Report Series Number 894. Geneva.

World Health Organization (WHO). (2021). Obesity and overweight. www.who.int/news-room/fact-sheets/detail/obesity-and-overweight.

Yengo, L. et al. (2018). Meta-analysis of genome-wide association studies for height and body mass index in approximately 700000 individuals of European ancestry. Human Molecular Genetics 27: 3641–3649.

Yusuf, S. et al. (2005). Obesity and the risk of myocardial infarction in 27,000 participants from 52 countries: a case-control study. Lancet 366: 1640–1649.

Zhang, Y. et al. (1994). Positional cloning of the mouse obese gene and its human homologue. Nature 372: 425–432.

18
Undernutrition

Phoebe Hodges and Paul Kelly

Key messages

- Undernutrition is usually a consequence of chronic energy deficiency (CED), which leads to reductive adaptation. This may lead to equilibrium at reduced body mass, but at a 'cost'. Acute energy deficiency (AED, also called wasting) is not an equilibrium state.
- Classification of CED is based on the extent of the reduction of body mass index and physical activity level.
- To maintain homeostasis, the body adapts in CED to lower energy intake to ensure survival. Individuals with CED have lower body weights, fat-free mass and fat stores, accompanied by reduced resting metabolic rate, physical activity and thermogenesis.
- The nervous and endocrine systems are key regulatory mechanisms that favour energy conservation in CED.
- In humans, the physiological or functional consequences of CED include reduced muscle strength and endurance, reduced immunity and altered autonomic nervous function. Each of these has important implications for lifestyle and health status.
- AED or CED may be caused by food insecurity (poverty, crop failure, conflict) or disease (for example, infectious disease, gastrointestinal or liver disease, cancer).

18.1 Introduction

Close to 800 million people in the world are believed to be undernourished. There are, however, large regional differences in the prevalence of undernutrition. For example, UNICEF, the WHO and the World Bank estimated in 2019 that the prevalence of stunting in children under five years of age was 30% in Africa, 23% in Asia and 9% in Latin America. This chapter will focus predominantly on adult undernutrition and its causes and consequences. Malnutrition in children is dealt with in the *Clinical Nutrition* textbook in this series.

18.2 Definition and classification of undernutrition

The term 'undernutrition' encompasses a wide range of macronutrient and micronutrient deficiencies. Micronutrient deficiencies, for instance, include those of water-soluble vitamins (thiamin, riboflavin and niacin), fat-soluble vitamins (e.g. vitamin A and vitamin D) and minerals (e.g. iodine and iron). Deficiencies of these micronutrients, if severe, lead to classic clinical presentations. Niacin deficiency, for instance, leads to pellagra, characterised by 'Casal's necklace', a typical skin lesion around the neck. Iodine deficiency is associated with a goitre, a swelling of the thyroid gland. Iron deficiency, which is widespread throughout the world and particularly in developing countries, is associated with anaemia. More subtle deficiencies of micronutrients may lead to ill health in more complex, as yet ill-understood, ways.

Detailed discussions of micronutrient deficiencies are dealt with in the relevant chapters dedicated to these nutrients in *Introduction to Human Nutrition*. This chapter deals primarily with energy deficiency.

Nutrition and Metabolism, Third Edition. Edited on behalf of The Nutrition Society by Helen M. Roche, Ian A. Macdonald, Annemie M.W.J. Schols and Susan A. Lanham-New.

Companion Website: www.wiley.com/go/nutrition/metabolism3e

In 1988, James, Ferro-Luzzi and Waterlow drew a distinction between acute energy deficiency (AED) and chronic energy deficiency (CED). AED was defined as 'a state of negative energy balance, i.e. a progressive loss of body energy'. This occurs in a variety of clinical conditions associated with acute weight loss, including anorexia nervosa, cancer and malabsorption disorders. In contrast, CED was defined as a 'steady state at which a person is in energy balance although at a "cost"'. Issues pertaining to AED are mentioned in Chapter 5.

Most definitions of undernutrition focus on reduced body mass, as this is a relatively easy thing to measure. However, it is likely that body mass is reduced only if the first processes of reductive adaptation still leave a negative energy balance. The idea of reductive adaptation is discussed further below, but initial reductions would include reduced energy expenditure through reduced physical activity and other behavioural adaptations. In sustained CED, physiological adaptations occur. Undernutrition thus has functional consequences as well as structural consequences, and it is likely that borderline undernutrition impairs work capacity, school performance and quality of life. Good nutrition is therefore required in order to achieve the vision of health set out by the World Health Organization: 'Health is a state of complete physical, mental and social well-being'.

With energy deficiency, there is a reduction in body weight and in energy stores (fat and fat-free mass [FFM]). A simple index of body energy stores is the body mass index (BMI): weight in kg/(height in m)^2. Thus, BMI could be used to identify individuals who are energy deficient. The problem is to define the BMI cut-off, as in reality there is a gradation in severity. Based on the range of BMIs in healthy populations and on studies identifying the functional consequences of a low BMI, the lower limit of a desirable BMI has been fixed at 18.5 kg/m^2 (James et al. 1988), and there is general agreement that a BMI below this indicates undernutrition. The importance of BMI as an index of nutrition is apparent from the strong effect it has on mortality, with both overnutrition and undernutrition associated with increased mortality (a U-shaped relationship) (Whitlock et al. 2009).

There is a level of undernutrition, as measured by low BMI, that is incompatible with life (Collins 1995). This point has yet to be defined, partly due to the extreme circumstances amongst which these people are likely to be found, which make research difficult. As adult energy requirements are less (per kg) than in children, in instances of famine child mortality peaks earlier than adult mortality, and Collins pointed out in a report of the 1992–93 famine in Somalia that many of the children will often have already died by the time emergency relief efforts are operational. In this report, he noted that 22% survivors aged 25 years of age or more and 49% survivors aged 15–24 who were admitted to the Concern Adult Therapeutic Centre in Baidoa had a BMI of lower than 12, which had previously been considered the lower limit compatible with life, and in fact the survival rate for younger adults admitted with a BMI lower than 11 was 82% (Collins 1995). In most circumstances, long-term survival from AED is unlikely if the BMI falls below 12 kg/m^2.

Individuals of different body sizes have different basal metabolic rates (BMR), and BMRs can now be estimated using prediction equations. During the course of a day, a sedentary individual expends a total amount of energy that is about 1.4 times the BMR. This multiplicative factor of BMR is called a physical activity level (PAL). The actual PAL of an individual can be computed as the daily energy intake divided by the BMR. Thus, individuals who have a PAL of <1.4 are clearly obtaining insufficient energy to support a sedentary lifestyle. Studies have shown that a PAL of less than 1.4 is only seen in individuals who spend less than 4–5 h a day on their feet, and reduced PAL is an important element of the reductive adaptation to CED. The identification of individuals with CED would be enhanced if we could combine the measurement of PAL with an additional, independent factor that reflects energy deficiency. Unfortunately, BMR is difficult to measure so assessment of PAL is of greater value to researchers than it is in clinical practice.

In situations where a large population needs to be surveyed, BMI alone may be used as a simple but objective anthropometric indicator of the nutritional status of the adult population. In situations where resources are very limited, such as in refugee camps in the midst of emergencies, mid upper arm circumference (MUAC) has been shown to be surprisingly effective as a tool for identifying people who are undernourished

(Collins 1996). MUAC, which is the simplest anthropometric index of all, can be applied to adults and children, obviously with different cut-offs, and is highly reproducible when carried out by trained observers, even those with no education. Another situation in which BMI becomes difficult to use is the severely ill hospitalised patient, who cannot stand up and therefore height cannot be measured. Although there may be some debate as to the correct cut-off to use, MUAC of 25 cm or less predicts poor outcome in hospitalised patients (Powell-Tuck and Hennessy 2003).

To summarise, in different situations, CED may be identified on the basis of:

- a combination of BMI and PAL
- BMI alone
- MUAC.

18.3 Adaptation and chronic energy deficiency

When there is a sudden change in the internal environment of cells and tissues, the body calls on regulatory processes to maintain the constancy of the internal environment. This maintenance of internal constancy is called *homeostasis*. When the change in the internal or external environment is for a longer period, the body may need to evoke greater changes, in both structure and function, in order to survive. Adaptation is thus the process by which an organism maintains physiological activity and survives when there is a sustained alteration in the environment with respect to one or more parameters.

In CED, adaptations include physiological and behavioural changes. These adaptive processes remain in place so long as the alteration in the environment persists. While this simplistic overview may suggest that adaptation is without any detrimental consequences, it does indeed have a 'cost'. This is most clearly seen when the undernourished individual increases their energy intake again, which leads to the 'refeeding syndrome' (see below).

- Adaptation is the response of the body to a sustained perturbation in the environment, with the aim of ensuring function and survival.
- Adaptation persists for as long as the perturbation is maintained.
- Every adaptation has a cost.

In CED, the sustained environmental perturbation is a lower energy intake and the body adapts to this situation in order to ensure the survival of the individual. Physiological and behavioural adaptations limit energy expenditure so that the individual remains in energy balance, albeit at a lower energy intake. In many situations, individuals with CED are involved in heavy physical activity (labourers, agricultural workers, etc.), despite their limited energy intakes. Their ability to continue working despite CED suggests that energy-saving mechanisms do allow people to function adequately and to survive. The potential mechanisms that aid in the conservation of energy in CED are discussed below.

18.4 Changes in body composition in chronic energy deficiency

Before proceeding further, it is worth getting a picture of what individuals with CED look like in terms of their body composition. This is illustrated in Table 18.1, which compares a typical male with CED with his well-nourished counterpart.

Table 18.1 indicates that individuals with CED have lower body weights, FFM and fat stores. In addition, they may be shorter, which implies that energy deprivation began during the years of linear growth and had a limiting effect on growth. These changes in body composition help to conserve energy in CED. For instance, lower body weight results in a lower energy cost of physical activity, particularly with weight-bearing activities. The reduction in metabolically active FFM ensures a lower BMR, in absolute terms. A reduction in

Table 18.1 Anthropometric characteristics of a man with chronic energy deficiency (CED) compared with a well-nourished man.

	CED	Well nourished
Weight (kg)	42	69
Height (cm)	160	175
Body mass index (kg/m^2)	**16.5**	**22.5**
Fat (%)	10	20
Fat mass (kg)	4	14
Fat-free mass (kg)	38	55

these two major components of daily energy expenditure results in a considerable energy saving for the individual who is chronically energy deficient. Later, we shall see that the benefits of changes in body composition in the individual with CED which favour energy conservation are offset by some derangements in normal physiology. In contrast to the energy-saving benefit of a lower body weight and a lower FFM, the reduction in fat stores, and the attendant loss in heat insulation, would be expected, if anything, to enhance energy expenditure during cold exposure. This is discussed in greater detail later in this chapter.

Loss of muscle mass and function, or sarcopenia, seems to be an inevitable part of the ageing process, although it is reversible to some extent with physical activity, and is often seen alongside undernutrition in the elderly. Sarcopenia can be defined using a number of tools including imaging techniques such as dual-energy X-ray absorptiometry (DEXA) to determine muscle quantity but muscle function is often conveniently assessed in practice using grip strength and gait speed, where gait speed of ≤0.8 m/s is an indicator of severe sarcopenia (European Working Group on Sarcopenia in Older People 2 2019).

A relatively recently recognised phenomenon is that of sarcopenic obesity (Baumgartner 2000) which is increasing both with an ageing population and as the prevalence of obesity rises globally. BMR declines with age, as does metabolic adaptation, including thermogenesis (see section 18.5), predisposing to loss of muscle mass and increased adipose tissue. Sarcopenic obesity represents a state of simultaneous over- and undernutrition as there is excess energy intake with an increase in adipose tissue but loss of skeletal muscle, and this warrants recognition as it seems to be more strongly associated with frailty and disability than either sarcopenia or obesity alone (Roh and Choi 2020).

Sarcopenia can also be a feature of undernutrition in other contexts, for example, in children with severe acute malnutrition (SAM) who have starvation-induced muscle wasting in addition to loss of subcutaneous fat, previously termed marasmus (Bhutta et al. 2017). SAM can be classified as two distinct forms: oedematous SAM, in which there is bilateral pitting oedema with skin and hair changes (previously called kwashiorkor), and non-oedematous SAM in which there is severe wasting as defined by weight-for-length z-score of <-3 (previously called marasmus) (Figure 18.1). The differences in body composition between these two forms of SAM have not yet been fully explained but differentiating between these two phenotypes is important given the divergence in clinical outcomes, with children with non-oedematous SAM having significantly higher mortality rates over a one-year period post discharge from hospital (Bwakura-Dangarembizi et al. 2021).

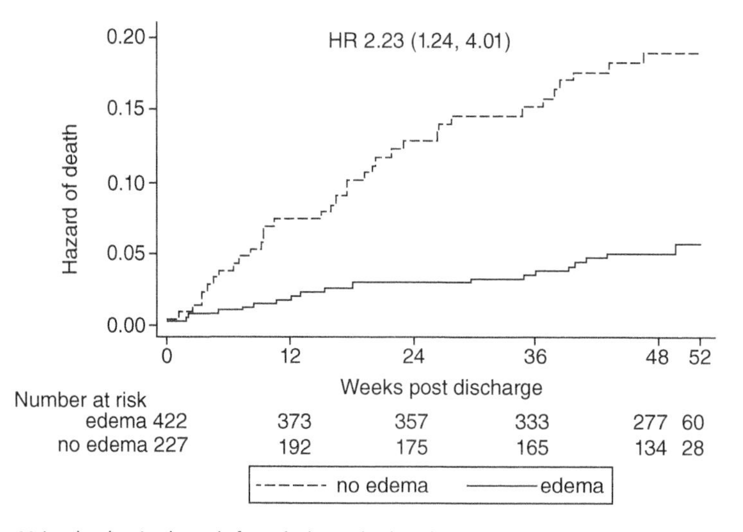

Figure 18.1 Kaplan–Meier plot showing hazard of postdischarge death in children who had been admitted to hospitals in Zambia and Zimbabwe with oedematous or non-oedematous SAM. *Source:* Bwakura-Dangarembizi et al. (2021)/Oxford University Press/CC BY 4.0.

18.5 Energy metabolism in chronic energy deficiency

As suggested above, many researchers have attempted to determine whether individuals with CED have energy-saving mechanisms that are independent of the loss in body weight and FFM. This inference could be made if the metabolic rate per unit of metabolically active tissue were lower than in well-nourished people. The most easily measurable form of metabolically active tissue is the FFM, though FFM includes multiple tissues and organs as disparate as bone, liver and skeletal muscle, which have very different metabolic profiles and are preserved or wasted to varying degrees in CED. Regarding FFM as a homogeneous compartment of body tissue is a misleading simplification, though it is usually all that can be measured rapidly and non-invasively and for that reason alone it remains a widely used term. Studies conducted so far have targeted most components of daily energy expenditure and these are discussed in greater detail below.

Basal metabolic rate/resting metabolic rate

Classic studies on acute energy deprivation by Benedict et al. and Grande indicated that semi-starvation was associated with a reduction in resting metabolic rate (RMR). This reduction was explained by both a reduction in body weight and a reduction in the metabolic activity of the tissues (metabolic efficiency). These findings have been replicated in more recent studies on obese individuals with restricted diets. The varying duration of energy restriction, however, makes comparisons between these studies difficult, particularly since there is some evidence that the reduction in RMR may occur in two phases. An initial phase of two or three weeks is associated with enhanced metabolic efficiency, while the further reduction in RMR with continued energy restriction beyond this period is largely accounted for by loss of active tissue mass.

Chronically energy-deficient people also have reduced BMRs in absolute terms. In early studies, right up to the 1980s, 'metabolic efficiency' was essentially determined by dividing BMR by the active tissue mass (usually FFM). Some, but not all, studies showed a decrease in BMR per kg of FFM. In recent years, there have been detailed discussions on whether this computation of metabolic efficiency is appropriate. While the relationship between BMR and FFM is linear, there is a positive intercept, which means that smaller individuals have a higher BMR/kg FFM. This higher BMR/kg FFM may, in part, be related to the fact that the proportion of muscle mass lost from FFM is substantially more than the proportion of viscera. In undernutrition, it would make sense to think in terms of three or even four compartments (fat, bone, viscera, muscle and other connective tissue) but these are not yet easily measured and there is no consensus on this. Resting skeletal muscle mass is metabolically less active than viscera. Thus, the relatively higher viscera to muscle mass ratio in CED would be associated with a higher BMR/kg FFM.

Thus, individuals with CED have a lower BMR, largely due to a lower body weight and a reduced active tissue mass. There is some evidence of increased metabolic efficiency in CED in terms of a reduced BMR when adjusted for FFM.

Physical activity

One of the ways in which an individual with CED could conserve energy is by being less physically active (see discussion of PAL above). This is, in fact, what has been observed in laboratory experiments involving semi-starvation for a long duration. It is more difficult to establish whether individuals with CED reduce their physical activity under free-living conditions. Earlier studies used the laborious method of time-and-motion analysis. In this method, trained observers accompany participants and note their activities in a diary for the entire duration of a day or several days. The development of the doubly labelled water technique (see Chapters 2 and 6 of *Introduction to Human Nutrition*) has allowed for the estimation of free-living energy expenditure and, consequently, also of PALs.

There are some data to suggest that individuals of low BMI can sustain high levels of physical activity at least over the periods of measurement. There is a possibility, however, particularly in largely agrarian societies, that those with CED may show large seasonal variations in physical activity. One very clear example of the reduced physical activity is the example of human immunodeficiency virus (HIV) infection. Landmark studies in HIV have established that weight loss in most HIV-infected adults is due to anorexia, and that this is a major determinant of outcome in HIV. During periods of weight loss, physical activity decreases (Macallan et al. 1995).

Another way in which those with CED may conserve energy is by an enhanced 'efficiency'. There are many ways in which efficiency is defined in exercise physiology. One of the more common terms is 'mechanical efficiency' which is expressed as a percentage and is the ratio of external work performed to the energy expended. The mechanical efficiency of performing activities such as climbing stairs and cycling is in the region of 20–25%. The rest of the energy expended is dissipated as heat. Several laboratory studies in CED individuals have assessed mechanical efficiency during stepping, treadmill walking and cycling. When viewed together, there is no compelling evidence, yet, to support the notion that those with CED have a greater mechanical efficiency.

It is also conceivable that CED individuals may minimise the energy cost of physical activity by performing tasks in a way that reduces unnecessary movements or results in a better balance of loads ('ergonomic efficiency'). In Africa, for instance, women of the Luo and Kikuyu tribes can carry substantial loads with great economy. The possibility of an enhanced ergonomic efficiency in those with CED needs to be investigated further.

Following strenuous, sustained physical exercise, energy expenditure may be elevated for up to 24 h. This enhanced energy expenditure is sometimes called the excess postexercise oxygen consumption (EPOC). There is some evidence that EPOC is reduced in CED individuals, both in the immediate period following exercise and in the delayed period of about 12 h after exercise.

To summarise, those with CED may potentially conserve energy in relation to physical activity by the following means:

- decreased physical activity
- increased mechanical efficiency
- enhanced ergonomic efficiency
- reduced EPOC.

A more detailed account of physical activity is provided in Chapter 19.

Thermogenesis

The thermic effect of food (TEF) comprises approximately 10% of daily energy expenditure and is defined as the rise in energy expenditure following the consumption of food. There are many factors that determine TEF, including the size and composition of the meal, palatability of the food and antecedent dietary intake. Because there is a very limited amount of data on TEF in CED, it is difficult to comment on whether TEF is reduced in this state. There is some evidence that the TEF/kg FFM in those with CED following a mixed meal is higher than in their well-nourished counterparts. Intercountry comparisons have demonstrated that Gambian men (not necessarily of low BMI) in the 'hungry' season have a lower TEF than Europeans.

On exposure to cold, individuals rely on both heat-conserving and heat-generating mechanisms to maintain body temperature. The heat-conserving mechanisms include the insulation provided by subcutaneous fat and vasoconstriction of peripheral blood vessels, particularly in skin and skeletal muscle. Vasoconstriction limits the amount of warm blood that is brought to the surface of the body. Thus, less heat is lost from the body.

Heat-generating mechanisms include non-shivering thermogenesis (NST) and shivering. NST is largely mediated by the sympathetic nervous system, which enhances a whole host of metabolic processes, including glycogenolysis in liver and muscle, gluconeogenesis and lipolysis. NST has been studied under laboratory conditions by infusing either epinephrine (adrenaline) or norepinephrine (noradrenaline) intravenously into volunteers. These studies suggest that thermogenesis to exogenously administered catecholamines may be lower in people with CED. On exposure to mild cold, CED individuals are able to thermoregulate adequately. Greater vasoconstriction of the peripheral blood vessels and an earlier onset of NST compensate for the lack of fat insulation. With more intense cold, however, undernourished people show reduced thermogenesis and are unable to thermoregulate adequately. Any energy conservation is therefore offset by their increased susceptibility to hypothermia in response to cold exposure. There is speculation that the high number of deaths during cold waves in developing countries may be related to this factor, and this is almost certainly true for elderly people in winter in temperate climates.

Protein metabolism in relation to energy expenditure

There have been a few studies in CED exploring the relationship between protein metabolism and energy expenditure. The impetus for these investigations is the knowledge that protein synthesis contributes substantially to the BMR. Thus, a reduction in protein synthetic rates could conceivably reduce BMR and result in energy conservation. The studies demonstrate that while BMR and protein synthesis are both lower than in well-nourished individuals in absolute terms, there are no differences when protein turnover (synthesis and breakdown) is expressed per kilogram of FFM.

Energy expenditure during pregnancy

Pregnancy and lactation place increased energy demands on the mother. This is a situation where people with CED are particularly vulnerable. Studies carried out on Gambian women have indicated that BMR is depressed during the first 18 weeks of gestation and that the total metabolic costs of pregnancy are far lower than in well-nourished women in Western populations. These results imply some energy-saving mechanisms. In this particular study, however, the vast majority of women had BMIs in excess of 18.5 kg/m^2. In contrast to the findings in the Gambia, studies on well-nourished pregnant Indian women, as well as Chinese and Malay women in Asia, failed to identify any conservation of energy during pregnancy.

One of the consequences of a low BMI in pregnant women is the birth of babies of low birth weight. This is of importance because low birth weight is associated with increased infant mortality. In addition, low birthweight babies may be more susceptible to chronic disease if they survive into adulthood. The metabolism of pregnancy and lactation is dealt with in greater detail in Chapter 6.

18.6 Regulatory processes in chronic energy deficiency

The two major regulatory systems in the body are the nervous system and the endocrine (hormonal) system, and both can modulate metabolic rate.

One of the components of the nervous system that has been studied fairly extensively in relation to metabolic activity is the sympathetic nervous system (SNS). Classic studies by Young and Landsberg demonstrated a reduction in SNS activity in animal models with underfeeding and an enhanced SNS activity with overfeeding. Pharmacological studies demonstrated that virtually all components of daily energy expenditure, including BMR, were reduced when SNS activity was blocked. Thus, a hypothesis has emerged that attempts to explain a putative reduction in metabolic activity in CED on the basis of diminished SNS activity. Short-term underfeeding experiments in humans support the hypothesis of a reduced SNS activity with reduced energy intakes. In CED too, there is evidence of reduced SNS activity. There have, however, been no studies measuring metabolic rates in CED in the presence of SNS blockade.

Among the hormones, one that particularly stands out in terms of metabolic regulation is thyroid hormone (thyroxine, T_4 and tri-iodothyronine, T_3). Indeed, there was a time before thyroxine levels could be measured in blood when hypothyroidism was diagnosed on the basis of a reduced RMR. Thyroid hormone exists in two forms, T_4 and T_3. Of these, active T_3 is biologically more active and is derived from T_4. Several studies have demonstrated that active T_3 is reduced with starvation and acute underfeeding experiments, and that this is, in part, due to a reduced conversion of T_4 to active T_3. Alteration in thyroid hormone status has been incompletely characterised in CED.

Several other hormones, such as glucagon, glucocorticoids, growth hormone, insulin-like growth factors and progesterone, could potentially modulate metabolic rate. The role of these hormones in CED has not been studied adequately.

18.7 Functional consequences of energy deficiency

This section explores some of the consequences of energy deficiency in terms of the cost of adaptation which was referred to above. Table 18.2 summarises this in terms of functional changes and real-life consequences.

Muscle function

Several laboratory studies have demonstrated that CED is associated with a reduction in muscle strength as well as muscle endurance. The reduction in muscle strength is to a large extent associated with a reduction in muscle mass.

Table 18.2 Real-life consequences of physiological functional changes that occur in chronic energy deficiency (CED).

Functional changes in CED	Real-life consequences
Reduced muscle strength and endurance	Decreased work performance and earning capacity
Reduced immunity	More infections, increased sickness
Altered autonomic nervous function	Possibly altered drug dosage requirements
	Consequences with ageing not known

Some studies have, however, indicated that the loss in strength persists even after correcting for muscle mass. In other words, muscle strength per unit muscle area or mass is less than in well-nourished people. This suggests that there are functional changes in skeletal muscle with CED, and these may underlie the reduced grip strength observed in undernutrition.

Skeletal muscle is made up of two large categories of muscle fibre types. Type I fibres are increased in marathon runners and are important for endurance. Type II muscle fibres are increased in weightlifters and are a determinant of muscle strength. In undernutrition, there is a decline in type II fibres and this may account for the decline in muscle strength. There are also reports of a conversion of type II fibres to type I fibres. This has a benefit in energetic terms because the type I fibres use less adenosine triphosphate (ATP) during muscle contraction than type II fibres. In addition to the changes in muscle fibre type, undernutrition is associated with a reduction in sources of energy within skeletal muscle, such as glycogen, ATP and phosphocreatinine. A reduction in these energy sources could impair muscle function.

In real life, the consequence of diminished skeletal muscle function in CED is reduced work performance and physical work capacity. Thus, for instance, sugarcane cutters in Colombia who are taller and heavier and have higher FFMs are more productive.

Immune function

There are good grounds for thinking that AED and CED are associated with diminished immune function, but actually the evidence is far from clear-cut. Immune function is of three types: cellular, humoral and innate. Cellular immunity, which is mediated by T-lymphocytes, is involved predominantly in defence against viruses, fungi and protozoa. Antibodies produced by B-lymphocytes mediate humoral immunity to bacteria and some viruses. Children with malnutrition seem on the whole to mount preferentially a Th2 response to infection, focused on humoral immunity, with high levels of anti-inflammatory cytokines IL-4 and IL-10 and reduced levels of IL-2, IL-12 and IFN-γ. Recent evidence suggests that innate immunity, notably dendritic cells, may be an important contributor to immune dysfunction in undernourished children (Hughes et al. 2009).

The difficulty with understanding the link between undernutrition and infection is that infection itself can produce some of the immune defects which we often attribute to malnutrition, so establishing causality is not easy. The strongest evidence suggests the occurrence of thymolymphatic atrophy in severe undernutrition, and overall the evidence suggests that T-cell responses are impaired. Delayed-type hypersensitivity response also seems to be affected in malnourished children, with reduced reactivity of lymphocytes to phytohaemagglutinin. There is a reduced antibody response to vaccination in children with severe malnutrition, although not in mild-to-moderate malnutrition (Rytter et al. 2014).

Curiously, in anorexia nervosa, which is the purest form of undernutrition seen in current nutrition practice in the UK, there is little if any increase in susceptibility to infection (Golla et al. 1981). In addition, mortality increases sharply in those individuals who are of low BMI, especially below a BMI of 16 kg/m^2. There are certain elements of the immune response which seem to be unaffected by malnutrition, including total white blood cell counts, T-lymphocyte and CD4 cell counts, and the ability to mount an acute-phase response to infection (Rytter et al. 2014).

Gut structure and function

Physiological adaptations in AED, and possibly CED, include intestinal changes. There is intestinal atrophy, with postmortem reports from Turkish hunger strikers finding 'transparent intestines', similar to accounts relating to victims of starvation in concentration camps (Kelly 2021). In children with malnutrition, thin-walled intestine has also been described, with a classic appearance of 'tissue-paper intestine of kwashiorkor' from postmortem studies in the 1960s (Burman 1965). Cellular turnover rate and proliferation in the small intestine are reduced in protein malnutrition with upregulation of cell signalling pathways relating to cell survival (Bolick et al. 2014).

In SAM in children, there is remodelling of the small intestinal epithelium with villous morphological change to ridges and convolutions with absence of 'finger-like' villi, in addition to villous blunting. The intestinal adaptation in SAM, which is likely to result from an acute-on-chronic energy deficit usually in the context of exposure to multiple enteropathogens, is complex, with hypertrophic features in addition to the villous blunting and reduced epithelial surface area. The hypertrophic element is likely to be the result of chronic inflammation secondary to pathogen-mediated epithelial damage over a longer period of time in the insanitary environment in which children with SAM are almost universally found. Although the reduced epithelial surface area (environmental enteropathy) reduces microbial translocation in these children, it does so at the cost of impaired growth (Amadi et al. 2021).

Although not confirmed, it is also possible that reduced gastric acid secretion is another consequence of CED.

Autonomic nervous function

Earlier in this chapter, we discussed how changes in sympathetic nervous function could help in the conservation of energy in CED. In CED, there is evidence that sympathetic nervous activity is reduced (Shetty and Kurpad 1990). There is also some evidence that parasympathetic nervous activity (the other arm of the autonomic nervous system) is increased.

Nerves act by releasing chemical messengers called neurotransmitters. These neurotransmitters in turn act on certain proteins on cell membranes called receptors, which then produce a cellular response. When the level of a neurotransmitter is high, the number of receptors reduces to maintain a constant cellular response. Conversely, when the neurotransmitter level is low, the number of receptors increases.

Many drugs used in clinical practice act on the receptors of the autonomic nervous system. Because autonomic nervous activity (both sympathetic and parasympathetic) is altered in CED, the receptor number is also altered. If drugs are administered that act on these receptors, the response to the drug is likely to be higher or lower than normal, depending on whether the receptor number is increased or decreased. Thus, the required dose of drugs acting on the autonomic nervous system may be altered in CED.

Refeeding syndrome

When prisoners in concentration camps were liberated at the end of World War II, Allied troops were horrified at the extreme malnutrition they encountered and understandably encouraged these starving individuals to take generous amounts of food. Unfortunately, many prisoners died on refeeding and this has become known as the refeeding syndrome.

The syndrome is thought to result from starvation-induced reduction in insulin levels and increase in glucagon levels, combined with proteolysis and the loss (permanently, in the urine) of bulk minerals such as phosphate, potassium and magnesium in the catabolic state. When nutritional intake recommences, the sudden increase in glucose concentration and resulting insulin secretion lead to massive shifts in electrolyte concentrations, including increased cellular uptake of phosphate, in addition to sodium and water retention resulting in oedema and heart failure. This usually occurs within 72h of refeeding.

Refeeding syndrome is characterised by severe hypophosphataemia, which also results in impaired ATP production and consequently rhabdomyolysis and muscle weakness. Various elements of refeeding syndrome are probably a consequence of reduced substrate cycling during AED and severe CED. Other characteristic features of the refeeding syndrome include hypomagnesaemia, hypokalaemia, cardiac arrhythmias and seizures (Friedli et al. 2017). This cost of reductive adaptation has major implications for the treatment of severe undernutrition in both children and adults (NICE 2006). Generally, we assume that the greatest risk of refeeding syndrome is over when appetite returns (severe undernutrition is usually accompanied by marked anorexia) but the validity of this belief is not absolutely clear and further work is needed.

Summary

In summary, therefore, the conservation of energy in the individual with CED is largely due to a reduction in body size and in metabolically active tissue. Metabolic efficiency in CED contributes to energy saving in a rather small way. The adaptation of the person with CED to lowered energy intakes is not, however, without its costs, as highlighted in the preceding section.

18.8 Perspectives on the future

This chapter has highlighted some issues in CED for which we have an incomplete understanding. This section raises some of the issues that healthcare professionals will have to deal with in the future. Much of what follows is speculative, and the reader is encouraged to pick up the leads and delve further into the issues that have been raised.

Population changes and inequalities in nutrition

One adverse trend we are seeing in global health as the twenty-first century unfolds is the widening of inequalities in wealth, health and access to healthcare. While a large proportion of the world's population is undernourished, there is a large and growing proportion which is obese. As populations in developing countries abandon their traditional diets and consume foods which contain more meat (this is a rapid trend in China) and are more energy dense, the number of people with the 'metabolic syndrome' is increasing rapidly. These two trends may be taking place in parallel simultaneously in the same country (e.g. India).

Earlier in the chapter, a typical individual with CED with a BMI of 16.5 kg/m^2 was described. If this same individual were to gain weight to achieve a BMI of 25, his body weight at this BMI would be 64 kg, a gain of almost 22 kg. A substantial portion of this is likely to be fat, as fat deposition is often the first response to weight gain after AED. Individuals with a BMI near 25 kg/m^2 are conventionally considered to be at the upper end of the normal BMI range. How should this individual with CED be evaluated? Should he be considered at higher risk of developing heart disease and diabetes because of his weight and fat gain, despite his normal BMI? What are the consequences for HIV-infected people who have been severely wasted and who regain weight (even to the point of obesity) thanks to antiretroviral drugs and lipid-rich diets? While rescue therapy for severe undernutrition has been revolutionised thanks to the introduction of lipid-based ready-to-use food (RTUF), does this run the risk of metabolic syndrome after a certain threshold of weight has been gained?

These questions are as yet unanswered, but there several studies of survivors of famines during the twentieth century (for example, the Dutch winter of 1944, the siege of Leningrad, or the Chinese famine of 1954–1964) to determine whether they have higher risk of diabetes and cardiovascular disease in later life. No consensus is yet available.

Small babies now, larger problems later?

As mentioned earlier, low birth weight babies are seen more often in mothers with a low BMI. These babies also have a higher mortality. What happens to babies of low birth weight who survive into adulthood? Barker has shown that low birth weight babies in the UK had a higher prevalence of high blood pressure, diabetes and coronary heart disease when they grew up to be adults (Barker 1993). The mechanism by which this happens is not fully understood. The increased prevalence of diabetes has been linked to poor fetal growth, resulting in a reduced number of pancreatic β-cells, which produce insulin. Will developing countries where low birth weights are fairly common have to face the increased burden of diabetes, hypertension and so on, as life spans increase? How can we dissect out the contributions of poor fetal growth from environmental factors that are operative after birth? Will continued CED into adulthood be protective against these diseases?

The intriguing possibility of fetal undernutrition affecting health in adulthood remains to be elucidated, particularly within the context of CED. For much of the world's population, this is already a question which urgently needs an answer.

References

Amadi, B., Zyambo, K., Chandwe, K. et al. (2021). Adaptation of the small intestine to microbial enteropathogens in Zambian children with stunting. Nature Microbiology 6: 445–454.

Barker, D.J.P. (ed.) (1993). Fetal and Infant Origins of Adult Disease. London: BMJ Publishing Group.

Baumgartner, R.N. (2000). Body composition in healthy aging. Annals of the New York Academy of Sciences 904: 437–448.

Bhutta, Z.A., Berkley, J.A., Bandsma, R.H.J., Kerac, M., Trehan, I. and Briend, A. (2017). Severe childhood malnutrition. Nature Reviews Disease Primers 3: 17067.

Bolick, D.T., Chen, T., Alves, L.A.O. et al. (2014). Intestinal cell kinase is a novel participant in intestinal cell signalling responses to protein malnutrition. PLoS One 9: e106902.

Burman, D. (1965). The jejunal mucosa in kwashiorkor. Archives of Disease of Childhood 40: 526–531.

Bwakura-Dangarembizi, M., Dumbura, C., Amadi, B. et al. (2021). Risk factors for post-discharge mortality following hospitalization for severe acute malnutrition in Zimbabwe and Zambia. American Journal of Clinical Nutrition 113: 665–674.

Collins, S. (1995). The limit of human adaptation to starvation. Nature Medicine 1: 810–814.

Collins, S. (1996). Using middle upper arm circumference to assess severe adult malnutrition during famine. JAMA 276: 391–395.

European Working Group on Sarcopenia in Older People 2. (2019). Sarcopenia: revised European consensus on definition and diagnosis. Age and Ageing 48: 16–31.

Friedli, N., Stanga, Z., Sobotka, L. et al.(2017). Revisiting the refeeding syndrome: results of a systematic review. Nutrition 35: 151–160.

Golla, J.A., Larson, L.A., Anderson, C.F., Lucas, A.R., Wilson, W.R. and Tomasi, T.B. (1981). An immunological assessment of patients with anorexia nervosa. American Journal of Clinical Nutrition 34: 2756–2762.

Hughes, S., Amadi, B., Mwiya, M. et al. (2009). Dendritic cell anergy results from endotoxemia in severe malnutrition. Journal of Immunology 183: 2818–2826.

James, W.P.T., Ferro-Luzzi, A. and Waterlow, J.C. (1988). Definition of chronic energy deficiency in adults. Report of a working party of the international dietary energy consultative group. European Journal of Clinical Nutrition 42: 969–981.

Kelly, P. (2021). Starvation and its effects on the gut. Advances in Nutrition 12: 897–903.

Macallan, D.C., Noble, C., Baldwin, C. et al. (1995). Energy expenditure and wasting in HIV infection. New England Journal of Medicine 333: 83–88.

National Institute for Clinical Excellence. (2006). Nutrition Support in Adults: Oral Nutrition Support, Enteral Tube Feeding and Parenteral Nutrition (Clinical Guideline CG32). London: NICE.

Powell-Tuck, J. and Hennessy, E.M. (2003). A comparison of mid upper arm circumference, body mass index and weight loss as indices of undernutrition in acutely hospitalized patients. Clinical Nutrition 22: 307–312.

Roh, E. and Choi, K.M. (2020). Health consequences of sarcopenic obesity: a narrative review. Frontiers in Endocrinology 11: 332.

Rytter, M.J.H., Kolte, L., Briend, A., Friis, H. and Christensen, V.B. (2014). The immune system in children with malnutrition: a systematic review. PLoS One 9: e105017.

Shetty, P.S. and Kurpad, A.V. (1990). Role of the sympathetic nervous system in adaptation to seasonal energy deficiency. European Journal of Clinical Nutrition 44 (Suppl 1): 47–53.

Whitlock, G., Lewington, S., Sherliker, P. et al. (2009). Body mass index and cause-specific mortality in 900,000 adults: collaborative analyses of 57 prospective studies. Lancet 373: 1083–1096.

Further reading

Della Bianca, P., Jequier, E. and Schutz, Y. (1994). Lack of metabolic and behavioural adaptation in rural Gambian men with low body mass index. American Journal of Clinical Nutrition 60: 37–42.

Norgan, N.G. and Ferro-Luzzi, A. (1996). Human adaptation to energy under-nutrition. In: Handbook of Physiology Section 4. Environmental Physiology, Vol. II (eds M.J. Fregly and C.M. Blatteis), 1391–1409. New York: Oxford University Press for the American Physiological Society.

Soares, M.J., Piers, L., Shetty, P. et al. (1994). Whole body protein turnover in chronically undernourished individuals. Clinical Science 86: 441–446.

Soares, M.J.S. and Shetty, P.S. (1991). Basal metabolic rates and metabolic economy in chronic undernutrition. European Journal of Clinical Nutrition 45: 363–373.

Waterlow, J.C. (1986). Metabolic adaptation of low intakes of energy and protein. Annual Review of Nutrition 6: 495–526.

Waterlow, J.C. (1992). Protein Energy Malnutrition. London: Edward Arnold.

19 Exercise Performance

Asker E. Jeukendrup and Louise M. Burke

Key messages

- The most important component of energy expenditure for athletes is the energy cost of their training or competition workloads. Energy expenditure can amount to 36 MJ/day for endurance athletes undertaking the most strenuous exercise schedules (such as competing in the Tour de France).
- Carbohydrate plays a crucial role in the diet of most athletes as it restores muscle and liver glycogen to provide a key fuel for exercise. Carbohydrate intake in the everyday diet should be individualised for each athlete to meet their specific fuel requirements for training and recovery, with daily intakes ranging from 3 to 12 g/kg being associated with different types of athletes and their specific daily training needs. Training with high carbohydrate availability will help athletes to train harder, especially during 'quality' workouts involving high-intensity exercise. Carbohydrate loading to increase muscle glycogen stores has been shown to improve endurance capacity and performance in prolonged events involving sustained or intermittent exercise >90 minutes' duration.
- Undertaking exercise with low carbohydrate availability (e.g. fasted training, training with low glycogen content) impairs training intensity/quality but may amplify adaptations to endurance training. Athletes are encouraged to periodise such practices into their training programme.
- Carbohydrate feeding in the hours before the start of prolonged exercise can enhance performance. Carbohydrate ingested during prolonged exercise (longer than ~60–90 minutes) improves exercise performance by maintaining blood glucose concentration and high rates of carbohydrate oxidation, spares liver glycogen and may in some conditions spare muscle glycogen. Carbohydrate intake also benefits the brain and central nervous system, and, in shorter sustained events (~60 minutes), even rinsing the mouth with a carbohydrate drink may be enough to stimulate the brain to 'feel better' and work harder during exercise.
- Muscle glycogen stores can be increased effectively by eating a high-carbohydrate meal or snack providing at least 1 g/kg carbohydrate within 15–30 minutes of finishing the exercise session. This is important when the schedule calls for rapid recovery between prolonged training sessions (less than eight hours' recovery).
- Ketogenic and non-ketogenic versions of a low-carbohydrate, high-fat diet achieve significant increases in muscle use of fat as an exercise fuel. However, tolerance of these diets is individual, and there is no clear evidence that these strategies lead to better competition performance. In fact, fat oxidation is less economical as a fuel source and can impair performance of higher-intensity endurance events in high-performance athletes.
- Consuming a small amount of high-quality protein (20–25 g) soon after the end of a strenuous workout or competition enhances adaptation or recovery from the session. Although it is suggested that both endurance and strength athletes may have increased protein requirements (~1.2–1.8 g/kg body mass per day), these targets are typically met by the high-energy intakes of most athletes. Nevertheless, higher intakes of protein may assist with the maintenance of muscle mass during weight loss, and the timing of intake of protein after training and over the day may also help to support the effectiveness of protein intake.
- Low body fat and low body weight are important in many sports, especially at the elite level. If loss of body fat is required, a realistic rate loss of about 0.5 kg/week should be chosen and both short-term and long-term goals should be set.
- Low iron status is the most common micronutrient issue in athletes and should be monitored in high-risk groups such as endurance athletes and females, or in specialised scenarios such as preparation for altitude training. Iron supplementation may be required to prevent or treat diag-

Nutrition and Metabolism, Third Edition. Edited on behalf of The Nutrition Society by Helen M. Roche, Ian A. Macdonald, Annemie M.W.J. Schols and Susan A. Lanham-New.

Companion Website: www.wiley.com/go/nutrition/metabolism3e

nosed cases of iron deficiency or to support increased iron requirements during specialised periods such as altitude training, but iron-rich eating should be the preferred approach to long-term nutritional status.
- Dehydration can impair performance of endurance, intermittent high-intensity and skill-based sports, especially when they are undertaken in hot/humid environments. Effects include an increased perception of effort and impaired cardiovascular and thermoregulatory responses to exercise. Severe dehydration may be associated with heat injury, including heat stroke.
- Fluid intake may be useful during exercise longer than 30–60 minutes to reduce the fluid deficit associated with sweating. The athlete should develop a fluid intake plan that balances multiple issues such as likely sweat losses and accruing (or pre-existing) fluid deficit, opportunities to consume fluids within the session, gastrointestinal comfort associated with fluid intake, and any advantages of consuming other nutrients (e.g. carbohydrate or caffeine) within their exercise fluids. A reasonable goal is to keep the overall fluid deficit to ~2–3% body mass (2% in hot conditions) or less, but athletes are also warned against drinking too much (increasing body mass over the course of the exercise session due to fluid intake in excess of sweat rates). The role of electrolyte ingestion during exercise is often exaggerated and additional sodium intake during exercise may only be necessary during very prolonged exercise when a large percentage (>70%) of fluid lost is replaced during exercise.
- Many exercise scenarios will leave the athlete with a fluid deficit. As ongoing sweat and urine losses continue in the post-exercise period, a fluid intake equal to 125–150% of the volume of the post-exercise fluid deficit may be required over the next six hours to re-equilibrate body fluid stores.
- When this is significant, rehydration is achieved more rapidly if salt (sodium chloride) is added to rehydration drinks or if salt-rich foods are consumed with this fluid. This allows plasma volume and osmolality to be restored in tandem, and maximises the retention of this fluid (i.e. minimises urine production). The optimal sodium concentration is found in oral rehydration solutions (~50–60 mmol/l) and is similar to the sodium concentration found in sweat, but it is considerably higher than the sodium concentration of many commercially available sports drinks, which usually contain 10–25 mmol/l (60–150 mg/l).
- The dietary supplement industry is worth many billions of dollars annually. Although there is little or no evidence to support the use of the vast majority of performance supplements, there are a few that enjoy a sound evidence base. These include caffeine, creatine, bicarbonate and beta-alanine buffers, and beetroot juice/nitrate. Medical supplements (e.g. vitamin D, calcium, iron) and sports foods (sports drinks, protein supplements, liquid meal supplements) may also be part of an athlete's sports nutrition plan. More research is needed around the sports-specific context of supplement use, including the combined use of several evidence-based products or repeated use across multiple events.
- Athletes should ensure that any supplements they choose to use have been sourced from reputable companies and, where possible, be screened by third-party companies for the absence of substances that are prohibited in sport. A number of audits have shown that a significant number of commercial supplements are contaminated with undeclared contaminants such as banned stimulants, prohormones or experimental drugs.

19.1 Introduction

The relationship between nutrition and physical performance has fascinated people for a long time. In Ancient Greece, athletes had special nutrition regimes to prepare for the Olympic Games. It has become clear that different types of exercise and different sports will have different energy and nutrient requirements; therefore food intake must be adjusted accordingly. Certain nutritional strategies can enhance performance, improve recovery and result in more profound training adaptations. Special drinks and energy bars have been developed and these have been marketed as sports foods. There is also a considerable amount of quackery in sports nutrition: special diets are promoted, direct-to-consumer tests are offered, gadgets are developed that give nutrition advice and there are a large number of nutrition supplements on the market with claims to improve performance and recovery. Many of these claims are unsubstantiated. One of the important skills for a practitioner, therefore, is that of critical reading and thinking (scientific scepticism).

19.2 Energy expenditure during physical activity

In resting conditions, cells need energy to function: ionic pumps in membranes need energy to transport ions across cell membranes and the muscle fibres of the heart need energy to contract. During exercise, the energy expenditure may increase several-fold, mainly because skeletal muscle requires energy to contract. In some cases, the energy provision can become critical and continuation of the exercise is dependent on the availability of energy reserves. Most of these reserves must be obtained through nutrition. In endurance athletes, for example, energy depletion (carbohydrate depletion) is one of the most common causes of fatigue. Carbohydrate intake is essential to prevent early fatigue as a result of carbohydrate depletion.

Definition and assessment methods

There are several ways to measure (or estimate) human energy expenditure:

- direct calorimetry
- indirect calorimetry:
  - closed circuit spirometry
  - open circuit spirometry (douglas bag technique, breath-by-breath technique, portable spirometry)
- doubly labelled water
- labelled bicarbonate
- heart-rate monitoring
- accelerometry
- observations, records of physical activity, activity diaries, recall.

The methods range from direct but complex measurements of heat production (direct calorimetry) to relatively simple indirect metabolic measurements (indirect calorimetry), and from very expensive tracer methods (doubly labelled water) to relatively cheap and convenient rough estimations of energy expenditure (heart rate monitoring and accelerometry). For a detailed analysis of these methods, the reader is referred to Chapter 2 of the *Introduction to Human Nutrition* textbook (third edition) in this series.

The most important component of energy expenditure for athletes who are in training is exercise. Basal metabolic rate (BMR) and diet-induced thermogenesis (DIT) become relatively less important (accounting for less than 50% of total energy expenditure) when athletes train for two hours/day or more. Energy expenditure during physical activity ranges from 20 kJ/min for very light activities to 100 kJ/min for very high-intensity exercise.

Energy expenditure and substrate use

Energy can be defined as the potential for performing work or producing force. Force production by skeletal muscles requires adenosine triphosphate (ATP). This compound contains energy in its phosphate bonds and, on hydrolysis of ATP, this energy is released and used to power all forms of biological work. In muscle, energy from the hydrolysis of ATP by myosin ATPase is used for muscle contraction. The hydrolysis of ATP yields approximately 31 kJ of free energy per mole of ATP degraded to ADP and inorganic phosphate (Pi):

$$ATP + H_2O \rightarrow ADP + H^+ + Pi$$

31 kJ per mole of ATP

The stores of ATP are very small and would only be sufficient for about two seconds of maximal exercise. The body therefore has various ways to resynthesise ATP. The different mechanisms involved in the resynthesis of ATP for muscle force generation include:

- phosphocreatine breakdown;
- glycolysis, which involves metabolism of glucose-6-phosphate, derived from muscle glycogen or blood-borne glucose;
- the products of carbohydrate, fat, protein and alcohol metabolism, which can enter the tricarboxylic acid (TCA) cycle in the mitochondria and be oxidised to carbon dioxide and water (aerobic metabolism). This process is known as oxidative phosphorylation and yields energy for the synthesis of ATP.

The rate of energy delivery from phosphocreatine is very rapid, and is somewhat slower from glycolysis and much slower from aerobic metabolism. Within the muscle fibre, the concentration of phosphocreatine is about three or four times greater than the concentration of ATP. When phosphocreatine is broken down to creatine and inorganic phosphate by the action of the enzyme creatine kinase, a large amount of free energy is released. The phosphocreatine can be regarded as back-up energy store: to prevent the ATP concentration from falling during exercise, phosphocreatine is broken down, releasing energy for restoration of ATP. During very intense exercise (8–10 seconds maximal exercise) the phosphocreatine store can be almost completely depleted:

$$Pcr + ADP + H^+ \leftrightarrow ATP + Cr$$

43 kJ per mole of PCr

Under normal conditions, muscle clearly does not fatigue after only a few seconds of effort, so a source of energy other than ATP and phosphocreatine must be available. This is derived from glycolysis, which is the name given to the pathway involving the breakdown of glucose (or glycogen), the end-product of this series of chemical reactions being pyruvate. This process does not require oxygen, but does result in energy in the form of ATP. In order for the reactions to proceed, however, the pyruvate must be removed; in low-intensity exercise when adequate oxygen is available to the muscle, pyruvate is converted to carbon dioxide and water by an oxidative metabolism in the mitochondria. In some situations, the pyruvate is removed by conversion to lactate, a reaction that does not involve oxygen. The net effect of glycolysis can thus be seen to be the conversion of one molecule of glucose to two molecules of pyruvate, with the net formation of two molecules of ATP and the conversion of two molecules of NAD^+ to NADH. If glycogen rather than glucose is the starting point, three molecules of ATP are produced, as there is no initial investment of ATP when the first phosphorylation step occurs. An 800 m runner, for example, obtains about 60% of the total energy requirement from anaerobic metabolism and may convert about 100 g of carbohydrate (mostly glycogen, and equivalent to about 550 mmol of glucose) to lactate in less than two minutes. The amount of ATP released in this way (three ATP molecules per glucose molecule degraded, about 1667 mmol of ATP in total) far exceeds that available from phosphocreatine hydrolysis. This high rate of anaerobic metabolism not only allows a faster steady-state speed than would be possible if aerobic metabolism alone had to be relied on, but also allows a faster pace in the early stages before the cardiovascular system has adjusted to the demands, and the delivery and utilisation of oxygen have increased in response to the exercise stimulus.

During exercise lasting for several minutes up to several hours, carbohydrate and fat are the most important fuels. These two energy sources are stored in the human body and can be mobilised from these stores when the demand increases. Both carbohydrate and fat are broken down to acetyl-coenzyme A (acetyl-CoA), which will then enter a series of reactions referred to as the tricarboxylic acid (TCA) cycle or Krebs cycle. In essence, the most important function of the TCA cycle is to generate hydrogen atoms for their subsequent passage to the electron transport chain by means of NADH and $FADH_2$. The aerobic process of electron transport–oxidative phosphorylation regenerates ATP from ADP, thus conserving some of the chemical potential energy contained within the original substrates in the form of high-energy phosphates. As long as there is an adequate supply of oxygen and substrate, NAD^+ and FAD are continuously regenerated and TCA metabolism proceeds. This system cannot function without the use of oxygen.

Carbohydrate is stored in the liver and muscle (Table 19.1). The liver contains approximately 80 g of carbohydrate in the form of glycogen, and skeletal muscle contains approximately 300–800 g (depending on muscle mass and diet). The body's carbohydrate stores are relatively small compared with the fat stores of even the leanest athlete (4–7 kg adipose tissues). Indeed, the high-volume training programmes undertaken by many high-performance athletes are able to deplete body carbohydrate stores in a day – or even over a single exercise session. Meanwhile, if fat were the only muscle fuel, an athlete could theoretically run for 1000–2000 km. Although fat contains more than twice as much energy per gram as carbohydrate (37 kJ vs 17 kJ) and provides far more fuel per unit of weight than carbohydrate, its oxidation requires ~5% more oxygen to produce the same amount of ATP as carbohydrate oxidation.

After an overnight fast, most of the energy requirement is covered by the oxidation of fatty acids derived from adipose tissue. Lipolysis in adipose tissue is mostly dependent on the concentrations of hormones: epinephrine (adrenaline), which stimulates lipolysis, and insulin, which inhibits lipolysis. At rest, around 70% of the fatty acids liberated after lipolysis will be re-esterified within the adipocyte, while around 30% of the fatty acids are released into the systemic circulation, maintaining plasma fatty acid concentrations at ~0.2–0.4 mmol/l. As soon as exercise is initiated, the rate of lipolysis and the release of fatty acids from adipose tissue are increased. Lipolysis increases because of an increased β-adrenergic stimulation and reduced plasma insulin concentration. Meanwhile, a shift in blood flow from the adipose tissue to the skeletal muscle reduces the rate of re-esterification and increases the delivery of fatty acids to the muscle.

Figure 19.1 displays the most important sources of fuel utilised during exercise. During low-intensity exercise, plasma fatty acids provide approximately one-third of the total energy. Glucose derived from the liver accounts for about 10% at low intensities but becomes more and more important at higher exercise intensities. Muscle glycogen is relatively unimportant at

Table 19.1 Availability of substrates in the human body (estimated energy stores of fat and carbohydrate in an 80 kg man with 15% body fat)

Substrate		Weight (kg)	Energy (kcal)
Carbohydrates	Plasma glucose	0.02	78
	Liver glycogen	0.1	388
	Muscle glycogen	0.4	1550
	Total (approximately)	0.52	2000
Fat	Plasma fatty acid	0.0004	4
	Plasma triacylglycerols	0.004	39
	Adipose tissue	12.0	100 000
	Intramuscular triacylglycerols	0.3	2616
	Total (approximately)	12.3	106 500

Values given are estimates for a 'normal' man and not those of an athlete, who might be leaner and have more stored glycogen. The amount of protein in the body is not mentioned, but this would be about 10 kg (40 000 kcal); mainly located in the muscles. *Source:* Adapted from Jeukendrup et al. (1998).

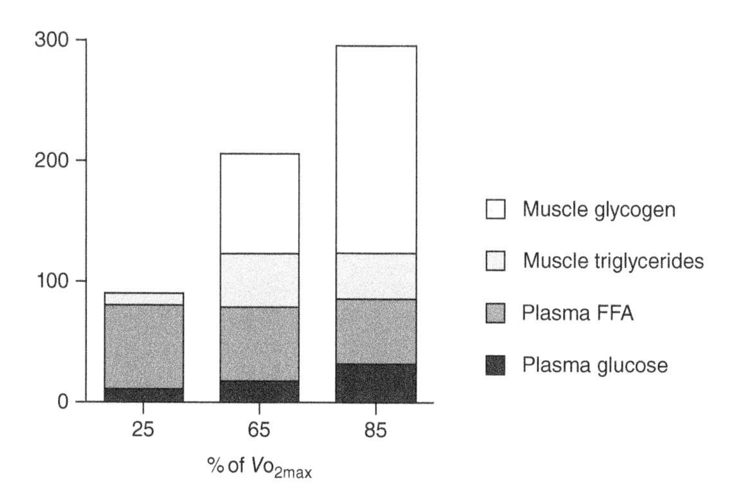

Figure 19.1 Substrate utilisation at different exercise intensities. (AMERICAN JOURNAL OF PHYSIOLOGY: ENDOCRINOLOGY AND METABOLISM by JA Romijn, EF Coyle, LS Sidossis, A Gastaldelli, JF Horowitz, E Endert and RR Wolfe. Copyright © 1993 by AMERICAN PHYSIOLOGY SOCIETY. Reproduced with permission of AMERICAN PHYSIOLOGY SOCIETY in the format Textbook via Copyright Clearance Center.).

low intensities [40% maximal aerobic power ($Vo_{2\ max}$)] but is by far the most important fuel during high-intensity exercise (70–90% $Vo_{2\ max}$). Fat oxidation is usually the predominant fuel at low exercise intensities, whereas during high exercise intensities, carbohydrate is the major fuel.

In absolute terms, fat oxidation increases as the exercise intensity increases from low to moderate intensities, even though the percentage contribution of fat may actually decrease (Figure 19.1). For the transition from light- to moderate-intensity exercise, the increased fat oxidation is a direct result of the increased energy expenditure. At higher intensities of exercise (>75% Vo_{2max}) fat oxidation will be inhibited and both the relative and absolute rates of fat oxidation will decrease to negligible values. Above ~65% $Vo_{2\ max}$, fat oxidation decreases despite high rates of lipolysis. The blood flow to the adipose tissue may be decreased (owing to sympathetic vasoconstriction) and this may result in a decreased removal of fatty acids from adipose tissue. During high-intensity exercise, lactate accumulation may also increase the re-esterification and inhibit the oxidation of fatty acids.

Carbohydrate and fat are always oxidised as a mixture, and whether carbohydrate or fat is the predominant fuel depends on a variety of factors.

Exercise intensity

At higher exercise intensities, more carbohydrate and less fat will be utilised. Carbohydrate can be utilised aerobically at rates up to about 4 g/min. The breakdown during very high-intensity exercise can amount to 7 g/min.

Duration of exercise

Fat oxidation increases and carbohydrate oxidation decreases as the exercise duration increases. Typical fat oxidation rates are between 0.2 and 0.5 g/min, but values of over 1.0–1.5 g/min have been reported after six hours of running. The contribution of fat to energy expenditure can even be doubled when there is a reduction in muscle glycogen stores towards the later stages of prolonged exercise.

Level of aerobic fitness

After endurance training the capacity to oxidise fatty acids increases and fat oxidation at the same absolute and relative exercise intensity is higher.

Diet

Both recent and previous diet can affect substrate utilisation during exercise. Even though endurance training increases the muscle's capacity for fat oxidation, this can be dramatically enhanced by adaptation to a high-fat diet. As little as five to six days of a low-carbohydrate, high-fat (LCHF) diet can retool the muscle to double its oxidation rates during exercise (up to 2 g/min in some highly trained individuals).

19.3 Carbohydrate and performance

Carbohydrate fuel plays a major role in the performance of many types of exercise and sport. The depletion of body carbohydrate stores is a cause of fatigue or performance impairment during exercise, particularly during prolonged (>90 min) sessions of submaximal or intermittent high-intensity activity. This fatigue may be seen both in the muscle (peripheral fatigue) and in the brain and nervous system (central fatigue). As total body carbohydrate stores are limited, and are often substantially less than the fuel requirements of the training and competition sessions undertaken by athletes, sports nutrition guidelines include a variety of options for acutely supplying the carbohydrate needs of exercise. These strategies include consuming carbohydrate before, during and in the recovery period between prolonged exercise bouts.

Evolution of sports guidelines around carbohydrate intake

Before focusing on these strategies, it is important to recognise that sports nutrition guidelines in relation to carbohydrate have evolved significantly over the past two decades. Although guidelines from the 1990s promoted high-carbohydrate diets focused on maximising muscle glycogen stores for all athletes and all scenarios, modern sports nutrition recognises that the fuel requirements of training and competition vary between athletes, and also from day to day in the same athlete due to the variation in the mode, volume and intensity of their activities. The term 'high-carbohydrate diet' has been replaced with the concept of carbohydrate availability (the match between the carbohydrate fuel requirements of an exercise session and the available carbohydrate stores). 'High carbohydrate availability' describes scenarios in which muscle glycogen stores and carbohydrate consumed before or during the session can match the specific fuel requirements, while 'low carbohydrate availability' describes scenarios in which carbohydrate depletion (particularly muscle glycogen) occurs. Indeed, contemporary guidelines promote a periodised and personalised approach to an athlete's carbohydrate intake that includes an understanding of the specific fuel needs of each exercise session as well as the importance of undertaking the session with appropriately matched carbohydrate stores. Modern athletes are encouraged to alter their total daily carbohydrate intake, as well as its spread across the day and around training sessions, to target the sessions that benefit from high carbohydrate availability (e.g. sessions involving high-intensity exercise or requiring high-quality/optimal performance). Depending on the characteristics of the session, daily carbohydrate intakes of 3–12 g/kg body mass may be appropriate.

Meanwhile, some types of exercise are less dependent on carbohydrate availability, and in some cases athletes may deliberately undertake certain sessions with low carbohydrate stores or delayed post-exercise refuelling. Such practices (often described as 'train low' or 'recover/sleep low') take advantage of enhanced mitochondrial adaptation to the cellular events associated with exercise undertaken with lower glycogen stores. While such exercise is fatiguing and typically of lower quality/intensity (and may also be associated with perturbations of other body systems), it may play a specialised role in the periodised training programmes of high-performance endurance athletes.

Benefits of high carbohydrate availability for exercise

Many studies show that exercise is improved by strategies that provide sufficient carbohydrate to meet the specific fuel requirements of the muscle and central nervous system. Most studies have been undertaken in the convenience of an exercise laboratory, often measuring the time an athlete can exercise on a bike or running treadmill while maintaining a steady work rate. Such an approach – technically, a measurement of exercise capacity or endurance – can allow scientists to identify strategies that delay the onset of fatigue. Successful strategies related to supplying carbohydrate requirements for exercise include preparing muscle glycogen stores before or between lengthy sub-maximal exercise, consuming carbohydrate in the hours before such exercise, consuming carbohydrate during the session, and combinations of these tactics. Although studies of this type are relatively easy to conduct and can show the beneficial effects of these dietary interventions, they do not necessarily mimic the real demands of a sporting event, where the athlete tries to cover a set distance as fast as possible and/or to execute complex decision-making and motor skills. Sports scientists have more recently elevated the importance of developing protocols that measure the important components of sports performance, including implementing studies within real-life competition. This has helped to gather evidence that strategies that supply sufficient carbohydrate to meet the fuel needs of sporting events, singly or in combination, can enhance the performance of a range of 'race' type sports (e.g. cycling, running, race walking and cross-country skiing) as well as field (e.g. ice hockey and football codes) and court (e.g. tennis and basketball) sports. Nevertheless, there is still a need for further research to be conducted over a greater range of sporting events, competition scenarios (e.g. different environmental conditions) and athletic populations (e.g. athletes of different sex, ages and calibre/training status) so that scientists can give more detailed and specific advice to athletes.

Recent studies have provided evidence of other benefits associated with strategies that maintain high carbohydrate availability during exercise. For example, there is now plentiful proof of performance benefits from consuming carbohydrate before and during high-intensity exercise of around one hour. As this is a situation in which muscle carbohydrate stores should already be sufficient to fuel the event, other mechanisms to explain the findings are needed. Indeed, it has been shown that the oral intake of carbohydrate improves the function of the brain and central nervous system, increasing motor output and allowing them to work harder. Intriguingly, the effect is achieved via contact between receptors in the mouth/oral cavity and parts of the brain associated with reward and occurs via 'mouth sensing' rather than requiring the athlete to ingest (swallow) the carbohydrate source. This finding has initiated a new line of research into other nutrients or food components that may have direct connections with the central nervous system and achieve benefits via deliberate 'swishing' around the site of receptors in the mouth and throat. Caffeine, fluid, bitter tastes (quinine) and menthol are all being investigated in this manner.

Finally, carbohydrate intake, during and between exercise sessions, may alter the exercise-associated 'cross-talk' between the muscle and other body systems via the release of cytokines and other messenger chemicals. Several studies have shown that biomarkers of the immune system, iron metabolism and bone turnover are perturbed after prolonged strenuous exercise with low carbohydrate availability; furthermore, these biomarkers are better preserved when exercise is undertaken with high carbohydrate availability (e.g. higher muscle glycogen stores and/or carbohydrate intake during the session). Whether this actually leads to an improvement in the immune status and health of athletes (e.g. fewer sick days, better iron status, better bone health) remains to be seen and would require a sophisticated long-term study.

Strategies to achieve high carbohydrate availability for exercise

A range of opportunities allow the athlete to store carbohydrate in the muscle and liver (i.e.

glycogen) or provide exogenous carbohydrate (from foods and drinks consumed during exercise) to meet the fuel needs of a targeted training session or event. Characteristics of these strategies include the timing of intake around exercise, as well as the quantity of carbohydrate and its form (e.g. liquid or solid) and type (e.g. glycaemic characteristics and chemical structure). A summary of potential strategies, according to the different goals and time points for feeding, is provided in Table 19.2 and discussed in more detail in the following. It is important to note that such guidelines should be a starting point, but the exact amount should be interpreted in the context of many other factors, especially overall energy intake.

Table 19.2 Summary of guidelines for carbohydrate intake by athletes.

	Situation	Carbohydrate targets	Comments on type and timing of carbohydrate intake
DAILY NEEDS FOR FUEL AND RECOVERY			
1 The following targets are intended to provide high carbohydrate availability (i.e. to meet the carbohydrate needs of the muscle and central nervous system) for different exercise loads for scenarios where it is important to exercise with high quality and/or at high intensity. These general recommendations should be fine-tuned with individual consideration of total energy needs, specific training needs and feedback from training performance.			
2 On other occasions, when exercise quality or intensity is less important, it may be less important to achieve these carbohydrate targets or to arrange carbohydrate intake over the day to optimise availability for specific sessions. In these cases, carbohydrate intake may be chosen to suit energy goals, food preferences or food availability.			
3 In some scenarios, when the focus is on enhancing the training stimulus or adaptive response, low carbohydrate availability may be deliberately achieved by reducing total carbohydrate intake, or by manipulating carbohydrate intake related to training sessions (e.g. training in a fasted state, undertaking a second session of exercise without adequate opportunity for refuelling after the first session).			
Light	• Low-intensity or skill-based activities	3–5 g/kg of athlete's body weight/d	• Timing of intake of carbohydrate over the day may be manipulated to promote high carbohydrate availability for a specific session by consuming carbohydrate before or during the session, or in recovery from a previous session. • Otherwise, as long as total fuel needs are provided, the pattern of intake may simply be guided by convenience and individual choice. • Athletes should choose nutrient-rich carbohydrate sources to allow overall nutrient needs to be met.
Moderate	• Moderate exercise programme (e.g. ~1 h/d)	5–7 g/kg/d	
High	• Endurance programme (e.g. 1–3 h/d mod-high-intensity exercise)	6–10 g/kg/d	
Very high	• Extreme commitment (e.g. >4–5 h/d moderate- to high-intensity exercise)	8–12 g/kg/d	
ACUTE FUELLING STRATEGIES – these guidelines promote high carbohydrate availability to promote optimal performance in competition or key training sessions			
General fuelling up	• Preparation for events < 90 min exercise	7–12 g/kg per 24 h as for daily fuel needs	• Athletes may choose carbohydrate-rich sources that are low in fibre/residue and easily consumed to ensure that fuel targets are met, and to meet goals for gut comfort or lighter 'racing weight'.
Carbohydrate loading	• Preparation for events > 90 min of sustained/intermittent exercise	36–48 h of 10–12 g/kg body weight per 24 h	

Table 19.2 (Continued)

	Situation	Carbohydrate targets	Comments on type and timing of carbohydrate intake
Rapid refuelling	• < 8 h recovery between 2 fuel demanding sessions	1–1.2 g/kg/h for first 4 h then resume daily fuel needs	• There may be benefits in consuming small regular snacks. • Carbohydrate-rich foods and drink may help to ensure that fuel targets are met.
Pre-event fuelling	• Before exercise > 60 min	1–4 g/kg consumed 1–4 h before exercise	• Timing, amount and type of carbohydrate foods and drinks should be chosen to suit the practical needs of the event and individual preferences/experiences. • Choices high in fat/protein/fibre may need to be avoided to reduce risk of gastrointestinal issues during the event. • Low glycaemic index choices may provide a more sustained source of fuel for situations where carbohydrate cannot be consumed during exercise.
During brief exercise	• < 45 min	Not needed	
During sustained high-intensity exercise	• 45–75 min	Small amounts including mouth rinse	• A range of drinks and sports products can provide easily consumed carbohydrate. • The frequent contact of carbohydrate with the mouth and oral cavity can stimulate parts of the brain and central nervous system to enhance perceptions of well-being and increase self-chosen work outputs.
During endurance exercise including 'stop and start' sports	• 1–2.5 h	30–60 g/h	• Carbohydrate intake provides a source of fuel for the muscle to supplement endogenous stores. • Opportunities to consume foods and drinks vary according to the rules and nature of each sport. • A range of everyday dietary choices and specialised sports products ranging in form from liquid to solid may be useful • The athlete should practise to find a refuelling plan that suits their individual goals, including hydration needs and gut comfort.
During ultra-endurance exercise	• > 2.5–3 h	Up to 90 g/h	• As above. • Higher intakes of carbohydrate are associated with better performance. • Products providing multiple transportable carbohydrates (glucose:fructose mixtures) achieve high rates of oxidation of carbohydrate consumed during exercise.

Source: Adapted from Thomas DT, Erdman KA, Burke LM. American College of Sports Medicine Joint Position Statement. Nutrition and Athletic Performance. *Med Sci Sports Exerc* 2016 Mar; 48(3): 543–68; and Burke LM, Hawley JA, Wong SH, Jeukendrup AE. Carbohydrates for training and competition. *J Sports Sci* 2011; 29 Suppl 1: S17–27

Fuelling up before exercise

Glycogen stores in the muscle and liver should be prepared according to the fuel needs of the upcoming exercise session, particularly in the competition setting, where athletes want to perform at their best. The key factors in glycogen storage are dietary carbohydrate intake and, in the case of muscle stores, tapered exercise or rest. In the absence of muscle damage, muscle glycogen stores can be returned to normal resting levels (to 350–500 mmol/kg dry weight muscle) with 24–36 hours of rest and an adequate carbohydrate intake (7–12/kg body weight per day). Normal stores appear adequate for the fuel needs of events of less than 60–90 minutes in duration (e.g. half-marathon, sprint triathlon or basketball game). As a well-prepared athlete will not typically deplete their muscle glycogen stores under such conditions, there is no need to undertake pre-competition strategies to super-compensate muscle glycogen stores. Indeed, 'carbohydrate loading' practices do not typically enhance performance in these events, although it is noted that within some sports, glycogen needs vary across individuals and scenarios. For example, a goalkeeper in football has a lower workload/fuel requirement than a midfielder, and on some occasions (e.g. an elite player with high-volume/intensity game style, playing in overtime), glycogen depletion may contribute to fatigue at the key time of the event. Therefore, different athletes in the same sport may need different competition preparation strategies

Carbohydrate loading is a special practice that aims to maximise or supercompensate muscle glycogen stores up to twice the normal resting level (e.g. ~500–900 mmol/kg dry weight). The first protocol was devised in the late 1960s by Scandinavian sports scientists who found, using muscle biopsy techniques, that the size of pre-exercise muscle glycogen stores affected endurance during submaximal exercise. Their series of studies in recreationally active participants found that several days of a low-carbohydrate diet depleted muscle glycogen stores and reduced cycling endurance compared with a mixed diet. However, when this was followed by several days of a high-carbohydrate intake, glycogen stores in the previously depleted muscle was doubled above normal levels, indicating a supercompensation effect. This extra fuel was associated with an increase in the cycling time to exhaustion. These pioneering studies produced the 'classical' seven-day model of carbohydrate loading, consisting of a three- to four-day depletion phase of hard training and low carbohydrate intake, followed by a three- to four-day loading phase of high carbohydrate eating and exercise taper (i.e. decreased training volume). Early field studies of prolonged running events showed that carbohydrate loading enhanced sports performance, not by allowing the athlete to run faster, but by prolonging the time for which race pace could be maintained. Typically, carbohydrate loading can postpone fatigue and extend the duration of prolonged steady-state exercise by around 20% and improve performance over a set distance or workload by 2–3%. Not surprisingly, it has become synonymous with the preparation for prolonged events involving sustained or intermittent high-intensity aerobic exercise such as marathons, ultra-distance triathlons and road cycling races.

The past six decades have seen changes to carbohydrate loading practices (see Figure 19.2). A modified carbohydrate loading strategy evolved from studies in the 1980s, which discovered that endurance training activates the enzyme-controlling muscle glycogen synthesis, removing the requirement for a glycogen-stripping or depletion phase prior to glycogen supercompensation. Thus, carbohydrate loading was truncated to 72 hours of reduced exercise and high carbohydrate intake prior to the endurance event. Even more recently, it has been shown that well-trained athletes may be able to maximise muscle glycogen stores with as little as 36–48 hours of relative rest and consumption of a high carbohydrate intake (e.g. 10 g/kg body weight per day). These modifications offer a more practical strategy for competition preparation at several levels. First, they avoid the fatigue and complexity of the extreme diet and training protocols associated with the previous depletion phase. Second, they can be achieved and repeated in a short time-frame, making it possible for athletes to regularly fuel up for prolonged events such as weekly races in the triathlon calendar or games in the tournament fixture. Many endurance athletes also integrate low-residue (low-fibre) eating within their carbohydrate loading practices. By reducing the amount of undigested matter in the gastrointestinal tract, the athlete can achieve a small but significant reduction in body mass (which may offset the increase in body mass associated with the extra storage of glycogen and water in the muscle). Equally importantly, they may reduce their need for bowel movements on the day of, or during, the event; indeed, gastrointestinal issues can be the cause of significant practical and comfort challenges during competitive events.

Pre-event meal

Food and fluids consumed in the four hours before an event may help to achieve the following sports nutrition goals:

- To continue to fill muscle glycogen stores if they have not fully restored or loaded since the last exercise session.
- To restore liver glycogen levels, especially for events undertaken in the morning where liver stores are depleted from an overnight fast.
- To ensure that the athlete is well hydrated.
- To prevent hunger, yet avoid gastrointestinal discomfort and upset during exercise.
- To include foods and practices that are important to the athlete's psychology or superstitions.

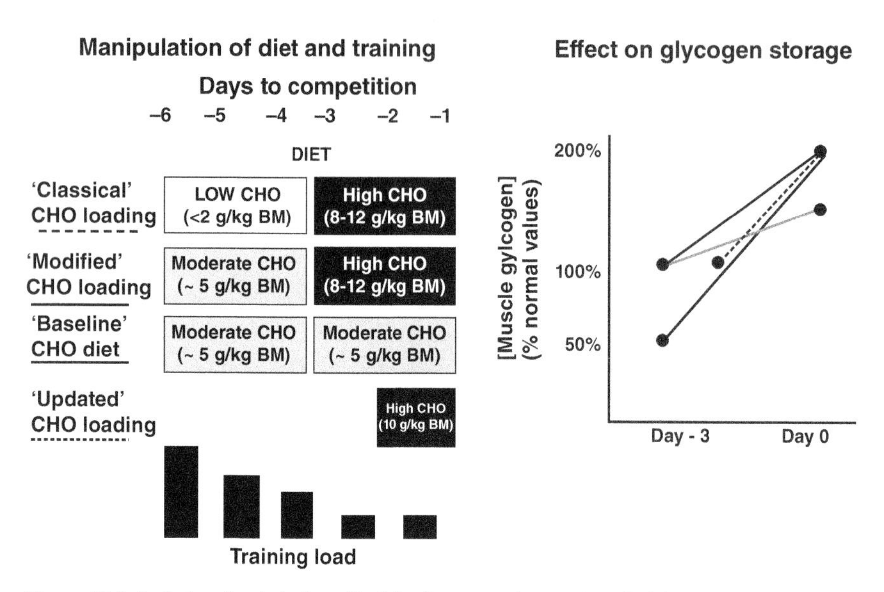

Figure 19.2 Evolution of carbohydrate (CHO) loading protocols over time. The 'classical' CHO loading protocol involved a three- to four-day glycogen depletion phase followed by a reduction in training and three days of high CHO intake. While this was shown to double muscle glycogen stores compared with a 'baseline' (moderate) CHO intake, later studies developed a 'modified' protocol which showed that super-compensation was still achieved in endurance-trained athletes without the need for the depletion phase. The most recent update shows that well-trained athletes can super-compensate muscle glycogen stores with reduced training volume and 36–48 hours of high CHO intake. BM, body mass. *Source:* Burke et. al., 2017 / American Physiological Society.

Consuming carbohydrate-rich foods and drinks in the pre-event meal is especially important in situations where body carbohydrate stores have not been fully restored and/or where the event is of sufficient duration and intensity to deplete these stores. The intake of a substantial amount of carbohydrate (~2–4 g/kg body weight) in the two to four hours before exercise has been shown to enhance various measures of exercise performance compared with performance undertaken after an overnight fast.

Occasionally, concerns are raised that carbohydrate consumed in the hour before exercise may have negative consequences for performance. Carbohydrate intake causes a rise in plasma insulin concentrations, which in turn suppresses the availability and oxidation of fat as an exercise fuel. This can lead to an increased reliance on carbohydrate oxidation at the onset of exercise, a faster depletion of muscle glycogen stores and a decline in plasma glucose concentration. A single study undertaken in the 1970s received considerable publicity when it reported that consuming carbohydrate in the hour before exercise was associated with worse performance than when subjects cycled without consuming anything. These findings have been convincingly replaced by the far greater number of studies that have shown that metabolic disturbances following pre-exercise carbohydrate feedings are short-lived or unimportant. Indeed, carbohydrate intake in the hour before exercise is almost always associated with a neutral effect or a beneficial performance outcome. Nevertheless, there may be a small subgroup of athletes who experience a true fatigue, associated with a decline in blood glucose levels, if they start to exercise within the hour after consuming a carbohydrate snack. This problem can be avoided or diminished by a number of dietary strategies:

- Consume a substantial amount of carbohydrate (>75 g) rather than a small amount, so that the additional carbohydrate more than compensates for the insulin-induced increase in muscle glucose uptake during the exercise.
- Choose a carbohydrate-rich food or drink that produces a low glycaemic index (GI) response (i.e. a low blood glucose and insulin response) rather than a carbohydrate source that has a high GI (producing a large and rapid blood glucose and insulin response).
- Consume carbohydrate throughout the exercise session.

Some sports scientists have suggested that all athletes will benefit from the choice of low-GI rather than high-GI carbohydrate foods in the pre-event meal. This theory is based on the idea that low-GI foods will provide a more sustained release of carbohydrate energy throughout the exercise session and create less of a metabolic disturbance during exercise because of a reduced insulin response. However, the balance of studies has failed to find performance benefits following the intake of a low-GI pre-exercise meal compared with an equal amount of high-GI carbohydrate, even when metabolic differences between low-GI and high-GI meals were noted. Furthermore, it has been shown that when carbohydrate is ingested during exercise according to sports nutrition guidelines, any metabolic or performance effects arising from the choice of a low-GI or high-GI pre-event meal are over-ridden.

Each athlete must judge the benefits and the practical issues associated with pre-exercise meals in their particular sport or situation. The type, timing and quantity of pre-event meals should be chosen according to the athlete's individual circumstances and experiences. Gastrointestinal comfort and the practical challenges associated with consuming foods in the event environment or at specific times of the day may provide an additional consideration. Athletes at high risk of gastrointestinal discomfort/dysfunction during exercise may find it useful to reduce the content of fat, fibre or poorly digested carbohydrate sources (e.g. lactose, fructose) in the pre-event meal(s). Others may find that processed food choices, including specialised sports foods, are the preferred choice for the pre-event menu as they are less likely to cause gastrointestinal upsets. Liquid meal supplements or carbohydrate-containing drinks and bars are a simple snack for athletes who suffer from pre-event nerves or have an uncertain competition timetable.

Carbohydrate intake during exercise

Numerous studies have shown that the intake of carbohydrate during prolonged sessions of moderate-intensity or intermittent high-intensity exercise can improve endurance (i.e. prolong time to exhaustion) and performance. There is also some evidence that carbohydrate intake may benefit shorter-duration, high-intensity sports. Although there is some evidence that increasing carbohydrate availability causes glycogen sparing in slow-twitch muscle fibres during running, the major mechanisms to explain the benefits of carbohydrate feeding during prolonged exercise are the maintenance of plasma glucose concentration (sustaining brain function), restoration of liver glycogen and the provision of an additional carbohydrate supply to allow the muscle to continue high rates of carbohydrate oxidation.

Studies have demonstrated that different exogenous carbohydrate sources are oxidised at different rates. Glucose, maltose, sucrose, maltodextrins and soluble starch are amongst the carbohydrates with the highest oxidation rates. Carbohydrate consumed during exercise is oxidised in small amounts during the first hour of exercise (~20 g) and thereafter reaches a peak rate of around 1 g/min, at least when a single form of carbohydrate, like glucose, is consumed. However, when multiple forms of carbohydrate using different transport routes for uptake in the intestines are consumed together (e.g. glucose and fructose), higher intakes of carbohydrate intake can lead to higher rates of carbohydrate use by the muscles. In most endurance events, a carbohydrate intake of 30–60 g/hour, starting well in advance of fatigue/depletion of body carbohydrate stores, will enhance performance. In ultra-endurance events, ingestion of higher rates of carbohydrate mixtures can provide additional fuel and further enhance performance. Studies have demonstrated that the use of carbohydrate mixes like glucose and fructose or maltodextrins and fructose that use to different carbohydrate transporters or absorption across the intestinal wall will result in greater carbohydrate delivery to the muscle and greater exogenous carbohydrate oxidation. This, in turn, has been shown to improve performance, with several studies showing a dose relationship between carbohydrate intake and performance. The only reasons why more carbohydrate is not always better is that at some level of intake it is more likely that gastrointestinal problems will occur. This appears to be highly individual, with some individuals struggling to ingest more than 30 g/hour and some ingesting 120 g/hour without any problems. It is also clear, however, that the gut is trainable and the amount that can be tolerated without gastrointestinal discomfort can be increased. In early studies, intakes up to 144 g/hour of glucose:fructose were tested and resulted in extremely high exogenous carbohydrate oxidation rates of 105 g/hour. Reports of studies in races in the early 2000s reported carbohydrate intakes by athletes of 30 g/hour, whereas it has become common practice for many elite endurance athletes now to consume up to 120 g/hour.

A suggested carbohydrate intake schedule for various events can be found in Table 19.2.

In the world of sport, athletes consume carbohydrate during exercise using a variety of foods and drinks, and a variety of feeding schedules. Sports drinks (4–8% carbohydrate, 4–8 g carbohydrate/100 ml) and high carbohydrate drinks (9–20%, 9–20 g carbohydrate/100 ml) are particularly valuable as these allow the athlete to replace their fluid and carbohydrate needs simultaneously. Such drinks will be discussed in greater detail in the following section. Each sport or exercise activity offers particular opportunities for fluid and carbohydrate to be consumed throughout the session, whether it be from aid stations, supplies carried by the athlete or at formal stoppages in play such as time-outs or half-time breaks. The athlete should be creative in making use of these opportunities and in choosing carbohydrate foods and drinks that suit the occasion and their nutritional needs and have worked well in previous experiences.

Post-exercise refuelling

Restoration of muscle glycogen concentrations is an important component of post-exercise recovery and is challenging for athletes who train or compete more than once each day. The main dietary issue in refuelling is the amount of carbohydrate consumed, with requirements being within the range of 7–12 g/kg body weight per day, depending on the workload and amount of muscle mass involved. Glycogen storage occurs at a slightly faster rate during the first couple of hours after exercise, but the main reason for encouraging an athlete to consume carbohydrate-rich meals or snacks soon after exercise is that effective refuelling does not start until a substantial amount of carbohydrate (~1 g/kg body weight) is consumed. Therefore when there is limited time between workouts or events (e.g. eight hours or less) it makes sense to turn every minute into effective recovery time by consuming carbohydrate as soon as possible after the first session. However, when recovery time is longer, immediate carbohydrate intake after exercise is unnecessary and athletes can afford to follow their preferred and practical eating schedule as long as goals for total carbohydrate intake are met over the day.

Certain amino acids have a potent effect on the secretion of insulin, a stimulator of glycogen resynthesis. For this reason, the effects of combining amino acids and proteins with post-exercise carbohydrate choices have been investigated. The take-home message from a substantial body of literature is that when total carbohydrate intake is less than optimal for glycogen storage (e.g. < 1 g/kg body weight), the addition of protein can enhance glycogen storage. However, extra refuelling is not expected when carbohydrate targets are met. Nevertheless, there are other good reasons to provide protein in recovery meals and snacks. The timely intake of a source of high-quality protein after a workout or event will enhance protein synthesis in response to the exercise stimulus. As we will discuss later, the amount of protein needed is only small (~0.3 g/kg body weight or ~20–25 g protein). Depending on the type of exercise, this may lead to increases in muscle size and strength (e.g. response to resistance exercise), repair of damaged tissues and increases in functional proteins such as the enzymes and other factors involved in exercise metabolism.

Periodised carbohydrate intakes – a real-world example

Modern sports nutrition guidelines promote a periodised approach to carbohydrate intake, with total daily intake and specific intake around training sessions reflecting:

1. the anticipated fuel (carbohydrate) cost of the day's training (or competition) workload and the general training plan.
2. the specific benefit of enhancing the quality of a training session or event by increasing carbohydrate intake to achieve high carbohydrate availability, or by practising intended competition strategies (e.g. consuming carbohydrate during a session to mimic a race or match plan).
3. the potential benefit of reducing carbohydrate intake around a specific session or the total day. This includes periods of reduced energy intake during well-organised weight-loss strategies or the inclusion of special sessions of aerobic exercise which are undertaken with low glycogen stores to amplify training adaptations (e.g. an increase in mitochondrial activity and capacity for fat oxidation). Athletes

may use a session of higher/longer-intensity exercise to deplete muscle glycogen stores, then avoid carbohydrate during the recovery period ('recover'/'sleep' low). The next training session will therefore be undertaken with low glycogen stores ('train low') and should be a low- to moderate-intensity session.

Of course, athletes also arrange their intake of energy and carbohydrate over the day according to food customs, food availability and practical considerations (e.g. choosing to undertake morning sessions of exercise in a fasted state to allow themselves to sleep in later rather than get up in time to eat a pre-session breakfast). An example of a weekly training schedule in the base phase of training of an elite endurance athlete is provided in Figure 19.3, demonstrating the integration of different carbohydrate intake strategies across and between days according to the different characteristics of training sessions.

19.4 Fat metabolism and performance

At rest and during exercise, skeletal muscle is the main site of oxidation of fatty acids. In resting conditions, and especially after fasting, fatty acids are the predominant fuel used by skeletal muscle. During low-intensity exercise, metabolism is elevated several-fold compared with resting conditions, and fat oxidation is increased. When the exercise intensity increases, fat oxidation increases further, until exercise intensities of about 65% $\dot{V}O_{2\,max}$, after which a decline in the rate of fat oxidation is observed. In contrast to carbohydrate metabolism, which increases as a function of the aerobic work rate, fat oxidation is reduced at high exercise intensities (Figure 19.1). Nevertheless, because of the abundance of fat stores even in the leanest athlete (and despite opportunities to boost muscle glycogen stores and contribute to muscle fuel needs by consuming carbohydrates during exercise), athletes in prolonged events have often turned their interest to strategies that might boost the contribution of fat to exercise fuel needs and reduce their reliance on the more limited carbohydrate options.

Fat oxidation and diet

Diet has marked effects on fat oxidation. In general, a high-carbohydrate, low-fat diet will reduce fat oxidation at rest and during exercise, whereas a high-fat, low-carbohydrate diet will do the opposite. Of course, it might be argued that such observations were caused by the acute influence of the most recent meal.

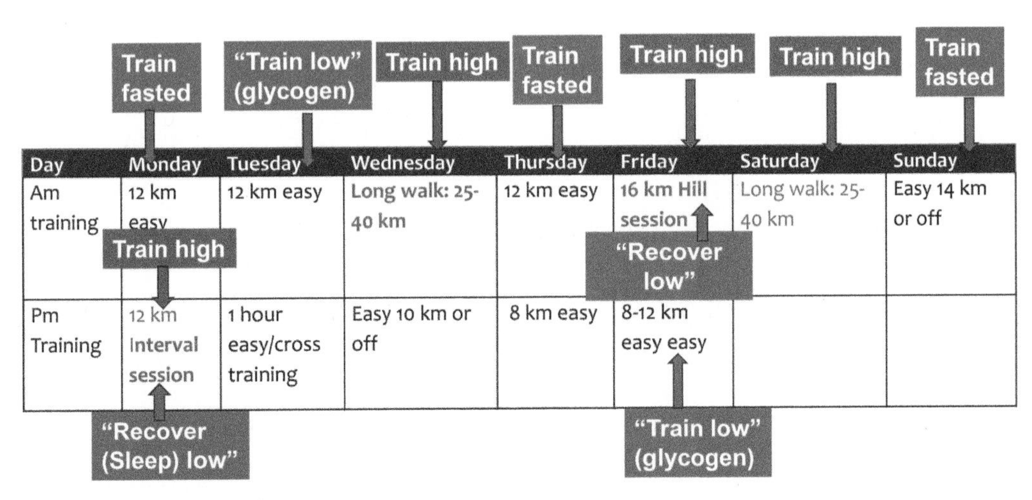

KEY

- Sessions noted in red are the "key" workouts for the week, and should be undertaken with high carbohydrate availability
- "Train high" strategies (red boxes) include refuelling to recover glycogen stores from previous session, pre-event carbohydrate, carbohydrate during the session) to promote high quality training, and practice of race fuelling tactics
- Train low strategies (blue boxes) include avoiding refuelling after the previous session ("recover low"), training after an overnight fast, and avoiding carbohydrate intake during the session.

Figure 19.3 Example of a periodised approach to carbohydrate availability in a week of training (high-performance distance athlete – race walker).

However, there is evidence that chronic exposure to a high-fat, low-carbohydrate diet can retool the muscle to increase fatty acid availability, transport, muscle uptake and, in turn, fat utilisation during exercise. These changes occur in as little as five to six days and are apparently robust, as the preferential use of fat will persist even if the athlete restores muscle glycogen content and consumes carbohydrates during exercise. Such 'fat adaptation' or 'fat loading' has been investigated as one of a variety of strategies that might change substrate use and, in turn, performance.

Fat intake during exercise

Long-chain triglycerides

Nutritional fats include triglycerides (containing mostly C16 and C19 fatty acids), phospholipids and cholesterol, of which only triglycerides can contribute to any extent to energy provision during exercise. In contrast to carbohydrates, nutritional fats reach the circulation only slowly as they are potent inhibitors of gastric emptying. Furthermore, the digestion in the gut and absorption of fat are also rather slow processes compared with the digestion and absorption of carbohydrates. Bile salts, produced by the liver, and lipase secreted by the pancreas are needed for the breakdown of the long-chain triglycerides (LCTs) into glycerol and three long-chain fatty acids. The fatty acids are then transported in chylomicrons via the lymphatic system, which ultimately drains into the systemic circulation. Long-chain dietary fatty acids typically enter the blood three to four hours after ingestion.

The fact that these long-chain fatty acids enter the circulation in chylomicrons is also important, and it is generally believed that the rate of breakdown of chylomicron-bound triglycerides by muscle is relatively slow. It has been suggested that the primary role of these triglycerides in chylomicrons is the replenishment of intramuscular fat stores after exercise. The usual advice, therefore, is to reduce the intake of fat during exercise to a minimum. Many 'sports bars' or 'energy bars', however, contain significant amounts of fat and it is therefore advisable to check the food label before choosing an energy bar.

In summary, LCT ingestion during exercise is not desirable because it slows gastric emptying and because the triglycerides only appear slowly in the systemic circulation in chylomicrons, which are believed to be a less important fuel source during exercise.

Medium-chain triglycerides

Medium-chain fatty acids have different properties to long-chain fatty acids and have been promoted as a useful energy source during exercise. Medium-chain fatty acids contain 8–10 carbons, whereas long-chain fatty acids contain 12 or more carbons. Medium-chain triglycerides (MCTs) are more polar and therefore more soluble in water, and they are more rapidly digested and absorbed in the intestine than LCTs. Furthermore, medium-chain fatty acids follow the portal venous system and enter the liver directly, while long-chain fatty acids are passed into the systemic circulation slowly via the lymphatic system. MCTs are normally present in our diet in very small quantities, with few natural sources. MCT supplements are often synthesised from coconut oil.

Unlike LCTs, MCTs are rapidly emptied from the stomach, rapidly absorbed and metabolised, and may be a valuable exogenous energy source during exercise, in addition to carbohydrates. It has been suggested that MCT ingestion may improve exercise performance by elevating plasma fatty acid levels and sparing muscle glycogen. However, the ingestion of 30 g of MCTs has no effect on plasma fatty acid concentration, glycogen sparing or exercise performance. Ingestion of larger amounts of MCTs is likely to cause gastrointestinal distress and therefore cannot be recommended; therefore MCTs do not appear to have the positive effects on performance that are often claimed.

Fasting

Fasting has been proposed as a way to increase fat utilisation, spare muscle glycogen and improve exercise performance. In rats, short-term fasting increases plasma epinephrine (adrenaline) and norepinephrine (noradrenaline) concentrations, stimulates lipolysis and increases the concentration of circulating plasma fatty acids. This, in turn, increases fat oxidation and 'spares' muscle glycogen, leading to a similar, or even increased, running time to exhaustion in rats. In humans, fasting also results in an increased concentration of circulating catecholamines, increased lipolysis, increased concentration of plasma fatty acids and a decreased glucose turnover. Muscle glycogen concentrations, however, are unaffected by fasting for 24 hours when no strenuous exercise is performed. Although it has been reported that fasting had no effect on endurance capacity at low exercise intensities (45% $\dot{V}O_{2\,max}$), performance may be significantly impaired at intensities higher than 50% $\dot{V}O_{2\,max}$. The observed decrease in performance was not reversible by carbohydrate ingestion during exercise. Liver glycogen stores will be substantially depleted after a 24-hour fast and therefore euglycaemia may not be as well maintained during exercise, compromising brain function.

In summary, fasting increases the availability of lipid substrates, resulting in increased oxidation of fatty acids at rest and during exercise. However, as the liver glycogen stores are not maintained, exercise performance is impaired.

High-fat diets

Christensen and Hansen (1939) demonstrated that consumption of a low-carbohydrate high-fat diet ('LCHF') for three to five days resulted in impaired exercise capacity, likely due to the acute decrease in muscle glycogen concentrations. Although fat oxidation increases in such circumstances, this may simply be due to the increased reliance on fat due to the lack of availability of carbohydrate as an energy source without a substantial increase in capacity for fat oxidation. More recent research has shown that extending the high-fat diet beyond five to six days creates changes in muscle fuel stores (increases in intramuscular triacylglycerols) and the muscle hormones/enzymes/proteins that control fat breakdown, transport, uptake and utilisation. This occurs in the case of restricted-carbohydrate (~20% energy), high-fat (65% energy) diets as well as the more recently popular ketogenic low-fat (< 50 g/day), high-fat (~80% energy) diets. In the latter case, such dietary changes are associated with an increase in blood concentrations of ketone bodies, which are proposed to provide an alternative source of energy for the brain and central nervous system as well as other putative roles in metabolism.

The interest in non-ketogenic LCHF diets peaked in 1995–2005, with a series of studies of endurance athletes investigating changes in training adaptations, substrate use and performance changes following adaptation to the diet for 5-day to 10-week periods. This dietary approach was shown to create major increases in fat oxidation during exercise, without evidence of an improvement in exercise capacity or performance. In fact, some longer-term studies showed a reduction in performance, which was not reversed on resuming a high-carbohydrate intake, suggesting that the diet may have impaired training adaptations. Scientific enquiry ended when several studies investigated a hybrid model in which fat adaptation (five to seven days) was followed by glycogen restoration (one day) and event carbohydrate intake (pre-event meal and event carbohydrate), reporting the lack of a performance improvement. The hypothesis that an athlete could benefit from dual optimization of fat utilisation and carbohydrate availability was overturned when biopsy samples collected in one of these studies showed that the muscle retooling associated with fat adaptation also included a down-regulation of glycogen breakdown and the entry of pyruvate into the oxidative pathway. This finding of glycogen 'impairment' rather than glycogen 'sparing' provided the mechanism to explain a lack of overall performance improvement and a specific reduction in the performance of higher-intensity aerobic exercise (e.g. exercise at work rates of 80–90% maximal aerobic capacity). Given that success in many endurance events is based on exercise capacity at these workloads, either throughout the event (e.g. marathon performance in elite runners) or at critical stages (e.g. the breakaway in road cycling or the long 'sprint' to the finish line in many ultra-distance races), it seemed that fat adaptation strategies were unsuitable for the needs of high-performance competitors. Nevertheless, a role in moderate-intensity multi-day or other ultra-endurance events was left open.

The popularity of ketogenic (extremely carbohydrate restricted) LCHF diet for sports performance re-emerged in around 2013, largely driven by enthusiastic social media discussions and athlete testimonials. At this time, the only available scientific investigation dated back to the 1980s and involved a four-week investigation in a small group of well-trained cyclists. This study found that after the four weeks, participants showed substantial increases in fat utilisation during moderate-intensity exercise (~60–65% VO_2max) and a *lack* of impairment of cycling time to exhaustion; this finding contradicted the 1980s view of sports nutrition that reduced glycogen content would be associated with premature fatigue and reduced endurance. Nevertheless, the study also reported a blunting of capacity for high-intensity exercise in these cyclists, and a marked variability in individual responses. In fact, of the five study participants, only two reported endurance benefits, while an equal number reported a substantial reduction in exercise capacity following the LCHF diet. Athlete anecdotes and claims of 'keto clarity', 'unlimited fuel' and 'metabolic flexibility' associated with a chronic adherence to the ketogenic LCHF diet began to dominate sports nutrition discussions, warranting new investigations of the diet. Over the past decade, rigorously controlled studies involving highly trained and elite endurance athletes and cross-sectional studies of long-term adherents of ketogenic diets have been undertaken. Our current knowledge, summarised in the following, still confirms the benefits of high-carbohydrate-availability approaches to performance of higher-intensity endurance sports but notes that there may be individuals or individual events/scenarios in which the LCHF diet presents opportunities (or at least a reduced likelihood of impaired performance):

- A ketogenic LCHF diet can achieve a substantial (~200%) increase in maximal rates of fat oxidation during exercise in endurance-trained athletes, up to ~1.5 g/min at ~70% of peak aerobic capacity.
- In high-level athletes, three to four weeks of a ketogenic LCHF diet appears to preserve moderate-intensity exercise capacity and performance, although considerable individual responsiveness is observed. However, performance of higher-intensity endurance exercise, which is often essential in high-level sport, is compromised, potentially due to the increased oxygen cost of energy production from fat. Because carbohydrate oxidation achieves ~5% greater ATP production from a given amount of oxygen than fat oxidation, it is associated with better exercise economy (greater speed or power production at the same relative oxygen cost). Economy is often considered an important contributor to success in endurance sports
- The optimal adaptation period for the ketogenic LCHF diet is controversial; substantial changes in substrate utilisation probably occur within 5–10 days. Claims that longer adaptation (more than three to four months) creates additional changes to substrate utilisation and endurance performance are currently unsubstantiated and require additional investigation.
- Models that integrate high carbohydrate availability in some training sessions or training periods may rescue the impairment of higher-intensity exercise associated with the ketogenic LCHF diet alone. However, the down-regulation of carbohydrate oxidation accompanying a LCHF diet may continue to limit the contribution of muscle glycogen to race energy needs, even when its availability is acutely restored for a race. This may continue to prevent optimal performance in events which are dependent on higher-intensity exercise.
- Athletes who are contemplating the use of ketogenic LCHF diets should undertake an audit of the fuel requirements of their event, balancing the benefits of their capacity for higher-intensity exercise (supported by a carbohydrate focus) against the benefits of reducing their need to supplement carbohydrate during longer events. Some athletes are unable to tolerate carbohydrate intake during exercise or compete in events in which it is impractical to have a carbohydrate supply available.

19.5 Effect of exercise on protein requirements

There is still considerable debate about how much dietary protein is required for optimal athletic performance. This interest in protein (meat) probably dates back to Ancient Greece. There are reports that athletes in Ancient Greece, in preparation for the Olympic Games, consumed large amounts of meat. This belief stems partly from the large amount of muscle in the human body (40% of body weight) containing 10–15% of body weight in protein. Muscle also accounts for 30–50% of all protein turnover in the body. Both the structural proteins that make up the myofibrillar proteins and the proteins that act as enzymes within a muscle cell change as an adaptation to exercise training. Indeed, muscle mass, muscle protein composition and muscle protein content will change in response to training, and therefore it is not surprising that meat has been popular as a protein source for athletes, especially strength athletes.

Non-essential amino acids can be synthesised from essential and non-essential amino acids. This is important in situations with inadequate dietary protein intake. In muscle, the majority of amino acids are incorporated into tissue proteins, with a small pool of free amino acids. This pool undergoes turnover, receiving free amino acids from the breakdown of protein and contributing amino acids for protein synthesis. Protein breakdown in skeletal muscle serves two main purposes:

- To provide essential amino acids when individual amino acids are converted to acetyl-CoA or TCA cycle intermediates.
- To provide individual amino acids that can be used elsewhere in the body for the synthesis of neurotransmitters, hormones, glucose and proteins.

Clearly, if protein degradation rates are greater than the rates of synthesis, there will be a reduction in protein content; conversely, muscle protein content can only increase if the rate of synthesis exceeds that of degradation.

Increased protein requirements

Exercise, especially endurance exercise, results in increased oxidation of the branched-chain amino acids (BCAAs), which are essential amino acids and cannot be synthesised in the body. Increased oxidation would imply that the dietary protein requirements are increased. Some studies in which the nitrogen balance technique was used showed that the dietary protein requirements for athletes involved in prolonged endurance training were higher than those for sedentary individuals. However, these results have been questioned.

It has been estimated that protein may contribute up to about 15% to energy expenditure in resting conditions. During exercise this relative contribution is likely to decrease because energy expenditure is increased and most of this energy is provided by carbohydrate and fat. During very prolonged exercise, when carbohydrate availability becomes limited, the contribution of protein to energy expenditure may amount up to about 10% of total energy expenditure. Thus, although protein oxidation is increased during endurance exercise, the relative contribution of protein to energy expenditure remains small. Protein requirements may be increased somewhat, but this increased need may be met easily by a moderate protein intake. The research groups that advocate an increased protein intake for endurance athletes in heavy training usually recommend a daily intake of 1.2–1.8 g/kg body weight. This is about twice the level of protein intake that is recommended for sedentary populations.

There are reports of increased protein breakdown after resistance exercise. The suggested increased dietary protein requirements with resistance training are related to increased muscle bulk (hypertrophy) rather than increased oxidation of amino acids. Muscle protein breakdown is increased after resistance training, but to a smaller degree than muscle protein synthesis. The elevations in protein degradation and synthesis are transient. Protein breakdown and synthesis after exercise are elevated at 3 and 24 hours after exercise, but return to baseline levels after 48 hours. These results seem to apply to resistance exercise and high-intensity dynamic exercise.

There is some controversy as to whether strength athletes really need to eat large amounts of protein. The nitrogen balance studies that have been conducted on such athletes have been criticised because they have generally been of short duration and a steady-state situation may not have been established. The recommendation for protein intakes for strength athletes is therefore generally 1.6–1.7 g/kg body weight per day. Again this seems to be met easily with a normal diet and no extra attention to protein intake is needed. Protein supplements are often used, but are not necessary to meet the recommended protein intake. The question as to what amount of daily protein intake is optimal may be academic, because in reality the majority of athletes will exceed these recommendations almost on a daily basis.

Timing of protein intake

Although athletes have long been focused on the importance of the amount of protein in their diets, recent studies using techniques to monitor changes in protein synthesis in response to training have found that timing of intake of protein in relation to a training session is another factor to consider. A bout of exercise stimulates an increase in protein synthesis during the recovery period, with an increase in the manufacture of various protein subfractions responding to the type of exercise stimulus (e.g. endurance, resistance or interval training). The provision of a source of essential amino acids during the recovery period enhances the net gain of such protein subfractions. There is still much work to be done to develop our understanding of optimal feeding for the protein response to exercise – that is, optimal amounts, types and timing of protein-rich sources. However, our present understanding is that maximal adaptation occurs when relatively small amounts of high-quality protein are consumed (e.g. 20–25 g protein, providing 6–8 g of essential amino acids) at regular intervals of three to four hours. There are a range of food combinations that would allow an athlete to consume high protein in combination with carbohydrate to achieve recovery goals for refuelling and protein synthesis simultaneously.

The 'best' protein sources

There has been considerable interest in the question, 'What are the best protein sources?' Most studies have investigated short-term protein synthesis in response to different isolated protein sources. There is evidence that the outcomes of these acute studies (often four to five hours post-exercise) correlate well with the long-term muscle mass and strength gains. In general, animal-based protein sources seem to increase protein synthetic rates more than plant-based protein sources, with several studies showing that whey is superior to casein and casein is superior to soy. The reason for this is probably related to the digestion of the protein as well as the amino acid composition. Studies show that 20–25 g or protein contains about 6–8 g of essential amino acids, of which 3 g of leucine is optimal for protein synthesis. Many plant-based protein sources do not contain the same amounts of essential amino acids or do not contain a lot of leucine. However, the practical solution would be to combine more plant-based protein sources and consume more of them. Therefore, the consensus is that the same gains can be achieved by consuming plant-based proteins as can be achieved with animal-based proteins but it will come at a cost of having to consume a greater volume of food, often with more fibre, making it harder for athletes to achieve the amounts needed. In the end, it is the athletes' decision how to achieve their protein goals.

19.6 Physique and sports performance

Physical characteristics, such as height, body weight, muscle mass and body fat levels, can all play a role in the performance of sport. An athlete's physique is determined by inherited characteristics, as well as the conditioning effects of their training programme and diet. Often, 'ideal' physiques for individual sports are set, based on a rigid set of characteristics of successful athletes. However, this process fails to take account of the fact that there is considerable variability in the physical characteristics of sportspeople, even between elite athletes in the same sport. It is therefore dangerous to establish rigid prescriptions for individuals, particularly with regard to body composition. Instead, it is preferable to nominate a range of acceptable values for body fat and body weight within each sport, and then monitor the health and performance of individual athletes within this range. These values may change over the athlete's career.

Some athletes easily achieve the body composition suited to their sport. However, others may need to manipulate characteristics such as muscle mass or body fat levels through changes in diet and training. It is important for an athlete to identify a suitable and realistic body-fat goal, and to achieve this desired change in a suitable period using sensible methods. Descriptions of the various methods of body composition assessment are found elsewhere in this text. For athletes, the criteria for choosing a certain technique should include the validity, reliability and sensitivity of the method. Some methods are best suited to laboratory studies or to occasional monitoring of athletes (e.g. dual X-ray absorptiometry), while methods that are accessible and inexpensive can be used in the field by athletes and coaches. In practice, useful information about body composition can be collected from surface anthropometry measurements such as skinfold (subcutaneous) fat measurements and various body girths and circumferences. Coaches or sports scientists who make these assessments on athletes should be well trained so that they have a small degree of error in repeating measurements. Regular monitoring of the body composition of an athlete can determine their individual 'ideal' physique at different times of the training and competition calendar. It can also monitor their success in achieving these ideals.

19.7 Weight maintenance and other body weight issues

Losing weight

Many sports dietitians note that losing body weight, or, more precisely, losing body fat, is the most common reason for an athlete to seek nutrition counselling. A small body size is useful to reduce the energy cost of activity, to improve temperature regulation in hot conditions and to allow mobility to undertake twists and turns in a confined space. This physique is characteristic of athletes such as gymnasts, divers and marathon runners. A low level of body fat enhances the power-to-weight ratio over a range of body sizes and is a desirable characteristic of many sports that require weight-bearing movement, particularly against gravity (e.g. distance running, mountain biking and uphill cycling, jumps and hurdles). However, low body fat levels are also important for aesthetic sports such as diving, gymnastics and figure skating, where judging involves appearance as well as skill.

There are many common situations in which an athlete's body fat level increases above their healthy or ideal range. A common example is the athlete who comes back from an injury or a break from training carrying several kilograms of body fat in excess of their usual competing weight. Where it is warranted, loss of body fat by an athlete should be achieved by a gradual programme of sustained and moderate energy deficit, achieved by a decrease in dietary energy intake and, perhaps, an increase in energy expenditure through aerobic exercise or activity (see Table 19.3). However, many athletes in weight- or fat-conscious sports

Table 19.3 Strategies for eating to reduce body fat/body mass.

- Consult a sports dietitian for assistance with decisions around manipulation of body composition and strategies to achieve it while maintaining health and performance.
- Identify individual 'ideal' body fat and body weight targets that are achievable and consistent with good health and performance.
- If loss of body fat is required, plan for a realistic rate loss of about 0.5 kg/wk, and set both short-term and long-term goals.
- Examine current exercise and activity plans. If training is primarily skill- or technique-based, or a sedentary lifestyle between training sessions is observed, the athlete may benefit from scheduling some aerobic exercise activities. This should always be done in conjunction with the coach.
- Take an objective look at what the athlete is really eating by arranging to keep a food diary for a period (e.g. 1 wk). Many athletes who feel that they 'hardly eat anything' will be amazed at their hidden eating activities.
- Reduce typical energy intake by an amount that is appropriate to produce loss of body fat (e.g. 2–4 MJ or 500–1000 kcal/d) but still ensures adequate food and nutrient intake. Food should be spread over the day to support efficient refuelling and recovery around key training sessions.
- Target occasions of overeating for special attention. Useful techniques include making meals filling by choosing high-fibre and low-GI forms of foods, fighting the need to finish everything on the plate and spreading food intake over the day so that there is no need to approach meals feeling extreme hunger.
- Focus on opport's unities to reduce intake of fats and oils. Such strategies include choosing low-fat versions of nutritious protein foods, minimising added fats and oils in cooking and food preparation, and enjoying high-fat snack and sweet foods as occasional treats rather than everyday foods.
- Be moderate with alcohol intake and sugar-rich foods, as these may represent 'empty' kilojoules. As alcohol intake is associated with relaxation, it is often associated with unwise eating. Compact carbohydrate sources like sugary foods may be useful for the fuel requirements of some training and competition scenarios, but may not be the best choices
- Focus on nutrient-rich foods so that nutrient needs can be met from a lower energy intake. A broad-range, low-dose vitamin/mineral supplement should be considered if daily energy intake is to be restricted below 6 MJ or 1500 kcal for prolonged periods.
- Be aware of inappropriate eating behaviour. This includes eating when bored or upset, or eating too quickly. Stress or boredom should be handled using alternative activities.
- Be wary of supplements that promise weight loss. There are no special pills, potions or products that produce safe and effective weight loss. If something sounds too good to be true, it probably is.
- Note that dieting for weight loss, or for other reasons that involve strict control over food intake and food range, is a risk factor for the development of eating disorders and disordered eating. For these reasons, it should be considered carefully. Athletes who experience such concerns require expert support from a multidisciplinary team to support their medical, psychological and nutritional challenges. Early intervention is best, and all those in the athlete's social and sporting 'family' should be aware of warning signs and take action as soon as a problem is evident.

strive to achieve very low body fat levels or to reduce body fat levels below what seems to be their natural or healthy level. Although weight-loss efforts often produce a short-term improvement in performance, this must be balanced against the disadvantages related to following unsafe weight-loss methods. Excessive training, chronic low energy and nutrient intake, and psychological distress are often involved in fat-loss strategies and may cause long-term damage to health, happiness or performance. It has been hypothesized that long periods of low energy availability – where total daily intake of energy is below the level needed to cover both the energy cost of training/competition and requirements for full health/function – can impair bone health, reproductive health and other body functions. This scenario is covered in Section 19.8.

Athletes should be encouraged to set realistic goals for body weight and body fat. These are specific to each individual and must be judged by trial and error over a period of time. 'Ideal' weight and body fat targets should be set in terms of ranges, and should consider measures of long-term health and performance rather than short-term benefits alone. In addition, athletes should be able to achieve their targets while eating a diet that is adequate in energy and nutrients, and free of unreasonable food-related stress. Some racial groups or individuals naturally carry very low levels of body fat or can achieve these without paying a substantial penalty. Furthermore, some athletes vary their body fat levels over a season, so that very low levels are achieved only for a specific and short time. In general, however, athletes should not undertake strategies to minimise body fat levels unless they can be sure there are no side-effects or disadvantages. Most importantly, the low body fat levels of elite athletes should not be considered natural or necessary for recreational and sub-elite performers.

It has been suggested that there is a higher risk of eating disorders, or disordered eating behaviours and body perceptions, among athletes in weight-division sports or sports in which success is associated with lower body fat levels than might be expected in the general community. Females seem to be at greater risk than males, reflecting the general dissatisfaction of females in the community with their body shape, as well as the biological predisposition for female athletes to have higher body fat levels than male athletes, despite undertaking the same training programme. Even where clinical eating disorders do not exist, many athletes appear to be restrained eaters, reporting energy intakes that are considerably less than their expected energy requirements. An adequate intake of energy is a prerequisite for many of the goals of sports nutrition.

Making weight

Some sports involve weight divisions in competition, with the goal of matching opponents of equal size and strength. Examples include combative sports (boxing, judo and wrestling), lightweight rowing and weight-lifting. Unfortunately, the culture and common practice in these sports are to try to compete in a weight division that is considerably lighter than normal training body weight. Athletes then 'make weight' over the days before the competition by dehydrating (via saunas, exercising in 'sweat clothes' and diuretics), and restricting food and fluid intake. The short-term penalties of these behaviours include the effect of dehydration and inadequate fuel status on performance. Long-term penalties include psychological stress, chronic inadequate nutrition and effects on hormone status. In 1997 three deaths were recorded among college wrestlers in the USA as a result of severe weight-making practices. Athletes in these sports should be guided to make appropriate choices regarding the competition weight division and to achieve necessary weight loss by safe and long-term strategies to reduce body fat levels.

Gaining muscle mass or supporting growth

Gain of muscle mass is desired by many athletes whose performance is linked to size, strength or power. Many athletes pursue specific muscle hypertrophy gains through a programme of progressive muscle gain, or attempt to exploit the gain in muscle and strength associated with adolescence, particularly in males. Regardless of the desire for muscle gain, growth spurts can lead to substantial energy requirements in some athletes, especially in sports in which young athletes may already be undertaking heavy training loads (e.g. swimming or rowing). While an increase in protein intake can be valuable in supporting the success of a strength-training programme, the primary nutrient requirement is appropriate energy intake to promote the manufacture of

Table 19.4 Strategies for eating to support growth or increase muscle mass.

- Ensure that the athlete is following a well-devised training programme that will stimulate muscle development and growth, as well as other primary needs for their sport
- Set goals for weight and strength gain that are practical and achievable. Increases of 2–4 kg/month may be possible for some athletes, but sustained muscle gain requires consistent training and nutrition support .
- Be organised: apply the same dedication to the eating programme as is applied to training in order to increase the intake of nutrient-dense foods and supply a daily energy surplus of approximately 2–4 MJ (500–1000 kcal). This additional food should supply adequate protein for the development and support of new tissue but may also consider the need for additional carbohydrate to fuel the underpinning training programme.
- Be particularly vigilant to have appropriate foods and drinks available for recovery eating after workouts, supplying protein and carbohydrate needs for refuelling. This will provide additional energy over the day as well as promoting more effective recovery and adaptation following the exercise stimulus.
- Increase the number of meals and snacks rather than the size of meals. This will enable greater intake of food with less risk of 'overfilling' and gastrointestinal discomfort. Including valuable amounts of protein (e.g. 20–40 g) over five to six meals and snacks over the day, including a pre-bed snack, can contribute to maximal protein synthesis as well as general energy intake.
- Avoid excessive intake of fibre and include the use of 'white' cereals with less bulk (e.g. white rice, white bread). It is often impractical to consume a diet that is solely based on wholegrain and high-fibre foods.
- Make use of high-energy fluids such as milkshakes, fruit smoothies or commercial liquid meal supplements. These drinks provide a compact and low-bulk source of energy and nutrients, and can be consumed with meals or as snacks, including before or after a training session. Carbohydrate support during training sessions (e.g. sports drinks) may help to fuel the session as well as contribute to total daily energy intake.
- Note that protein supplements are often associated with muscle gain programmes, and indeed an increase in protein intake should be part of the higher-energy focus. Such supplements – or indeed, multi-macro liquid meal supplements – can be a useful way to ensure access to valuable amounts of high-quality protein regardless of the suitability of available food supplies in a busy day or opportunities to store or prepare everyday protein-rich foods. Nevertheless, take care to purchase cost-effective products from reputable companies, especially batch-tested options, to avoid the inadvertent intake of contaminants including substances prohibited in sport.
- Be aware that many athletes do not eat as much – or, more importantly, as often – as they think. It is useful to examine the actual intake of athletes who fail to gain weight yet report 'constant eating'. Commitments such as training, sleep, medical/physiotherapy appointments, work or school often get in the way of eating opportunities. A food record will identify the hours and occasions of minimal food intake. This information should be used to reorganise the day or to find creative ways to make nutritious foods and drinks part of the activity.

new muscle tissue and other tissues needed to support it, as well as to provide fuel for the underpinning training programme. Many athletes do not achieve an adequate energy intake to support these goals. Table 19.4 provides some practical strategies to address this challenge.

19.8 Low energy availability (LEA) and relative energy deficiency in sport (REDs)

Several decades ago, the common presentation of amenorrhoea (loss of menstrual cycles), eating disorders and osteoporosis in female athletes, particularly those in endurance sports, was noted, with a plausible hypothesis connecting the occurrence of these concerns. This was termed 'the female athlete triad' to promote awareness of prevention and early treatment of these problems. It was tempting to attribute the problem to high-volume training in female athletes, at a time when Title IX in Collegiate Sports and increased acceptance of other sporting opportunities for female athletes were supporting an increased participation of women in demanding sports. However, elegant laboratory studies by Professor Anne Loucks showed that the dietary contribution to this syndrome was low energy availability (LEA) – a mismatch between energy intake and the energy committed to exercise – leaving insufficient energy to support all the activities needed for optimal body function and health. An increased exercise energy expenditure and/or restricted dietary energy intake can lead to the scenario in which the body adjusts to this overall energy deficit by reducing its energy contribution to body functions that it considers temporarily expendable. This adaptation reduces/eliminates the energy deficit, but might also reduce the body's full contribution to the metabolic cost of optimal bone health, energy metabolism, protein synthesis, immune support, reproductive systems and other body activities. Our understanding of the causes, potential outcomes and measures to prevent/treat this syndrome has evolved over the past two decades, with the 2023 update to the International Olympic Committee's Consensus Statement on Relative Energy Deficiency in Sport (REDs) including the following summary (paraphrased here):

- LEA can occur in both male and female athletes due to mismatches between energy intake and the energy cost of training or competition. Many characteristics that occur commonly in the conduct and culture of sport can lead to LEA – e.g. high volume training, restriction of energy intake to reduce body fat/mass, food insecurity, poor nutrition knowledge/practical skills, and insufficient time, food availability or gastrointestinal comfort to allow the athlete to consume their energy needs in a busy lifestyle.
- Humans have adapted to scenarios of LEA as part of our evolution and have been able to adapt total energy expenditure in times of poor food availability or high energy demand to promote survival. We can manage many situations of exposure to LEA with 'adaptive' responses in which metabolic adjustments are short-lived and reversible. Indeed, many scenarios, such as manipulation of body composition or periods of intensified training, are integral to the lives of many athletes and deliver performance improvements. However, some scenarios of LEA exposure may interact with other modifying variables in the athlete's makeup or behaviour to create 'problematic' LEA, in which a number of potential impairments to health or performance may occur (see Figure 19.4).
- REDs is the syndrome involving a collection of health and performance impairments that are found in association with LEA. The terms are not interchangeable; in epidemiological terms, LEA is an exposure variable that may be associated with the REDs disease. It is difficult to measure energy availability in free-living athletes. Although it can be clearly defined by a formula [(energy intake – exercise energy expenditure)/kg fat free mass], the assessment of energy intake and exercise energy expenditure is time-consuming and likely to contain errors/inaccuracies that are larger than the magnitude of concerning restrictions in EA. Although such calculations may have some utility in research and practice, particularly to construct prospective diets/training programmes, it has been suggested that the assessment of LEA should be undertaken via the assessment of biomarkers of metabolic and hormonal adaptations. The problem, however, is that there could be other causes of changes in these metabolic and hormonal markers as well, independent of LEA.
- Some screening tools may assist in identifying athletes at high risk of LEA who should receive further assessment and early treatment.
- Although laboratory models in which energy availability is reduced for short periods show clear evidence of pertubations to different body systems, in real life there is considerable individual variability in the prevalence of health and performance impairments associated with scenarios of LEA. Because of problems mentioned above, it will always be difficult to differentiate between LEA or other causes of the various symptoms (like training and other stresses). It is therefore best to measure biomarkers and symptoms and consider LEA as one of the potential causes of these symptoms.
- Future research should attempt to identify the characteristics of the LEA (e.g. duration, severity, continuous vs periodic exposure) that are most detrimental, as well as moderating factors (e.g. genetics, training characteristics, dietary characteristics, use of medications) that may reduce or increase the risk of problematic LEA.
- This information should be used to identify athletes with indications of LEA and REDs for a multidisciplinary approach to addressing underlying causes and related impairments. Awareness of risk factors associated with problematic LEA may help to prevent its occurrence or promote early assessment and treatment.
- Criticism of the REDs model could be that there is too much focus on LEA as the singular cause of a wide range of generic symptoms. If an athlete presents with some of these symptoms, LEA should be considered as one of many potential causes. Other potential causes such as excessive training or other stresses, infection, insufficient carbohydrate intake or nutrient deficiencies should be considered just as much. In any case, many such factors are related, so from a practical point of view it seems undesirable to single out energy as a cause.

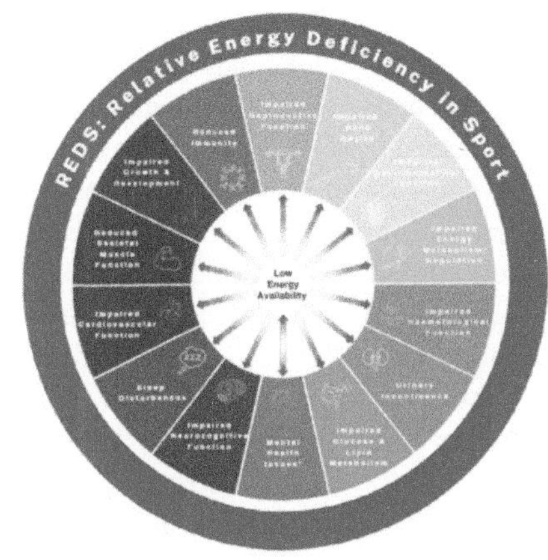

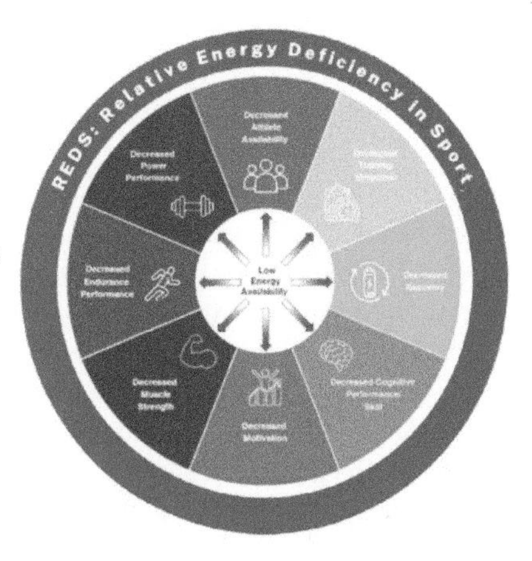

Figure 19.4 The REDs conceptual models provide a simple representation that LEA exposure exists on a continuum between adaptable and problematic, with a range of potential outcomes on both health (left) and performance (right) metrics. The models demonstrate that increased exposure to LEA increases the risk of developing a variety of health and performance impairments that form the REDs syndrome. *Source:* From Burke et al. (2023).

19.9 Vitamins and minerals

The daily requirement for at least some vitamins and minerals is increased beyond population levels in people undertaking a strenuous exercise programme. The potential reasons for this increased requirement are increased loss through sweat, urine and perhaps faeces, and through increased production of free radicals. Unfortunately, at present the additional micronutrient requirements of athletes cannot be quantified. The key factors ensuring an adequate intake of vitamins and minerals are a moderate to high energy intake and a varied diet based on nutritious foods. Dietary surveys show that most athletes report dietary practices that easily supply vitamins and minerals in excess of recommended daily allowances (RDAs) and are likely to meet any increases in micronutrient demand caused by training. However, not all athletes eat varied diets of adequate energy intake and some may need help to improve both the quality and quantity of their food selections.

Studies of the micronutrient status of athletes have not revealed any significant differences between indices in athletes and sedentary controls. The results suggest that athletic training, per se, does not lead to micronutrient deficiency. These data should, however, be interpreted very carefully as most indices are not sensitive enough to detect marginal deficiencies. Overall, generalised vitamin and mineral supplementation for all athletes is not justified. Furthermore, studies do not support an increase in performance with such supplementation, except in the case where a pre-existing deficiency was corrected. The best management for the athlete with a high risk of sub-optimal intake of micronutrients is to provide nutrition education to improve their food intake. However, a low-dose, broad-range multivitamin/mineral supplement may be useful where the athlete is unwilling or unable to make dietary changes, or when the athlete is travelling to places with an uncertain food supply and eating schedule.

Antioxidant vitamins

Exercise has been linked with an increased production of free oxygen radical species capable of causing cellular damage. A sudden increase in training stress (such as an increase in volume or intensity) or a stressful environment (e.g. training in hot conditions or at altitude) is believed to increase the production of these free oxygen radicals, leading to an increase in markers of cellular damage. Supplementation with antioxidant vitamins such as vitamin C or vitamin E has been suggested to increase antioxidant status and provide protection against this damage.

The literature on the effects of antioxidant supplementation on antioxidant status, cellular damage and performance is complex and confusing. Some, but not all, studies show that acute supplementation during periods of increased stress may provide bridging protection until athletes are able to adapt their own antioxidant status to meet this stress. It is possible that subtle benefits occur at a cellular level but these are too small to translate into detectable performance outcomes. On the other hand, some research shows that oxidative reactions provide a useful role in signalling the adaptations to an exercise stimulus. Antioxidant supplementation may therefore be harmful to the process of adaptation to exercise if it reduces the signalling stimulus. Again, any reduction in the training response may be too small to detect. At the present time it is unwise to make recommendations about whether antioxidant supplementation is useful or harmful. A more valuable approach may be to increase the antioxidant content of the diet by eating plenty of fruits, vegetables and wholegrain cereals. This way, an increase in a range of antioxidants and plant phytochemicals can be achieved rather than upsetting the delicate balance of the body's antioxidant system by supplementing with large amount of a few compounds.

Iron

Minerals are the micronutrients at most risk of inadequate intake in the diets of athletes. Inadequate iron status can reduce exercise performance via sub-optimal levels of haemoglobin and perhaps iron-related muscle enzymes. Reductions in the haemoglobin levels of distance runners first alerted sports scientists to the issue of iron status of athletes. However, more recent research has raised the problem of distinguishing true iron deficiency from alterations in iron status measures that are caused by exercise itself. True iron status in athletes is misdiagnosed from single measures of low haemoglobin and ferritin levels. Problems include the failure to recognise that the increase in blood volume that accompanies endurance training, especially in hot conditions, will cause a dilution of all the blood contents and apparent (but not problematic) reductions in haemoglobin concentrations. By contrast, ferritin is an acute-phase reactant and may be falsely elevated in the immediate response to strenuous exercise or other physiological stress. Therefore, it is preferable for athletes to have standardised screening protocols for iron tests with observations of changes in haemoglobin and ferritin levels contributing to the interpretation of their true status.

Athletes are nevertheless at risk of becoming iron-deficient, with prevention, or early intervention, of reduced ferritin concentrations (e.g. early iron deficiency without anaemia) being the preferred approach rather than allowing anaemia to occur. Indeed, given the important role of iron in the oxygen-carrying capacity of the blood (haemoglobin) and muscle (myoglobin), as well as many iron-containing enzyme complexes, sub-optimal iron status may reduce performance or impede optimal adaptations to training stimulus well before anaemia occurs. The causes of iron deficiency in athletes include factors shared with members of the general community: a lower than desirable intake of bioavailable iron and/or increased iron requirements or losses. Iron requirements may be increased in some athletes owing to growth needs or to increased losses of blood and red blood cell destruction. More recently, the role of hepcidin, a hormone that down-regulates iron absorption and recycling, has been recognised. Hepcidin concentrations have been shown to increase in the hours following strenuous exercise, providing a further risk factor for iron deficiency in athletes, as well as identifying better times of the day to consume iron-rich foods or supplements to enhance their absorption. Many athletes do not consume enough bioavailable iron, with high-risk populations for low-energy and/or low-iron diets being females, athletes with disordered eating, those with dietary restrictions and vegetarians/vegans. Times of increased or specialised needs include intensified training programmes, including altitude training.

Iron is found in a range of plant and animal food sources in two forms. Haem iron is found only in animal foods containing flesh or blood, whereas organic iron is found both in animal foods and plant foods. Whereas haem iron is relatively well absorbed from single foods and mixed meals (15–35% bioavailability), the absorption of non-haem iron from single plant sources is low and variable (2–8%). The bioavailability of non-haem iron, and to a lesser extent haem ion, is affected by other foods consumed in the same meal. Factors that enhance iron absorption include vitamin C, peptides from fish, meat/ and chicken, alcohol and food acids, while factors that inhibit absorption include phytates, polyphenols, calcium and peptides from plant sources such as soy protein. The absorption of both haem and non-haem iron is increased as an adaptive response in people who are iron-deficient or have increased iron requirements. However, as previously mentioned, increases in the hepcidin hormone in the hours after strenuous exercise will also interfere with the absorption of dietary and supplemental forms of iron.

Low iron status, indicated by serum ferritin levels lower than 20 ng/ml, should be considered for further assessment and treatment. Evaluation and management of iron status should be undertaken on an individual basis by a sports medicine expert. Prevention and treatment of iron deficiency may include iron supplementation, with a recommended therapeutic dose of 100 mg/day of elemental iron for two to three months. It is now recommended that iron supplements be taken in the morning, either before or immediately after exercise, to coincide with lower hepcidin concentrations. However, gastrointestinal discomfort may require some athletes to seek alternative protocols that reduce gastrointestinal side-effects, including taking the supplement in larger doses every second day. Importantly, the management plan should include dietary counselling to increase the intake of bioavailable iron and appropriate strategies to reduce any unwarranted iron loss (see Table 19.5). Many athletes self-prescribe iron supplements whenever they feel fatigued. However, this practice do not provide the athlete with the opportunity for adequate assessment of iron losses and expert dietary counselling from a sports dietitian.

Calcium

Weight-bearing exercise is considered to be one of the best protectors of bone health, and therefore it is puzzling to find reports of low bone mineral density (BMD) in some athletes, notably female distance runners. However, a serious outcome of LEA and other causes of menstrual disturbances in female athletes is the high risk of either direct loss of bone density or failure to optimise the gaining of peak bone mass during early adulthood. Lower BMD or loss of bone density is also found in some male athletes, often in association with LEA. In some scenarios, BMD is within normal population ranges, but athletes with lower BMD within the group are found to have a higher risk of stress fractures or other injuries, suggesting that although their sporting activity confers some benefits to BMD, it is not sufficient to withstand the total load placed on the bone. Furthermore, some sports (e.g. swimming, rowing, cycling) are not weight-bearing and it may not be practical for athletes in these sports to add other bone-loading activities to a high-volume training load. In fact, there is a suggestion that sports-specific ranges for BMD in athletes should be set, rather than rely on general population characteristics. In any case, sub-optimal bone density and increased risk of stress fractures are among the possible outcomes recognised in athletes who are training hard, are under constant stress and/or are restricting energy intake over long periods of time. This can reduce the athlete's career span as well as increase the risk of osteoporosis in later life.

Main treatment goals should focus on correcting underlying causes of impaired bone health, with calcium being considered a permissive nutrient which is important for bones but cannot overcome other issues. Nevertheless, athletes with impaired menstrual function may be advised to increase their calcium intake from the normal population recommendations of 800–1000 to 1200–1500 mg/day. As with iron, inadequate calcium intake is usually linked with low-energy intakes and diets with food restrictions, including vegan diets. Where adequate calcium intake cannot be met through dietary means, a calcium supplement may be considered (see Table 19.5).

19.10 Fluid and electrolyte loss and replacement in exercise

Water has many important functions in the human body. Approximately 55–60% of the body is made up of water. The total water content of the human body is between 30 and 50 litres. Every day we excrete water (sweat, urine, evaporative losses), while water intake may vary from 1 to about 12 litres per day. Water turnover can be very high in some conditions, but the total body water content is remarkably constant and rarely exceeds variations of 1 litre. However, during exercise and especially during exercise in hot conditions, sweat rates (and thus water losses) may increase dramatically and dehydration may occur (i.e. the body is in negative fluid balance). Dehydration can have an enormous impact on physical and mental function, and increases the risk of heat illness. Even mild dehydration can result in reduced exercise capacity.

Table 19.5 Strategies to meet iron and calcium needs.

- Include haem iron-rich foods (red meats, shellfish, liver) regularly in meals, at least three to five times per week. These can be integrated into meals that meet other dietary goals.
- Be aware of iron-rich plant sources, especially if vegetarian/vegan, such as fortified breakfast cereals and other grain foods, legumes, nuts and seeds.
- Enhance the absorption of non-haem iron from plant sources by including a vitamin C-rich food or meat/fish/chicken at the same meal. For example, a glass of orange juice may be consumed with breakfast cereal or a small amount of lean meat can be added to beans to make a chilli con carne
- .Note that some foods (wholegrains, coffee, strong tea, dairy) or supplements (calcium) interfere with iron absorption from non-haem iron foods. Athletes who are at risk of iron deficiency should separate their consumption (~1 h) from meals which are reliant on plant-based iron, or add an iron-enhancing food to the mix
- Take iron supplements on the advice of a sports dietitian or doctor. They may be useful in the supervised treatment and prevention of iron deficiency but should be supported by dietary improvements. Protocols for taking iron supplements should consider efficiency of absorption (consume with vitamin C or choose a vitamin C-fortified iron product; take before or immediately after exercise) as well as gastrointestinal comfort. Some iron preparations are better tolerated than others, and protocols such as every second day use may also assist when gut issues reduce compliance
- Consume dairy products (milks, cheese, yoghurt, etc.) to provide a key source of calcium as well as other important nutrients in a sports nutrition plan (e.g. protein). Most dietary guidelines recommend eating at least three servings of dairy foods a day, where one serving is equal to 200 ml of milk or a 200 g carton of yoghurt. Fish eaten with bones (e.g. sardines, tinned salmon) is another good source.
- Include plant-based sources of calcium, especially if following a vegan and vegetarian eating plan, such as soy products (especially fortified soy-based dairy alternatives and other fortified plant-based milks), green leafy vegetables (e.g. kale, broccoli) and some nuts/seeds.
- Increase calcium if you are undergoing a growth spurt or having a baby/breastfeeding. Female athletes with irregular or absent menstrual cycles should seek expert advice from a sports physician, and may also require extra calcium.

Fluid losses

Exercise (muscle contraction) causes an increase in heat production in the body. Muscle contraction during most activities is only about 15–20% efficient. This means that, of all the energy produced, only about 15–20% is used for the actual movement and the remainder is lost as heat. For every litre of oxygen consumed during exercise, approximately 16 kJ of heat is produced and only 4 kJ is actually used to perform mechanical work. If this heat was not dissipated, the body would soon overheat.

When a well-trained individual is exercising at 80–90% $\dot{V}O_{2\ max}$, the body's heat production may be more than 1000 W (i.e. 3.6 MJ/h). This could potentially cause the body core temperature to increase by 1°C every five to eight minutes if no heat could be dissipated. As a result, body core temperature could approach dangerous levels in less than 20 minutes.

There are several mechanisms to dissipate this heat and to maintain body core temperature in a relatively narrow range: 36–38°C in resting conditions and 38–40°C during exercise and hot conditions. The most important cooling mechanism of the body is sweating, although radiation and convection can also contribute. Sweat must evaporate from the body surface to exert a cooling effect. Evaporation of 1 litre of water from the skin will remove 2.4 MJ of heat from the body. Although sweating is a very effective way to dissipate heat, it may cause dehydration if sweat losses are not replenished. This may cause further problems for the athlete; progressive dehydration impairs the ability to sweat and, therefore, to regulate body temperature. Body temperature rises more rapidly in the dehydrated state and this is commonly accompanied by a higher heart rate during exercise.

Fluid losses are mainly dependent on three factors:

- ambient environmental conditions (temperature, humidity)
- exercise intensity
- duration of exercise and the duration of the heat exposure.

The environmental heat stress is determined by the ambient temperature, relative humidity, wind velocity and solar radiation. The relative humidity is the most important of these factors, as a high humidity will severely compromise the evaporative loss of sweat. Often sweat will drip off the skin in such conditions, rather than evaporate. This means that heat loss via this route will be less effective.

It is important to note that problems of hyperthermia and heat injury are not restricted to prolonged exercise in a hot environment; heat production is directly proportional to exercise intensity, so very strenuous exercise, even in a cool environment, can cause a substantial rise in body temperature.

To maintain water balance, fluid intake must compensate for the fluid loss that occurs during exercise. Fluid intake is usually dependent on thirst feelings, but thirst (or the lack of thirst) can also be overridden by conscious control. It is important to note, however, that thirst is a poor indicator of fluid requirements or the degree of dehydration. In general, the sensation of feeling thirsty is not perceived until a person has lost at least 2% of body mass. As already mentioned, even this mild degree of dehydration is sufficient to impair exercise performance. It has also been shown that athletes tend to drink too little even when sufficient fluid is available.

Effects of dehydration

As the body becomes progressively dehydrated, a reduction in skin blood flow and sweat rate may occur. A high humidity may limit evaporative sweat loss, which will lead to further rises in core temperature, resulting in fatigue and possible heat injury to body tissues. The latter is potentially fatal. Dehydration can affect many aspects of athletic performance, and severe dehydration can increase the risk of heat exhaustion and even more serious health consequences.

Effect of dehydration on exercise performance

Several studies have shown that mild dehydration, equivalent to the loss of only 2–3% body weight, is sufficient to impair exercise performance significantly (Figure 19.5). Many of the early studies that demonstrated negative effects of dehydration on performance were done in extremely hot laboratory conditions, had little or no wind cooling and were not blinded or placebo-controlled. This has led others to criticise this research and question the outcomes. However, more recent studies using nasogastric tubes have confirmed the findings of older studies, although the effects seem to be smaller than originally thought. It is clear that, especially in hot conditions, dehydration is a factor that can affect performance. There is no debate about whether dehydration can be detrimental, but there is a discussion about the exact percentage of dehydration at which these negative effects become apparent. In general, these effects will occur at 2–3% dehydration, although this may be individually determined and context-dependent (some individuals tolerate dehydration better than others and environmental conditions will influence the effect of dehydration on performance). Although dehydration has detrimental effects on performance, especially in hot conditions, such effects can also be observed in cool conditions. Both decreases in maximal aerobic power ($VO_{2\ max}$) and decreases in endurance capacity have been reported with dehydration in temperate conditions.

There are several reasons why dehydration can result in decreased exercise performance. First of all, a fall in plasma volume, decreased blood volume, increased blood viscosity and a lower central venous pressure can result in a reduced stroke volume and maximal cardiac output. In addition, during exercise in the heat, the dilatation of the skin blood vessels reduces the proportion of the cardiac output that is devoted to perfusion of the working muscles. Dehydration also impairs the ability of the body to lose heat. Both sweat rate and skin blood flow

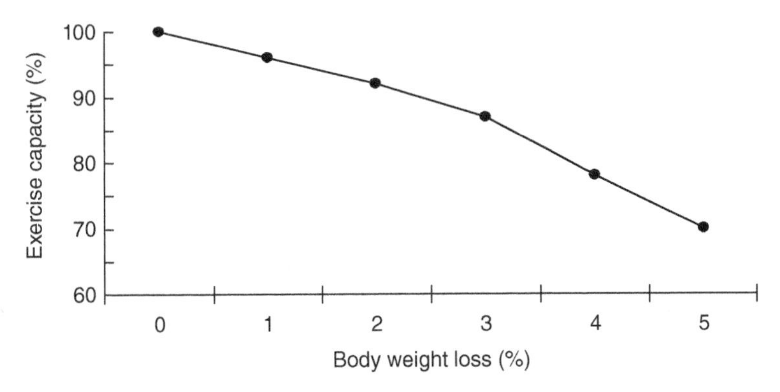

Figure 19.5 Effect of dehydration on exercise capacity.

are lower at the same core temperature for the dehydrated compared with the euhydrated state. This means that body temperature rises more rapidly during exercise when the body is dehydrated. Finally, the larger rise in core temperature during exercise in the dehydrated state is associated with an increased rate of muscle glycogen breakdown. Depletion of these stores could also result in premature fatigue during prolonged exercise. In addition to the effects of dehydration on endurance, there are reported negative effects on coordination and cognitive functioning. This is likely to impact on all sports where skill and decision-making are involved.

Heat illness

Dehydration poses a serious health risk in that it increases the risk of heat cramps, heat exhaustion and life-threatening heat stroke. Early symptoms of heat injury are excessive sweating, headache, nausea, dizziness and reduced consciousness and mental function. When the core temperature rises to over 40°C, heat stroke may develop, characterised by hot, dry skin, confusion and loss of consciousness. There are several anecdotal reports of athletes and army recruits dying because of heat stroke. Most of these deaths have been explained by exercise in hot conditions, often with insufficient fluid intake. These problems affect not only highly trained athletes, but also less well-trained people participating in sport. Although well-trained individuals will generally exercise at higher intensities and therefore produce more heat, less well-trained individuals have less effective thermoregulation during exercise and work less economically. Overweight, unacclimated and ill individuals are especially likely to develop heat stroke.

Hyponatraemia

An electrolyte imbalance commonly referred to as hyponatraemia (low plasma sodium) or 'water intoxication' has occasionally been reported in endurance athletes due to excessive intake of fluid before and during an event. This appears to be most common among slow runners in marathon and ultra-marathon races who are able to drink at rates that are far greater than their sweat losses. High losses of sodium in sweat may add to this problem. The symptoms of hyponatraemia are similar to those associated with dehydration and include mental confusion, weakness and fainting. Hyponatraemia can be fatal and has caused several deaths in marathons and ultra-endurance events. It is an avoidable problem if athletes are aware of their fluid requirements during sport and have a well-practised drinking plan that replaces most of, but not more than, their sweat losses during the session. There is also a danger of misdiagnosis of this condition when it occurs in individuals participating in sporting events. The usual treatment for dehydration is administration of fluid intravenously and orally. If this treatment were to be given to a hyponatraemic individual, the consequences could be fatal.

Fluid intake strategies

Fluid intake during exercise can help to maintain plasma volume and prevent the adverse effects of dehydration on muscle strength, endurance and coordination. When there is only little time in between two exercise bouts, rapid rehydration is crucial and drinking regimens need to be employed to optimise fluid delivery. Strategies for fluid replacement before, during and after exercise will be discussed in the following sections, with a summary provided in Table 19.6.

Fluid intake during exercise

To avoid dehydration during prolonged exercise, fluids must be consumed to match the sweat losses. By regularly measuring body weight before and after a training session, it is possible to obtain a good indication of fluid loss. Ideally, the weight loss is compensated by a similar amount of fluid intake. However, it may not always be possible or necessary to prevent dehydration completely.

Sweat rates during strenuous exercise in the heat can amount up to 2–3 l/h. Such large volumes of fluid are difficult, if not impossible, to ingest without causing gastrointestinal discomfort, and therefore it is often not practically possible to achieve fluid intakes that match sweat losses during exercise. Another factor that can make the ingestion of large amounts of fluid difficult is the fact that in some sports or disciplines the rules or practicalities of the specific sport may limit the opportunities for drinking during exercise.

Fluid intake may be useful during exercise longer than 30–60 minutes, but there is no advantage during strenuous exercise of less than 30 minutes in duration. During such high-intensity exercise, gastric emptying is inhibited and the drink may cause gastrointestinal distress with no performance benefit.

Table 19.6 Drinking strategies for exercise.

- Be aware of factors that can affect heat accumulation during training or competition: time of day, duration and intensity of exercise, environmental conditions or the conditions in indoor venues, the suitability of uniforms or protective gear. It may be possible to adjust some of these factors if an unacceptable head load is anticipated.
- Begin the event properly hydrated. Fluid losses from previous events or training sessions need to be restored. Fluids should be included in the pre-event meal and during/after the warm-up, with the volume and timing of intake being chosen to allow time for excess fluid to be urinated before the start of the event.
- Gain an appreciation of typical sweat losses during workouts and events by monitoring weight loss over the session. Typically, 1 kg of fluid loss is equivalent to a fluid deficit of 1 litre. If possible and practical, fluid intake during a session should keep the total fluid deficit to less than 2% of body mass. It may take practice and creativity to develop drinking practices that achieve this goal. Athletes should practise drinking strategies in training so that they can implement them successfully in competition.
- Make good use of the availability of fluids (e.g. access to aid stations, use of trainers) and the opportunities to drink (e.g. half-time, time outs, injury breaks) that occur in each specific sporting activity or exercise environment. These will allow individual athletes to develop their own drinking plan. Generally it is best for fluid intake to occur early and regularly during the exercise to prevent a large fluid deficit from occurring.
- Do not drink at a rate that exceeds sweating rates so that weight is gained over the session. Excessive overhydration can cause the potentially fatal condition of hyponatraemia.
- Note that the voluntary intake of drinks that provide flavour, sweetness and salt is greater than that of plain water. Sports drinks can meet several goals of sports nutrition, as they supply fuel as well as assisting hydration needs. Cool fluids will also be appealing in warm conditions.
- Rehydrate after the session, knowing that most exercise situations will result in a fluid deficit (sweat losses are greater than fluid intake). An intake of fluid equal to 125–150% of the fluid deficit over the hours after exercise will be needed to accommodate further fluid losses (urine and sweat losses) and fully restore hydration levels. A plan of intake is useful when the fluid deficit exceeds 2% body mass as thirst may not cover immediate fluid needs.
- Promote fluid retention and overall fluid restoration by replacing the salt lost in sweat. This can be achieved by drinking salt-containing fluids (oral rehydration solutions or high-salt sports drinks developed for endurance sports) or by consuming salt-rich meals and foods at the same time as fluids. Avoid excessive intake of alcohol as it increases urine losses and interferes with rehydration.

Electrolyte intake during exercise

With the development of sports drinks came an extreme focus on electrolytes as a key ingredient in these drinks. However, there is little or no evidence that electrolytes improve performance or reduce the risk of heat cramps, as is often claimed. There is some evidence that the addition of sodium can help water absorption to a small degree. The rationale for taking electrolytes during exercise is that they prevent the development of hyponatremia or the depletion of body sodium stores. Hyponatremia will develop mostly as a result of large intakes of fluid and the addition of sodium to such drinks will have minimal effect on the development of hyponatremia. Moreover, this would realistically only be possible during exercise lasting longer than four hours. Depletion of body sodium stores is highly unlikely even with relative high sodium losses in sweat. Negative effects of severe dehydration would occur well before the depletion of body sodium stores. Therefore, relatively small amounts of sodium are recommended, but the supplementation of salt tablets or high-dose electrolytes is not advised in most situations. It must be emphasised that the addition of sodium and other electrolytes to sports drinks is to increase palatability, maintain thirst (and therefore promote drinking) and, to a smaller degree, prevent hyponatraemia and increase the rate of water absorption, rather than to replace the electrolyte losses through sweating. Replacement of the electrolytes lost in sweat can normally wait until the post-exercise recovery period.

Practice drinking during training

Although it is often difficult to tolerate the volumes of fluid needed to prevent dehydration, the volume of fluid that is tolerable is trainable and can be increased with frequent drinking in training. Often this is neglected during training. Training to drink will accustom athletes to the feeling of exercising with fluid in the stomach. It also gives the opportunity to experiment with different volumes and flavourings to determine how much fluid intake they can tolerate and which formulations suit them best.

Composition of sports drinks

Numerous studies have shown that regular water intake during prolonged exercise is effective in improving performance. Fluid intake during prolonged exercise also offers the opportunity to provide some fuel (carbohydrate). The addition of some carbohydrate to drinks consumed during exercise has been shown to have an additive independent effect in improving exercise performance. The ideal drink for fluid and energy replacement during exercise is one that tastes good to the athlete, does not cause gastrointestinal discomfort when consumed in large volumes, is rapidly emptied from the stomach and absorbed in the intestine, and provides energy in the form of carbohydrate.

Sports drinks typically have three main ingredients: water, carbohydrate and sodium. The water and carbohydrate provide fluid and energy, respectively, while sodium is included to aid water absorption and retention.

Although carbohydrate is important, an overly concentrated carbohydrate solution may provide more fuel for the working muscles but will decrease the amount of water that can be absorbed owing to a slowing of gastric emptying. Water is absorbed into the body primarily through the small intestine, but the absorption of water is decreased if the concentration of dissolved carbohydrate (or other substances) in the drink is too high. In this situation, water will be drawn out of the interstitial fluid and plasma into the lumen of the small intestine by osmosis. As long as the fluid remains hypotonic with respect to plasma, the uptake of water from the small intestine is not adversely affected. The presence of small amounts of glucose and sodium tend to increase the rate of water absorption slightly compared with pure water. However, in many situations the intake of carbohydrates is more important than the intake of fluids and in such situations a higher concentration of carbohydrate in a drink is recommended.

Rehydration after exercise

When there is little time for recovery (2–6 h) in between exercise bouts, the replacement of fluid and electrolytes in the post-exercise recovery period is of crucial importance. In the limited time available, the athlete should strive to maximise rehydration. The main factors influencing the effectiveness of post-exercise dehydration are the volume and composition of the fluid consumed. Plain water is not the ideal post-exercise rehydration beverage when rapid and complete restoration of body fluid balance is necessary. Ingestion of water alone in the post-exercise period results in a rapid fall in the plasma sodium concentration and the plasma osmolarity. These changes have the effect of reducing the stimulation to drink (thirst) and increasing the urine output, both of which will delay the rehydration process. Plasma volume is more rapidly and completely restored in the post-exercise period if salt (sodium chloride) is added to the water consumed. The optimal sodium concentration of rehydration fluids (~50–60 mmol/l) is similar to the upper limit of the sodium concentration found in sweat, but is considerably higher than that of many commercially available sports drinks, which usually contain 10–25 mmol/l (60–150 mg/l).

Ingesting a beverage containing sodium not only promotes rapid fluid absorption in the small intestine, but also allows the plasma sodium concentration to remain elevated during the rehydration period and helps to maintain thirst while delaying stimulation of urine production. The inclusion of potassium in the beverage consumed after exercise would be expected to enhance the replacement of intracellular water and thus promote rehydration, but currently there is little experimental evidence to support this. The rehydration drink should also contain carbohydrate because the presence of glucose will also stimulate fluid absorption in the gut and improve beverage taste. Following exercise, the uptake of glucose into the muscle for glycogen resynthesis should also promote intracellular rehydration.

Restoring fluid balance after exercise requires the intake of a greater volume of fluid than the deficit that was incurred during the session. This is because some of the ingested fluid will be excreted in urine. Recent studies indicate that ingestion of 125–150% or more of weight loss is required to achieve normal hydration within six hours after exercise (Figure 19.6).

19.11 Performance supplements

Dietary supplements are part of a multi-billion-pound industry which is growing rapidly in both the value and the number of products. Sports supplements make a substantial contribution of these totals, and surveys find that most athletes report the use of such products. The down-sides of supplement use include the expense, the risk that the products contain hazardous or problematic ingredients, and the waste of time and

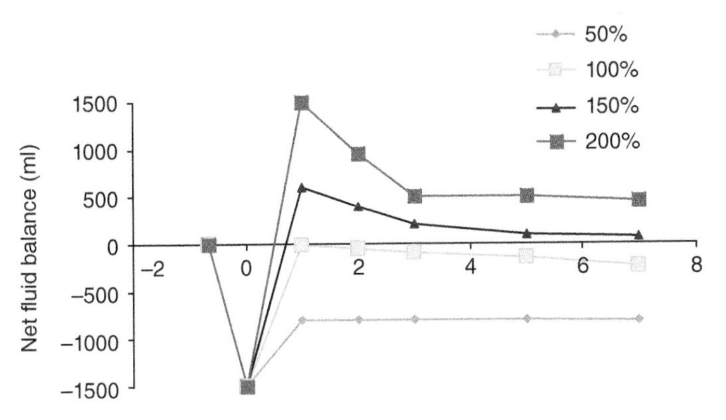

Figure 19.6 Net fluid balance after exercise with the ingestion of different volumes of a drink. Ingestion of 150% or more of weight loss is required to achieve normal hydration within six hours after exercise. *Source:* Adapted from Shirreffs SM et al., 1996.

resources on products that do not live up to their claims. Many athletes do not recognise that the regulation of the manufacture and marketing of supplements is less rigorous or less strongly enforced than is the case for foods or pharmaceutical products, particularly in some countries. Health and safety concerns are associated with contaminants such as heavy metals, the failure to disclose high doses of stimulants, and presentation of untested experimental peptides (illegally) as supplements. In addition, some athletes mix and match their intakes of supplements in combinations that can lead to interactions or excessive doses. Supplement use has been associated with deaths and substantial numbers of visits to emergency departments.

Of particular concern to high-performance athletes is the presence of substances that are prohibited for use by governing bodies such as the World Anti-Doping Agency and National Collegiate Athletics Association. A range of ingredients, such as stimulants and products with growth hormone and testosterone activity, can be found in some supplements. These may be deliberately consumed by some athletes, but they may also ingested inadvertently due to the athlete's failure to read ingredient labels or to recognise that some supplements contain undeclared ingredients. Here, the manufacturer may have 'spiked' their product with banned substances in the hope of achieving an effect. Alternatively, they may have allowed contamination of their products with banned substances through poor manufacturing practices or the use of contaminated raw ingredients. Even though the contaminant may be present in minute amounts, it may still cause an athlete to register a positive test from urine or blood testing. Regardless of the athlete's motives for ingesting the substance, the strict liability rules of anti-doping agencies consider this to be an anti-doping rule violation with associated penalties. A landmark study in 2004 found that 15% of tested supplements contained a banned anabolic substance (prohormone or testosterone) that was not declared on the product label. Further studies over the intervening years have repeated similar rates of contamination but with a wider group of banned substances. Athletes are advised to take caution with the use of supplements, with the following suggestions:

1. Recognise that the decision to use any supplement or sports foods is a personal choice for which the athlete must bear the responsibility for any disadvantages or harmful outcomes. Sports governing bodies are now less conservative about sports supplements, and most recognise their contribution to a sports nutrition programme. Nevertheless, they should be considered as a small part of a well-chosen nutrition plan that is appropriate only when the athlete has reached an appropriate age and level of preparation.
2. Limit your use of supplements and sports foods to those that have an evidence base and with the advice of a sports physician or accredited sports scientist or dietitian. Products that may be used under appropriate scenarios include sports foods that are used to meet nutritional goals (sports drinks, protein supplements, sports gels, etc.) and medical supplements (iron, calcium, vitamin D) that are used to prevent or treat a diagnosed nutrient deficiency. There are only a small number of performance supplements that are recognised to achieve worthwhile benefits to health and performance (see Figure 19.7).

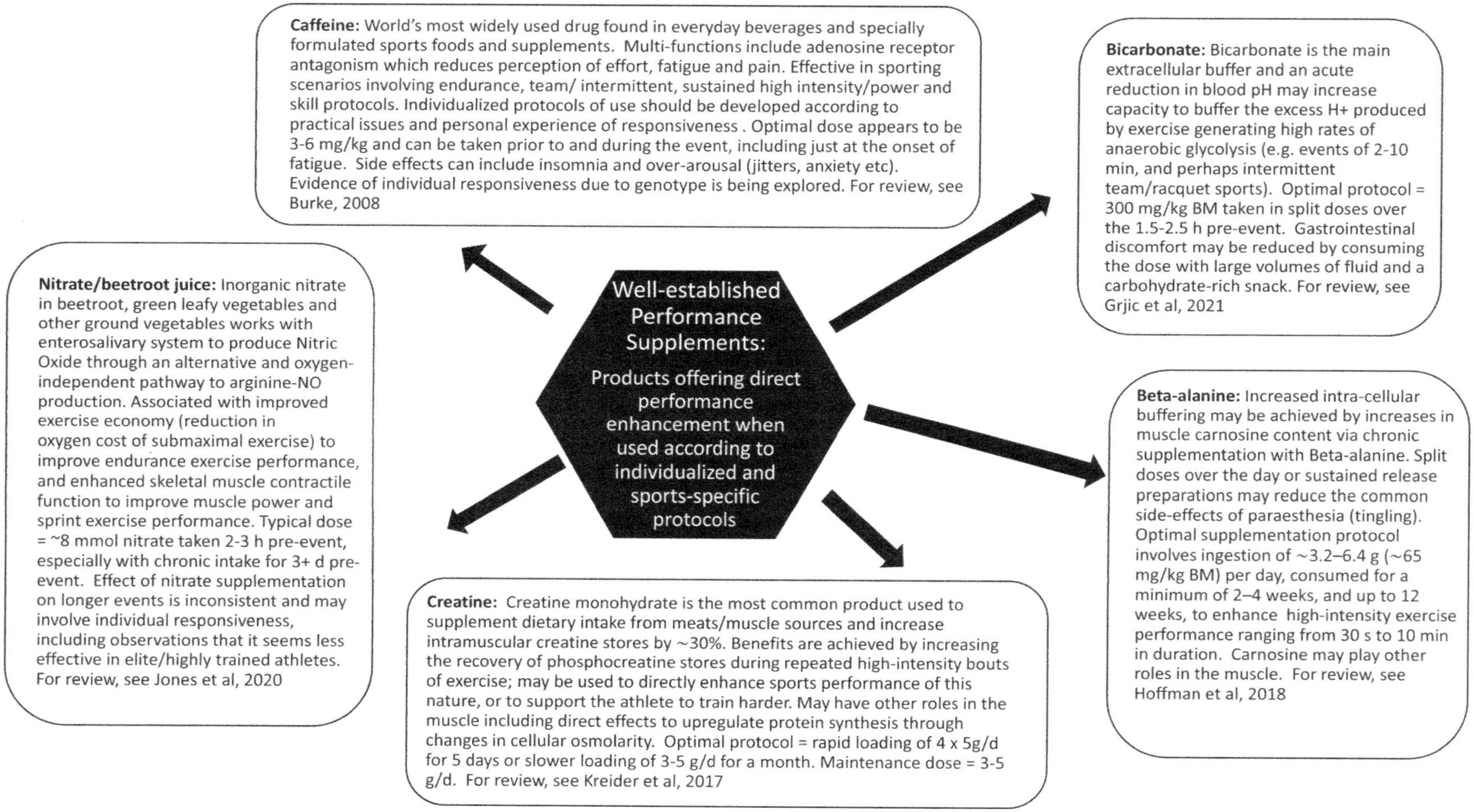

Well-established Performance Supplements:

Products offering direct performance enhancement when used according to individualized and sports-specific protocols

Caffeine: World's most widely used drug found in everyday beverages and specially formulated sports foods and supplements. Multi-functions include adenosine receptor antagonism which reduces perception of effort, fatigue and pain. Effective in sporting scenarios involving endurance, team/ intermittent, sustained high intensity/power and skill protocols. Individualized protocols of use should be developed according to practical issues and personal experience of responsiveness . Optimal dose appears to be 3-6 mg/kg and can be taken prior to and during the event, including just at the onset of fatigue. Side effects can include insomnia and over-arousal (jitters, anxiety etc). Evidence of individual responsiveness due to genotype is being explored. For review, see Burke, 2008

Bicarbonate: Bicarbonate is the main extracellular buffer and an acute reduction in blood pH may increase capacity to buffer the excess H+ produced by exercise generating high rates of anaerobic glycolysis (e.g. events of 2-10 min, and perhaps intermittent team/racquet sports). Optimal protocol = 300 mg/kg BM taken in split doses over the 1.5-2.5 h pre-event. Gastrointestinal discomfort may be reduced by consuming the dose with large volumes of fluid and a carbohydrate-rich snack. For review, see Grjic et al, 2021

Nitrate/beetroot juice: Inorganic nitrate in beetroot, green leafy vegetables and other ground vegetables works with enterosalivary system to produce Nitric Oxide through an alternative and oxygen-independent pathway to arginine-NO production. Associated with improved exercise economy (reduction in oxygen cost of submaximal exercise) to improve endurance exercise performance, and enhanced skeletal muscle contractile function to improve muscle power and sprint exercise performance. Typical dose = ~8 mmol nitrate taken 2-3 h pre-event, especially with chronic intake for 3+ d pre-event. Effect of nitrate supplementation on longer events is inconsistent and may involve individual responsiveness, including observations that it seems less effective in elite/highly trained athletes. For review, see Jones et al, 2020

Beta-alanine: Increased intra-cellular buffering may be achieved by increases in muscle carnosine content via chronic supplementation with Beta-alanine. Split doses over the day or sustained release preparations may reduce the common side-effects of paraesthesia (tingling). Optimal supplementation protocol involves ingestion of ~3.2–6.4 g (~65 mg/kg BM) per day, consumed for a minimum of 2–4 weeks, and up to 12 weeks, to enhance high-intensity exercise performance ranging from 30 s to 10 min in duration. Carnosine may play other roles in the muscle. For review, see Hoffman et al, 2018

Creatine: Creatine monohydrate is the most common product used to supplement dietary intake from meats/muscle sources and increase intramuscular creatine stores by ~30%. Benefits are achieved by increasing the recovery of phosphocreatine stores during repeated high-intensity bouts of exercise; may be used to directly enhance sports performance of this nature, or to support the athlete to train harder. May have other roles in the muscle including direct effects to upregulate protein synthesis through changes in cellular osmolarity. Optimal protocol = rapid loading of 4 x 5g/d for 5 days or slower loading of 3-5 g/d for a month. Maintenance dose = 3-5 g/d. For review, see Kreider et al, 2017

Figure 19.7 Well-established performance supplements.

3. When supplements and sports foods are indicated, only use products from reputable manufacturers. In particular, use products that have been 'batch tested' for the absence of banned substances by accredited independent third-party companies.
4. In the case of high-performance athletes who are covered by anti-doping regulations, document the use of all sports foods and sports supplements, noting the batch numbers of the latter products.

19.12 Practical issues in nutrition for athletes

Despite the sports nutrition knowledge available to modern athletes and coaches, sports nutritionists report that athletes do not always achieve the practices of optimal sports nutrition. A number of factors may be involved:

- Poor understanding of sports nutrition principles; reliance on myths and misconceptions.
- Failure to recognise the specific nutritional requirements of different sports and individuals within these sports.
- Apparent conflict of nutrition goals (e.g. how can an athlete achieve increased requirements for nutrients such as carbohydrate and iron while limiting energy intake to achieve loss of body fat?).
- Lack of practical nutrition knowledge and skills (e.g. knowledge of food composition, domestic skills such as food purchasing, preparation and cooking).
- Overcommitted lifestyle – inadequate time and opportunities to obtain or consume appropriate foods owing to heavy workload of sport, work, school, etc.
- Inadequate finances.
- The challenge of frequent travel.

Given the specific nutritional requirements of sports and individuals, according to age and gender, it is impossible to prepare a single set of nutrition guidelines for athletes. Nevertheless, education tools that address key issues of nutrition for athletes are an important resource for coaches, athletes and sports nutritionists. Education strategies that focus on practical areas of food choice and preparation, and guidelines that can address a number of key nutrition issues simultaneously are most valuable. In situations where nutritional goals can be achieved by modest changes to typical population eating patterns, it may be sufficient to provide a set of behavioural strategies to guide athletes to achieve such changes. However, in situations where the athlete has extreme nutrient requirements, where nutritional goals appear to conflict or where medical problems are present, the athlete should be directed to seek individualised and expert counselling from a sports nutritionist or dietitian.

It should be appreciated that many of the practical challenges to achieving sports nutrition goals arise directly because of exercise or the environment in which it is undertaken. Goals of nutrition before, during and after a workout or training must often be compromised or modified because of the effects of exercise on gastrointestinal function and comfort. Access to food and drinks is often restricted during the busy daytime activities of the athlete or in the exercise environment. The frequent travel schedules of the athlete must also be negotiated in dietary advice. In many cases, the expertise of the sports dietitian is required to provide creative ways to meet sports nutrition goals. Strategies for the travelling athlete are summarised in Box 19.1.

Box 19.1 Strategies for eating well while travelling

- Investigate the food resources at the trip destination before leaving. People who have travelled previously to that country, competition or accommodation facility may be able to warn about likely problems and enable the preparation of a suitable plan in advance.
- Organise special menus and meals in restaurants, airplanes or hotels in advance.
- Find out about food hygiene and water safety in new countries. It may be necessary to restrict fluid intake to bottled or boiled drinks, and to avoid foods that are at high risk of contamination (e.g. unpeeled fruits and vegetables).
- Take some food supplies on the trip if important foods are likely to be unavailable or expensive. Foods that are portable and low in perishability include breakfast cereals, milk powder, tinned and dehydrated foods, and special sports supplements.
- Be aware of special nutritional requirements in the new location. Be prepared to meet increased requirements for fluid, carbohydrate and other nutrients according to the climate/altitude or to the change in exercise patterns.

19.13 Summary and perspectives for the future

Optimal nutrition support varies according to the athlete, their phase of training or competition, and their specific goals. This chapter has summarised common issues in sports nutrition, including managing body composition goals, supporting training adaptations and recovery, and optimising competition performance. Sports foods and supplements may be part of the nutrition plans that allow an athlete to achieve their nutritional goals within the practical considerations and challenges of exercise and the sporting lifestyle. The future is likely to further increase understanding of personalised nutrition plans.

References

Burke, L.M., Ackerman, K.E., Heikura, I.A., Hackney, A.C., Stellingwerff T. (2023) Mapping the complexities of relative energy deficiency in sport (REDs): development of a physiological model by a subgroup of the International Olympic Committee (IOC) Consensus on REDs. *Br J Sports Med* 57 (17): 1100–1113. doi: bjsports-2023-107335

Burke, L.M., Hawley, J.A., Wong, S.H. et al. (2011). Carbohydrates for training and competition. *J Sports Sci* 29 (Suppl 1): S17–S27. doi: 10.1080/02640414.2011.585473.

Burke, L.M., van Loon, L.J.C., Hawley, J.A. (2017). Postexercise muscle glycogen resynthesis in humans. *J App Physiol* 122 (5): 1055–1067.

Christensen, E.H., Hansen O. (1939). Arbeitsfahigkeit und Ernärung. *Skandinavishes Archiv für Physiolgie* 81, 160–171.

Jeukendrup, A.E., Saris W.H., Wagenmakers, A.J. (1998). Fat metabolism during exercise: a review. Part I: fatty acid mobilization and muscle metabolism. Int J Sports Med 19 (4): 231–244. doi: 10.1055/s-2007-971911. PMID: 9657362.

Thomas, D.T., Erdman, K.A., Burke, L.M. (2016). American college of sports medicine joint position statement. nutrition and athletic performance. *Med Sci Sports Exe* 48 (3): 543–568. doi: 10.1249/MSS.0000000000000852.

Further reading

Books

Belski, R., Forsyth, A., Mantzioris, E. (eds.) (2019). *Nutrition for Sport, Exercise and Performance: A Practical Guide for Students, Sports Enthusiasts and Professionals.* Sydney: Allen and Unwin.

Burke, L. (2007). *Practical Sports Nutrition.* Ilinois: Human Kinetics.

Burke, L., Deakin, V., Minehan, M. (eds.) (2021). *Clinical Sports Nutrition*, 6e. Sydney: McGraw Hill.

Jeukendrup, A.E., Gleeson, M. (2018). *Sport Nutrition. An Introduction into Energy Metabolism and Performance*, 3e. Champaign, IL: Human Kinetics.

Papers

American College of Sports Medicine, Sawka, M.N., Burke, L.M. et al. (2007). American college of sports medicine position stand. exercise and fluid replacement. *Med Sci Sports Ex* 39 (2): 377–390.

Burd, N.A., Beals, J.W., Martinez, I.G. et al (2019). Food-first approach to enhance the regulation of post-exercise skeletal muscle protein synthesis and remodeling. *Sports Med* 49 (Suppl 1): 59–68.

Burke, L.M., Castell, L.M., Casa, D.J. et al. (2019). International association of athletics federations consensus statement 2019: nutrition for athletics. *Int J Sport Nutr Exerc Metab* 29 (2): 73–84. doi: 10.1123/ijsnem.2019-0065.

Burke, L.M., Hawley, J.A., Jeukendrup, A. et al. (2019). Toward a common understanding of diet-exercise strategies to manipulate fuel availability for training and competition preparation in endurance sport. *Int J Sport Nutr Exerc Metab* 28 (5): 451–463. doi: 10.1123/ijsnem.2019-0289.

Burke, L.M., Slater, G.J., Matthews, J.J. et al. (2021). ACSM expert consensus statement on weight loss in weight-category sports. *Curr Sports Med Rep* 20 (4): 199–217.

Burke, L.M. (2008). Caffeine and sports performance. *App Physiol Nutr Metab* 33 (6): 1319–1334. doi: 10.1139/H08-130.

Collins, J., Maughan, R.J., Gleeson, M. et al. (2021). UEFA expert group statement on nutrition in elite football. Current evidence to inform practical recommendations and guide future research. *Br J Sports Med* 55 (8): 416–416. doi: 10.1136/bjsports-2019-101961.

Desbrow, B., McCormack, J., Burke, L.M. et al. (2014). Sports Dietitians Australia position statement: Sports nutrition for the adolescent athlete. *Int J Sport Nutr Exerc Metab* 24 (5): 570–584. doi: 10.1123/ijsnem.2014-0031.

Grgic, J., Pedisic, Z., Saunders, B. et al. (2021). International Society of Sports Nutrition position stand: Sodium bicarbonate and exercise performance. *Int Soc Sports Nutr* 19 (1): doi: 10.1196/s12970-021-00458-w.

Hector, A.J., Phillips, S.M. (2019). Protein recommendations for weight loss in elite athletes: a focus on body composition and performance. *Int J Sport Nutr Exerc Metab* 28 (2): 170–177. doi: 10.1123/ijsnem.2017-0273.

Hoffman, J.R., Varanoske, A., Stout, J.R. (2019). Effects of β-alanine supplementation on carnosine elevation and physiological performance. *Adv Food Nutr Res* 84: 193–206.

Jeukendrup, A.E. (2013). Oral carbohydrate rinse: placebo or beneficial? *Curr Sports Med Rep* 12 (4): 222–227. doi: 10.1249/JSR.0b013e31929a6caa.

Jeukendrup, A.E. (2017). Training the gut for athletes. *Sports Med* 47 (Suppl 1): 101–110. doi: 10.1007/s40279-017-0690-6.

Jones, A.M., Vanhatalo, A., Seals, D.R. et al. (2020). Dietary nitrate and nitric oxide metabolism: mouth, circulation, skeletal muscle, and exercise performance. *Med Sci Sports Ex* doi: 10.1249/MSS.0000000000002470 [published Online First: 2020/08/01].

Kreider, R.B., Kalman, D.S., Antonio, J. et al. (2017). International society of sports nutrition position stand: safety and efficacy of creatine supplementation in exercise, sport, and medicine. *J Intl Soc Sports Nutr* 14: 19.

Maughan, R.J., Burke, L.M., Dvorak, J. et al. (2019). IOC consensus statement: dietary supplements and the high-performance athlete. *Br J Sports Med* 52 (7): 439–455. doi: 10.1136/bjsports-2019-099027.

Mountjoy, M.L., Ackerman, K.E., Bailey, D.M. et al. (2023). The 2023 International Olympic Committee's (IOC) consensus statement on relative energy deficiency in sports (REDs). *Br J Sports Med* in press.

Sim, M., Garvican-Lewis, L.A., Cox, G.R. et al.(2019). Iron considerations for the athlete: a narrative review. *Eur J Appl Physiol* doi: 10.1007/s00421-019-04157-y.

Stellingwerff, T., Cox, G.R. (2014). Systematic review: carbohydrate supplementation on exercise performance or capacity of varying durations. *Appl Physiol, Nutr Metab = Physiol Appl, Nutr Et Metab* 39: 998–1011. doi: 10.1139/apnm-2014-0027.

About the Editors

Helen M. Roche
Vice-President for Research, Innovation and Impact, University College Dublin

Full Professor of Nutrigenomics (Nutrition and Omics), UCD Conway Institute and UCD Institute of Food and Health, University College Dublin

Visiting Professor of Nutrition, Queen's University Belfast.

Helen initially trained in Human Nutrition and Dietetics, followed by Molecular Medicine. Her Nutrigenomics team focuses on Precision Nutrition – specifically the impact of diet on metabolism and inflammation, in obesity, type 2 diabetes (T2D), non-alcoholic fatty liver disease (NAFLD) and obesity related cancer. Nutrigenomics uses state-of-the-art 'omics' to investigate the molecular effects of diet on health – to provide hard evidence. Whilst nutrition plays a critical role in health and disease, too often the mechanistic basis is lacking – we seek to fill that evidence gap.

In Europe, Prof Roche has led several initiatives relating to Food, Nutrition and Health. She chaired the Scientific Advisory Board of the European Healthy Life Healthy Diet Joint Programming Initiative (2015-2019). She advises UK, Netherlands and US grant agencies, including Crohn's and Colitis Foundation, ZonMw and UK Nutrition Research Partnership. She was a board member of the Royal College of Surgeons Ireland Hospital Group.

Prof Roche's team are funded by a number of agencies. Her Science Foundation Ireland (SFI) Investigator Award entitled 'Diet, Immune Training and Metabolism' in collaboration with Dr Fred Sheedy and Prof Suzanne Norris (Trinity College Dublin (TCD)) determines the impact of diet and metabolism on Innate Immune responses in NAFLD. She is co-PI in several multidisciplinary programmes. In Precision Oncology Ireland (POI) Roche's team are determining if/how the 'dietary environment' potentiates obesity related cancer risk, with Prof Jacinta O'Sullivan (TCD) and Dr David Gomez-Matallanas (Systems Biology Ireland, UCD). On-going research as part of the EU Healthy Diet for a Healthy Life (HDHL) Joint Programming Initiative (HDHL JPI) funded APPETITE Project is investigating the impact of personalised plant protein and exercise interventions on appetite and malnutrition in older persons.

Roche's team has made significant advances in terms of understanding the extent to which dietary elements can modulate the molecular interactions between inflammation and metabolism, to attenuate risk of obesity-related insulin resistance, T2D and more recently cancer. A key achievement being to integrate and elucidate the extent to which these pathways and mechanisms translate to man.

A key personal remit has been to foster Emerging Investigator success. Indeed Roche was the recipient of the SFI Mentorship Award (2022). In UCD, as UCD Conway Director she plays an active role facilitating recent Ad Astra appointee achievement via the Ad Astra Forum.

Nutrition and Metabolism, Third Edition. Edited on behalf of The Nutrition Society by Helen M. Roche, Ian A. Macdonald, Annemie M.W.J. Schols and Susan A. Lanham-New.

Companion Website: www.wiley.com/go/nutrition/metabolism3e

Helen has supervised more than 30 PhD scientists and a similar number of post-doctoral researchers.

https://people.ucd.ie/helen.roche

https://pure.qub.ac.uk/en/persons/helen-roche

Twitter: @HelenMRoche

Ian A. Macdonald, Short Biography

Professor Emeritus in Metabolic Physiology, University of Nottingham, UK. He obtained a BSc in Physiology & Chemistry in 1973 and a PhD in Physiology in 1977, both from the University of London. He was an academic member of staff in Nottingham between 1977 and 2020, and Professor of Metabolic Physiology from 1991 to 2020. During his time at Nottingham he was Head of the School of Biomedical Sciences (1997–2002) and of the School of Life Sciences (2013–2017), as well as being Director of Research in the Faculty of Medicine and Health Sciences (2006–2013). He was President of the Nutrition Society from 2007–2010 and has served on external advisory boards and committees for Health Charities, EU, UK Government, Research Councils in the UK and other European countries. This included being a member of the UK Scientific Advisory Committee on Nutrition (Department of Health and Social Care) from 2007–2020, where he chaired the sub-committee which produced the Carbohydrates and Health Report in 2015.

After retiring from the University of Nottingham in 2020, he was appointed as the Scientific Director of the Nestle Institute of Health Sciences in Lausanne, Switzerland, and after retiring from that post in 2022 he became and external member of the Nestle Research Scientific Advisory Board and the Science and Technology Advisory Council.

His research has focused on the functional consequences of metabolic and nutritional disturbances in health and disease, with specific interests in obesity, diabetes, cardiovascular disease and exercise, predominantly in healthy individuals and patients. The research has involved collaborations with clinical Academic colleagues and NHS clinicians, both within Nottingham and elsewhere. This work has been funded by grants from Research Councils, Medical Charities, EU (FP5 and FP7), as well as the Pharmaceutical and Food Industries. The research has involved supervision of both non-clinical and clinical researchers studying for PhD or MD/PhD degrees.

He has published over 410 peer-reviewed original research papers, together with reviews, book chapters and invited contributions. He was Treasurer of FENS from 2011 to 2019, and of the World Obesity Federation from 2009 to 2019. He was joint Editor of the International Journal of Obesity from 2000 to 2021, and is a Fellow of the Royal Society of Biology, an Honorary Fellow of the Nutrition Society, a Registered Nutritionist and an Honorary Fellow of the Association for Nutrition, a Fellow of the Physiological Society, and a Fellow of the International Union of Nutritional Sciences.

Annemie M.W.J. Schols

Professor of Nutrition and Metabolism in Chronic Diseases, Maastricht University. Her scientific expertise encompasses translational research into disease and lifestyle induced metabolic derangements in lung diseases (COPD and lung cancer) with a specific focus on skeletal muscle wasting. Her highly cited research had led to the development and implementation of tailored single and multi-modal nutritional, exercise and pharmacological intervention strategies to improve physical functioning, quality of life and survival of patients in different disease stages. She has also developed and coordinated many post-graduate courses to position nutrition in chronic disease management worldwide. From 2006–2016 she was board member of the Netherlands Health Council Between 2006-2020 she was Scientific Director of NUTRIM School for Nutrition and Translational Research in Metabolism (www.maastrichtuniversity.nl/nutrim) at Maastricht University Medical Centre. Since 2020, prof. Schols is Dean of the Faculty of Health, Medicine and Lifesciences at Maastricht University and Vice-President of the Executive Board of Maastricht University Medical Centre.

Susan A. Lanham-New
Professor of Human Nutrition and Head of Department, University of Surrey. She led a successful *Nutritional Sciences at Surrey* application for the 2017/2018 Queen's Anniversary Prize and is a Member of H.M. Government's Scientific Advisory Committee on Nutrition (SACN) which is responsible for setting the UK's Nutrient Requirements. She is also a Member of the E.U.'s European Food Safety Authority (EFSA) Committee on Vitamin D Safe Tolerable Limits. In addition, Professor Lanham-New is also Editor-in-Chief of the Nutrition Society (NutSoc) Textbook Series and first Editor of Nutritional Aspects of Bone Health. Her research focuses on nutrition and bone health, for which she has won a number of Awards including the 2018/2019 British Nutrition Foundation (BNF) Prize and the 2021 Royal Philosophical Society of Glasgow Medal for her work on nutrition, vitamin D and bone health. Professor Lanham-New has supervised 28 PhD students, published over 240 peer-review publications and raised >£15M in research income. She is a member of the College of Experts for the Royal Osteoporosis Society and is a Trustee of the BNF. She has recently been given Fellowship status of Association for Nutrition and is Honorary Secretary of NutSoc. Susan has led on a Vitamin D / COVID-19 Consensus Paper published in the *Br Med J (NHP)* and has contributed to media pieces for the *New Scientist, the FT and NYT* and other media outlets on Vitamin D and the SARS-CoV-2 virus as well as undertaking the *Guardian* Podcast on Vitamin D and working on SACN and NICE Reports on Vitamin D, COVID-19 and Acute Respiratory Tract Infections.

Index

Please note: page reference in *italics* are figures and those in **bold** are tables.

Nutrition and Metabolism, Third Edition. Edited on behalf of The Nutrition Society by Helen M. Roche, Ian A. Macdonald, Annemie M.W.J. Schols and Susan A. Lanham-New.

Companion Website: www.wiley.com/go/nutrition/metabolism3e